Handbook for
Chemical Technicians

OTHER McGRAW-HILL HANDBOOKS OF INTEREST

AMERICAN INSTITUTE OF PHYSICS *American Institute of Physics Handbook*

AMERICAN SOCIETY OF MECHANICAL ENGINEERS *ASME Handbooks:*
 Engineering Tables   *Metals Engineering—Processes*
 Metals Engineering—Design *Metals Properties*

BAUMEISTER AND MARKS *Standard Handbook for Mechanical Engineers*

BRADY *Materials Handbook*

CAGLE *Handbook of Adhesive Bonding*

CONDON AND ODISHAW *Handbook of Physics*

CONSIDINE *Chemical and Process Technology Encyclopedia*

CONSIDINE *Encyclopedia of Instrumentation and Control*

CONSIDINE *Instruments and Controls Handbook*

CONSIDINE AND ROSS *Handbook of Applied Instrumentation*

DEAN *Lange's Handbook of Chemistry*

FACTORY MUTUAL ENGINEERING DIVISION *Handbook of Industrial Loss Prevention*

FINK AND CARROLL *Standard Handbook for Electrical Engineers*

GARTMANN *De Laval Engineering Handbook*

GRANT *Hackh's Chemical Dictionary*

HEYEL *The Foreman's Handbook*

JURAN *Quality Control Handbook*

KALLEN *Handbook of Instrumentation and Controls*

KAUFMAN AND SEIDMAN *Handbook for Electronics Engineering Technicians*

KLERER AND KORN *Digital Computer User's Handbook*

KORN AND KORN *Mathematical Handbook for Scientists and Engineers*

LUND *Industrial Pollution Control Handbook*

MAGILL, HOLDEN, AND ACKLEY *Air Pollution Handbook*

MANTELL *Engineering Materials Handbook*

MAYNARD *Industrial Engineering Handbook*

MEITES *Handbook of Analytical Chemistry*

PERRY *Engineering Manual*

RAŽNJEVIĆ *Handbook of Thermodynamic Tables and Charts*

ROHSENOW AND HARTNETT *Handbook of Heat Transfer*

SHAND *Glass Engineering Handbook*

SHUGAR *Chemical Technician's Ready Reference Handbook*

STOUT AND KAUFMAN *Handbook of Operational Amplifier Circuit Design*

TUMA *Technology Mathematics Handbook*

Handbook for Chemical Technicians

HOWARD J. STRAUSS, Ph.D.

Director, Market and Technology Development,
Gould Inc.

Edited by
MILTON KAUFMAN

President, Electronic Writers and Editors, Inc.

McGRAW-HILL BOOK COMPANY

New York   St. Louis   San Francisco   Auckland   Düsseldorf
Johannesburg   Kuala Lumpur   London   Mexico   Montreal
New Delhi   Panama   Paris   São Paulo   Singapore
Sydney   Tokyo   Toronto

Library of Congress Cataloging in Publication Data

Strauss, Howard J
    Handbook for chemical technicians.

    1. Chemistry, Technical—Handbooks, manuals,
etc. I. Title.
TP151.S89   542   76-10459
ISBN 0-07-062164-0

1234567890   KPKP   785432109876

The editors for this book were Harold B. Crawford and Lester Strong,
the designer was Naomi Auerbach, and the
production supervisor was George Oechsner. It was set in
Century Schoolbook by York Graphic Services, Inc.

Printed and bound by The Kingsport Press.

Contents

Index follows Chapter 10.

Preface

The keystone in preparing the *Handbook for Chemical Technicians* was utility. It was felt that this would result in a reference book which, it was hoped, would become dog-eared from frequent use by students, laboratory people, chemical plant operators, and, in fact, anyone who basically works with chemicals. To do this not only required that a large store of fundamental data be concisely and clearly presented, but also some information as to the theory behind the various tables and how they should be used. Consequently, the organization of each subject is such as to include a short exposition of the pertinent theory and examples illustrating the use of the tabular or graphic information.

It was also recognized that sometimes years may elapse between first learning a particular item and the need to use it for some practical purpose, and that this situation may be true even for basic concepts. While the expository material was kept as brief as possible, consistent with clarity, each subject was covered sufficiently to refresh the user's memory, or serve as supplementary information for those who may be in the initial learning process.

In keeping with the principle of utility, a sizeable problem developed from the fact that several values are frequently reported for ostensibly the same data, or that the reported data are not for standard or normal conditions. Where multiple values were reported, an effort was made to select the best one, that is, the one which was judged to be the most practical. Failing this, an average value was used. For data which were not at standard or normal conditions, corrections were applied to bring them to such conditions. In a few instances, this may have resulted in some slight loss of accuracy, but it was felt that this was more than compensated by the increased utility of the handbook, which was, after all, the guiding objective.

Howard J. Strauss

Milton Kaufman

Handbook for
Chemical Technicians

Chapter 1

Units and Measurements

1-1 PRIMARY UNITS AND MEASUREMENTS

Primary and Derived Units

Modern chemical technology is highly quantitative. That is, descriptions of materials are given in terms of how much matter is involved (mass) and how much space it occupies (volume). For descriptions of reactions involving materials—i.e., physical or chemical changes—additional quantities are frequently used to indicate how long (time) or how hot (temperature). While qualitative (general) descriptions are still important, wherever possible the preference in both commercial and academic work is almost always for quantitative details.

In general scientific work, there are three primary measurements: mass, length, and time. In chemical work, temperature is also considered to be a primary measurement. While temperature can in fact be related to mass, length, and time, the relationship tends to be distant. All other measurements can be readily expressed in terms of the primary quantities and are therefore called *derived measurements*. For example, the liter is a unit of volume measure but is derived from length measurements, since a liter is the space occupied by a cube that is 10 cm long, 10 cm wide, and 10 cm high. In general, volume = length × length × length, or $V = L^3$. In a more complex relationship, the calorie, a unit of heat measure, is derived from mass and temperature, since a calorie is the heat necessary to raise the temperature of one gram of water one Celsius degree. In general, heat = mass × temperature, or $H = M \times t$. In a similar manner, other measurements, including those involving electrical and thermal quantities, can be related back to the primary $M, L, T,$ and t quantities, although the derivation can be fairly complex. Nevertheless, an appreciation of this can be important, as we shall see, in setting up calculations involving measurements of related quantities, and particularly in handling measurements in more than one system.

English and Metric Systems

One of the daily problems with which chemical technicians must contend routinely is that two fundamental systems for measurements of primary and derived quantities have developed and are in common usage. These are the *metric* system, generally used in countries on the European continent and in those countries whose histories have been influenced by continental European nations; and the *English* system, until recently in general use in the British Empire and in those countries, such as the United States, whose backgrounds have been significantly influenced by England. While the metric system is generally the system of science, the English system is generally the system of commerce; and it is extremely important for the chemical technician to be able to work in either system, to be able to move from one system to the other with ease, and even to handle hybrid systems that come up from time to time.

Mass

The basic unit of mass in the metric system is the kilogram (kg) or 1,000 grams. In the English system, it is the pound (lb). Fundamentally, all metric mass measurements are referred to a 1-kg bar of platinum maintained at the International Bureau of Weights and Measures in Paris. In the United States, a similar bar at the Bureau of Standards in Washington, D.C., serves a similar purpose. The Paris bar is the primary standard, and the Washington bar, and others like it in other countries, are secondary standards. In principle, all weights are traceable to the Paris primary standard.

In recent years, the behavior of particular atoms subjected to known magnetic and electrostatic force fields has served as a basis for even more precise standards of mass.

A factor that often results in some confusion, and which should be resolved even at the outset, is the general interchangeability between mass and weight in common usage. Strictly speaking, mass is a measure of the quantity of matter present in a body, while weight is a measure of the gravitational attraction of the earth on a body. Thus, weight is in fact a force, and that force depends on the mass of the body and the gravitational attraction it experiences. The interchangeability and confusion have arisen from the fact that forces can also be expressed in mass units such as kilograms and pounds. On earth, a mass of 1 kg experiences a gravitational attraction (force) of 1 kg on the surface of the earth at sea level, and is said to weigh 1 kg. Similarly, a 1-lb mass experiences a force of 1 lb attracting it to the earth when at sea level, and is said to weigh 1 lb. On the moon, the 1-kg mass would weigh only 167 g, since the gravitational attraction of the moon is about one-sixth that of the earth. The 1-lb mass would similarly weigh only $2\frac{2}{3}$ oz on the moon. Since the force of gravity diminishes as the body recedes from the center of the earth, the weight of the body decreases as its elevation is increased. This is a minor factor for most work, but where elevation is of significance, weights are conveniently standardized at the gravitational force that exists at sea level.

In chemical work, weights are fundamentally measured by comparing the gravitational force experienced by an unknown mass against a known one on a balance. When the balance is level, the effect of gravity on the two masses must therefore be equal. The measurement of weight and mass is illustrated in Fig. 1-1. A force of 1 kg—or a weight of 1 kg—is thus conveniently defined as the force exerted by the earth on a 1-kg mass. In the English system, a weight of 1 lb is defined as the force exerted by the earth on a 1-lb mass. In common usage, the word "force" is understood, and weights are expressed simply in kilograms and pounds. Recognizing that there is a fundamental difference, for practical purposes in chemical work, mass and weight can be and are used interchangeably. Table 1-1 gives the most common weight units used in the metric and English systems and the equivalents which relate the various units.

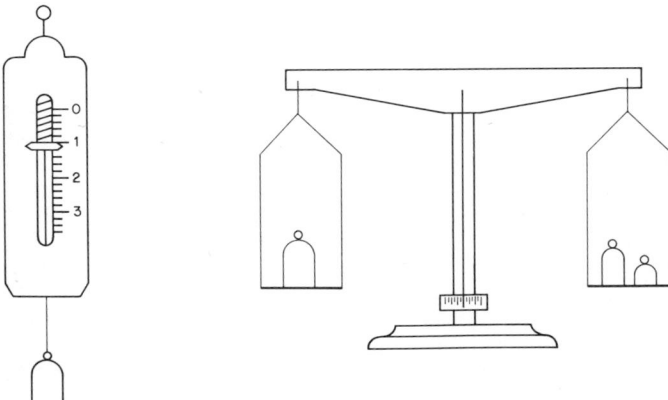

Fig. 1-1 Measurement of weight and mass. Spring scale measures force of gravity on body, i.e., its weight. Balance equalizes effects of gravity and therefore compares masses.

TABLE 1-1   Common Weight Units and Equivalents

1 kilogram (kg) = 1,000 grams
1 hectogram (hg) = 100 grams
1 dekagram (dakg) = 10 grams
1 gram (g) = 1,000 milligrams (mg)
1 decigram (dg) = 100 milligrams
1 centigram (cg) = 10 milligrams
1 pound (lb) = 16 ounces
1 pound (lb) = 453.6 grams
1 ounce (oz) = 28.35 grams
1 ton (short) = 2,000 pounds
1 ton (long) = 2,240 pounds

Note: The common pound-ounce units used in the United States are referred to as the avoirdupois (avdp) system. Commonly accepted abbreviations are shown in parentheses.

TABLE 1-2   Special Weight Units and Equivalents

1 grain (gr) = 0.0648 g
1 carat (c) = 0.2 g
1 dram, apothecaries (drap) = 60 gr
1 dram, avoirdupois (dravdp) = 27.34 gr
1 ton, metric = 2,204.6 lb
1 ounce, troy (ozt) = 31.10 g
1 pound, troy (lbt) = 373.2 g

Note: Apothecaries and troy weights are equivalent.

Example 1-1   How many grams are there in 0.75 lb?
solution   Since there are 453.6 g in 1 lb (Table 1-1), there are 0.75 × 453.6 or 340.2 g in 0.75 lb.

Example 1-2   How many ounces are there in 196 g?
solution   There are 28.35 g to an ounce (Table 1-1), and therefore there are 196 ÷ 28.35 or 6.92 oz in 196 g.

Example 1-3   How many ounces are there in 1.5 kg?
solution   There are 1.5 × 1,000 or 1,500 g in 1.5 kg (Table 1-1). Since there are 28.35 g in an ounce (Table 1-1), there are 1,500 ÷ 28.35 or 52.9 oz in 1,500 g or in 1.5 kg.

While Table 1-1 shows the most commonly used weight measures, there are others of importance in chemical work. These are usually associated with particular industries, such as troy weight used in conjunction with precious metals. Table 1-2 summarizes several of the more important special weight systems and gives the relationship to the common weight units.

Example 1-4   Which is heavier, a troy ounce or an avoirdupois ounce?
solution   From Table 1-2, a troy ounce is equal to 31.10 g. From Table 1-1, an avoirdupois ounce is equal to 28.35 g. The troy ounce is therefore heavier.

Example 1-5   How many troy ounces are there in a troy pound?
solution   From Table 1-2, there are 31.10 g in a troy ounce, and 373.2 g in a troy pound. Therefore, there are 373.2/31.10 = 12 troy ounces in a troy pound.

Length

The basic unit of length in the metric system is the meter (m). In the English system, it is the yard (yd). Fundamentally, all metric length measurements are referenced to a bar of platinum maintained at the International Bureau of Weights and Measures in Paris, the length of which is taken as the standard meter. A similar bar at the U.S. Bureau of Standards in Washington, D.C., serves as the standard yard. In principle, any device which measures length can be standardized against the standard meter or standard yard or, as is more practical, against calibrated meter or yard bars whose length has been carefully referred back to the standard bars in Paris or Washington. Because of the physical difficulties in accomplishing this, and the problems of compensating for changes that occur in the standard bars because of environmental conditions, particularly temperature, length measurements can now be made against new international standards, such as the wavelength of the orange-red radiation emitted by suitably excited krypton.

For practical measurements, the meter is frequently either too large or too small. Consequently, a number of derived units are now in use. For measuring small objects, it is usually convenient to use the *centimeter*, which is $1/_{100}$, or the *millimeter*, which is $1/_{1000}$ of a meter. For even smaller units, such as might be used for microscopic and submicroscopic units, the *angstrom* (Å) is used. An angstrom unit is $1/_{100}$ millionth of a centimeter $(1 \times 10^{-8}$ cm). Where this unit is too small, the *micrometer* is used. A micrometer is $1/_{10,000}$ cm. For measurement of large lengths, the *kilometer* is used. A kilometer is 1,000 m.

In the English system, small measurements are expressed in *feet* ($\frac{1}{3}$ yd) and *inches* ($\frac{1}{36}$ yd). Although inches can be broken down in a geometric sequence ($\frac{1}{2}$, $\frac{1}{4}$, $\frac{1}{8}$, $\frac{1}{16}$, $\frac{1}{32}$, and $\frac{1}{64}$), the handling of such fractions can become cumbersome, and for measurements smaller than $\frac{1}{64}$ in it is common to go over to a metric subdivision, i.e., $\frac{1}{1,000}$ in. For large measurements the English system jumps right to the *mile*, or 1,760 yd.

The fundamental relationship between the metric and English systems is that there are 1.0936 yd to 1 m. From this fundamental relationship, the factors for converting all other units of length from the English system to the metric system, or vice versa, can be derived. These are given in Table 1-3.

In measuring areas, secondary units are used. In the metric system, the unit of area is the *square meter* (i.e., the area enclosed by a square each side of which is 1 m in length). Similarly, the *square yard* is the unit of surface measurement in the English system. In some instances, these areas are too large for convenience, as would be the case for most laboratory measurements, or they may be too small, as would be the case for land measurements. Consequently, fractions of these measurements, or multiples thereof, are used for convenience. Thus, the metric system uses square centimeters and square kilometers, and the English system has square inches and square miles. The common area measurements and the equivalents in the English and metric systems are given in Table 1-4.

Still another secondary measurement derived from the primary length measurement is that of volume. In the metric system, the fundamental unit is the *cubic meter* or the volume enclosed by a cube each edge of which is 1 m in length. In the English system, the basic unit of volume is the *cubic yard*, or the volume enclosed by a cube each edge of which is 1 yd in length. As in area measurement, fractions and multiples of the fundamental volume units are used where convenient. The common units of volume and their various equivalents are given in Table 1-5.

TABLE 1-3 Common Length Units and Equivalents

1 meter (m) = 39.37 inches (in)
1 kilometer (km) = 1,000 meters
1 meter = 100 centimeters (cm)
1 centimeter = 10 millimeters (mm)
1 mile (mi) = 5,280 feet (ft)
1 mile = 1,609.3 meters
1 inch (in) = 2.54 centimeters
1 foot (ft) = 12 inches
1 yard (yd) = 3 feet
1 yard = 36 inches

TABLE 1-4 Common Area Units and Equivalents

1 acre (acre) = 4,840 square yards (yd²)
1 yd² = 0.836 square meter (m²)
1 m² = 10.76 ft²
1 square inch (in²) = 6.45 square centimeters (cm²)
1 square kilometer (km², = 0.386 square mile (mi²)
1 yd² = 9 ft²
1 ft² = 144 in²

TABLE 1-5 Common Volume Units and Equivalents

1 cubic meter (m³) = 1,000,000 cubic centimeters (cm³)
1 cubic meter = 1,308 cubic yards (yd³)
1 liter (l) = 1,000 cm³
1 gallon (gal) = 231 cubic inches (in³)
1 gal = 3.785 liters
1 cubic foot (ft³) = 7.48 gal
1 fluid ounce (fl oz) = 1.805 in³
1 gal = 8 pints (pt)
1 fl oz = 29.57 cm³
1 quart (qt) = 0.946 liter
1 milliliter (ml) = 1 cm³

Example 1-6   How many centimeters are there in 3.25 m?
solution   From Table 1-3, there are 100 cm in 1 m; therefore, there are 3.25 × 100 or 325 cm in 3.25 m.

Example 1-7   How much larger is a mile than a kilometer?
solution   From Table 1-3, a mile is equivalent to 1,609.3 m, or 1.6093 km. A mile is therefore 0.6093 km larger than a kilometer (or 60.93 percent larger).

Example 1-8   How many centimeters are there in 1 ft?
solution   From Table 1-3, there are 2.54 cm to each inch, and therefore, there are 12 × 2.54 or 30.48 cm in 1 ft.

Example 1-9   How many square meters are there in 1 acre?
solution   From Table 1-4, there are 4,840 yd^2 in 1 acre, or 4,840 × 9 = 43,560 ft^2. Also shown is that there are 10.76 ft^2 in 1 m^2, and therefore, there are 43,560 ÷ 10.76 or 4,060 m^2 in 1 acre.

Example 1-10   How many square yards are equivalent to 1 m^2?
solution   From Table 1-4, there are 10.76 ft^2 in 1 m^2, and therefore, there are 10.76 ÷ 9 or 1.196 yd^2 in 1 m^2.

Example 1-11   How many square inches are there in 10.33 cm^2?
solution   From Table 1-4, there are 6.45 cm^2 in 1 in^2. Therefore, there are 10.33 ÷ 6.45 or 1.60 in^2.

Example 1-12   How many gallons are there in 1 m^3?
solution   From Table 1-5, there are 3.785 l, or 3,785 cm^3, in 1 gal, and there are 1,000,000 cm^3 in 1 m^3. Therefore, there are 1,000,000 ÷ 3,785 or 264 gal.

Example 1-13   How many fluid ounces are there in 1 gal?
solution   From Table 1-5, there are 1.805 in^3 in 1 fl oz, and 231 in^3 in 1 gal. Therefore, there are 231 ÷ 1.805 or 128 fl oz in 1 gal.

In the tables we have already covered, as well as in all subsequent ones, only the most important (in terms of general utility) quantities are presented so as to make them easy and convenient to use. On the relatively rare occasions where the desired equivalents do not appear in the tables, they can be readily derived, as demonstrated in Example 1-13. While the most commonly used length-area-volume units are given in Tables 1-3 to 1-5, there are others of importance in chemical work. As in the case of weight measures, these special length-area-volume units are associated with particular industries. The more important special length-area-volume units and equivalents are given in Tables 1-6 to 1-8.

Example 1-14   How many angstroms are there in 1 μm?
solution   From Table 1-6, we see that there are 1,000 μm in 1 mm (i.e., 1 μm = 0.001 mm), and since there are 10 mm in 1 cm (from Table 1-3), there are 10,000 μm in 1 cm. From Table 1-6 again, we get that there are 100,000,000 Å in 1 cm (i.e., 1 Å = 1 × 10^{-8} cm). Therefore, there are 100,000,000 ÷ 10,000 or 10,000 Å in 1 μm.

TABLE 1-6   Special Length Units and Equivalents

1 angstrom (Å) = 1 × 10^{-8} cm
1 fathom (fath) = 6 ft
1 furlong (fur) = 220 yd
1 nautical mile (nmi) = 1.151 mi
1 rod (rd) = 16.5 ft
1 micrometer (μm) = 0.001 mm
1 μm = 1,000 nanometers (nm)

TABLE 1-7   Special Area Units and Equivalents for Wire

1 circular millimeter (cmm) = 0.7854 mm^2
1 circular mil (cmil) = 0.7854 mil^2
1 cmm = 0.007854 cm^2
1 cmil = 7.854 × 10^{-7} in^2
1 rd^2 = 30.25 yd^2

TABLE 1-8   Special Volume Units and Equivalents

1 bushel (bu) = 32 qt
1 fluid dram (fl dr) = 0.125 fl oz
1 pint (pt) = 0.5 qt
1 minim = 0.00208 fl oz
1 peck (pk) = 2 gal
1 gill = 4 fl oz

Example 1-15 How many minims are there in 1 in³?
solution   Table 1-8 indicates that there are $1 \div 0.00208$ or 481 minims in 1 fl oz (i.e., 1 minim = 0.00208 fl oz). From Table 1-5, 1 fl oz = 1.805 in³. Therefore, there are 0.554×481 or 266 minims in 1 in³.

Special Mathematical Manipulations

In converting from one set of units to another, and particularly when several conversions have to be carried out in sequence (as was the case in the last two examples), some confusion can arise as to whether to multiply or divide by the equivalents given in the tables. A quick check on the computations can be made by the method of *cancellation of units*. This method depends upon the fact that units can be manipulated algebraically just as if they were algebraic symbols. In addition, the answer resulting from any computation must have the appropriate units. The cancellation-of-units method is based on the fact that any number can be multiplied by 1 without changing its value. Furthermore, if any two quantities are equal, the fraction formed by making one of these quantities the numerator and the other the denominator is equal to 1 and therefore may be used to multiply any other quantity without changing the latter's value. Thus, in Table 1-1, 1 lb = 453.6 g. Therefore, the fraction 1 lb/453.6 g is equal to 1, or the fraction 453.6 g/1 lb is also equal to 1. Suppose now we wanted to use this information to convert 1.37 lb to grams. We would simply multiply 1.37 lb by 453.6 g/1 lb and cancel the units just as if they were algebraic quantities. This then leaves us with 620 g: $1.37 \text{ lb} \times 453.6 \text{ g}/1 \text{ lb} = 620 \text{ g}$. Notice that the final answer contains only the appropriate units. It is important in using the cancellation-of-units method to include all the quantities throughout the calculation. A few examples will serve to illustrate the utility of this method.

Example 1-16 Convert 3.20 nmi into yards.
solution

$$3.20 \text{ nmi} \times \frac{1.151 \text{ mi}}{\text{nmi}} \times \frac{5,280 \text{ ft}}{\text{mi}} \times \frac{\text{yd}}{3 \text{ ft}} = 6,480 \text{ yd}$$

Example 1-17 How many grains are there in 1.80 oz?
solution

$$1.80 \text{ oz} \times \frac{28.35 \text{ g}}{\text{oz}} \times \frac{\text{gr}}{0.0648 \text{ g}} = 788 \text{ gr}$$

In Example 1-16, the number 1 has been omitted before the units nmi, mi, and yd. This is because the units are considered to be algebraic quantities for the purposes of the calculation, and therefore, 1 nmi can be written simply as nmi. In a similar manner, in Example 1-17, 1 oz is written simply oz, and 1 gr is written simply gr. As the conversion of units becomes more complicated owing to the introduction of derived units later on, it will be found that the method of cancellation of units turns out to be indeed very useful. In Example 1-17, the fraction 28.35 g/oz (meaning of course that 28.35 g is equivalent to 1 oz) is usually read as 28.35 grams per ounce. The word "per" is used to indicate a fraction in which the denominator has a numerical value of 1. Thus, in Example 1-17 we can read that there are 0.0648 gram per grain. It should be noticed that the latter fraction is used in the solution of Example 1-17 in its inverted form. In its inverted form, the fraction still has a value of 1, so that multiplying by the inverted form does not change the value of the quantities. In this particular case, the inverted form is used, since we are thereby really dividing. Notice that the units of the answer are appropriate, as is required by the method of cancellation of units.

In carrying out a sequence of unit conversions, it is also sometimes difficult to locate the decimal point in the answer, particularly when the sequential calculations are carried out on a slide rule. The decimal point can be conveniently located by a method of approximation. In the following sequential calculation, the answer is given without its appropriate decimal point:

$$3,790 \times 0.016 \times \frac{1}{3} \times 6.2 = 1,253$$

To locate the decimal, perform simple approximations and cancellations:

$$4{,}000 \times 0.015 \times \frac{1}{\cancel{3}} \times \cancel{6.2}^{2}$$

$$4{,}000 \times 0.03$$

$$40 \times 3 = 120$$

The approximate answer is 120, indicating that the decimal point is after the 5. The actual answer is therefore 125.3.

Example 1-18 Perform the indicated calculations and locate the decimal point in the answer by the method of approximation:

$$0.032 \times \frac{6{,}392{,}000}{52} \times 60 \times 60 \times 0.8$$

solution Performing the multiplications and divisions yields 1,132. The location of the decimal point can be determined by the following approximations:

$$0.03 \times \frac{6{,}000{,}000}{50} \times 50 \times 50 \times 1$$

$$6{,}000{,}000 \times 1.5 = 9{,}000{,}000$$

Notice that most of the approximations are smaller than the original quantities, i.e., 0.03 for 0.32, 6,000,000 for 6,392,000, etc. We would therefore expect the approximate answer to be 10,500,000, again indicating the actual answer is 11,320,000.

Sometimes it is inconvenient to use very large or very small numbers. In this case, exponential numerics can be helpful. In Example 1-18, several of the factors can be written in exponential form:

$$3.2 \times 10^{-2} \times \frac{6.392 \times 10^{6}}{5.2 \times 10} \times 6 \times 10^{1} \times 6 \times 10^{1} \times 8 \times 10^{-1}$$

Multiplying only the 10 factors by adding exponents, we get

$$3.2 \times \frac{6.392}{5.2} \times 6 \times 6 \times 8 \times 10^{4}$$

Using approximate substitutions:

$$3 \times 1 \times 35 \times 10 \times 10^{4}$$

Notice that 6.392/5.2, 6 × 6, and 8 appear as 1, 35, and 10, respectively, in the approximate form. Thus, the approximate answer is 105×10^{5}. Notice also that there is no set rule for substituting approximate numbers for the actual ones, except that cancellations should be convenient, meaning that multiples of 5 and 10 are usually preferred. Because of their convenience in many calculations, the exponential forms of expressing numbers and the rules for carrying out calculations with numbers in exponential form are given in Table 1-9.

1-2 SIGNIFICANT FIGURES

In arithmetic manipulations, each number has a very precise value. The number 6, for example, refers to exactly 6 units, and not almost 6 units, or about 6 units. Measurements, on the other hand, are never perfectly accurate, since as we have seen, they involve a comparison and sometimes a long sequence of comparisons with certain essentially arbitrary standards. Just how precise actual measurements really are depends on how carefully they are made and, almost always, on the quality of the instruments on which the measurements are made. We may say that an object weighs 6 g, but this in itself needs some indication as to how accurate a measurement is involved. If it is measured on a spring balance, the 6 g measurement may be closer to perhaps 5.8 or 6.4 g. If measured on a precision balance, the 6 g measurement actually may lie between 5.9990 and 6.0010 g.

Just how accurate a measurement should be depends on the circumstances and the use to which the measurement will be put. A gasoline tank of an automobile may have a volume measured at 20 gal. It would normally serve very little purpose to measure the volume of the tank down to the last pint. In special circumstances, perhaps a check on gasoline consumption, the tank volume may have to be known to the accuracy of the nearest pint.

To measure the tank volume much more precisely may not even be practical. If we wanted the tank volume accurate to a fluid dram, we would have to take into serious consideration the fact that the tank volume actually changes because of temperature, fullness, seams, and many other factors. In other words, the volume may actually be continuously changing to the extent of several fluid drams, and the last few fluid drams would essentially be meaningless. The precision with which measurements are made should therefore reflect the physical nature of the object being measured and the use to which the measurement will be put.

In chemical work in particular, it is important to note that measurements become more costly as they become more precise, not only in terms of the time necessary to make the measurement but also in terms of equipment. Any measurement should therefore be a careful balance of precision, need, utility, and speed.

Since the precision of a measurement is so important, some means is needed of having a measurement indicate its precision. This is done with the concept of *significant figures*. Suppose a group of 10 people were given micrometers and asked to measure the outside diameter of a 1-in standard pipe. Their measurements might be reported as follows:

1.310	1.300
1.304	1.305
1.300	1.310
1.308	1.303
1.305	1.301

It can be seen that for any one measurement, some uncertainty is associated with the last digit. Since the variation among these measurements is 0.01 unit (i.e., the difference between the highest and the lowest measurements) it is apparent that the last digit can vary in any one of the measurements. The last digit in any one of the above measurements is therefore not significant. To indicate the reliability of his measurement, each observer should therefore have reported only in three-digit numbers, i.e., in numbers having three significant digits:

1.31	1.30
1.30	1.31
1.30	1.31
1.31	1.30
1.31	1.30

The position of the decimal point has no influence in establishing significance. What is important is that the significance of a digit conveys the information that it is reasonably reliable.

TABLE 1-9   Exponential Numbers and Manipulations

Ordinary form	Exponential form
1	1×10^0
10	1×10 (or 1×10^1)
100	1×10^2
1,000	1×10^3
10,000	1×10^4
0.1	1×10^{-1}
0.01	1×10^{-2}
0.001	1×10^{-3}
0.001	1×10^{-4}

To multiply numbers in exponential form, add the exponents:

$$(1 \times 10^2) \times (1 \times 10^3) = 1 \times 10^5 = 100,000$$
$$(1 \times 10^2) \times (1 \times 10^{-3}) = 1 \times 10^{-1} = 0.1$$

To divide, subtract exponents:

$$(1 \times 10^2) \div (1 \times 10^3) = 1 \times 10^{-1} = 0.1$$
$$1 \div (1 \times 10^2) = (1 \times 10^0) \div (1 \times 10^2) = 1 \times 10^{-2}$$
$$= 0.01$$

Notice that $\dfrac{1}{1 \times 10^2} = 1 \times 10^{-2}$

Since each of the above (original) measurements is independent, their average should be more reliable than any one measurement. Since the original measurements disagree in the last two digits, we cannot give weight to these digits. The average, 1.3046, being more reliable than any one measurement, can be assigned one more (but no more than one more) significant digit. The average should thus be reported as 1.305.

It should be noted that we had to drop one digit from the average as not being significant. This action is known as *rounding off* the number. By convention, if the digits to be rounded off are halfway or more to the next higher significant number, the number is rounded off to the next higher significant number. If it is less than halfway, the number is rounded off to the next lower significant number. A few typical cases are illustrated below:

51.4233 may be rounded off to 51.42
3.09   may be rounded off to 3.1
0.6371 may be rounded off to 0.64
11.3853 may be rounded off to 11.39
11.395 may be rounded off to 11.40

The digit zero needs special consideration in determining whether or not it is significant. Final zeros after decimal points are significant digits and indicate to which decimal place the measurements are reliable. A weight of 3.20 g indicates that the measurement is reliable to 0.01 g, whereas a weight of 3.200 g indicates reliability to 0.001 g.

When a number has no decimal suffix, final zeros may or may not be significant. In this case, the context in which the number is used will usually indicate the number of significant digits. If the distance between two points is given as 10 mi, the final zero may or may not be significant, and it is necessary to know more about how the measurement was made. If the distance is given as 1,100 mi, the final two zeros are probably not significant. It should be noted that final zeros are not significant but are important in that they still serve to locate the decimal point.

In some instances where a number ends in several zeros, the number of significant digits can be conveniently indicated by expressing the number in exponential form and having the appropriate number of significant digits in the preexponential factor. Thus, expressing 1,100 mi in exponential form as 1.1×10^3 mi indicates that the figure is good for two significant digits.

For numbers less than 1 expressed in decimal form, the zeros after the decimal point and preceding other digits are not significant. Thus, 0.0081 g has two significant digits, 8 and 1. Since 0.0081 g equals 8.1 mg, and we still have the same significant digits, the number of significant digits should not be changed in converting a measurement from conventional to exponential form.

Example 1-19 How many significant figures are there in each of the following numbers? Identify them: 0.0386; 38,600; 386; 0.038600; and 3.86?
solution 0.0386 has 3 significant digits: 3, 8, and 6. 38,600 has 5 significant figures if we mean exactly 38,600: 3,8,6,0, and 0. If we mean about 38,600, then it has 3 significant figures: 3,8, and 6. 386 has 3 significant figures: 3,8, and 6. 0.038600 has 5 significant figures: 3,8,6,0, and 0. 3.86 has 3 significant figures: 3,8, and 6.

Example 1-20 Express 3.600 oz in grams.
solution 3.600 oz $\times$ 28.35 g/oz = 102.06000 g. However, since the original number 3.600 has only four significant figures (3,6,0, and 0), the same measurement expressed in grams should have four significant figures. The proper answer is thus 102.1 g (noting that the last digit was rounded off).

An important rule to note is that when carrying out any sequence of multiplications or divisions, the answer should have no more significant digits than the smallest number of significant figures present in any of the numbers involved.

Example 1-21 Carry out the following indicated operations: $0.0321 \times 6 \times 3.11 \times 1.235$.
solution By slide rule, we read 4850. Locating the decimal point by the method of approximation, we get $0.03 \times 5 \times 3 \times 1 = 0.45$, so that the answer is 0.4850. However, since the factor 6 has only 1 significant digit, the answer should have only 1 significant figure. The proper answer (rounded off) would then be 0.5.

Example 1-22 In 1 min, 5,037 drops of oil issue from a special nozzle under a particular set of conditions. How many drops would issue in 1 h?

solution

$$\frac{5{,}037 \text{ drops}}{\text{min}} \times \frac{60 \text{ min}}{\text{h}} = 302{,}220 \text{ drops/h}$$

In this case, the number 60 is exact and should really be written as 60.00000. . . , in effect having an infinite number of significant figures. By common usage, and convenience, it is written simply as 60. Since 5,037 has four significant digits, the answer should be given as 302,200 drops/h.

In addition and subtraction, the significant digits of the sum or difference are determined by the least significant-digit place of any of the numbers.

Example 1-23 What is the sum of 362.47, 10.2, and 0.9?
solution
$$
\begin{array}{r}
362.47 \\
10.2 \\
0.9 \\
\hline
373.57
\end{array}
$$

However, since the lowest significant-digit place is the tenths place in 362.4, the answer should have significance only to the tenths place. The answer is thus 373.6, after rounding off 0.57 to 0.6.

Example 1-24 Add 26,300 and 47.1.
solution
$$
\begin{array}{r}
26{,}300 \\
47.1 \\
\hline
26{,}347.1
\end{array}
$$

Since 26,300 is significant only to the hundreds place, the answer must be no more precise. The answer, therefore, is 26,300 (after rounding off 347.1 to 300).

In Example 1-24, it is interesting to note that compared with 26,300, the 47.1 has no significance, since 26,300 in itself could be as high as 26,400 and as low as 26,200.

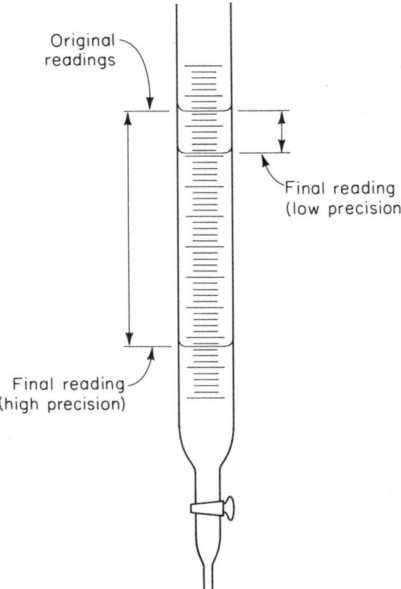

Original readings

Final reading (low precision)

Final reading (high precision)

Fig. 1-2 Accuracy of measurements by differences. The volume determination on the left is more precise than the one on the right even though all readings of the burette are taken with the same care.

Example 1-25 A dried filter and the solids it has removed from a liquid weigh 3.0743 g. The filter itself weighs 3.0308 g. What is the weight of the solids?
solution
$$
\begin{array}{r}
3.0743 \text{ g} \\
-3.0308 \text{ g} \\
\hline
0.0435 \text{ g}
\end{array}
$$

Example 1-25 also brings out an important fact, namely, that the difference between two numbers of similar magnitude has fewer significant figures than the numbers. In the example, 3.0743 and 3.0308 have five significant figures, whereas their difference has only three. Where a quantity is measured by the difference between two other measurements, especially when the two measurements are similar, as in the example above, the precision of the difference must be carefully considered. Figure 1-2 illustrates a practical application of this in which a volume of liquid is determined by the difference between an initial volume and a final volume in a calibrated tube called a *buret*. When the two readings are too similar, as on the right in the illustration, the accuracy is less than when the two readings are substantially dissimilar, as on the left.

1-3   PRECISION OF MEASUREMENTS

Distribution of Errors

The concept of significant figures infers that the last significant digit is useful but not entirely reliable. The question therefore arises as to just how reliable the last figure really is and how we can make it more reliable. One concept that tries to answer these questions involves taking a large number of observations and averaging them, on the theory that the various measurements will have both positive and negative errors, so that on the average the errors will cancel each other out, and the average value will be close to the true value. If we take a large number of observations and plot the frequency with which any single value appears, we get a curve showing the distribution of the observations. When we take observations that are not prejudiced by any intentional or unintentional biasing, we find that a graph of the frequency with which each measurement occurs against the actual measurement produces a bell-shaped curve which is symmetrical around the true value. This is called a *normal-distribution curve,* and is illustrated in Fig. 1-3.

One of the properties of a normal-distribution curve is that a vertical line drawn through the peak of the curve is equal to the average value of the observations $\bar{X}$. When the number of observations is very large, the average is represented by the Greek letter mu (μ), as in Figs. 1-3 and 1-4, i.e., $\mu = \bar{X}$. Naturally, those observations which are closest to the true value will occur more frequently than those which are somewhat farther away, and because the errors can be on either side of the true value, the curve is symmetrical about the true value.

When the measurements are made with great care using instruments that are as precise as possible, we can expect that most of the observations will lie close to the true value, whereas those which are somewhat remote from the true value will occur only a very small number of times. Thus a curve of observations made with great care is likely to look like that in Fig. 1-3, whereas a curve of observations made with less care or with less precise instruments is more likely to look like the curve shown in Fig. 1-4. Two important methods are used to describe the breadth of a normal-distribution curve and thereby the precision and reliability with which the measurements are made. These two devices are known as the average deviation and the standard deviation.

Average Deviation

The difference between any measured value and the average value is known as the deviation for that particular value. If $\bar{X}$ is the average value and X is any measurement, then $X - \bar{X}$ is the deviation for that particular measurement. Since deviations show how far a particular value lies from the average, there is no particular significance in whether a deviation is

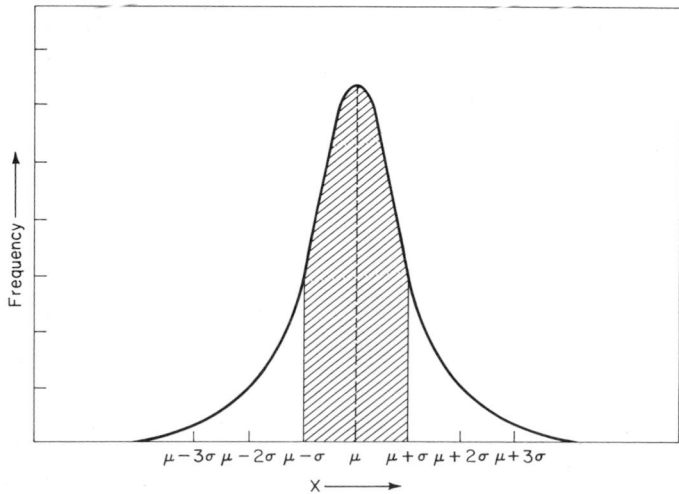

Fig. 1-3 Normal-distribution curve with small standard deviation.

positive or negative, and we are interested only in the numerical value of the deviation. The average deviation is then simply the average of all the absolute values of the individual deviations, as shown in Eq. (1-1):

$$\text{Average deviation} = \frac{(X_1 - \bar{X}) + (X_2 - \bar{X}) + (X_3 - \bar{X}) + \cdots + (X_n - \bar{X})}{n} \qquad (1\text{-}1)$$

It should be remembered that where $X_1 - \bar{X}$ is negative (i.e., $\bar{X}$ is larger than X_1), the positive value of the difference is used in calculating the average deviation. The subscripts 1,2,3, and so on, represent individual observations: the subscript n refers to the nth or last observation. Occasions of this type, involving the sum of a number of terms, can be more simply written with the Greek letter sigma (Σ), which in mathematical usage means "the sum of." The average value $\bar{X}$ can thus be expressed in simplified form as in Eq. (1-2), which can be read as "the average value, $\bar{X}$, is the sum of all the values of X divided by the number of values."

$$\bar{X} = \frac{\Sigma X}{n} \qquad (1\text{-}2)$$

In a similar way, the average deviation can be written as in Eq. (1-3).

$$\text{Average deviation} = \frac{\Sigma(X - \bar{X})}{n} \qquad (1\text{-}3)$$

Example 1-26 Six individual measurements were made of the period of a pendulum. These were 6.87, 6.84, 6.88, 6.88, 6.85, and 6.83 s. What are the average value and average deviation of each measurement?

solution

Measurement (X)	Deviation ($X - \bar{X}$)
6.87 s	0.01 s
6.84	0.02
6.88	0.02
6.88	0.02
6.85	0.01
6.83	0.03
$X = \overline{41.15}$	$(X - \bar{X}) = \overline{0.11}$

$$\bar{X} = \frac{\Sigma X}{n} = \frac{41.15}{6} = 6.86 \text{ s}$$

$$\text{Average deviation} = \frac{\Sigma(X - \bar{X})}{n} = \frac{0.11}{6} = 0.02 \text{ s}$$

In Example 1-26, it is interesting to note that the average is closer to 6.858 s, but this was rounded off to 6.86 in order to indicate the number of significant figures properly.

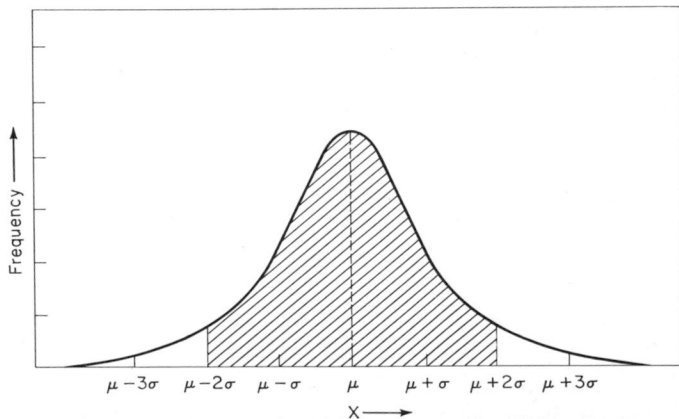

Fig. 1-4 Normal-distribution curve with large standard deviation.

Since there were only a limited number of observations, the average is probably not the true value. If a large number of observations were made, the average would, of course, be much closer to the true value; and in the limit for an infinite number of measurements, the average will be exactly equal to the true value. A report of the average value and the average deviation gives a fairly good description of the reliability of the measurements together with an indication of the care and precision with which they were carried out.

Standard Deviation

Perhaps an even better way of estimating and reporting the reliability of a set of measurements is given by the *standard deviation*. The standard deviation S is defined as the square root of the variance of the set of observations. In turn, the variance is defined by Eq. (1-4).

$$\text{Variance} = \frac{(X_1 - \bar{X})^2 + (X_2 - \bar{X})^2 + (X_3 - \bar{X})^2 + \cdots + (X_n - \bar{X})^2}{n - 1} \tag{1-4}$$

Equation (1-4) can, of course, be written in short form as in Eq. (1-5):

$$\text{Variance} = \frac{\Sigma(X - \bar{X})^2}{n - 1} \tag{1-5}$$

The variance is almost the same as the average of the squares of the deviations of each individual measurement, but it is not quite the same, since the denominator in Eqs. (1-4) or (1-5) is $n - 1$ instead of n. This has the effect of indicating that a small number of measurements is considerably less accurate than a large number. For example, if there are only 6 measurements, the value of $n - 1$ is 5, and this is substantially different from 6; but if there are a large number of measurements, say 100, the value of $n - 1$ is 99, and this is not very far different from 100. Thus the denominator in Eqs. (1-4) or (1-5) has the effect of giving greater credibility to large numbers of observations. Since the standard deviation is the square root of the variance, the standard deviation is given by Eq. (1-6).

$$\text{Standard deviation} = S = \sqrt{\frac{\Sigma(X - \bar{X})^2}{n - 1}} \tag{1-6}$$

Example 1-27 Calculate the standard deviation of the observations given in Example 1-26.
solution From the deviations, we can write $(X - \bar{X})^2$ directly;

$$
\begin{array}{l}
0.0001 \text{ s}^2 \\
0.0004 \\
0.0004 \\
0.0004 \\
0.0001 \\
0.0009 \\
\hline
0.0023 = \Sigma(X - \bar{X})^2
\end{array}
$$

$$\text{Standard deviation} = S = \sqrt{\frac{\Sigma(X - \bar{X})^2}{n - 1}} = \sqrt{\frac{0.0023}{6 - 1}} = \sqrt{\frac{0.0023}{5}} = 0.00021 \text{ s}$$

In order to make the value of S more useful in certain calculations, as will be shown shortly, it is customary to write S to one more significant figure than those of the measurements from which it was derived.

The usefulness of the standard deviation is derived from the mathematical properties of a normal-distribution curve. The generalized equation of such a curve, called a gaussian-distribution equation, is shown below.

$$F = \frac{1}{\sigma(2\pi)^{1/2}} e^{-(X-\mu)^2/2\sigma^2} \tag{1-7}$$

In this equation, F is the frequency of any particular observation, and the lowercase Greek letter sigma (σ) is the standard deviation. The larger the number of observations n, the more precisely will the equation describe the distribution.

At the peak of the curve, when $X = \mu$,

$$F = \frac{1}{\sigma(2\pi)^{1/2}}$$

for which we see that the maximum height is determined only by the value of σ (since 2π is a constant). If σ is small, because the errors are large, F for values near μ will be small, and the normal-distribution curve will be broad, as in Fig. 1-4.

The value of σ is also related to the *confidence level* of the observations. The confidence level relates to the fraction of observations that lie within a given distance from μ. It turns out that 68.3 percent of all the observations have values of X which lie between $\mu - \sigma$ and $\mu + \sigma$. This is represented by the shaded area under the curve in Fig. 1-3. For observations that follow a normal-distribution curve, the confidence level that the observations will fall between $\mu - \sigma$ and $\mu + \sigma$ is therefore 0.683. For the interval $\mu - 2\sigma$ and $\mu + 2\sigma$, the confidence level is 0.954, as illustrated by the shaded area under the curve in Fig. 1-4; and for the interval $\mu - 3\sigma$ and $\mu + 3\sigma$ it is 0.997. Thus, for the latter interval, 99.7 percent of the observations will lie between $\mu - 3\sigma$ and $\mu + 3\sigma$.

Most practical observations are limited in number, so that some measure of the confidence level is needed when only a limited number of observations is available. This has been worked out mathematically and summarized in t tables of the type given in Table 1-10. Here the percent confidence that a given observation lies within the interval $\mu - t\sigma$ and $\mu + t\sigma$ is shown for various sample sizes or observations. Note that as the sample size gets large, the confidence level approaches values derived from the mathematical equation of the normal-distribution curve.

Example 1-28 What is the confidence interval for a set of 20 measurements for which a confidence level of 80 percent is desired, given that the true value is 827 and the standard deviation is 4?
solution From Table 1-10, we select the value of t from the column headed 80 percent and the row corresponding to $n = 20$. This value is 1.328, and the confidence interval is 827 ± 1.328 (4) $= 827 \pm 5$; i.e., 80 percent of the observations, at the σ given, will lie between 822 and 832.

1-4 DENSITY AND SPECIFIC GRAVITY

Although a large variety of derived units and measurements are in use, those important in chemical technology are relatively few in number. In describing materials physically, density or specific gravity is one of the most important characteristics. The density of a substance is defined as its mass per unit volume. The density of a substance can be obtained, therefore, by simply measuring the mass and volume of a representative sample and applying the formula $D = M/V$. In the case of a regularly shaped material, the required information can be obtained rather easily; but in the case of an irregular object, indirect measurements are often more accurate. To obtain the volume of an irregular object, the volume of a liquid it displaces is commonly measured. In this procedure, the volume of a liquid is observed in a graduated cylinder before and after submersion of the object in the liquid. The difference between the two readings gives the volume of the object.

Example 1-29 A cylindrical rod of polyethylene 2.54 in in diameter and 12.70 cm long weighs 61.8 g. What are the density and specific gravity of the polyethylene?

TABLE 1-10   *t* Values for Various Sample Sizes and Confidence Levels

Sample size	% confidence level						
	50	60	70	80	90	95	99
2	1.000	1.376	1.963	3.078	6.314	12.706	63.657
3	0.816	1.061	1.386	1.886	2.920	4.303	9.925
4	0.765	0.978	1.250	1.638	2.353	3.182	5.841
5	0.741	0.941	1.190	1.533	2.132	2.776	4.604
6	0.727	0.920	1.150	1.476	2.015	2.571	4.032
7	0.718	0.906	1.134	1.440	1.943	2.447	3.707
8	0.711	0.896	1.119	1.415	1.895	2.365	3.499
9	0.706	0.889	1.108	1.397	1.860	2.306	3.355
10	0.703	0.883	1.100	1.383	1.833	2.262	3.250
20	0.688	0.861	1.066	1.328	1.729	2.093	2.861
30	0.683	0.854	1.055	1.311	1.699	2.045	2.756
40	0.681	0.851	1.050	1.303	1.684	2.021	2.704
50	0.680	0.849	1.048	1.299	1.676	2.008	2.678
60	0.679	0.848	1.046	1.296	1.671	2.000	2.660
120	0.677	0.845	1.041	1.289	1.658	1.980	2.617
∞	0.674	0.842	1.036	1.282	1.645	1.968	2.576

solution

$$\text{Volume} = \frac{\pi d^2}{4} \times h = \frac{\pi(2.54)^2 \text{ cm}^2}{4} \times 12.70 \text{ cm} = 64.4 \text{ cm}^3$$

$$\text{Density} = \frac{61.8 \text{ g}}{64.4 \text{ cm}^3} = 0.96 \text{ g/cm}^3$$

$$\text{Specific gravity} = \frac{\text{density of sample}}{\text{density of water}} = \frac{0.96 \text{ g}}{\text{cm}^3} \times \frac{\text{cm}^3}{1.0 \text{ g}} = 0.96$$

Example 1-30 A graduated cylinder is partially filled with n-propyl alcohol having a density of 0.78 g/cm³. An irregular price of sulfur weighing 113.91 g is then immersed in the liquid, causing its level to rise. If the original and final liquid levels are 19.2 and 74.2 cm³, respectively, what is the density of sulfur?

solution

$$\text{Volume of sample} = 74.2 - 19.2 = 55.0 \text{ cm}^3$$

$$\text{Density} = \frac{113.91 \text{ g}}{55.0 \text{ cm}^3} = 2.07 \text{ g/cm}^3$$

(Note that even though the weight is given to five significant figures, the volume is known to only three significant figures and the density must therefore be reported to only three significant figures.)

In this particular method, care must be taken to have the object completely wetted by the liquid in which it is immersed. If any air bubbles or voids in the object are not displaced or filled by the liquid, an erroneously high value will be obtained for the volume of the object. For this reason, liquids with a very low surface tension, such as kerosene or alcohol, are frequently used for this kind of measurement.

The measurement of the volume displaced by a solid by observing the rise of a liquid level in a graduated cylinder can sometimes be rather inaccurate, since it may depend upon the difference between two similar numbers. The accuracy of such measurements can be improved by having the liquid displaced overflow into another receptacle in which its volume can be determined more accurately. Alternatively, the overflow can be weighed, and by knowing the density of the liquid, it is possible to calculate back to the volume displaced. These methods are illustrated in Fig. 1-5.

One of the most accurate methods of determining the volume of an irregular object is to weigh it both in air and submerged in a liquid of known density. By Archimedes' principle, the apparent loss in weight of the object in air and submerged is equal to the weight of the displaced liquid. Again, knowing the density of the liquid used, it is possible

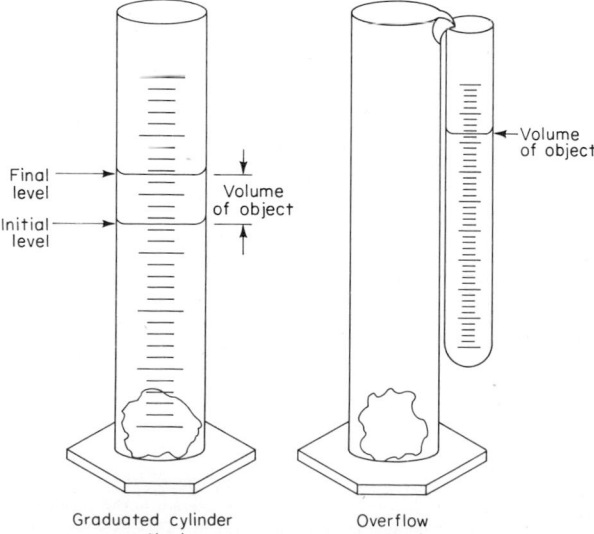

Fig. 1-5 Measurement of volume of an irregular object.

to calculate the volume of the object with great accuracy. A typical determination of this type is shown in Fig. 1-6.

Example 1-31 A piece of sulfur weighs 6.88 g in air and 3.56 g when submerged in water. What is the density of sulfur?

solution The loss in weight when submerged, 6.88 g − 3.56 g = 3.32 g, is equal to the weight of water displaced. Since the density of water is 1 g/cm³, this is also equal to the volume of water displaced, which in turn is equal to the volume of the sulfur. The density of sulfur is then 6.88 g/3.32 cm³ = 2.07 g/cm³.

Example 1-32 A glass test tube having a volume of 25.61 cm³ is weighted with lead shot so that it floats upright in toluene. Fine sand is then slowly and carefully added to the test tube until toluene spills over the top and the test tube sinks. If the weighted test tube weighs 13.38 g and the weight of the added sand is 8.62 g, what is the density of toluene?

solution The weight of the test tube plus sand is equal to the weight of toluene displaced: 13.38 g + 8.62 g = 22.00 g. The volume displaced is 25.61 cm³, and the density of toluene is therefore 22.00 g/25.61 cm³ = 0.8669 g/cm³.

The measurement of the density of a liquid may involve some special techniques. Among the most accurate methods, the use of a *pycnometer* is preferred for simplicity and convenience. A pycnometer is a glass vessel of carefully measured volume which is fitted with a capillary overflow tube. To make a determination with a pycnometer, the weight of the empty vessel is obtained, and then the vessel is filled with the liquid. When the stopper is placed in position, some of the liquid spills out through the capillary. After any material on the outside is cleaned off, the pycnometer is weighed again. The fact that the spill-off occurred through the capillary tube provides a very accurate means of reproducing the volume of a liquid, since the volume of the capillary is extremely small compared with the total volume of the pycnometer. Thus any small variations in the level of the liquid in the capillary do not have any significant influence on the reproducibility of the volume of the liquid. A conventional pycnometer with an integral thermometer is shown in Fig. 1-7. Since liquids expand or contract with a rise or fall in temperature, their densities will vary with temperature, and measuring the temperature at which the density is determined is of considerable importance.

Example 1-33 A pycnometer having a volume of 25.90 cm³ weighs 32.44 g empty. When filled with road oil, it weighs 55.06 g. What is the density of the road oil?

solution The weight of road oil is 55.06 g − 32.44 g = 22.62 g. Its density is therefore 22.62/25.90 = 0.875 g/cm³.

Another convenient way of measuring the density of a liquid is a modification of the method of determining the volume of an irregular object by Archimedes' principle. In

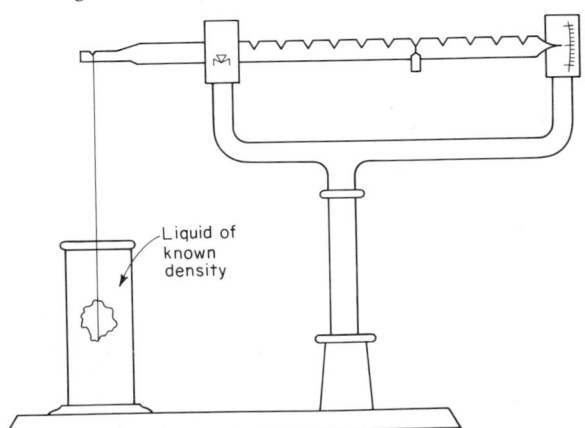

Fig. 1-6 Volume of an irregular object by displacement method. The difference between the weight of the object as measured in air and when submerged is equal to the weight of the liquid displaced. The volume of the object is therefore equal to the weight of the liquid displaced divided by the density of the liquid.

this particular case, a reverse procedure is used in that an object of known volume is immersed in the liquid whose density is desired. The loss in weight of the object is, as before, related to the density of the liquid and the volume of the object, again providing a very accurate means for determining the density of the liquid.

Archimedes' principle also affords one of the most convenient means of measuring the density of a liquid. For a floating body, the weight of liquid displaced is equal to the weight of the body. If the body is floating in a low-density liquid, a greater volume of liquid will have to be displaced than when the body is floating in a high-density liquid. Thus a greater portion of a given body will be submerged in a low-density liquid than in a high-density liquid. A device designed to indicate readily how much of a body is submerged when it is floating in a liquid is called a *hydrometer*, a typical form of which is shown in Fig. 1-8. Because the upper stem is thin, small differences in the density of the test liquid will cause a considerable change in the amount of the stem that is submerged, so that a hydrometer can be made not only to read directly in the density of the liquid in which it is floating but to do so with a high degree of accuracy.

The density of gases is somewhat more difficult to measure because the weight of a reasonable volume of gas is extremely small. The most direct method of measuring the density of gas is by the use of the Dumas, or gas-density, flask shown in Fig. 1-9. This is a rather large flask having a small capillary-bore stopcock as a means of entry into the flask. In use, the flask is weighed after evacuation and again after filling with the gas whose density is desired. The volume of the flask is usually obtained by filling with a liquid of known density. Other methods of measuring gas density are less direct and depend upon measurements of rates of flow through capillaries and other similar procedures. These will be discussed elsewhere, as they depend on more advanced information.

Where the gas can be liquefied, however, other direct methods are available. In this case, a weighed quantity of liquid is allowed to vaporize in a confined space, and the volume of the resultant vapor can thereby be obtained. This procedure is known as the *Hofmann method*. A modification of the Hofmann apparatus, shown in Fig. 1-10, is known as *Victor Meyer's apparatus*. In the Victor Meyer procedure, the volume of the vapor is not measured

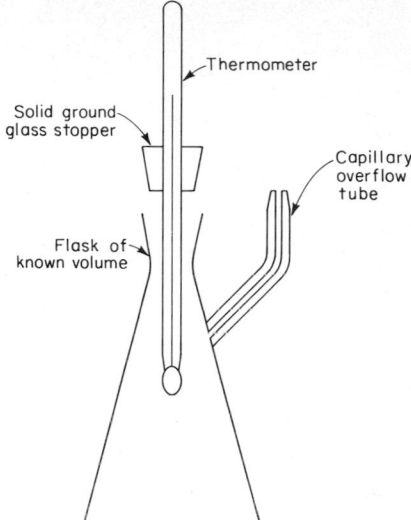

Fig. 1-7 Pycnometer for measurement of density of liquids.

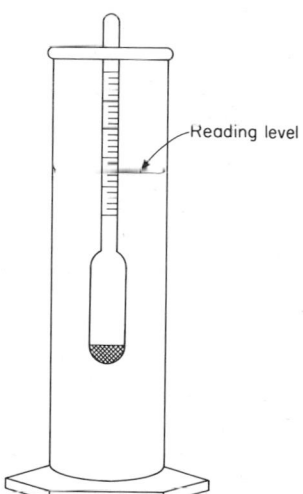

Fig. 1-8 Hydrometer method for measurement of liquid density. As long as the hydrometer floats, the weight of liquid displaced is equal to the weight of the hydrometer. Therefore, the volume of displaced liquid need not be as great for a dense liquid as for a light one and the hydrometer will ride higher in the liquid. The scale showing how high the hydrometer rides in the liquid can therefore be calibrated to read density directly.

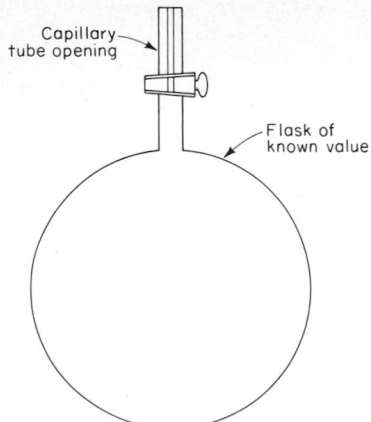

Fig. 1-9   Gas-density flask.

directly, but rather causes a displacement of an equal volume of air, and it is the volume of this air that is actually measured. This method is more convenient than the Hofmann procedure and is used where extreme accuracy is not required.

In all measurements of gas and vapor densities, the pressure and temperature of the gas when the measurements are taken are extremely important, since the volume of a gas is strongly affected by its temperature and pressure. Consequently, in all determinations of gas densities, the temperature and pressure of the gas when its density measurement is made must also be carefully measured. Since the range of temperatures and pressures under which such measurements can be made is so great, the density of a gas is usually reported at a particular set of conditions known as

normal temperature and pressure (NTP) or *standard temperature and pressure* (STP). Normal temperature and pressure are taken as 22°C and 760 mm of mercury (i.e. 1 atmosphere pressure). Standard temperature and pressure are taken as 0°C and 760 mm of mercury. Knowing the density of a gas at either NTP or STP, its density can be calculated at any other temperature and pressure using the gas laws to be discussed later.

In comparing the densities of two substances, it is frequently convenient to use the concept of specific gravity. The specific gravity of a solid or liquid is the ratio of the density of that substance to the density of water. Since liquids, like gases, are affected by temperature and pressure (although not to nearly the same degree), the density of water for most specific-gravity measurements is taken as 1 g/cm³. This is the maximum density of water under normal atmospheric pressures and occurs at a temperature of 4°C. It can be seen, therefore, that the specific gravity of a solid or liquid is also the ratio of the weight of the substance to the weight of an equal volume of water.

In the case of the specific gravity of a gas, this ratio is extremely small, and consequently the specific gravity of a gas is compared with that of an equal volume of air. Since all gases are not affected by temperature and pressure in precisely the same way, the specific gravity of the gas is usually referred to air at STP or NPT. Generally speaking, however, for normal ambients the density of air and most other gases will be affected by changes in temperature and pressure in about the same way, so that the specific gravity of a gas will not change very much with either pressure or temperature. However, variations in the density of a gas compared with that of air can become extremely significant at high temperatures and high pressures, and gas specific gravities should be used with caution at such conditions.

Tables 1-11 and 1-12 give the densities and specific gravities of common solids, liquids, and gases.

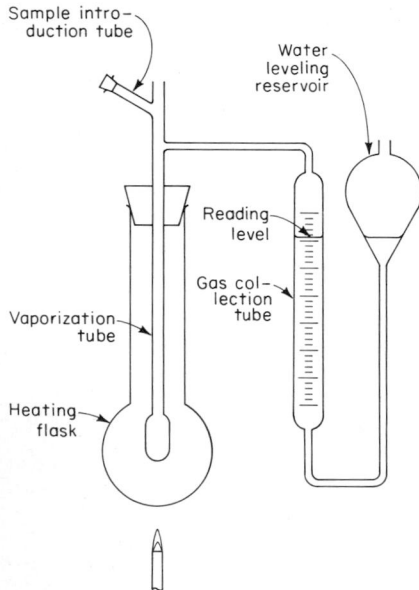

Fig. 1-10   Measurement of density of a vapor by Victor Meyer method. A weighed sample of liquid is introduced into the vaporization tube, where it is gently vaporized by heat from the heating flask. The volume of air displaced by the vapor is transferred to the gas-collection tube by raising or lowering the water-level reservoir and its volume is read on the calibrated gas-collection tube.

TABLE 1-11   Physical Constants of Elements and Compounds

Abbreviations: A Reference specific gravity air = 1; D Reference specific gravity hydrogen = 1; anh.—anhydrous; d.—decompose; expl.—explodes; ign.—ignition; rhb.—rhombic; subl.—sublime; tr.—transition; vac.—vacuum.

INORGANIC

Name	Formula	Formula weight	Specific gravity	Melting point, °C.	Boiling point, °C.
Aluminum................................	Al	26.97	2.70	660	2056
acetate, normal..................	Al(C₂H₃O₂)₃	204.10	....	d. 200	....
acetate, basic....................	Al(OH)(C₂H₃O₂)₂	162.07	....	d.	....
bromide.............................	AlBr₃	266.72	3.01	97.5	268
bromide.............................	AlBr₃.6H₂O	374.82	....	d. 100	....
carbide..............................	Al₄C₃	143.91	2.95	d. >2200	....
chloride.............................	AlCl₃	133.34	2.44	194⁵·²ᵃᵗᵐ·	182.7 subl. 178
chloride.............................	AlCl₃.6H₂O	241.44	....	d.	....
fluoride (fluellite)...............	AlF₃.H₂O	101.99	2.17	d.	....
fluoride.............................	Al₂F₆.7H₂O	294.05	....	−4H₂O, 120 −2H₂O, 300	−6H₂O, 250
hydroxide..........................	Al(OH)₃	77.99	2.42		
nitrate..............................	Al(NO₃)₃.9H₂O	375.14	....	73	d. 134
nitride..............................	Al₂N₂	81.96	3.05	2150⁴ᵃᵗᵐ·	d. >1400
oxide................................	Al₂O₃	101.94	3.99	1999 to 2032	
oxide (corundum)................	Al₂O₃	101.94	4.00	1999 to 2032	2210
phosphate.........................	AlPO₄	121.95	2.59	....	....
potassium silicate (muscovite)........	3Al₂O₃.K₂O.6SiO₂.2H₂O	796.40	2.9	d.	
potassium silicate (orthoclase)........	Al₂O₃.K₂O.6SiO₂	556.49	2.56	1450 (1150)	....
Aluminum potassium tartrate..........	AlK(C₄H₄O₆)₂	362.21	....	1000	
sodium fluoride (cryolite)............	AlF₃.3NaF	209.96	2.90	1000	
sodium silicate....................	Al₂O₃.Na₂O.6SiO₂	524.29	2.61	1100	
sulfate..............................	Al₂(SO₄)₃	342.12	2.71	d. 770	
Alum, ammonium (tschermigite)........	Al₂(SO₄)₃.(NH₄)₂SO₄.24H₂O	906.64	1.64	93.5	−20H₂O, 120; −24H₂O, 200
ammonium chrome..................	Cr₂(SO₄)₃.(NH₄)₂SO₄.24H₂O	956.72	1.72	100 d.	
ammonium iron.......................	Fe₂(SO₄)₃.(NH₄)₂SO₄.24H₂O	964.40	1.71	40	....
potassium (kalinite)...............	Al₂(SO₄)₃.K₂SO₄.24H₂O	948.76	1.76	92	−18H₂O, 64.5
potassium chrome..................	Cr₂(SO₄)₃.K₂SO₄.24H₂O	998.84	1.83	89	....
sodium..............................	Al₂(SO₄)₃.Na₂SO₄.24H₂O	916.56	1.675	61	....
Ammonia..............................	NH₃	17.03	0.817	−77.7	−33.4
Ammonium acetate...................	NH₄C₂H₃O₂	77.08	1.073	114	d.
auricyanide........................	NH₄CN.Au(CN)₃.H₂O	337.33	....	d. 200	....
bicarbonate.......................	NH₄HCO₃	79.06	1.573	d. 35–60	....
bromide.............................	NH₄Br	97.96	2.327	subl. 542	....
carbonate..........................	(NH₄)₂CO₃.H₂O	114.11	....	d. 58	....
carbonate, carbamate.............	NH₄HCO₃.NH₂CO₂NH₄	157.11	....	subl.	....
carbonate, sesqui-.................	(NH₄)₂CO₃.2NH₄HCO₃.H₂O	272.22	....	d.	....
chloride (salammoniac)...........	NH₄Cl	53.50	1.53	d. 350	subl. 520
chloroplatinate...................	(NH₄)₂PtCl₆	444.05	3.065	d.	....
chloroplatinite...................	(NH₄)₂PtCl₄	373.14	....	d.	....
chlorostannate....................	(NH₄)₂SnCl₆	367.52	2.4		
chromate...........................	(NH₄)₂CrO₄	152.09	1.917	d. 180	....
cyanide.............................	NH₄CN	44.06	....	36	....
dichromate.........................	(NH₄)₂Cr₂O₇	252.10	2.15	d. 185	....
ferrocyanide.......................	(NH₄)₄Fe(CN)₆.6H₂O	392.21	....	d.	....
fluoride.............................	NH₄F	37.04	....		....
fluoride, acid.....................	NH₄F.HF	57.05	2.21	....	....
formate.............................	HCO₂NH₄	63.06	1.266	114–116	d. 180; subl. in vac.
hydrosulfide.......................	NH₄HS	51.11	....	d.	subl. 120
hydroxide..........................	NH₄OH	35.05	....		
molybdate..........................	(NH₄)₂MoO₄	196.03	2.27	d.	....
molybdate, hepta-.................	(NH₄)₆Mo₇O₂₄.4H₂O	1235.95	....	....	....
nitrate (α), stable −16° to 32°.......	NH₄NO₃	80.05	1.66	169.6	d. 210
nitrate (β), stable 32° to 84°........	NH₄NO₃	80.05	1.725	....	d. 210
nitrite..............................	NH₄NO₂	64.05	1.69	expl.	....
osmochleride......................	(NH₄)₂OsCl₆	439.02	2.93		
oxalate.............................	(NH₄)₂C₂O₄.H₂O	142.12	1.501	d.	....
oxalate, acid......................	NH₄HC₂O₄.H₂O	125.08	1.556		....

Adapted from Robert H. Perry and Cecil H. Chilton (eds.), "Chemical Engineers Handbook," 5th ed., McGraw-Hill Book Company, New York, 1973. Melting points and boiling points shown are at normal pressures and temperatures. Other pressures or temperatures shown as superscripts.

TABLE 1-11 Physical Constants of Elements and Compounds—(Continued)

Name	Formula	Formula weight	Specific gravity	Melting point, °C	Boiling point, °C
Ammonium perchlorate...............	NH_4ClO_4	117.50	1.95	d.	....
persulfate........................	$(NH_4)_2S_2O_8$	228.20	1.98	d. 120	....
phosphate, monobasic...............	$NH_4H_2PO_4$	115.04	1.803	....	....
phosphate, dibasic.................	$(NH_4)_2HPO_4$	132.07	1.619	....	....
phosphate, meta-...................	$(NH_4)_4P_4O_{12}$	388.08	2.21	....	....
phosphomolybdate..................	$(NH_4)_3PO_4.12MoO_3.3H_2O$	1930.55	....	d.	....
silicofluoride.....................	$(NH_4)_2SiF_6$	178.14	2.01	....	subl.
sulfamate.........................	$NH_4.SO_3NH_2$	114.12	....	132	d. 160
sulfate (mascagnite)...............	$(NH_4)_2SO_4$	132.14	1.769	513 d.	....
sulfate, acid.....................	NH_4HSO_4	115.11	1.78	146.9	490
sulfide...........................	$(NH_4)_2S$	68.14	....	d.	....
sulfide, penta-...................	$(NH_4)_2S_5$	196.38	....	....	....
sulfite...........................	$(NH_4)_2SO_3.H_2O$	134.16	1.41	d.	....
sulfite, acid.....................	NH_4HSO_3	99.11	2.03	d.	....
tartrate..........................	$(NH_4)_2C_4H_4O_6$	184.15	1.60	d.	....
thiocyanate.......................	NH_4CNS	76.12	1.305	149.6	d. 170
vanadate, meta-...................	NH_4VO_3	116.99	2.326	d.	....
Antimony..........................	Sb	121.76	6.684	630.5	1380
chloride, tri- (butter of antimony).....	$SbCl_3$	228.13	3.14	73.4	220.2
oxide, tri- (valentinite)..........	Sb_2O_3	291.52	5.67	656	1570
oxide, tri- (senarmontite).........	Sb_2O_3	291.52	5.2	652	....
sulfide, tri- (stibnite)...........	Sb_2S_3	339.70	4.64	550	....
sulfide, penta-...................	Sb_2S_5	403.82	$4.120^{0°}$	d. 135	....
telluride, tri-...................	Sb_2Te_3	626.35	....	629	....
Antimonyl potassium tartrate (tartar emetic)....................	$(SbO)KC_4H_4O_6.\frac{1}{2}H_2O$	333.94	2.60	$-\frac{1}{2}H_2O$, 100	....
sulfate, normal..................	$(SbO)_2SO_4$	371.58	4.89	....	....
sulfate, basic...................	$(SbO)_2SO_4.Sb_2(OH)_4$	683.13	....	....	....
Argon.............................	A	39.94	$1.65^{-233°}$; $1.402^{-185.7°}$; 1.38 (A)	-189.2	-185.7
Arsenic (crystalline)(α).........	As_4	299.64	5.727	$814^{36atm.}$	subl. 615
Arsenic (black)(β).............	As_4	299.64	4.7	....	....
Arsenic (yellow)(γ)............	As_4	299.64	2.0	d. 358	....
acid, ortho-......................	$H_3AsO_4.\frac{1}{2}H_2O$	150.94	2.0–2.5	35.5	$-H_2O$, 160
acid, meta-.......................	$HAsO_3$	123.92	....	d.	....
acid, pyro-.......................	$H_4As_2O_7$	265.85	....	d. 206	....
pentoxide.........................	As_2O_5	229.82	4.086	....	d.
sulfide, di- (realgar)............	As_2S_2	213.94	$(\alpha)3.506$ $(\beta)3.254$	$(\alpha)tr. 267$; $(\beta)307$	565
sulfide, penta-...................	As_2S_5	310.12	....	....	d. 500
Arsenious chloride (butter of arsenic).....	$AsCl_3$	181.28	lq. 2.163	-18	130
hydride (arsine)..................	AsH_3	77.93	2.695 (A)	-113.5	-55; d. 230
oxide (arsenolite)................	As_2O_3	197.82	3.865	subl.	....
oxide (claudetite)................	As_2O_3	197.82	3.85	subl.	....
oxide.............................	As_2O_3	197.82	3.738	315	....
Auric chloride....................	$AuCl_3.2H_2O$	339.60	....	d.	....
cyanide...........................	$Au(CN)_3.6H_2O$	383.35	....	d. 50	....
Aurous chloride...................	$AuCl$	232.66	7.4	d. 170	d. 290
cyanide...........................	$AuCN$	223.22	....	d.	....
Cf. also under *Gold*					
Barium............................	Ba	137.36	3.5	850	1140
acetate...........................	$Ba(C_2H_3O_2)_2$	255.45	2.468	....	....
acetate...........................	$Ba(C_2H_3O_2)_2.H_2O$	273.46	2.19	$-H_2O$, 41	....
bromide...........................	$BaBr_2$	297.19	4.781	847	d.
bromide...........................	$BaBr_2.2H_2O$	333.22	3.69	$-2H_2O$, 100	d.
carbonate (witherite).............	$BaCO_3$	197.37	4.29	tr. 811 to α	d. 1450
carbonate (α)...............	$BaCO_3$	197.37	....	tr. 982 to β	....
carbonate (β)...............	$BaCO_3$	197.37	....	1740^{90atm}	....
chlorate..........................	$Ba(ClO_3)_2$	304.27	....	414	....
chlorate..........................	$Ba(ClO_3)_2.H_2O$	322.29	3.179	d. 120	....
chloride..........................	$BaCl_2$	208.27	3.856	tr. 925	1560
chloride..........................	$BaCl_2$	208.27	....	962	1560
chloride..........................	$BaCl_2.2H_2O$†	244.31	3.097	$-2H_2O$, 100	....
hydroxide.........................	$Ba(OH)_2$	171.38	4.495	....	....
hydroxide.........................	$Ba(OH)_2.8H_2O$	315.50	2.188	77.9	$-8H_2O$, 550
nitrate (nitrobarite).............	$Ba(NO_3)_2$	261.38	3.244	592	d.
oxalate...........................	BaC_2O_4	225.38	2.658	....	....
oxide.............................	BaO	153.36	5.72	1923	2000
peroxide..........................	BaO_2	169.36	4.958	d. 800	....
peroxide..........................	$BaO_2.8H_2O$	313.49	....	$-8H_2O$, 100	....
phosphate, monobasic..............	$BaH_4(PO_4)_2$	331.35	$2.94^°$	....	....
phosphate, dibasic................	$BaHPO_4$	233.35	4.165	....	....
phosphate, tribasic...............	$Ba_3(PO_4)_2$	602.04	4.1	....	....
phosphate, pyro-..................	$Ba_2P_2O_7$	448.68	3.9	....	....
silicofluoride....................	$BaSiF_6$	279.42	4.279	....	....
sulfate (barite, barytes).........	$BaSO_4$	233.42	4.499	1580 d.	....

TABLE 1-11 Physical Constants of Elements and Compounds—(Continued)

Name	Formula	Formula weight	Specific gravity	Melting point, °C.	Boiling point, °C.
Barium sulfide, mono-..................	BaS	169.42	4.25	....	....
sulfide, tri-............................	BaS₃	233.54		d. 400	
sulfide, tetra-.........................	BaS₄.2H₂O	301.63	2.988	d. 200	
Beryllium (glucinum)....................	Be(Gl)	9.02	1.816	1284	2767
Bismuth	Bi	209.00	9.80	271	1450
carbonate, sub-.......................	Bi₂O₃.CO₂.H₂O	528.03	6.86	d.	
chloride, di-...........................	BiCl₂	279.91	4.86	163	d. 300
chloride, tri-..........................	BiCl₃	315.37	4.75	230	447
nitrate................................	Bi(NO₃)₃.5H₂O	485.10	2.82	d. 30	−5H₂O, 80
nitrate, sub-..........................	BiONO₃.H₂O	305.02	4.928	d. 260	
oxide, tri-.............................	Bi₂O₃	466.00	8.9	820	1900
oxide, tri-.............................	Bi₂O₃	466.00	8.55	860	
oxide, tri-.............................	Bi₂O₃	466.00	8.20	tr. 704	
oxychloride............................	BiOCl	260.46	7.72	....	
Boric acid...............................	H₃BO₃	61.84	1.435	185 d.	
Boron....................................	B	10.82	2.32	2300	2550
carbide................................	B₄C	55.29	2.54	2450	>3500
oxide.................................	B₂O₃	69.64	1.85	577	>1500
oxide (sassolite)......................	B₂O₃.3H₂O	123.69	1.49	d.	
Bromic acid..............................	HBrO₃	128.92		d. 100	
Bromine..................................	Br₂	159.83	3.119 5.87 (Λ)	−7.2	58.78
hydrate................................	Br₂.10H₂O	339.99		d. 6.8	
Cadmium..................................	Cd	112.41	8.65	320.9	767
acetate................................	Cd(C₂H₃O₂)₂	230.50	2.341	256	d.
acetate................................	Cd(C₂H₃O₂)₂.2H₂O	266.53	2.01	−H₂O, 130	
carbonate..............................	CdCO₃	172.42	4.258	d. <500	
chloride...............................	CdCl₂	183.32	4.047	568	960
chloride...............................	CdCl₂.2½H₂O	228.36	3.327	tr. 34	
cyanide................................	Cd(CN)₂	164.45	....	d. >200	
hydroxide..............................	Cd(OH)₂	146.43	4.79	d. 300	
nitrate................................	Cd(NO₃)₂	236.43		350	
nitrate................................	Cd(NO₃)₂.4H₂O	308.49	2.455	59.4	132
oxide..................................	CdO	128.41	8.15		
oxide..................................	CdO	128.41	6.95	d. 900–1000	
oxide, sub-............................	Cd₂O	240.82	8.192	d.	
Cadmium sulfate.......................	CdSO₄	208.47	4.691	1000	
sulfate................................	CdSO₄.H₂O	226.49	3.786	tr. 108	
sulfate................................	3CdSO₄.8H₂O	769.54	3.09	tr. 41.5	
sulfate................................	CdSO₄.4H₂O	280.53	3.05	....	
sulfate................................	CdSO₄.7H₂O	334.50	2.40	tr. 4	
sulfide (greenockite)..................	CdS	144.47	4.58	1750¹⁰⁰ᵃᵗᵐ	subl. in N₂, 980
Calcium..................................	Ca	40.08	1.55	810	1200
acetate................................	Ca(C₂H₃O₂)₂.H₂O	176.18		d.	
aluminate..............................	Ca(AlO₂)₂	158.02	3.67	1600	
aluminum silicate (anorthite).........	CaO.Al₂O₃.2SiO₂	278.14	2.765	1551	
arsenate...............................	Ca₃(AsO₄)₂	398.06			
bromide................................	CaBr₂	199.91	3.353	760	810
carbonate (aragonite).................	CaCO₃	100.09	2.93	d. 825	
carbonate (calcite)...................	CaCO₃	100.09	2.711	1339¹⁰³ᵃᵗᵐ	
chloride (hydrophilite)...............	CaCl₂	110.99	2.152	772	>1600
chloride...............................	CaCl₂.H₂O	129.01			
chloride...............................	CaCl₂.6H₂O	219.09	1.68	29.92	−6H₂O, 200
citrate................................	Ca₃(C₆H₅O₇)₂.4H₂O	570.50		−2H₂O, 130	−4H₂O, 185
cyanamid...............................	CaCN₂	80.11		....	
ferrocyanide...........................	Ca₂Fe(CN)₆.12H₂O	508.31	1.7		
fluoride (fluorite)....................	CaF₂	78.08	3.180	1330	
formate................................	Ca(HCO₂)₂	130.12	2.015	d.	
hydride................................	CaH₂	42.10	1.7	d. 675	
hydroxide..............................	Ca(OH)₂	74.10	2.2	−H₂O, 580	
hypochlorite...........................	Ca(ClO)₂.4H₂O	215.06		d.	
hypophosphate..........................	Ca₂P₂O₆.2H₂O	274.15		−2H₂O, 200	
lactate................................	Ca(C₃H₅O₃)₂.5H₂O	308.30		−3H₂O, 100	
magnesium carbonate (dolomite).......	CaO.MgO.2CO₂	184.42	2.872	d. 730–760	
magnesium silicate (diopside)........	CaO.MgO.2SiO₂	216.52	3.3	1391	
nitrate (nitrocalcite)................	Ca(NO₃)₂	164.10	2.36	561	
nitrate................................	Ca(NO₃)₂.4H₂O	236.16	1.82	42.7	
nitride................................	Ca₃N₂	148.26	2.63	900	
nitrite................................	Ca(NO₂)₂.H₂O	150.11	2.23		
oxalate................................	CaC₂O₄	128.10	2.2	d.	
oxalate................................	CaC₂O₄.H₂O	146.12	2.2	−H₂O, 200	

TABLE 1-11   Physical Constants of Elements and Compounds—(Continued)

Name	Formula	Formula weight	Specific gravity	Melting point, °C.	Boiling point, °C.
Calcium oxide................	CaO	56.08	3.32	2570	
peroxide................	CaO₂.8H₂O	216.21	....	−8H₂O, 100	expl. 275
phosphate, monobasic...............	CaH₄(PO₄)₂.H₂O	252.09	2.220	−H₂O, 100	d. 200
phosphate, dibasic..............	CaHPO₄.2H₂O	172.10	2.306	d.	....
phosphate, tribasic...............	Ca₃(PO₄)₂	310.20	3.14	1670	....
phosphate, meta-............	Ca(PO₃)₂	198.04	2.82	975	....
phosphate, pyro-............	Ca₂P₂O₇	254.12	3.09	1230	....
phosphate, pyro- (brushite)..........	Ca₂P₂O₇.5H₂O	344.20	2.25	....	....
phosphide................	Ca₃P₂	182.20	2.51	>1600]	....
silicate (α) (pseudowollastonite)........	CaSiO₃	116.14	2.905	1540	
silicate (β) (wollastonite).............	CaSiO₃	116.14	2.915	tr. 1190 to α	
sulfate (anhydrite)................	CaSO₄	136.14	2.96		tr. 1193 to rhb.
sulfate (gypsum).................	CaSO₄.2H₂O	172.17	2.32	−1½H₂O, 128	−2H₂O, 163
sulfhydrate	Ca(SH)₂.6H₂O	214.31	....	d. 15	
sulfide (oldhamite)................	CaS	72.14	2.8	....	....
sulfite................	CaSO₃.2H₂O	156.17	....	−2H₂O, 100	d. 650
tartrate................	CaC₄H₄O₆.4H₂O	260.22	....	d.	....
thiocyanate................	Ca(CNS)₂.3H₂O	210.28	....		
thiosulfate................	CaS₂O₃.6H₂O	260.30	1.873	d.	....
tungstate (scheelite)...	CaWO₄	288.00	6.06		
Carbon, Cf. table of organic compounds					
Carbon, amorphous................	C	12.01	1.8–2.1	>3500	4200
Carbon, diamond................	C	12.01	3.51	>3500	4200
Carbon, graphite................	C	12.01	2.26	>3500	4200
dioxide................	CO₂	44.01	lq. 1.101⁻³⁷°; 1.53 (A); solid 1.56⁻⁷⁹°	−56.6⁵·²ᵃᵗᵐ.	subl. −78.5
disulfide................	CS₂	76.13	1.261	−108.6	46.3
monoxide................	CO	28.01	lq. 0.814⁻¹⁹⁵°; 0.968 (A)	−207	−192
oxychloride (phosgene)...............	COCl₂	98.92	1.392	−104	8.2
oxysulfide................	COS	60.07	lq. 1.24⁻⁸⁷°;	−138.2	−50.2
suboxide................	C₃O₂	68.03	lq. 1.114⁰°	−107	7
thionyl chloride................	CSCl₂	114.98	1.509	....	73.5
Ceric hydroxide................	2CeO₂.3H₂O	398.31	....		
hydroxynitrate................	Ce(OH)(NO₃)₃.3H₂O	397.21	....		
oxide................	CeO₂	172.13	7.3	1950	....
sulfate................	Ce(SO₄)₂.4H₂O	404.31	3.91	....	....
Cerium................	Ce	140.13	6.9 cb.; 6.7 hex.	645	1400
Cerous sulfate................	Ce₂(SO₄)₃	568.44	3.91	....	....
sulfate................	Ce₂(SO₄)₃.8H₂O	712.57	2.886	−8H₂O, 630	
Cesium................	Ce	132.91	1.90	28.5	670
Chloric acid................	HClO₃.7H₂O	210.58	1.282	<−20	d. 40
Chlorine................	Cl₂	70.91	lq. 1.56⁻³³·⁶°; 2.49⁰° (A)	−101.6	−34.6
hydrate................	Cl₂.8H₂O	215.04	1.23	d. 9.6	
Chloroplatinic acid................	H₂PtCl₆.6H₂O	518.08	2.431	60	....
Chlorostannic acid................	H₂SnCl₆.6H₂O	441.55	1.971	19.2	....
Chlorosulfonic acid................	HO.SO₂.Cl	116.52	1.787	−80	151.5
Chromic acetate................	Cr₂(C₂H₃O₂)₆.2H₂O	494.32	....		
chloride................	CrCl₃	158.38	2.757		1200–1500 d.
chloride................	CrCl₃.6H₂O	266.48	1.835	subl. 83	....
fluoride................	CrF₃	109.01	3.8	>1000	d.
hydroxide................	Cr(OH)₃	103.03	....	....	....
hydroxide................	Cr(OH)₃.2H₂O	139.07	....	−2H₂O, 100	....
nitrate................	Cr(NO₃)₃.9H₂O	400.18	....	36.5	d. 100
nitrate................	Cr(NO₃)₃.7½H₂O	373.15	....	100	d.
oxide................	Cr₂O₃	152.02	5.21	1900	....
sulfate................	Cr₂(SO₄)₃	392.20	3.012	....	....
sulfate................	Cr₂(SO₄)₃.5H₂O	482.28	....	....	....
sulfate................	Cr₂(SO₄)₃.15H₂O	662.44	1.867	100	−10H₂O, 100
sulfate................	Cr₂(SO₄)₃.18H₂O	716.49	1.7		−12H₂O, 100
sulfide................	Cr₂S₃	200.02	3.77	d. 1350	
Chromium................	Cr	52.01	7.1	1615	2200
trioxide (chromic acid)...............	CrO₃	100.01	2.70	197 d.	....
Chromous chloride................	CrCl₂	122.92	2.75		....
hydroxide................	Cr(OH)₂	86.03	....	d.	....
oxide................	CrO	68.01	....		....
sulfate................	CrSO₄.7H₂O	274.18	....		....
sulfide (daubrelite)................	CrS	84.07	3.97	1550	....
Chromyl chloride................	CrO₂Cl₂	154.92	1.92	−96.5	117.6

TABLE 1-11   Physical Constants of Elements and Compounds—(Continued)

Name	Formula	Formula weight	Specific gravity	Melting point, °C.	Boiling point, °C.
Cobalt	Co	58.94	8.9	1480	2900
carbonyl	Co(CO)₄	170.98	1.73	51	d. 52
sulfide, di-	CoS₂	123.06	4.269		
Cobaltic chloride	CoCl₃	165.31	2.94	subl.	
chloride, dichro	Co(NH₃)₃Cl₃.H₂O	234.42			
chloride, luteo	Co(NH₃)₆Cl₃	267.50	1.7016		
chloride, praseo	Co(NH₃)₄Cl₃.H₂O	251.46	1.847		
Cobaltic chloride, purpureo	Co(NH₃)₅Cl₃	250.47	1.819		
chloride, roseo	Co(NH₃)₅Cl₃.H₂O	268.49		d. 100	
hydroxide	Co(OH)₃	109.96		−1½H₂O, 100	
oxide	Co₂O₃	165.88	5.18	d. 900	
sulfate	Co₂(SO₄)₃	406.06			
sulfide	Co₂S₃	214.06	4.8		
Cobalto-cobaltic oxide	Co₃O₄	240.82	6.07		
Cobaltous acetate	Co(C₂H₃O₂)₂.4H₂O	249.09	1.7053	−4H₂O, 140	
chloride	CoCl₂	129.85	3.356	subl.	1049
chloride	CoCl₂.6H₂O	237.95	1.924	86	−6H₂O, 110
nitrate	Co(NO₃)₂.6H₂O	291.05	1.883	<100	d.
oxide	CoO	74.94	5.68	d. 1800	
sulfate	CoSO₄	155.00	3.710	d. 880	
sulfate	CoSO₄.H₂O	173.02	3.13	d.	
sulfate (bieberite)	CoSO₄.7H₂O	281.11	1.948	96.8	−7H₂O, 420
sulfide (syeporite)	CoS	91.00	5.45	>1100	
Copper	Cu	63.57	8.92	1083	2300
Cupric acetate	Cu(C₂H₃O₂)₂	181.66	1.930		
acetate	Cu(C₂H₃O₂)₂.H₂O	199.67	1.882	115	240 d.
aceto-arsenite (Paris green)	(CuOAs₂O₃)₃. Cu(C₂H₃O₂)₂	1013.83			
ammonium chloride	CuCl₂.2NH₄Cl.2H₂O	277.51	1.90	d. 110	
ammonium sulfate	CuSO₄.4NH₃.H₂O	245.77	1.81	d. 150	
carbonate, basic (azurite)	2CuCO₃.Cu(OH)₂	344.75	3.88	d. 220	
carbonate, basic (malachite)	CuCO₃.Cu(OH)₂	221.17	3.9	d.	
chloride (eriochalcite)	CuCl₂	134.48	3.054	498	Forms Cu₂Cl₂ 993
chloride	CuCl₂.2H₂O	170.52	2.39	−2H₂O, 110	d.
chromate, basic	CuCrO₄.2CuO.2H₂O	574.75		2H₂O, 260	
cyanide	Cu(CN)₂	115.61		d.	
dichromate	CuCr₂O₇.2H₂O	315.62	2.286	−2H₂O, 100	
ferricyanide	Cu₃[Fe(CN)₆]₂	614.63			
ferrocyanide	Cu₂Fe(CN)₆.7H₂O	465.21			
formate	Cu(HCO₂)₂	153.61	1.831		
hydroxide	Cu(OH)₂	97.59	3.368	H₂O	
lactate	Cu(C₃H₅O₃)₂.2H₂O	277.74			
nitrate	Cu(NO₃)₂.3H₂O	241.63	2.047	114.5	−HNO₃, 170
nitrate	Cu(NO₃)₂.6H₂O	295.68	2.074	−3H₂O, 26.4	
oxide (paramelaconite)	CuO	79.57	6.40	d. 1026	
oxide (tenorite)	CuO	79.57	6.45	d. 1026	
oxychloride	CuCl₂.2CuO.4H₂O	365.69		−3H₂O, 140	
phosphide	Cu₃P₂	252.67	6.35	d.	
sulfate (hydrocyanite)	CuSO₄	159.63	3.606	d. >600	Forms CuO, 650
sulfate (blue vitriol or chalcanthite)	CuSO₄.5H₂O*	249.71	2.286	−4H₂O, 110	−5H₂O, 250
sulfide (covellite)	CuS	95.63	4.6	tr. 103	d. 220
tartate	CuC₄H₄O₆.3H₂O	265.69		d.	
Cuprous ammonium iodide	CuI.NH₄I.H₂O	353.47			
carbonate	Cu₂CO₃	187.15	4.4	d.	
chloride (nantokite)	Cu₂Cl₂	198.05	3.53	422	1366
cyanide	Cu₂(CN)₂	179.16	2.9	474.5	d.
ferricyanide	Cu₃Fe(CN)₆	402.67			
ferrocyanide	Cu₄Fe(CN)₆	466.24			
fluoride	Cu₂F₂	165.14		908	subl. 1100
hydroxide	CuOH	80.58	3.4	−½H₂O, 360	
oxide (cuprite)	Cu₂O	143.14	6.0	1235	d. 1800
Cuprous phosphide	Cu₆P₂	443.38	6.4 to 6.8		
sulfide (chalcocite)	Cu₂S	159.20	5.6	1100	
sulfide	Cu₂S	159.20	5.80	1130	
Cyanogen	C₂N₂	52.02	lq. 0.866⁻¹⁷·²°; 1.806 (A)	−34.4	−20.5
Cyanogen compounds, *Cf.* table of organic compounds					
Ferric acetate, basic	Fe(OH)(C₂H₃O₂)₂	190.95			
ammonium sulfate, *Cf.* Alum					
chloride (molysite)	FeCl₃	162.22	2.804	282	315
chloride	FeCl₃.6H₂O	270.32		37	280
ferrocyanide (Prussian blue)	Fe₄[Fe(CN)₆]₃	859.27		d.	
hydroxide	Fe(OH)₃	106.87	3.4 to 3.9	−1½H₂O, 500	
lactate	Fe(C₃H₅O₃)₃	323.06			
nitrate	Fe(NO₃)₃.6H₂O	349.97	1.684	35	d.

TABLE 1-11 Physical Constants of Elements and Compounds—(Continued)

Name	Formula	Formula weight	Specific gravity	Melting point, °C.	Boiling point, °C.
Ferric acetate, oxide (hematite)	Fe_2O_3	159.70	5.12	1560 d.	....
sulfate	$Fe_2(SO_4)_3$	399.88	3.097	d. 480	....
sulfate (coquimbite)	$Fe_2(SO_4)_3.9H_2O$	562.02	2.1	....	....
Ferroso-ferric chloride	$FeCl_2.2FeCl_3.18H_2O$	775.49	....	d. 50	....
ferricyanide (Prussian green)	$Fe'''_4Fe''_3[Fe(CN)_6]_6$	1662.70	....	d. 180	....
oxide (magnetite; magnetic iron oxide)	Fe_3O_4	231.55	5.2	1538 d.	....
oxide, hydrated	$Fe_3O_4.4H_2O$	303.61	....	d.	....
Ferrous ammonium sulfate	$FeSO_4.(NH_4)_2SO_4.6H_2O$	392.15	1.864	d.	....
chloride (lawrencite)	$FeCl_2$	126.76	2.7	....	....
chloroplatinate	$FePtCl_6.6H_2O$	571.92	2.714	....	....
ferricyanide (Turnbull's blue)	$Fe_3[Fe(CN)_6]_2$	591.47	....	d.	....
ferrocyanide	$Fe_2Fe(CN)_6$	323.66	....	....	....
formate	$Fe(HCO_2)_2.2H_2O$	181.92	....	d.	....
hydroxide	$Fe(OH)_2$	89.87	3.4	....	....
nitrate	$Fe(NO_3)_2.6H_2O$	287.96	....	60.5	....
oxide	FeO	71.85	5.7	1420	....
phosphate (vivianite)	$Fe_3(PO_4)_2.8H_2O$	501.64	2.58	....	....
silicate	$FeSiO_3$	131.91	3.5	1550	....
sulfate (siderotilate)	$FeSO_4.5H_2O$	241.99	2.2	....	$-5H_2O$, 300
sulfate (copperas)	$FeSO_4.7H_2O$	278.02	1.899	64	$-7H_2O$, 300
sulfide	FeS	87.91	4.84	1193	d.
Cf. also under iron					
Fluoboric acid	HBF_4	87.83	....	....	130 d.
Fluorine	F_2	38.00	lq. $1.51^{-187°}$; 1.31 (A)	-223	-187
Fluosilicic acid	H_2SiF_6	144.08	....	....	....
Gadolinium	Gd	156.9	....	....	....
Gallium bromide	$GaBr_3$	309.47	....	....	....
Glucinum *Cf.* Beryllium					
Gold	Au	197.20	19.3	1063	2600
Gold, colloidal	Au	197.20	....	....	....
Gold salts *Cf.* under Auric and Aurous					
Hafnium	Hf	178.6	12.1	>1700	>3200
Helium	He	4.00	0.1368 (A)	<-272.2	-268.9
Hydrazine	N_2H_4	32.05	1.011	1.4	113.5
formate	$N_2H_4.2HCO_2H$	124.10	....	128	....
hydrate	$N_2H_4.H_2O$	50.06	1.03	-40	118.5
hydrochloride	$N_2H_4.HCl$	68.51	....	....	....
hydrochloride, di-	$N_2H_4.2HCl$	104.98	1.42	198	....
nitrate	$N_2H_4.HNO_3$	95.06	....	70.7	subl. 140
nitrate, di-	$N_2H_4.2HNO_3$	158.08	....	104	d.
sulfate	$N_2H_4.\frac{1}{2}H_2SO_4$	81.09	....	85	....
sulfate	$N_2H_4.H_2SO_4$	130.12	1.378	254	....
Hydrazoic acid (azoimide)	HN_3	43.03	....	-80	37
Hydriodic acid	HI	127.93	$4.4^{0°}$ (A)	-50.8	-35.5
Hydriodic acid	$HI.H_2O$	145.94	1.7	....	127
Hydriodic acid	$HI.2H_2O$	163.96	....	-43	....
Hydriodic acid	$HI.3H_2O$	181.98	....	-48	....
Hydriodic acid	$HI.4H_2O$	199.99	....	-36.5	....
Hydrobromic acid	HBr	80.92	$2.71^{0°}$ (A)	-86	-67
Hydrobromic acid	$HBr.H_2O$	98.94	1.78	....	....
Hydrobromic acid (47.8% in H_2O)	HBr	80.92	1.486	....	126
Hydrobromic acid	$HBr.2H_2O$	116.96	$2.11^{-15°}$	-11	....
Hydrochloric acid	HCl	36.47	$1.268^{0°}$ (A)	-111	-85
Hydrochloric acid (45.2% in H_2O)	HCl	36.47	1.48	-15.35	....
Hydrochloric acid	$HCl.2H_2O$	72.50	1.46^{-18}	0	d.
Hydrochloric acid	$HCl.3H_2O$	90.51	....	-24.4	d.
Hydrocyanic acid (prussic acid)	HCN	27.03	0.697	-14	26
Hydrofluoric acid	HF	20.01	0.988	-83	19.4
Hydrofluoric acid (35.35% in H_2O)	HF	20.01	1.15	-35	120
Hydrogen	H_2	2.016	lq. $0.0709^{-252.7°}$ 0.06948 (A)	-259.1	-252.7
peroxide	H_2O_2	34.02	1.438	-0.89	151.4
selenide	H_2Se	81.22	$2.12^{-42°}$	-64	-42
sulfide	H_2S	34.08	1.1895 (A)	-82.9	-59.6
Hydroxylamine	NH_2OH	33.03	1.35	34	56.5^{22mm}
hydrochloride	$NH_2OH.HCl$	69.50	1.67	151	d.
nitrate	$NH_2OH.HNO_3$	96.05	....	48	d. <100
sulfate	$NH_2OH.\frac{1}{2}H_2SO_4$	82.07	....	170 d.	....
Hypobromous acid	$HBrO$	96.92	....	....	40^{50mm}
Illinium	Il	146(?)	....	....	....
Indium	In	114.76	7.3	155	1450
Iodic acid	HIO_3	175.93	4.629	110 d.	....
Iodine	I_2	253.84	4.93	113.5	184.35
oxide, penta-	I_2O_5	333.84	4.799	d. 300	....
Iodoplatinic acid	$H_2PtI_6.9H_2O$	1120.91	....	....	....
Iridium	Ir	193.10	22.4	2350	>4800

TABLE 1-11 Physical Constants of Elements and Compounds—(Continued)

Name	Formula	Formula weight	Specific gravity	Melting point, °C.	Boiling point, °C.
Iron, cast.............................	Fe	55.85	7.03	1275	....
pure.........................	Fe	55.85	7.86	1535	3000
steel...........................	Fe	55.85	7.6 to 7.8	1375	....
white pig....................	Fe	55.85	7.6 to 7.8	1075	....
wrought......................	Fe	55.85	7.86	1505	....
carbide (cementite)...............	Fe_3C	179.56	7.4	1837	....
carbonyl........................	$Fe(CO)_5$	195.90	1.457	−21	102.5
nitride........................	Fe_2N	125.71	6.35	d. >560	....
silicide........................	FeSi	83.91	6.1	....	....
sulfide, di- (marcasite).............	FeS_2	119.97	4.87	tr. 450	d.
sulfide, di- (pyrite)...............	FeS_2	119.97	5.0	1171	d.
sulfide (pyrrhotite)...............	Fe_7S_8	647.43	4.6	d. >700	....
Cf. also under ferric and ferrous					
Krypton..............................	Kr	83.70	2.818 (A)	−169	−151.8
Lanthanum............................	La	138.92	6.15	826	1800
Lead................................	Pb	207.21	11.337	327.5	1620
acetate........................	$Pb(C_2H_3O_2)_2$	325.30	3.251	280	....
acetate (sugar of lead)...............	$Pb(C_2H_3O_2)_2.3H_2O$	379.35	2.55	−3H_2O, 75	....
acetate........................	$Pb(C_2H_3O_2)_2.10H_2O$	505.46	1.689	22	....
acetate, basic..................	$Pb_2(C_2H_3O_2)_3OH$	608.56	....	....	....
acetate, basic..................	$Pb(C_2H_3O_2)_2.$ $Pb(OH)_2.H_2O$	584.54	....	....	....
acetate, basic..................	$Pb(C_2H_3O_2)_2.$ $2Pb(OH)_2$	807.75	....	....	....
arsenate, monobasic..............	$PbH_4(AsO_4)_2$	489.06	4.46	d. 140	....
arsenate, dibasic (schultenite).........	$PbHAsO_4$	347.13	5.94	d. >200	−H_2O, 280
arsenate, meta-..................	$Pb(AsO_3)_2$	453.03	6.42	....	....
arsenate, pyro-...................	$Pb_2As_2O_7$	676.24	6.85	802	....
Lead azide.........................	PbN_6	291.26	....	expl. 350	....
bromide.....................	$PbBr_2$	367.05	6.66	373	918
carbonate (cerussite)...............	$PbCO_3$	267.22	6.6	d. 315	....
carbonate, basic (hydrocerussite; white lead).....	$2PbCO_3.Pb(OH)_2$	775.67	6.14	d. 400	....
chloride (cotunnite).................	$PbCl_2$	278.12	5.80	501	954
chromate (crocoite)................	$PbCrO_4$	323.22	6.12	844	d.
chromate, basic.................	$PbCrO_4.PbO$	546.43	....	....	....
formate........................	$Pb(HCO_2)_2$	297.25	4.56	d. 190	....
hydroxide......................	$3PbO.H_2O$	687.65	7.592	−H_2O, 130	....
nitrate........................	$Pb(NO_3)_2$	331.23	4.53	d. 470	....
oxide, sub-....................	Pb_2O	430.42	8.34	d.	....
oxide, mono- (litharge)...........	PbO	223.21	9.53	888	....
oxide, mono (massicotite)...........	PbO	223.21	8.0	....	....
oxide, mono-..................	PbO	223.21	9.2 to 9.5	....	....
oxide, red (minium)................	Pb_3O_4	685.63	9.1	d. 500	....
oxide, sesqui-..................	Pb_2O_3	462.42	....	d. 360	....
oxide, di- (plattnerite)...........	PbO_2	239.21	9.375	d. 290	....
silicate.......................	$PbSiO_3$	283.27	6.49	766	....
sulfate (anglesite)................	$PbSO_4$	303.27	6.2	1170	....
sulfate, acid...................	$Pb(HSO_4)_2.H_2O$	419.36	....	d.	....
sulfate, basic (lanarkite)...........	$PbSO_4.PbO$	526.48	6.92	977	....
sulfide (galena).................	PbS	239.27	7.5	1120	....
thiocyanate.....................	$Pb(CNS)_2$	323.37	3.82	d. 190	....
Lithium............................	Li	6.94	0.53	186	1336
benzoate.......................	$LiC_7H_5O_2$	128.05	....	....	....
bromide.......................	LiBr	86.86	3.464	547	1265
bromide.......................	$LiBr.2H_2O$	122.89	....	44	....
carbonate......................	Li_2CO_3	73.89	2.11	618	d.
chloride.......................	LiCl	42.40	2.068	614	1360
citrate........................	$Li_3C_6H_5O_7.4H_2O$	281.98	....	d.	....
fluoride.......................	LiF	25.94	2.295	870	1670
formate.......................	$LiHCO_2.H_2O$	69.97	1.46	−H_2O, 94	....
hydride.......................	LiH	7.95	0.820	680	....
hydroxide......................	LiOH	23.95	2.54	445	925
hydroxide......................	$LiOH.H_2O$	41.96	1.83	....	d.
nitrate........................	$LiNO_3$	68.95	2.38	261	....
nitrate........................	$LiNO_3.3H_2O$	123.00	....	29.88	....
oxide.........................	Li_2O	29.88	2.013	....	subl. <1000
phosphate, monobasic.............	LiH_2PO_4	103.94	2.461	>100	....
phosphate, tribasic..............	Li_3PO_4	115.80	2.537	837	....
phosphate, tribasic..............	$Li_3PO_4.12H_2O$	331.99	1.645	100	....
salicylate......................	$LiC_7H_5O_3$	144.05	....	d.	....
sulfate........................	Li_2SO_4	109.94	2.22	860	....
sulfate........................	$Li_2SO_4.H_2O$	127.96	2.06	−H_2O, 130	....
sulfate, acid...................	$LiHSO_4$	104.01	2.123	170.5	....
Lutecium............................	Lu	174.99	....	....	....

TABLE 1-11 Physical Constants of Elements and Compounds—(Continued)

Name	Formula	Formula weight	Specific gravity	Melting point, °C.	Boiling point, °C.
Magnesium	Mg	24.32	1.74	651	1110
acetate	Mg(C₂H₃O₂)₂	142.41	1.42	323	....
acetate	Mg(C₂H₃O₂)₂.4H₂O	214.47	1.454	80	....
aluminate (spinel)	MgO.Al₂O₃	142.26	3.6	2135	....
ammonium chloride	MgCl₂.NH₄Cl.6H₂O	256.83	1.456	−4H₂O, 195	....
ammonium phosphate (struvite)	MgNH₄PO₄.6H₂O	245.44	1.715	d. 100	....
ammonium sulfate (boussingaultite)	MgSO₄.(NH₄)₂SO₄.6H₂O	360.62	1.72	>120	....
benzoate	Mg(C₇H₅O₂)₂.3H₂O	320.59	....	−3H₂O, 110]	....
carbonate (magnesite)	MgCO₃	83.43	3.037	d. 350	....
carbonate (nesquehonite)	MgCO₃.3H₂O	138.38	1.852	−H₂O, 100	....
carbonate, basic (hydromagnesite)	3MgCO₃.Mg(OH)₂.3H₂O	365.37	2.16	d.	....
Magnesium chloride (chloromagnesite)	MgCl₂	95.23	2.325	712	1412
chloride (bischofite)	MgCl₂.6H₂O	203.33	1.56	118 d.	d.
hydroxide (brucite)	Mg(OH)₂	58.34	2.4	d.	....
nitride	Mg₃N₂	100.98	....	d.	....
oxide (magnesia; periclase)	MgO	40.32	3.65	2800	3600
perchlorate	Mg(ClO₄)₂	223.23	2.60	d.	....
peroxide	MgO₂	56.32	....	expl. 275	....
phosphate, pyro-	Mg₂P₂O₇	222.60	2.598	1383	....
phosphate, pyro-	Mg₂P₂O₇.3H₂O	276.65	2.56	−3H₂O, 100	....
potassium chloride (carnallite)	MgCl₂.KCl.6H₂O	277.88	1.60	265	....
potassium sulfate (picromerite)	MgSO₄.K₂SO₄.6H₂O	402.73	2.15	d. 72	....
silicofluoride	MgSiF₆.6H₂O	274.48	1.788	d.	....
sodium chloride	MgCl₂.NaCl.H₂O	171.70	....	....	....
sulfate	MgSO₄	120.38	2.66	1185	....
sulfate (epsom salt; epsomite)	MgSO₄.7H₂O	246.49	1.68	70 d.	....
Manganese	Mn	54.93	7.2	1260	1900
acetate	Mn(C₂H₃O₂)₂	173.02	1.74	....	....
acetate	Mn(C₂H₃O₂)₂.4H₂O	245.08	1.589	....	....
carbonate (rhodocrosite)	MnCO₃	114.94	3.125	d.	....
chloride (scacchite)	MnCl₂	125.84	2.977	650	1190
chloride	MnCl₂.4H₂O	197.91	2.01	58.0	−H₂O, 106; −4H₂O, 200
chloride, per-	MnCl₄	196.76	....	....	....
hydroxide (ous) (pyrochroite)	Mn(OH)₂	88.95	3.258	d.	....
hydroxide (ic) (manganite)	Mn₂O₃.H₂O	175.88	3.258	d.	....
nitrate	Mn(NO₃)₂.6H₂O	287.04	1.82	25.8	129.5
oxide (ous) (manganosite)	MnO	70.93	5.18	1650	....
oxide (ic)	Mn₂O₃	157.86	4.81	d. 1080	....
oxide, di- (pyrolusite; polianite)	MnO₂	86.93	5.026	d. >230	....
sulfate (ous)	MnSO₄	150.99	3.235	700	d. 850
sulfate (ous) (szmikite)	MnSO₄.H₂O	169.01	2.87	Stable 57 to 117	....
sulfate (ous)	MnSO₄.2H₂O	187.02	2.526	Stable 40 to 57	....
sulfate (ous)	MnSO₄.3H₂O	205.04	2.356	Stable 30 to 40	....
sulfate (ous)	MnSO₄.4H₂O	223.05	2.107	Stable 18 to 30	−4H₂O, 450
sulfate (ous)	MnSO₄.5H₂O	241.07	2.103	Stable 8 to 18	....
sulfate (ous)	MnSO₄.6H₂O	259.09	....	Stable −5 to +8	....
sulfate (ous)	MnSO₄.7H₂O	277.10	2.092	Stable −10 to −5; 19 d.	−7H₂O, 280
sulfate (ic)	Mn₂(SO₄)₃	398.04	3.24	d. 160	....
Masurium	Ma	98–99.5	11.5	2300	....
Mercuric acetate	Hg(C₂H₃O₂)₂	318.70	3.270	d.	....
bromide	HgBr₂	360.44	6.053	237	322
carbonate, basic	HgCO₃.2HgO	693.84	....	....	....
chloride (corrosive sublimate)	HgCl₂	271.52	5.44	277	304
fulminate	Hg(CNO)₂	284.65	4.42	expl.	....
hydroxide	Hg(OH)₂	234.63	....	−H₂O, 175	....
oxide (montroydite)	HgO	216.61	11.14	d. 100	....
oxychloride (kleinite)	HgCl₂.3HgO	921.35	7.93	d. 260	....
silicofluoride, basic	HgSiF₆.HgO.3H₂O	613.33	....	....	....
sulfate	HgSO₄	296.67	6.47	d.	....
sulfate, basic (turpeth)	HgSO₄.2HgO	729.89	6.44	....	....
Mercurous acetate	HgC₂H₃O₂	259.65	....	d.	....
bromide	HgBr	280.53	7.307	subl. 345	....
carbonate	Hg₂CO₃	461.23	....	d. 130	....
chloride (calomel)	HgCl	236.07	7.150	302	383.7
iodide	HgI	327.53	7.70	290 d.	subl. 140; 310d.
nitrate	HgNO₃.H₂O	280.63	4.785	70	expl.
Mercurous oxide	Hg₂O	417.22	9.8	d. 100	....
sulfate	Hg₂SO₄	497.28	7.56	d.	....
Mercury	Hg	200.61	13.546	−38.87	356.9
Molybdenum	Mo	95.95	10.2	2620	3700
chloride, di-	MoCl₂	166.85	3.714	d.	....

TABLE 1-11   Physical Constants of Elements and Compounds—(Continued)

Name	Formula	Formula weight	Specific gravity	Melting point, °C	Boiling point, °C
Molybdenum chloride, tri-.............	MoCl₃	202.32	3.578	d.	....
chloride, tetra-........................	MoCl₄	237.78	....	....	d.
chloride, penta-.......................	MoCl₅	273.24	2.928	194	268
oxide, tri- (molybdite)...............	MoO₃	143.95	4.50	795	subl.
sulfide, di- (molybdenite)............	MoS₂	160.07	4.801	1185	....
sulfide, tri-..........................	MoS₃	192.13	....	d.	....
sulfide, tetra-.......................	MoS₄	224.19	....	d.	....
Molybdic acid.........................	H₂MoO₄	161.97	....	d. 115	....
Molybdic acid.........................	H₂MoO₄.H₂O	179.98	3.124	−H₂O, 70	−2H₂O, 200
Neodymium............................	Nd	144.27	6.9	840	....
Neon..................................	Ne	20.18	lq. 1.204⁻²⁴⁵·⁹° 0.674 (A)	−248.67	−245.9
Neptunium............................	Np²³⁹	239	....	....	....
Nickel................................	Ni	58.69	8.90	1452	2900
acetate...............................	Ni(C₂H₃O₂)₂	176.78	1.798	d.	....
ammonium chloride...................	NiCl₂.NH₄Cl.6H₂O	291.20	1.645	....	....
ammonium sulfate...................	NiSO₄.(NH₄)₂SO₄. 6H₂O	394.99	1.923	....	....
bromate...............................	Ni(BrO₃)₂.6H₂O	422.62	2.575	d.	....
bromide...............................	NiBr₂	218.52	4.64	d.	....
bromide..............................	NiBr₂.3H₂O	272.57	....	−3H₂O, 200	....
bromide, ammonia....................	NiBr₂.6NH₃	320.71	1.837	....	....
bromoplatinate......................	NiPtBr₆.6H₂O	841.51	3.715	....	....
carbonate............................	NiCO₃	118.70	....	d.	....
carbonate, basic.....................	2NiCO₃.3Ni(OH)₂. 4H₂O	587.58	....	d.	....
carbonyl..............................	Ni(CO)₄	170.73	1.31	−25	43
chloride...............................	NiCl₂	129.60	5.544	subl.	973
chloride...............................	NiCl₂.6H₂O	237.70	....	....	....
chloride, ammonia...................	NiCl₂.6NH₃	231.80	....	....	....
cyanide...............................	Ni(CN)₂.4H₂O	182.79	....	−4H₂O, 200	d.
dimethylglyoxime....................	NiC₈H₁₄O₄N₄	288.61	....	subl. 250	....
formate...............................	Ni(HCO₂)₂.2H₂O	184.76	2.154	d.	....
hydroxide (ic)........................	Ni(OH)₃	109.71	....	d.	....
hydroxide (ous)......................	Ni(OH)₂.¼H₂O	97.21	4.36	d.	....
nitrate................................	Ni(NO₃)₂.6H₂O	290.80	2.05	56.7	136.7
nitrate, ammonia....................	Ni(NO₃)₂.4NH₃.2H₂O	286.87	....	....	....
oxide, mono- (bunsenite)............	NiO	74.69	7.45	Forms Ni₂O₃ at 400	
potassium cyanide...................	Ni(CN)₂.2KCN.H₂O	258.97	1.875	−H₂O, 100	....
sulfate................................	NiSO₄	154.75	3.68	d. 840	....
sulfate................................	NiSO₄.6H₂O	262.85	2.07	tr. 53.3	−6H₂O, 280
sulfate (morenosite).................	NiSO₄.7H₂O	280.86	1.948	98–100	−6H₂O, 103
Nitric acid...........................	HNO₃	63.02	1.502	−42	86
Nitric acid...........................	HNO₃.H₂O	81.03	....	−38	....
Nitric acid...........................	HNO₃.3H₂O	117.06	....	−18.5	....
Nitro acid sulfite....................	NO₂HSO₃	127.08	....	73 d.	....
Nitrogen..............................	N₂	28.02	1.026⁻²⁵²·⁵° 0.808⁻¹⁹⁵·⁸° 12.5° (D)	−209.86	−195.8
Nitrogen oxide, mono- (ous).........	N₂O	44.02	lq. 1.226⁻⁸⁹° 1.530 (A)	−102.3	−90.7
oxide, di- (ic)........................	NO or (NO)₂	30.01 (60.02)	lq. 1.269⁻¹⁵⁰·²° 1.0367 (A)	−161	−151
oxide, tri-............................	N₂O₃	76.02	1.447²°	−102	3.5
oxide, tetra- (per- or di-)............	NO₂ or (NO₂)₂	46.01 (92.02)	1.448	−9.3	21.3
oxide, penta-.........................	N₂O₅	108.02	1.63	30	47
oxybromide...........................	NOBr	109.92	>1.0	−55.5	−2
oxychloride...........................	NOCl	65.47	1.417⁻¹²° 2.31 (A)	−64.5	−5.5
Nitroxyl chloride.....................	NO₂Cl	81.47	lq. 1.32¹⁴°	<−30	5
Osmium...............................	Os	190.2	22.48	2700	>5300
chloride, di-..........................	OsCl₂	261.11	....	....	....
chloride, tri-.........................	OsCl₃	296.57	....	d. 560–600	....
chloride, tetra-......................	OsCl₄	332.03	....	....	....
Oxygen................................	O₂	32.00	1.14⁻¹⁸³° 1.426⁻²⁵²·⁵° 1.1053 (A)	−218.4	−183
Ozone.................................	O₃	48.00	1.71⁻¹⁸³° 3.03⁻⁸⁰° 1.658 (A)	−251	−112
Palladium.............................	Pd	106.70	12.0 11¹⁵⁵⁰°	1555	2200
bromide (ous)........................	PdBr₂	266.53	....	....	....
chloride..............................	PdCl₂	177.61	....	500 d.	....
chloride..............................	PdCl₂.2H₂O	213.65	....	....	....
cyanide..............................	Pd(CN)₂	158.74	....	d.	....
hydride...............................	Pd₂H	214.41	11.06	d.	....

TABLE 1-11 Physical Constants of Elements and Compounds—(Continued)

Name	Formula	Formula weight	Specific gravity	Melting point, °C.	Boiling point, °C.
Palladous dichlorodiammine.............	$Pd(NH_3)_2Cl_2$	211.68	2.5	....	....
Perchloric acid...................	$HClO_4$	100.46	1.768	−112	16¹⁸ᵐᵐ
Perchloric acid...................	$HClO_4.H_2O$	118.48	1.88	50	d.
Perchloric acid...................	$HClO_4.2H_2O$ 73.6% anh.	136.50	1.71	−17.8	200
Periodic acid....................	HIO_4	191.93	....	d. 138	subl. 110
Periodic acid....................	$HIO_4.2H_2O$	227.96	....	d. 110	....
Permanganic acid................	$HMnO_4$	119.94	....	....	....
Permolybdic acid................	$HMoO_4.2H_2O$	196.99	....	....	....
Persulfuric acid.................	$H_2S_2O_8$	194.14	....	<60	....
Phosphamic acid.................	$PONH_2.(OH)_2$	97.02	....	d.	....
Phosphatomolybdic acid..............	$H_7P(Mo_2O_7)_6.28H_2O$	2365.88	....	78	−25H_2O, 140
Phosphine.......................	PH_3	34.00	lq. 0.746⁻⁹⁰° 1.146 (A)	−132.5	−85
Phosphonium chloride.............	PH_4Cl	70.47	....	28⁴⁶ᵃᵗᵐ	subl.
Phosphoric acid, hypo-...........	$H_4P_2O_6$	161.99	....	55	d. 70
Phosphoric acid, meta-...........	HPO_3	79.99	2.2–2.5	subl.	....
Phosphoric acid, ortho-..........	H_3PO_4	98.00	1.834	42.35	−½H_2O, 213
Phosphoric acid, pyro-...........	$H_4P_2O_7$	177.99	....	61	....
Phosphorous acid, hypo-..........	H_3PO_2	66.00	1.493	26.5	d.
Phosphorous acid, ortho-.........	H_3PO_3	82.00	1.651	74	d. 200
Phosphorous acid, pyro-..........	$H_4P_2O_5$	145.99	....	38	d. 130
Phosphorus, black...............	P_4	123.92	2.69	....	ign. in air, 400
Phosphorus, red.................	P_4	123.92	2.20	590⁴³ᵃᵗᵐ	ign. in air, 725
Phosphorus, yellow..............	P_4	123.92	1.82; lq. 1.745⁴⁴·⁵°	44.1; ign. 34	280
chloride, tri-...................	PCl_3	137.35	1.574	−111.8	75.95
chloride, penta-................	PCl_5	208.27	1.6	148	subl. 160
oxide, penta-...................	P_2O_5	141.96	2.387	subl. 250	....
oxychloride....................	$POCl_3$	153.35	1.675	2	107.2
Phosphotungstic acid..............	$P_2O_5.2WO_3.42H_2O$	3681.67	....	....	....
Platinum........................	Pt	195.23	21.45	1755	4300
chloride (ic)...................	$PtCl_4$	337.06	....	d. 370	....
chloride (ous).................	$PtCl_2$	266.14	5.87	d. 581	....
chloride (ic)..................	$PtCl_4.8H_2O$	481.19	2.43	−4H_2O, 100	....
cyanide (ous).................	$Pt(CN)_2$	247.27	....	....	....
Plutonium.......................	Pu	238	....	....	....
Plutonium.......................	Pu	239	....	....	....
Potassium.......................	K	39.10	0.86; lq. 0.83⁶²°	62.3	760
acetate........................	$KC_2H_3O_2$	98.14	1.8	292	....
acetate, acid..................	$KH(C_2H_3O_2)_2$	158.19	....	148	d. 200
aluminate......................	$K_2(AlO_2)_2.3H_2O$	250.18	....	....	....
amide.........................	KNH_2	55.12	....	338	subl. 400
arsenate (monobasic)...........	KH_2AsO_4	180.02	2.867	288	....
auricyanide...................	$KAu(CN)_4.1.5H_2O$	367.39	....	d. 200	....
aurocyanide...................	$KAu(CN)_2$	288.33	....	....	....
bicarbonate....................	$KHCO_3$	100.11	2.17	d. 100–200	....
bisulfate......................	$KHSO_4$	136.16	2.35	210	d.
bromate.......................	$KBrO_3$	167.01	3.27	370 d.	....
bromide.......................	KBr	119.01	2.75	730	1380
carbonate.....................	K_2CO_3	138.20	2.29	891	d.
carbonate.....................	$K_2CO_3.2H_2O$	174.23	2.043	....	....
carbonate.....................	$2K_2CO_3.3H_2O$	330.45	2.13	....	....
chlorate......................	$KClO_3$	122.56	2.32	368	d. 400
chloride (sylvite).............	KCl	74.56	1.988	790	1500
chloroplatinate...............	K_2PtCl_6	486.16	3.499	d. 250	....
chromate (tarapacaite).........	K_2CrO_4	194.20	2.732	975	....
cyanate.......................	$KCNO$	81.11	2.048	....	....
cyanide.......................	KCN	65.11	1.52	634.5	....
dichromate....................	$K_2Cr_2O_7$	294.21	2.69	398	d.
ferricyanide..................	$K_3Fe(CN)_6$	329.25	1.84	d.	....
ferrocyanide..................	$K_4Fe(CN)_6.3H_2O$	422.39	1.853	−3HO_2, 70	....
formate.......................	$KHCO_2$	84.11	1.91	167.5	d.
hydride.......................	KH	40.10	0.80	d.	....
hydrosulfide..................	KHS	72.16	2.0	455	....
hydroxide.....................	KOH	56.10	2.044	380	1320
iodate........................	KIO_3	214.02	3.89	560	....
iodide........................	KI	166.02	3.13	723	1330
iodide, tri-...................	KI_3	419.86	3.498	45	d. 225
iodoplatinate.................	K_2PtI_6	1034.94	5.18	....	....
manganate.....................	K_2MnO_4	197.12	....	d. 190	....
metabisulfite.................	$K_2S_2O_5$	222.31	....	d. 150	....
nitrate (saltpeter)...........	KNO_3	101.10	2.11	tr. 129; 333	d. 400
nitrite.......................	KNO_2	85.10	1.915	297	d. 350
oxalate.......................	$K_2C_2O_4.H_2O$	184.23	2.13	d.	....
oxalate, acid.................	KHC_2O_4	128.12	2.0	d.	....
oxalate, acid.................	$KHC_2O_4.½H_2O$	137.13	....	d.	....
oxide.........................	K_2O	94.19	2.32	....	....
perchlorate...................	$KClO_4$	138.55	2.524	d. 400	....

TABLE 1-11 Physical Constants of Elements and Compounds—(*Continued*)

Name	Formula	Formula weight	Specific gravity	Melting point, °C.	Boiling point, °C.
Potassium permanganate	KMnO₄	158.03	2.703	d. <240	....
persulfate	K₂S₂O₈	270.31	....	d. <100	....
phosphate, monobasic	KH₂PO₄	136.09	2.338	256	....
phosphate, dibasic	K₂HPO₄	174.18	....	d.	....
phosphate, tribasic	K₃PO₄	212.27	2.564	1340	....
phosphate, meta-	KPO₃	118.08	2.258	tr. 450; 798	1320
phosphate, meta-	K₄P₄O₁₂.2H₂O	508.34	2.264	−2H₂O, 100	d.
phosphate, pyro-	K₄P₂O₇.3H₂O	384.39	2.33	−2H₂O, 180	−3H₂O, 300
phthalate, acid	KHC₈H₄O₄	204.22	1.63	d.	....
platinocyanide	K₂Pt(CN)₄.3H₂O	431.54	2.45	....	....
silicate	K₂SiO₃	154.25	....	976	....
silicate, tetra-	K₂Si₄O₉.H₂O	352.45	2.417	d. 400	....
sulfate (arcanite)	K₂SO₄	174.25	2.662	tr. 588	....
Potassium sulfate, pyro-	K₂S₂O₇	254.31	2.277	300	....
sulfide, mono-	K₂S.5H₂O	200.33	....	60	−3H₂O, 150
sulfite	K₂SO₃.2H₂O	194.28	....	d.	....
sulfite, acid	KHSO₃	120.16	....	d. 190	....
tartrate	K₂C₄H₄O₆.½H₂O	235.27	1.98	....	d.
tartrate, acid	KHC₄H₄O₆	188.18	1.956	....	....
thiocyanate	KCNS	97.17	1.886	172.3	d. 500
thiosulfate	K₂S₂O₃	190.31	....	d. 400	....
thiosulfate	3K₂S₂O₃.H₂O	588.95	2.23	−H₂O, 180	d.
Praseodymium	Pr	140.92	6.5	940	....
Radium	Ra	226.05	5	960	1140
bromide	RaBr₂	385.88	5.79	728	subl. 900
Radon (Niton)	Rn	222.0	lq. 5.5	−71	−62
Rhenium	Re	186.31	....	3440	....
Rhodium	Rh	102.91	12.5	1955	>2500
chloride	RhCl₃	209.28	....	d. 450	subl. 800±
chloride	RhCl₃.4H₂O	281.35	....	....	....
Rubidium	Rb	85.48	1.53	38.5	700
Ruthenium	Ru	101.70	8.6	>1950	....
Ruthenium	Ru	101.70	12.2	2450	>2700
Samarium	Sm (also Sa)	150.43	7.7	>1300	....
Scandium	Sc	45.10	2.5	1200	2400
Selenic acid	H₂SeO₄	144.98	2.950	58	260
Selenic acid	H₂SeO₄.H₂O	162.99	2.627	26	205
Selenium	Se₈	631.68	4.26	50	688
Selenium	Se₈	631.68	4.80; 4.50	220	688
Selenium	Se₈	631.68	4.8	217	688
Selenous acid	H₂SeO₃	128.98	3.004	d.	....
Silicic acid, meta-	H₂SiO₃	78.08	2.1–2.3	....	....
Silicic acid, ortho-	H₄SiO₄	96.09	1.576	....	....
Silicon, crystalline	Si	28.06	2.4	1420	2600
Silicon, graphitic	Si	28.06	2.0–2.5	....	2600
Silicon, amorphous	Si	28.06	2	....	2600
carbide	SiC	40.07	3.17	>2700	subl. 2200
chloride, tri-	Si₂Cl₆	268.86	1.58	−1	144
chloride, tetra-	SiCl₄	169.89	1.50	−70	57.6
fluoride	SiF₄	104.06	3.57 (A)	−95.7	−65¹⁸¹⁰mm
hydride (silane)	SiH₄	32.09	lq. 0.68⁻¹⁸⁸°	−185	−112
oxide, di- (opal)	SiO₂.xH₂O	......	2.2	1600–1750	subl. 1750
oxide, di- (cristobalite)	SiO₂	60.06	2.32	1710	2230
oxide, di- (lechatelierite)	SiO₂	60.06	2.20	....	2230
oxide, di- (quartz)	SiO₂	60.06	2.650	tr. <1425	2230
oxide, di- (tridymite)	SiO₂	60.06	2.26	tr. 1670	2230
Silver	Ag	107.88	10.5	960.5	1950
bromide (bromyrite)	AgBr	187.80	6.473	434	d. 700]
carbonate	Ag₂CO₃	275.77	6.077	218 d.	....
chloride (cerargyrite)	AgCl	143.34	5.56	455	1550
cyanide	AgCN	133.90	3.95	d. 320	....
nitrate (lunar caustic)	AgNO₃	169.89	4.352	212	444 d.
Sodium	Na	22.997	0.97	97.5	880
acetate	NaC₂H₃O₂	82.04	1.528	324	....
acetate	NaC₂H₃O₂.3H₂O	136.09	1.45	58	−3H₂O, 120
aluminate	NaAlO₂	81.97	....	1650	....
amide	NaNH₂	39.02	....	210	400
ammonium phosphate	NaNH₄HPO₄.4H₂O	209.09	1.574	79 d.	....
antimonate, meta-	2NaSbO₃.7H₂O	511.63	....	....	....
arsenate	Na₃AsO₄.12H₂O	424.09	1.759	86.3	....
arsenate, acid (monobasic)	NaH₂AsO₄.H₂O	181.94	2.535	d. 100	....
arsenate, acid (dibasic)	Na₂HAsO₄.7H₂O	312.02	1.871	125	−7H₂O, 100
arsenate, acid (dibasic)	Na₂HAsO₄.12H₂O	402.10	1.72	28	−12H₂O, 100

TABLE 1-11   Physical Constants of Elements and Compounds—(Continued)

Name	Formula	Formula weight	Specific gravity	Melting point, °C.	Boiling point, °C.
Sodium arsenite, acid	Na₂HAsO₃	169.91	1.87		
benzoate	NaC₇H₅O₂	144.11			
bicarbonate	NaHCO₃	84.01	2.20	d. 270	
bifluoride	NaHF₂	62.00		d.	
bisulfate	NaHSO₄	120.06	2.742	>315	d., −H₂O
bisulfite	NaHSO₃	104.06	1.48	d.	
borate, tetra-	Na₂B₄O₇	201.27	2.367	741	
borate, tetra-	Na₂B₄O₇.5H₂O	291.35	1.815		
borate, tetra- (borax)	Na₂B₄O₇.10H₂O	381.43	1.73	75	−10H₂O, 200
bromate	NaBrO₃	150.91	3.339	381	
bromide	NaBr	102.91	3.205	755	1390
bromide	NaBr.2H₂O	138.95	2.176	50.7	
carbonate (soda ash)	Na₂CO₃	106.00	2.533	851	d.
carbonate	Na₂CO₃.H₂O	124.02	1.55	−H₂O, 100	
carbonate	Na₂CO₃.7H₂O	232.12	1.51	d. 35.1	
carbonate (sal soda)	Na₂CO₃.10H₂O	286.16	1.46		
carbonate, sesqui- (trona)	Na₃H(CO₃)₂.2H₂O	226.05	2.112	d.	
chlorate	NaClO₃	106.45	2.490	248	d.
chloride	NaCl	58.45	2.163	800.4	1413
chromate	Na₂CrO₄	162.00	2.723	392	
chromate	Na₂CrO₄.10H₂O	342.16	1.483	19.9	
citrate	2Na₃C₆H₅O₇.11H₂O	714.36	1.857	−11H₂O, 150	d.
cyanide	NaCN	49.02		563.7	1496
dichromate	Na₂Cr₂O₇.2H₂O	298.05	2.52	−2H₂O, 84.6; 356 (anh.)	d. 400
ferricyanide	Na₃Fe(CN)₆.H₂O	298.97			
ferrocyanide	Na₄Fe(CN)₆.10H₂O	484.11	1.458		
fluoride (villiaumite)	NaF	42.00	2.79	992	
formate	NaHCO₂	68.01	1.919	253	
hydride	NaH	24.005	0.92	d. 800	
hydrosulfide	NaSH.2H₂O	92.10		d.	
hydrosulfide	NaSH.3H₂O	110.11		22	d.
hydrosulfite	Na₂S₂O₄.2H₂O	210.15		d.	
hydroxide	NaOH	40.00	2.130	318.4	1390
hydroxide	NaOH.3½H₂O	103.06		15.5	
hypochlorite	NaOCl	74.45		d.	
iodide	NaI	149.92	3.667	651	1300
iodide	NaI.2H₂O	185.95	2.448		
lactate	NaC₃H₅O₃	112.07		d.	
nitrate (soda niter)	NaNO₃	85.01	2.257	308	d. 380
nitrite	NaNO₂	69.01	2.168	271	d. 320
oxide	Na₂O	61.99	2.27	subl.	
perborate	NaBO₃.H₂O	99.83		d. 40	
perchlorate	NaClO₄	122.45		482 d.	
perchlorate	NaClO₄.H₂O	140.47	2.02	d. 130	
peroxide	Na₂O₂	77.99	2.805	d.	
peroxide	Na₂O₂.8H₂O	222.12		d. 30	
phosphate, monobasic	NaH₂PO₄.H₂O	138.01	2.040	−H₂O, 100	d. 200
phosphate, monobasic	NaH₂PO₄.2H₂O	156.03	1.91	60	
phosphate, dibasic	Na₂HPO₄.7H₂O	268.09	1.679	d.	
phosphate, dibasic	Na₂HPO₄.12H₂O	358.17	1.52	34.6	−12H₂O, 180
phosphate, tribasic	Na₃PO₄	163.97	2.537	1340	
phosphate, tribasic	Na₃PO₄.12H₂O	380.16	1.62	73.4	−11H₂O, 100
phosphate, meta-	Na₄P₄O₁₂	407.91	2.476	616 d.	
phosphate, pyro-	Na₄P₂O₇	265.95	2.45	988	
phosphate, pyro-	Na₄P₂O₇.10H₂O	446.11	1.82	d.	
phosphate (pyrodisodium)	Na₂H₂P₂O₇	221.97	1.862	d. 220	
phosphate (pyrodisodium)	Na₂H₂P₂O₇.6H₂O	330.07	1.848		
potassium tartrate	NaKC₄H₄O₆.4H₂O	282.23	1.790	70 to 80	−4H₂O, 215
silicate, meta-	Na₂SiO₃	122.05		1088	
silicate, meta-	Na₂SiO₃.9H₂O	284.20		47	−6H₂O, 100
silicate, ortho-	Na₄SiO₄	184.05		1018	
silicofluoride	Na₂SiF₆	188.05	2.679	d.	
stannate	Na₂SnO₃.3H₂O	266.74		d. 140	
sulfate (thenardite)	Na₂SO₄	142.05	2.698	tr. 100	
sulfate	Na₂SO₄	142.05		884	
sulfate	Na₂SO₄.7H₂O	268.17			
sulfate (Glauber's salt)	Na₂SO₄.10H₂O	322.21	1.464	32.4	−10H₂O, 100
sulfide, mono-	Na₂S	78.05	1.856		
sulfide, tetra-	Na₂S₄	174.23		275	
sulfide, penta-	Na₂S₅	206.29		251.8	
sulfite	Na₂SO₃	126.05	2.633	d.	
sulfite	Na₂SO₃.7H₂O	252.17	1.561	−7H₂O, 150	d.
tartrate	Na₂C₄H₄O₆.2H₂O	230.10	1.818		
thiocyanate	NaCNS	81.08		287	
thiosulfate	Na₂S₂O₃	158.11	1.667		
thiosulfate (hypo)	Na₂S₂O₃.5H₂O	248.19	1.685	d. 48.0	
tungstate	Na₂WO₄	293.91	4.179	692	
tungstate	Na₂WO₄.2H₂O	329.95	3.245	−2H₂O, 100	
tungstate, para-	Na₆W₇O₂₄.16H₂O	2097.68	3.987	−16H₂O, 300	
uranate	Na₂UO₄	348.06			

TABLE 1-11   Physical Constants of Elements and Compounds—(Continued)

Name	Formula	Formula weight	Specific gravity	Melting point, °C.	Boiling point, °C.
Sodium vanadate	$Na_3VO_4.16H_2O$	472.20	....	866 (anh.)	....
vanadate, pyro-	$Na_4V_2O_7$	305.89	....	654	....
Stannic chloride	$SnCl_4$	260.53	2.226	−30.2	114.1
oxide (cassiterite)	SnO_2	150.70	7.0	1127	....
sulfate	$Sn(SO_4)_2.2H_2O$	346.85	....	....	....
Stannous bromide	$SnBr_2$	278.53	5.12	215.5	620
chloride	$SnCl_2$	189.61	....	246.8	623
chloride (tin salt)	$SnCl_2.2H_2O$	225.65	2.71	37.7	d.
sulfate	$SnSO_4$	214.76	....	d. 360	....
Strontium	Sr	87.63	2.6	800	1150
acetate	$Sr(C_2H_3O_2)_2$	205.72	2.099	....	d.
carbonate (strontianite)	$SrCO_3$	147.64	3.70	1497^{60atm}	d. 1350
chloride	$SrCl_2$	158.54	3.052	873	....
chloride	$SrCl_2.6H_2O$	266.64	1.933	−4H_2O, 61	−6H_2O, 100
hydroxide	$Sr(OH)_2$	121.65	3.625	375	....
hydroxide	$Sr(OH)_2.8H_2O$	265.77	1.90	−7H_2O in dry air	....
nitrate	$Sr(NO_3)_2$	211.65	2.986	570	....
nitrate	$Sr(NO_3)_2.4H_2O$	283.71	2.2	....	....
oxide (strontia)	SrO	103.63	4.7	2430	....
peroxide	SrO_2	119.63	....	d.	....
peroxide	$SrO_2.8H_2O$	263.76	....	−8H_2O, 100	d.
sulfate (celestite)	$SrSO_4$	183.69	3.96	1580 d.	....
sulfate, acid	$Sr(HSO_4)_2$	281.77	....	d.	....
Sulfamic acid	NH_2SO_3H	97.09	2.05	205 d.	....
Sulfur, amorphous	S	32.06	2.046	120	444.6
Sulfur, monoclinic	S_8	256.48	1.96	119.0	444.6
Sulfur, rhombic	S_8	256.48	2.07	112.8	444.6
Sulfur bromide, mono-	S_2Br_2	223.95	2.635	−46	54$^{0.18mm}$
chloride, mono-	S_2Cl_2	135.03	1.687	−80	138
chloride, di-	SCl_2	102.97	1.621	−78	59
chloride, tetra-	SCl_4	173.89	....	−30	d. > −20
oxide, di-	SO_2	64.06	lq., 1.434$^{0°}$; 2.264 (A)	−75.5	−10.0
oxide, tri-(α)	SO_3	80.06	lq., 1.923; 2.75 (A)	16.83	44.6
oxide, tri-(β)	$(SO_3)_2$	160.12	1.97	50	....
Sulfuric acid	H_2SO_4	98.08	1.834	10.49	d. 340
Sulfuric acid	$H_2SO_4.H_2O$	116.09	1.842	8.62	290
Sulfuric acid	$H_2SO_4.2H_2O$	134.11	1.650	−38.9	167
Sulfuric acid, pyro-	$H_2S_2O_7$	178.14	1.920°	35	d.
Sulfuric oxychloride	SO_2Cl_2	134.97	1.667	−54.1	69.1
Sulfurous oxybromide	$SOBr_2$	207.89	2.68	−50	68^{40mm}
oxychloride	$SOCl_2$	118.97	1.638	−104.5	78.8
Tantalum	Ta	180.88	16.6	2850	>4100
Tellurium	Te	127.61	(α) 6.24; (β) 6.00	452	1390
Terbium	Tb	159.20	....	....	....
Thallium	Tl	204.39	11.85	303.5	1650
acetate	$TlC_2H_3O_2$	263.43	3.68	110	....
chloride, mono-	$TlCl$	239.85	7.00	430	806
chloride, sesqui-	Tl_2Cl_3	515.15	5.9	400–500	d.
chloride, tri-	$TlCl_3$	310.76	....	25	d.
chloride, tri-	$TlCl_3.4H_2O$	382.83	....	37	−4H_2O, 100
sulfate (ic)	$Tl_2(SO_4)_3.7H_2O$	823.07	....	−6H_2O, 200	d.
sulfate (ous)	Tl_2SO_4	504.84	6.77	632	d.
sulfate, acid	$TlHSO_4$	301.46	....	115 d.	....
Thio, Cf. sulfo or sulfur					
Thorium	Th	232.12	11.2	1845	>3000
oxide, di- (thorianite)	ThO_2	264.12	9.69	>2800	4400
sulfate	$Th(SO_4)_2$	424.24	4.225	....	....
sulfate	$Th(SO_4)_2.9H_2O$	586.38	2.77	−9H_2O, 400	....
Thulium	Tm	169.40	....	....	....
Tin	Sn	118.70	7.31	231.85	2260
Tin	Sn	118.70	5.750	Stable −163 to +18	2260
Tin salts, Cf. stannic and stannous					
Titanic acid	H_2TiO_3	97.92	....	....	....
Titanium	Ti	47.90	4.50	1800	>3000
chloride, di-	$TiCl_2$	118.81	....	Unstable in air	....
chloride, tri-	$TiCl_3$	154.27	....	d. 440	....
chloride, tetra-	$TiCl_4$	189.73	lq., 1.726	−30	136.4

TABLE 1-11   Physical Constants of Elements and Compounds—(Continued)

Name	Formula	Formula weight	Specific gravity	Melting point, °C.	Boiling point, °C.
Titanium oxide, di- (anatase)..........	TiO_2	79.90	3.84	....	....
oxide, di- (brookite).................	TiO_2	79.90	4.17	....	....
oxide, di- (rutile).................	TiO_2	79.90	4.26	1640 d.	<3000
Tungsten...........................	W	183.92	19.3	3370	5900
carbide...........................	WC	195.93	15.7	2777	6000
carbide...........................	W_2C	379.85	16.06	2877	6000
oxide, tri-........................	WO_3	231.92	7.16	>2130	....
Tungstic acid (tungstite)...............	H_2WO_4	249.94	5.5	$-\frac{1}{2}H_2O$, 100; 1473	....
Uranic acid........................	H_2UO_4	304.09	5.926	$-H_2O$, 250 to 300	....
Uranium...........................	U	238.07	18.485	1133	3500
carbide...........................	U_2C_3	512.14	11.28	2400	....
oxide, di- (uraninite)...............	UO_2	270.07	10.9	2176	....
oxide (pitchblende)................	U_3O_8	842.21	7.31	d.	....
sulfate (ous)......................	$U(SO_4)_2.4H_2O$	502.25	....	$-4H_2O$, 300	....
Uranyl acetate........................	$UO_2(C_2H_3O_2)_2.2H_2O$	424.19	2.89	$-2H_2O$, 110	....
carbonate (rutherfordine)...........	UO_2CO_3	330.08	5.6	....	....
nitrate...........................	$UO_2(NO_3)_2.6H_2O$	502.18	2.807	60.2	118
sulfate...........................	$UO_2SO_4.3H_2O$	420.18	3.28	d. 100	....
Vanadic acid, meta-...................	HVO_3	99.96	....	....	....
Vanadic acid, pyro-..................	$H_4V_2O_7$	217.93	....	....	....
Vanadium...........................	V	50.95	5.96	1710	3000
chloride, di-......................	VCl_2	121.86	3.23	....	....
chloride, tri-.....................	VCl_3	157.23·	3.00	d.	....
chloride, tetra-...................	VCl_4	192.78	1.816	-109	148.5
oxide, di-.........................	V_2O_2	133.90	3.64	ign.	....
oxide, tri-........................	V_2O_3	149.90	4.87	1970	....
oxide, tetra-..'..................	V_2O_4	165.90	4.399	1967	....
oxide, penta-......................	V_2O_5	181.90	3.357	800	d. 1750
oxychloride, mono-.................	$VOCl$	102.41	2.824	....	....
Vanadyl chloride....................	$(VO)_2Cl$	169.36	3.64	d. in air	....
chloride, di-......................	$VOCl_2$	137.86	2.88	....	....
chloride, tri-.....................	$VOCl_3$	173.32	1.829	<-15	127.19
Water	H_2O	18.016	1.00 (lq.); 0.915$^{0°}$ (ice)	0	100
Water, heavy.......................	D_2O	20.029	1.107	3.82	101.42
Xenon..............................	Xe	131.30	lq., $3.06^{-109.1}$ $2.7^{-140°}$ 4.53 (A)	-140	-109.1
Ytterbium..........................	Yb	173.04	....	....	....
Yttrium............................	Y	88.92	5.51	1490	2500
Zinc..............................	Zn	65.38	7.140	419.4	907
acetate...........................	$Zn(C_2H_3O_2)_2$	183.47	1.840	242	subl. in vac.
acetate...........................	$Zn(C_2H_3O_2)_2.2H_2O$	219.50	1.735	237	$-2H_2O$, 100
bromide...........................	$ZnBr_2$	225.21	4.219	394	650
carbonate.........................	$ZnCO_3$	125.39	4.42	d. 300	....
chloride..........................	$ZnCl_2$	136.29	2.91	283	732
cyanide...........................	$Zn(CN)_2$	117.42	....	d. 80	....
hydroxide.........................	$Zn(OH)_2$	99.40	3.053	d. 125	....
iodide............................	ZnI_2	319.22	4.666	446	624
nitrate...........................	$Zn(NO_3)_2.6H_2O$	297.49	2.065	36.4	$-6H_2O$, 105
oxide (zincite)...................	ZnO	81.38	5.606	>1800	....
oxide.............................	ZnO	81.38	5.47	>1800	....
peroxide..........................	ZnO_2	97.38	1.571	expl. 212	....
phosphide.........................	Zn_3P_2	258.10	4.55	>420	1100
silicate..........................	$ZnSiO_3$	141.44	3.52	1437	....
sulfate (zincosite).................	$ZnSO_4$	161.44	3.74	d. 740	....
sulfate...........................	$ZnSO_4.H_2O$	179.46	3.28	d. 238	....
sulfate...........................	$ZnSO_4.6H_2O$	269.54	2.072	$-5H_2O$, 70	....
sulfate (goslarite).................	$ZnSO_4.7H_2O$	287.55	1.966	tr. 39	$-7H_2O$, 280
sulfide (α) (wurtzite)........	ZnS	97.44	4.087	1850^{150atm}	subl. 1185
sulfide (β) (sphalerite).......	ZnS	97.44	4.102	tr. 1020	....
sulfide (blende)...................	ZnS	97.44	4.04	....	....
sulfite...........................	$ZnSO_3.2\frac{1}{2}H_2O$	190.48	....	$-2\frac{1}{2}H_2O$, 100	d. 200
Zirconium..........................	Zr	91.22	6.4	1700	>2900
oxide, di- (baddeleyite).............	ZrO_2	123.22	5.49	2700	....
oxide, di- (free from Hf)...........	ZrO_2	123.22	5.73	....	4300

TABLE 1-11   Physical Constants of Elements and Compounds—(*Continued*)

ORGANIC

Name	Formula	Formula weight	Specific gravity	Melting point, °C	Boiling point, °C
Abietic acid	$C_{20}H_{30}O_2$	302.44		182	
Acenaphthene	$C_{10}H_6(CH_2)_2$	154.20	1.069	95	278–9
Acetal	$CH_3CH(OC_2H_5)_2$	118.17	0.821		102.2
Acet-aldehyde-	CH_3CHO	44.05	0.783	−123.5	20.2
-aldehyde, par-	$(C_2H_4O)_3$	132.16	0.994	10.5–12	124.4
-aldehyde ammonia	$CH_3CHOHNH_2$	61.08		97	100–10 d.
-amide	CH_3CONH_2	59.07	1.159	81	222
-anilide	$C_6H_5NHCOCH_3$	135.16	1.21	113–4	305
-phenetidide (o-)	$CH_3CONHC_6H_4OC_2H_5$	179.21		79	>250
(m-)	$CH_3CONHC_6H_4OC_2H_5$	179.21		96–7	
-toluidide (o-)	$CH_3C_6H_4NHCOCH_3$	149.19	1.168	110	296
(p-)	$CH_3C_6H_4NHCOCH_3$	149.19	1.212	153	306–7
Acetic acid	CH_3CO_2H	60.05	1.049	16.7	118.1
anhydride	$(CH_3CO)_2O$	102.09	1.082	−73	139.6
nitrile	CH_3CN	41.05	0.783	−41	81.6–2.0
Acetone	CH_3COCH_3	58.08	0.792	−94.6	56.5
Acetonyl urea	$<NHCONHCOC>(CH_3)_2$	128.13		175	subl.
Acetophenone benzoyl hydride	$CH_3COC_6H_5$	120.14	1.033	20.5	202.3
Acetyl-chloride	CH_3COCl	78.50	1.105	−112.0	51–2
-phenylenediamine (-p)	$C_2H_3ONHC_6H_4NH_2$	150.18		162	
Acetylene	$HC:CH$	26.04	(A) 0.906	−81.5	−84
dichloride (cis)	$CHCl:CHCl$	96.95	1.291	−80.5	60.3
(trans)	$CHCl:CHCl$	96.95	1.265	−50	48.4
Aconitic acid	$C_3H_3(CO_2H)_3$	174.11		192 d.	
Acridine	$C_6H_4 < (CH)(N) > C_6H_4$	179.21		110–1	346
Acrolein ethylene aldehyde	$CH_2:CH.CHO$	56.06	0.841	−87.7	52.5
Acrylic acid	$CH_2:CH.CO_2H$	72.06	1.062	12–13	141–2
nitrile	$CH_2:CH.CN$	53.06	0.811	83	78.9
Adipic acid	$(CH_2CH_2CO_2H)_2$	146.14	1.360	151–3	265 10 mm
amide	$(CH_2CH_2CONH_2)_2$	144.17		226–7	
nitrile	$(CH_2CH_2CN)_2$	108.14	0.951	1	295
Adrenaline (1-) (3,4,1)	$C_6H_3(OH)_2(CHOHCH_2NHCH_3)$	183.20		d. 207–11	
Alanine (α) (dl-)	$CH_3CH(NH_2)CO_2H$	89.09		295 d.	subl. >200
Aldol acetaldol	$CH_3CH(OH)CH_2CO_2H$	88.10	1.103		83 200 mm
Alizarin	$C_6H_4(CO)_2C_6H_2(OH)_2$	240.20		289–90	430
Allyl alcohol	$CH_2:CH.CH_2OH$	58.08	0.854	−129	96.6
bromide	$CH_2:CH.CH_2Br$	120.99	1.398	−119.4	70–1
chloride	$CH_2:CH.CH_2Cl$	76.53	0.938	−136.4	44.6
thiocyanate (i)	$CH_2:CH.CH_2NCS$	99.15	1.013	−80	152
thiourea	$CH_2:CH.CH_2NHCSNH_2$	116.18	1.219	77–8	
Aluminum ethoxide	$Al(OCH_2CH_3)_3$	164.15	1.142	150–60	200–510 mm
Amino-anthraquinone (α)	$C_6H_4(CO)_2C_6H_3NH_2$	223.22		256	subl.
(β)	$C_6H_4(CO)_2C_6H_3NH_2$	223.22		302	subl.
-azobenzene	$C_6H_5N:N.C_6H_4NH_2$	197.23		126–7	225 120 mm
-benzoic acid (m-)	$H_2N.C_6H_4CO_2H$	137.13	1.511	173–4	
(p-)	$H_2N.C_6H_4CO_2H$	137.13		187–8	
Amino-diphenylamine (p-)	$H_2N.C_6H_4NH.C_6H_5$	184.23		67	354
-G-acid (2-)(6-,8-), Na₂ salt	$C_{10}H_5(NH_2)(SO_3Na)_2$	347.28			
-mono-potassium salt	$C_{10}H_5(NH_2)S_2O_6HK$	341.39			
-sodium salt	$C_{10}H_5(NH_2)S_2O_6HNa$	325.29			
-J-acid (2-)(5-,7-)	$C_{10}H_5(NH_2)(SO_3H)_2$	303.30			
-mono-potassium salt	$C_{10}H_5(NH_2)S_2O_6HK$	341.39			
-naphthol sulfonic (1-,2-,4-)(α-)	$C_{10}H_5ONHNH_2SO_3H\frac{1}{2}H_2O$	248.25			
(1-,8-,4-)	$NH_2(OH)C_{10}H_5SO_3H$	239.24			
-phenol (o-)	$H_2N.C_6H_4.OH$	109.12		173	subl.
(m-)	$H_2N.C_6H_4.OH$	109.12		122–3	
(p-)	$H_2N.C_6H_4.OH$	109.12		184–6 d.	subl.
-toluene sulfonic acid (1-,2-,3-)	$C_6H_3(CH_3)(NH_2)SO_3H$	187.21			
(1-,4-,2-)	$C_6H_3(CH_3)(NH_2)SO_3H.H_2O$	205.23		d.	
(1-,4-,3-)	$C_6H_3(CH_3)(NH_2)SO_3H.\frac{1}{2}H_2O$	196.22			
(1-,2-,5-)	$C_6H_3(CH_3)(NH_2)SO_3H.H_2O$	205.23		−H₂O, 120	
Amyl acetate (n-)	$CH_3CO_2CH_2(CH_2)_3CH_3$	130.18	0.879	−70.8	148.4
(i-)	$CH_3CO_2CH_2CH_2CH(CH_3)_2$	130.18	0.876		142
(s-)	$CH_3CO_2CH_2CH(CH_3)C_2H_5$	130.18	0.880		141–2
(s-)	$CH_3CO_2CH(CH_3)CH_2C_2H_5$	130.18	0.922⁰		133.5
(s-)	$CH_3CO_2CH(C_2H_5)_2$	130.18	0.871		133
(t-)	$CH_3CO_2C(CH_3)_2C_2H_5$	130.18	0.874		124.5
alcohol (n-) fusel oil	$CH_3(CH_2)_3CH_2OH$	88.15	0.817	−78.5	137.9
(s-,n-) methyl-propyl carbinol	$CH_3CH_2CH(OH)CH_3$	88.15	0.810		119.5
(prim.-,i-) isobutyl carbinol	$(CH_3)_2CHCH_2CH_2OH$	88.15	0.813	−117.2	132.0
	$(C_2H_5)_2CHOH$	88.15	0.815		115.6
(s-,i-)	$(CH_3)_2CHCH(OH)CH_3$	88.15	0.819		113–4
(t-)	$(CH_3)_2C(OH)C_2H_5$	88.15	0.809	−11.9	102
	$(CH_3)_3CCH_2OH$	88.15		52–3	113–4
(d-)	$C_2H_5CH(CH_3)CH_2OH$	88.15	0.816		128
-amine (n-)	$CH_3(CH_2)_4NH_2$	87.16	0.766	−55	103–4
(s-,n-)	$(C_3H_7)(CH_3)CHNH_2$	87.16	0.749		91–2
(i-)	$(CH_3)_2CH(CH_2)_2NH_2$	87.16	0.751		95
(t-)	$(C_2H_5)(CH_3)_2CNH_2$	87.16	0.731	−105	77–8
	$C_2H_5CH(CH_3)CH_2NH_2$	87.16	0.755		95–6
	$(C_2H_5)_2CHNH_2$	87.16	0.749		90–1
	$(CH_3)_2CHCH(CH_3)NH_2$	87.16	0.757		83–4
aniline (i-)	$C_6H_5NHC_5H_{11}$	163.25	0.928		254.5
benzoate (i-)	$C_6H_5CO_2C_5H_{11}$	192.25	0.992		261
bromide (n-)	$CH_3(CH_2)_3CH_2Br$	151.05	1.218	−95	129.7
(i-)	$(CH_3)_2CH(CH_2)_2Br$	151.05	1.220		120

TABLE 1-11 Physical Constants of Elements and Compounds—(*Continued*)

Name	Formula	Formula weight	Specific gravity	Melting point, °C.	Boiling point, °C.
Amyl bromide (t-)................	$(CH_3)_2C(Br)C_2H_5$	151.05	1.216	....	108
n-butyrate (n-)...............	$C_2H_5CH_2CO_2(CH_2)_4CH_3$	158.23	0.871	−73.2	186.4
(i-)................	$C_2H_5CH_2CO_2.C_5H_{11}$	158.23	0.866	....	178.6
(t-)................	$C_3H_7CO_2C(CH_3)_2C_2H_5$	158.23	0.865	....	164
i-butyrate (i-)...............	$(CH_3)_2CHCO_2C_5H_{11}$	158.23	0.876	....	168.8
chloride (n-)................	$CH_3(CH_2)_3CH_2Cl$	106.60	0.878	−99	108.4
(s-)................	$C_2H_5CH_2CHClCH_3$	106.60	0.870	....	96.7
(s-)................	$(C_2H_5)_2CHCl$	106.60	0.895	....	97.3
(i-)................	$(CH_3)_2CH(CH_2)_2Cl$	106.60	0.893	....	99.7
(s-,i-)............	$(CH_3)CHCHClCH_3$	106.60	0.883^0	....	91
(t-)................	$(CH_3)_2CClC_2H_5$	106.60	0.871	−72.9	85.7
	$(CH_3)(C_2H_5)CHCH_2Cl$	106.60	0.881	....	98–9
i-cyanide (i-)................	$(CH_3)_2CH(CH_2)_2NC$	97.16	....	....	137–9
formate (n-)................	$HCO_2CH_2(CH_2)_3CH_3$	116.16	0.902^0	−73.5	132
(i-)................	$HCO_2CH_2CH_2CH(CH_3)_2$	116.16	0.882	−93.5	123.5
iodide (n-)................	$CH_3(CH_2)_3CH_2I$	198.06	1.510	−86	157.0
(i-)................	$(CH_3)_2CHCH_2CH_2I$	198.06	1.515	....	147
(s-,n-)............	$C_2H_5CH_2CHICH_3$	198.06	1.507	....	144–5
(t-)................	$(CH_3)_2CIC_2H_5$	198.06	1.471	....	127
	$C_2H_5CH(CH_3)CH_2I$	198.06	1.524	....	148
mercaptan (n-)...............	$CH_3(CH_2)_3CH_2SH$	104.21	0.857	....	126
(i-)................	$(C_2H_5)_2CHSH$	104.21	....	....	105
(i-)................	$(CH_3)_2CH(CH_2)_2SH$	104.21	0.835	....	120
phenol (t-)(p-)...............	$C_5H_{11}.C_6H_4OH$	164.24	....	93	265–7
propionate (n-)...............	$C_2H_5CO_2(CH_2)_4CH_3$	144.21	0.876	−73.1	168.7
(i-)................	$C_2H_5CO_2(CH_2)_2CH(CH_3)_2$	144.21	0.870	....	160.2
(act.)............	$C_2H_5CO_2C_5H_{11}$	144.21	0.866	....	58^{16mm}
salicylate (n-)...............	$HOC_6H_4CO_2C_5H_{11}$	208.25	1.065	....	265
Amyl i-valerate (i)...............	$C_4H_9CO_2C_5H_{11}$	172.26	0.858	....	194
(t-)................	$C_4H_9CO_2C_5H_{11}$	172.26	0.861	....	173–4
Amylene (n-)(α-)...............	$C_2H_5CH_2CH:CH_2$	70.13	0.644	....	30–1
(i-)................	$(CH_3)_2CHCH:CH_2$	70.13	0.632	−135	20.5
(α-)................	$(C_2H_5)(CH_3)C:CH_2$	70.13	0.667^0	....	31–2
(-n)(β-)............	$C_2H_5CH:CHCH_3$	70.13	0.650	−139	36.4
(i-)(β-)............	$(CH_3)_2C:CHCH_3$	70.13	0.663	−124	37–8
Anethole (p-)..................	$CH_3CH:CH.C_6H_4OCH_3$	148.20	0.991	22.5	235.3
Anhydroformald-aniline............	$(CH_2NC_6H_5)_3$	315.40	....	143	185
Aniline.........................	$C_6H_5NH_2$	93.12	1.022	−6.2	184.4
hydrochloride.............	$C_6H_5NH_2.HCl$	129.59	1.222^4	198	245
nitrate....................	$C_6H_5NH_2.HNO_3$	156.14	1.356^4	d. 190	....
sulfate....................	$(C_6H_5NH_2)_2.H_2SO_4$	284.32	1.377^4	d.	....
Anisal-acetone (p-).............	$CH_3OC_6H_4CH:CHCOCH_3$	176.22	....	73–4	....
Anisic acid (p-)...............	$CH_3OC_6H_4CO_2H$	152.14	1.385^4	184.2	275–80
aldehyde (p-).............	$CH_3OC_6H_4CHO$	136.14	1.123	2.5	247–8
Anisidine (o-)...............	$CH_3OC_6H_4NH_2$	123.15	1.098	5.2	225
(m-)...............	$CH_3OC_6H_4NH_2$	123.15	1.096	<−12	251
(p-)...............	$CH_3OC_6H_4NH_2$	123.15	1.089	57.2	243
Anisole........................	$CH_3OC_6H_5$	108.13	0.990	−37.3	154–5
Anthracene.....................	$C_6H_4:(CH)_2:C_6H_4$	178.22	1.25	217–8	340–2
Anthramine (α)................	$C_6H_4:(CH)_2:C_6H_3NH_2$	193.24	....	130	....
(β)................	$C_6H_4:(CH)_2:C_6H_3NH_2$	193.24	....	238	subl.
Anthranil......................	$C_6H_4:(NH)CO$	119.12	1.187	<−18	d. >215
Anthranilic acid (o-)............	$H_2NC_6H_4CO_2H$	137.13	....	144–5	subl.
Anthrapurpurin (1-,2-,7-).........	$C_{14}H_5O_2(OH)_3$	256.20	....	369	462
Anthraquinone..................	$C_6H_4:(CO)_2:C_6H_4$	208.20	1.438	286	379–81
disulfonate Na₂ (1-,5-).....	$C_{14}H_6O_2(SO_3Na)_2.5H_2O$	502.38	....	....	....
(1-,8-).....	$C_{14}H_6O_2(SO_3Na)_2.4H_2O$	484.37	....	....	....
(2-,6-).....	$C_{14}H_6O_2(SO_3Na)_2.7H_2O$	538.41	....	....	....
(2-,7-).....	$C_{14}H_6O_2(SO_3Na)_2.4H_2O$	484.37	....	....	....
sulfonate Na (1-)...........	$C_{14}H_7O_2SO_3Na$	310.25	....	....	....
(2-)...........	$C_{14}H_7O_2SO_3Na$	310.25	....	....	....
Anthrarufin (1-,5-)............	$C_{14}H_6O_3(OH)_2$	240.20	....	280	subl.
Antipyrene.....................	$C_{11}H_{12}ON_2$	188.22	1.088^{113}	113	319^{174mm}
Apiole.........................	$C_{12}H_{14}O_2$	222.23	1.02	30	294
Arabinose (α)(d- or l-).........	$CH_2OH(CHOH)_3CHO$	150.13	1.585	159.5	....
(dl-)............	$CH_2OH(CHOH)_3CHO$	150.13	....	164.5	....
Arachidic acid..................	$CH_3(CH_2)_{18}CO_2H$	312.52	....	77	328
Arsanilic acid (p-)..............	$H_2N.C_6H_4.AsO_3H_2$	217.04	....	232	....
Asparagine (l-).................	$HO_2C.C_2H_3(NH_2).CONH_2$	132.12	1.543	227–35	d. 235
Aspirin (o-)....................	$CH_3CO_2.C_6H_4.OH$	180.15	....	135–6	....
Atropic acid...................	$C_6H_5C(:CH_2).CO_2H$	148.15	....	106–7	267 d.
Auramine......................	$[(CH_3)_2NC_6H_4]_2C:NH$	267.36	....	136	....
Aurine, coralline (4-,4'-).......	$(HOC_6H_4)_2C:C_6H_4:O$	290.30	....	310 d.	....
Azo-anisole (2-,2'-)............	$(CH_3O.C_6H_4N:)_2$	242.27	....	153	....
benzene...................	$C_6H_5N:N.C_6H_5$	182.22	1.203	68	297
Azoxybenzene..................	$(C_6H_5)_2N_2O$	198.22	1.248	36	d.
Barbituric acid.................	$CO:(NHCO)_2:CH_2.2H_2O$	164.12	....	d. 245	....
Benzal acetone.................	$C_6H_5CH:CHCOCH_3$	146.18	1.035	41–2	260–2
Benzaldehyde..................	C_6H_5CHO	106.12	1.046	−26	179
Benzamide.....................	$C_6H_5CONH_2$	121.13	1.341	130	290
Benzanilide....................	$C_6H_5CONHC_6H_5$	197.23	1.31^4	163	117–910^{mm}
Benzene.......................	C_6H_6	78.11	0.879	5.5	80.1
sulfinic acid..............	$C_6H_5SO_2H$	142.17	....	83–4	d. >100
sulfonic acid.............	$C_6H_5SO_3H$	158.17	....	65–6	d.
sulfonic amide...........	$C_6H_5SO_2NH_2$	157.18	....	156	....
sulfonic chloride.........	$C_6H_5SO_2Cl$	176.62	1.384	14.5	251.5

TABLE 1-11   Physical Constants of Elements and Compounds—(Continued)

Name	Formula	Formula weight	Specific gravity	Melting point, °C.	Boiling point, °C.
Benzidine (4-,4'-)	$NH_2.C_6H_4.C_6H_4.NH_2$	184.23	....	128–9	400
disulfonic acid (2-,2'-)	$(.C_6H_3(NH_2)SO_3H)_2.3H_2O$	398.40	....	d. >175	....
(3-,3'-)	$(.C_6H_3(NH_2)SO_3H)_2$	344.35	....	....	....
Benzil	$C_6H_5CO.COC_6H_5$	210 22	1.23	95	348 d.
Benzoic acid	$C_6H_5CO_2H$	122.12	1.266	121.7	249.2
anhydride	$(C_6H_5CO)_2O$	226.22	1.199	42	360
nitrile	C_6H_5CN	103.12	1.001	−12.9	190.7
Benzoin (dl-)	$C_6H_5CO.CHOHC_6H_5$	212.24	....	133–7	344
Benzophenone	$C_6H_5COC_6H_5$	182.21	1.083⁵⁴	48.5	305.4
Benzotrichloride	$C_6H_5CCl_3$	195.48	1.380	−4.75	220.7
Benzoyl-benzoic acid (o-)	$C_6H_5COC_6H_4CO_2H.H_2O$	244.24	....	93(128)	....
-chloride	C_6H_5COCl	140.57	1.212	−0.5	197.2
-peroxide	$(C_6H_5CO)_2O_2$	242.22	....	108 d.	expl.
Benzyl acetate	$CH_3CO_2CH_2C_6H_5$	150.17	1.057	−51.5	213.5
alcohol	$C_6H_5CH_2OH$	108.13	1.043	−15.3	204.7
amine	$C_6H_5CH_2NH_2$	107.15	0.982	....	184.5
aniline	$C_6H_5CH_2NHC_6H_5$	183.24	1.065	37–8	306
benzoate	$C_6H_5CO_2CH_2C_6H_5$	212.24	1.12	21	323–4
butyrate	$C_2H_5CH_2CO_2CH_2C_6H_5$	178.22	1.016	....	238–40
chloride	$C_6H_5CH_2Cl$	126.58	1.100	−39	179.4
ether	$(C_6H_5CH_2)_2O$	198.25	1.036	....	295–8
formate	$HCO_2CH_2C_6H_5$	136.14	1.081	3.6	202–3
propionate	$C_2H_5CO_2CH_2C_6H_5$	164.20	1.036	....	220–2
Berberonic acid (2-,4-,5-)	$C_5H_2N(CO_2H)_3.2H_2O$	247.16	....	243	....
Biuret	$NH(CONH_2)_2$	103.08	....	192–3 d.	....
Borneol (dl-)	$C_{10}H_{17}OH$	154.24	1.011	210.5	subl.
(d- or l-)	$C_{10}H_{17}OH$	154.24	1.011	208–9	212–3
(iso-)	$C_{10}H_{17}OH$	154.24	....	212	....
Bornyl acetate (d-)	$CH_3CO_2C_{10}H_{17}$	196.28	0.991	29	226–7
Bromo-aniline (p-)	$BrC_6H_4NH_2$	172.03	1.8	65–4	....
-benzene	C_6H_5Br	157.02	1.495	−30.6	156.2
-camphor (3-)(d-)	$BrC_{10}H_{15}O$	231.14	1.449	77–8	274
-diphenyl (p-)	$BrC_6H_4.C_6H_5$	233.11	....	90–1	310
-naphthalene (α-)	$C_{10}H_7Br$	207.07	1.482	3–4	281.1
(β-)	$C_{10}H_7Br$	207.07	1.605⁵⁰	59	281–2
-phenol (o-)	BrC_6H_4OH	173.02	1.553³⁰	5.6	194–5
(m-)	BrC_6H_4OH	173.02	....	32–3	236–7
(p-)	BrC_6H_4OH	173.02	1.588⁸⁰	63.5	238
-styrene (ω)(1)	$C_6H_5CH:CHBr$	183.05	1.422	....	221
(2)	$C_6H_5CH:CHBr$	183.05	1.427	7	108²⁶ᵐᵐ
-toluene (o-)	$CH_3.C_6H_4Br$	171.04	1.422	−7.5	181.8
(m-)	$CH_3.C_6H_4Br$	171.04	1.410	−28	183.7
(p-)	$CH_3.C_6H_4Br$	171.04	1.390	28.5	184–5
Bromoform	$CHBr_3$	252.77	2.890	8–9	150.5
Butadiene (1-,2-)	$CH_2CH:C:CH_2$	54.09	....	....	18–9
(1-,3-)	$CH_2:CHCH:CH_2$	54.09	0.621	−108.9	−4.41
Butadienyl acetylene	$CH_2:(CH)_2:CH.C:CH$	78.11	0.773	....	83–6
Butane	$CH_3CH_2CH_2CH_3$	58.12	0.60⁰	−135	−0.6
(i-)	$(CH_3)_2CHCH_3$	58.12	0.60⁰	−145	−10
Butyl acetate (n-)	$CH_3CO_2(CH_2)_2C_2H_5$	116.16	0.882	−76.3	125
(s-)	$CH_3CO_2CH(CH_3)C_2H_5$	116.16	0.865	....	112
(i-)	$CH_3CO_2CH_2CH(CH_3)_2$	116.16	0.871	−98.9	118
(tert-)	$CH_3CO_2C(CH_3)_3$	116.16	0.866	....	95–6
alcohol (n-)	$C_2H_5CH_2CH_2OH$	74.12	0.810	−79.9	117
(s-)	$C_2H_5CH(OH)CH_3$	74.12	0.808	−114.7	99.5
(i-)	$(CH_3)_2CHCH_2OH$	74.12	0.805	−108	107–8
(tert-)	$(CH_3)_3COH$	74.12	0.779	25.5	82.9
amine (n-)	$C_2H_5CH_2CH_2NH_2$	73.14	0.739	−50	77.8
(s-)	$C_2H_5CH(NH_2)CH_3$	73.14	0.724	−104	66
(i-)	$(CH_3)_2CHCH_2NH_2$	73.14	0.732	−85	68–9
(t-)	$(CH_3)_3CNH_2$	73.14	0.698	−67.5	45.2
p-aminophenol (N)(n)	$C_4H_9NH.C_6H_4.OH$	165.23	....	71	....
(N)(i-)	$C_4H_9NH.C_6H_4.OH$	165.23	....	79	....
aniline (n-)	$C_4H_9NHC_6H_5$	149.23	....	....	235
(i-)	$C_4H_9NHC_6H_5$	149.23	0.940	....	231–2
arsonic acid (n-)	$C_4H_9AsO(OH)_2$	182.04	....	158–9	....
benzoate (n-)	$C_6H_5CO_2C_4H_9$	178.22	1.005	−22	249–50
(i-)	$C_6H_5CO_2C_4H_9$	178.22	0.997	....	241.5
bromide (n-)	$C_2H_5CH_2CH_2Br$	137.03	1.277	−112.4	101.6
(s-)	$C_2H_5CH(Br)CH_3$	137.03	1.251	−112	91.3
(i-)	$(CH_3)_2CHCH_2Br$	137.03	1.258	−118.5	91.5
(t2)	$(CH_3)_3CBr$	137.03	1.211	−16.2	73.3
butyrate (n-)(n-)	$C_2H_5CH_2CO_2CH_2CH_2C_2H_5$	144.21	0.872	....	165.7
(n-)(i-)	$C_2H_5CH_2CO_2CH_2CH(CH_3)_2$	144.21	0.863	....	156.9
(i-)(i-)	$(CH_3)_2CHCO_2CH_2CH(CH_3)_2$	144.21	0.875 ⁰	−80.7	148–9
caproate	$CH_3(CH_2)_4CO_2C_4H_9$	172.26	0.882⁰	....	204.3
carbamate (i-)	$NH_2CO_2CH_2CH(CH_3)_2$	117.15	0.956⁷⁶	65	206–7
cellosolve (n-)	$C_4H_9OCH_2CH_2OH$	118.17	0.903	....	171.2
chloride (n-)	$C_2H_5CH_2CH_2Cl$	92.57	0.887	−123.1	77.9
(s-)	$C_2H_5.CHCl.CH_3$	92.57	0.871	−131	67.8
(i-)	$(CH_3)_2CHCH_2Cl$	92.57	0.884	−131.2	68.9
(t-)	$(CH_3)_3CCl$	92.57	0.847	−26.5	51–2
dimethylbenzene (t-)(1-,3-,5-)	$(CH_3)_3C.C_6H_3:(CH_3)_2$	162.26	....	....	200–2¹⁴⁷ᵐᵐ
formate (n-)	$HCO_2CH_2CH_2C_2H_5$	102.13	0.911¹⁰	....	106.9
(s-)	$HCO_2CH(CH_3)C_2H_5$	102.13	0.882	....	97
(i-)	$HCO_2CH_2CH(CH_3)_2$	102.13	0.885	−95.3	98.2

TABLE 1-11   Physical Constants of Elements and Compounds—(Continued)

Name	Formula	Formula weight	Specific gravity	Melting point, °C.	Boiling point, °C.
Butyl furoate (n-)	$OC_4H_3CO_2C_4H_9$	168.19	1.056		$118–20^{25\,mm}$
iodide (n-)	$C_2H_5CH_2CH_2I$	184.03	1.617	−103.5	129.9
(s-)	$C_2H_5CHICH_3$	184.03	1.595	−104	118–9
(i-)	$(CH_3)_2CHCH_2I$	184.03	1.606	−90.7	120
(t-)	$(CH_3)_3CI$	184.03	1.370	−34	99
lactate (n-)	$CH_3CH(OH)CO_2C_4H_9$	146.18	0.968	....	$75–6^{8\,mm}$
mercaptan (n-)	$C_2H_5CH_2CH_2SH$	90.18	0.837	−116	97–8
(i-)	$(CH_3)_2CHCH_2SH$	90.18	0.836	<−79	88
(t-)	$(CH_3)_3CSH$	90.18			65–7
methacrylate (n-)	$CH_2{:}C(CH_3)CO_2C_4H_9$	142.19	0.889		155
(i-)	$CH_2{:}C(CH_3)CO_2C_4H_9$	142.19	0.889		155
phenol (p-)(t-)	$(CH_3)_3C.C_6H4.OH$	150.21	0.908^{112}	99	236–8
propionate (n-)	$C_2H_5CO_2C_4H_9$	130.18	0.883	−89.55	146
(s-)	$C_2H_5CO_2C_4H_9$	130.18	0.866	....	132.5
(i-)	$C_2H_5CO_2C_4H_9$	130.18	0.888^0	−71.4	136.8
stearate (n-)	$CH_3(CH_2)_{16}CO_2C_4H_9$	340.57	0.855	27.5	$220–5^{25\,mm}$
(i-)	$CH_3(CH_2)_{16}CO_2C_4H_9$	340.57		25	
iso-thiocyanate (n-)	$C_2H_5CH_2CH_2.N{:}CS$	115.19	0.956	....	165
(i-)	$(CH_3)_2CHCH_2.N{:}CS$	115.19	0.964	....	162
(s-)(d-)	$C_4H_9.N{:}CS$	115.19	0.943	....	159–63
(t-)	$(CH_3)_3C.N{:}CS$	115.19	0.919	10.5	140
valerate (n-)(n-)	$CH_3(CH_2)_3CO_2(CH_2)_3CH_3$	158.23	0.870	−93	186
(i-)(n-)	$(CH_3)_2CHCH_2CO_2(CH_2)_3CH_3$	158.23	0.862	....	168.8
(i-)(s-)	$(CH_3)_2CHCH_2CO_2C_4H_9$	158.23	0.848	....	163–4
(i-)(i-)	$C_4H_9CO_2C_4H_9$	158.23	0.874^0	....	168.7
Butylene (α-)	$C_2H_5CH{:}CH_2$	56.10	0.6^9	−130	−5
(β-)	$CH_3CH{:}CHCH_3$	56.10		−127	3
Butyraldehyde (n-)	$CH_3CH_2CH_2CHO$	72.10	0.817	−99	75.7
(i-)	$(CH_3)_2CHCHO$	72.10	0.794	−65.9	64
Butyric acid (n-)	$C_2H_5CH_2CO_2H$	88.10	0.964	−4.7	163.5
(i-)	$(CH_3)_2CHCO_2H$	88.10	0.949	−47	154.5
amide (n-)	$C_2H_5CH_2CONH_2$	87.12	1.032	115–6	216
(i-)	$(CH_3)_2CHCONH_2$	87.12	1.013	129–30	216–20
anhydride (n-)	$(C_2H_5CH_2CO)_2O$	158.19	0.968	−75	199.5
(i-)	$[(CH_3)_2CHCO]_2O$	158.19	0.950	−53.5	181.5
anilide (n-)	$C_3H_7CONHC_6H_5$	163.21	1.134	92	$189^{15\,mm}$
Caffeic acid (3-,4-)	$(HO)_2C_6H_3C_2H_2CO_2H$	180.15		195–213	d.
Caffeine	$C_8H_{10}O_2N_4.H_2O$	212.21	1.23	237	subl.
Camphene (dl-)	$C_{10}H_{16}$	136.23	0.822^{78}	50	160
(d- or l-)	$C_{10}H_{16}$	136.23	0.845^{50}	42.7	159.6
Camphor (d-)	$C_{10}H_{16}O$	152.23	0.999	178–9	209.1
Camphoric acid (d-)	$C_8H_{14}(CO_2H)_2$	200.23	1.186	187	....
Cantharidine	$C_{10}H_{12}O_4$	196.20		212	
Capric acid	$CH_3(CH_2)_8CO_2H$	172.26	0.889	31.5	268–70
Caproic acid (n-)	$CH_3(CH_2)_4CO_2H$	116.16	0.922	−1.5	202
(i-)	$(CH_3)_2CH(CH_2)_2.CO_2H$	116.16	0.925	−35	207.7
Caprylic acid (n-)	$CH_3(CH_2)_6CO_2H$	144.21	0.910	16	237.5
Carbazole	$(C_6H_4)_2NH$	167.20		244.8	354.8
Carbitol	$C_2H_5O(CH_2)_2O(CH_2)_2OH$	134.17	0.990		201.9
Carbon disulfide	CS_2	76.13	1.263	−108.6	46.3
monoxide	CO	28.01	0.81^{-195}	−207	−192
suboxide	$OC{:}C{:}CO$	68.03	1.114^0	−107	7
tetrabromide	CBr_4	331.67	3.42	90.1	189.5
tetrachloride	CCl_4	153.84	1.595	−22.6	76.8
tetrafluoride	CF_4	88.01			−128
Carbonyl sulfide	COS	60.07	1.24^{-87}	−138.2	−50.2
Carminic acid	$C_{22}H_{20}O_{13}$	492.40		d. 136	....
Carvacrol (1-,2-,4-)	$CH_3C_6H_3(OH)CH(CH_3)_2$	150.21	0.977	0.5	238
Carvacrylamine (2-,1-,4-)	$H_2NC_6H_3(CH_3)C_3H_7$	149.23	0.994	−16	241
Carvone (d-)	$C_{10}H_{14}O$	150.21	0.961		230
Cellosolve	$C_2H_5O(CH_2)_2OH$	90.12	0.931	−70	135.1
acetate	$CH_3CO_2CH_2CH_2OC_2H_5$	132.16	0.975		156.3
Cellulose	$(C_6H_{10}O_5)x$	162.14	1.3–1.4		
Cetyl acetate	$CH_3CO_2(CH_2)_{15}CH_3$	284.47	0.858	22–3	$200^{15\,mm}$
alcohol	$CH_3(CH_2)_{14}CH_2OH$	242.43	0.818^{50}	49–50	$189.5^{15\,mm}$
Chloral	$CCl_3.CHO$	147.40	1.505	−57	97.6
hydrate	$CCl_3.CH(OH)_2$	165.42	1.619	51.7	d. 98
Chloranil	$OC{:}(CCl.CCl)_2{:}CO$	245.89		290	subl.
Chloretone	$Cl_3C.C(OH)(CH_3)_2$	177.47		97	167
Chloro-acetanilide (p-)	$CH_3CO_2NHC_6H_4Cl$	169.61	1.385	175–6	....
-acetic acid	$ClCH_2CO_2H$	94.50	1.58	61.2	189.5
-acetone	CH_3COCH_2Cl	92.53	1.162	−44.5	121
-acetophenone (ω-)	$C_6H_5COCH_2Cl$	154.59	1.324	58–9	245–7
-acetyl chloride	$ClCH_2COCl$	112.95	1.498		105
-aniline (o-)	$ClC_6H_4NH_2$	127.57	1.213	0	210.5
(m-)	$ClC_6H_4NH_2$	127.57	1.216	−10.4	230
(p-)	$ClC_6H_4NH_2$	127.57	1.427	70–1	230–1
-anthraquinone (1-)	$C_6H_4(CO)_2C_6H_3Cl$	242.65		162	subl.
(2-)	$C_6H_4(CO)_2C_6H_3Cl$	242.65		208–9	
-benzaldehyde (o-)	ClC_6H_4CHO	140.57	1.29	11	208
(m-)	ClC_6H_4CHO	140.57	1.250	17–8	213–4
(p-)	ClC_6H_4CHO	140.57	1.196^{61}	47.8	213
-benzene	C_6H_5Cl	112.56	1.107	−45.2	132.1
-benzoic acid (o-)	$ClC_6H_4CO_2H$	156.57	1.544	141–2	....
(m-)	$ClC_6H_4CO_2H$	156.57	1.496	158	....
(p-)	$ClC_6H_4CO_2H$	156.57	1.541	242–3	subl.

TABLE 1-11   Physical Constants of Elements and Compounds—(Continued)

Name	Formula	Formula weight	Specific gravity	Melting point, °C.	Boiling point, °C.
Chloro-buta-1,3-diene (2-)............	$CH_2:CCl.CH:CH_2$	88.54	0.958	....	59.4
(1-)...............	$CH_2:CH.CH:CHCl$	88.54	0.965	....	69
-buta-1,2-diene (4-)...............	$CH_2:C:CH.CH_2Cl$	88.54	0.991	....	88
-dimethylhydantoin.............	$-C(CH_3)_2N(Cl)CON(Cl)CO-$	197.03	1.5	130	....
-dinitrobenzene (α)(1-,2-)(4-)....	$ClC_6H_3(NO_2)_2$	202.56	....	39	315 d.
(α)(1-,3-)(4-)....	$ClC_6H_3(NO_2)_2$	202.56	1.697	53	315 d.
-diphenyl (o-)............	$C_6H_5.C_6H_4Cl$	188.65	....	34	267–8
(m-)............	$C_6H_5.C_6H_4Cl$	188.65	....	89	284–5
(p-)............	$C_6H_5.C_6H_4Cl$	188.65	....	77.5	282
-hydroquinone.............	$ClC_6H_3(OH)_2$	144.56	....	106	263 sl. d.
-naphthalene (α-)..........	$C_{10}H_7Cl$	162.61	1.194	−20	259.3
(β-)..........	$C_{10}H_7Cl$	162.61	1.266	56–7	264
-nitrobenzene (o-)...........	$ClC_6H_4NO_2$	157.56	1.305⁸⁰	32.5	245.5
(m-)...........	$ClC_6H_4NO_2$	157.56	1.343⁵⁰	44.4	235.6
(p-)...........	$ClC_6H_4NO_2$	157.56	1.298⁹¹	83–4	242
-nitrotoluene (2-,4-)...........	$CH_3C_6H_3(NO_2)(Cl)$	171.56	1.256⁸⁰	38.2	240
(2-,6-)...........	$CH_3C_6H_3(NO_2)(Cl)$	171.56	....	37.5	238
-phenol (o-)...............	ClC_6H_4OH	128.56	1.241	7	175–6
(m-)...............	ClC_6H_4OH	128.56	1.268	32–3	214
(p-)...............	ClC_6H_4OH	128.56	1.306	41–3	217
-propionic acid (α)(dl-)..........	$CH_3.CHCl.CO_2H$	108.53	1.306	< −20	186
-toluene (o-)...............	$CH_3.C_6H_4Cl$	126.58	1.082	−34	159.5
(m-)...............	$CH_3.C_6H_4Cl$	126.58	1.072	−47.8	161.6
(p-)...............	$CH_3.C_6H_4Cl$	126.58	1.070	7.5	162.2
Chloroform..................	$CHCl_3$	119.39	1.489	−63.5	61.2
Chlorophyll (α-)...........	$C_{55}H_{72}O_5N_4Mg$	893.48	....	....	d.
Chloropicrin.................	Cl_3CNO_2	164.39	1.651	−64	112.3
Cholesterol.................	$C_{27}H_{45}OH.H_2O$	404.65	1.067	149–51	subl.
Chrysene..................	$C_{18}H_{12}$	228.28	....	253–4	448
Chrysoidine (2-,4-)...........	$C_6H_5.N:N.C_6H_3(NH_2)_2$	212.25	....	117.5	....
Chrysophanic acid...........	$C_{14}H_5(OH)_2(CH_3)O_2$	254.23	....	195	subl.
Cinchomeronic acid (3-,4-)......	$C_5H_3N(CO_2H)_2$	167.12	....	258–9 d.	subl. d.
Cineole, eucalyptole..........	$C_{10}H_{18}O$	154.24	0.927	1.5	176–7
Cinnamic acid (cis-)...........	$C_6H_5CH:CHCO_2H$	148.15	1.284	68	125
(trans-)..........	$C_6H_5CH:CHCO_2H$	148.15	1.245	133	300
aldehyde........	$C_6H_5CH:CHCHO$	132.15	1.110	−7.5	252 sl. d.
Cinnamyl alcohol..............	$C_6H_5CH:CHCH_2OH$	134.17	1.040	33	257.5
cinnamate...........	$C_8H_7O_2C_9H_9$	264.31	1.085	44	....
Citraconic acid (cis-)..........	$CH_3C(CO_2H):CHCO_2H$	130.10	1.617	92–3	....
Citral (α)............	$C_9H_{16}CHO$	152.23	0.890	....	229
Citric acid..................	$C_3H_4(OH)(CO_2H)_3$	192.12	1.542	153	d.
Citronellal (d-)...............	$C_9H_{17}.CHO$	154.24	0.855	....	204–8
Citronellol (d-)...............	$C_{10}H_{20}O$	156.26	0.848	....	224–5
Coniine (d-)(2-)..............	$C_8H_7.C_6H_{10}N$	127.22	0.847	−2	166–7
Coumaric acid (o-)............	$HOC_6H_4CH:CHCO_2H$	164.15	....	207–8	subl.
(p-)............	$HOC_6H_4CH:CHCO_2H$	164.15	....	206–7 d.	....
Coumarin..................	$C_9H_6O_2$	146.14	0.935	70	290–1
Coumarone..................	C_8H_6O	118.13	1.078	< −18	173–4
Creatine..................	$C_4H_9N_3O_2.H_2O$	149.15	....	295	....
Creatinine.................	$C_4H_7N_3O$	113.12	....	260 d.	....
Creosol (3-,1-,4-)............	$CH_3.C_6H_3(CH_3)OH$	138.16	1.092	5.5	221–2
Cresidine (1-,2-,4-)...........	$CH_3(NH_2)C_6H_3.OCH_3$	137.18	....	93–4	235
Cresol (o-).................	$CH_3C_6H_4OH$	108.13	1.048	30.8	190.8
(m-).................	$CH_3C_6H_4OH$	108.13	1.034	10.9	202.8
(p-).................	$CH_3C_6H_4OH$	108.13	1.035	35–6	202
Cresyl benzoate (o-)...........	$C_6H_5CO_2C_6H_4CH_3$	212.24	....	....	308
(m-)...........	$C_6H_5CO_2C_6H_4CH_3$	212.24	....	55	314
(p-)...........	$C_6H_5CO_2C_6H_4CH_3$	212.24	....	71.5	316
Crotonic acid (α-,1-,4-)........	$CH_3CH:CHCO_2H$	86.09	0.964⁷⁹·⁷	72	189
acid (β-)(cis-).........	$CH_3CH:CHCO_2H$	86.09	1.031	15.5	170–1 d.
aldehyde (α)...........	$CH_3CH:CHCHO$	70.09	0.853	−69	102.2
Cumene..................	$C_6H_5CH(CH_3)_2$	120.19	0.862	−96.9	152.5
Cumic acid (p-).............	$(CH_3)_2CH.C_6H_4CO_2H$	164.20	1.162⁴	116–7	subl.
Cumidine (p-)...............	$(CH_3)_2CH.C_6H_4NH_2$	135.20	0.953	< −20	225
Cyanamide.................	$H_2N.CN$	42.04	1.073⁴⁸	44–5	140¹⁹ᵐᵐ
Cyanic acid................	$HOCN$ or $HNCO$	43.03	1.140⁰	−80	−64⁰ᵐᵐ
Cyanoacetic acid.............	$CH_2(CN)CO_2H$	85.06	....	65–6	108⁰·²ᵐᵐ
Cyanogen..................	$(CN)_2$	52.04	0.866	−34.4	−21
bromide................	$BrCN$	105.93	2.015	52	61.3
chloride................	$ClCN$	61.48	1.222⁰	−6.5	12.5–13
Cyanuric acid..............	$C_3H_3O_3N_3.2H_2O$	165.11	1.768⁰	> 360	d.
Cyclo-butane...............	$CH_2 < (CH_2)_2 > CH_2$	56.10	0.703³⁰	−50	11–12
-heptane..............	$CH_2 < (CH_2CH_2CH_2)_2 >$	98.18	0.810	−12	118–20
-hexane..............	$CH_2 < (CH_2CH_2)_2 > CH_2$	84.16	0.779	6.5	80–1
-hexanol..............	$CH_2 < (CH_2CH_2)_2 > CHOH$	100.16	0.962	23.9	160–1
-hexanone.............	$CH_2 < (CH_2CH_2)_2 > CO$	98.14	0.947	−45	155–6
-hexene..............	$(.CH_2.CH_2CH:)_2$	82.14	0.810	−103.7	83.3
-hexyl acetate...........	$CH_3CO_2C_6H_{11}$	142.19	0.985⁰	....	174
amine.............	$CH_2 < (CH_2CH_2)_2 > CHNH_2$	99.17	0.865	....	134
bromide............	$CH_2 < (CH_2CH_2)_2 > CHBr$	163.06	1.324	....	165
chloride...........	$CH_2 < (CH_2CH_2)_2 > CHCl$	118.61	0.977	−43.9	142
-pentadiene (1-,3-)...........	$CH_2 < (CH:CH)_2 >$	66.10	0.805	−85	41–2
-pentane.................	$CH_2 < (CH_2CH_2)_2 >$	70.13	0.745	−93.3	49–50
-pentanone...............	$< (CH_2CH_2)_2 > CO$	84.11	0.948	−58.2	129–30
-propane.................	$< CH_2CH_2CH_2 >$	42.08	0.720⁻⁷⁹	−126.6	−34

TABLE 1-11 Physical Constants of Elements and Compounds—(Continued)

Name	Formula	Formula weight	Specific gravity	Melting point, °C.	Boiling point, °C.
Cymene (o-)	$CH_3.C_6H_3CH(CH_3)_2$	134.21	0.875	......	177
(m-)	$CH_3.C_6H_4CH(CH_3)_2$	134.21	0.862	<-26	175-6
(p-)	$CH_3.C_6H_4CH(CH_3)_2$	134.21	0.857	-73.5	176-7
Cystine (l-)	$[.SCH_2CH(NH_2)CO_2H]_2$	240.29	......	d. 258-61	......
Dambose	$C_6H_6(OH)_6$	180.16	1.752	253	319^{15mm}
Decahydronaphthalene (cis-)	$C_{10}H_{18}$	138.24	0.895	-51	193.3
(trans-)	$C_{10}H_{18}$	138.24	0.872	-32	185.3
Decane (n-)	$CH_3(CH_2)_8CH_3$	142.28	0.730^2	-29.7	174.0
Decyl alcohol	$CH_3(CH_2)_8CH_2OH$	158.28	0.830	7	232.9
Dextrin	$(C_6H_{10}O_5)x$	162.14	1.038	......	......
Diacetone alcohol	$(CH_3)_2C(OH).CH_2COCH_3$	116.16	0.931	-47	167.9
Diamino-benzophenone (4-,4'-)	$H_2NC_6H_4COC_6H_4NH_2$	212.24	......	237-9	......
-diphenylamine (4,4'-)	$H_2NC_6H_4NHC_6H_4NH_2$	199.25	......	158	d.
-diphenylmethane (4,4'-)	$H_2NC_6H_4CH_2C_6H_4NH_2$	198.26	......	93-4	$249-53^{15mm}$
-diphenylurea (4,4'-)	$(H_2NC_6H_4NH)_2CO$	242.28	......	subl. 310	......
Diamyl-amine (i-)	$[(CH_3)_2CHCH_2CH_2]_2NH$	157.29	0.767	-44	188-90
ether (n-)	$(C_2H_5CH_2CH_2CH_2)_2O$	158.28	0.774	-69	190
(i-)	$[(CH_3)_2CH(CH_2)_2]_2O$	158.28	0.777	......	173.4
Diamyl ketone (i-)	$[(CH_3)_2CHCH_2CH_2]_2CO$	170.29	0.821	14.6	228
phthalate (n-)	$C_6H_4(CO_2C_5H_{11})_2$	306.39	......	......	$204-6^{11mm}$
(i-)	$C_6H_4(CO_2C_5H_{11})_2$	306.39	1.03	......	225^{40mm}
tartrate (i-)	$(HOCH.CO_2C_5H_{11})_2$	290.35	1.063	......	195^{16mm}
Dianisidine (o-)(4-,3-)₂	$[NH_2(OCH_3)C_6H_3.]_2$	244.28	......	131.5	......
Diazo-aminobenzene	$C_6H_5N:N.NHC_6H_5$	197.23	......	96-8	expl.
-aminotoluene (2-,2'-)	$C_7H_7N:N.NHC_7H_7$	225.28	......	51	......
-methane	$CH_2:N_2$	42.04	......	-145	-23
Dibenzothiazyl-disulfide (2-,2'-)	$(C_6H_4NSC)_2S_2$	232.46	1.50	180	d.
Dibenzoyl methane	$(C_6H_5CO)_2CH_2$	224.25	......	78	$219-21^{18mm}$
Dibenzyl-amine	$(C_6H_5CH_2)_2NH$	197.27	1.028	-26	$268-7^{250mm}$
-aniline	$C_6H_5N(CH_2C_6H_5)_2$	273.36	......	70-1	>300
ketone	$(C_6H_5CH_2)_2CO$	210.26	......	34-5	330.6
phthalate (o-)	$C_6H_4(CO_2CH_2C_6H_5)_2$	346.36	......	42-3	274^{12mm}
succinate	$(.CH_2CO_2CH_2C_6H_5)_2$	298.32	......	45-6	238^{14mm}
Dibromo-benzene (o-)	$C_6H_4Br_2$	235.92	1.956	1.8	221-2
(m-)	$C_6H_4Br_2$	235.92	1.952	-6.9	219
(p-)	$C_6H_4Br_2$	235.92	2.261	87-8	218.6
-diphenyl (4,4'-)	$BrC_6H_4.C_6H_4Br$	312.02	1.897	164-5	355-60
Dibutyl-adipate (n-)	$(.CH_2CH_2CO_2C_4H_9)_2$	258.35	0.965	-38	183^{14mm}
(i-)	$(.CH_2CH_2CO_2C_4H_9)_2$	258.35	0.950	-20	278-80
-amine (n-)	$(C_2H_5CH_2CH_2)_2NH$	129.24	0.768	......	159
(i-)	$[(CH_3)_2CHCH_2]_2NH$	129.24	0.741	-70	139-40
-p-aminophenol (s-)	$(C_4H_9)_2N.C_6H_4OH$	221.33	......	......	170^{10mm}
-aniline (n-)	$C_6H_5N(C_4H_9)_2$	205.33	......	......	262.8
carbonate (n-)	$CO(OC_4H_9)_2$	174.23	0.924	......	207
(i-)	$CO(OC_4H_9)_2$	174.23	0.919	......	190
(s-)	$CO(OC_4H_9)_2$	174.23	......	......	178-80
ether (n-)	$(C_2H_5CH_2CH_2)_2O$	130.22	0.769	-98	142.4
(i-)	$[(CH_3)_2CHCH_2]_2O$	130.22	0.762	......	122.5
(s-)	$[C_2H_5(CH_3)CH]_2O$	130.22	0.756	......	121
ketone (n-)	$(C_2H_5CH_2CH_2)_2CO$	142.23	0.827	-5.9	187.7
(i-)	$[(CH_3)_2CHCH_2]_2CO$	142.23	0.805	......	168.1
malate (l-)(n-)	$C_2H_4O(CO_2C_4H_9)_2$	246.30	1.038	......	$170-1^{13mm}$
oxalate (n-)	$(.CO_2C_4H_9)_2$	202.24	0.986	-29.6	245.5
phthalate (n-)	$C_6H_4(CO_2C_4H_9)_2$	278.34	1.045	......	340
tartrate (d-)(n-)	$(CHOHCO_2C_4H_9)_2$	262.30	1.098	22-2.5	$200-3^{18mm}$
(d-)(i-)	$(CHOHCO_2C_4H_9)_2$	262.30	1.03^{75}	73-4	323-5
Dichloro-acetic acid	$Cl_2CH.CO_2H$	128.95	1.560	9.7	194.4
-acetone (αα-)	$Cl_2CHCOCH_3$	126.98	1.234	......	120
-aniline (2,5-)	$Cl_2C_6H_3NH_2$	162.02	......	50	251
-anthraquinone (1,3-)	$C_6H_4:(CO)_2:C_6H_2Cl_2$	277.10	......	208-9	......
(1,4-)	$C_6H_4:(CO)_2:C_6H_2Cl_2$	277.10	......	187.5	......
(1,5-)	$C_6H_3Cl:(CO)_2:C_6H_3Cl$	277.10	......	251	......
(1,6-)	$C_6H_3Cl:(CO)_2:C_6H_3Cl$	277.10	......	203-4	......
(1,8-)	$C_6H_3Cl:(CO)_2:C_6H_3Cl$	277.10	......	202-3	......
(2,3-)	$C_6H_4:(CO)_2:C_6H_2Cl_2$	277.10	......	268-70	......
(2,6-)	$C_6H_3Cl:(CO)_2:C_6H_3Cl$	277.10	......	282	......
(2,7-)	$C_6H_3Cl:(CO)_2:C_6H_3Cl$	277.10	......	210-11	......
-benzene (o-)	$C_6H_4Cl_2$	147.01	1.305	-17.6	179
(m-)	$C_6H_4Cl_2$	147.01	1.288	-24.8	172
(p-)	$C_6H_4Cl_2$	147.01	1.458	53	174
-butane (n-)(1,4-)	$ClCH_2(CH_2)_2CH_2Cl$	127.02	......	-38.7	161-3
-diphenyl (4,4'-)	$ClC_6H_4.C_6H_4Cl$	223.10	1.442^{20}	148	315-9
-ethane (1,2-)	$ClCH_2.CH_2Cl$	98.97	1.256	-35.3	83.7
-naphthalene (β-)(1,4-)	$C_{10}H_6Cl_2$	197.06	1.300^{76}	67-8	286-7
(γ-)(1,5-)	$C_{10}H_6Cl_2$	197.06	......	107	subl.
-nitrobenzene (2,5-)	$Cl_2C_6H_3NO_2$	192.01	1.669	54.6	266
-pentane (1,5-)	$ClCH(CH_2)_3CH_2Cl$	141.04	1.094	......	180-1
-phenol (2,4-)	$Cl_2C_6H_3OH$	163.01	1.383^{60}	45	209-10
Dichloramine T (p-)	$CH_3C_6H_4SO_2NCl_2$	240.11	......	83	......
Dicyandiamide	$H_2N.C(:NH).NH.CN$	84.08	1.40	207-8	d.
Diethanolamine	$HN(CH_2CH_2OH)_2$	105.14	1.097	28	270
Diethyl adipate	$(.CH_2CH_2CO_2C_2H_5)_2$	202.24	1.009	-21	239-41
-amine	$(C_2H_5)_2NH$	73.14	0.712	-38.9	55.5
-aminophenol (m-)	$(C_2H_5)_2N.C_6H_4.OH$	165.23	......	78	276-80
-aniline	$(C_2H_5)_2NC_6H_5$	149.23	0.934	-34.4	216
sulfonic acid (m-)	$(C_2H_5)_2NC_6H_4SO_3H$	229.29	......	270 d.	......
carbonate	$OC(OC_2H_5)_2$	118.13	0.975	-43	126
diethyl malonate	$(C_2H_5)_2C(CO_2C_2H_5)_2$	216.27	0.985	......	230

TABLE 1-11   Physical Constants of Elements and Compounds—(Continued)

Name	Formula	Formula weight	Specific gravity	Melting point, °C.	Boiling point, °C.
Diethyl dimethyl malonate..........	$(CH_3)_2C(CO_2C_2H_5)_2$	188.22	0.994		196.7
glutarate.....................	$CH_2(CH_2CO_2C_2H_5)_2$	188.22	1.025	−24	237
ketone.......................	$(C_2H_5)_2CO$	86.13	0.816	−42	101.7
malonate.....................	$CH_2(CO_2C_2H_5)_2$	160.17	1.055	−49.8	198.9
-malonic acid.................	$(C_2H_5)_2C(CO_2H)_2$	160.17		125	d. 170–80
-naphthylamine (α-)........	$C_{10}H_7N(C_2H_5)_2$	199.28	1.005	....	285–90
(β-)........	$C_{10}H_7N(C_2H_5)_2$	199.28	1.026	....	318
oxalate......................	$(.CO_2C_2H_5)_2$	146.14	1.079	−40.6	186
phthalate (o-)................	$C_6H_4(CO_2C_2H_5)_2$	222.23	1.121	....	298–9
sulfate......................	$O_2S(OC_2H_5)_2$	154.18	1.172	−25	210
sulfide......................	$(C_2H_5)_2S$	90.18	0.837	−99.5	92–3
tartrate (d-).................	$(CHOH.CO_2C_2H_5)_2$	206.19	1.204	17	280
-toluidine (o-)...............	$CH_3.C_6H_4.N(C_2H_5)_2$	163.25		....	208–9
(m-)...............	$CH_3.C_6H_4.N(C_2H_5)_2$	163.25	....	....	231–2
(p-)...............	$CH_3.C_6H_4.N(C_2H_5)_2$	163.25	0.924	....	228–9
Diethyleneglycol dinitrate..........	$O(CH_2CH_2ONO_2)_2$	196.12	1.377	−11.3	....
Difluorodichloromethane............	F_2CCl_2	120.92	1.486^{-80}	−155	−29.2
Diglycerol........................	$[(HO)_2C_3H_5]_2O$	166.17		....	$220–30^{10 mm}$
Dihydroxy-dinaphthyl (α-)........	$(HO.C_{10}H_6.)_2$	286.31	....	300	....
(-2,-2',-1,-1')..	$(HO.C_{10}H_6.)_2$	286.31	....	218	subl.
-diphenyl (4-,4'-)............	$(HO.C_6H_4.)_2$	186.20	1.25	270–2	subl.
-ethyl formal (β-).........	$CH_2(OCH_2CH_2OH)_2$	136.15	1.154	−5.3	264
-naphthalene (1-,5-).........	$C_{10}H_6(OH)_2$	160.16	....	258–60	d.
(1-,8-).........	$C_{10}H_6(OH)_2$	160.16	....	140	
Dimethoxy-benzene (p-)........	$(CH_3O)_2C_6H_4$	138.16	1.053	56	212.6
-diphenylamine (4-,4'-)......	$HN(C_6H_4OCH_3)_2$	229.26		103	
-ethyl adipate...............	$(CH_2)_4(CO_2C_2H_4OCH_3)_2$	262.30	1.075		$145–50^{2 mm}$
Dimethyl adipate...................	$[(CH_2)_2CO_2CH_3]_2$	174.19	1.063	10–1	$115^{18 mm}$
-amine.......................	$(CH_3)_2NH$	45.08	0.680^0	−96	7.4
-aminoazobenzene (p-)........	$C_6H_5N:N.C_6H_4N(CH_3)_2$	225.28		116–7	d.
-aminoethanol...............	$(CH_3)_2NCH_2CH_2OH$	89.14	0.887		135
-aminophenol (m-)............	$(CH_3)_2NC_6H_4OH$	137.18		85	265–8
-aniline.....................	$(CH_3)_2NC_6H_5$	121.18	0.956	2.5	193
sulfonic acid (m-)........	$(CH_3)_2NC_6H_4SO_3H$	201.24		d. 266	
(p-)........	$(CH_3)_2NC_6H_4SO_3H.H_2O$	219.25		257	....
carbonate....................	$OC(OCH_3)_2$	90.08	1.070	0.5	89–90
ether........................	CH_3OCH_3	46.07		−138.5	−23.7
formamide...................	$HCON(CH_3)_2$	73.09	0.945	−58.3	152.8
fumarate.....................	$(:CHCO_2CH_3)_2$	144.12		102	192
glutarate....................	$(CH_2)_3(CO_2CH_3)_2$	160.17	1.089	−37	$130^{60 mm}$
glyoxime.....................	$(CH_3.C:NOH)_2$	116.12		240–6	....
-naphthalene (1-,4-).........	$C_{10}H_6(CH_3)_2$	156.22	1.016	<−18	264–6
(2-,3-).........	$C_{10}H_6(CH_3)_2$	156.22		104	265
-naphthylamine (α-)........	$C_{10}H_7N(CH_3)_2$	171.23	1.042		274.5
(β-)........	$C_{10}H_7N(CH_3)_2$	171.23	1.039^{70}	46	304–5
oxalate......................	$(.CO_2CH_3)_2$	118.09	1.148^{54}	54	163.3
phthalate (o-)...............	$C_6H_4(CO_2CH_3)_2$	194.18	1.189		280
sulfate......................	$(CH_3)_2SO_4$	126.13	1.352^0	−26.8	188.3
sulfide......................	$(CH_3)_2S$	62.13	0.846	−83.2	37.3
tartrate (d-).................	$(CHOH.CO_2CH_3)_2$	178.14	1.328	61.5	280
-vinyl-ethenyl carbinol......	$(CH_3)_2COH.C:C.OH:CH_2$	110.15	0.887		150
Dinaphthyl ($\alpha\alpha$-)...............	$C_{10}H_7.C_{10}H_7$	254.31	....	160	$240–412 mm$
-methane ($\alpha\alpha'$-)...........	$(C_{10}H_7)_2CH_2$	268.34	....	109	>360
(β,β'-)...........	$(C_{10}H_7)_2CH_2$	268.34	....	92	....
Dinitro-anisole (1-)(2,4-)...........	$CH_3OC_6H_3(NO_2)_2$	198.13	1.341	94–5	....
-benzene (o-)................	$C_6H_4(NO_2)_2$	168.11	1.59	117–8	319
(m-)................	$C_6H_4(NO_2)_2$	168.11	1.575	89.8	300–2
(p-)................	$C_6H_4(NO_2)_2$	168.11	1.625	173–4	299
sulfonic acid (2,4-)(1-)..	$(NO_2)_2C_6H_3SO_3H.3H_2O$	302.22	....	106–8	
-benzoic acid (2,4-).........	$(NO_2)_2C_6H_3CO_2H$	212.12	....	179–80	
(3,5-).........	$(NO_2)_2C_6H_3CO_2H$	212.12	....	204–5	subl.
-benzophenone (4,4'-)........	$(NO_2C_6H_4)_2CO$	272.21	....	189	....
-diphenyl (4,4'-)............	$(NO_2C_6H_4)_2$	244.20	1.445	233	....
(2,4'-)............	$(NO_2C_6H_4)_2$	244.20	1.474	93.5	....
-naphthalene (1-,5-).........	$C_{10}H_6(NO_2)_2$	218.16	....	216	subl.
(1-,8-).........	$C_{10}H_6(NO_2)_2$	218.16	....	170–2	d.
Dinitro-phenol (2-,3-)...............	$(NO_2)_2C_6H_3OH$	184.11	1.681	144–5	....
(2-,4-)...............	$(NO_2)_2C_6H_3OH$	184.11	1.683	114–5	subl.
(2-,6-)...............	$(NO_2)_2C_6H_3OH$	184.11	....	63–4	
-salicylic acid (3-,5-).......	$(NO_2)_2C_6H_2(OH)CO_2H.H_2O$	246.13	....	173 d.	
-stilbene (4,4'-)............	$(NO_2C_6H_4CH:)_2$	270.24	....	210–6	
-toluene (2,4-)..............	$(NO_2)_2C_6H_3CH_3$	182.13	1.321^{71}	70	300
(3,4-)..............	$(NO_2)_2C_6H_3CH_3$	182.13	1.259^{111}	60–1	
(3,5-)..............	$(NO_2)_2C_6H_3CH_3$	182.13	1.277^{111}	92–3	subl.
Dioxane..........................	$O<(CH_2.CH_2)_2>O$	88.10	1.033	9.5–10.5	101.1
Dipentene........................	$C_{12}H_{16}$	136.23	0.865		178
Diphenyl..........................	$C_6H_5.C_6H_5$	154.20	0.992^{73}	69–70	254.9
-amine........................	$C_6H_5NHC_6H_5$	169.22	1.160	52.9	302
carbonate....................	$O(COC_6H_5)_2$	214.21	1.272	80	302–6
-chloroarsine................	$(C_6H_5)_2AsCl$	264.57	1.583	43–4	d. 327
-ethane......................	$(C_6H_5CH_2)_2$	182.25	0.978	52–3	284
ether........................	$C_6H_5OC_6H_5$	170.20	1.073	27	259
guanidine....................	$(C_6H_5NH)_2C:NH$	211.26	....	147–8	d. >170
-methane.....................	$(C_6H_5)_2CH_2$	168.23	1.001	26–7	265
phenylenediamine (p-)........	$(C_6H_5NH)_2C_6H_4$	260.32	....	152	....
succinate....................	$(.CH_2CO_2C_6H_5)_2$	270.27	....	122–3	330
sulfide......................	$(C_6H_5)_2S$	186.26	1.119	<−40	296–7

TABLE 1-11   Physical Constants of Elements and Compounds—(Continued)

Name	Formula	Formula weight	Specific gravity	Melting point, °C.	Boiling point, °C.
Diphenyl sulfone	$(C_6H_5)_2SO_2$	218.26	1.248	128–9	379
urea (uns.)	$(C_6H_5)_2NCONH_2$	212.24	1.276	189	....
Diphenylene oxide	$< (C_6H_4)_2O$	168.18		86–7	287–8
Dipropyl adipate (n-)	$(.CH_2CH_2CO_2C_3H_7)_2$	230.30	0.979	−20.3	143–5 ^{10mm}
-amine (n-)	$(C_2H_5CH_2)_2NH$	101.19	0.739	−39.6	110–1
(i-)	$[(CH_3)_2CH]_2NH$	101.19	0.722	−61	83.5
aniline (n-)	$C_6H_5N(C_3H_7)_2$	177.28	0.910	....	245.4
carbonate (n-)	$O(COCH_2C_2H_5)_2$	146.18	0.968	....	168.2
ether (n-)	$(C_2H_5CH_2)_2O$	102.17	0.744	−122	91
(i-)	$[(CH_3)_2CH]_2O$	102.17	0.725	−60	69
ketone (n-)	$(C_2H_5CH_2)_2CO$	114.18	0.822	−32.6	144.2
(i-)	$[(CH_3)_2CH]_2CO$	114.18	0.806		123.7
oxalate (n-)	$(CO_2CH_2C_2H_5)_2$	174.19	1.038^0	−51.7	213.5
(i-)	$[CO_2CH(CH_3)_2]_2$	174.19			190
Disalicylal ethylenediamine	$[HOC_6H_4CH{:}NCH_2.]_2$	268.30	1.34	125–6	....
Ditolyl guanidine (o-)	$(C_7H_7NH)_2C{:}NH$	239.31	1.10	178–9	....
Divinyl acetylene	$(H_2C{:}CH.C{:})_2$	78.11	0.776		85
Docosane (n-)	$CH_3(CH_2)_{20}CH_3$	310.59	0.778	44.5	224.5 ^{15mm}
Dodecane (n-)	$CH_3(CH_2)_{10}CH_3$	170.33	0.751	−9.6	214.5
Dulcitol	$CH_2OH(CHOH)_4CH_2OH$	182.17	1.466	189	290–5 ^{3mm}
Durene (1-,2-,4-,5-)	$(CH_3)_4C_6H_2$	134.21	0.838^{81}	79–80	193–5
Elaidic acid	$C_8H_{17}CH{:}CH(CH_2)_7CO_2H$	282.45	0.851^{79}	51–2	288 ^{100mm}
Eosine	$C_{20}H_8O_5Br_4$	647.93	....	....	....
Ephedrine (l-)	$C_6H_5CHOHCH(CH_3)NHCH_3$	165.23	....	40	255
Epichlorhydrin (α-)	$C_3H_5O.CH_2Cl$	92.53	1.183	−25.6	117
Epidichlorohydrin (α-)	$CH_2{:}CCl.CH_2Cl$	110.98	1.204		94
Erythritol (dl-)	$CH_2OH(CHOH)_2CH_2OH$	122.12	1.451	126	329–31
tetranitrate	$C_4H_6(ONO_2)_4$	302.12		61	expl.
Ethane	CH_3CH_3	30.07	0.546^{-88}	−172	−88.6
Ethanol-amine	$HOCH_2CH_2NH_2$	61.08	1.022	10.5	171
formamide	$HCONHCH_2CH_2OH$	89.09	1.169	<-40	d.
Ether	$(CH_3CH_2)_2O$	74.12	0.708	−116.3	34.6
Ethyl abietate	$C_{19}H_{29}CO_2C_2H_5$	330.49	1.020	....	200 ^{4mm}
acetate	$CH_3CO_2C_2H_5$	88.10	0.901	−82.4	77.1
acetoacetate	$CH_3COCH_2CO_2C_2H_5$	130.14	1.025	−45	180
alcohol	CH_3CH_2OH	46.07	0.789	−112	78.4
-amine	$C_2H_5NH_2$	45.08	0.689	−80.6	16.6
hydrochloride	$C_2H_5NH_2.HCl$	81.55	1.216	108–9	....
aniline	$C_6H_5NHC_2H_5$	121.18	0.963	−63.5	204
sulfonic acid (m-)	$C_2H_5NHC_6H_4SO_3H$	201.24		d. 294	....
anisate (p-)	$CH_3OC_6H_4CO_2C_2H_5$	180.20	1.103	7–8	269–70
anthranilate (o-)	$NH_2C_6H_4CO_2C_2H_5$	165.19	1.117	13	266–8
benzene	$C_6H_5.C_2H_5$	106.16	0.867	−94.4	136.2
benzoate	$C_6H_5CO_2C_2H_5$	150.17	1.052	−34.6	211–2
-benzyl-aniline	$C_6H_5N(C_2H_5)CH_2C_6H_5$	211.29	1.034		285 ^{10mm}
bromide	C_2H_5Br	108.98	1.431	−117.8	38.4
butyrate (n-)	$C_2H_5CH_2CO_2C_2H_5$	116.16	0.879	−93.3	120–1
(i-)	$(CH_3)_2CHCO_2C_2H_5$	116.16	0.871	−88.2	110–1
caprate (n-)	$CH_3(CH_2)_8CO_2C_2H_5$	200.31	0.859	−20	244.6
Ethyl caproate (n-)	$CH_3(CH_2)_4CO_2C_2H_5$	144.21	0.873	−67.5	165–6
caprylate (n-)	$CH_3(CH_2)_6CO_2C_2H_5$	172.26	0.878	−45	207–8
chloride	CH_3CH_2Cl	64.52	0.917^6	−139	13
chloroacetate	$ClCH_2CO_2C_2H_5$	122.55	1.159	−26	144
chlorocarbonate	$ClCO_2C_2H_5$	108.53	1.138	−80.6	94–5
cinnamate (trans-)	$C_6H_5CH{:}CHCO_2H$	176.21	1.049	12	271
cyanoacetate	$CH_2(CN)CO_2C_2H_5$	113.11	1.062	−22.5	208
formate	$HCO_2CH_2CH_3$	74.08	0.923	−79	54
furoate (α)	$OC_4H_3CO_2C_2H_5$	140.13	1.117	34	195
heptoate	$CH_3(CH_2)_5CO_2C_2H_5$	158.23	0.872	−66.1	187–8
hypochlorite	$ClOCH_2CH_3$	80.52	1.013^{-6}	expl.	36
iodide	CH_3CH_2I	155.98	1.933	−105	72.4
lactate	$CH_3CH(OH)CO_2C_2H_5$	118.13	1.030	....	155
laurate	$CH_3(CH_2)_{10}CO_2C_2H_5$	228.36	0.868	−10.7	269
mercaptan	CH_3CH_2SH	62.13	0.839	−121	36–7
methacrylate	$CH_2{:}C(CH_3)CO_2C_2H_5$	114.14	0.913	....	118
naphthylamine (α-)	$C_{10}H_7NHC_2H_5$	171.23	1.060		303
naphthyl ether (α-)	$C_{10}H_7OC_2H_5$	172.22	1.061	5.5	276.4
nitrate	$C_2H_5ONO_2$	91.07	1.100	−102	87–8
nitrite	C_2H_5ONO	75.07	0.900		17
oleate	$C_{17}H_{33}CO_2C_2H_5$	310.50	0.867	<-15	216–8 ^{15mm}
palmitate	$CH_3(CH_2)_{14}CO_2C_2H_5$	284.47	0.858	24–5	191 ^{10mm}
pelargonate	$CH_3(CH_2)_7CO_2C_2H_5$	186.29	0.866	−44.5	227–8
propionate	$CH_3CH_2CO_2C_2H_5$	102.13	0.891	−72.6	99.1
salicylate (o-)	$HOC_6H_4CO_2C_2H_5$	166.17	1.136	1.3	233–4
stearate	$CH_3(CH_2)_{16}CO_2C_2H_5$	312.52	0.848	33.4	201 ^{10mm}
toluate (o-)	$CH_3.C_6H_4CO_2C_2H_5$	164.20	1.032	<-10	227
(m-)	$CH_3.C_6H_4CO_2C_2H_5$	164.20	1.030		231
toluene sulfonate (p-)	$CH_3.C_6H_4SO_3C_2H_5$	200.25	1.166^{48}	33–4	221.3
toluidine (o-)	$CH_3.C_6H_4NHC_2H_5$	135.20	0.948	<-15	215–6
(p-)	$CH_3.C_6H_4NHC_2H_5$	135.20	0.942		217
urea	$C_2H_5NH.CO.NH_2$	88.11	1.213	92	....
valerate (n-)	$CH_3(CH_2)_3CO_2C_2H_5$	130.18	0.877	−91.2	145.5
(i-)	$(CH_3)_2CH(CH_2)_2CO_2C_2H_5$	130.18	0.867	−99.3	135
Ethylal	$CH_2(OC_2H_5)_2$	104.15	0.824	−66.5	89
Ethylene	$H_2C{:}CH_2$	28.05	0.57^{-102}	−169	−103.9
bromide	$BrCH_2.CH_2Br$	187.88	2.180	10	131.5
bromohydrin	$BrCH_2.CH_2OH$	124.98	1.772	....	150.3

TABLE 1-11   Physical Constants of Elements and Compounds—(Continued)

Name	Formula	Formula weight	Specific gravity	Melting point, °C	Boiling point, °C
Ethylene chlorobromide............	$ClCH_2.CH_2Br$	143.43	1.689	−16.6	106.7
chlorohydrin...................	$ClCH_2.CH_2OH$	80.52	1.213	−69	128.8
diamine....................	$H_2NCH_2.CH_2NH_2$	60.10	0.900	8.5	117.2
oxide.....................	$< (CH_2)_2 > O$	44.05	0.887	−111.3	13.5
Ethylidene diacetate..............	$CH_3CH(O_2CCH_3)_2$	146.14	1.061	18.85	168
Eugenol (1-,4-,3-)............	$C_3H_5.C_6H_3(OH)OCH_3$	164.20	1.070	10.3	253.5
(i-)(1-,3-,4-)............	$C_3H_5.C_6H_3(OCH_3)OH$	164.20	1.091	−10	267.5
Fenchyl alcohol (dl-)............	$C_{10}H_{17}OH$	154.24	0.935	35	201
(d-)(α-).........	$C_{10}H_{17}OH$	154.24	0.964	45–7	201–2
(i-)(l-).........	$C_{10}H_{17}OH$	154.24	0.961	61–2	201–2
Ferric dimethyl-dithiocarbamate....	$Fe[SSCN(CH_3)_2]_3$	416.47	....	d. 100–30	ign. >150
Fluorene......................	$(C_6H_4)_2 > CH_2$	166.21	1.203	115–6	293–5
Fluorescein....................	$C_{20}H_{12}O_5$	332.30	....	d. >290	....
Fluoro-dichloromethane..........	$FCHCl_2$	102.93	1.426^0	−127	14.5
-trichloromethane...........	Cl_3CF	137.38	1.494	....	24.9
Formaldehyde.................	$HOHO$	30.03	0.815^{-20}	−92	−21
(m-)......................	$(CH_2O)_3$	90.08	1.17^{65}	64	114.5
(p-)......................	$(CH_2O)_x.xH_2O$	(30.03)	....	150–60	subl.
Formamide...................	$HCONH_2$	45.04	1.139	2	193
Formanilide..................	$HCONHC_6H_5$	121.13	1.147	47	216^{120mm}
Formic acid..................	HCO_2H	46.03	1.220	8.6	100.8
Fructose.....................	$CH_2OH(CHOH)_3COCH_2OH$	180.16	1.669	95–105	....
Fuchsin......................	$C_{20}H_{19}N_3HCl$	337.84	1.22	d. >200	....
Fulminic acid.................	$C:NOH$	43.03			
Fumaric acid (trans-).............	$HO_2CCH:CHCO_2H$	116.07	1.635	286–7	290
Furfural.....................	$C_4H_3O.CHO$	96.08	1.159	−38.7	161.7
Furfuran.....................	C_4H_4O	68.07	0.937	....	31–2
Furfuryl acetate...............	$CH_3CO_2CH_2C_4H_3O$	140.13	1.118	....	175–7
alcohol...................	$C_4H_3O.CH_2OH$	98.10	1.129	....	169.5
butyrate..................	$C_3H_7CO_2CH_2.C_4H_3O$	168.19	1.053	....	212–3
propionate................	$C_2H_5CO_2CH_2.C_4H_3O$	154.16	1.109	....	195–6
Furoic acid..................	$C_4H_3O.CO_2H$	112.08	....	133–4	230–2
G-acid, K salt (2-)(6-,8-).......	$HOC_{10}H_5(SO_3K)_2$	380.46	....		
Na salt (2-)(6-,8-).......	$HOC_{10}H_5(SO_3Na)_2$	348.26	....		
Galactose (d-)...............	$C_5H_{11}O_5.CHO$	180.16	....	165.5	.
Gallic acid (3-,4-,5-)...........	$(HO)_3C_6H_2CO_2H.H_2O$	188.13	1.694	d. 220	
Gamma acid (2-,8-,6-)...........	$C_{10}H_5(NH_2)(OH)SO_3H$	239.24			
Geraniol....................	$C_9H_{16}CH_2OH$	154.24	0.883	<−15	230
Glucose (d-)(α-)...............	$C_5H_{11}O_5.CHO$	180.16	1.544	146	
(d-)(β-)...............	$C_6H_{12}O_6.H_2O$	198.17	1.562	150	
Glucuronic acid..............	$CHO(CHOH)_4CO_2H$	194.14	....	154	d.
Glutam(in)ic acid (dl-)...........	$[.CHNH_2(CH_2)_2.](CO_2H)_2$	147.13	1.460	199 d.	
Glutaric acid................	$CH_2(CH_2CO_2H)_2$	132.11	1.429	97.5	200^{20mm}
Glycerol....................	$CH_2OH.CHOH.CH_2OH$	92.09	1.260^{50}	17.9	290
acetate (mono-)..............	$C_5H_{10}O_4$	134.13	1.20	....	158^{168mm}
(di-)...............	$(CH_3CO_2)_2C_3H_5OH$	176.17	1.178	40	$175–6^{40mm}$
nitrate (mono-) (α-)...	$CH_2OH.CHOH.CH_2NO_3$	137.09	1.40	58–9	155–60
(β-)...	$CH_2OH.CHNO_3.CH_2OH$	137.09	1.40	54	155–60
dinitrate (1-,3-).............	$CHOH(CH_2ONO_2)_2$	182.09	1.47	<−30	$146–8^{15mm}$
Glyceryl triacetate.............	$(CH_3CO_2)_3C_3H_5$	218.20	1.161	−78	258–9
tribenzoate.................	$(C_6H_5CO_2)_3C_3H_5$	404.40	1.228	75–6	d.
tributyrate.................	$(C_3H_7CH_2CO_2)_3C_3H_5$	302.36	1.032	<−75	305–9
tricaprate.................	$[CH_3(CH_2)_8CO_2]_3C_3H_5$	554.83	0.921	31	....
tricaproate................	$[CH_3(CH_2)_4CO_2]_3C_3H_5$	386.51	0.987	−25	....
tricaprylate...............	$[CH_3(CH_2)_6CO_2]_3C_3H_5$	470.67	0.954	8.3	....
trilaurate.................	$[CH_3(CH_2)_{10}CO_2]_3C_3H_5$	638.98	0.894^{60}	45–6	....
trimyristate...............	$[CH_3(CH_2)_{12}CO_2]_3C_3H_5$	723.14	0.885^{60}	56.5	....
trinitrate.................	$CH_2NO_3.CHNO_3.CH_2NO_3$	227.09	1.601	13.3	160^{15mm}
trinitrite.................	$CH_2NO_2.CHNO_2.CH_2NO_2$	179.09	1.291	....	150 d.
trioleate.................	$(C_{17}H_{33}CO_2)_3C_3H_5$	885.40	0.915	−4	240^{18mm}
tripalmitate................	$[CH_3(CH_2)_{14}CO_2]_3C_3H_5$	807.29	0.866^{80}	65.1	$310–20^{0.1mm}$
tristearate................	$[CH_3(CH_2)_{16}CO_2]_3C_3H_5$	891.45	0.862^{80}	70.8	....
Glycide....................	$C_2H_3O.CH_2OH$	74.08	1.114	....	166 d.
Glycine, Glycocoll...........	$NH_2CH_2.CO_2H$	75.07	1.161	232–6 d.	....
Glycol.....................	$CH_2OH.CH_2OH$	62.07	1.113	−15.6	197.4
diacetate.................	$(CH_3CO_2CH_2.)_2$	146.14	1.109	−31	190.5
dibenzoate...............	$(C_6H_5CO_2CH_2.)_2$	270.27	....	73–4	>360
dibutyrate................	$(C_3H_7CO_2CH_2.)_2$	202.24	1.024^{0}	....	240
dicaprylate...............	$(C_7H_{15}CO_2CH_2.)_2$	314.45	....	22	....
diformate................	$(HCO_2CH_2.)_2$	118.09	....	....	174
dilaurate.................	$(C_{11}H_{23}CO_2CH_2.)_2$	426.66	....	52–4	188^{20mm}
dinitrate.................	$(O_2NO.CH_2.)_2$	152.07	1.482	−20	expl. 114
dinitrite.................	$(ONO.CH_2.)_2$	120.07	1.216^{0}	<−15	96–8
dipalmitate...............	$(C_{15}H_{31}CO_2CH_2.)_2$	538.87	....	71–2	$260^{0.1mm}$
dipropionate..............	$(C_2H_5CO_2CH_2.)_2$	174.19	1.045	....	211–2
ether....................	$(HO.CH_2CH_2)_2O$	106.12	1.118	−10.5	244.8
formal...................	$< O.CH_2CH_2OCH_2 >$	74.08	1.060	....	75–6
formate (mono-)............	$HCO_2CH_2CH_2OH$	90.08	1.199	....	180
Glycolic acid................	$HOCH_2CO_2H$	76.05	....	79	d.
Guaiacol (o-)...............	$CH_3O.C_6H_4OH$	124.13	1.140	28.3	205
Guanidine...................	$NH:C(NH_2)_2$	59.07	....	50	....
H-acid, Na salt (1-,8-,3-,6-)........	$C_{10}H_5O_7NS_2Na.1\frac{1}{2}H_2O$	368.31			
Heptacosane (n-)............	$CH_3(CH_2)_{25}CH_3$	380.72	0.780^{60}	59.5	270^{15mm}
Heptane (n-)................	$CH_3(CH_2)_5CH_3$	100.20	0.684	−90.6	98.4
(i-)................	$(CH_3)_2CH(CH_2)_3CH_3$	100.20	0.679	−118.2	90.0
	$C_3H_7.CH(CH_3).C_2H_5$	100.20	0.687	−119.4	91.8
	$(CH_3)_3C.CH_2.C_2H_5$	100.20	0.674	−125	79.1

TABLE 1-11   Physical Constants of Elements and Compounds—(Continued)

Name	Formula	Formula weight	Specific gravity	Melting point, °C.	Boiling point, °C.
	$[(CH_3)_2CH]_2CH_2$	100.20	0.675	−119.4	80.8
	$(CH_3)_2C(C_2H_5)_2$	100.20	0.693	−135.0	86.0
	$(C_2H_5)_3CH$	100.20	0.698	−118.7	93.5
	$(CH_3)_3C.CH(CH_3)_2$	100.20	0.690	−25	80.8
Heptoic acid	$CH_3(CH_2)_5CO_2H$	130.18	0.918	−10	221–2
aldehyde	$CH_3(CH_2)_5CHO$	114.18	0.850	−42	155
Heptyl acetate (n-)	$CH_3CO_2CH_2(CH_2)_5CH_3$	158.24	0.874	....	191.5
alcohol (n-)	$CH_3(CH_2)_6CH_2OH$	116.20	0.824	34.6	175
	$[(CH_3)_2CH]_2CHOH$	116.20	0.829	....	140
	$(C_2H_5.CH_2.CH_2)_2CHOH$	116.20	0.820	−37	156
mercaptan	$CH_3CH(SH).C_5H_{11}$	132.26	0.835	....	174–5
Hexachloro-benzene	C_6Cl_6	284.80	2.044	228–31	309
-ethane	$CCl_3.CCl_3$	236.76	2.091	186–7	186
Hexacosane (n-)	$CH_3(CH_2)_{24}CH_3$	366.69	0.77^{967}	56.6	262^{15mm}
Hexadecane (n-)	$CH_3(CH_2)_{14}CH_3$	226.43	0.774	18.5	287.5
Hexaethylbenzene	$C_6(C_2H_5)_6$	246.42	0.831^{130}	130	298.3
Hexamethylbenzene	$C_6(CH_3)_6$	162.26	....	166	265
Hexamethylene-diamine	$NH_2(CH_2)_6NH_2$	116.20	....	42	204–5
-diisocyanate	$OCN(CH_2)_6NCO$	168.19	1.04	....	$143–4^{20mm}$
-glycol	$HO(CH_2)_6OH$	118.17	....	42	250
tetramine	$(CH_2)_6N_4$	140.19	....	subl.	....
Hexane (n-)	$CH_3(CH_2)_4CH_3$	86.17	0.659	−94	69
(i-)	$(CH_3)_2CH(CH_2)_2CH_3$	86.17	0.654	−153.7	60.2
(neo-)	$(CH_3)_3C.C_2H_5$	86.17	0.649	−98.2	49.7
	$(CH_3)_2CH.CH(CH_3)_2$	86.17	0.662	−129.8	58.0
	$(C_2H_5)_2CHCH_3$	86.17	0.664	−118	63.2
Hexyl acetate (n-)	$CH_3CO_2(CH_2)_5CH_3$	144.21	0.890^0	....	169.2
alcohol (n-)	$CH_3(CH_2)_4CH_2OH$	102.17	0.820	−51.6	157.2
	$(CH_3)_2CH.C(CH_3)_2OH$	102.17	0.821	−14	120–1
	$(CH_3)_2COH.CH_2C_2H_5$	102.17	0.809	−107	123
formate (n-)	$HCO_2CH_2(CH_2)_4CH_3$	130.18	0.898^0	....	153.6
resorcinol (2-,4-)	$CH_3(CH_2)_5C_6H_3(OH)_2$	194.26	....	68–70	179^{7mm}
Hippuric acid	$C_6H_5CONHCH_2CO_2H$	179.17	1.371	187–8	d.
Histidine (l-)	$C_6H_9O_2N_3$	155.16	....	d. 287	....
Homophthalic acid (o-)	$HO_2C.C_6H_4.CH_2CO_2H$	180.15	....	175–80	....
Hydracrylic acid	$HOCH_2CH_2CO_2H$	90.08	....	....	d.
Hydro-cyanic acid	HCN	27.03	0.697	−12	25–6
-quinone (p-)	$C_6H_4(OH)_2$	110.11	1.332	170.3	285
Hydroxy-benzaldehyde (p-)	$HO.C_6H_4.CHO$	122.12	1.129^{130}	116–7	subl.
-benzanilide (o-)	$HO.C_6H_4.CONHC_6H_5$	213.23	....	135	d.
-quinoline (2-)(α-)	$C_9H_6N.OH$	145.15	....	199–200	subl.
(8-)(o-)	$C_9H_6N.OH$	145.15	....	75–6	266.6
Indigo	$[C_6H_4(CO)(NH)C:]_2$	262.26	1.35	390–2	subl.
White	$C_{16}H_{12}O_2N_2$	264.27	....	....	....
Indole	C_8H_7N	117.14	....	52	253–4
Indoxyl	C_8H_6NOH	133.14	....	85	110
Iodo-benzene	C_6H_5I	204.02	1.824	−28.5	188.6
-phenol (p-)	IC_6H_4OH	220.02	1.857^{112}	93–4	d.
Iodoform	HCI_3	393.78	4.008	119	subl.
Ionone (α-)	$C_{10}H_{16}:CHCOCH_3$	192.29	0.930	....	136.1^{17mm}
(β-)	$C_{10}H_{16}:CHCOCH_3$	192.29	0.944	....	140^{18mm}
Irone (β-)	$C_{14}H_{22}O$	206.32	0.939	....	144^{16mm}
Isatin	$C_6H_4 < (CO)(N) > COH$	147.13	....	200–1	subl.
Isoprene	$CH_2:CH.C(CH_3):CH_2$	68.11	0.681	−120	34
Ketene	$H_2C:CO$	42.04	....	−151	−56
Koch acid (1-)(3-,6-,8-)	$C_{10}H_4(NH_2)S_3O_9HNa_2$	427.34	....	....	....
Lactic acid (dl-)	$CH_3CH(OH)CO_2H$	90.08	1.249	16.8	122^{14mm}
anhydride	$C_6H_{10}O_5$	162.14	....	....	d. 250
Lactide (dl-)	$C_6H_8C_4$	144.12	0.862	124.5	255
Lactose	$C_{12}H_{22}O_{11}.H_2O$	360.31	1.525	202	d.
Lauric acid	$CH_3(CH_2)_{10}CO_2H$	200.31	0.869^{50}	48	225
Laurone	$[CH_3(CH_2)_{10}]_2CO$	338.60	0.809^{69}	69–70	....
Lauryl alcohol	$CH_3(CH_2)_{10}CH_2OH$	186.33	0.831	24	255–9
Lead tetraethyl	$Pb(C_2H_5)_4$	323.45	1.659	−136	152^{291mm}
tetramethyl	$Pb(CH_3)_4$	267.35	1.995	−27.5	110
Lecithin (protagon)	$C_{42}H_{48}O_9PN$	778.08	....	150–200 d.	....
Lepidine (py-4-)	$C_9H_6N.CH_3$	143.18	1.086	9–10	261–3
Leucine (l-)	$(CH_3)_2CHCH_2CH(NH_2)CO_2H$	131.17	1.293	295	subl.
Levulinic acid	$CH_3CO(CH_2)_2CO_2H$	116.11	1.140	33.5	245–6
Limonene (d- or l-)	$C_{10}H_{16}$	136.23	0.842	−96.9	177
Linalool (d- or l-)	$C_{10}H_{17}OH$	154.24	0.868	....	198–200
Linalyl acetate	$CH_3CO_2C_{10}H_{17}$	196.28	0.895	....	220 d.
Linoleic acid	$C_{17}H_{31}CO_2H$	280.44	0.903	−9.5	$229–30^{16mm}$
Maleic acid	$HO_2C.CH:CH.CO_2H$	116.07	1.609	130.5	135 d.
anhydride	$< (.CHCO)_2 > O$	98.06	1.5	57–60	202
Malic acid (dl-)	$HO_2CCH_2CH(OH)CO_2H$	134.09	1.601	128–9	150 d.
(d- or l-)	$HO_2CCH_2CH(OH)CO_2H$	134.09	1.595	99–100	140 d.
Malonic acid	$H_2C(CO_2H)_2$	104.06	1.631	130–5 d.	....
Maltose	$C_{12}H_{22}O_{11}.H_2O$	360.31	1.540	d.	....
Mandelic acid (dl-)	$C_6H_5CH(OH)CO_2H$	152.14	1.300	118.1	d.
Mannitol (d-)	$CH_2OH(CHOH)_4CH_2OH$	182.17	1.489	166	$290–5^{3mm}$
Mannose (d-)	$CH_2OH(CHOH)_4CHO$	180.16	1.539	132	d.
Margaric acid	$CH_3(CH_2)_{15}CO_2H$	270.44	0.853^{60}	60–1	227^{100mm}
Mellitic acid	$C_6(CO_2H)_6$	342.17	....	286–8	d.
Menthol (l-)(α-)	$C_{10}H_{19}OH$	156.26	0.890	42–3	212
Mercapto-benzothiazole (2-)	$< C_6H_4N:C(SH)S >$	167.24	1.42	179	d.
-thiazoline (2-)	$< CH_2N:C(SH)SCH_2 >$	119.20	1.50	106	....

TABLE 1-11 Physical Constants of Elements and Compounds—(Continued)

Name	Formula	Formula weight	Specific gravity	Melting point, °C.	Boiling point, °C.
Mercuric cyanide	$Hg(CN)_2$	252.65	4.003	d. 320	....
fulminate	$Hg(ONC)_2.\frac{1}{2}H_2O$	293.65	4.4	expl.	....
Mesityl oxide	$(CH_3)_2C:CHCOCH_3$	98.14	0.858	−59	130
Mesitylene (1-,3-,5-)	$C_6H_3(CH_3)_3$	120.19	0.865	−45	164.8
Metanilic acid (m-)	$H_2NC_6H_4SO_3H$	173.18	d.		
Methane	CH_4	16.04	0.415^{-164}	−182.6	−161.4
Methoxy-methoxyethanol	$CH_3(OCH_2)_2CH_2OH$	106.12	1.038	<-70	167.5
Methyl acetate	$CH_3CO_2CH_3$	74.08	0.924	−98.7	57.1
acrylic acid (α-)	$CH_2:C(CH_3)CO_2H$	86.09	1.015	15–16	161–3
alcohol	CH_3OH	32.04	0.792	−97–8	64.7
-amine	CH_3NH_2	31.06	0.699^{-11}	−92.5	−6.7
-amine hydrochloride	$CH_3NH_2.HCl$	67.52	1.23	226–8	230^{15} mm
aniline	$C_6H_5NHCH_3$	107.15	0.989	−57	195.5
anthracene (α-)	$C_6H_4:(CH)_2:C_6H_3CH_3$	192.25	$1.047^{99.4}$	86	....
(β-)	$C_6H_4:(CH)_2:C_6H_3CH_3$	192.25	1.181^0	207	....
anthranilate (o)	$NH_2O_6H_4CO_2CH_3$	151.16	1.168	24	135.5^{15} mm
anthraquinone (2-)	$C_6H_4:(CO)_2:C_6H_3CH_3$	222.23		176–7	subl.
benzoate	$C_6H_5CO_2CH_3$	136.14	1.087	−12.5	198–9
benzylaniline	$C_6H_5N(CH_3)CH_2C_6H_5$	197.27		9.2	305–6
bromide	CH_3Br	94.95	1.732^0	−93	4.5
butyrate (n-)	$CH_3(CH_2)_2CO_2CH_4$	102.13	0.898	<-95	102.3
(i-)	$(CH_3)_2CHCO_2CH_3$	102.13	0.891	−84.7	92.6
caprate (n-)	$CH_3(CH_2)_8CO_2CH_3$	186.29		−18	223–4
caproate (n-)	$CH_3(CH_2)_4CO_2CH_3$	130.18	0.904^0		149.5
caprylate	$CH_3(CH_2)_6CO_2CH_3$	158.23	0.887	−40	192–4
cellosolve	$CH_3OCH_2CH_2OH$	76.09	0.965		174–5
chloride	CH_3Cl	50.49	0.952^0	−97.7	−24
chloroacetate	$ClCH_2CO_2CH_3$	108.53	1.236	−32.7	130
chloroformate	$ClCO_2CH_3$	94.50	1.236		71–2
cinnamate	$C_6H_5CH:CHCO_2CH_3$	162.18	1.042	33.4	263
cyclohexane	$CH_2 < (CH_2CH_2)_2 > CHCH_3$	98.18	0.769	−126.3	101
ethyl carbonate	$CH_3O.CO.OC_2H_5$	104.10	1.002	−14.5	109.2
ethyl ketone	$CH_3.CO.C_2H_5$	72.10	0.805	85.9	79.6
ethyl oxalate	$CH_3OCO.CO_2C_2H_5$	132.11	1.156^0		173.7
formate	HCO_2CH_3	60.05	0.974	−99.8	32
furoate	$C_4H_3O.CO_2CH_3$	126.11	1.179		181.3
glucamine	$CH_2OH(CHOH)_4CH_2NHCH_3$	195.21			
glycolate	$HOCH_2CO_2CH_3$	90.08	1.168		151.2
heptoate	$CH_3(CH_2)_5CO_2CH_3$	144.21	0.881		172–3
hypochlorite	$ClOCH_3$	66.49			12
iodide	CH_3I	141.95	2.279	−64.4	42.4
lactate	$CH_3CH(OH)CO_2CH_3$	104.10	1.090		144.8
laurate	$CH_3(CH_2)_{10}CO_2CH_3$	214.34		5	148^{18} mm
mercaptan	CH_3SH	48.10	0.896^0	−121	5.8
methacrylate	$CH_2:C(CH_3)CO_2CH_3$	100.11	0.950	−48	100.3
myristate	$CH_3(CH_2)_{12}CO_2CH_3$	242.39		18–9	295
naphthalene (α-)	$C_{10}H_7CHO$	142.19	1.025	−19	244.6
(β-)	$C_{10}H_7CH_3$	142.19	0.994^{40}	35–6	241–2
nitrate	CH_3ONO_2	77.04	1.203	expl.	65
nitrite	CH_3ONO	61.04	0.991		−12
nonyl ketone (n-)	$CH_3(CH_2)_8COCH_3$	170.29	0.828	13.5	228
oleate	$C_{17}H_{33}CO_2CH_3$	296.48	0.879		$190–1^{10}$ mm
orange	$(CH_3)_2NC_6H_4N_2C_6H_4SO_3Na$	327.33			
palmitate	$CH_3(CH_2)_{14}CO_2CH_3$	270.44		30–1	196^{15} mm
phosphine	CH_3PH_2	48.03			−14
propionate	$CH_3CH_2CO_2CH_3$	88.10	0.915	−87.5	79.7
propyl ketone (n-)	$CH_3COCH_2CH_2CH_3$	86.13	0.812	−77.8	102
salicylate (o-)	$HO.C_6H_4CO_2CH_3$	152.14	1.182	−8.3	222.2
stearate	$CH_3(CH_2)_{16}CO_2CH_3$	298.49		38–9	215^{15} mm
toluate (o-)	$CH_3.C_6H_4CO_2CH_3$	150.17	1.073	<-50	213
(m-)	$CH_3.C_6H_4CO_2CH_3$	150.17	1.066		215
(p-)	$CH_3.C_6H_4CO_2CH_3$	150.17		33–4	217
Methyl toluidine (o-)	$CH_3.C_6H_4NHCH_3$	121.18	0.973		206–7
(m-)	$CH_3.C_6H_4NHCH_3$	121.18			206–7
(p-)	$CH_3.C_6H_4NHCH_3$	121.18	0.935^{55}		211
valerate (n-)	$CH_3(CH_2)_3CO_2CH_3$	116.16	0.895	−91	127.3
(i-)	$(CH_3)_2CHCH_2CO_2CH_3$	116.16	0.881		116–7
vinyl ketone	$CH_3COCH:CH_2$	70.09	0.836		81
Methylal	$HCH(OCH_3)_2$	76.09	0.866	−104.8	42–3
Methylene-bis-(phenyl-4-isocyanate)	$(OCN.C_6H_4)_2CH_2$	250.25	1.222		$210–2^{13}$ mm
bromide	CH_2Br_2	173.86	2.495	−52.8	98.5
chloride	CH_2Cl_2	84.94	1.336	−96.7	40–1
dianiline	$(C_6H_5NH)_2CH_2$	198.26		65	208–9 d.
iodide	CH_2I_2	267.87	3.325	5.7	180 d.
Michler's hydrol (p-,p'-)	$[(CH_3)_2NC_6H_4]_2CHOH$	270.36		96–7	
ketone	$[(CH_3)_2NC_6H_4]_2CO$	268.35		174	>360 d.
Morphine	$C_{17}H_{19}O_3N.H_2O$	303.35	1.317	254 d.	
Mucic acid	$(.CHOHCHOHCO_2H)_2$	210.14		206–14	
Mustard gas	$(ClCH_2.CH_2)_2S$	159.08	1.275	13–4	217
Myricyl alcohol	$C_{31}H_{63}OH(?)$	452.82	0.777^{95}	88	
Myristic acid	$CH_3(CH_2)_{12}CO_2H$	228.36	0.853^{70}	57–8	250.5^{100} mm
Myristyl alcohol	$CH_3(CH_2)_{12}CH_2OH$	214.38	0.824	38	167^{15} mm
Naphthalene	$C_{10}H_8$	128.16	1.145	80.2	217.9
disulfonic acid (1-,5-)	$C_{10}H_6(SO_3H)_2$	288.28		d.	
(1-,6-)	$C_{10}H_6(SO_3H)_2$	288.28		d. 125	
sulfonic acid (α-)	$C_{10}H_7SO_3H.2H_2O$	244.26		90	
(β-)	$C_{10}H_7SO_3H.H_2O$	226.24		125	

TABLE 1-11 Physical Constants of Elements and Compounds—(Continued)

Name	Formula	Formula weight	Specific gravity	Melting point, °C.	Boiling point, °C.
Naphthasultam (1-,8-)	$C_{10}H_7O_2NS$	205.22	....	177–8	....
disulfonate Na (1-,8-)	$C_{10}H_5O_8NS_3Na_2.2H_2O$	445.35	....	....	....
(2-,4-)	$C_{10}H_4O_8NS_3Na_3.8\frac{1}{2}H_2O$	584.45	....	....	....
Naphthoic acid (α-)	$C_{10}H_7CO_2H$	172.17	....	160–1	300
(β-)	$C_{10}H_7CO_2H$	172.17	1.077^{100}	184	>300
Naphthol (α-)	$C_{10}H_7OH$	144.16	1.224	96	278–80
(β-)	$C_{10}H_7OH$	144.16	1.217	122–3	285–6
sulfonic acid (α-)(1-,2-)	$HO.C_{10}H_6SO_3H$	224.22	....	>250	....
(β-)(2-,6-)	$HO.C_{10}H_6SO_3H$	224.22	....	125	....
Naphthyl acetate (α-)	$CH_3CO_2C_{10}H_7$	186.20	....	46–9	....
(β-)	$CH_3CO_2C_{10}H_7$	186.20	....	69–70	....
amine (α-)	$C_{10}H_7NH_2$	143.18	1.123	50	300.8
(β-)	$C_{10}H_7NH_2$	143.18	1.061^{98}	111–2	306.1
amine hydrochloride (α-)	$C_{10}H_7NH_2.HCl$	179.65	....	....	subl.
(β-)	$C_{10}H_7NH_2.HCl$	179.65	....	....	....
amine sulfonic acid (1-,4-)	$NH_2.C_{10}H_6.SO_3H$	223.24	....	d.	....
(1-,5-)	$NH_2.C_{10}H_6.SO_3H.H_2O$	241.26	....	....	....
(1-,7-)	$NH_2.C_{10}H_6.SO_3H.H_2O$	241.26	....	....	....
(1-,8-)	$NH_2.C_{10}H_6.SO_3H.H_2O$	241.26	....	....	....
(2-,5-)	$NH_2.C_{10}H_6.SO_3H$	223.24	....	....	....
(2-,6-)	$NH_2.C_{10}H_6.SO_3H.H_2O$	241.26	....	....	....
(2-,7-)	$NH_2.C_{10}H_6.SO_3H.H_2O$	241.26	....	....	....
isocyanate (α-)	$C_{10}H_7N{:}CO$	169.17	1.18	....	269–70
Nicotine	$C_{10}H_{14}N_2$	162.23	1.009	<−80	246
Nicotinic acid (3-)	$C_5H_4NCO_2H$	123.11	....	235.2	subl.
(i-)(4-)	$C_6H_4NCO_2H$	123.11	....	317	d.
Nitro-acetanilide (p-)	$CH_3CONHC_6H_4NO_2$	180.16	....	215–6	....
-acetophenone (m-)	$CH_3COC_6H_4NO_2$	165.14	....	80–1	202
-aminoanisole (4-,1-,2-)	$NO_2.C_6H_3(OCH_3)NH_2$	168.15	1.207^{156}	118	....
(5-,1-,2-)	$NO_2.C_6H_3(OCH_3)NH_2$	168.15	1.211^{156}	139–40	....
(3-,1-,4-)	$NO_2.C_6H_3(OCH_3)NH_2$	168.15	....	123	....
-aminophenol (4-,2-,1-)	$NO_2.C_6H_3(NH_2)OH$	154.12	....	142–3	....
-aniline (o-)	$NO_2.C_6H_4NH_2$	138.12	1.442	71.5	284.1
(m-)	$NO_2.C_6H_4NH_2$	138.12	1.43	114	306.4
(p-)	$NO_2.C_6H_4NH_2$	138.12	1.437	146–7	331.7
-anisole (o-)	$CH_3OC_6H_4NO_2$	153.13	1.254	9.4	272–3
(p-)	$CH_3OC_6H_4NO_2$	153.13	1.233	54	274
-anthraquinone (α-)	$C_6H_4{:}(CO)_2{:}C_6H_3NO_2$	253.20	....	230	207^{7mm}
-anthraquinone sulfonic acid (1-,5-)	$NO_2.C_{14}H_6O_2.SO_3H$	333.26	....	65	....
-benzal chloride (m-)	$NO_2.C_6H_4CHCl_2$	206.03	....	65	....
-benzaldehyde (m-)	$NO_2.C_6H_4CHO$	151.12	....	58	164^{23mm}
-benzene	$C_6H_5NO_2$	123.11	1.205	5.7	210.9
-benzidine (2-)	$NH_2C_6H_4C_6H_3(NH_2)NO_2$	229.23	....	143	....
-benzoic acid (o-)	$NO_2.C_6H_4.CO_2H$	167.12	1.575	147.5	....
(m-)	$NO_2.C_6H_4.CO_2H$	167.12	1.494	140–1	....
(p-)	$NO_2.C_6H_4.CO_2H$	167.12	1.550	240–2	subl.
-benzyl alcohol (m-)	$NO_2.C_6H_4.CH_2OH$	153.13	....	27	$175–80^{3mm}$
-benzyl bromide (p-)	$NO_2.C_6H_4CH_2Br$	216.04	....	99–100	....
-chlorotoluene (1-,2-,6-)	$CH_3.C_6H_3(NO_2)Cl$	171.58	....	37.5	238
-cresol (1-,3-,4-)	$CH_3.C_6H_3(NO_2)OH$	153.13	1.240	32	125^{22mm}
-cymene (1-,2-,4-)	$CH_3.C_6H_3(NO_2)CH(CH_3)_2$	179.21	1.067	....	152^{15mm}
-dimethylaniline (o-)	$NO_2.C_6H_4NHCH_3$	166.18	1.179	....	$151–3^{20mm}$
(m-)	$NO_2.C_6H_4NHCH_3$	166.18	1.313	60–1	280–5
(p-)	$NO_2.C_6H_4NHCH_3$	166.18	....	163–4	....
-diphenyl (o-)	$C_6H_5.C_6H_4NO_2$	199.20	1.44	37	320
(p-)	$C_6H_5.C_6H_4NO_2$	199.20	....	113–4	340
-diphenylamine (o-)	$C_6H_5.NH.C_6H_4NO_2$	214.22	....	75–6	....
-guanidine	$H_2NC(NH)NHNO_2$	104.07	....	246–7	....
-naphthalene (α-)	$C_{10}H_7NO_2$	173.16	1.223^{82}	59–60	304
(β-)	$C_{10}H_7NO_2$	173.16	....	79	165^{15mm}
-phenol (o-)	$NO_2.C_6H_4.OH$	139.11	1.295^{45}	44–5	214.5
(m-)	$NO_2.C_6H_4.OH$	139.11	1.485	96–7	194^{70mm}
(p-)	$NO_2.C_6H_4.OH$	139.11	1.479	113–4	subl.
-phenol sulfonic acid (1-,4-,2-)	$HO.C_6H_3(NO_2)SO_3H.3H_2O$	273.22	....	d. 110	....
(1-,2-,4-)	$HO.C_6H_3(NO_2)SO_3H.3H_2O$	273.22	....	51.5	....
-phthalic acid (3-)	$NO_2.C_6H_3(CO_2H)_2$	211.13	....	222	....
(4-)	$NO_2.C_6H_3(CO_2H)_2$	211.13	....	164–5	....
-toluene (o-)	$CH_3.C_6H_4NO_2$	137.13	1.163	−4.1	222.3
(m-)	$CH_3.C_6H_4NO_2$	137.13	1.160	15–16	230–1
(p-)	$CH_3.C_6H_4NO_2$	137.13	1.139^{55}	51.9	237.7
-toluene sulfonic acid (1-,4-,2-)	$CH_3.C_6H_3(NO_2)SO_3H.2H_2O$	253.23	....	130	....
-toluidine (4-,1-,2-)	$NO_2.C_6H_3(CH_3)NH_2$	152.15	1.365	105–7	....
(3-,1-,4-)	$NO_2.C_6H_3(CH_3)NH_2$	152.15	1.312	116–7	....
Nitron	$C_{20}H_{16}N_4$	312.36	....	189–90 d.	....
Nitroso-dimethylaniline (p-)	$ON.C_6H_4N(CH_3)_2$	150.18	....	86–7	....
-naphthol (β-)(1-)	$ON.C_{10}H_6OH$	173.16	....	109.5	....
Nonadecane (n-)	$CH_3(CH_2)_{17}CH_3$	268.51	0.777	32	330
Nonane (n-)	$CH_3(CH_2)_7CH_3$	128.25	0.718	−53.7	150.5
Octadecane (n-)	$CH_3(CH_2)_{16}CH_3$	254.48	0.775	28	317
Octane (n-)	$CH_3(CH_2)_6CH_3$	114.22	0.703	−56.5	125.7
(iso-)	$(CH_3)_3CCH_2CH(CH_3)_2$	114.22	0.692	−107.4	99.3
Octyl acetate (n-)	$CH_3CO_2CH_2(CH_2)_6CH_3$	172.26	0.885^0	−38.5	210
(sec-)	$CH_3CO_2CH(CH_3)C_6H_{13}$	172.26	0.863	....	195
alcohol (n-)	$CH_3(CH_2)_6CH_2OH$	130.22	0.827	−16	194–5
(sec-)	$CH_3(CH_2)_5CH(OH)CH_3$	130.22	0.822	−38.6	179–80
Octylene (n-)	$CH_3(CH_2)_5CH{:}CH_2$	112.21	0.721	....	126
Oleic acid	$C_8H_{17}CH{:}CH(CH_2)_7CO_2H$	282.45	0.854^{78}	14	$285–6^{100mm}$
Orcinol (1-,3-,5-)	$(HO)_2C_6H_3.CH_3$	124.13	1.290^4	107–8	287–90
Oxalic acid	$HO_2C.CO_2H.2H_2O$	126.07	1.653	101.5	subl.

TABLE 1-11   Physical Constants of Elements and Compounds—(Continued)

Name	Formula	Formula weight	Specific gravity	Melting point, °C.	Boiling point, °C.
Palmitic acid	$CH_3(CH_2)_{14}CO_2H$	256.42	0.849^{70}	63–4	$271.5^{100 \text{ mm}}$
Pelargonic acid	$CH_3(CH_2)_7CO_2H$	158.23	0.906	12.5	253–4
Penta-chloroethane	$CHCl_2.CCl_3$	202.31	1.671	--22	162
-decane (n-)	$CH_3(CH_2)_{13}CH_3$	212.41	0.770	10	270.5
-erythritol	$C(CH_2OH)_4$	136.15		262	$276^{80 \text{ mm}}$
Pentandiol	$HOCH_2(CH_2)_3CH_2OH$	104.15	0.994		239.4
Pentane (n-)	$CH_3(CH_2)_3CH_3$	72.15	0.630	−129.7	36.3
(i-)	$(CH_3)_2CHCH_2CH_3$	72.15	0.621	−160.0	27.95
(neo-)	$(CH_3)_2C(CH_3)_2$	72.15	0.613	−20	9.5
Phenacetin	$C_2H_5O.C_6H_4NHCOCH_3$	179.21		134–5	d.
Phenanthrene	$< (C_6H_4CH_2) >$	178.22	1.179	99–100	340
Phenetidine (o-)	$C_2H_5O.C_6H_4.NH_2$	137.18		<-21	228–9
(p-)	$C_2H_5O.C_6H_4.NH_2$	137.18	1.061	3–4	254–5
Phenetole	$C_2H_5O.C_6H_5$	122.16	0.967	−30.2	172
Phenol	C_6H_5OH	94.11	1.071	42–3	181.4
-phthalein	$C_{20}H_{14}O_4$	318.31	1.299	261–2	
-sulfonic acid (o-)	$HO.C_6H_4SO_3H.\frac{3}{4}H_2O$	187.68		50 d.	
Phenyl acetaldehyde	$C_6H_5CH_2CHO$	120.14	1.025		193–4
acetic acid	$C_6H_5CH_2CO_2H$	136.14	1.081^{80}	76–7	265.5
-acetylene	$C_6H_5C:CH$	102.13	0.930	43	142–3
aniline (o-)	$C_6H_5.C_6H_4.NH_2$	169.22		45–6	299
(p-)	$C_6H_5.C_6H_4.NH_2$	169.22		50–2	302
Phenyl-ethyl alcohol	$C_6H_5CH_2CH_2OH$	122.16	1.023		219–21
-glycine	$C_6H_5NHCH_2CO_2H$	151.16		127	
-hydrazine	$C_6H_5NH.NH_2$	108.14	1.097	19.6	243.5
-hydrazine sulfonic acid (p-)	$H_2NNHC_6H_4SO_3H$	100.20		288	
isocyanate	$C_6H_5N:CO$	119.12	1.096		166
-methylpyrazolone (3-)(N-)	$C_4H_5ON_2.C_6H_5$	174.20		128	$191^{17 \text{ mm}}$
-mustard oil	$C_6H_5N:CS$	135.18	1.138	−21	219–20
naphthalene (α-)	$C_{10}H_7.C_6H_5$	204.26		45	336–7
(β-)	$C_{10}H_7.C_6H_5$	204.26		102.5	345–6
naphthylamine (α-)	$C_{10}H_7NHC_6H_5$	219.27	1.17	62	$335^{268 \text{ mm}}$
(β-)	$C_{10}H_7NHC_6H_5$	219.27	1.18	107–8	399.5
phenol (o-)	$C_6H_5.C_6H_4OH$	170.20		56–7	275
(p-)	$C_6H_5.C_6H_4OH$	170.20		164–5	305–8
propyl alcohol (γ-)	$C_6H_5(CH_2)_3OH$	136.19	1.008	<-18	235–7
quinoline (2-)(α-)	$C_6H_5.C_9H_6N$	205.25		86	363
(8-)(U-)	$C_6H_5.C_9H_6N$	205.25			$283^{187 \text{ mm}}$
salicylate, salol	$HO.C_6H_4CO_2C_6H_5$	214.21	1.250	42–3	$172–3^{12 \text{ mm}}$
stearate	$CH_3(CH_2)_{16}CO_2C_6H_5$	360.56		52	$267^{16 \text{ mm}}$
urethane	$C_6H_5NHCO_2C_2H_5$	165.19	1.106	52–3	237–8
Phenylene-diamine (o-)	$C_6H_4(NH_2)_2$	108.14		103–4	256–8
(m-)	$C_6H_4(NH_2)_2$	108.14		62.8	284–7
(p-)	$C_6H_4(NH_2)_2$	108.14	1.139	140	267
Phloroglucinol (1-,3-,5-)	$C_6H_3(OH)_3.2H_2O$	162.14		117	subl.
Phorone	$[(CH_3)_2C:CH]_2CO$	138.20	0.885	28	197.2
Phosgene	$OCCl_2$	98.92	1.392	−104	8.2
Phthalic acid (o-)	$C_6H_4(CO_2H)_2$	166.13	1.593	208	d.
(m-)(iso-)	$C_6H_4(CO_2H)_2$	166.13		330	subl.
anhydride (o-)	$C_6H_4 < (CO)_2 > O$	148.11	1.527^4	130.8	284.5
nitrile (o-)	$C_6H_4(CN)_2$	128.13		141	
Phthalide	$C_6H_4(CH_2)(CO) > O$	134.13	1.164^{99}	73	290
Phthalimide (o-)	$C_6H_4 < (CO)_2 > NH$	147.13		238	subl.
Picoline (α-)	$C_5H_4N.CH_3$	93.12	0.950	−70	128.8
(β-)	$C_5H_4N.CH_3$	93.12	0.961		143.5
(γ-)	$C_5H_4N.CH_3$	93.12	0.957		143.1
Picramic acid (1-,2-,4-,6-)	$HO.C_6H_2(NH_2)(NO_2)_2$	199.12		169	
Picric acid (2-,4-,6-)	$HO.C_6H_2(NO_2)_3$	229.11	1.763	121.8	expl.
Picryl chloride (2-,4-,6-)	$ClC_6H_2(NO_2)_3$	247.56	1.797	83	d.
Pinacol	$[(CH_3)_2C.OH]_2$	118.17	0.967	43	171–2
Pinacoline	$CH_3COC(CH_3)_3$	100.16	0.800	−52.5	106.2
Pinene (α-)(dl-)	$C_{10}H_{16}$	136.23	0.878	−55	154–6
hydrochloride	$C_{10}H_{17}Cl$	172.69		131–2	207–8
Pinol (dl-)	$C_{10}H_{16}O$	152.23	0.953		183–4
Piperidine	$CH_2 < (CH_2CH_2)_2 > NH$	85.15	0.860	−9	106
carboxylic acid (α-)(dl-)	$HO_2C.CH < (CH_2CH_2)_2 > NH$	129.16		264	
Piperidinium pentamethylene dithiocarbamate	$(CH_2)_5C.S_2H.HN(CH_2)_5$	232.41	1.13	175	
Propane	$CH_3CH_2CH_3$	44.09	0.585^{-45}	−187.1	−42.2
Propionic acid	$CH_3CH_2CO_2H$	74.08	0.992	−22	141.1
aldehyde	CH_3CH_2CHO	58.08	0.807	−81	49.5
anhydride	$(CH_3CH_2CO)_2O$	130.14	1.012	−45	168.8
Propyl acetate (n-)	$CH_3COCH_2CH_2CH_3$	102.13	0.886	−92.5	101.6
(i-)	$CH_3CO_2CH(CH_3)_2$	103.13	0.874	−73.4	88.4
alcohol (n-)	$CH_3CH_2CH_2OH$	60.09	0.804	−127	97.8
(i-)	$(CH_3)_2CHOH$	60.09	0.789	−85.8	82.5
amine (n-)	$CH_3CH_2CH_2NH_2$	59.11	0.718	−83	49–50
(i-)	$(CH_3)_2CHNH_2$	59.11	0.694	−101	33–4
aniline (n-)	$C_6H_5NHCH_2CH_2CH_3$	135.20	0.949		222
benzoate (n-)	$C_6H_5CO_2CH_2CH_2CH_3$	164.20	1.021	−51.6	231
(i-)	$C_6H_5CO_2CH(CH_3)_2$	164.20	1.010		218.5
bromide (n-)	$CH_3CH_2CH_2Br$	123.00	1.353	−109.9	70.8
(i-)	$(CH_3)_2CHBr$	123.00	1.310	−89	60
n-butyrate (n-)	$C_2H_5CH_2CO_2CH_2C_2H_5$	130.18	0.879	−95.2	142.7
i-butyrate (n-)	$(CH_3)_2CHCO_2CH_2C_2H_5$	130.18	0.884^{0}		134–5
n-butyrate (i-)	$C_2H_5CH_2CO_2CH(CH_3)_2$	130.18	0.865		128
i-butyrate (i-)	$(CH_3)_2CHCO_2CH(CH_3)_2$	130.18	0.869^{0}		120.8
chloride (n-)	$CH_3CH_2CH_2Cl$	78.54	0.890	−122.8	46.4
(i-)	$(CH_3)_2CHCl$	78.54	0.859	−117	36.5

TABLE 1-11　Physical Constants of Elements and Compounds—(Continued)

Name	Formula	Formula weight	Specific gravity	Melting point, °C.	Boiling point, °C.
Propyl formate (n-)	HCO₂CH₂CH₂CH₃	88.10	0.901	−92.9	81.3
(i-)	HCO₂CH(CH₃)₂	88.10	0.873	....	68–71
furoate (n-)	C₄H₃O.CO₂C₃H₇	154.16	1.075	....	211
lactate (n-)	CH₃CH(OH)CO₂CH₂C₂H₅	132.16	....	....	122–3¹⁵⁰ ᵐᵐ
(i-)	CH₃CH(OH)CO₂CH(CH₃)₂	132.16	....	....	167.5
mercaptan (n-)	CH₃CH₂CH₂SH	76.15	0.836	−112	67–8
(i-)	(CH₃)₂CHSH	76.15	0.809	−130.7	58–60
propionate (n-)	C₂H₅CO₂CH₂C₂H₅	116.16	0.883	−76	122–3
(i-)	C₂H₅CO₂CH(CH₃)₂	116.16	0.893³⁰	....	109–11
thiocyanate (i-)	(CH₃)₂CH.CNS	101.16	0.963	....	152–3
n-valerate (n-)	CH₃(CH₂)₃CO₂CH₂C₂H₅	144.21	0.874	−70.7	67.5
i-valerate (n-)	(CH₃)₂CHCH₂CO₂C₃H₇	144.21	0.863	....	155.9
i-valerate (i-)	(CH₃)₂CHCH₂CO₂C₃H₇	144.21	0.854	....	142
Propylene	CH₃CH:CH₂	42.08	0.609⁻⁴⁷	−185	−48
bromide	CH₃CHBrCH₂Br	201.91	1.933	−55.5	141.6
chlorohydrin	CH₃CHClCH₂OH	94.54	1.103	....	133–4
chloride	CH₃CHClCH₂Cl	112.99	1.159	<−70	96.8
glycol	CH₃CH(OH)CH₂OH	76.09	1.040	....	188–9
oxide	CH₃(CHCH₂)O	58.08	0.831	....	35
Protocatechuic acid (3-,4-)	(HO)₂C₆H₃CO₂H.H₂O	172.13	1.542	199 d.	....
Pulegol (iso-)(d-)	C₁₀H₁₇OH	154.24	0.911	....	86–9¹⁰ ᵐᵐ
Pulegone	C₁₀H₁₆O	152.23	0.932	....	224
Pyrazole	—NH.N:CH.CH:CH—	68.08	....	70	186–8
Pyrazoline	—NH.N:CH.CH₂CH₂—	70.09	....	....	144
Pyrazolone	—NH.CO.CH₂CH:N—	84.08	....	165	subl. d.
Pyrene	C₁₆H₁₀	202.24	1.277⁰	149–50	>360
Pyridazine	N₂ < (CHCH)₂ >	80.09	1.107	−8	208
Pyridine	CH < (CHCH)₂ > N	79.10	0.982	−42	115–6
Pyrocatechol (o-)	C₆H₄(OH)₂	110.11	1.344	104–5	240–5
Pyrogallol (1-,2-,3-)	C₆H₃(OH)₃	126.11	1.453	133–4	309
Pyrone	CO < (CHCH)₂ > O	96.08	1.190⁴⁰·³	32.5	215–7
Pyrrole	< (CH:CH)₂ > NH	67.09	0.948	....	131
Pyrrolidine	< (CH₂.CH₂)₂ > NH	71.12	0.852	....	87–8
Pyrroline	< (CH.CH₂)₂ > NH	69.10	0.910	....	90–1
Pyruvic acid	CH₃COCO₂H	88.06	1.267	13.6	165
Quercitrin	C₂₁H₂₀O₁₁.2H₂O	484.40	....	182–5	....
Quinaldine (py-2)	CH₃.C₉H₆N	143.18	1.059	−1	244–5
Quinoline	C₉H₇N	129.15	1.095	−15	237.1
(iso-)	C₉H₇N	129.15	1.099	24.6	240.5
-diol (1-,3-)	—C₆H₄CH:C(OH)N:C(OH)—	161.15	....	237	....
Quinone (p-)	CO < (CHCH)₂ > CO	108.09	1.318	115.7	subl.
R-acid Ca salt (2-)(3-,6-)	HOC₁₀H₅(SO₃)₂Ca	342.35	....	....	....
K salt	HOC₁₀H₅(SO₃K)₂	380.46	....	....	....
Na salt	HOC₁₀H₅(SO₃Na)₂	348.26	....	....	....
Raffinose	C₁₈H₃₂O₁₆.5H₂O	594.52	1.465⁰	119	d. 130
Resorcinol (m-)	C₆H₄(OH)₂	110.11	1.272	110.7	276.5
Retene	C₁₈H₁₈	234.32	1.13	98–9	390–4
Rhamnose (β-)	CH₃(CHOH)₄CHO.H₂O	182.17	1.471	126	....
Ricinoleic acid	C₁₇H₃₂(OH)CO₂H	298.45	0.954	4–5	226–8¹⁰ ᵐᵐ
Rosaniline	C₂₀H₂₁ON₃	319.39	....	186 d.	....
Rosolic acid	C₂₀H₁₆O₃	304.33	....	308–10 d.	....
Saccharin	C₆H₄(CO)(SO₂) > NH	183.18	....	225–8	subl.
Safrole (1-,3-,4-)	CH₂:CHCH₂.C₆H₃O₂CH₂	162.18	1.100	11.2	233–4
(iso-)(1-,3-,4-)	CH₃.CH:CH.C₆H₃O₂CH₂	162.18	1.122	6–7	252–3
Salicylic acid (o-)	HO.C₆H₄.CO₂H	138.12	1.443	159	21¹²⁰ ᵐᵐ
aldehyde (o-)	HO.C₆H₄.CHO	122.12	1.153	~7	196.5
Saligenin	HO.C₆H₄.CH₂OH	124.13	1.161	86–7	subl.
Schaeffer's salt, Ca	(HOC₁₀H₆SO₃)₂Ca.5H₂O	576.59	....	....	....
K	HOC₁₀H₆SO₃K	262.31	....	....	....
Na	HOC₁₀H₆SO₃Na	246.21	....	....	....
Semicarbazide	NH₂.CO.NH.NH₂	75.07	....	96	....
hydrochloride	NH₂.CO.NH.NH₃Cl	111.54	....	173 d.	....
Skatole (3-)	CH₃.C₈H₆N	131.17	....	95	265–6
Sodium methylate	CH₃ONa	54.03	....	d. 300	....
Sorbitol	[CH₂OH(CHOH)]₂	182.17	....	110–2	....
Sorbose (d- or l-)	C₆H₁₂O₆	180.16	1.65	165	....
Starch	(C₆H₁₀O₅)x	162.14	1.50	d.	....
Stearic acid	CH₃(CH₂)₁₆CO₂H	284.47	0.847⁶⁹·³	70–1	291¹¹⁰ ᵐᵐ
amide	CH₃(CH₂)₁₆CONH₂	283.48	....	108–9	251¹² ᵐᵐ
Styrene	C₆H₅CH:CH₂	104.14	0.903	−31	145–6
Suberic acid	HO₂C(CH₂)₆CO₂H	174.19	1.266	140–4	279¹⁰⁰ ᵐᵐ
Succinic acid	HO₂C(CH₂)₂CO₂H	118.09	1.572	189–90	235 d.
Sucrose	C₁₂H₂₂O₁₁	342.30	1.588	170–86 d.	....
Sulfanilic acid (p-)	H₂N.C₆H₄.SO₃H	173.18	....	d. >280	....
Sylvestrene (d-)	C₁₀H₁₆	136.23	0.863	....	176–7
Tartaric acid (meso-)	(CHOHCO₂H)₂	150.09	1.737	159–60	....
(racemic)	(CHOHCO₂H)₂.H₂O	168.10	1.697	205–6	d.
(d- or l-)	(CHOHCO₂H)₂	150.09	1.760	168–70	d.
Tartronic acid	CH(OH)(CO₂H)₂.½H₂O	129.07	....	d. 155–8	subl.
Terephthalic acid (p-)	C₆H₄(CO₂H)₂	166.13	1.510	subl.	....
Terpin hydrate (cis-)	C₁₀H₂₀O₂.H₂O	190.28	....	117	d.
Terpineol (α-)(d- or l-)	C₁₀H₁₈O	154.24	0.935	38–40	219–21
(dl-)	C₁₀H₁₈O	154.24	0.935	35	218–9
Terpinyl acetate (α-)(dl-)	CH₃CO₂.C₁₀H₁₇	196.28	0.966	<−50	220 d.
Tetrabromo-ethane (sym)	Br₂CH.CHBr₂	345.70	2.964	−1.0	15¹⁵⁴ ᵐᵐ
(uns)	Br₃C.CH₂Br	345.70	2.875	0	104¹³ ᵐᵐ
Tetrachloro-ethane (sym)	Cl₂CH.CHCl₂	167.86	1.600	−36	146.3
(uns)	Cl₃C.CH₂Cl	167.86	1.588	....	129–30
-ethylene	Cl₂C:CCl₂	165.85	1.624	−19	120.8

TABLE 1-11 Physical Constants of Elements and Compounds—(*Continued*)

Name	Formula	Formula weight	Specific gravity	Melting point, °C.	Boiling point, °C.
Tetracosane (n-)	$CH_3(CH_2)_{22}CH_3$	338.64	0.779^{81}	51.1	324
Tetradecane (n-)	$CH_3(CH_2)_{12}CH_3$	198.38	0.765	5.5	252.5
Tetraethyl-thiuram disulfide	$[(C_2H_5)_2NCS]_2S_2$	296.52	1.17	70	
Tetrafluoro-ethylene	$F_2C:CF_2$	100.02	1.58^{-78}	−142.5	−76.3
Tetrahydro-furan	$—CH_2(CH_2)_2CH_2O—$	72.10	0.888	−65	65-6
-furfuryl alcohol	$C_4H_7O.CH_2OH$	102.13	1.050	....	177-8
-pyran	$—CH_2(CH_2)_3CH_2O—$	86.13	0.881	....	88
Tetralin	$—C_6H_4CH(CH_2)_2CH_2—$	132.20	0.973	−31	206
Tetramethyl-thiuram disulfide	$[(CH_3)_2NCS]_2S_2$	240.41	1.29	155-6	
Tetryl (2-,4-,6-)	$(NO_2)_3C_6H_2.N(CH_3)NO_2$	287.15	1.57	130.5	expl.
Theobromine	$C_7H_8O_2N_4$	180.17		330	
Thio-acetic acid	$CH_3.CO.SH$	76.11	1.074	<−17	93
-aniline (4-, 4′-)	$(NH_2.C_6H_4)_2S$	216.29		108	
-carbanilide	$(C_6H_5.NH)_2CS$	228.30	1.3	154	d.
-naphthol (β-)	$C_{10}H_7.SH$	160.22		81	286-8
phenol	$C_6H_5.SH$	110.17	1.074		168-9
-salicylic acid (o-)	$HS.C_6H_4.CO_2H$	154.18		164	subl.
-urea	$NH_2.CS.NH_2$	76.12	1.405	180-2	d.
Thiophene	$< (CH:CH)_2 > S$	84.13	1.070	−30	84
Thymol (5-,2-,1-)	$(CH_3)(C_3H_7)C_6H_3OH$	150.21	0.972	51.5	232
Tolidine (0)(3,3′,4,4′)	$[CH_3(NH_2)C_6H_3]_2$	212.28		128-9	
Toluene	$C_6H_5.CH_3$	92.13	0.866	−95	110.8
sulfonic acid (o-)	$CH_3.C_6H_4SO_3H.2H_2O$	208.23		d.	128.8$_0$ mm
(p-)	$CH_3.C_6H_4SO_3H.H_2O$	190.21		104-5	146-7$_0$ mm
sulfonic amide (p-)	$CH_3.C_6H_4SO_2NH_2$	171.21		137	
sulfonic chloride (p)	$CH_3.C_6H_4.SO_2Cl$	190.64		64	134.5 10 mm
Toluic acid (o-)	$CH_3.C_6H_4.CO_2H$	136.14	1.062^{115}	104-5	259
(m-)	$CH_3.C_6H_4.CO_2H$	136.14	1.054^{112}	110-1	263
(p-)	$CH_3.C_6H_4.CO_2H$	136.14		179-80	274-5
Toluidine (o-)	$CH_3.C_6H_4.NH_2$	107.15	0.999	−16.3	199.7
(m-)	$CH_3.C_6H_4.NH_2$	107.15	0.989	−31.5	203.3
(p-)	$CH_3.C_6H_4.NH_2$	107.15	1.046	44-5	200.3
hydrochloride (o-)	$CH_3.C_6H_4.NH_2.Cl$	143.62		218-20	242
sulfonic acid (1-,2-,3-)	$CH_3(NH_2)C_6H_3SO_3H$	187.21			
Toluylenediamine (1-,2-,4-)	$CH_3.C_6H_3(NH_2)_2$	122.17		99	283-5
Tolylene diisocyanate (1-,2-,4-)	$CH_3.C_6H_3(NCO)_2$	174.15	1.23		134.5 20 mm
Trehalose	$C_{12}H_{22}O_{11}.2H_2O$	378.33		97	
Triamylamine (n)	$[CH_3)_3CH_2]_3N$	227.42			240-5
(i-)	$[(CH_3)_2CH(CH_2)_2]_3N$	227.42	0.786		235
Tributyl-amine (n-)	$[CH_3(CH_2)_2CH_2]_3N$	185.34	0.778		216.5
phosphite	$[CH_3(CH_2)_3O]_3P$	250.32	0.925		122-312 mm
Trichloro-acetic acid	$Cl_3C.CO_2H$	163.40	1.617^{46}	58	195.5
-benzene (s-)(1-,3-,5-)	$C_6H_3Cl_3$	181.46		63.5	208.5
-ethane (1-,1-,1-)	$Cl_3C.CH_3$	133.42	1.325		74.1
-ethylene	$Cl_2C:CHCl$	131.40	1.466	−73	87.2
-phenol	$Cl_3C_6H_2OH$	197.46	1.490^{75}	68-9	246
Tricosane (n-)	$CH_3(CH_2)_{21}CH_3$	324.61	0.779^{48}	47.7	234 15 mm
Tricresyl phosphate (o-)	$OP(OC_6H_4CH_3)_3$	368.36			
Tridecane (n-)	$CH_3(CH_2)_{11}CH_3$	184.35	0.757	−6.2	234
Triethanol amine	$(HOCH_2CH_2)_3N$	149.19	1.126	20-1	277-9 150 mm
Triethyl amine	$(CH_3CH_2)_3N$	101.19	0.729	−114.8	89.4
-benzene (1-,3-,5-)	$(C_2H_5)_3C_6H_3$	162.26	0.861		215
(1-,2-,4-)	$(C_2H_5)_3C_6H_3$	162.26	0.882		217-8
borate	$B(OCH_2CH_3)_3$	146.00	0.864		120
citrate	$HOC_3H_4(CO_2C_2H_5)_3$	276.28	1.137		294
Triethylene glycol	$(.CH_2OCH_2CH_2OH)_2$	150.17	1.125	−5	290
Trifluoro-chloromethane	CF_3Cl	104.47	1.726^{-130}	−182	−80
chloroethylene	$F_2C:CFCl$	116.48		−157.5	−27.9
-trichloroethane	$Cl_2CF.CClF_2$	187.39	1.576	−35	47.6
Trimethoxybutane (1-,3-,3-)	$CH_2(OCH_3)CH_2C(OCH_3)_2CH_3$	148.20	0.932		63-5 25 mm
Trimethylamine	$(CH_3)_3N$	59.11	0.662^{-5}	−124	3.5
Trimethylene bromide	$BrCH_2CH_2CH_2Br$	201.91	1.987	−34.4	167.5
chloride	$ClCH_2CH_2CH_2Cl$	112.99	1.201		123-5
glycol	$HOCH_2CH_2CH_2OH$	76.09	1.060		214
Trinitro-benzene (1-,3-,5-)	$C_6H_3(NO_2)_3$	213.11	1.688	121	d.
-benzoic acid (2-,4-,6-)	$(NO_2)_3C_6H_2CO_2H$	257.12		210-20 d.	....
-tert-butylxylene	$(NO_2)_3C_6(CH_3)_2C_4H_9$	297.26		110	....
-naphthalene (α-)(1-,3-,5-)	$C_{10}H_5(NO_2)_3$	263.16		122-3	....
(β-)(1-,3-,8-)	$C_{10}H_5(NO_2)_3$	263.16		218-9	....
(γ-)(1-,4-,5-)	$C_{10}H_5(NO_2)_3$	263.16		148-9	....
-phenol (2-,3-,6-)	$(NO_2)_3C_6H_2OH$	229.11		117-8	....
-toluene (β-)(2-,3-,4-)	$CH_3C_6H_2(NO_2)_3$	227.13	1.620	112	expl.
(γ-)(2-,4-,5-)	$CH_3C_6H_2(NO_2)_3$	227.13	1.620	104	expl.
(α-)(2-,4-,6-)	$CH_3C_6H_2(NO_2)_3$	227.13	1.654	80.8	expl.
Trional	$C_2H_5(CH_3)C(SO_2C_2H_5)_3$	242.34	1.199^{85}	76	d.
Triphenyl-arsine	$(C_6H_5)_3As$	306.21	1.306	59-60	>360
carbinol	$(C_6H_5)_3COH$	260.32	1.188	162.5	>360
guanidine (α-)	$C_6H_5N:C(NHC_6H_5)_2$	287.35	1.13	144-5	d.
methane	$(C_6H_5)_3CH$	244.32	1.014^{99}	93.4	359
methyl	$(C_6H_5)_3C$ . . .	243.31		145-7	d.
phosphate	$OP(OC_6H_5)_3$	326.28	1.206^{68}	49-50	245 11 mm
Tripropylamine (n-)	$(CH_3CH_2CH_2)_3N$	143.27	0.757	−93.5	156.5
Undecane (n-)	$CH_3(CH_2)_9CH_3$	156.30	0.741	−25.6	194.5
Urea	$H_2N.CO.NH_2$	60.06	1.335	132.7	d.
nitrate	$CO(NH_2)_2.HNO_3$	123.07		152 d.	....
Uric acid	$C_5H_4O_3N_4$	168.11	1.893	d.	....

TABLE 1-11 Physical Constants of Elements and Compounds—(Continued)

Name	Formula	Formula weight	Specific gravity	Melting point, °C.	Boiling point, °C.
Valeric acid (n-)	$C_2H_5CH_2CH_2CO_2H$	102.13	0.939	−34.5	187
(i-)	$(CH_3)_2CHCH_2CO_2H$	102.13	0.931	−37.6	176
aldehyde (n-)	$C_2H_5CH_2CH_2CHO$	86.13	0.819	−92	103.4
(i-)	$(CH_3)_2CHCH_2CHO$	86.13	0.803	−51	92.5
amide (n-)	$C_2H_5CH_2CH_2CONH_2$	101.15	1.023	106	....
(i-)	$(CH_3)_2CHCH_2CONH_2$	101.15	0.965	135–7	232
Vanillic acid (3-,4-,1-)	$CH_3O(OH)C_6H_3CO_2H$	168.14	....	207	subl.
alcohol (3-,4-,1-)	$CH_3O(OH)C_6H_3CH_2OH$	154.16	....	115	d.
Vanillin (3-,4-,1-)	$CH_3O(OH)C_6H_3CHO$	152.14	1.056	81–2	285
Veratrole (o-)	$C_6H_4(OCH_3)_2$	138.16	1.091	22.5	207.1
Vinyl acetate	$CH_3CO_2CH{:}CH_2$	86.09	0.932	<−60	72–3
(poly-)	$(CH_3CO_2CH{:}CH_2)x$	(86.09)	1.19	100–25	....
acetic acid	$CH_2{:}CH.CH_2CO_2H$	86.09	1.013	−39	163
acetylene	$CH_2{:}CH.C{:}CH$	52.07	$0.705^{1.5}$	....	5.5
alcohol	$CH_2{:}CHOH$	44.06			
(poly-)	$(CH_2{:}CHOH)x$	(44.06)	1.3	d. >200	....
chloride	$CH_2{:}CHCl$	62.50	0.908	−160	−12
propionate	$C_2H_5CO_2CH{:}CH_2$	100.11	....		93–5
Xylene (o-)	$C_6H_4(CH_3)_2$	106.16	0.881	−25	144
(m-)	$C_6H_4(CH_3)_2$	106.16	0.867	−47.4	139.3
(p-)	$C_6H_4(CH_3)_2$	106.16	0.861	13.2	138.5
sulfonic acid (1-,4-,2-)	$(CH_3)_2C_6H_3SO_3H.2H_2O$	222.25		86	$149^{0.1\,mm}$
Xylidine (1:2)(3-)	$(CH_3)_2C_6H_3NH_2$	121.18	0.991	<−15	223
(1:2)(4-)	$(CH_3)_2C_6H_3NH_2$	121.18	1.076	49–50	224–6
(1:3)(2-)	$(CH_3)_2C_6H_3NH_2$	121.18	0.980		216–7
(1:3)(4-)	$(CH_3)_2C_6H_3NH_2$	121.18	0.978		213–4
(1:3)(5-)	$(CH_3)_2C_6H_3NH_2$	121.18	0.972		221–2
(1:4)(2-)	$(CH_3)_2C_6H_3NH_2$	121.18	0.979	15.5	215
Xylose (l-)(+)	$CH_2OH(CHOH)_3CHO$	150.13	1.535^0	153–4	....
Xylylene dichloride (p-)	$C_6H_4(CH_2Cl)_2$	175.06	1.417^0	100.5	240–5 d.
Zinc diethyl	$Zn(CH_2CH_3)_2$	123.50	1.182	−28	118
dimethyl	$Zn(CH_3)_2$	95.45	1.386	−40	46
dimethyl-dithiocarbamate	$Zn[S_2CN(CH_3)_2]_2$	305.79	2.00^{40}	248–50	....

TABLE 1-12 Densities of Common Materials

Material	Density, lb/ft^3	Density, g/cm^3 or sp.gr.
ALLOYS		
Aluminum-copper (Al/Cu, 10/90)	480	7.69
(Al/Cu, 5/95)	523	8.37
Aluminum-zinc (Al/Zn, 91/9)	175	2.80
Bell metal (Cu/Zn, 78/22)	543	8.70
Bismuth-lead-tin (Bi/Pb/Sn, 53/40/7)	659	10.56
Brass, yellow (Cu/Zn, 70/30)	534	8.56
red (Cu/Zn, 90/10)	537	8.60
white (Cu/Zn, 50/50)	511	8.20
Bronze (gun metal Cu/Sn, 90/10)	548	8.78
(Cu/Sn, 80/20)	546	8.74
(Cu/Sn, 75/25)	551	8.83
Cadmium-tin (Cd/Sn, 32/68)	481	7.70
Constantan (Cu/Ni, 60/40)	554	8.88
German silver (Cu/Zn/Ni, 26/37/37)	518	8.30
(Cu/Zn/Ni, 52/26/22)	528	8.45
(Cu/Zn/Ni, 63/30/6)	518	8.30
Gold-copper (Au/Cu, 98/2)	1,176	18.84
(Au/Cu, 92/8)	1,094	17.52
(Au/Cu, 86/14)	1,028	16.47
Invar (Fe/Ni/C, 63.9/35.9/0.2)	499	8.00
Lead-tin (Pb/Sn, 88/12)	662	10.60
(Pb/Sn, 78/22)	627	10.05
(Pb/Sn, 64/36)	589	9.43
(Pb/Sn, 30/70)	514	8.24
Magnalium (Al/Mg, 90/10)	156	2.50
(Al/Mg, 70/30)	125	2.00
Manganese bronze (Cu/Mn, 95/5)	549	8.80
Manganin (Cu/Mn/Ni, 84/12/4)	531	8.50
Monel (Ni/Cu/Fe, 71/27/2)	556	8.90
Phosphor bronze (Cu/Sn/Sb/P, 80/10/9/1)	549	8.80

TABLE 1-12   Densities of Common Materials (Continued)

Material	Density, lb/ft³	Density, g/cm³ or sp.gr.
ALLOYS (continued)		
Platinum-iridium (Pt/Ir, 90/10)	1,350	21.62
(Pt/Ir, 67/33)	1,365	21.87
(Pt/Ir, 5/95)	1,397	22.38
Speculum (Cu/Sn, 67/33)	537	8.60
Steel (Fe/C, 99/1)	489	7.83
(Fe/Mn/C, 86/13/1)	488	7.81
Wood's metal (Be/Pb/Cd/Sn, 50/25/13/12)	659	10.56
VARIOUS SOLIDS		
Agate	162	2.6
Amber	67	1.1
Asbestos	150	2.4
Asphalt	82	1.3
Basalt	170	2.8
Beeswax	61	0.97
Beryl	169	2.7
Biotite	180	2.9
Bone	116	1.9
Brick	112	1.8
Butter	54	0.87
Calamine	268	4.3
Calspar	169	2.7
Camphor	62	0.99
Caoutchouc	60	0.96
Cardboard	43	0.7
Celluloid	87	1.4
Cement (set)	180	2.9
Chalk	147	2.4
Charcoal, oak	35	0.57
pine	23	0.36
Cinnabar	507	8.1
Clay	137	2.2
Coal, anthracite	125	1.6
bituminous	85	1.4
Cocoa butter	57	0.90
Coke	84	1.4
Cork	15	0.24
Corundum	248	4.0
Diamond	204	3.3
Dolomite	177	2.8
Ebonite	72	1.2
Emery	250	4.0
Feldspar	166	2.7
Flint	164	2.6
Fluorite	198	3.2
Galena	465	7.5
Garnet	235	3.8
Gelatin	79	1.3
Glass, common	163	2.6
flint	275	4.4
Glue	79	1.3
Granite	169	2.7
Graphite	157	2.5
Gum arabic	84	1.4
Gypsum	145	2.3
Hematite	318	5.1
Hornblende	187	3.0
Ice	57	0.92
Ivory	117	1.9
Lampblack	117	1.9
Leather	55	0.9
Lime, slaked	84	1.4

TABLE 1-12   Densities of Common Materials (*Continued*)

Material	Density, lb/ft^3	Density, g/cm^3 or sp.gr.
VARIOUS SOLIDS (continued)		
Limestone	169	2.7
Linoleum	74	1.2
Magnetite	315	5.1
Malachite	244	3.9
Marble	169	2.7
Mica	183	2.9
Muscovite	180	2.8
Ocher	218	3.5
Opal	137	2.2
Paper	58	0.95
Paraffin	56	0.90
Pitch	67	1.1
Porcelain	150	2.4
Porphyry	172	2.8
Pressed board (pulp)	12	0.2
Pyrite	314	5.0
Quartz	165	2.7
Resin	67	1.1
Rock salt	136	2.2
Rubber, hard	74	1.2
soft	69	1.1
gum	58	0.92
Sandstone	141	2.3
Serpentine	161	2.6
Silica	134	2.1
Slag	183	3.0
Slate	184	3.0
Soapstone	169	2.7
Spermaceti	59	0.95
Starch	95	1.53
Sugar	99	1.59
Talc	171	2.8
Tallow	59	0.94
Tar	66	1.02
Topaz	221	3.6
Tourmaline	195	3.1
Wax, sealing	112	1.8
WOODS		
Apple	47	0.75
Ash	47	0.75
Balsa	8	0.13
Bamboo	22	0.36
Basswood	29	0.46
Beech	50	0.80
Birch	40	0.64
Box	66	1.06
Cedar	33	0.53
Cherry	50	0.80
Dogwood	47	0.76
Ebony	76	1.22
Elm	36	0.57
Hickory	48	0.77
Locust	43	0.69
Mahogany, Honduras	41	0.66
Spanish	58	0.85
Maple	43	0.69
Oak	47	0.76
Pear	42	0.67
Pine, white	27	0.43
yellow	30	0.49

TABLE 1-12 Densities of Common Materials (Continued)

Material	Density, lb/ft^3	Density, g/cm^3 or sp.gr.
Woods (continued)		
Plum	45	0.72
Poplar	27	0.43
Spruce	37	0.59
Sycamore	31	0.50
Teak, Indian	48	0.76
African	61	0.98
Walnut	42	0.67
Willow	31	0.50
Common Liquids		
Acetone	49.4	0.79
Alcohol, ethyl	49.4	0.79
methyl	50.5	0.81
Benzene	56.1	0.89
Carbon disulfide	80.7	1.29
tetrachloride	99.6	1.60
Chloroform	93.0	1.49
Ether	45.9	0.74
Gasoline	42.0	0.68
Glycerin	78.6	1.26
Kerosene	51.2	0.82
Mercury	849.0	13.6
Milk	64.4	1.03
Naphtha, petroleum	41.5	0.67
wood	51.7	0.83
Oil, castor	60.5	0.97
coconut	57.7	0.93
cottonseed	57.8	0.93
creosote	66.8	1.07
linseed, boiled	58.8	0.94
olive	57.3	0.92
Turpentine	54.3	0.87
Water, fresh	62.4	1.00
sea	64.0	1.03

1-5 ADDITIONAL PHYSICAL PROPERTIES OF MATTER

Elasticity and Young's Modulus

The property of *elasticity* in a body is its ability to return to its original condition after the force which caused a distortion is removed. We do not normally think of glass or steel as being highly elastic bodies, but nevertheless, they are more elastic than such materials as rubber, since they return more perfectly to their original shape once a distorting force has been removed. When a force, generally referred to as a stress, is applied to an elastic body, the distortion, or strain, which it causes is directly proportional to the stress. Conversely, if an elastic body is strained, an internal stress will be developed. The fact that the stress and strain are directly proportional is known as *Hooke's law*. Expressing this mathematically: k = stress/strain. The value of k is a constant for any given material and is known as the *modulus of elasticity*.

If too great a stress is applied to a body, it can be permanently distorted even after removal of the stress, indicating that there is a limit beyond which the proportionality expressed by Hooke's law does not hold. This stress is known as the *elastic limit* of the material. If a stress is applied to a body beyond the elastic limit, eventually the body will rupture. The stress at which this occurs is known as the *breaking stress*.

The type of elasticity which is usually of greatest interest is the change in length of a body under a tensile or compressive load. When the stress is expressed as force per unit area and the strain as change in length per unit length, a special form of the elastic modulus results which is highly characteristic of the given material up to its elastic limit. This is known as its *Young's modulus*. This is expressed mathematically by the following equation, the components of which are illustrated in Fig. 1-11. [The uppercase Greek letter delta (Δ) indicates a change, so that the symbol ΔL means the change in length.]

$$Y = \frac{\text{stress}}{\text{strain}} = \frac{F/A}{\Delta L/L}$$

In English units, force is expressed in pounds, area in square inches, and length in inches; therefore, the unit for Young's modulus is lb/in².

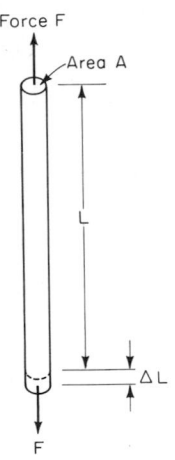

Force F

Area A

L

ΔL

F

Fig. 1-11 Determination of Young's modulus.

$$Y = \frac{\text{lb/in}^2}{\text{in/in}} = \frac{\text{lb}}{\text{in}^2}$$

Table 1-13 presents the Young's moduli for a number of common materials.

Example 1-34 A copper wire 0.016 in in diameter and 36 in long is placed under a tensile force of 1.65 lb. This causes an elongation of 0.017 in. What is Young's modulus for copper?

solution Cross-sectional area of wire $= \dfrac{\pi d^2}{4} = \dfrac{\pi (0.016 \text{ in})^2}{4}$

$$= 0.000206 \text{ in}^2$$

$$Y = \frac{F/A}{\Delta L/L} = \frac{1.65 \text{ lb}/0.000206 \text{ in}^2}{0.017 \text{ in}/36 \text{ in}} = \frac{8{,}000 \text{ lb/in}^2}{0.00047} = 17{,}000{,}000 \text{ lb/in}^2$$

In Young's modulus calculations, it is frequently easier to locate the decimal point using exponential numerics:

$$Y = \frac{1.65 \text{ lb}/0.206 \times 10^{-3} \text{ in}^2}{17 \times 10^{-3}/36} = \frac{8.0 \times 10^3 \text{ lb/in}^2}{0.47 \times 10^{-3}}$$

$$= 17 \times 10^6 \text{ lb/in}^2$$

TABLE 1-13   Elastic (Young's) Modulus for Common Materials

METALS AND ALLOYS

Material	Young's modulus	
	lb/in² $\times 10^6$	dyn/cm² $\times 10^{11}$
Aluminum, cast	9.50	6.65
rolled	9.85	6.91
Aluminum-magnesium (6% Mg)	8.96	6.18
-manganese (8% Mn)	9.39	6.47
Antimony	11.31	7.80
Brass, cold-rolled	13.09	9.02
Bronze, commercial (90% Cu, 10% Zn)	17.0	11.6
phospor (10% Sn)	16.0	11.0
phospor (5% Sn)	16.0	11.0
Cadmium	10.06	6.93
Constantan	22.05	15.20
Copper, rolled	18.06	12.46
wire	15.80	11.10
Duralumin	10.00	6.89
Gold	11.38	7.85
Gunmetal	10.0	7.0
Iron, cast	13.0	9.1
wrought	27.5	19.4
Lead, rolled	2.28	1.57
antimonial (6% Sb)	3.0	2.1
Magnesium, drawn	6.06	4.18
Magnesium-aluminum (8% Al)	6.06	4.18
Monel	24.96	17.22

TABLE 1-13   Elastic (Young's) Modulus for Common Materials (Continued)

METALS AND ALLOYS (continued)

Material	Young's modulus	
	lb/in² × 10⁶	dyn/cm² × 10¹¹
Nickel	30.01	20.61
Nickel-iron (5% Ni)	30.86	21.28
(1.8% Ni)	25.17	17.36
Palladium, drawn	17.07	11.77
Platinum, drawn	24.18	16.67
Silver, drawn	11.24	7.75
Steel, drawn (0.08% C)	27.88	19.22
annealed (0.4% C)	29.01	20.01
annealed (0.7% C)	28.45	19.61
Tantalum	27.0	18.6
Tin, rolled	6.76	4.66
Tungsten, drawn	51.49	35.50
Zinc, rolled	13.10	9.0
PLASTICS		
Acrylonitrile butadiene styrene (ABS)	0.28	0.196
Acetal	0.41	0.29
Acrylic	0.40	0.28
Fluorocarbon, trifluorochlorethylene	0.25	0.18
tetrafluoroethylene	0.052	0.037
Polyamide, nylon 66	0.41	0.29
nylon 6	0.29	0.20
nylon 11	0.185	0.130
Polycarbonate	0.32	0.22
Phenolic, molded	1.10	0.77
cast	0.45	0.32
filled, glass fabric	3.4	2.39
filled, asbestos	5.5	3.86
Polyester	0.25	0.18
Polystyrene	0.50	0.35
Polyethylene, low-density	0.024	0.017
Polypropylene	0.15	0.11
Polyvinyl butyral	0.38	0.27
Polyvinyl dichloride	0.040	0.28
Polyvinyl chloride, rigid	0.38	0.27
flexible	0.002	0.0014
Polyvinylidene chloride (Saran)	0.06	0.042
Urea, molded	1.45	1.02

Example 1-35   If the breaking stress of the wire in Example 1-34 is 32,000 lb/in², under what weight will the wire break?

solution

$$\text{Breaking stress} = \frac{\text{force}}{\text{area}}$$

$$\text{Force} = \text{breaking stress} \times \text{area}$$
$$= 32{,}000 \text{ lb/in}^2 \times 0.000206 \text{ in}^2 = 6.6 \text{ lb}$$

Compressibility and Bulk Modulus

Solids can, of course, be placed under tensile or compressive stresses, but there would be little meaning in everyday experience in applying a tensile force to a liquid. Nevertheless, liquids display a high degree of elasticity when subjected to a compressive stress. When pressure is applied to a given volume of liquid, the volume tends to become smaller; and like Hooke's law for solids, the ratio of the change in pressure to the change in volume per unit volume is a constant and is known as the *bulk modulus*. The bulk modulus is defined by the following equation:

$$B = -\frac{\Delta p}{\Delta V / V}$$

in which the minus sign appears because an increase in pressure causes a decrease in the volume of the liquid. Table 1-14 lists the bulk modulus for a variety of common liquids. As in the case of Young's modulus, the units for bulk modulus are lb/in².

Example 1-36 A vessel holding 1,000 gal of water is subjected to a pressure of 2,000 lb/in². At this pressure, what is the volume of the water?

solution From Table 1-14, the bulk modulus of water is given as 31×10^4 lb/in².

$$B = -\frac{\Delta p}{\Delta V/V}$$

$$\frac{\Delta V}{V} = -\frac{\Delta p}{B} = -\frac{2,000 \text{ lb/in}^2}{31 \times 10^4 \text{ lb/in}^2} = -0.0065$$

$$\Delta V = -0.0065 V = -0.0065 (1,000 \text{ gal}) = -6.5 \text{ gal}$$

The final volume is therefore $V - \Delta V = 994$ gal

TABLE 1-14   Bulk Modulus for a Variety of Liquids at 20°C

Liquid	Bulk modulus	
	lb/in² $\times 10^5$	dyn/cm² $\times 10^{10}$
Acetic acid	1.60	1.10
Acetone	1.15	0.79
Aniline	3.21	2.21
Benzene	1.54	1.06
Butyl alcohol	1.74	1.20
Carbon disulfide	1.57	1.08
Carbon tetrachloride	1.41	0.97
Chloroform	1.47	1.01
Ethyl alcohol	1.31	0.90
Ethylene chloride	1.82	1.25
Ethyl ether	0.78	0.54
Glycol	3.92	2.70
Isopropyl alcohol	1.29	0.89
Mercury	3.53	2.43
Methanol	1.19	0.82
n-Propyl alcohol	1.60	1.10
Toluene	1.63	1.12
Water	3.17	2.18

Bulk modulus of water = 3.17×10^5 lb/in². Compressibility is the reciprocal of bulk modulus; i.e., compressibility of toluene = 0.61×10^{-5} in²/lb.

Example 1-37 A volume of mercury contained in a suitable vessel is found to be 96.37 cm³. The mercury is subjected to a pressure of 250 atm and its volume is found to be 96.28 cm³. What is the bulk modulus of mercury?

solution

$$B = -\frac{\Delta p}{\Delta V/V} = -\frac{250 \text{ atm} \times 14.7 \text{ lb/atm} - \text{in}^2}{(96.28 - 96.37) \text{ cm}^3/96.37 \text{ cm}^3}$$

$$B = -\frac{250 (14.7) \text{ lb/in}^2}{-0.09/96.37} = 3.9 \times 10^6 \text{ lb/in}^2$$

Hardness

Hardness is another property of matter which frequently has to be described in chemical work. The *hardness* of a substance is fundamentally a measure of its ability to abrade or indent another substance. An arbitrary scale of hardness, based on the ability of 10 selected minerals to scratch each other, is known as the Mohs' scale. The Mohs' scale is based on the following minerals, with talc the softest and diamond the hardest.

1. Talc
2. Rock salt or gypsum
3. Calcite
4. Fluorite
5. Apatite
6. Feldspar
7. Quartz
8. Topaz
9. Corundum
10. Diamond

The hardness of various materials can also be obtained based on the ability of one to scratch the other. These relative hardnesses are summarized in Table 1-15 for a number of important chemical materials and minerals.

For metals and plastic materials, special systems for measuring hardness have been developed. These will be considered in later chapters.

TABLE 1-15   Average Hardness of Various Materials (Mohs' Scale)

Agate	6.5	Hematite	6.0
Alabaster	1.7	Hornblende	5.5
Alum	2.3	Indium	1.2
Aluminum	2.5	Iridium	6.3
Alundum	9.0	Iridosmium	7.0
Amber	2.3	Iron	4.5
Andalusite	7.5	Kaolinite	2.3
Anthracite	2.2	Lead	1.5
Antimony	3.2	Lithium	0.6
Apatite	5.0	Magnesium	2.0
Aragonite	3.5	Magnetite	6.0
Arsenic	3.5	Manganese	5.0
Asbestos	5.0	Marble	3.5
Asphalt	1.5	Mica	2.8
Augite	6.0	Opal	5.0
Barite	3.3	Orthoclase	6.0
Bell metal	4.0	Osmium	7.0
Beryl	7.8	Palladium	4.8
Bismuth	2.5	Phosphorous	0.5
Boric acid	3.0	Phosphor bronze	4.0
Boron	9.5	Platinum	4.3
Brass	3.5	Potassium	0.5
Cadmium	2.0	Pumice	6.0
Calamine	5.0	Pyrite	6.3
Calcite	3.0	Quartz	7.0
Calcium	1.5	Rock salt	2.0
Carbon (diamond)	10.0	Rubidium	0.3
Carborundum	9.5	Ruthenium	6.5
Cesium	0.2	Selenium	2.0
Chromium	9.0	Serpentine	3.5
Copper	2.8	Silicon	7.0
Corundum	9.0	Silver	2.6
Diamond	10.0	Silver chloride	1.3
Diatomaceous earth	1.3	Sodium	0.4
Dolomite	3.8	Steel	6.7
Emery	8.0	Stibnite	2.0
Feldspar	6.0	Strontium	1.8
Flint	7.0	Sulfur	2.0
Fluorite	4.0	Talc	1.0
Galena	2.5	Tellurium	2.3
Gallium	1.5	Tin	1.7
Garnet	6.8	Topaz	8.0
Glass	5.5	Tourmaline	7.3
Gold	2.8	Wood's metal	3.0
Graphite	0.8	Zinc	2.5
Gypsum	1.8		

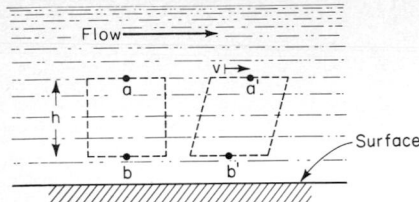

Fig. 1-12 Shear in a flowing liquid. During the time interval in which point a moves to a', point b moves only to b', putting the liquid between a and b in shear.

Viscosity and Fluidity

The resistance of a liquid to flow or to a change of form is called its *viscosity*. From common experience, some liquids appear to be much more mobile than others. As an example, water will flow down over a tilted surface much more rapidly than oil. When a liquid flows over a surface, the liquid particles immediately adjacent to the surface tend to adhere to the surface and essentially remain stationary. The next layer of liquid moves over the first layer, and the third over the second, and so on, the speed of each layer increasing with its distance from the solid surface. This distribution of speed causes a portion of the liquid that was cubical at one point in time to become rhomboidal at a later time, as shown in Fig. 1-12. The shearing stress set up in the cubical portion of liquid shown in Fig. 1-12 is measured by the force per unit area on the upper face, while the rate of shear is the difference in velocity v between the upper and lower faces divided by the distance h between the two faces. The shear force per unit area for practically all liquids over a wide range of conditions is proportional to the rate of shear, as shown by the equation $F/A \propto v/h$. This proportionality can be rewritten as an equation in which a constant, the lowercase Greek letter eta (η) is the proportionality constant, so that $F/A = \eta\,(v/h)$. This constant is called the *coefficient of viscosity* and is characteristic of a given liquid at a given temperature. Thus the coefficient of viscosity of a liquid is defined as $\eta = (F/A)/(v/h)$. In the metric system, if a 1-cm^2 plate is drawn through a liquid at a speed of 1 cm/s in a plane 1 cm from a stationary surface, the force in dynes necessary to move the plate is numerically equal to the coefficient of viscosity of the liquid. (The dyne, a unit of force used extensively in scientific work, is described in detail in the next chapter.) The coefficient of viscosity is expressed in poise (P), 1 poise being equal to 1 dyne-s/cm^2:

$$1 \text{ poise} = 1\,\frac{\text{dyne}}{\text{cm}^2} \times \frac{\text{s}}{\text{cm}} \times \text{cm} = 1\,\frac{\text{dyne} - \text{s}}{\text{cm}^2} = 1\,\frac{\text{g}}{\text{s} - \text{cm}}$$

The coefficients of viscosity for a number of common liquids are given in Table 1-16.

TABLE 1-16 Coefficient of Viscosity of Common Liquids

Liquid	Temp (°C)	Viscosity (cP)
Acetic acid	20	1.30
	40	1.00
	100	0.43
Acetone	0	0.40
	20	0.30
	40	0.28
Ammonia	−34	0.27
Aniline	0	8.65
	20	4.40
	40	2.41
Benzene	0	0.91
	20	0.65
	40	0.50
Butyl alcohol	0	5.19
	20	2.95
	100	0.54
Carbon disulfide	−10	0.50
	0	0.43
	20	0.38
Carbon tetrachloride	0	1.35
	20	0.98
	40	0.75
Chloroform	0	0.71
	20	0.57
	40	0.47
Ethyl acetate	20	0.46

**TABLE 1-16 Coefficient of Viscosity
of Common Liquids (Continued)**

Liquid	Temp (°C)	Viscosity (cP)
	40	0.37
	60	0.30
Ethyl alcohol	0	1.77
	20	1.20
	40	0.83
	60	0.59
Ethylene chloride	0	1.13
	20	0.84
	40	0.65
Glycerin	0	4,230
	20	833
Glycol	0	2.18
Mercury	−20	1.85
	0	1.68
	20	1.55
	50	1.39
	100	1.21
	200	1.01
Methyl alcohol	0	0.82
	20	0.60
	40	0.46
Oil, castor	20	986
	100	16.9
linseed	20	25.1
olive	20	84.0
	40	36.3
rape	20	163
soybean	20	30.3
sperm	20	54.0
Pentane	0	0.29
	20	0.24
Propyl acetate	20	0.59
	40	0.44
alcohol	0	3.88
	20	2.26
	40	1.41
Sulfur	200	50,000
	300	2,400
Sulfur dioxide	−10	0.43
	0	0.38
Toluene	0	0.77
	20	0.59
	40	0.47
Turpentine	20	1.49
	40	1.07
Water	0	1.79
	20	1.01
	40	0.65
	80	0.36
	100	0.28
Xylene, meta	0	0.81
	20	0.62
	40	0.50
ortho	0	1.11
	20	0.81
	40	0.63
para	20	0.65
	40	0.51

To convert centipoises to	Multiply by
lb/(ft)(s)	0.672×10^{-3}
lb/(ft)(h)	2.42
kg/(m)(s)	0.001

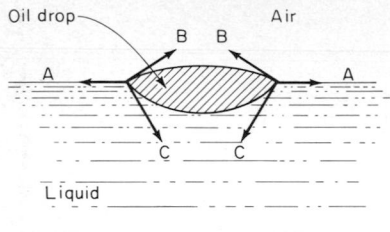

Fig. 1-13 Surface-tension forces on an oil drop. Force *A* results from the air-liquid interface, *B* from the oil-air interface; and *C* from the oil-liquid interface. The oil drop takes the shape of intersecting spheres as a result of these forces.

Since viscosity is essentially a measure of the internal friction of a liquid, it can be expected that the viscosity will appear in a great many equations describing the manner in which a liquid flows. This is actually the case, and in fact, the ratio of viscosity to density appears so often that the special name *kinematic viscosity* has been given to this ratio. The unit of kinematic viscosity is the stoke (St), equal to 1 dyne $-$ s $-$ cm/g $=$ cm^2/s.

Fluidity, defined simply as the reciprocal of viscosity, is most frequently used in qualitative descriptions; i.e., alcohol is more fluid than oil, meaning that is has a lower viscosity and flows more easily.

Surface Tension

The surface of a liquid tends to assume the smallest possible area, acting very much as if it were under tension. Any portion of the liquid, of course, exerts an attractive pull upon all adjacent portions of the liquid, or upon other objects with which it is in contact. Within the bulk of the liquid, these forces are exerted equally in all directions and thereby neutralize each other, but, on the surface, the forces are not balanced and result in a net force which acts in the plane of the surface. This force per unit length of the surface boundary is known as the *surface tension*. One way of visualizing the concept of surface tension is to consider an imaginary line on the surface of a liquid and tensile forces being exerted on the line in both directions perpendicular to the line. The surface tension of the liquid is the force that counteracts the forces pulling on the line so that the surface remains undisturbed.

The surface tension of a liquid manifests itself only at the free surface of the liquid, or at the boundaries between the liquid and other materials, which could be a solid, gas, or another liquid. As an example, the forces acting on a drop of oil floating on water are shown in Fig. 1-13.

Perhaps the most convenient way of measuring the surface tension of a liquid is to observe its rise in a capillary tube, as illustrated in Fig. 1-14. A liquid will rise in a capillary tube owing to the upward pull of surface-tension forces. The liquid will rise to the height necessary to counteract these forces by the weight of the column of liquid. Thus the height is given by the equation.

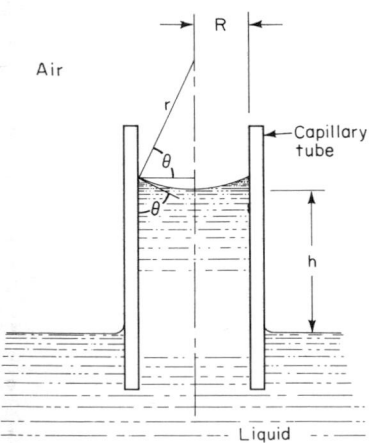

Fig. 1-14 Rise of a liquid in a capillary tube. The liquid, which contacts the tube at an angle of θ, is pulled up by surface tension to such a height that the weight of the column of liquid exactly balances the surface-tension forces.

$$h = \frac{2T \cos \theta}{Rdg}$$

where the lowercase Greek letter theta (θ) is the contact angle the liquid makes with the capillary tube as shown in Fig. 1-14. T is the surface tension in dynes per centimeter and g the gravitational constant in metric units, 980 cm/s^2. When the contact angle is larger than 90°, as is the case for mercury, the column of liquid in the capillary tube will be depressed below the free surface of the liquid, and h will be negative. (Mathematically this is consistent, since cos θ for obtuse angles is negative.) From the dimensions of the capillary, the rise (or fall) of the liquid in the capillary, and the density of the liquid, it is possible to calculate the surface tension of the liquid. The surface tensions of a variety of common liquids important in chemical work are given in Table 1-17.

The surface tension of a liquid is important in determining whether the liquid will "wet" a solid. When the contact angle is more than 90°, the liquid tends to "ball up," i.e., form droplets on the solid, much like mercury on glass, and does not wet, or spread over, the surface. A material with a low surface tension tends to wet solids, and this is important in detergents (where the dirt particles must be wetted before the liquid can wash them away) and in emulsifiers (where the particles of one liquid must be wetted by the other liquid before the former can be dispersed in the latter). In surface-tension analyses, it must be remembered that it is only the nature of the surface itself that is important. Thus, water will not spread on a piece of glass that has even an extremely thin layer of oil on it, whereas it will spread readily on a clean glass surface.

Many materials, such as soaps and alcohols, have the property of lowering the surface tension of a liquid. Materials have been developed which show this property to a pronounced degree. Since they have such an effect on the surface properties of a liquid, as a group they are known as surface-active materials.

TABLE 1-17 Surface Tension of Common Liquids at 20°C

Liquid	In contact with	Surface tension (dyn/cm)
Acetone	Air/vapor	23.7
Aniline	Vapor	42.9
Benzene	Air	28.9
n-Butyl alcohol	Air/vapor	29.2
Carbon disulfide	Vapor	32.3
Carbon tetrachloride	Vapor	25.4
Ethyl alcohol	Air	21.2
Ethyl ether	Vapor	17.0
Glycerin	Air	63.4
Glycol	Air/vapor	47.7
Isobutyl alcohol	Vapor	23.0
Isopropyl alcohol	Air/vapor	21.7
Mercury	Air	470
Methyl alcohol	Air	22.6
Methyl ethyl ketone	Air/vapor	24.6
Nitrobenzene	Air/vapor	43.9
n-Octyl alcohol	Air	27.5
n-Propyl acetate	Air/vapor	24.3
n-Propyl alcohol	Vapor	23.8
Pyridine	Air	38.0
Tetrachloroethylene	Vapor	31.7
Toluene	Vapor	28.5
Water	Air	72.9
m-Xylene	Vapor	28.9
o-Xylene	Air	30.1
p-Xylene	Vapor	28.4

Index of Refraction

A property of liquids which has developed into the basis of important analytical procedures is refractive index. *Refraction* occurs when light undergoes an abrupt change of direction when passing obliquely from one medium into another. The ratio of the speed of light in two contacting media V_1/V_2 is known as the *refractive index* of the second medium relative to the first, and is represented by the lowercase Greek letter mu (μ). Frequently, the subscript 1-2 is used to indicate the direction in which the light is traveling, i.e., from medium 1 to medium 2. Thus, the index of refraction is defined by the following equation:

$$\mu_{1\text{-}2} = \frac{V_1}{V_2} = \frac{\sin i}{\sin r}$$

where i and r are the angles of incidence and refraction, as illustrated in Fig. 1-15.

Where medium 1 is air, medium 2 would simply be a liquid (or another gas) under observation. Normally, tables giving indices of refraction simply refer to the material, it being understood that the other medium is air. Table 1-18 gives the refractive indices of a variety of gases, liquids, and solids of importance in chemical work. It should be noted that since the index of refraction is the ratio of two velocities, the index of refraction itself is dimensionless. Also, since different wavelengths have slightly different velocities in any given medium, the index of refraction is affected by the wavelength of the light used to make the measurements. In extremely accurate scientific work,

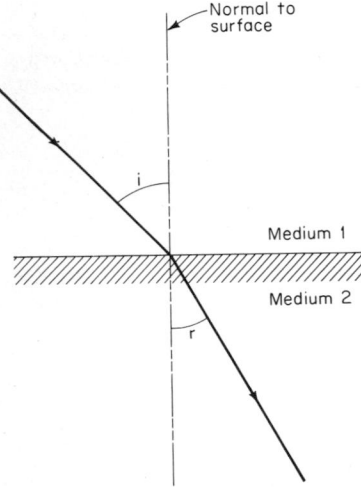

Fig. 1-15 Refraction of light at Interface of two transparent media. If medium 2 is denser than medium 1, the ray of light will be bent toward the normal to the surface; i.e., angle *r* will be less than angle *i*. The reverse is true if medium 1 is denser than medium 2.

index-of-refraction measurements are usually made with a sodium-vapor lamp which emits light of a very narrow band of wavelengths of known value. However, such precision is not usually needed, and most index-of-refraction measurements are made with ordinary white light.

TABLE 1-18 Index of Refraction of Common Materials

Gases and Vapors:	
Air	1.00029
Ammonia	1.00038
Argon	1.00028
Bromine	1.00113
Carbon dioxide	1.00045
Carbon monoxide	1.00034
Chlorine	1.00077
Ethyl alcohol	1.00087
Ethyl ether	1.00152
Helium	1.00004
Hydrogen	1.00013
Methane	1.00044
Methyl alcohol	1.00055
Nitrogen	1.00030
Oxygen	1.00027
Sulfur dioxide	1.00069
Water	1.00025
Liquids:	
Benzene	1.50
Carbon disulfide	1.63
Chlorobenzene	1.52
Ethyl alcohol	1.36
Glycerin	1.47
Water	1.34
Plastics:	
Cellulose acetate	1.48
butyrate	1.47
nitrate	1.50
Melamine	1.59
Methyl methacrylate	1.49
Nylon	1.53
Phenolic, cast	1.59
molded	1.60
Polyethylene	1.52
Polypropylene	1.52
Polystyrene	1.59
Urea	1.55
Polyvinylchloride	1.48
Vinylidene	1.61
Various Solids:	
Amber	1.55
Borax, fused	1.46
Fluorite	1.43
Glass, crown	1.53
flint	1.59
Rock salt	1.53
Quartz	1.56

Chapter 2

Thermal, Electrical and Mechanical Units

2-1 THERMAL PROPERTIES OF MATTER

Heat and Temperature

The internal energy of a body, resulting from the incessant movement (kinetic energy) of its molecules, is called heat. Heat is therefore a form of energy. Heat will flow from a hot body to a cold one, meaning that some of the internal energy of the hot body is transferred to the cold body. The transfer of heat from one body to another is governed by their temperatures. As a matter of fact, temperature can best be defined as the thermal condition that determines the direction in which the transfer of heat from one body to another will occur. For example, if a hot forging is submerged into cold oil, heat is transferred from the forging to the oil; i.e., the forging gets cooler while the oil gets hotter, the process continuing until equilibrium is reached and the forging and the oil have the same temperature. Various scales have been devised for describing the temperature of a body quantitatively. These involve the selection of two *fixed points,* i.e., readily reproducible fixed temperatures, and establishing the number of divisions, or degrees (°), between the fixed points. Such scales can then be extended above and below the fixed points as needed. For most temperature scales, the two fixed points are the boiling point of water and the melting point of ice. The four most important temperature scales are compared below:

Fixed points	Fahrenheit (°F)	Celsius (°C)	Kelvin (K)	Rankine (°R)
Boiling point of water	212	100	373	672
Melting point of ice	32	0	273	492
Divisions between these points	180	100	100	180

Notice that a Fahrenheit degree is the same as a Rankine degree, and that a Celsius degree is the same as a Kelvin degree.

In converting from one system to another, it should be noted that 180 Fahrenheit degrees are equivalent to 100 Celsius degrees, and that the fixed points are assigned different numerical values in the various systems.

Example 2-1 What does a temperature of 60°F correspond to on the Celsius scale?
solution The degrees above the melting point of ice, 60 − 32, are multiplied by 100/180. Thus,

$$60°F = (60 - 32)(100/180) = 15.6°C$$

The general relation between Fahrenheit and Celsius readings can be expressed by the following equation:

$$\frac{°F - 32}{180} = \frac{°C}{100} \tag{2-1}$$

Example 2-2 What does a temperature of $-20°C$ correspond to on the Fahrenheit scale?
solution 20 Celsius degrees are equal to 20 (180/100) or 36 Fahrenheit degrees. Therefore $-20°C$ is 36 Fahrenheit degrees below the melting point of ice, which is 32° on the Fahrenheit scale. Thus, $-20°C = 32 - 36 = -4°F$.

Example 2-3 What does a temperature of $-20°C$ correspond to on the Kelvin scale?
solution Since the Celsius and Kelvin degrees are equal, $-20°C$ corresponds to 20° below the melting point of ice, which on the Kelvin scale is 273°. Thus, $-20°C = 273 - 20 = 253$ K.

To facilitate conversion of temperatures from one system to another, Table 2-1 presents equivalent temperatures in the Celsius and Fahrenheit systems.

TABLE 2-1   Temperature-Conversion Chart °C to °F

FOR TEMPERATURES BELOW 0°C

Temp (°C)	0	5	10	15	20	25	30	35	40	45
0	+32	+23	+14	+5	−4	−13	−22	−31	−40	−49
−50	−58	−67	−76	−85	−94	−103	−112	−121	−130	−139
−100	−148	−157	−166	−175	−184	−193	−202	−211	−220	−229
−150	−238	−247	−256	−265	−274	−283	−292	−301	−310	−319
−200	−328	−337	−346	−355	−364	−373	−382	−391	−400	−409
−250	−418	−427	−436	−445	−454					

FOR TEMPERATURES FROM 0 TO 500°C

	0	5	10	15	20	25	30	35	40	45
0	32	41	50	59	68	77	86	95	104	113
50	122	131	140	149	158	167	176	185	194	203
100	212	221	230	239	248	257	266	275	284	293
150	302	311	320	329	338	347	356	365	374	383
200	392	401	410	419	428	437	446	455	464	473
250	482	491	500	509	518	527	536	545	554	563
300	572	581	590	599	608	617	626	635	644	653
350	662	671	680	689	698	707	716	725	734	743
400	752	761	770	779	788	797	806	815	824	833
450	842	851	860	869	878	887	896	905	914	923

FOR TEMPERATURES ABOVE 500°C

Temp (°C)	0	10	20	30	40	50	60	70	80	90
500	932	950	968	986	1004	1022	1040	1058	1076	1094
600	1112	1130	1148	1166	1184	1202	1220	1238	1256	1274
700	1292	1320	1328	1346	1364	1382	1400	1418	1436	1454
800	1472	1490	1508	1526	1544	1562	1580	1598	1616	1634
900	1652	1670	1688	1706	1724	1742	1760	1778	1796	1814
1000	1832	1850	1868	1886	1904	1922	1940	1958	1976	1994
1100	2012	2030	2048	2066	2084	2102	2120	2138	2156	2174
1200	2192	2210	2228	2246	2264	2282	2300	2318	2336	2354
1300	2372	2390	2408	2426	2444	2462	2480	2498	2516	2534
1400	2552	2570	2588	2606	2624	2642	2660	2678	2696	2714
1500	2732	2750	2768	2786	2804	2822	2840	2858	2876	2894
1600	2912	2930	2948	2966	2984	3002	3020	3038	3056	3074
1700	3092	3110	3128	3146	3164	3182	3200	3218	3236	3254
1800	3272	3290	3308	3326	3344	3362	3380	3398	3416	3434
1900	3452	3470	3488	3506	3524	3542	3560	3578	3596	3614
2000	3632	3650	3668	3686	3704	3722	3740	3758	3776	3794

Example 2-4 Using Table 2-1, what are the Fahrenheit equivalents of the following Celsius temperatures?

$$-165°$$
$$-57°$$
$$235°$$
$$1030°$$
$$1846°$$

solution On the first section of the table (i.e., for temperatures below 0°C) locate $-100°$ in the left-hand column; then move horizontally to the right until under the column headed 15 (i.e., $115 = 100 + 15$). For $165°C$, read $-175°F$.

To determine the equivalent of $-57°C$, Table 2-1 reads $-58°F$ for $-50°C$ and $-76°F$ for $60°C$. Interpolating (i.e., proportioning) the difference between -50 and $-60°C$ to get $-57°C$:

$$\frac{-57 + 50}{-60 + 50} = \frac{X + 58}{-76 + 58}$$

$$\frac{-7}{-10} = \frac{X + 58}{-18}$$

$$X + 58 = 0.7 \, (-18) = -12.6$$

$$X = -70.6°F$$

The equivalents of 235°C and 1030°C can be read directly from the second and third sections of Table 2-1:

$$235°C = 455°F$$
$$1030°C = 1886°F$$

To get the equivalent for 1846°C, however, requires interpolation between 1840 and 1850°C:

$$\frac{1846 - 1840}{1850 - 1840} = \frac{X - 3344}{3362 - 3344}$$

$$0.6 = \frac{X - 3344}{18}$$

$$10.8 = X - 3346$$

$$X = 3356.8 \text{ or, rounding off, } 3357°F$$

Note that to convert from Fahrenheit to Celsius, Table 2-1 is simply used in the reverse direction. Thus 2264°F is equivalent to 1200 + 40 or 1240°F.

Heat Units

If heat is applied to a substance, its temperature is raised; and conversely, if heat is removed, the temperature of the body is lowered. The unit of heat in the metric system is the calorie (cal), which is defined as the quantity of heat required to raise the temperature of 1 g of water 1°C. This amount of heat varies somewhat with the temperature of the water, and consequently, in very precise work, the temperature at which the measurement is made must be stated. The *normal calorie* is measured at 15°C. A convenient method of measuring a calorie is to determine the heat required to raise the temperature of 1 g of water from its freezing point, 0°C, to its boiling point, 100°C. The *mean calorie* is defined as 1/100th of this amount of heat. A *large calorie*, or kilocalorie, is equal to 1,000 cal and is abbreviated kcal.

In the English system, the unit of heat is the British thermal unit (Btu), which is defined as the amount of heat required to raise the temperature of 1 lb of water 1°F. The conversion factor for Btu into calories can be calculated as follows:

$$1 \text{ Btu} = \text{lb} - °F \times \frac{453 \text{ g}}{\text{lb}} \times \frac{5°C}{9°F} \times \frac{\text{cal}}{\text{g} - °C} = 252 \text{ cal}$$

Example 2-5 How much heat is required to raise the temperature of 1 gal of water from 65 to 80°F?
solution

$$\text{Heat} = 1 \text{ gal} \times \frac{8.35 \text{ lb}}{\text{gal}} \times (80 - 65)°F \times \frac{\text{Btu}}{\text{lb} - °F} = 125 \text{ Btu}$$

Heat Capacity and Specific Heat

The *heat capacity* of a substance is the quantity of heat required to raise the temperature of that substance 1°. In the metric system, it is expressed in calories per degree Celsius and in the English system, as British thermal units per degree Fahrenheit. Still another way of expressing the temperature-heat relationships of a substance is by *specific heat*. Specific heat is the ratio of the heat capacity of a solid, liquid, or gas to that of water. As in the case of specific gravity, specific heat, being the ratio of two numbers having the same units, is a pure number; i.e., it is dimensionless. The specific heats for a number of common solids, liquids, and gases are given in Table 2-2.

The measurement of specific heats or heat capacities is carried out in an apparatus known as a *calorimeter*. The essential features of a calorimeter are shown in Fig. 2-1. It measures the temperature rise (or fall) of a substance as a result of the addition (or withdrawal) of a known quantity of heat. One of the most convenient ways of introducing heat into a calorimeter is electrically, since the electrical input can be completely converted into heat in accordance with very exact relationships (as will be seen later in this chapter). An essential feature of the design of a calorimeter is to provide good thermal insulation in order for heat to be neither gained from nor lost to the external surroundings so that all the heat conveyed to or removed from the test material must result in an appropriate change in its temperature.

An important operation in chemical work is to heat or cool a material by conveying heat to or from the material by means of a second material. In other words, heating and cooling are accomplished by transferring heat from one material to another. Heat-transfer problems can be readily handled by recognizing that the heat gained by one substance is equal to that which is lost by the other (assuming, of course, that no heat is gained from or lost to the surroundings). When heat is gained or lost to the surroundings, the problem really becomes one of transfer among several materials, but the same principles still apply.

Example 2-6 How much heat is required to raise the temperature of a 24-g tube of copper from 45 to 87°C?

solution From Table 2-2, the specific heat of copper is 0.0931, meaning that it takes 0.0931 cal to raise the temperature of 1 g of Cu 1°C. Thus, the total heat required to raise the temperature of the copper tube will be

$$\frac{0.0931 \text{ cal}}{\cancel{g} - °\cancel{C}} \times 24 \cancel{g} \times (87 - 45)°\cancel{C} = 94 \text{ cal}$$

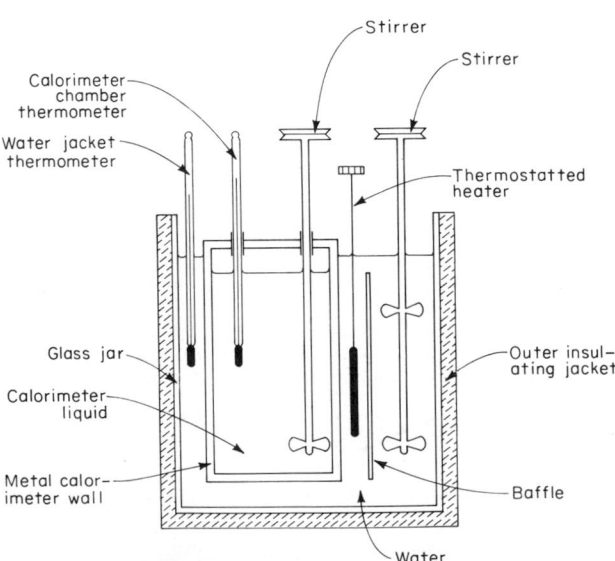

Fig. 2-1 Essentials of a typical calorimeter.

TABLE 2-2 Specific Heats of Various Solids, Liquids, and Gases

ELEMENTS

Element	Temp (°C)	Sp ht	Element	Temp (°C)	Sp ht
Aluminum	−250	0.0039	Calcium (continued)	100	0.1625
	−240.6	0.0092		300	0.1832
	−233	0.0165		600	0.188
	−200	0.076	Carbon, charcoal	0–24	0.165
	−150	0.1367	Diamond	−233	0.0005
	−100	0.1676		−185	0.0025
	−50	0.1914		−188 to −78	0.019
	0	0.2079		−78 to +18	0.079
	20	0.214		0	0.1044
	100	0.225		20	0.12
	300	0.0248		140	0.222
	600	0.277		223	0.264
Liquid	660	0.25		247	0.303
Antimony	−207.1	0.0322		606	0.441
	−150	0.0412		823	0.428
	−100	0.0448	Gas carbon	24–68	0.204
	−50	0.0476	Graphite	−243	0.005
	0	0.0494		−203	0.0175
	20–100	0.0504		−191 to −79	0.057
	100	0.0513		−66	0.053
	200	0.0520		20	0.17
	300	0.0537		85	0.177
	500	0.054		100	0.234
Argon, solid	−223	0.155		642	0.445
Liquid	−190	0.124		890	0.454
Arsenic	−216	0.032	Cerium	−253 to −196	0.033
	−117.6	0.0666		0–100	0.0423
	18	0.078		20–100	0.0511
Gray, crystal	0–100	0.0822	Cesium, solid	20	0.052
Black, amorphous	0–100	0.0861		0–26	0.0482
Barium	185 to +20	0.068	Liquid	50	0.058
Beryllium	−202	0.017	Chlorine	−113	0.19
	0–46	0.397	Liquid	0–24	0.226
	0–100	0.425	Chromium	−150	0.0599
	0–300	0.505		−100	0.0797
Bismuth	−150	0.0264		−50	0.0941
	−100	0.0273		0	0.1044
	−50	0.0282		20	0.11
	0	0.291		18–100	0.111
	20	0.0294		100	0.112
	100	0.0304		400	0.133
Liquid	297	0.0292		500	0.150
	400	0.035		600	0.187
Boron	−191 to −78	0.071	Cobalt	−150	0.0672
	−76 to 0	0.168		−100	0.0809
	0–100	0.307		−50	0.0914
	100	0.287		0	0.1028
	500	0.472		20	0.1001
	900	0.510		100	0.1067
Bromine, solid	−253.1	0.205		200	0.1134
	−173.1	0.659		300	0.121
	−73.1	0.080		508	0.135
	−13.1	0.088		800	0.160
Bromine, liquid	13–45	0.107		1000	0.184
Cadmium	−263	0.0019		1112	0.220
	−203.1	0.0415	Copper	−253	0.0031
	−103.1	0.0518		−189	0.0506
	27.9	0.0552		−150	0.0674
	107.9	0.0569		−100	0.0783
	277	0.060		−50	0.0862
Liquid	321	0.077		0	0.0910
Calcium	−185 to +20	0.157		20	0.0921
	0–20	0.145		15–100	0.09305
	24	0.168		100	0.0939

From C. D. Hodgman (ed.), "Handbook of Chemistry and Physics," The Chemical Rubber Co., Cleveland, 1942.

TABLE 2-2 Specific Heats of Various Solids, Liquids, and Gases—(Continued)

ELEMENTS

Element	Temp (°C)	Sp ht	Element	Temp (°C)	Sp ht
Copper (continued)	200	0.0963	Lithium (continued)	−100	0.600
	900	0.1259		0	0.079
	18–100	0.0928		50	0.96
Liquid	1084	0.101		100	1.0407
Gallium	−258.1	0.0049		190	1.374
	−213.1	0.044	Magnesium	−150	0.1767
	−73.1	0.084		−100	0.2025
	12–23	0.079		−50	0.2228
Liquid	13–110	0.080		0	0.2316
	119	0.079		20	0.246
Germanium	0–100	0.074		100	0.257
Gold	−258.1	0.0018		300	0.279
	−209.5	0.0211		600	0.311
	−150	0.0266	Liquid	650–775	0.284
	−100	0.0281	Manganese	−188 to −79	0.0820
	−50	0.0293		−100	0.0979
	0	0.0302		0	0.1072
	18	0.0312		20–100	0.1211
	0–100	0.0316		60	0.1211
	100	0.0314		325	0.1783
Liquid	1100	0.0327	Mercury, solid	−263.3	0.00552
Hydrogen, solid	−260.6	0.57		−259.8	0.00783
Liquid	−252	0.231		−245.6	0.0172
Indium	−186 to −79	0.0263		−220.2	0.0255
	−79 to +18	0.0303		−163.7	0.0298
	0–100	0.057		−81.4	0.0324
Iodine	−263.2	0.0037		−43.1	0.0337
	−255.9	0.0118	Liquid	−33.1	0.0338
	−221.1	0.353		0	0.03346
	−90 to +17	0.0485		20	0.03325
	20	0.0523		40	0.03308
Liquid	107–180	0.108		60	0.03294
Iridium	−186 to +18	0.0282		100	0.03269
	18–100	0.0323		200	0.0323
	0–900	0.0371		250	0.0321
Iron, cast	20–100	0.1189	Molybdenum	−257	0.0004
Wrought	15–100	0.1152		−239.1	0.0034
Hard-drawn	20–100	0.1146		−181.5	0.0300
Pure	−256.2	0.00067		−152.7	0.0399
	−214.0	0.0194		−34.5	0.0561
	−172.6	0.0512		0	0.0589
	−67.5	0.0939		20–100	0.065
	0	0.1043		250	0.0632
	20	0.107		475	0.0750
	100	0.115	Neodymium	0–100	0.045
α, β, γ	500	0.163	Nickel	−258	0.0008
	760	0.320		−247.9	0.0024
	1000	0.162		−201.2	0.0363
γ	100	0.127		−150	0.0660
	700	0.157		−100	0.0817
	1000	0.162		−50	0.0940
Lanthanum	0–100	0.0448		0	0.1032
Lead	−270	0.00001		20	0.105
	−267	0.00086		100	0.1146
	−259	0.0073		500	0.1270
	−150	0.0279		800	0.1413
	−100	0.0283	Liquid	1452	0.13
	−50	0.0289	Nitrogen, solid	−212	0.39
	0	0.0297	Liquid	−200	0.474
	20	0.0306	Osmium	19–98	0.0311
	100	0.0320	Oxygen, solid	−221.8	0.336
	300	0.0356	Liquid	−200	0.394
Liquid	360	0.0375	Palladium	−180 to +18	0.0528
	500	0.0370		0	0.0538
Lithium	−183	0.3		100	0.0564

ELEMENTS

Element	Temp (°C)	Sp ht	Element	Temp (°C)	Sp ht
Palladium	500	0.0653	Sulfur	−188 to +18	0.137
(continued)	900	0.0717	Rhombic	15–96	0.176
	1500	0.0766	Monoclinic	0–52	0.181
Phosphorus, yel-			Liquid	115–160	0.220
low	−136	0.124	Tantalum	−201.7	0.0205
	−40	0.165		20	0.036
	9	0.189		380	0.035
Red	−136	0.107		900	0.036
	−40	0.182		1100	0.043
	9	0.190		1400	0.044
Platinum	−255.6	0.00123	Tellurium	−188 to +18	0.047
	−237.7	0.0073	Crystalline	15–200	0.0483
	−191.7	0.0211		15–200	0.0487
	−152.1	0.0261	Thallium	−185 to +20	0.038
	−64.8	0.0307		28	0.0311
	0	0.03162		20–100	0.0326
	20	0.0324	Thorium	−253 to −196	0.0197
	500	0.0349		0–100	0.0276
	750	0.0365	Tin	−186 to −79	0.0486
	1000	0.0381		−186.7	0.0422
	1300	0.0400		−150	0.0450
Potassium	−258.4	0.032		−100	0.0483
	−255.8	0.045		−50	0.0512
	−201.3	0.140		0	0.0536
	−53.1	0.172		18	0.0542
	14	0.18		100	0.0577
	22–56	0.192		1100	0.0758
Liquid	63	0.18	Liquid		
	78–100	0.217	Gray	20	0.515
	90	0.200	Titanium	−185 to +20	0.082
	181	0.196		0–100	0.1125
Praseodymium	0–100	0.046	Tungsten	−247.1	0.0012
Rhenium	0–20	0.035		−218.4	0.0098
Rhodium	10–97	0.058		−173.1	0.0205
Rubidium, solid	0	0.0802		−73.1	0.0288
Liquid	50	0.0908		20–100	0.034
Ruthenium	0–100	0.0611		100	0.0320
Selenium	−188 to +18	0.068		500	0.0344
	3	0.072		1000	0.0367
	20.5	0.077		1500	0.0390
	29.5	0.085	Uranium	0–98	0.0280
	32	0.127	Vanadium	0–100	0.1153
	38	0.131	Zinc	−252.4	0.0071
Silicon	−212	0.029		−201.3	0.0573
	−143.3	0.087		−150	0.0740
	−86.2	0.126		−100	0.0814
	13.9	0.168		−50	0.0871
	18.2–99.1	0.181		0	0.0913
	18.0–900.6	0.210		0–100	0.095
Silver	−238	0.0146		20	0.0925
	−150	0.0461		100	0.0957
	−100	0.0505		300	0.1043
	−50	0.0537		400	0.1089
	0	0.0557	Zirconium	0–100	0.068
	20	0.0558			
	100	0.0564			
	500	0.0581			
	800	0.076			
Liquid	900	0.0685			
Sodium	−256.1	0.026			
	−238.5	0.108			
	−155.5	0.245			
	−40	0.279			
	20	0.295			
Liquid	100	0.32			

SOLID INORGANIC COMPOUNDS

Name	Formula	Temp (°C)	Sp ht
Aluminum chloride	$AlCl_3$ (α)	93	0.468
Chloride	$AlCl_3$ (β)	0	0.196
Chloride	$AlCl_36H_2O$	35	0.313
Fluoride	AlF_3	35	0.229
Fluoride	$2AlF_3 \cdot 7H_2O$	35	0.432
Hydroxide	$Al(OH)_3$	0	0.177
		50	0.202
Oxide	Al_2O_3	0	0.174
		50	0.198
Sulfate	$Al_2(SO_4)_3$	50	0.184
Sulfate	$Al_2(SO_4)_3 \cdot 18H_2O$	34	0.354
Ammonia	NH_3	-103 to -188	0.502
Ammonium bromide	NH_4Br	20	0.210
Chloride	NH_4Cl	-200	0.121
		-100	0.263
		0	0.357
		50	0.389
Iodide	NH_4I	0	0.111
		50	0.118
Nitrate	NH_4NO_3	-100	0.306
		0	0.397
		100	0.428
Sulfate	$(NH_4)_2SO_4$	-100	0.283
		0	0.337
		50	0.345
Antimony trisulfide	Sb_2S_3	0	0.0829
Arsenous oxide	As_2O_3	0	0.117
Barium carbonate	$BaCO_3$	0	0.0999
		100	0.110
Chlorate	$Ba(ClO_3)_2 \cdot H_2O$	32	0.158
Chloride	$BaCl_2 \cdot 2H_2O$	0	0.140
Nitrate	$Ba(NO_3)_2$	47	0.148
Sulfate	$BaSO_4$	0	0.111
Thiosulfate	BaS_2O_3	58	0.162
Beryllium oxide	BeO	50	0.260
Sulfate	$BeSO_4$	50	0.198
Bismuth sulfide	Bi_2S_3	50	0.0600
Trioxide	Bi_2O_3	50	0.0569
Cadmium nitrate	$Cd(NO_3)_2 \cdot 4H_2O$	40	0.260
Sulfate	$3CdSO_4 \cdot 8H_2O$	0	0.195
		20	0.200
Sulfide	CdS	0	0.0882
		50	0.0922
Calcium carbonate	$CaCO_3$	0	0.203
		100	0.214
Chloride	$CaCl_2$	61	0.164
Chloride	$CaCl_2 \cdot 6H_2O$	0	0.320
Fluoride	CaF_2	0	0.204
		40	0.212
Formate	$Ca(HCO_2)_2$	0	0.238
Hydroxide	$Ca(OH)_2$	0	0.260
		50	0.288
Molybdate	$CaMoO_4$	15	0.165
Oxide	CaO	0	0.177
		100	0.197
Sulfate	$CaSO_4 \cdot 2H_2O$	36	0.265
Tungstate	$CaWO_4$	15	0.104
Carbon dioxide, solid	CO_2	-225	0.124
Monoxide, solid	CO	-220	0.417
		-206	0.457
Ceric oxide	CeO_2	0	0.0870
		50	0.0946
Sulfate	$Ce(SO_4)_2$	50	0.117
Sulfate	$CeSO_4 \cdot 5H_2O$	50	0.201
Chromic oxide	Cr_2O_3	0	0.168
		50	0.189

SOLID INORGANIC COMPOUNDS

Name	Formula	Temp (°C)	Sp ht
Sulphate	$Cr_2(SO_4)_3$	50	0.172
Sulfate	$Cr_2(SO_4)_3 \cdot 5H_2O$	50	0.200
Cobaltous nitrate	$Co(NO_3)_2 \cdot 6H_2O$	32	0.373
Sulfate	$CoSO_4 \cdot 7H_2O$	48	0.342
Columbium pentoxide	Cb_2O_5	50	0.101
Copper ammonium sulfate	$CuSO_4 \cdot (NH_4)_2SO_4 \cdot 6H_2O$	0	0.256
Copper sulfate	$CuSO_4 \cdot H_2O$	0	0.172
		50	0.191
Sulfate	$CuSO_4 \cdot 3H_2O$	9	0.228
Sulfate	$CuSO_4 \cdot 5H_2O$	0	0.253
		50	0.287
Cupric carbonate	$2CuO \cdot CO_2 \cdot H_2O$	57	0.177
Chloride	$CuCl_2$	58	0.139
Oxide	CuO	0	0.125
		100	0.144
Sulfide	CuS	0	0.129
		100	0.151
Cuprous iodide	CuI	0	0.0658
		50	0.0671
Oxide	Cu_2O	0	0.110
		100	0.116
Selenide	Cu_2Se	60	0.104
Sulfide	Cu_2S	0	0.148
		50	0.166
Erbium oxide	Er_2O_3	50	0.0650
Ferric oxide	Fe_2O_3	0	0.148
		100	0.182
Ferrosoferric oxide (magnetite)	Fe_3O_4	0	0.151
		100	0.179
Ferrous carbonate	$FeCO_3$	54	0.194
Sulfate	$FeSO_4$	45	0.167
Sulfate	$FeSO_4 \cdot 4H_2O$	9	0.284
Sulfate	$FeSO_4 \cdot 7H_2O$	0	0.325
		10	0.337
Sulfide	FeS	0	0.135
Gallium sesquioxide	Ga_2O_3	50	0.105
Gold iodide	AuI	0	0.0404
		50	0.0432
Hydrogen peroxide	H_2O_2	−25	0.471
Indium sesquioxide	In_2O_3	50	0.0808
Iron diarsenide	$FeAs_2$	50	0.0860
Disulfide	FeS_2	0	0.118
		50	0.128
Lanthanum sesquioxide	La_2O_3	50	0.0750
Lead ammonium chloride	$2PbCl_2 \cdot NH_4Cl$	10	0.0865
Lead borate	PbB_2O_4	57	0.0903
Lead bromide	$PbBr_2$	0	0.0502
		50	0.0530
Carbonate	$PbCO_3$	32	0.0800
Chloride	$PbCl_2$	0	0.0649
		100	0.0681
Chromate	$PbCrO_4$	35	0.0908
Dioxide	PbO_2	0	0.0619
		50	0.0650
Iodide	PbI_2	0	0.0417
		100	0.0437
Molybdate	$PbMoO_4$	15	0.100
Monoxide	PbO	0	0.0483
		50	0.0509
Nitrate	$Pb(NO_3)_2$	45	0.115
Pyrophosphate	$Pb_2P_2O_7$	55	0.0820
Silicate	$PbSiO_3$	60	0.0779
Sulfate	$PbSO_4$	45	0.0839
Sulfide	PbS	0	0.0502
		100	0.0511

TABLE 2-2 Specific Heats of Various Solids, Liquids, and Gases—(Continued)

SOLID INORGANIC COMPOUNDS

Name	Formula	Temp (°C)	Sp ht
Thiosulfate	PbS_2O_3	58	0.0918
Tungstate	$PbWO_4$	15	0.0769
Lithium chloride	LiCl	55	0.282
Fluoride	LiF	10	0.373
Hydride	LiH	0	0.980
		50	1.07
Hydroxide	LiOH	0	0.327
		50	0.356
Nitrate	$LiNO_3$	210	0.387
Thiosulfate	$Li_2S_2O_3$	58	0.0920
Magnesium carbonate	$MgCO_3$	25	0.200
Chloride	$MgCl_2 \cdot 6H_2O$	44	0.378
Chloride	$MgCl_2$	48	0.194
Nitrate	$Mg(NO_3)_2 \cdot 6H_2O$	55	0.887
Oxide	MgO	0	0.209
		50	0.232
Sulfate	$MgSO_4 \cdot 7H_2O$	12	0.361
Sulfate	$MgSO_4 \cdot 6H_2O$	9	0.349
Sulfate	$MgSO_4H_2O$	9	0.239
Sulfate	$MgSO_4$	61	0.222
Manganese dioxide	MnO_2	0	0.152
		50	0.163
Nitrate	$Mn(NO_3)_2 \cdot 6H_2O$	47	0.373
Manganic oxide	Mn_2O_3	58	0.162
Oxide	$Mn_2O_3 \cdot 3H_2O$	38	0.177
Manganous oxide	MnO	58	0.158
Sulfate	$MnSO_4 \cdot 5H_2O$	32	0.323
Sulfate	$MnSO_4$	61	0.182
Mercuric chloride	$HgCl_2$	0	0.0640
		100	0.0669
Cyanide	$Hg(CN)_2$	29	0.100
Iodide	HgI_2 (red)	0	0.0404
		50	0.0413
Oxide	HgO	0	0.0485
		50	0.0521
Sulfide	HgS	0	0.0506
		50	0.0520
Mercurous chloride	HgCl	0	0.0499
		50	0.0512
Sulfate	Hg_2SO_4	0	0.0616
		50	0.0680
Molybdenum trioxide	MoO_3	54	0.134
Nickel nitrate	$Ni(NO_3)_2 \cdot 6H_2O$	80	0.473
Sulfate	$NiSO_4 \cdot 6H_2O$	35	0.313
Sulfate	$NiSO_4$	58	0.225
Sulfide	NiS	0	0.116
		100	0.128
Nitrogen pentoxide	N_2O_4	−80 to −5	0.239
Potassium acetate	$KC_2H_3O_2$	20	0.272
	$KHSO_4$	35	0.244
Aluminum sulfate, (alum)	$K_2SO_4Al_2(SO_4)_3 \cdot 24H_2O$	0	0.324
		50	0.360
Potassium arsenate, acid	KH_2AsO_4	31	0.174
Potassium bromide	KBr	0	0.104
		100	0.108
	K_2CO_3	47	0.210
Chlorate	$KClO_3$	0	0.191
		50	0.205
Chloride	KCl	0	0.162
		100	0.168
Chloroplatinate	K_2PtCl_6	30	0.112
Chromate	K_2CrO_4	46	0.186
Dichromate	$K_2Cr_2O_7$	0	0.178
Ferricyanide	$K_3Fe(CN)_6$	26	0.232
Ferrocyanide	$K_4Fe(CN)_6$	0	0.210
		50	0.225

TABLE 2-2 Specific Heats of Various Solids, Liquids, and Gases—(Continued)

SOLID INORGANIC COMPOUNDS

Name	Formula	Temp (°C)	Sp ht
Ferrocyanide	$K_4Fe(CN)_6 \cdot 3H_2O$	0	0.267
		50	0.285
Fluoride	KF	0	0.199
		50	0.204
Metaborate	$K_2B_2O_4$	57	0.225
Nitrate	KNO_3	0	0.214
		100	0.240
Perchlorate	$KClO_4$	30	0.189
Phosphate, dihydrogen	KH_2PO_4	33	0.208
Pyrophosphate	$K_4P_2O_7$	58	0.191
Thiosulfate	$K_2S_2O_3$	60	0.196
Silicon carbide	SiC	0	0.143
		100	0.194
Silver bromide	AgBr	0	0.0695
		100	0.0734
Chloride	AgCl	0	0.0848
		50	0.0906
Cyanate	AgCNO	40	0.124
Iodide	AgI	0	0.0548
		100	0.0593
Nitrate	$AgNO_3$	50	0.146
Selenide	Ag_2Se	37–187	0.0693
Sulfide	Ag_2S	0	0.0719
		50	0.0748
Sodium acetate	$NaC_2H_3O_2$	38	0.339
Acetate	$NaC_2H_3O_2 \cdot 3H_2O$	0	0.344
		40	0.602
Bromide	NaBr	0	0.118
		100	0.124
Carbonate	Na_2CO_3	45	0.256
Chloride	NaCl	0	0.204
		100	0.217
Fluoride	NaF	0	0.258
		100	0.279
Formate	$NaHCO_2$	46	0.306
Iodide	NaI	0	0.0829
		50	0.0848
Metaborate	$Na_2B_2O_4$	57	0.253
Nitrate	$NaNO_3$	0	0.247
		50	0.270
Phosphate, di-	$Na_2HPO_4 \cdot 12H_2O$	0	0.404
		50	0.464
Phosphate, di-	$Na_2HPO_4 \cdot 7H_2O$	0	0.351
		50	0.406
Pyrophosphate	$Na_4P_2O_7$	50	0.227
Sulfate	Na_2SO_4	0	0.202
		100	0.220
Tetraborate	$Na_2B_4O_7$	45	0.234
Tetraborate (borax)	$Na_2B_4O_7 \cdot 10H_2O$	35	0.385
Thiosulfate	$Na_2S_2O_3 \cdot 5H_2O$	21	0.346
Thiosulfate	$Na_2S_2O_3$	9	0.220
Stannic oxide	SnO_2	45	0.0898
Sulfide	SnS_2	54	0.119
Stannous chloride	$SnCl_2$	60	0.102
Sulfide	SnS	56	0.0839
Strontium molybdate	$SrMoO_4$	15	0.148
Nitrate	$Sr(NO_3)_2$	32	0.182
Sulfate	$SrSO_4$	48	0.143
Sulfuric acid	H_2SO_4	−30	0.239
		0	0.270
Sulfur dioxide	SO_2	−185 to −103	0.229
Thallium monochloride	TlCl	0	0.0520
		100	0.0542
Thorium chloride	ThCl	30	0.406
Dioxide	ThO_2	0	0.0571
		50	0.0589

SOLID INORGANIC COMPOUNDS

Name	Formula	Temp (°C)	Sp ht
Sulfate	$Th(SO_4)_2$	50	0.0980
Tin (see under Stannous and Stannic)			
Titanium dioxide	TiO_2	0	0.168
Tungsten trioxide	WO_3	0	0.0743
		50	0.0832
Uranium oxide (ous-ic)	U_3O_8	0	0.0671
		50	0.0750
Water, solid	H_2O	−250	0.0361
		−200	0.156
		−150	0.246
		−100	0.332
		−40	0.435
		0	0.492
Yttrium oxide	Y_2O_3	57	0.112
Zinc chloride	$ZnCl_2$	60	0.136
Zinc nitrate	$Zn(NO_3)_2 \cdot 6H_2O$	30	0.318
Oxide	ZnO	0	0.114
		100	0.129
Sulfate	$ZnSO_4 \cdot 7H_2O$	0	0.322
Sulfate	$ZnSO_4 \cdot 6H_2O$	9	0.299
Sulfate	$ZnSO_4 \cdot H_2O$	9	0.194
Sulfate	$ZnSO_4$	50	0.174
Sulfide	ZnS	0	0.116
		100	0.118
Zirconium dioxide	ZrO_2	0	0.103

LIQUID INORGANIC COMPOUNDS

Name	Formula	Temp (°C)	Sp ht
Ammonia	NH_3	−60	1.047
		0	1.098
		20	1.125
		100	1.48
Calcium chloride	$CaCl_2 \cdot 6H_2O$	33–99	0.552
Hydrogen peroxide	H_2O_2	0	0.578
Lead bromide	$PbBr_2$	550	0.0779
Chloride	$PbCl_2$	540	0.121
Lithium nitrate	$LiNO_3$	280	0.390
Potassium dichromate	$K_2Cr_2O_7$	397	0.0335
Nitrate	KNO_3	380	0.0332
Silver bromide	$AgBr$	500	0.0760
Chloride	$AgCl$	490	0.129
Nitrate	$AgNO_3$	250	0.187
Sodium acetate	$NaC_2H_3O_2$	61.8	0.846
Chlorate	$NaClO_3$	280	0.325
Nitrate	$NaNO_3$	350	0.430
Thiosulfate	$Na_2S_2O_3 \cdot 5H_2O$	13–98	0.570
Stannic chloride	$SnCl_4$	14–98	0.148
Sulfur dioxide	SO_2	−20	0.313
		0	0.318
		20	0.327
		100	0.418
Sulfuric acid	H_2SO_4	10	0.339
Acid pyro-	$H_2S_2O_7$	35	0.334
Water	H_2O		

SOLID ORGANIC COMPOUNDS

Name	Formula	Temp (°C)	Sp ht
Acetic acid	CH_3CO_2H	0	0.487
Acetone	$(CH_3)_2CO$	−210	0.540
o-Aminobenzoic acid	$H_2NC_6H_4CO_2H$	85	0.254
m-Aminobenzoic acid		120	0.253
p-Aminobenzoic acid		128	0.287
Aniline	$C_6H_6NH_2$	?	0.741
Anthracene	$C_{14}H_{10}$	50	0.308
		100	0.350

SOLID ORGANIC COMPOUNDS

Name	Formula	Temp (°C)	Sp ht
Anthraquinone	$(C_6H_4)_2(CO)_2$	0	0.258
Azobenzene	$(C_6H_5N)_2$	28	0.330
Benzene	C_6H_6	−250	0.0399
		−200	0.124
		−100	0.227
		−50	0.290
Benzoic acid	$C_6H_5CO_2H$	20	0.287
Benzophenone	$(C_6H_5)_2CO$	−150	0.115
		−50	0.220
		0	0.275
		20	0.303
Betol	$HOC_6H^4CO_2C_{10}H_7$	−150	0.129
		−100	0.167
o-Bromochorobenzene	C_6H_4BrCl	−34	0.192
m-Bromochlorobenzene		−52	0.150
p-Bromochlorobenzene		−40	0.150
		0	0.170
o-Bromoiodobenzene	C_6H_4BrI	−50	0.143
m-Bromoiodobenzene		−75 to −15	0.143
p-Bromoiodobenzene		−40	0.116
β-Bromonaphthalene	$C_{10}H_7Br$	41	0.260
Bromophenol	HOC_6H_4Br	32	0.263
Camphene	$C_{10}H_{16}$	35	0.380
Capric acid	$CH_3(CH_2)_8CO_2H$	8	0.695
Caprylic acid	$CH_3(CH_2)_6CO_2H$	2	0.628
Carbon tetrachloride	CCl_4	−200	0.0812
		−80	0.182
		−40	0.201
Catechol	$C_6H_4(OH)_2$	163	0.278
Chloral alcoholate	$CCl_3CHO \cdot C_2H_5OH$	78	0.509
Hydrate	$CCl_3CHO \cdot H_2O$	32	0.213
Chloroacetic acid	CH_2ClCO_2H	60	0.363
p-Chlorobenzoic acid	$ClC_6H_4CO_2H$	80	0.228
m-Chlorobenzoic acid		94	0.232
p-Chlorobenzoic acid		180	0.242
Crotonic acid	$CH_3CHCHCO_2H$	38	0.520
Cyamelide	$C_3H_3O_3N_3$	40	0.263
Cyanuric acid	$(HNCO)_3$	40	0.318
Dextrose	$C_6H_{12}O_6$	−250	0.0155
		0	0.277
		20	0.275
Dextrin	$(C_6H_{10}O_5)x$	0–90	0.292
o-Dibromobenzene	$C_6H_4Br_2$	−36	0.249
m-Dibromobenzene		−25	0.134
p-Dibromobenzene		−50	0.139
Dichloroacetic acid	$CHCl_2CO_2H$	Solid	0.406
o-Dichlorobenzene	$C_6H_4Cl_2$	−48.5	0.185
m-Dichlorobenzene		−52	0.186
p-Dichlorobenzene		−50	0.219
Dicyandiamide	$C_2H_4N_4$	0–204	0.456
Dulcitol	$C_6H_8(OH)_6$	20	0.282
m-Diiodobenzene	$C_6H_4I_2$	−52	0.100
p-Diiodobenzene		−50	0.101
Dibenzyl	$(C_6H_5CH_2)_2$	28	0.363
Dimethyl oxalate	$(CO_2CH_3)_2$	10	0.212
Dimethylpyrone	$(CH_3)_2C_5H_2O_2$	50	0.368
o-Dinitrobenzene	$C_6H_4(NO_2)_2$	−160	0.252
m-Dinitrobenzene		−160	0.248
p-Dinitrobenzene		119	0.259
Diphenyl	$(C_6H_5)_2$	40	0.385
Diphenylamine	$(C_6H_5)_2NH$	26	0.337
Ethyl alcohol (crystalline)	C_2H_5OH	−190	0.232
(Vitreous)		−190	0.260
Erythritol	$(CHOHCH_2OH)_2$	60	0.351
Formic acid	HCO_2H	−22	0.387
		0	0.430

SOLID ORGANIC COMPOUNDS

Name	Formula	Temp (°C)	Sp ht
Glutaric acid	$(CH_2)_3(CO_2H)_2$	20	0.299
Glycerol	$C_3H_6(OH)_3$	-250	0.0471
		-200	0.115
		-100	0.217
		0	0.330
Glycol	$(CH_2OH)_2$	40	0.528
Hexadecane	$C_{16}H_{34}$	19	0.495
Iodobenzene	C_6H_5I	40	0.191
Lactose	$C_{12}H_{22}O_{11}$	20	0.287
	$C_{12}H_{22}O_{11} \cdot H_2O$	20	0.299
Lauric acid	$C_{11}H_{23}CO_2H$	-30	0.430
Levulose	$C_6H_{12}O_6$	20	0.275
Malonic acid	$CH_2(CO_2H)_2$	20	0.275
Maltose	$C_{12}H_{22}O_{11}$	20	0.320
Mannitol	$C_6H_8(OH)_6$	0	0.313
Melamine	$C_3H_6N_6$	40	0.351
Myristic acid	$C_{13}H_{27}CO_2H$	0	0.381
Naphthalene	$C_{10}H_8$	-130	0.281
α-Naphthol	$C_{10}H_7OH$	50	0.240
β-Naphthol		61	0.252
α-Naphthylamine	$C_{10}H_7NH_2$	0	0.270
m-Nitroaniline	$H_2NC_6H_4NO_2$	-160	0.275
o-Nitroaniline		-160	0.269
p-Nitroaniline		-160	0.276
Nitrobenzene	$C_6H_5NO_2$	20	0.349
		100	0.356
o-Nitrobenzoic acid	$NO_2C_6H_4CO_2H$	-163	0.256
m-Nitrobenzoic acid		-160	0.247
Nitronaphthalene	$C_{10}H_7NO_2$	0	0.236
Oxalic acid	$(CO_2H)_2 \cdot 2H_2O$	0	0.338
		50	0.385
Palmitic acid	$C_{15}H_{31}CO_2H$	-180	0.167
		-100	0.251
		-50	0.306
		0	0.382
		20	0.430
Picric acid	$HOC_6H_2(NO_2)_3$	-100	0.165
		0	0.240
		50	0.263
Phthalic acid	$C_6H_4(CO_2H)_2$	20	0.232
Propionic acid	$C_2H_5CO_2H$	-33	0.726
n-Propyl alcohol	C_3H_7OH	-200	0.170
		-130	0.497
iso-Propyl alcohol	C_3H_7OH	-200	0.0507
Pyrotartaric acid	$C_5H_8O_4$	20	0.301
Quinhydrone	$C_{12}H_{10}O_4$	-250	0.0165
		-200	0.0980
		0	0.256
Quinol	$C_6H_4(OH)_2$	-250	0.0246
		-150	0.268
Quinone	$C_6H_4O_2$	-250	0.0311
		-200	0.113
		-150	0.282
Resorcinol	$C_6H_4(OH)_2$	-160	0.269
Salol	$HOC_6H_4CO_2C_6H_5$	32	0.289
Stearic acid	$C_{17}H_{35}CO_2H$	15	0.399
Succinic acid	$(CH_2CO_2H)_2$	0	0.248
Sucrose	$C_{12}H_{22}O_{11}$	20	0.299
Tartaric acid	$H_2C_4H_4O$	36	0.287
		0	0.308
		50	0.366
Thymol	$C_{10}H_{14}O$	0	0.315
Trichloroacetic acid	CCl_3CO_2H	solid	0.459
Trimethyl carbinol	$(CH_3)_3COH$	-4	0.559
Trinitrotoluene	$CH_3C_6H_2(NO_2)_3$	-100	0.170
		0	0.311
		100	0.385

TABLE 2-2 Specific Heats of Various Solids, Liquids, and Gases—(*Continued*)

Name	Formula	Temp (°C)	Sp ht
SOLID ORGANIC COMPOUNDS			
Trinitroxylene	$(CH_3)_2C_6H(NO_2)_3$	20–50	0.423
Triphenylmethane	$(C_6H_5)_3CH$	0	0.189
o-Toluic acid	$CH_3C_6H_4CO_2H$	54	0.277
m-Toluic acid		54	0.239
p-Toluic acid		130	0.271
p-Toluidine	$CH_3C_6H_4NH_2$	0	0.337
		20	0.387
Urea	$(NH_2)_2CO$	20	0.320
LIQUID ORGANIC COMPOUNDS			
Acetic acid	CH_3COOH	0	0.468
Acetone	$(CH_3)_2CO$	0	0.506
		20	0.528
Acetonitrile	CH_3CN	21–76	0.541
Acetophenone	$C_6H_5COCH_3$	20–100	0.474
Acetyl chloride	CH_3COCl	0	0.339
Allyl acetate	$CH_3CO_2C_3H_5$	0	0.431
Alcohol	C_3H_5OH	21–96	0.665
Benzoate	$C_6H_5CO_2C_3H_5$	20	0.388
Butyrate	$C_3H_7CO_2C_3H_5$	20	0.451
Chloride	CH_2CHCH_2Cl	0	0.313
Isobutyrate	$C_3H_7CO_2C_3H_5$	20	0.448
Propionate	$C_2H_5CO_2C_3H_5$	20	0.451
Valerate	$C_4H_9CO_2C_3H_5$	20	0.451
o-Aminobenzoic acid	$H_2NC_6H_4CO_2H$	145	0.435
m-Aminobenzoic acid		174	0.435
p-Aminobenzoic acid		186	0.444
iso-Amyl acetate	$CH_3CO_2C_5H_{11}$	20	0.459
Alcohol	$C_5H_{11}OH$	0	0.502
		20	0.535
		75.5	0.688
d-prim.-Amyl alcohol	$C_5H_{11}OH$	22–125	0.712
tert.-Amyl alcohol		20–99	0.753
iso-Amyl butyrate	$C_3H_7CO_2C_5H_{11}$	20	0.459
Formate	$HCO_2C_5H_{11}$	20	0.459
Isobutyrate	$C_3H_7CO_2C_5H_{11}$	20	0.459
Propionate	$C_2H_5CO_2C_5H_{11}$	20	0.459
Succinate	$(CH_2CO_2C_5H_{11})_2$	0	0.449
Valerate	$C_4H_9CO_2C_5H_{11}$	20	0.459
iso-Amylamine	$C_5H_{11}NH_2$	22–91	0.614
Amylene	C_5H_{10}	0	0.282
Anethol	$C_9H_9OCH_3$	22.48	0.551
Aniline	$C_6H_5 \cdot NH_2$	0	0.478
		50	0.521
		100	0.547
Anisol	$C_6H_5OCH_3$	20–152	0.483
Benzaldehyde	C_6H_5CHO	22–172	0.428
Benzene	C_6H_6	5	0.389
		20	0.406
		60	0.444
		90	0.473
Benzoic acid	$C_6H_5CO_2H$	0	0.424
Benzonitrile	C_6H_5CN	22–186	0.441
β-Benzophenone	$(C_6H_5)_2CO$	3–40	0.383
Benzyl alcohol	$C_6H_5CH_2OH$	20–100	0.511
Chloride	$C_6H_5CH_2Cl$	0	0.323
Betol	$HOC_6H_4CO_2C_{10}H_7$	19–63	0.356
Bromobenzene	C_6H_5Br	20	0.231
o-Bromochlorobenzene	C_6H_4BrCl	0	0.215
m-Bromochlorobenzene		0	0.212
o-Bromoiodobenzene	C_6H_4BrI	5–100	0.160
m-Bromoidobenzene		5–100	0.158
Bromophenol	HOC_6H_4Br	18–77	0.316
n-Butane	C_4H_{10}	0	0.550
iso-Butane		0	0.550

LIQUID ORGANIC COMPOUNDS

Name	Formula	Temp (°C)	Sp ht
iso-Butyl acetate	$CH_3CO_2C_4H_9$	20	0.459
n-Butyl alcohol	C_4H_9OH	2.3	0.526
		19.2	0.563
iso-Butyl alcohol		21–109	0.716
Butyl butyrate	$C_3H_7CO_2C_4H_9$	20	0.459
iso-Butyl butyrate		20	0.459
n-Butyl chloride	C_4H_9Cl	20	0.451
Formate	$HCO_2C_4H_9$	20	0.459
Butyl propionate	$C_2H_5CO_2C_4H_9$	20	0.459
iso-Butyl succinate	$(CH_2CO_2C_4H_9)_2$	0	0.442
Butyl valerate	$C_4H_9CO_2C_4H_9$	20	0.459
n-Butyric acid	$C_3H_7CO_2H$	20–100	0.515
iso-Butyric acid	$C_4H_8O_2$	20	0.450
n-Butyronitrile	C_3H_7CN	21–113	0.547
Caproic acid	$C_5H_{11}CO_2H$	29–105	0.533
Capronitrile	$C_5H_{11}CN$	18–156	0.542
Carbon tetrachloride	CCl_4	0	0.198
		20	0.201
Carvacrol	$C_9H_{13}OH$	24–233	0.577
Catechol	$C_6H_4(OH)_2$	0	0.462
Chloral	CCl_3CHO	17–53	0.250
Hydrate	$CCl_3CHO \cdot H_2O$	55–88	0.470
Chlorobenzene	C_6H_5Cl	20	0.309
o-Chlorobenzoic acid	$ClC_6H_4CO_2H$	0	0.392
m-Chlorobenzoic acid		0	0.266
p-Chlorobenzoic acid		226	0.547
Chloroform	$CHCl_3$	0	0.232
		15	0.226
		20	0.234
o-Chlorophenol	HOC_6H_4Cl	0–20	0.401
Chlorotoluene	$CH_3C_6H_4Cl$	0	0.316
o-Cresol	$CH_3C_6H_4OH$	0–20	0.499
m-Cresol		0–20	0.479
p-Cresyl methyl ether	$CH_3C_6H_4OCH_3$	0	0.405
Crotonic acid	$C_3H_5CO_2H$	71.4	0.500
Cyclohexanol	$C_6H_{11}OH$	15–18	0.417
Cyclohexanone	$C_6H_{10}O$	15–18	0.433
o-Cymene	$C_3H_7C_6H_4CH_3$	0	0.400
Decylene-2	$C_{10}H_{20}$	0–50	0.469
Diallyl oxalate	$(CO_2H_5)_2$	20	0.426
Succinate	$(CH_2CO_2C_3H_5)_2$	20	0.452
Diamylene	$C_{10}H_{20}$	20–130	0.545
o-Dibromobenzene	$C_6H_4Br_2$	0	0.180
m-Dibromobenzene		0	0.175
Dibutyl oxalate	$(CO_2C_4H_9)_2$	20	0.441
Dichloroacetic acid	Cl_2CHCO_2H	21–106	0.350
o-Dichlorobenzene	$C_6H_4Cl_2$	0	0.270
m-Dichlorobenzene		0	0.270
p-Dichlorobenzene		53–99	0.298
Diethylamine	$(C_2H_5)_2NH$	22.5	0.518
Diethylaniline	$C_6H_6N(C_2H_5)_2$	20	0.452
Diethyl carbonate	$CO(OC_2H_5)_2$	20–100	0.464
Ketone	$(C_2H_5)_2CO$	20–98.5	0.557
Malate	$HOC_2H_3(CO_2C_2H_6)_2$	24–186	0.475
Malonate	$CH_2(CO_2C_2H_5)_2$	20	0.433
Oxalate	$(CO_2C_2H_5)_2$	20	0.433
Succinate	$(CH_2CO_2C_2H_5)_2$	20	0.452
o-Diiodobenzene	$C_6H_4I_2$	0	0.136
m-Diiodobenzene		34.2–99.6	0.140
Diisoamyl	$C_{10}H_{22}$	21.5–155	0.590
Oxalate	$(CO_2C_5H_{11})_2$	20	0.449
Diisobutylamine	$(C_4H_9)_2NH$	22–130	0.571
Dimethylaniline	$C_6H_5N(CH_3)_2$	0–20	0.418
Dimethyl carbonate	$CO(OCH_3)_2$	19.8–88	0.452
o-Dinitrobenzene	$C_6H_4(NO_2)_2$	0	0.349
m-Dinitrobenzene		90	0.405

LIQUID ORGANIC COMPOUNDS

Name	Formula	Temp (°C)	Sp ht
p-Dinitrobenzene		0	0.279
Diphenylamine	$(C_6H_5)_2NH$	53	0.464
Diphenyl oxide	$(C_6H_5)_2O$	30	0.399
Dipropylamine	$(C_3H_7)_2NH$	22–100	0.597
Dipropyl ketone	$(C_3H_7)_2CO$	20–140	0.552
Malonate	$CH_2(CO_2C_3H_7)_2$	20	0.433
Succinate	$(CH_2CO_2C_3H_7)_2$	20	0.452
Di-n-propyl oxalate	$(CO_2C_3H_7)_2$	20	0.433
Dodecane	$C_{12}H_{26}$	0–50	0.500
Dodecylene	$C_{12}H_{24}$	0–50	0.457
Ether	$(C_2H_5)_2O$	−50	0.517
		0	0.529
		30	0.547
		120	0.803
		180	1.041
Ethyl acetate	$CH_3CO_2C_2H_5$	20	0.459
Acetoacetate	$CH_3COCH_2CO_2C_2H_5$	20–100	0.477
Alcohol	C_2H_5OH	−100	0.450
		0	0.535
		25	0.581
		100	0.824
Benzene	$C_6H_5C_2H_5$	30	0.409
Benzoate	$C_6H_5CO_2C_2H_5$	20	0.389
Bromide	C_2H_5Br	5–10	0.216
		15–20	0.215
Butyrate	$C_3H_7CO_2C_2H_5$	20	0.459
Chloride	C_2H_5Cl	0	0.368
Chloroacetate	$ClCH_2CO_2C_2H_5$	9–138	0.418
Dichloroacetate	$Cl_2CHCO_2C_2H_5$	20	0.329
Formate	$HCO_2C_2H_5$	14–49	0.510
Iodide	C_2H_5I	0	0.162
Isobutyrate	$C_3H_7CO_2C_2H_5$	20	0.459
Propionate	$C_2H_5CO_2C_2H_5$	20	0.459
Sulfide	$(C_2H_5)_2S$	0	0.470
		15–20	0.477
Trichloroacetate	$CCl_3CO_2C_2H_5$	10–81	0.295
Valerate	$C_4H_9CO_2C_2H_5$	20	0.459
Ethylene bromide	$(CH_2Br)_2$	20	0.174
Chloride	$(CH_2Cl)_2$	20	0.301
		60	0.319
Formamide	$HCONH_2$	19	0.551
Formic acid	HCO_2H	0	0.437
		15.5	0.511
		20–100	0.526
Furfural	$(C_4H_3O)CHO$	20–100	0.418
Glycerol (glycerin)	$HOCH_2 \cdot CHOH.-$		
	CH_2OH	0	0.540
		50	0.600
		100	0.669
Glycol	$(CH_2OH)_2$	0	0.544
		14.9	0.571
Heptaldehyde	$C_6H_{13}CHO$	0	0.365
n-Haptane (B. P. 98°)	C_7H_{16}	20	0.490
iso-Heptane		0–50	0.501
Heptylene (B. P., 98°)	C_7H_{14}	0–50	0.488
Heptylic acid	$C_6H_{13}CO_2H$	9	0.558
n-Hexadecane (B. P., 275°)	$C_{16}H_{34}$	0–50	0.496
1, 5-Hexadiene	C_6H_{10}	0	0.407
o-Hexahydrocresol	$CH_3C_6H_{10}OH$	15–18	0.418
m-Hexahydrocresol		15–18	0.422
p-Hexahydrocresol		15–18	0.423
n-Hexane	C_6H_{14}	20–100	0.600
Hexylene	C_6H_{12}	0–50	0.506
Lauric acid	$C_{11}H_{23}CO_2H$	57	0.515
Mesitylene	$C_6H_3(CH_3)_3$	0	0.393
Mesityl oxide	$C_6H_{10}O$	21–121	0.521

TABLE 2-2 Specific Heats of Various Solids, Liquids, and Gases—(Continued)

Name	Formula	Temp (°C)	Sp ht
Methyl acetate	$CH_3CO_2CH_3$	15	0.468
Alcohol	CH_3OH	0	0.566
		20	0.600
Methyl aniline	$C_6H_5NHCH_3$	20–197	0.513
Benzoate	$C_6H_5CO_2CH_3$	0	0.363
Butyl ketone	$CH_3COC_4H_9$	21–127	0.553
p-Butyrate	$C_3H_7CO_2CH_3$	20	0.459
Chloroacetate	$ClCH_2CO_2CH_3$	20	0.382
Dichloroacetate	$Cl_2CHCO_2CH_3$	20	0.311
Ethyl ketone	$CH_3COC_2H_5$	20–78	0.549
Ethyl ketoxime	$(CH_3)(C_2H_5)CNOH$	22–152	0.650
Formate	HCO_2CH_3	13–29	0.516
Hexyl ketone	$CH_3COC_6H_{13}$	22–168	0.552
Isobutyl ketone	$CH_3COC_4H_9$	20	0.459
Isopropyl ketone	$CH_3COC_3H_7$	20–91	0.525
Propionate	$C_3H_7CO_2CH_3$	20	0.459
Trichloroacetate	$Cl_3CCO_2CH_3$	20	0.267
Valerate	$C_4H_9CO_2CH_3$	20	0.459
o-Methylcyclohexanone	$C_7H_{12}O$	15–18	0.436
m-Methycyclohexanone		15–18	0.441
p-Methylcyclohexanone		15–18	0.441
Methylene chloride	CH_2Cl_2	15–40	0.288
Myristic acid	$C_{13}H_{27}CO_2H$	56–100	0.539
Naphthalene	$C_{10}H_8$	0	0.313
α-Naphthol	$C_{10}H_7OH$	0	0.389
β-Naphthol		0	0.403
α-Naphthylamine	$C_{10}H_7NH_2$	53.2	0.475
o-Nitraniline	$H_2NC_6H_4NO_2$	0	0.400
m-Nitraniline		0	0.392
p-Nitraniline		0	0.427
Nitrobenzene	$C_6H_5NO_2$	30	0.339
		120	0.394
o-Nitrobenzoic acid	$O_2NC_6H_4CO_2H$	0	0.314
m-Nitrobenzoic acid		0	0.405
p-Nitrobenzoic acid		238	0.449
Nitromethane	CH_3NO_2	17	0.412
α-Nitronaphthalene	$C_{10}H_7NO_2$	58.6	0.365
Nonane	C_9H_{20}	0–50	0.503
Nonylene	C_9H_{18}	0–50	0.485
n-Octane	C_8H_{18}	20–123	0.578
Octylene	C_8H_{16}	0–50	0.486
Olive oil		6.6	0.471
Palmitic acid	$C_{15}H_{31}CO_2H$	65–104	0.653
Paraldehyde	$(CH_3CHO)_3$	0	0.436
Pentadecane	$C_{15}H_{32}$	0–50	0.497
Pentadecylene	$C_{15}H_{30}$	0–50	0.471
iso-Pentane	C_5H_{12}	8	0.527
Petroleum		21–58	0.511
Phenetole	$C_6H_5OC_2H_5$	20	0.446
Phenol	C_6H_5OH	14–26	0.561
Piperidine	$C_5H_{11}N$	20–98	0.523
Propane	C_3H_8	0	0.576
Propionaldehyde	C_2H_5CHO	0	0.522
Propionic acid	$C_2H_5CO_2H$	20–137	0.560
Propionitrile	C_2H_5CN	19–95	0.538
n-Propyl acetate	$CH_3CO_2C_3H_7$	20	0.459
Propyl alcohol	$C_3H_7 \cdot OH$	−100	0.435
		0	0.526
		25	0.586
Benzene	$C_6H_5C_3H_7$	0	0.400
Propyl benzoate	$C_6H_5CO_2C_3H_7$	20	0.398
Butyrate	$C_3H_7CO_2C_3H_7$	20	0.459
Chloroacetate	$CH_2ClCO_2C_3H_7$	20	0.414
n-Propyl formate	$HCO_2C_3H_7$	20	0.459
Propyl isobutyrate	$C_3H_7CO_2C_3H_7$	20	0.459
Phenyl ether	$C_6H_5OC_3H_7$	0	0.429

TABLE 2-2 Specific Heats of Various Solids, Liquids, and Gases—(Continued)

LIQUID ORGANIC COMPOUNDS

Name	Formula	Temp (°C)	Sp ht
Propionate	$C_2H_5CO_2C_3H_7$	20	0.459
Valerate	$C_4H_9CO_2C_3H_7$	20	0.459
Pseudocumene	$C_6H_3(CH_3)_3$	20	0.414
Pyridine	C_5H_5N	21–108	0.431
Quinol	$C_6H_4(OH)_2$	0	0.492
Quinoline	C_9H_7N	0–20	0.352
Quinone	$C_6H_4O_2$	0	0.324
Resorcinol	$C_6H_4(OH)_2$	0	0.452
Salicylaldehyde	HOC_6H_4CHO	18	0.382
Salol	$HOC_6H_4CO_2C_6H_5$	44.1	0.391
Stearic acid	$C_{17}H_{35}CO_2H$	74–137	0.550
Tetrachloroethylene	C_2Cl_4	20	0.211
Tetradecane	$C_{14}H_{30}$	0–50	0.497
Tetradecylene	$C_{14}H_{28}$	0–50	0.453
m-Thymol	$C_9H_{11}OH$	50	0.567
Toluene	$C_6H_5CH_3$	0	0.386
		50	0.421
		100	0.470
o-Toluic acid	$CH_3C_6H_4CO_2H$	0	0.422
m-Toluic acid		0	0.503
p-Toluic acid		0	0.316
o-Toluidine	$CH_3C_6H_4NH_2$	22–195	0.524
p-Toluidine		43	0.508
Trichloroethylene	C_2HCl_3	20	0.223
Tridecane	$C_{13}H_{28}$	0–50	0.499
Tridecylene	$C_{13}H_{26}$	0–50	0.457
Trinitrotoluene (2, 4, 6)	$CH_3C_6H_2(NO_2)_3$	?	0.335
Turpentine, oil		0	0.411
Undecane	$C_{11}H_{24}$	0–50	0.501
Undecylene	$C_{11}H_{22}$	0–50	0.482
Valeronitrile	C_4H_9CN	23–121	0.520
iso Valeric acid	$C_4H_9CO_2H$	23–93	0.590
o-Xylene	$(CH_3)_2C_6H_4$	30	0.411
m-Xylene		16–35	0.387
p-Xylene		30	0.397

ALLOYS AND VARIOUS SOLIDS

Substance	Temp (°C)	Sp ht
Alloys:		
Aluminum bronze, 88.7 Cu, 11.3 Al	20–100	0.104
Antimony bismuth tin, 21.6Sb, 36.7Bi, 41.7Sn	22–99	0.046
Antimony lead, 37.1Sb, 62.9Pb	10–98	0.0388
Bell metal, 80Cu, 20Sn	14–98	0.0862
Bismuth tin, 63.8Bi, 36.2Sn	20–99	0.0400
46.9Bi, 53.1Sn	20–99	0.0450
56.9Bi, 43.1Sn	17–99	0.0450
Brass, 60Cu, 40Zn	−186 to −79	0.0743
	−79 to +18	0.0873
	20–100	0.0917
72Cu, 28Zn	14–98	0.094
Bronze, 80Cu, 20Sn	15–98	0.086
88Cu, 12Sn, 0.94P	20–100	0.0874
Constantan	0	0.098
	100	0.102
German silver	0	0.094
	100	0.095
Invar, 64Fe, 36Ni	−182 to +15	0.095
	15–100	0.120
	15–600	0.126
Lead bismuth, 39.9Pb, 60.1Bi	16–99	0.0317
Lead bismuth tin:		
32.5Pb, 49.0Bi, 18.5Sn	14–80	0.0600n
31.8Pb, 32.0Bi, 36.2Sn	11–98	0.0448

ALLOYS AND VARIOUS SOLIDS		
Substance	Temp (°C)	Sp ht
Lead tin, 63.7Pb, 36.3Sn	12–99	0.0407
46.7Pb, 53.3Sn	10–99	0.451
Lipowitz alloy, 24.97Pb, 10.13Cd, 50.66Bi, 14.24Sn	5–50	0.0345
Manganin	0	0.097
	100	0.095
Platinum iridium, 90Pt, 10Ir	20–100	0.0323
Rose alloy, 27.5Pb, 48.9Bi, 23.6Sn	20–89	0.0552
Solder, *see* Lead tin		
Steel, ordinary (.004C)	20	0.107
	100	0.117
Wood's alloy, 25.85Pb, 6.99Cd, 52.43Bi, 14.73Sn	5–50	0.0352
Amalgams:		
50.8Pb, 49.2Hg	23–99	0.0383
78.3Pb, 37.1Sn, 62.9Hg	22–99	0.0729
54.1Sn, 45.9Hg	25–99	0.0659
Asbestos	20–98	0.195
Basalt	20–100	0.20
Calcspar	0–100	0.2005
Carborundum	3–44	0.162
Cellulose, dry		0.37
Cement, powder	200–10	0.20
Chalk	20–99	0.214
Charcoal	10	0.16
Clay, dry	20–100	0.22
Ebonite	20–100	0.40
Glass, normal thermometer	19–100	0.1988
Crown	10–50	0.161
Flint	10–50	0.117
Carboloy		0.052
Granite	12–100	0.192
Ice	−200	0.168
	−180	0.199
	−160	0.230
	−140	0.262
	−100	0.325
	− 60	0.392
	− 20	0.480
	− 10	0.530
India rubber (Para)	100	0.481
Leather, dry		0.36
Marble	0–100	0.21
Mica (Mg)	20–98	0.2061
Paraffin	0–20	0.6939
Porcelain	15–950	0.26
Quartz	12–100	0.188
Rock salt	13–45	0.219
Sugar	20	0.274
Vulcanite	20–100	0.3312
Wood		0.42

GASES						

Gas or vapor	Temp (°C)	Sp ht	Gas or vapor	Temp (°C)	Sp ht
Acetic acid, $C_2H_4O_2$	118–140	1.50	Hydrochloric acid, HCl	10–190	0.185
	140–180	1.27		15	0.1939
Acetone, C_3H_6O	26–110	0.3470		−181	2.64
	130–230	0.4119		−76	3.15
Acetylene, C_2H_2	−71	0.3509		+15	3.389
	+15	0.3832		100	3.429
Air	−120 (10 atm)	0.2719		200	3.463
	(20 atm)	0.3221		400	3.533
	(40 atm)	0.4791		600	3.602
	(70 atm)	0.7771		800	3.672
	−50 (10 atm)	0.2440		1000	3.741
	(20 atm)	0.2521		2000	4.088
	(40 atm)	0.2741	Hydrogen sulfide, H_2S	−57	0.292
	(70 atm)	0.3121		−45	0.279
	+50 (20 atm)	0.2480		+10–190	0.243
	(100 atm)	0.2719		15	0.2533
	(220 atm)	0.2961	Iodine, I	206–077	0.034
	100 (1 atm)	0.2404	Methane, CH_4	−115	0.4502
	(20 atm)	0.2471		−80	0.5038
	(100 atm)	0.2600		−74	0.4979
	(220 atm)	0.2841		−10–200	0.5031
	400	0.2430		15	0.5284
	1000	0.2570	Methyl alcohol, CH_4O	77	0.390
	1400	0.2699		100–223	0.4581
	1800	0.2850	Nitric oxide, NO	−80	0.2445
Ammonia, NH_3	15	0.5232		−45	0.2389
Amylene, C_5H_{10}	ca. 210	0.631		−10–180	0.232
Argon, A	−180	0.133		15	0.2329
	+15	0.1253	Nitrogen, N_2	−181	0.256
Benzene, C_6H_6	80	0.260		+15	0.2477
	34–115	0.301	Nitrogen peroxide, NO_2	27–67	1.620
	120–220	0.370	Nitrous oxide N_2O	−70	0.1900
Bromine, Br	19–388	0.055		−30	0.1998
Carbon dioxide, CO_2	−75	0.184		+15	0.2004
	+15	0.1989		25–100	0.212
Carbon disulfide, CS_2	80–190	0.157	Oxygen, O_2	−181	0.2285
Carbon monoxide, CO	−180	0.259		−76	0.2143
	+15	0.2478		+15	0.2178
Carbon tetrachloride, CCl_4	0	0.140		100	0.2181
	30	0.132		200	0.2187
	70	0.115		400	0.2213
Chlorine, Cl_2	15	0.1149		600	0.2241
Chloroform, $CHCl_3$	27–118	0.145		800	0.2278
	120–230	0.157		1000	0.2325
Cyanogen, CN	15	0.4095		2000	0.2669
Ethane, C_2H_6	−82	0.3475	Phosphorus trichloride, PCl_3	110–250	0.135
	+15	0.3861	Silicon tetrachloride, $SiCl_4$	90–230	0.132
Ethyl acetate, $C_4H_8O_2$	35–189	0.3711	Stannic chloride, $SnCl_4$	149–273	0.094
Ethyl alcohol, C_2H_6O	90	0.406	Sulfur dioxide, SO_2	10–190	0.134
	100–223	0.454		15	0.1516
Ethyl bromide, C_2H_5Br	28–116	0.161	Water, H_2O	100	0.4820
Ethyl chloride, C_2H_5Cl	10–170	0.2750		120	0.4769
Ethyl cyanide, C_3H_5N	114–223	0.4260		140	0.4741
Ethyl ether, $C_4H_{10}O$	27–189	0.4619		160	0.4719
	35	0.4449		180	0.4710
Ethylene, C_2H_4	−91	0.3086		200	0.4710
	+15	0.3592		300	0.4769
	15–100	0.399		400	0.4901
	25–200	0.430		500 (1 atm)	0.5071
Ethylene chloride, $C_2H_4Cl_2$	111–221	0.23		(10 atm)	0.5159
Helium, He	−180	1.25		(20 atm)	0.5259
Hydrobromic acid, HBr	+11–100	0.082			

Example 2-7 A copper rod weighing 8.3 lb and at a room temperature of 68°F is dropped into 12.5 lb of water at 86°F. What is the final temperature of the combination?

solution The heat lost by the water is equal to the heat gained by the copper. If the final temperature of the combination is t°F, the heat lost by the water is

$$\frac{1 \text{ Btu}}{\cancel{lb} - °\cancel{F}} \times 12.5 \cancel{lb} \times (86 - t)°\cancel{F} = 1074 - 12.5t \text{ Btu}$$

Similarly, the heat gained by the copper is

$$\frac{0.0931 \text{ Btu}}{\cancel{lb} - °\cancel{F}} \times 8.3 \cancel{lb} \times (t - 68)°\cancel{F} = 0.77t - 52.5 \text{ Btu}$$

Thus,

$$0.77t - 52.5 = 1074 - 12.5t$$
$$13.3t = 1127$$
$$t = 83.8°F$$

Example 2-8 Benzene entering a heat exchanger at a temperature of 42°F and at a rate of 750 gal/h is to be heated to 86°F by means of hot water. If the water enters the heat exchanger at 135°F and leaves at 113°F, how much water is needed?

solution From Table 2-2, the specific heat of benzene is 0.41. Since the density of benzene is 0.879, the amount of heat that must be supplied to the benzene will be

$$\frac{750 \text{ } \cancel{gal}}{h} \times \frac{8.34 \text{ lb}}{\cancel{gal}} \times \frac{0.41 \text{ Btu}}{\cancel{lb} - °\cancel{F}} \times (86 - 42)°F = 113,000 \frac{\text{Btu}}{h}$$

This heat comes from the water:

$$\frac{113,000 \text{ } \cancel{Btu}}{h} \times \frac{\cancel{lb} - °\cancel{F}}{1 \text{ } \cancel{Btu}} \times \frac{1}{(135 - 113)°\cancel{F}} \times \frac{\text{gal}}{8.34 \cancel{lb}} = 616 \text{ gal/h}$$

Heat of Vaporization

The gain or loss of heat which results in a change in temperature of a substance is called *sensible heat*. When a change in the state (or phase) of a material accompanies a change in its heat content, such as occurs in boiling (in which the phase change is from a liquid to a gas), the temperature remains constant. From a theoretical view, for example, the addition of heat to a liquid causes an increase in the degree of agitation of its molecules, and at first this manifests itself as a rise in temperature. However, when the molecules of the liquid are already in a state of high agitation, the further addition of heat provides some molecules with sufficient energy to break through the surface, i.e., to enter the gaseous state, in preference to raising the temperature of the liquid still further. This process, known as boiling, uses the additional heat to form vapor and not to raise the temperature of the system. At the boiling point, therefore, as more heat is added, more vaporization occurs, until all the liquid has been vaporized. At this point, the further addition of heat goes into raising the temperature of the vapor, thereby again becoming sensible heat. Boiling is therefore a constant-temperature process, and the heat required to vaporize a liquid is known as its *heat of vaporization*. Usually, the heat of vaporization of a substance is large in comparison with its sensible heat.

In a similar manner, when a vapor is cooled, it gives up its heat of vaporization and, in a constant-temperature process, reverts back to a liquid. Further cooling then lowers the temperature of the liquid. These processes are illustrated in Fig. 2-2, which shows what happens to the temperature of 1 g of water as heat is added or removed. Table 2-3 gives the heats of vaporization for a number of common liquids.

Example 2-9 How much heat is required to vaporize 12 g of methanol (methyl alcohol) at its boiling point?

solution From Table 1-10, the boiling point is found to be 64.7°C. At this temperature, the heat of vaporization is given in Table 2-3 as 262.8 cal/g.

$$\text{Heat required} = 12 (262.8) = 3154 \text{ cal}$$

Heat of Fusion

When heat is added to a solid, its temperature rises until it begins to melt. In a manner analogous to boiling, the addition of heat to a substance at its melting point causes the substance to change phase, that is, to liquefy, but does not raise its temperature. After the material has completely liquefied, the further addition of heat results in a temperature rise of the liquid. Melting, like boiling, thus also occurs at constant temperature, and the heat required to liquefy a solid at its melting point is known as its *heat of fusion*. (Sometimes the word "latent" is used in connection with heat of vaporization or heat of fusion to indicate that the heat involved in these processes does not manifest itself as a change in temperature.) The heat of fusion of a variety of solids is given in Table 2-4. Again, in a manner analogous to condensation, when a liquid solidifies (freezes), it releases its heat of fusion without causing a temperature change. These processes are also illustrated in Fig. 2-2.

Example 2-10 What is the heat of fusion of ice in the English system; i.e., how much heat is required to melt 1 lb of ice?

solution From Table 2-4, the heat of fusion of ice is 79.7 cal/g.

$$\text{Heat required} = \frac{79.7 \text{ cal}}{\text{g}} \times \frac{453 \text{ g}}{\text{lb}} \times 1 \text{ lb} \times \frac{\text{Btu}}{252 \text{ cal}} = 143.3 \text{ Btu}$$

The calorimeter also provides a convenient apparatus for measuring heats of fusion and heats of vaporization.

Example 2-11 Starting with a flask containing 250 g of benzene, how much heat is required to boil off 150 g of the benzene?

solution From Table 1-10, the boiling point of benzene is 80.1°C, and from Table 2-2, its average specific heat between 20 and 80°C is 0.435. (This, the specific heat at 80°C, is obtained by extrapolating between the closest temperatures given, 60 and 90°C, and then averaging the values between 20 and 80°C.) The heat necessary to raise the temperature of the benzene to its boiling point is

$$250 \text{ g} \times (80.1 - 20)°C \times \frac{0.435 \text{ cal}}{\text{g} - °C} = 6{,}530 \text{ cal}$$

From Table 2-3, the heat of vaporization of benzene is 94.3 cal/g. The heat required to vaporize 150 g of benzene at its boiling point is therefore

$$150 \text{ g} \times \frac{94.3 \text{ cal}}{\text{g}} = 14{,}200 \text{ cal}$$

The total process, therefore, requires 6,530 + 14,200, or 20,700 cal.

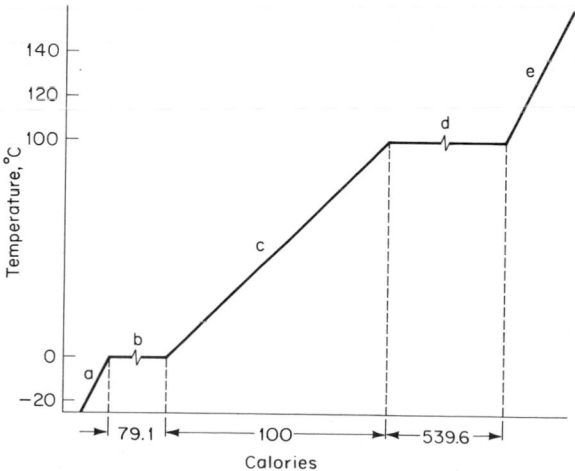

Fig. 2-2 Heat input–temperature relation for 1 g of water. Segment *a* represents the heating of ice to its melting point. The slope of *a* is the specific heat of ice. Segment *b* is for a mixture of ice and water. Segment *c* represents the heating of water, and its slope is the specific heat of water. Segment *d* represents the boiling of the water and is a mixture of water and water vapor. Segment *e* represents the heating of water vapor with its slope equal to the specific heat of water vapor. The heat of fusion of 1 g of ice is 79.1 cal; the heat of vaporization of 1 g of water is 539.6 cal.

TABLE 2-3 Heats of Vaporization of Various Liquids

Name	Formula	Temp (°C)	Heat of vaporization (cal/g)
ELEMENTS AND INORGANIC COMPOUNDS			
Air			50.97
Ammonia	NH_3	−33.4	327.1
		−20	317.6
		−10	309.7
		0	301.6
Ammonium chloride	NH_4Cl (solid)	350	78.9
Argon	A	−186	37.6
Boron chloride	BCl_3	10	38.2
Bromine	Br	63	43.7
Carbon dioxide	CO_2	−60	87.2
		−50	83.4
		−40	79.6
		−30	71.4
		−20	66.9
		−10	61.4
		0	55.0
		10	46.6
		20	35.1
		30	11.9
Carbon monoxide	CO	−192	50.4
Chlorosulfonic acid	$ClSO_3H$	151	110.2
Helium	He	−268.6	6
Hydriodic acid	HI	−37.2	33.9
Hydrobromic acid	HBr	−69.9	48.7
Hydrochloric acid	HCl	−84.3	98.7
Hydrofluoric acid	HF	17	360.8
Hydrogen	H_2	−252.8	108
Sulfide	H_2S	−61.4	131.9
Iodine		184	23.95
Mercury		357	65
Nitric acid	HNO_3	86.0	114.9
Nitrogen	N_2	−195.55	47.6
Tetroxide	N_2O_4	18	93.4
Oxygen	O_2	−182.9	50.9
Phosphorus	P	287	130
Phosphorus trichloride	PCl_3	78	51.4
Silicon tetrachloride	$SiCl_4$	57	36.1
Stannic chloride	$SnCl_4$	112	30.3
Sulfur chloride	S_2Cl_2	138	49.5
Dioxide	SO_2	−10.08	94.9
		0	91.3
		10	87.7
		20	84.1
		30	80.8
		40	71.2
		50	73.8
		60	70.3
Pentoxydichloride	$S_2O_5Cl_2$	140	61.2
Trioxide	SO_3	53	118.5
Sulfuric acid	H_2SO_4	326	122.1
Oxychloride	SO_2Cl_2	69.1	49.4
Sulfurous oxychloride	$SOCl_2$	82	54.5
Water	H_2O	0	595.9
		10	590.4
		20	584.9
		30	579.5
Water	H_2O	40	574.0
		50	568.5
		60	563.2
		70	557.5

From C. D. Hodgman (ed.), "Handbook of Chemistry and Physics," The Chemical Rubber Co., Cleveland, 1942.

TABLE 2-3   Heats of Vaporization of Various Liquids—(Continued)

Name	Formula	Temp (°C)	Heat of Vaporization (cal/g)
ELEMENTS AND INORGANIC COMPOUNDS			
Water (continued)	H_2O	80	551.7
		90	545.8
		100	539.55
		110	532.9
		120	525.7
		130	518.5
		140	511.1
		150	503.5
		160	495.6
		170	487.2
		180	478.6
ORGANIC COMPOUNDS			
Acetaldehyde	CH_3CHO	21	136
Acetic acid	CH_3CO_2H	118.3	96.8
Anhydride	$(CH_3CO)_2O$	137	66.2
Acetone	$(CH_3)_2CO$	56.1	124.5
Acetonitrile	CH_3CN	80	174
Acetyl chloride	CH_3COCl	51	78.9
Allyl alcohol	C_3H_5OH	96	163
n-Amyl alcohol	$C_5H_{11}OH$	131	120.2
iso-Amyl alcohol		130.2	119.8
n-Amyl bromide	$C_5H_{11}Br$	129	48.3
n-Amyl ether	$(C_5H_{11})_2O$	170	69.5
n-Amyl iodide	$C_5H_{11}I$	155	47.6
iso-Amyl isobutyrate	$C_3H_7CO_2C_5H_{11}$	168	57.6
iso-Amyl n-valerate	$C_4H_9CO_2C_5H_{11}$	187	56.2
Amylene	C_5H_{10}	12.5	76.0
n-Anethole	$C_3H_5OCH_3$	232	71.4
Benzene	C_6H_6	80.2	94.3
Butane	C_4H_{10}	0	91.5
iso-Butane		10	82.4
		−10	87.5
n-Butyl alcohol	C_4H_9OH	116.8	141
iso-Butyl alcohol		106.9	138
sec-Butyl alcohol		98.1	134
tert-Butyl alcohol		83	130.5
n-Butyl formate	$HCO_2C_4H_9$	105.1	86.8
iso-Butyl formate		97.0	78.5
n-Butyl iodide	C_4H_9I	129.5	459
iso-Butyl n-valerate	$C_4H_9CO_2C_4H_9$	169	57.8
iso-Butyl isovalerate		169	60.5
n-Butyric acid	$C_3H_7CO_2H$	163.5	114.0
iso-Butyric acid	$C_3H_7CO_2H$	154	111.6
n-Butyronitrile	C_3H_7CN	117.4	114.9
Carbon disulfide	CS_2	46.25	84.1
Tetrachloride	CCl_4	76.75	46.4
Carvacrol	$C_{10}H_{13}OH$	237	68.1
Chloral	CCl_3CHO		54.0
Hydrate	$Cl_3CCHO \cdot H_2O$	96	132
Chloroform	$CHCl_3$	61.5	59.0
Cyanogen	$(CN)_2$	0	10.3
Chloride	$ClCn$	13	135
p-Cymene	$C_{10}H_{14}$	176	67.6
Dichloroacetic acid	Cl_2CHCO_2H	194.4	77.2
n-Decane	$C_{10}H_{22}$	160	60.2
Diethyl carbonate	$CO(OC_2H_6)_2$	126	73.1
Ketone	$(C_2H_5)_2CO$	101	90.8
Diethylamine	$(C_2H_6)_2NH$	58	91.0
Dimethyl carbonate	$CO(OCH_3)_2$	90	88.2
Ethane	C_2H_6	0	75.0
		−20	87.0
		−40	97.5
		−90	127

TABLE 2-3   Heats of Vaporization of Various Liquids—(Continued)

Name	Formula	Temp (°C)	Heat of Vapori- zation (cal/g)
ORGANIC COMPOUNDS			
Ethyl acetate	$CH_3CO_2C_2H_5$	0.0	102.0
Alcohol	C_2H_5OH	78.3	204
Bromide	C_2H_5Br	38.4	59.9
Caprylate	$C_7H_{15}CO_2C_2H_6$	207	60.5
Chloride	C_2H_5Cl	4.7	92.95
		15.0	92.5
		20.0	92.2
		25.0	92.0
Ether	$(C_2H_5)_2O$	34.6	83.9
Formate	$HCO_2C_2H_5$	53.3	97.2
Iodide	C_2H_5I	71.2	45.6
Nonylate	$C_8H_{17}CO_2C_2H_5$	227	58.1
Propionate	$C_2H_5CO_2C_2H_5$	97.6	80.1
Ethylene bromide	$(CH_2Br)_2$	130.8	46.2
Chloride	$(CH_2Cl)_2$	0.0	85.3
		82.3	77.3
Oxide	$(CH_2)_2O$	13	139
Ethylamine	$C_2H_5NH_2$	15	14.6
Ethylidene chloride	CH_3CHCl_2	0.0	76.7
		60	67.1
Formic acid	HCO_2H	101	120.0
Furane	$(CH)_4O$	31	95.3
Furfural	C_4H_3OCHO	160.5	107.5
Glycol	$(CH_2OH)_2$	197	191
Hydrocyanic acid	HCN	20	210
Methane	CH_4	−159	138
Methyl acetate	$CH_3CO_2CH_3$	0.0	114.0
		56.3	98.1
Alcohol	CH_3OH	64.7	262.8
n-Butyrate	$C_3H_7CO_2CH_3$	102.6	79.8
Chloride	CH_3Cl	−23.8	102.3
		20.0	95.3
Ethyl ketone	$CH_3COC_2H_5$	78.2	106.0
Ethyl ketoxime	C_4H_8NOH	182	115.9
Formate	HCO_2CH_3	31.3	112.4
Iodide	CH_3I	42	45.9
Isobutyrate	$C_3H_7CO_2CH_3$	91.1	78.1
Isopropyl ketone	$C_4H_{10}CO$	92	89.8
Propionate	$C_2H_5CO_2CH_3$	79.0	87.6
Methylene chloride	CH_2Cl_2	40.5	78.6
Naphthalene	$C_{10}H_8$	218	75.5
Nitromethane	CH_3NO_2	99.9	135
iso-Pentane	C_5H_{12}	13	88.7
Piperidine	$C_5H_{11}N$	106	89.4
Propane	C_3H_8	20	83.4
		0	89.6
		−20	95.3
		−30	98.0
Propionic acid	$C_2H_5CO_2H$	139.3	98.8
Propionitrile	C_2H_5CN	97	134
n-Propyl acetate	$CH_3CO_2C_3H_7$	100.4	80.3
n-Propyl alcohol	C_3H_7OH	97.2	164
iso-Propyl alcohol		82.3	159
n-Propyl formate	$HCO_2C_3H_7$	80.0	88.1
Pyridine	C_5H_6N	114.1	107.4
Tetrachloroethane-1, 1, 2; 2	$(CHCl_2)_2$	145.0	55.1
Tetrachloroethylene	$(CCl_2)_2$	120.7	50.1
Trichloroethylene	C_2HCl_3	85.7	57.3
Turpentine	$C_{10}H_{16}$	156	68.6
n-Valeric acid	$C_4H_9CO_2H$	184.6	103.2
iso-Valeric acid		176.3	101.1
n-Valeronitrile	C_4H_9CN	129	96.3

TABLE 2-4 Heat of Fusion of Various Solids

Name	Formula	Temp (°C)	Heat of fusion (cal/g)
ELEMENTS AND INORGANIC COMPOUNDS			
Aluminum	Al	658	76.8
Ammonia	NH_3	−75	108.1
		−77.6	83.9
Antimony bromide	$SbBr_3$	94	9.76
Trichloride	$SbCl_3$	73.2	13.3
Trisulfide	Sb_2S_3	540	17.6
Argon	A	−190	6.71
Arsenous bromide	$AsBr_3$	31	8.94
Barium chloride	$BaCl_2$	958.9	27.5
Bismuth	Bi	268	12.64
Bromine	Br	−7.32	16.2
Cadmium	Cd	320.7	13.66
Nitrate	$Cd(NO_3)_2 \cdot 4H_2O$	59.5	25.3
Caesium hydroxide	CsOH	272.3	10.8
Calcium chloride	$CaCl_2$	773.9	54.3
Chloride	$CaCl_2 \cdot 6H_2O$	29	40.7
Nitrate	$Ca(NO_3)_2 \cdot 4H_2O$	42.1	34.0
Carbon dioxide	CO_2	−56.2	45.3
Monoxide	CO	−206	8.00
Chlorine	Cl	−103.5	23.0
Cobalt nitrate	$Co(NO_3)_2 \cdot 6H_2O$		30.2
Copper	Cu	1083	42
Cupric nitrate	$Cu(NO_3)_2 \cdot 6H_2O$	24.4	29.4
Gold	Au	1064	15.8
Hydriodic acid	HI	−53	5.68
Hydrobromic acid	HBr	−86	7.67
Hydrochloric acid	HCl	−114	13.9
Acid hydrate	$HCl \cdot 2H_2O$	−18.5	34.6
Hydrogen	H_2		14.0
Peroxide	H_2O_2	−1.7	74.1
Iodine	I		11.71
Iron, gray cast	Fe		5.50
White cast			7.89
Slag			11.9
Lead	Pb	327	5.86
Bromide	$PbBr_2$	490	12.3
Chloride	$PbCl_2$	485	20.9
Iodide	PbI_2	375	11.5
Lithium nitrate	$LiNO_3$	250	88.5
Silicate	Li_2SiO_3		80.2
Silicate	$Li_2SiO_3 \cdot Li_2O$		62.1
Magnesium chloride	$MgCl_2 \cdot 6H_2O$	116.7	41.2
Nitrate	$Mg(NO_3)_2 \cdot 6H_2O$	90	38.2
Manganese nitrate	$Mn(NO_3)_2 \cdot 6H_2O$	25.8	28.8
Mercuric bromide	$HgBr_2$	235	12.8
Iodide	HgI_2	250	9.80
Mercury	Hg	−39	2.82
Nickel	Ni	1435	73.8
Nitrate	$Ni(NO_3)_2 \cdot 6H_2O$	56.7	36.4
Nitric acid	HNO_3	−47	9.55
Nitrogen	N	−210	6.09
Dioxide	NO	−163	18.4
Pentoxide	N_2O_5	29.5	76.7
Tetroxide	N_2O_4	−10.14	34.81
Oxygen	O	−219	3.30
Palladium	Pd	1545	36.3
Platinum	Pt	1755	27.2
Potassium	K	62	15.7
Chloride	KCl	772.3	74.1
Dichromate	$K_2Cr_2O_7$	397	29.7

From C. D. Hodgman (ed.), "Handbook of Chemistry and Physics," The Chemical Rubber Co., Cleveland, 1942.

TABLE 2-4   Heat of Fusion of Various Solids—(Continued)

Name	Formula	Temp (°C)	Heat of fusion (cal/g)
ELEMENTS AND INORGANIC COMPOUNDS			
Potassium (continued)			
Fluoride	KF	859.9	108
Hydroxide	KOH	360.4	28.6
Nitrate	KNO_3	308	25.4
Phosphorous acid, hypo	H_3PO_2	17.4	35.0
Phosphorus	P	44.2	5.03
Oxychloride	$POCl_3$	2	19.8
Rubidium chloride	RbCl		38.0
Hydroxide	RbOH	301	15.8
Silicon tetrachloride	$SiCl_4$	70.3	10.9
Silver	Ag	961	21.07
Bromide	AgBr	430	12.5
Chloride	AgCl	451	30.7
		455	21.3
Nitrate	$AgNO_3$	208	17.7
Sodium	Na	97	31.7
Chlorate	$NaClO_3$	255	49.0
Chloride	NaCl	804.3	124
Chromate	$Na_2CrO_4 \cdot 10H_2O$	23	39.2
Fluoride	NaF	992.2	186
Hydroxide	NaOH	318.4	40.0
Nitrate	$NaNO_3$	333	45.3
Phosphate, dibasic	$Na_2HPO_4 \cdot 12H_2O$	36.1	66.8
Sulfate	$Na_2SO_4 \cdot 10H_2O$	31	51.3
Thiosulfate	$Na_2S_2O_3 \cdot 5H_2O$		47.8
Stannic bromide	$SnBr_4$	25.5	6.26
Chloride	$SnCl_4$	−33	8.40
Strontium chloride	$SrCl_2$	872.3	25.4
Sulfur	S	119	13.2
Trioxide	SO_3	−30	24.0
Sulfuric acid	$H_2SO_4 \cdot H_2O$	8.56	39.1
Acid	H_2SO_4	10.352	24.0
acid, pyro-	$H_2S_2O_7$	35	17.9
Thallium bromide	TlBr	460	12.7
Monochloride	TlCl	427	16.6
Tin	Sn	232	14.0
Titanium tetrachloride	$TiCl_4$	−25	11.8
Water	H_2O	0	79.71
ice from seawater	H_2O	−8.7	54.0
Zinc	Zn	419	28.13
Nitrate	$Zn(NO_3)_2 \cdot 6H_2O$	36.4	31.1
ORGANIC COMPOUNDS			
Acetic acid	CH_3CO_2H	16.7	43.2
Acetone	$(CH_3)_2CO$	−95.5	23.4
Acrylic acid	$C_2H_3CO_2H$	13	37.0
Allocinnamic acid	$C_6H_5C_2H_2CO_2H$	58	27.4
o-Aminobenzoic acid	$H_2NC_6H_4CO_2H$	145	35.5
m-Aminobenzoic acid		180	38.0
p-Aminobenzoic acid		188.5	36.5
tert-Amyl alcohol	$C_5H_{11}OH$		12.5
Anethole	$C_3H_5C_6H_4OCH_3$	21.5	25.8
Aniline	$C_6H_5NH_2$	−7.03	21.0
Anthracene	$C_{14}H_{10}$	216.55	38.7
Anthraquinone	$(C_6H_4)_2(CO)_2$	282	37.5
Azobenzene	$(C_6H_5N)_2$	69.1	28.9
Azoxybenzene	$(C_6H_5)_2ON_2$	34.6	21.6
Benzene	C_6H_6	5.42	30.3

TABLE 2-4   Heat of Fusion of Various Solids—(Continued)

Name	Formula	Temp (°C)	Heat of fusion (cal/g)
ORGANIC COMPOUNDS			
Benzil	$(C_6H_5CO)_2$	94.94	22.2
Benzoic acid	$C_6H_5CO_2H$	121.8	33.9
Benzophenone	$(C_6H_5)_2CO$	48.25	23.5
Benzylaniline	$C_6H_5NHC_7H_7$	36	21.9
Bromal hydrate	$CBr_3CHO \cdot H_2O$	46	16.9
Bromocamphor	$C_{10}H_{15}BrO$		41.6
o-Bromochlorobenzene	C_6H_4BrCl	−12.6	15.4
m-Bromochlorobenzene		−21.2	15.3
p-Bromochlorobenzene		64.6	23.4
o-Bromoiodobenzene	C_6H_4BrI	21	12.2
m-Bromoiodobenzene		−9.3	10.3
p-Bromoiodobenzene		90.1	16.6
p-Bromophenol	HOC_6H_4Br	64	20.5
p-Bromotoluene	$CH_3C_6H_4Br$	27.6	20.9
n-Butyl alcohol	C_4H_9OH	−89.2	29.9
tert-Butyl alcohol	C_4H_9OH	25.45	21.0
n-Butyric acid	$C_3H_7CO_2H$	−5.7	30.1
n-Capric acid	$C_9H_{19}CO_2H$	31.2	38.9
n-Caprylic acid	$C_7H_{15}CO_2H$	16.34	35.4
Carbazole	$C_{12}H_9N$	236	42.1
Carbon tetrachloride	CCl_4	−24	4.16
Carvoxime (d)	$C_{10}H_{14}NOH$	71.5	23.3
Carvoxime (l)		71	23.4
Carvoxime (dl)		91	24.6
Catechol	$C_6H_4(OH)_2$	104.3	49.4
Cetyl alcohol	$C_{16}H_{33}OH$	47	33.8
Cinnamic acid	$C_6H_5C_2H_2CO_2H$	133	36.5
Anhydride	$(C_6H_5C_2H_4CO)_2O$	48	28.1
Chloral alcoholate	$CCl_3CHO \cdot C_2H_5OH$	9	24.0
Hydrate	$CCl_3CHO \cdot H_2O$		33.2
Chloroacetic acid (α)	$ClCH_2CO_2H$	61.2	31.1
Acid (β)		56	35.1
p-Chloroaniline	$H_2NC_6H_4Cl$	69	37.2
o-Chlorobenzoic acid	$ClC_6H_4CO_2H$	140.2	39.3
m-Chlorobenzoic acid		154.25	36.4
p-Chlorobenzoic acid		239.7	49.2
m-Chloronitrobenzene	$ClC_6H_4NO_2$	43.8	29.4
p-Chloronitrobenzene		82	21.4
p-Cresol	$CH_3C_6H_4OH$	34	26.3
Cyanamide	H_2NCN	42.9	49.8
Cyclohexanol	$C_6H_{11}(OH)$	23.2	4.19
Dibenzyl	$(C_6H_5CH_2)_2$	51	31.0
o-Dibromobenzene	$C_6H_4Br_2$	18	12.8
m-Dibromobenzene		−6.9	13.4
p-Dibromobenzene		86	20.5
Dibromophenol (2, 4)	$HOC_6H_3Br_2$	12	14.0
Dichloroacetic acid	Cl_2CHCO_2H	10.8	14.2
o-Dichlorobenzene	$C_6H_4Cl_2$	−17.5	21.0
m-Dichlorobenzene		−24.4	20.5
p-Dichlorobenzene		52.7	29.7
o-Diiodobenzene	$C_6H_4I_2$	23.4	10.2
m-Diiodobenzene		34.2	11.6
p-Diiodobenzene		129	16.2
Dimethyl tartrate (d)	$(CHOH)_2(CO_2CH_3)_2$	49	21.5
Dimethyl tartrate (dl)		87	35.1
o-Dinitrobenzene	$C_6H_4(NO_2)_2$	116.93	32.3
m-Dinitrobenzene		90.08	24.7
p-Dinitrobenzene		173.5	40.0
Dinitrotoluene (2, 4)	$CH_3C_6H_3(NO_2)_3$	70	26.4
Diphenyl	$(C_6H_5)_2$	71	26.1

TABLE 2-4   Heat of Fusion of Various Solids—(Continued)

Name	Formula	Temp (°C)	Heat of fusion (cal/g)
ORGANIC COMPOUNDS			
Diphenylamine	$(C_6H_5)_2NH$	53.4	25.2
Diphenylmethane	$(C_6H_5)_2CH_2$	26.3	25.2
Ethyl alcohol	C_2H_5OH	−114.4	24.9
Ethylene dibromide	$(CH_2Br)_2$	9.55	13.5
Elaidic acid	$C_{17}H_{33}CO_2H$	47	52.1
Formic acid	HCO_2H	8.0	58.9
Glutaric acid	$(CH_2)_3(CO_2H)_2$	99.3	37.4
Glycerol	$C_3H_5(OH)_3$	18	47.5
Glycol	$(CH_2OH)_2$	−11.5	43.3
Hydrazobenzene	$(C_6H_5NH)_2$	134	22.9
Hydrocinnamic acid	$C_6H_5C_2H_4CO_2H$	48	28.1
p-Iodotoluene	$IC_6H_4CO_2H$	34	18.8
n-Lauric acid	$C_{11}H_{23}CO_2H$	43.85	43.7
Levulinic acid	$CH_3CO(CH_2)_2CO_2H$	33	19.0
α-Menthol (l)	$C_{10}H_{19}OH$	42	18.6
Methane	CH_4	−182.6	14.5
Methyl alcohol	CH_3OH	−97	16.4
Cinnamate	$C_6H_5C_2H_2CO_2CH_3$	34.5	26.5
Fumarate	$(CHCO_2CH_3)_2$	102	57.9
Oxalate	$(CO_2CH_3)_2$	49.5	42.7
Phenylpropiolate	$C_6H_5C_2CO_2CH_3$	18	22.9
Succinate	$(CH_2CO_2CH_3)_2$	18	35.7
Myristic acid	$C_{13}H_{27}CO_2H$		47.5
Naphthalene	$C_{10}H_8$	79.9	35.6
α-Naphthol	$C_{10}H_7OH$	95	38.9
β-Naphthol		120.6	31.3
α-Naphthylamine	$C_{10}H_7NH_2.$	47.5	22.3
o-Nitroaniline	$H_2NC_6H_4NO_2$	69.3	27.9
m-Nitroaniline	$H_2NC_6H_4NO_2$	111.8	41.0
p-Nitroaniline		147.5	36.5
Nitrobenzene	$C_6H_5NO_2$	5.72	22.5
o-Nitrobenzoic acid	$O_2NC_6H_4CO_2H$	145.8	40.1
m-Nitrobenzoic acid		141.1	27.6
p-Nitrobenzoic acid		239.2	52.8
α-Nitronaphthalene	$C_{10}H_7NO_2$	56	25.4
o-Nitrophenol	$HOC_6H_4NO_2$	42.8	26.8
Palmitic acid	$C_{15}H_{31}CO_2H$	55	39.2
Paraffin		52.40	35.10
Paraldehyde	$(CH_3CHO)_3$	12.6	25.0
Phenanthrene	$C_{14}H_{10}$	98.2	24.3
Phenol	C_6H_5OH	25.37	29.0
Phenylacetic acid	$C_6H_5CH_2CO_2H$	74.9	25.4
Phenylhydrazine	$C_6H_5N_2H_2$	22.1	36.3
iso-Propyl alcohol	C_3H_7OH	−88.5	21.0
Quinol	$C_6H_4(OH)_2$	172.3	58.8
Quinone	$C_6H_4O_2$	112.85	40.9
Resorcinol	$C_6H_4(OH)_2$	109.65	46.2
Spermaceti		43.9	36.98
Stearic acid	$C_{17}H_{35}CO_2H$	64	47.6
Stilbene	$(C_6H_5CH)_2$	124	39.9
Succinic anhydride	$(CH_2CO)_2O$	119	48.7
Succinonitrile	$(CH_2CN)_2$	54.5	11.7
Thymol	$C_{10}H_{13}OH$	48.5	27.5
Tolane	$(C_6H_5C)_2$	60	28.7
o-Toluic acid	$CH_3C_6H_4CO_2H$	103.7	35.4
m-Toluic acid		108.75	27.6
p-Toluic acid		179.6	39.9
p-Toluidine	$CH_3C_6H_4NH_2$	40.01	39.9
Tribromoaniline (2, 4, 6)	$H_2NC_6H_2Br_3$	122	16.8
Tribromophenol (2, 4, 6)	$HOC_6H_2Br_3$	93	13.4

TABLE 2-4   Heat of Fusion of Various Solids—(Continued)

Name	Formula	Temp (°C)	Heat of fusion (cal/g)
ORGANIC COMPOUNDS			
Trichloroacetic acid	CCl_3CO_2H	59.1	8.6
Trinitroglycerol	$C_3H_5(NO_3)_3$	12.3	23.0
Trinitrotoluene (TNT)			
(2, 4, 6)	$CH_3C_6H_2(NO_2)_3$	79	22.3
Triphenylmethane	$(C_6H_5)_3CH$	92.3	17.8
Tristearin	$(C_{17}H_{35}CO_2)_3C_3H_5$	56	45.6
n-Undecylic acid (α)	$C_{10}H_{21}CO_2H$	28.25	32.2
n-Undecylic acid (β)			42.9
Urethane	$H_2NCO_2C_2H_5$	48.7	40.9
Veratrol (1, 2)	$C_6H_4(OCH_3)_2$	22.7	27.5
Wax (bees')		61.8	42.3
p-Xylene	$C_6H_4(CH_3)_2$	16	39.3

Example 2-12 How much heat (in Btu) is absorbed by a refrigerator in changing 2.5 lb of water at 58°F to ice at 32°F?

solution The heat absorbed in cooling the water to its freezing point is

$$2.5 \text{ lb} \times (58 - 32)°F \times \frac{1 \text{ Btu}}{\text{lb} - °F} = 65 \text{ Btu}$$

From Table 2-4, the heat of fusion of ice (which is also the heat of freezing of water) is 79.7 cal/g.

$$79.7 \frac{\text{cal}}{g} \times \frac{\text{Btu}}{252 \text{ cal}} \times \frac{453 g}{\text{lb}} = 143 \frac{\text{Btu}}{\text{lb}}$$

$$2.5 \text{ lb} \times \frac{143 \text{ Btu}}{\text{lb}} = 358 \text{ Btu}$$

The total heat required is therefore $65 + 358 = 423$ Btu.

Under proper conditions of temperature and pressure, a substance will pass directly from the vapor phase without going through a liquid stage. This process is known as *sublimation*. Examples of materials which sublimate under ordinary conditions are iodine, naphthalene, and dry ice (solid carbon dioxide).

Thermal Conductivity

The thermal agitation of the molecules of a section of a body at a high temperature will, by impact with its neighbors, soon impart an increased thermal agitation to the molecules of a cooler section of the body, this process continuing until the body is at a uniform temperature. The heat of the hotter parts of the body is thus transferred by the process of *conduction* to the cooler sections. The rate of heat transfer by conduction is given by the equation

$$q = kAT\frac{\Delta t}{d} \qquad (2-2)$$

where q is the amount of heat flowing across an area A in time T as a result of a temperature difference of Δt between two points separated by a distance d. Frequently, the term *temperature gradient* is used to refer to the quantity $\Delta t/d$. These conditions are illustrated in Fig. 2-3. The proportionality constant k is called the thermal conductivity and, as can be seen from Eq. (2-2), has dimensions of $cal/(s)(cm)(°C)$ in the metric system and $Btu/(h)(ft)(°F)$ in the English system. The thermal conductivities of common materials are given in Table 2-5.

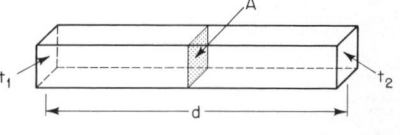

$(t_1 > t_2)$
$\Delta t = (t_1 - t_2)$
Direction of heat flow, q: ⟶

Fig. 2-3 Conductivity of heat.

Example 2-13 The thermal conductivity of brass is given in Table 2-5 as 0.26 cal/(s)(cm)(°C). (*Example 2-13 continued on p. 2-35.*)

TABLE 2-5   Thermal Conductivities of Common Materials

METALS

Substance	Temp (°C)	Conductivity*	Substance	Temp (°C)	Conductivity*
Aluminum	−160	0.514	Cast	18	0.109
	18	0.492		100	0.108
	100	0.49		54	0.114
	200	0.55		102	0.111
	300	0.64	Steel	−160	0.113
	400	0.76		18	0.115
	600	1.01		100	0.107
Antimony	0	0.0442	Lead	−160	0.092
	100	0.040		18	0.083
	0–30	0.042		100	0.082
Bismuth	−186	0.025	Magnesium	0–100	0.376
	0	0.0177	Manganin	18	0.15186
	18	0.0194	(84Cu, 4Ni, 12Mn)	100	0.06310
	100	0.0161		−160	0.035
Brass (70Cu + 30Zn)	−160	0.181	Mercury	0	0.0148
(70Cu + 30Zn)	17	0.260		50	0.0189
Yellow	0	0.204		17	0.0197
Red	0	0.246	Molybdenum	17	0.346
Bronze, aluminum		0.18	Nickel	−160	0.129
(90Cu, 10Al)				18	0.142
Cadmium	−160	0.239		100	0.138
	0	0.220		300	0.126
	18	0.222		600	0.088
	100	0.216		800	0.068
Constantan	18	0.054		1200	0.058
(60Cu, 40Ni)	100	0.064	Palladium	18	0.1683
Copper, pure	−160	1.097		100	0.182
	13	1.00	Platinum	18	0.1664
	18	0.918		100	0.1733
Copper, pure	100	0.908	Platinum-iridium	17	0.074
	100–197	1.043	10% Ir		
	100–268	0.969	Platinum-rhodium	17	0.072
	100–370	0.931	10% Rh		
	100–541	0.902	Platinoid	18	0.060
	100–837	0.858	Rhodium	17	0.210
German silver	0	0.070	Silver, pure	−160	0.998
	100	0.089		18	1.006
		0.10		100	0.992
(52Cu, 26Zn, 22Ni)			Tin	−160	0.192
Gold	18	0.700		0	0.1528
	100	0.703		18	0.155
Iridium	17	0.141		100	0.145
Iron, pure	18	0.161	Tantalum	17	0.130
	100	0.151	Tungsten	17	0.41
	100–727	0.202	Wood's alloy		0.0319
	100–1245	0.191	Zinc	−160	0.278
Wrought	−160	0.152		18	0.2653
	18	0.144		100	0.2619
	100	0.143			

VARIOUS SOLIDS

Substance	Conductivity	Substance	Conductivity
Asbestos fiber, 500°C	0.00019	Brick, 150–1200°	0.032–0.027
Paper	0.0006	Cardboard	0.0005
Basalt	0.0052	Cement, portland	0.00071
Brick, common red	0.0015	Chalk	0.0020
Blotting paper	0.00015	Concrete, cinder	0.00081
Carbon	0.01	Stone	0.0022
Carborundum	0.0005	Cork	0.0004
			0.00013

TABLE 2-5   Thermal Conductivities of Common Materials—(Continued)

VARIOUS SOLIDS

Substance	Conductivity	Substance	Conductivity
Cotton wool	0.000043	Lime	0.00029
Felted	0.000033	Linen	0.00021
Diatomaceous earth	0.00013	Magnesia, MgO	0.00016-0.00045
Earth's crust, avg	0.004	Brick, 50-1130°	0.0027-0.0072
Ebonite	0.0003	Magnesium carbonate:	
Eiderdown, $d = 0.109$	0.000046	100°	0.00023
Felt	0.000087	300°	0.00025
Fiber, red	0.0011	Marble	0.0071
Firebrick	0.0007	Mica, perpendicular to	0.0018
Flannel	0.00023	cleavage plane	
Gas carbon, 20°	0.0085	Paper	0.0003
100°	0.0095	Paraffin	0.0006
Glass:		Plaster of paris	0.00070
Crown (window)	0.0025	Porcelain	0.0025
Flint	0.002	165-1055°	0.0039-0.0047
Jena	0.001-0.002	Quartz, parallel to axis	0.030
Soda, 20°	0.0017	Perpendicular to axis	0.16
100°	0.0018	Rubber, para	0.00045
Granite, 100°	0.0045-0.0050	Sand, dry	0.00093
500°	0.0040	Sandstone	0.0055
Graphite	0.012	Sawdust	0.00012
Graphite brick, 300-700°	0.24	Silica, fused, 20°	0.00237
Gutta-percha	0.00048	100°	0.00255
Gypsum	0.0031	Silica brick, 100-1000°C	0.002-0.003
Haircloth, felt	0.000042	Silk	0.000095
Ice	0.004	Slate	0.004700
Infusorial earth, 100°	0.00034	Snow, compact	0.00051
300°	0.00040	Soil, dry	0.00033
Pressed bricks, 100°	0.00030	Wax, bees'	0.00009
Lampblack, 100	0.00007	Wood, fir ‖ to axis	0.0003
Leather, cowhide	0.00042	⊥ to axis	0.00009
Chamois	0.00015		

LIQUIDS

Liquid	Conductivity	Liquid	Conductivity
Acetic acid	0.00047	Castor	0.000425
Amyl alcohol	0.000328	Petroleum, 13°	0.000355
Aniline, 12°	0.00041	Turpentine	0.000325
Benzole, 5°	0.000333	Petroleum jelly, 25°	0.00044
Carbon disulfide, 9-15°	0.000343	Water:	
Chloroform, 9-15°	0.000288	4°	0.00138
Ether, 9-15°	0.000303	0°	0.00120
Ethyl alcohol	0.000423	17°	0.00131
Glycerin, 9-15°	0.000637	20°	0.00143
Methyl alcohol	0.000495		
Oils:			
Olive	0.000395		

GASES

Gas	Conductivity	Gas	Conductivity
Air, 0°	0.0000568	Hydrogen, 0°	0.000327
Ammonia gas, 0°	0.0000458	100°	0.000369
Argon, 0°	0.0000389	Methane, 8°	0.0000647
Carbon dioxide, 0°	0.0000307	Nitric oxide, NO, 8°	0.0000460
Monoxide	0.0000499	Nitrogen, 8°	0.0000524
Ethylene	0.0000395	Nitrous oxide, N_2O	0.0000350
Helium, 0°	0.000339	Oxygen, 8°	0.0000563

From C. D. Hodgman (ed.), "Handbook of Chemistry and Physics," The Chemical Rubber Co., Cleveland, 1942.

*Quantity of heat in calories transmitted per second through a plate 1 cm thick across an area of 1 cm² when the temperature difference is 1°C, i.e., kcal-cm/(s)(cm²)(°C).

TABLE 2-5 Thermal Conductivities of Common Materials—(Continued)

INSULATING AND BUILDING MATERIALS

Material	Density (lb/ft^3)	Conductivity [Btu-in/(h)(ft^2)(°F)]
Soft flexible materials in sheet form:		
Dry zero, kapok between burlap or paper	1.0	0.24
	2.0	0.25
Cabot's quilt, eelgrass between kraft paper	3.4	0.25
	4.6	0.26
Hair felt, felted cattle hair	11.0	0.26
	13.0	0.26
Balsam wool, chemically treated wood fiber	2.2	0.27
Hairinsul:		
75% hair, 25% jute	6.3	0.27
50% hair, 50% jute	6.1	0.26
Linofelt, flax fibers between paper	4.9	0.28
Thermofelt:		
Jute and asbestos fibers, felted	10.0	0.37
Hair and asbestos fibers, felted	7.8	0.28
Loose materials:		
Rock wool, fibrous material made from rock, also made in sheet form, felted and confined with wire netting	6.0	0.26
	10.0	0.27
	14.0	0.28
	18.0	0.29
Glass wool, Pyrex glass, curled	4.0	0.29
	10.0	0.29
Sil-O-Cel, powdered diatomaceous earth	10.6	0.31
Regranulated cork:		
Fine particles	9.4	0.30
About ³/₁₆-in particles	8.1	0.31
Thermofill, gypsum in powdered form	26.	0.52
	34.	0.60
Sawdust:		
Various	12.0	0.41
Redwood	10.9	0.42
Shavings, various, from planer	8.8	0.41
Charcoal, from maple, beech, and birch, coarse	13.2	0.36
6 mesh	15.2	0.37
20 mesh	19.2	0.39
Semiflexible materials in sheet form:		
Flaxlinum, flax fiber	13.0	0.31
Fibrofelt, flax and rye fiber	13.6	0.32
Semirigid materials in board form:		
Corkboard, no added binder; very low density	5.4	0.25
Corkboard, no added binder; low density	7.0	0.27
Corkboard, no added binder; medium density	10.6	0.30
Corkboard, no added binder; high density	14.0	0.34
Eureka, corkboard with asphaltic binder	14.5	0.32
Rock cork, rock-wool block with binder	14.5	0.326
Lith, board containing rock wool, flax, and straw pulp	14.3	0.40
Stiff fibrous materials in sheet form:		
Insulite, wood pulp	16.2	0.34
	16.9	0.34
Celotex, sugarcane fiber	13.2	0.34
	14.8	0.34
Masonite		0.33
Inso-board		0.33
Maizewood		0.33–0.39
Cornstalk pith board		0.24–0.30
Malftex		0.34
Cellular gypsum, Insulex or Pyrocell	8	0.35
	12	0.44
	18	0.59
	24	0.77
	30	1.00

TABLE 2-5   Thermal Conductivities of Common Materials—(Continued)

INSULATING AND BUILDING MATERIALS

Material	Density (lb/ft³)	Conductivity [Btu-in/(h)(ft²)(°F)]
Woods (across grain):		
Balsa	7.3	0.33
	8.8	0.38
	20	0.58
Cypress	29	0.67
White pine	32	0.78
Mahogany	34	0.90
Virginia pine	34	0.98
Oak	38	1.02
Maple	44	1.10
Miscellaneous building materials:		
Cinder concrete		2–3
Building gypsum		About 3
Plaster		2–5
Building brick		3–6
Glass		5–6
Limestone		4–9
Concrete		6–9
Sandstone		8–16
Marble		14–20
Granite		13–28

How much heat is transferred across a 4×4-in brass plate in 1 h if it is 2 mm thick and its faces are maintained at 100 and 65°C?

solution

$$q = \frac{0.26 \text{ cal}}{(\text{g})(\text{cm})(°\text{C})} \times 16 \text{ in}^2 \times 1 \text{ h} \times \frac{(100-65)°\text{C}}{2 \text{ mm}} \times \frac{3,600 \text{ s}}{\text{h}} \times \frac{6.45 \text{ cm}^2}{\text{in}^2} \times \frac{10 \text{ mm}}{\text{cm}}$$

$$q = \frac{0.26 \ (16)(35)(3,600)(6.45)}{2} \text{ cal} = 1,690,000 \text{ cal (or 1,690 kcal)}$$

Note that this example requires careful attention to units.

Example 2-14   How many pounds of water at 212°F can be vaporized in 1 h by the heat transmitted through a submerged steel tube 12 in long and having an ID of 15/16 in and an OD of 1 in? The tube will be heated by rapidly flowing oil so that its inner surface will essentially be maintained at 300°F.

solution   For heat being transferred radially through the walls of a pipe, the area across which the heat is flowing on the inside of the tube is different from that on the outside. This particular geometry, as well as other special cases, has received special mathematical analysis. However, for most practical cases, especially where the pipe walls are thin, the average area A can be used.

$$A_{\text{inside}} = \pi(0.938)(12) = 35.4 \text{ in}^2$$
$$A_{\text{outside}} = \pi(1)(12) = 37.7 \text{ in}^2$$
$$A_{\text{average}} = \frac{35.4 + 37.7}{2} = 36.6 \text{ in}^2$$

From Table 2-5, k for steel is 0.11 cal/(s)(cm)(°C).

$$q = \frac{0.11 \text{ cal}}{(\text{g})(\text{cm})(°\text{C})} \times 36.6 \text{ in}^2 \times 1 \text{ hr} \times \frac{88°\text{F}}{0.0625 \text{ in}} \times \frac{2.54 \text{ cm}}{\text{in}} \times \frac{3,600 \text{ s}}{\text{h}} \times \frac{°\text{C}}{1.8°\text{F}}$$

$$q = \frac{0.11 \ (36.6)(88)(2.54)(3,600)}{0.0625 \ (1.8)} = 28,800,000 \text{ cal}$$

Table 2-3 gives the heat of vaporization of water at its boiling point as 540 cal/g.

$$\text{Amount of water vaporized} = 28,800,000 \text{ cal} \times \frac{\text{g}}{540 \text{ cal}} = 53,300 \text{ g or } \frac{53,300}{454} = 118 \text{ lb}$$

Convection and Radiation

The process of transferring heat from one part of a body to another, or directly from one body to another with which it is in contact, by the process of molecular impacts, is called *conduction.* Another way of transmitting heat is by *convection,* a process in which heat is moved from one point in a fluid to another by actual physical movement of the fluid itself.

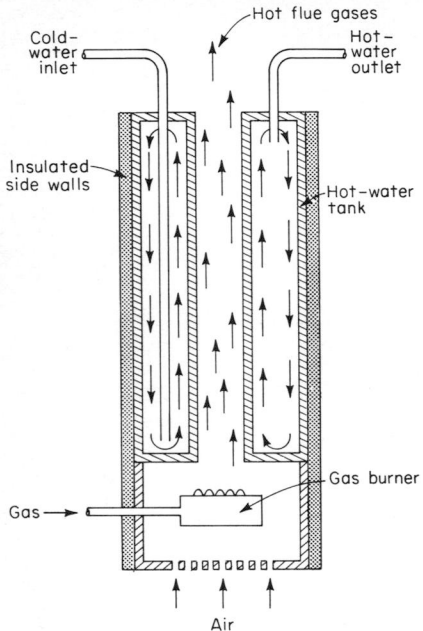

In a water heater, as illustrated in Fig. 2-4, the heat of the hot combustion gases is transferred through the tank walls by conduction and distributed by the water nearest the walls to the rest of the water by convection. Although some generalized equations describing heat transfer by convection have been worked out, they have limited applicability, and many convection problems are most readily solved by empirical methods.

A third means of transferring heat is by the radiant energy emitted from a hot body being absorbed and converted back into heat by a cooler body, a process known as *radiation.* While the laws of radiant-energy transfer are precise, their application involves detailed descriptions of the geometry of the system. The equations for several generalized systems have been developed, but these have only limited utility. As with convection, many problems yield most easily to empirical methods.

2-2 ELECTRICAL UNITS AND MEASUREMENTS

Voltage

In a manner somewhat analogous to heat conduction, when a higher concentration of free electrons exists in one part of a conducting wire than in another part, the electrons will flow from the point of high concentration to the point of low concentration. This flow, which constitutes an *electric current,* will continue until the free-electron concentrations at the two points become equal. If a steady electric current is to be set up, the difference in electron concentrations must be maintained by a source of energy capable of producing a difference in electron concentrations, such as batteries or generators. The electron concentration at a point is called its *potential,* and it is the difference between the potentials at two points, or the *electromotive force* (abbreviated emf) that results in an electric current. Potential differences are measured in *volts,* but before voltage can be defined, other concepts must be introduced.

Fig. 2-4 Heat transfer in a water heater. The walls of the water tank get hot by contact with the hot flue gases. The heat is then transferred to the inner wall of the tank, where contact with the water heats the water nearest the tank walls. The rest of the water is then heated by convection.

Current

The flow of electricity from the source of energy through its associated wires, and to the device using the energy, or load, is called the electric circuit. Since it is actually electrons that are flowing, they can flow (except for very short times) only in a closed loop, or circuit. In this respect, the analogy to water flowing in a pipe, as shown in Fig. 2-5, is useful. The main effects produced by electricity are heat, electrolysis, and magnetism, and electrical quantities can be defined and measured in terms of these associated effects. Current, for example, is defined as the rate at which electrical charges are flowing across a given cross section in a circuit.

Coulombs and Faradays

While current can be defined in many fully equivalent ways, for chemical technicians perhaps the best definition is in terms of a given quantity of electricity called a *coulomb* (C). In electrolysis, the passage of 96,500 coulombs of electricity oxidizes or reduces a precise weight of a given chemical called an *equivalent weight.* This will be discussed in detail later, but this particular quantity of electricity has special importance and is called a *faraday*. The unit of electric current, the *ampere* (A), corresponds to a flow of 1 coulomb per second. The relationship defining current is thus

$$I = \frac{Q}{t} \qquad (2\text{-}3)$$

where I is in the current in amperes, Q the quantity of electricity in coulombs, and t the time in seconds. Current is measured on an ammeter, a direct-reading device which senses the magnetic effects of the current.

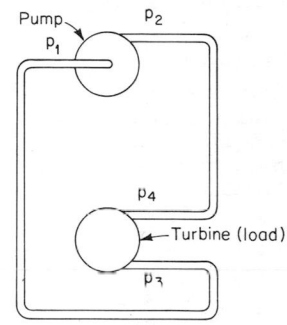

Example 2-15 How much electricity flows through a circuit carrying a current of 500 mA for 30 min?
solution

$$Q = 500 \text{ mA} \times \frac{A}{1,000 \text{ mA}} \times 30 \text{ min} \times \frac{60 \text{ s}}{\text{min}}$$

$$Q = 900 \text{ C}$$

An electric current flowing in a circuit is capable of performing work. A potential difference of 1 volt is needed to set up a current such that 1 coulomb of electricity will do a particular amount of work called a *joule* (J). (The joule is discussed later in this chapter.) The relationship defining voltage is thus

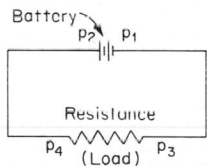

$$E = \frac{W}{Q} \qquad (2\text{-}4)$$

where E is the potential in volts, W the work done in joules, and Q the quantity of electricity in coulombs involved in producing the work. Voltage is measured on a *voltmeter*, a direct-reading device which senses the magnetic effects of a current resulting from the voltage. Other aspects of current and voltage will be considered in later chapters.

Referring to Eq. (2-4), and noting that $Q = It$ [from Eq. (2-3)], we obtain

Fig. 2-5 Analogy between electrical and hydraulic circuits. For both examples, the driving potential is $p_2\text{-}p_1$. The load causes a drop in potential of $p_3\text{-}p_4$. For the hydraulic system, the p's are expressed as pressures; for the electrical system, the p's are expressed in volts.

$$W = QE = EIt \qquad (2\text{-}5)$$

which gives the work done by a current flowing in an electric circuit. When E is in volts, I in amperes, and t in hours, the work is expressed in *watthours* (Wh).

Example 2-16 How much energy is used to operate a toaster drawing 10 A at 110 V for a period of 5 min?
solution The answer to this problem is obtained directly by using Eq. (2-5). However, it is important to use the units stipulated in the text.

$$t = 5 \text{ min} = \frac{5}{60} \text{ or } 0.08 \text{ h}$$

$$W = 110 \times 10 \times 0.08 = 92 \text{ Wh}$$

Example 2-17 How many joules are there in 1 Wh?
solution For 1 A flowing for 1 h, the quantity of electricity is

$$Q = 1 \text{ A} \times 1 \text{ h} \times \frac{3,600 \text{ s}}{\text{h}} = 3,600 \text{ C}$$

At 1 V,

$$W = 1 \times 3,600 = 3,600 \text{ J}$$

Since power is the rate at which work is done, from Eq. (2-5) we get

$$P = \frac{W}{t} = EI \qquad (2\text{-}6)$$

When E is in volts and I in amperes, the power available in an electric circuit is expressed in *watts* (W).

Example 2-18 How much power does it take to operate an electric motor that draws 5 A at 115 V?
solution $P = EI = 115 \times 5 = 575$ W

Resistance

Ohm's law is a fundamental law of flow which simply states that the flow is proportional to the driving force which gives rise to the flow. The best-known form of Ohm's law is in terms of electrical quantities. It states that the current between two points in a circuit is proportional to the voltage between the two points, i.e., $I \propto E$. In mathematical form, this becomes

$$\frac{E}{I} = R \tag{2-7}$$

in which R is the proportionality constant and is called the *resistance*. When a current of 1 A is flowing through an electrical device as a result of a potential difference across its terminals of 1 V, the device is said to have a resistance of 1 *ohm* (Ω). A means of measuring resistance is schematically shown in Fig. 2-6.

Example 2-19 What is the value of the resistance shown in Fig. 2-6 if the voltmeter reads 0.95 V when a current of 0.28 A is flowing through the circuit?
solution

$$R = \frac{E}{I} = \frac{0.95}{0.28} = 3.39 \ \Omega$$

Conductivity is defined as the reciprocal of resistance, i.e., $C = 1/R$, and is expressed in reciprocal ohms (Ω^{-1} or mho).

Example 2-20 What is the conductivity of the resistor described in Example 2-19?
solution

$$C = \frac{1}{R} = \frac{1}{3.39 \ \Omega} = 0.30 \ \Omega^{-1} \text{ or } 0.30 \text{ mho}$$

The resistance of a conductor to the passage of an electric current can be expected to be directly proportional to its length and inversely proportional to its cross-sectional area. Further, two conductors having the same physical dimensions, but made of different materials, can be expected to have different resistances. This is expressed mathematically by the equation

$$R = \rho \frac{L}{A} \tag{2-8}$$

in which the proportionality constant, the Greek letter rho (ρ), is called the *resistivity*, and is a characteristic of the substance involved.

In determining the resistance of conductors, two systems of units are in general use. In the metric system, the length is expressed in centimeters and the area in square centimeters. Resistivity, from Eq. (2-8), therefore has units of Ω-cm:

$$\rho = R \times \frac{A}{L} = \Omega \times \frac{\text{cm}^2}{\text{cm}} = \Omega\text{-cm}$$

In the English system, the length is expressed in feet and the cross-sectional area in circular mils. [One circular mil (1 cmil) is the area of a circle 0.0001 in in diameter. See Table 1-7.] The unit of resistivity in the English system is therefore

$$\rho = R \times \frac{A}{L} = \Omega \times \frac{\text{cmil}}{\text{ft}} = \frac{\Omega\text{-cmil}}{\text{ft}}$$

These unit systems are illustrated in Fig. 2-7. Table 2-6 gives the resistivities of common conductors.

Example 2-21   A constantan wire 10 cm long and having a cross-sectional area of 1 mm² is found to have a resistance of 49,000 Ω at room temperature. What is the resistivity of constantan?
solution

$$\rho = R \times \frac{A}{L} = 49{,}000\ \Omega \times \frac{1\ \text{mm}^2}{10\ \text{cm}} \times \frac{\text{cm}^2}{100\ \text{mm}^2} = 49\ \Omega\text{-cm}$$

Example 2-22   Table 2-6 gives the resistivity of aluminum as 2.83×10^{-6} Ω-cm. Express the resistivity of aluminum in units of the English system.
solution   A piece of aluminum 1 cm long and having a cross-sectional area of 1 cm² has a resistance of $R = \rho(L/A) = 2.83 \times 10^{-6}$ Ω-cm $\times$ (1 cm/1 cm²) $= 2.83 \times 10^{-6}$ Ω. The length of this aluminum in feet is

$$L = 1\ \text{cm} \times \frac{1\ \text{in}}{2.54\ \text{cm}} \times \frac{1\ \text{ft}}{12\ \text{in}} = 0.0328\ \text{ft}$$

Its area in circular mils is

$$A = 1\ \text{cm}^2 \times \frac{\text{in}^2}{(2.54)^2\ \text{cm}^2} \div \frac{\pi(0.001)^2\ \text{in}^2}{4} \times \frac{1}{\text{cmil}}$$

$$= 0.155\ \text{in}^2 \times \frac{\text{cmil}}{0.785 \times 10^{-6}\ \text{in}^2} = 0.198 \times 10^6\ \text{cmil}$$

$$\rho - 2.83 \times 10^{-6}\ \Omega \times \frac{0.198 \times 10^6\ \text{cmil}}{0.0328\ \text{ft}} = 17.1\ \frac{\Omega\text{-cmil}}{\text{ft}}$$

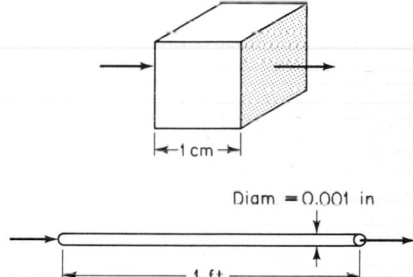

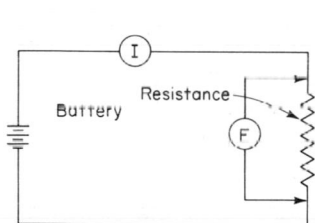

Fig. 2-6   Measuring resistance with an ammeter and voltmeter.

Fig. 2-7   Conductor dimensions for resistivity values.

Example 2-23   What is the resistance of an aluminum wire 1 mile long and ⅜ in in diameter?
solution   From Example 2-22, the resistivity of aluminum is 17.1 Ω-cmil/ft. The area of the wire in circular mils can be readily calculated.

$$A = \frac{\pi(0.375)^2}{4} \div \frac{\pi(0.001)^2}{4} = \frac{\pi(0.375)^2}{4} \times \frac{4}{\pi(0.001)^2} = \frac{0.375^2}{0.001^2} = (375)^2 = 141 \times 10^3\ \text{cmil}$$

$$R = \frac{17.1\ \Omega\text{-cmil}}{\text{ft}} \times \frac{5280\ \text{ft}}{141 \times 10^3\ \text{cmil}} = 0.64\ \Omega$$

The resistivity of a substance is affected by temperature changes, the resistance of most (but not all) materials increasing as their temperature increases. The relation between temperature and resistance may be expressed as follows:

$$R_t = R_0(1 + \alpha t) \tag{2-9}$$

where R_t is the resistance at $t°$ above (or below) a given temperature at which the conductor has a known resistance R_o. The empirical factor α (lowercase Greek letter alpha) is called the *temperature coefficient of resistance*. The temperature coefficient of resistance has units of $°\text{C}^{-1}$, and also varies with temperature, as shown in Table 2-7, which gives the temperature coefficient of resistance for common conductors at various final temperatures.

Example 2-24   What is the resistance of a constantan wire at 500°C and at 0°C if its resistance at 100°C is 25,000 Ω?   (*Example 2-24 continued on p. 2-45.*)

TABLE 2-6 Resistivities of Metals and Alloys

Material	Temp (°C)	Resistivity (Ω-cm) (×10⁻⁶)
Advance	0	48
Aluminum:		
Commercial	20	2.828
Al 99.57, Si 0.29, Fe 0.14		
Pure	−189	0.64
	−100	1.53
	0	2.63
	+100	3.86
	400	8.0
Aluminum bronze	0	12–13
Cu 97, Al 3	0	8.26
Cu 90, Al 10	0	12.6
Cu 6, Al 94	0	3.1
Antimony	20	41.7
	−190	10.5
Liquid	+860	120
Argentan, Cu 56, Ni 26	15	42
Arsenic	0	35
Bismuth	18	119.0
	100	160.2
	−200	34.8
	−100	75.6
	+100	156.5
	200	214.5
Liquid	300	128.9
	500	139.9
	700	150.8
Brass:		
Various	0	7.4
Hard-drawn Cu 70.2, Zn 29.8	0	8.2
Annealed	0	7.0
Bronze:		
Cu 88, Sn 12	20	18
Cu 89, Sn 6, Zn 4	15	13.5
Cadmium, drawn	18	7.54
	100	9.82
	−252.9	0.17
	−200	1.66
	−100	4.80
	+300	16.50
Liquid	400	33.70
	500	35.12
	700	35.78
Calcium, Ca 99.57%	20	4.6
Calido	0	110
Carboloy	20	19.6
Carbon	0	3,500
	500	2,700
	1000	2,100
	2000	1,100
	2500	900
Cesium	0	19
	−187	5.25
	27	22.2
Liquid	30	36.6
Chromium	0	2.6
Climax,	20	87
Cobalt, Co 99.8%	20	9.7
Constantan, Cu 60, Ni 40	20	49
	−200	42.4
	−150	43.0
	−100	43.5
	−50	43.9
	0	44.1
	+100	44.6
	400	44.8

TABLE 2-6   Resistivities of Metals and Alloys—(Continued)

Material	Temp (°C)	Resistivity (Ω-cm) (×10⁻⁶)
Copper, commercial:	20	1.7241
Annealed	20	1.77
Hard-drawn	20	1.692
Pure, annealed	−258.6	0.014
	−206.6	0.163
	−150	0.567
	−100	0.904
	+100	2.28
	200	2.96
	500	5.08
	1000	9.42
	1500	24.62
Liquid		
Copper-manganese:	0	4.83
Mn 0.98	0	6.66
Mn 1.49	20	17.9
Mn 4.2	20	19.7
Mn 7.4	20	50
Mn 15		
Copper-manganese-iron:	0	20
Cu 91, Mn 7.1, Fe 1.9	0	77
Cu 70.6, Mn 23.2, Fe 6.2	0	48
Copper-manganese-nickel, Cu 73, Mn 24, Ni 3	0	47
Eureka	20	92
Excello	0	53
Gallium	20	33
German silver, Ni 18%	−200	27.9
Cu 60.16, Zn 25.37, Ni 14.03, Fe 0.3, Co and Mn,	100	29.3
trace	+100	33.1
	20	2.44
Gold, pure, drawn	−252.8	0.018
	−200	0.601
	−183	0.68
99.9 pure	−150	0.997
	−100	1.400
	+100	2.97
Gold, 99.9 pure	200	3.83
	500	6.62
	1000	12.52
	1500	3.70
Gold-copper-silver:		
Au 58.3, Cu 26.5, Ag 15.2	0	13.2
Au 66.5, Cu 15.4, Ag 18.1	0	14.6
Au 7.4, Cu 78.3, Ag 14.3	0	3.6
Gold-silver:		
Au 90, Ag 10	0	6.3
Au 67, Ag 33	0	10.8
Graphite	0	800
	500	830
	1000	870
	2000	1,000
	2500	1,100
Ideal (see Constantan)		
Illinium		91.61
Indium	0	8.37
Invar (see Steel)		
Iridium	−186	1.92
	0	6.10
	100	3.30
Iron 99.98% pure	20	10
	−252.7	0.011
	−205.3	.652
	−200	2.27
	−192.5	.844

TABLE 2-6 Resistivities of Metals and Alloys—(Continued)

Material	Temp (°C)	Resistivity (Ω-cm) ($\times 10^{-6}$)
Iron 99.98% pure (continued)	−100	5.92
	+100	16.61
	200	24.50
	400	43.29
(*See also under* Steel)		
Lead	20	22
	−252.9	0.59
	−203	4.42
	−192.8	5.22
	−103	11.8
	+100	27.8
	200	38
	319	50
Liquid	333	95.0
	400	98.3
	600	107.2
	800	116.2
Cold-pressed	−183	6.02
	−78	14.1
	0	20.4
	90.4	28.0
	196.1	36.9
Lithium	−187	1.34
	0	8.55
	99.3	12.7
Lithium liquid	230	45.2
Magnesium	20	4.6
Zn free	−183	1.00
	−78	2.97
	0	4.35
	98.5	5.99
Pure	400	11.9
Manganese		5.0
Manganese-copper, Mn 30, Cu 70	0	100
Manganin, Cu 84, Mn 12, Ni 4	20	44
	22.5	45
	−200	37.8
	−100	38.5
	−50	38.7
	0	38.8
	100	38.9
	400	38.3
Mercury	20	95.783
Solid	−183.5	6.97
	−102.9	15.04
	−50.3	12.3
	−39.2	25.5
Liquid	−36.1	80.6
	0	94.07
	50	98.50
	100	103.25
	200	114.27
	350	135.5
	100	103.1
	200	114.0
	300	127.0
Molybdenum, drawn	20	5.7
Monel metal	20	42
Nichrome	20	100
Nickel	20	7.8
Pure	−182.5	1.44
	−78.2	4.31
	0	6.93
	94.9	11.1
	400	60.2
Nickel-copper, Ni 40, Cu 60	0	50

TABLE 2-6   Resistivities of Metals and Alloys—(Continued)

Material	Temp (°C)	Resistivity (Ω-cm) (×10⁻⁶)
Nickel-copper-zinc, Ni 12.84, Cu 30.59, Zn 6.57 by vol.	0	20.3
Nickelin, Ni 18.46, Cu 61.63, Zn 19.67, Fe 0.24, Co 0.19, Mn 0.18	0	33
Osmium	20	60.2
Palladium	20	11
	−183	2.78
	−78	7.17
	0	10.21
	98.5	13.79
Patent nickel, Ni 25.1, Cu 74.41, Fe 0.42, Zn 0.23, Mn 0.13, Co trace	0	34
Phosphor bronze:		
Sn 5.08, P 0.01		10.5
Sn 2	0	5.5
Platinoid, Cu 62, Ni 15, Zn 22	−160	32.5
	18	34.4
Platinum	20	10
	−203.1	2.44
	−97.5	6.87
	0	10.96
	+100	14.85
	400	26
	−265	0.10
	−253	0.15
	−233	0.54
	−150	4.18
	73	7.82
	0	11.05
	+100	14.1
	200	17.9
	400	25.4
	800	40.3
	1000	47.0
	1200	52.7
	1400	58.0
	1600	63.0
Platinum-iridium:		
P 90, Ir 10	0	24
P 80, Ir 20	0	31
Platinum-rhodium	−200	14.49
Pt 90, Rh 10	−100	18.05
	0	21.14
	+100	24.20
Platinum-silver, Pt 67, Ag 33	0	24.2
Platinite, nickel steel, Ni 46–48%	0	45
Potassium	−200	1.72
	−100	3.72
	−75	4.0
	0	6.1
	+55	8.4
Liquid	100	15.31
Rheotan, Cu 53.28, Ni 25.31, Zn 16.80, Fe 4.46, Mn 0.37	0	53
Rhodium	−186	0.7
	−78.3	3.09
	0	4.69
	+100	6.60
Rose metal, Bi 49, Pb 28, Sm 23	0	64
Rubidium	−190	2.5
	0	11.6
	+35	13.4
Liquid	40	19.6
Silicium (silicon)	20	58
Silicium bronze	0	2.4

TABLE 2-6 Resistivities of Metals and Alloys—(Continued)

Material	Temp (°C)	Resistivity (Ω-cm) ($\times 10^{-6}$)
Silver 99.98%	18	1.629
Electrolytic	−183	0.390
	−78	1.021
	0	1.468
	+98.15	2.062
	192.1	2.608
	−258.6	0.009
	−200	0.357
	−100	0.916
	0	1.506
	+100	2.15
	200	2.80
	400	3.46
	750	6.65
Liquid	1000	11.3
	1500	15.3
Sodium	−180	1.0
	−75	2.8
	0	4.3
	55	5.4
Liquid	116	10.2
	−200	0.605
	140	10.34
Sodium-amalgam, Hg 98, Na 2	0	95
Steel:		
Aluminum	20	64
Al 5, C 0.2		
Al 15, C 09	20	88
Chromium	20	60
Cr 13, C oz		
Cr 40, C 08	20	71
Invar		
35% Ni	20	81
Manganese	20	70
Nickel		
Ni 10, C 0.1	20	29
Ni 25, C 0.1	20	39
Ni 80, C 0.1	20	82
Piano wire	0	11.8
Siemens-martin	20	18
Silicon, Si 25%	20	45
Si 4%	20	62
Tempered glass, hard		45.7
Tempered yellow		27
Tempered blue		20.5
Tempered soft		15.9
Titanium Ti 2.5, C 0.15	20	16
Tungsten:		
W 5, C 0.2	20	20
W 20, C 0.2	20	24
Vanadium, V 5, C 1.1	20	121
Strontium	20	24.8
Tantalium	20	15.5
Tellurium	19.6	200,000
Thallium, pure	−183	4.08
	−78	11.8
	0	17.60
	+98.5	24.7
Therlo	20	47
Thorium	15	40.1
	20	18
Tin	20	11.5
	−184	3.40
	−78	8.8
	0	13.0

TABLE 2-6   Resistivities of Metals and Alloys—(Continued)

Material	Temp (°C)	Resistivity (Ω-cm) (×10⁻⁶)
Tin (continued)	+91.45	18.2
	200	20.30
	225	22.00
Liquid	235	47.60
	750	61.22
Tin-bismuth:		
Sn 90.5, Bi 9.5	12	16
Sn 2, Bi 98	0	244
Tin-lead:		
Sn 90, Pb 10	15	13.5
Sn 33.3, Pb 66.7	15	16
Titanium		3.2
Tungsten	20	5.51
	727	25.3
	1227	41.4
	1727	59.4
	2727	98.9
	3237	118
Wood's metal, Bi 56, Pb 14, Sn 14	0	52
Zinc	−183	1.62
	−78	3.34
	0	5.75
	+92.5	8.00
	181.5	10.37
Liquid	440	37.2
	100	7.95
	300	13.25
	415	17.00
	427	37.30
	500	36.60
	600	35.90
	700	35.60
	800	36.60
	850	35.74
Alloy 193	0	91.5
Alumel	0	33.3
Chromel	0	90
Copel	0	49.5
Dowmetal	0	15
Duralumin	0	3.35
Nichrome II	0	110
Nichrome III	0	93.5
Nichrome IV	0	101

From C. D. Hodgman (ed.), "Handbook of Chemistry and Physics," The Chemical Rubber Co., Cleveland, 1942.

solution From Table 2-7, the temperature coefficient of resistance of constantan at a final temperature of 500°C is 0.000027/°C.

$$t = 500 - 100 = 400°C$$

$$R_{500} = 25,000\ \Omega\ \left(1 + \frac{0.000027}{\cancel{C}} \times 400°\cancel{C}\right) = 25,000\ \Omega\ (1 + 0.0108)$$

$$= 25,000\ (1.0108)\ \Omega = 25,270\ \Omega$$

For 0°C, t is negative: $t = 0 - 100 = -100°C$. Table 2-7 gives α as 0.000008 at 12°C, which is close enough to 0°C, since α does not change very much for small changes in temperature.

$$R_{0°} = 25,000\ \Omega\ \left(1 - \frac{0.000008}{°C} \times 100°C\right) = 25,000\ (0.999992) = 24,980\ \Omega$$

The use of the temperature coefficients of resistance in this manner may be somewhat inconvenient, and an alternate procedure is frequently used based on the equation:

TABLE 2-7   Temperature Coefficient of Resistance of Metals

<table>
<tr><td colspan="4" align="center">ELEMENTS</td></tr>
<tr><td>Material</td><td>α^*</td><td>Material</td><td>α</td></tr>
<tr><td>Aluminum</td><td>0.0044</td><td>Neptunium</td><td>0.0010</td></tr>
<tr><td>Antimony</td><td>0.0045</td><td>Nickel</td><td>0.0062</td></tr>
<tr><td>Arsenic</td><td>0.0052</td><td>Niobium</td><td>0.0034</td></tr>
<tr><td>Barium</td><td>0.0038</td><td>Osmium</td><td>0.0042</td></tr>
<tr><td>Beryllium</td><td>0.0091</td><td>Palladium</td><td>0.0040</td></tr>
<tr><td>Bismuth</td><td>0.0048</td><td>Platinum</td><td>0.0039</td></tr>
<tr><td>Cadmium</td><td>0.0036</td><td>Plutonium</td><td>−0.0003</td></tr>
<tr><td>Calcium</td><td>0.0040</td><td>Polonium</td><td>0.0043</td></tr>
<tr><td>Cerium</td><td>0.0012</td><td>Potassium</td><td>0.0052</td></tr>
<tr><td>Cesium</td><td>0.0049</td><td>Praseodymium</td><td>0.0021</td></tr>
<tr><td>Chromium</td><td>0.0030</td><td>Rhenium</td><td>0.0046</td></tr>
<tr><td>Cobalt</td><td>0.0057</td><td>Rhodium</td><td>0.0044</td></tr>
<tr><td>Copper</td><td>0.0044</td><td>Rubidium</td><td>0.0051</td></tr>
<tr><td>Dysprosium</td><td>0.0018</td><td>Ruthenium</td><td>0.0046</td></tr>
<tr><td>Erbium</td><td>0.0024</td><td>Samarium</td><td>0.0019</td></tr>
<tr><td>Europium</td><td>0.0016</td><td>Scandium</td><td>0.0041</td></tr>
<tr><td>Gadolinium</td><td>0.0024</td><td>Silver</td><td>0.0043</td></tr>
<tr><td>Gallium</td><td>0.0040</td><td>Sodium</td><td>0.0049</td></tr>
<tr><td>Gold</td><td>0.0043</td><td>Strontium</td><td>0.0039</td></tr>
<tr><td>Hafnium</td><td>0.0042</td><td>Tantalum</td><td>0.0038</td></tr>
<tr><td>Holmium</td><td>0.0020</td><td>Terbium</td><td>0.0008</td></tr>
<tr><td>Indium</td><td>0.0043</td><td>Thallium</td><td>0.0042</td></tr>
<tr><td>Iridium</td><td>0.0041</td><td>Thorium</td><td>0.0032</td></tr>
<tr><td>Iron</td><td>0.0057</td><td>Thulium</td><td>0.0031</td></tr>
<tr><td>Lanthanum</td><td>0.0024</td><td>Tin</td><td>0.0041</td></tr>
<tr><td>Lead</td><td>0.0040</td><td>Titanium</td><td>0.0048</td></tr>
<tr><td>Lithium</td><td>0.0044</td><td>Tungsten</td><td>0.0048</td></tr>
<tr><td>Luticium</td><td>0.0037</td><td>Uranium</td><td>0.0030</td></tr>
<tr><td>Magnesium</td><td>0.0042</td><td>Vanadium</td><td>0.0040</td></tr>
<tr><td>Manganese</td><td>0.0003</td><td>Ytterbium</td><td>0.0016</td></tr>
<tr><td>Mercury (l)</td><td>0.0009</td><td>Yttrium</td><td>0.0041</td></tr>
<tr><td>Molybdenum</td><td>0.0046</td><td>Zinc</td><td>0.0039</td></tr>
<tr><td>Neodymium</td><td>0.0020</td><td>Zirconium</td><td>0.0045</td></tr>
<tr><td colspan="4" align="center">ALLOYS</td></tr>
<tr><td>90 Pt/10 Rh</td><td>0.0017</td><td>Chromel</td><td>0.0004</td></tr>
<tr><td>87 Pt/13 Rh</td><td>0.0016</td><td>Alumel</td><td>0.0024</td></tr>
<tr><td>80 Ni/20 Cr</td><td>0.0002</td><td>Constantan</td><td>−0.0001</td></tr>
<tr><td>60 Ni/24 Fe/16 Cr</td><td>0.0003</td><td>Manganin</td><td>0.0001</td></tr>
<tr><td>50 Fe/30 Ni/20 Cr</td><td>0.0004</td><td></td><td></td></tr>
</table>

$^*R_t = R_0 (1 + \alpha t)$.
R_t = resistance at temperature $t°C$ (ohms)
R_0 = resistance at 0°C (ohms)
α = coefficient of electrical resistance (1/°C)

From C. D. Hodgman (ed.), "Handbook of Chemistry and Physics," The Chemical Rubber Co., Cleveland, 1942.

$$\frac{R_2}{R_1} = \frac{T + t_2}{T + t_1} \tag{2-10}$$

Table 2-8 gives the T values in the above equation for a number of common conductors.

Example 2-25   What is the resistance of an iron wire at 550°C if its resistance at 20°C is 750 Ω?
solution   From Table 2-8, the T value for iron is 162.

$$\frac{R_2}{R_1} = \frac{T + t_2}{T + t_1}$$

$$\frac{R_2}{750} = \frac{162 + 550}{162 + 20} = \frac{712}{182} = 3.91$$

$$R_2 = 3.91 \times 750 = 2,930 \text{ }\Omega$$

It should be particularly noted that the T values given in Table 2-8 are for t temperatures in degrees Celsius only.

In chemical work, the fact that the resistance of a substance varies with temperature frequently affords a convenient and accurate means of measuring temperatures. This is known as *resistance thermometry* and involves calibrating the resistance of a wire at various temperatures, and thereafter measuring its resistance. Metals such as nickel and iron are frequently used for this purpose because of their high temperature coefficients of resistance. The method is particularly useful for remote-reading and continuous-recording applications, or where space limitations or other physical conditions would make the use of ordinary thermometers difficult.

Example 2-26 An iron wire having a resistance of 750 Ω at 68°F is placed in a furnace. After it has come to thermal equilibrium with the furnace, its resistance is found to be 1,850 Ω. What is the temperature of the furnace?
solution The temperatures must first be expressed in Celsius degrees to be compatible with the T values given in Table 2-8. From Table 2-1, 68°F = 20°C.

$$\frac{1850}{750} = \frac{162 + t}{162 + 20}$$

$$2.47 = \frac{162 + t}{182}$$

$$448 = 162 + t$$

$$t = 286°C$$

Interpolating between 285 and 290°C in Table 2-1, 286°C is equivalent to 457°F.

TABLE 2-8 T Values for Common Conductors

Material	T^*
Aluminum	236
Brass	480
Copper	235
Iron, commercial	162
Lead	236
Mercury	1,100
Nichrome	2,250
Silver	243
Steel, hard	600
Tin	220
Tungsten	176
Zinc	230

* For use in $\dfrac{R_2}{R_1} = \dfrac{T + t_2}{T + t_1}$

Thermoelectricity

Another electrical method of measuring temperatures is based on the fact that an electrical potential is set up between the junctions of two dissimilar wires when the junctions are at different temperatures. The device used, called a *thermocouple*, is of considerable importance in chemical work. The potential of a thermocouple depends on the composition of the dissimilar wires, and the temperature difference between the two junctions. For accurate work, one junction is immersed in ice water (i.e., to maintain it at 0°C). As the other junction is heated, the emf between the two junctions may rise to a maximum, called the *neutral point*, and then start falling, frequently reversing polarity at the *inversion* point. This is illustrated in Fig. 2-8, which shows the thermo-emf diagram for the iron-copper couple. The slope of the thermo-emf diagram at any point is called the *thermoelectric power* of the couple. Strictly speaking, this is not a power at all, but merely indicates the ability of the couple to be used as a thermocouple. A high slope (either positive or negative), or high thermoelectric power, indicates that a relatively small junction temperature difference will produce a large thermo-emf. Figure 2-9 gives the thermoelectric power of various metals against lead. This figure can be used for any combination of metals by comparing them individually with lead. Figure 2-10 illustrates the principle involved in such a calculation.

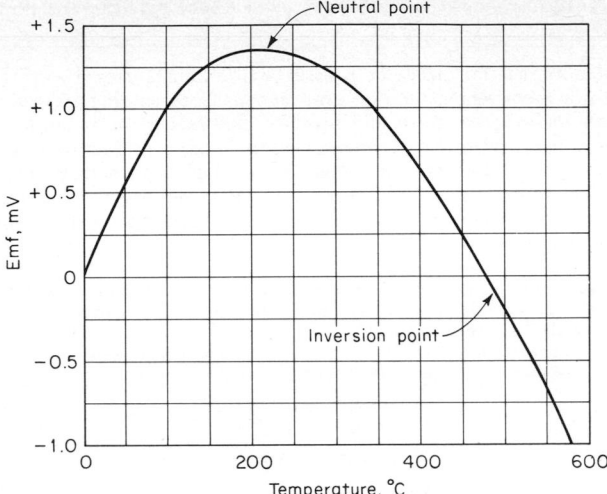

Fig. 2-8 Thermo-emf curve of iron-copper couple.

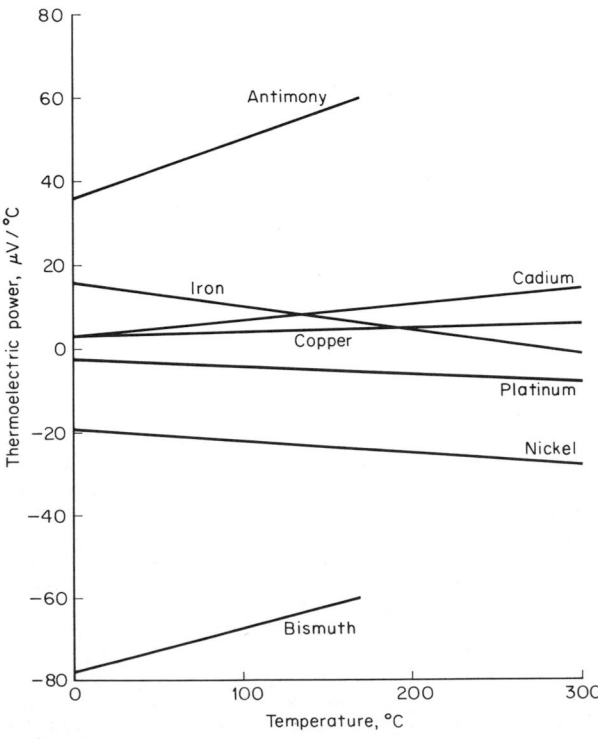

Fig. 2-9 Thermoelectric power of various metals against lead.

Example 2-27 What is the emf generated by an iron-nickel couple if the hot junction is at 250°C and the cold junction is at room temperature (20°C)?

solution From Fig. 2-9 the thermoelectric power of the iron-lead couple at 250°C is 1.6 μV/°C, meaning that the emf of this couple (with its cold junction at 0°C) is $1.6 \times 250 = 400$ μV. The thermoelectric power of this couple at 20°C is 15 μV/°C, meaning that its emf is $15 \times 20 \times 300$ μV. Thus the net emf for junctions of 20 and 250°C is $400 - 300$ or 100 μV.

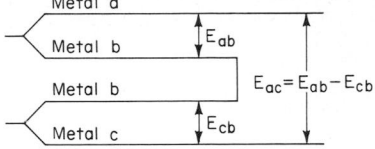

Similarly, for the nickel-lead couple that has thermoelectric powers of -21.5 and -19.5 at 250 and 20°C, respectively:

emf (0 to 250°) $= -21.5 (250) = -538$ μV
emf (0 to 20°) $= -19.5 (20) = -39$ μV
emf (20 to 250°) $= -538 + 39 = -499$ μV

The emf of the iron-nickel couple with junctions at 20 and 250°C is therefore

$$\text{emf} = 100 - (-499) = 599 \ \mu V$$

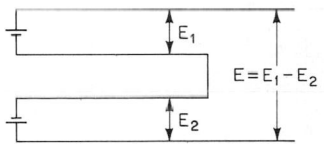

Fig. 2-10 Use of Fig. 2-9 to calculate the thermoelectric power of any couple. The thermoelectric power of any couple is the difference of the thermoelectric powers of each of the metals against a common third metal.

2-3 ENERGY AND POWER

Force and Pressure

Scientific terms, such as force, pressure, power, and work, which have found their way into common usage, are frequently used with only casual attention to their precise technical meanings. This tends to lead to confusion, and the chemical technician is well advised to fix the precise use of these terms clearly in mind. *Force* can be described as a push or a pull acting on a body. As simple examples, a man pushing a wheelbarrow or a boy pulling a sled exerts a force on these objects, and there is some mechanical contact between the person and the object. Mechanical contact, however, is not a necessary condition, since forces can be exerted on bodies by gravity, magnetic fields, or electrostatic fields. A force can influence a body to alter its motion; that is, it can cause a body at rest to start moving or, if moving, to stop. Any change in the speed or direction of a body must be caused by the action of a force. *Friction*, for example, refers to a force which resists the motion of one body past or through another. To overcome inertia, the tendency of a body to move at a constant speed in a given direction, requires the application of a force to the body.

It is important to realize that when the application of force does not result in a change in the motion of a body, other forces must be present to counteract the applied force. Thus, if the sled the boy is pulling moves at a constant velocity (i.e., constant speed and constant direction), the force he is exerting is exactly counterbalanced by the frictional force between the runners and the snow.

As described in Chap. 1, there is also confusion in the units used to measure forces. However, as pointed out, since the weight of a body is the force of gravity acting upon the body, forces can be measured in the same units used for weight, i.e., kilograms and grams, etc., in the metric system; pounds and ounces, etc., in the English system. While these are the units in general use in scientific work, the so-called "absolute units" are also in common usage. These are based on the fundamental law of motion:

$$F = ma = \frac{w}{g}a \qquad (2\text{-}11)$$

where F is the force on a body of mass m experiencing an acceleration a, and w is its weight. In the metric system, the force necessary to give a mass of 1 g an acceleration of 1 cm/s² is

$$F = 1 \text{ g} \times \frac{1 \text{ cm}}{\text{s}^2} = 1 \frac{\text{g} - \text{cm}}{\text{s}^2}$$

For convenience, this unit of force is called a *dyne* (dyn). Similarly, in the English system, the force necessary to give a mass of 1 lb an acceleration of 1 ft/s² is given by

$$F = 1 \text{ lb} \times \frac{1 \text{ ft}}{s^2} = 1 \frac{\text{lb} - \text{ft}}{s^2}$$

Again for convenience, this unit of force is called a *poundal* (pdl).

Example 2-28   A body having a mass of 2 kg is accelerating at 4 cm/s². What force is it experiencing?

solution

$$F = ma = 2 \cancel{kg} \times \frac{1,000 \text{ g}}{\cancel{kg}} \times \frac{4 \text{ cm}}{s^2} = \frac{8,000 \text{ g} - \text{cm}}{s^2} = 8,000 \text{ dyn}$$

Example 2-29   What acceleration will a force of 3 lb produce in a body weighing 12 lb?

solution $F = ma = \dfrac{w}{g} a = 12 \text{ lb} \times \dfrac{s^2}{32 \text{ ft}} \times a$

$$3 \text{ lb} = 12 \text{ lb} \times \frac{s^2}{32 \text{ ft}} \times a$$

$$a = \frac{32 \text{ ft}}{s^2} \times 3 \cancel{lb} \times \frac{1}{12 \cancel{lb}} = 8 \text{ ft/s}^2$$

Note that (unlike Example 2-28) both force and weight in this problem are expressed in the same units—pounds. Strictly speaking, this is incorrect, but it reflects common usage. Recalling (from Chap. 1) that weight is the force of gravity on a mass, the weight of 12 lb is converted to a "mass" by dividing by the acceleration of gravity *g*.

Example 2-30   What force is needed to accelerate an automobile weighing 3,000 lb at 2 mi/(h)(s)?
solution   In this particular problem, there is some ambiguity as to whether the 3,000 lb is a weight or a mass. Considering it as a mass:

$$F = ma = 3,000 \text{ lb} \times \frac{2 \cancel{mi}}{(\cancel{h})(s)} \times \frac{5,280 \text{ ft}}{\cancel{mi}} \times \frac{\cancel{h}}{3,600 \text{ s}}$$

$$= 8,790 \frac{\text{lb} - \text{ft}}{s^2} = 8,790 \text{ pdl}$$

Considering the 3,000 lb as a weight,

$$F = \frac{w}{g} a = \frac{3,000 \text{ lb}}{32 \cancel{ft}} \times \cancel{s^2} \times \frac{2 \cancel{mi}}{(\cancel{h})(\cancel{s})} \times \frac{5,280 \cancel{ft}}{\cancel{mi}} \times \frac{h}{3,600 \cancel{s}}$$

$$= 27.5 \text{ lb}$$

The former solution would reflect scientific usage, the latter common and engineering usage. Incidentally, a force of 27.5 lb hardly seems adequate to accelerate a 3,000-lb automobile 2 mi/(h)(s) or 2.94 ft/s². However, it must be recalled that this does not cover the force necessary to overcome friction.

Force divided by the area on which it acts, i.e., force per unit area, is called *pressure*. Generally speaking, in mechanical systems, forces are applied at a point on a body, or over a very small part of the body, such as a man pushing on a wall, in which case pressure does not have any great usefulness. Where a mechanical force acts fairly uniformly across a known area, as in the case of a steel bar under compression or tension, the force per unit area is referred to as the *stress*. The term pressure is usually used in connection with fluid systems, i.e., when the force is exerted by or on a liquid or gas. Fluid pressure is exerted in all directions on the enclosing vessel, and for a closed system at rest is exerted equally, except for the weight of the fluid, as shown in Fig. 2-11. In chemical work, the effect of the weight of the fluid can be very important. The pressure due to the weight of a fluid depends only on the height of the column of fluid and its density. In mathematical terms, this fluid pressure is given by the equation

$$P = \rho h \tag{2-12}$$

where *P* is the pressure exerted by a column of fluid of density ρ and height *h*.

In terms of units, pressure can be expressed in any force and area terms, i.e., lb/in², g/cm², dyn/cm², pdl/ft², etc.

Example 2-31 What is the pressure at the bottom of a 125-ft water tower when filled to capacity?
solution From Table 1-12, the density of water is 62.4 lb/ft³.

$$P = \rho h = \frac{62.4 \text{ lb}}{\text{ft}^3} \times 125 \text{ ft} \times \frac{\text{ft}^2}{144 \text{ in}^2} = 54.1 \frac{\text{lb}}{\text{in}^2}$$

Example 2-32 What is the stress on a wire 0.01 in in diameter supporting a weight of 10 lb?
solution

$$\text{Diameter of wire} = \frac{\pi (0.01)^2 \text{ in}^2}{4} = 0.0000785 \text{ in}^2$$

$$\text{Stress} = \frac{10 \text{ lb}}{0.0000785 \text{ in}^2} = 1{,}270{,}000 \text{ lb/in}^2$$

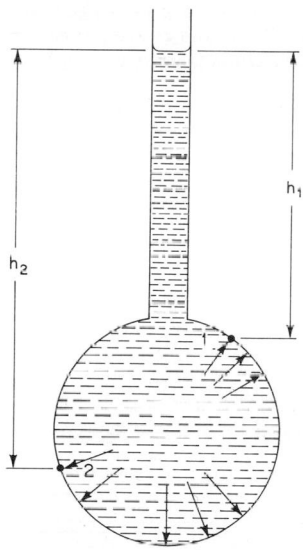

Fig. 2-11 Fluid pressure is exerted in all directions. The pressure depends only on the height and density of the fluid. The pressure is greater at 2 than at 1.

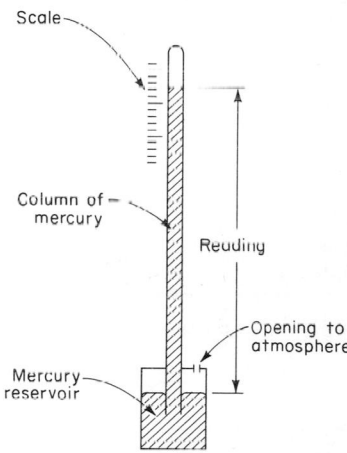

Fig. 2-12 Measurement of atmospheric pressure by means of a barometer.

Because of its general importance, special pressure units have been developed, usually to provide convenience within special areas of use. Under general average conditions at sea level and 32°F, i.e., "standard" conditions, the pressure of the atmosphere is 14.7 lb/in². This pressure is said to be "1 atmosphere" (atm), and gas pressures are sometimes reported in atmospheres. Thus, a pressure of 10 atm corresponds to 147 lb/in².

The fundamental means of measuring atmospheric pressures is with a *barometer*. This device provides means for measuring the height of a column of liquid supported by atmospheric pressure, as illustrated in Fig. 2-12. Under standard conditions, for mercury this turns out to be 760 mm (or 29.9 in). Pressures of fluids can also be expressed, therefore, as the equivalent column of a given liquid. Thus, a pressure of 1 atm can be expressed as 760 mm of Hg or, if the column of liquid is water, as 33.9 ft of water.

Example 2-33 What pressure corresponds to 723 mm of mercury?
solution $P = \dfrac{723}{760} \times 14.7 \dfrac{\text{lb}}{\text{in}^2} = 14.0 \dfrac{\text{lb}}{\text{in}^2}$

In recent years the *millibar* (mbar) has come into general use in connection with low-pressure and meteorological work. The bar, or 1,000 mbar, is the atmospheric pressure at sea level and 20°C. (Note that because of the temperature difference, the bar is slightly smaller than 1 standard atmosphere; i.e., 1 bar = 0.987 atm).

Example 2-34   What pressure in millibars corresponds to 723 mm of Hg?
solution

$$P = \frac{723}{760} \text{ atm} \times \frac{1 \text{ bar}}{0.987 \text{ atm}} \times \frac{1{,}000 \text{ mbar}}{1 \text{ bar}} = 964 \text{ mbar}$$

Three general types of devices are used to measure pressures; bourdon tube, aneroid, and manometer. A bourdon tube is a curved, flattened tube shaped so that when pressure is applied to a fluid inside the tube, the force on the outer section is greater than that on the inner section, since the area of the latter is smaller. This force difference causes the tube to straighten out from its rest position, and this movement, magnified by mechanical linkages, is transmitted to a pointer reading on a calibrated scale, as shown in Fig. 2-13. In general, bourdon gages are used to measure pressures (or vacuums) of 5 lb/in² and greater, since the instrument tends to stick and is somewhat inaccurate for smaller readings. For very high pressures, bourdon gages are used almost exclusively.

For more accurate work in the low (5 to 50 lb/in²) pressure range, an aneroid gage is generally used. This device, illustrated in Fig. 2-14, utilizes a closed, elastic vessel capable of a physical distortion as a result of a change in the external (or internal) pressure. This physical distortion, as in the case of a bourdon gage, is magnified by mechanical linkages as it is transmitted to a pointer reading on a calibrated scale.

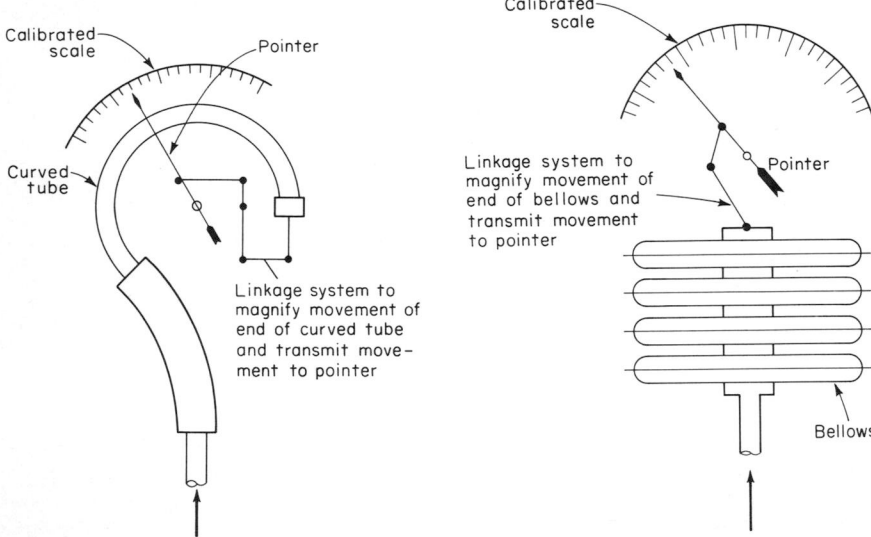

Fig. 2-13   Bourdon gage.   **Fig. 2-14**   Aneroid gage.

For extremely accurate work, particularly at very low to moderate pressures, manometers are used. There are a large variety of such devices, all of which are extensively used in chemical work. Basically, manometers balance the unknown pressure against a known pressure and a column of liquid in a suitably connected tube. Common forms of manometers are shown in Fig. 2-15.

Example 2-35   In the simple manometer shown in Fig. 2-15, what is the pressure differential if $h = 12$ cm and the manometer liquid is carbon tetrachloride?
solution   From Table 1-12, the density of carbon tetrachloride is 1.60 g/cm³.

$$p_2 = p_1 + \rho h$$

$$p_2 - p_1 = \rho h = \frac{1.60 \text{ g}}{(\text{cm})(\text{cm})(\text{cm})} \times 12 \text{ cm} = 19.2 \text{ g/cm}^2$$

Example 2-36 For the absolute manometer shown in Fig. 2-15, what is the pressure if $h = 12$ cm and the manometer liquid is mercury?

solution From Table 1-12, the density of mercury is 13.6 g/cm³.

$$p_2 = p_1 + \rho h$$

but since $p_1 = 0, p_2 = \rho h = \dfrac{13.6\text{ g}}{(\cancel{cm})(cm)(cm)} \times 12\ \cancel{cm} = 163\ \dfrac{g}{cm^2}\ \text{abs}$

Example 2-37 In the inclined-tube manometer shown in Fig. 2-15, what is the pressure if $L = 12$ cm, $\theta = 30°$, and the manometer liquid is a red-tinted oil having a density of 0.88 g/cm³? (sin 30° − 0.50)

solution $h = L \sin \theta = 12$ cm $\times 0.50 = 6$ cm

$$p_2 - p_1 = \rho h = \dfrac{0.88\text{ g}}{(\cancel{cm})(cm)(cm)} \times 6\ \cancel{cm} = 5.28\ g/cm^2$$

The pressure-measuring devices shown in Figs. 2-13 to 2-15 read pressures relative to atmospheric pressure, but as we have seen, the atmospheric pressure itself has an actual value of 14.7 lb/in² or 760 mm of mercury. Thus, to get the *absolute pressure*, the atmospheric pressure must be added to the gage reading. Pressure readings should be designated as absolute or gage, as for example, pounds per square inch absolute or pounds per square

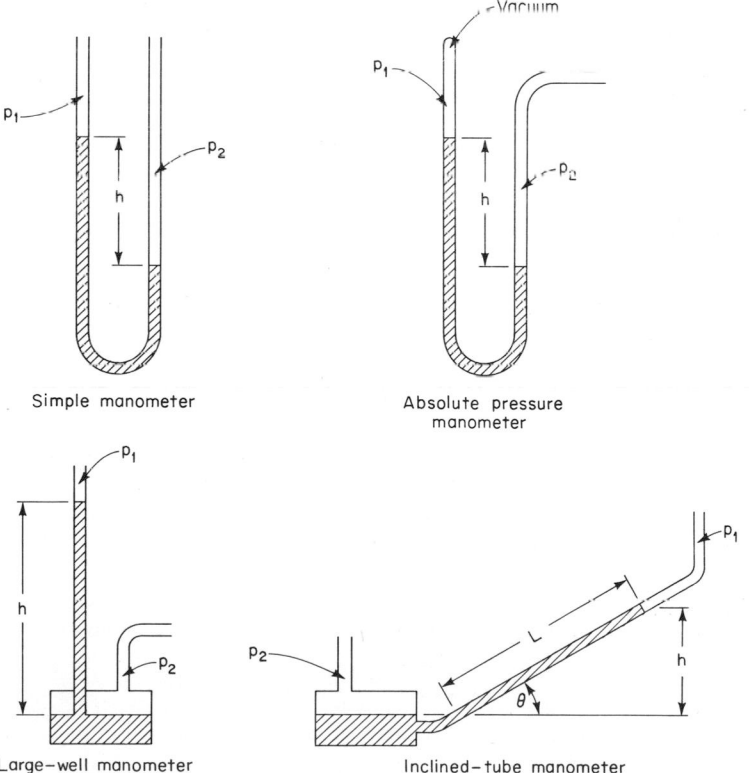

Simple manometer

Absolute pressure manometer

Large-well manometer

Inclined-tube manometer

Fig. 2-15 Types of manometers. The pressure reading is always the difference in heights of the liquid in each leg multiplied by the density of the liquid. The absolute-pressure manometer reads directly in absolute pressure since the pressure on the liquid in the vacuum leg is zero. The advantage of the large-well manometer is that the liquid level in one leg (i.e., the large well) does not change very much, even with large changes in the height of liquid in the tube. The inclined-tube manometer magnifies the height of liquid in the tube with a large movement in the L direction for a small movement in the h direction. $h = L \sin \theta$.

inch gage. Usually, however, since we frequently deal in pressure differentials, it is more convenient to work in gage pressures, and most pressure information, unless otherwise stated, refers to gage pressures.

Work

In scientific terms, work occurs when a force causes a displacement of an object *in the direction of the force*. To illustrate this, a boy pushing a box across a floor or lifting it from the floor to a table performs work on the box. The amount of work done is the product of the force and the displacement while the force is being applied, or mathematically:

$$W = Fd \tag{2-13}$$

where W is the work performed by application of force F for a distance d. In the metric system, the unit of work is the erg, which is the work done by a force of 1 dyn exerted on a body while it is displaced 1 cm. In the English system, where force is generally expressed in pounds, the common work unit is the foot-pound (ft-lb).

A frequently used unit of work is the joule. A joule is the work done by a force of 1 newton (N) operating through a displacement of 1 meter. [A newton is a metric unit of force. A force of 1 N will cause a 1-kg mass to accelerate 1 m/s²; i.e., $F = 1 \text{ kg} \times 1 \text{ m/s}^2$.]

Example 2-38 How much work is done on a body weighing 8 lb when it is lifted to a height of 12 ft?
solution To lift a body weighing 8 lb requires a force of 8 lb (to overcome gravity). Therefore,

$$W = Fd = 8 \text{ lb} \times 12 \text{ ft} = 96 \text{ ft-lb}$$

TABLE 2-9   Work-Energy Equivalents

1 Btu	= 0.293 Wh
1 Btu	= 778 ft-lb
1 erg	= 2.39×10^{-8} cal
1 ft-lb	= 3.77×10^{-4} Wh
1 ft-lb	= 0.324 cal
1 cal	= 0.00116 Wh
1 cal	= 4.19 J
1 hp-h	= 746 Wh
1 hp-h	= 2,545 Btu
1 J	= 2.78×10^{-4} Wh
1 J	= 1 Ws
1 Wh	= 3.41 Btu

A body possesses energy if it is capable of doing work. When such a body does work, its energy is decreased by an amount exactly equivalent to the work done. There are many forms of energy, such as an extended spring (mechanical), a raised weight (gravitational), fuel (chemical), a hot body (thermal), or a charged capacitor (electrical). All forms of energy are convertible from one form to another through the use of suitable equipment, and the appropriate equivalents are given in Table 2-9.

Example 2-39 If the weight in Example 2-38 were to be lifted by an electric motor at an efficiency of 80 percent (as a result of motor loses, friction, etc.), how much electrical energy would be needed?
solution

$$\text{Total work} = \frac{96 \text{ ft-lb}}{0.80} = 120 \text{ ft-lb}$$

As derived from Table 2-9, 1 Wh = 2.66×10^3 ft-lb:

$$\text{Equivalent electrical energy} = 120 \text{ ft-lb} \times \frac{\text{Wh}}{2.66 \times 10^3 \text{ ft-lb}}$$

$$= 0.045 \text{ Wh}$$

Power

When work has to be done, both the amount of work and the time in which it is to be done are important. As illustrated in Example 2-38, if a body is lifted against gravity, the work done on the body is dependent only on its weight and the height to which it is raised. However, if the body is to be lifted quickly, a more powerful motor, and stronger ropes, pulleys, etc., are needed than if the object were lifted slowly. The rate at which work is done is called power, or mathematically:

$$P = \frac{W}{t} \tag{2-14}$$

where P is the power required to do work W in time t.

In the metric system, power is usually expressed in ergs per second, and in the English system in foot-pounds per second. In addition to these units, many special units of power have been developed for convenience in particular areas of application, and since work or energy can appear in many equivalent forms, there are many forms in which power equivalents can appear. These are summarized in Table 2-10.

Example 2-40 Calculate the power required by a pump that discharges 125 gal/min of oil (of specific gravity 0.85) into a tank 50 ft above the intake.

solution From Table 1-12, the density of water is 62.4 lb/ft³. The oil, therefore, has a density of 0.85 (62.4) or 53 lb/ft³. As derived from Table 1-5, 1 ft³ = 7.48 gal. Thus,

$$\frac{125 \text{ gal}}{\text{min}} \times \frac{\text{ft}^3}{7.48 \text{ gal}} \times \frac{53 \text{ lb}}{\text{ft}^3} = 886 \frac{\text{lb}}{\text{min}}$$

$$P = \frac{W}{t} = \frac{886 \text{ lb}}{\text{min}} \times 50 \text{ ft} = 44,300 \text{ ft-lb/min}$$

TABLE 2-10 Power Equivalents

1 Btu/min	= 0.0236 hp
1 Btu/min	= 17.58 W
1 erg/s	= 1 × 10⁻⁷ W
1 erg/s	= 5.69 × 10⁻⁶ Btu
1 ft-lb/min	= 0.0226 W
1 hp	= 0.707 Btu/s
1 hp	= 746 W
1 hp	= 550 ft-lb/s
1 W	= 0.00134 hp
1 W	= 0.0569 Btu/min
1 W	= 1 J/s

TABLE 2-11 Mechanical Equivalent of Heat

1 cal	= 4.186 × 10⁷ erg
1 cal	= 4.186 J
1 Btu	= 778 ft-lb
1 J	= 0.239 cal

From Table 2-10, 1 hp = 33,000 ft-lb/min

$$P = \frac{44,300 \text{ ft-lb}}{\text{min}} \times \frac{\text{hp}}{33,000 \text{ ft-lb}} = 1.34 \text{ hp}$$

Example 2-41 At an overall efficiency of 45 percent, what power rating is required of the motor that operates the pump in Example 2-40?

solution Total power = $\dfrac{1.34}{0.45}$ = 2.98 hp

From Table 2-10, 1 hp = 746 W

Power required = $2.98 \text{ hp} \times \dfrac{746 \text{ W}}{\text{hp}}$ = 2,220 W·

Thermal Equivalents

In chemical technology, the relation between heat and work is particularly important. The numerical relation between heat and mechanical energy is determined by performing a known amount of mechanical work (as by descending weights) on paddles immersed in water in an insulated vessel. The mechanical energy so generated is transformed into heat by the friction of the paddles in the water. Measurements show that the heat developed in the water is directly proportional to the mechanical work supplied. The proportionality constant, known as the *mechanical equivalent of heat*, has the values given in Table 2-11.

Example 2-42   A motor supplies 0.4 hp to a paddle immersed in 1.2 gal of water. Assuming that all the work goes into heat, how much will the temperature of the water rise in 18 min?
solution   From Table 2-11, 1 Btu = 778 ft-lb, and from Table 2-10, 1 hp = 33,000 ft-lb/min. Thus,

$$\frac{1 \text{ Btu}}{\text{min}} = \frac{778 \text{ ft-lb}}{\text{min}} \times \frac{\text{hp-min}}{33,000 \text{ ft-lb}} = 0.24 \text{ hp}$$

or 1 Btu = 0.24 hp-min.

In 18 min, the energy supplied by the motor is 0.4 (18) or 7.2 hp-min.

$$7.2 \text{ hp-min} \times \frac{\text{Btu}}{0.24 \text{ hp-min}} = 30 \text{ Btu}$$

From Tables 1-12 and 1-5,

$$1.2 \text{ gal} \times \frac{\text{ft}^3}{7.48 \text{ gal}} \times \frac{62.4 \text{ lb}}{\text{ft}^3} = 10.0 \text{ lb}$$

$$\text{Temperature rise} = 30 \text{ Btu} \times \frac{\text{lb-F}}{\text{Btu}} \times \frac{1}{10.0 \text{ lb}} = 3°\text{F}$$

Electrical Equivalents

When an electrical charge circulates around an electric circuit, it does work, as indicated by the evolution of heat, the generation of mechanical work by a motor, or the evolution of chemical materials in an electrochemical cell. The electrical energy associated with the flow of electricity is given by the product of the charge and the electromotive force which causes the flow, or mathematically:

$$J = QE \tag{2-15}$$

where J is the work done by a charge Q flowing between points having a voltage difference E. Since, as we have seen, $Q = It$,

$$J = EIt \tag{2-16}$$

In Eq. (2-16), the work J is given in joules when E is in volts, I in amperes, and t in seconds. Since power is the rate of doing work, in electrical terms,

$$P = \frac{J}{t} = \frac{EIt}{t} = EI \tag{2-17}$$

When E is in volts and I in amperes, P has units of joules per second, which, because of its importance, is given a special unit called the *watt* (W).

Example 2-43   A potential of 110 V causes a current of 5.0 A to flow in a resistor. What power is being absorbed by the resistor?
solution $P = EI = 110 \text{ V} \times 5.0 \text{ A} = 550 \text{ W}$

We can now backtrack to Eq. (2-16) and note that

$$Wt = EIt = Pt \tag{2-18}$$

Example 2-44   If the resistor in Example 2-43 is allowed to run for 30 min, how much energy is consumed?

solution $Wt = Pt = 550 \text{ W} \times 30 \text{ min} \times \dfrac{\text{h}}{60 \text{ min}}$

$$= 225 \text{ Wh}$$

If P is in watts and t in hours, Wt will be in watthours, the conventional unit of electrical energy. Recalling that from Ohm's law $E = IR$, Eq. (2-18) can be rewritten

$$Wt = I^2 Rt \tag{2-19}$$

Example 2-45   What is the resistance of the load described in Example 2-43?
solution $Wt = I^2 Rt$

$$R = \frac{Wt}{I^2 t} = \frac{225 \text{ wh}}{(5.0)^2 \text{ A}^2 \times 0.5 \text{ h}}$$

$$R = 18 \frac{\text{W}}{\text{A}^2} = \frac{18 \text{ V-A}}{\text{A}^2} = 18 \frac{\text{V}}{\text{A}} = \frac{18 \text{ V}}{\text{A}} \times \frac{\Omega\text{-A}}{\text{V}} = 18 \ \Omega$$

It should be noted [as can be seen from Eqs. (2-16) and (2-17)] that $1 \text{ Ws} = 1 \text{ J}$. Since Table 2-9 indicates that $1 \text{ J} = 0.239 \text{ cal}$, Eq. (2-19) can be extended:

$$H = 0.239 \, I^2 \, Rt \qquad\qquad (2\text{-}20)$$

Equation (2-20) gives the heat generated in calories by the passage of a current of I A through a resistance of $R \, \Omega$ for t s.

Example 2-46 How much heat is generated in the resistor described in Example 2-44?
solution $H = 0.239 \, I^2 \, Rt$
$\qquad\quad = 0.239 \, (5.0)^2 \, (18) \, (1,800)$
$\qquad\quad = 194,000 \text{ cal} = 194 \text{ kcal}$

Because of its importance—and the fact that many people confuse them—the difference between energy and power should be clearly fixed in mind, together with the proper units for each.

Chapter 3

Review of Chemical Fundamentals (Part I)

3-1 GASES

Gas Laws

If a gas is contained in a vessel whose volume can be altered (such as in a cylinder closed by a movable piston), it will be found that the pressure of the gas increases as the volume is decreased, as illustrated in Fig. 3-1a. This can be expressed mathematically by the equation

$$PV = K_1 \tag{3-1}$$

which indicates that the product of pressure and volume is equal to a constant. In this equation, P is the absolute pressure of a gas contained in volume V. This relationship is known as *Boyle's law*.

Example 3-1 A quantity of gas under a pressure of 50 lb/in² occupies a volume of 0.85 ft³. If it is to be compressed to 0.66 ft³, how much must the pressure be increased?
solution From Eq. (3-1),

$$PV = P_2 V_2$$
$$P_2 = \frac{P_1 V_1}{V_2} = \frac{50 \text{ lb}}{\text{in}^2} \times \frac{0.85 \text{ ft}^3}{0.66 \text{ ft}^3}$$
$$P_2 = 64 \text{ lb/in}^2$$
$$P_2 - P_1 = 64 - 50 = 14 \text{ lb/in}^2$$

In an analogous manner, if a gas is heated at a constant pressure, its volume will increase; and it has been found that its volume is directly proportional to its absolute temperature:

$$V = K_2 T \tag{3-2}$$

In Eq. (3-2), V is the volume of the gas at absolute temperature T. This is known as *Charles' law*, and is illustrated in Fig. 3-1b.

Example 3-2 The volume of a quantity of gas at 20°C is 22.4 l. What will the volume of the gas be if it is heated to 100°C?
solution From Eq. (3-2),

$$\frac{V_1}{T_1} = \frac{V_2}{T_2}$$
$$V_2 = V_1 \frac{T_2}{T_1}$$

In using Eq. (3-2), the temperatures must be expressed in absolute scales, in this case in K.

$$T_1 = 273 + 20 = 293 \text{ K}$$
$$T_2 = 273 + 100 = 373 \text{ K}$$
$$V_2 = 22.4 \, l \times \frac{373 \text{ K}}{293 \text{ K}} = 28.5 \, l$$

Finally, if a gas is kept at a constant volume, its pressure is found to be proportional to its absolute temperature:

$$P = K_3 T \tag{3-3}$$

This is illustrated in Fig. 3-1c.

Example 3-3   A gas at room temperature (78°F) is contained in a cylinder having a volume of 2.8 ft³. Its pressure is found to be 84 lb/in². If the cylinder is moved to a warehouse whose temperature is 16°F, what will the pressure of the gas in the cylinder be?

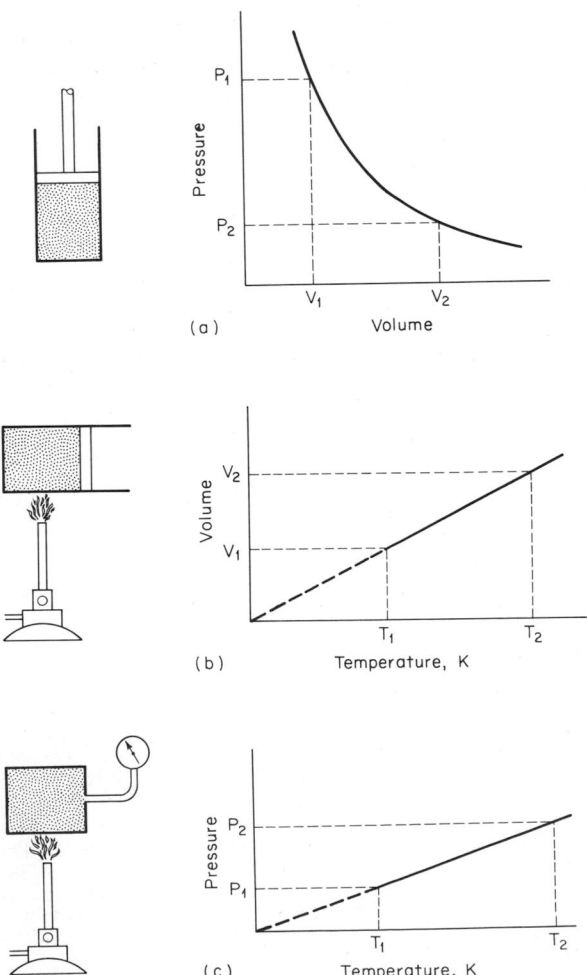

Fig. 3-1   Pressure-volume-temperature relations for gases. (a) Pressure-volume at constant temperature. (b) Volume-temperature at constant pressure. (c) Pressure-temperature at constant volume.

solution   From Eq. (3-3),

$$\frac{P_1}{T_1} = \frac{P_2}{T_2}$$

$$P_2 = P_1 \frac{T_2}{T_1}$$

Again, the temperatures must be expressed in absolute values, in this case °R.

$$T_1 = 492 + (78 - 32) = 538°\text{R}$$
$$T_2 = 492 + (16 - 32) = 476°\text{R}$$

$$P_2 = \frac{84\ \text{lb}}{\text{in}^2} \times \frac{476°\cancel{\text{R}}}{538°\cancel{\text{R}}} = 74\ \text{lb/in}^2$$

The three equations above can be multiplied algebraically as follows:

$$PV \frac{V}{T} \frac{P}{T} = K_1 K_2 K_3 \tag{3-4}$$

$$\frac{P^2 V^2}{T^2} = K_1 K_2 K_3 \tag{3-5}$$

$$\frac{PV}{T} = \sqrt{K_1 K_2 K_3} \tag{3-6}$$

Since the product of several constants is also a constant, Eq. (3-6) can be simplified to

$$\frac{PV}{T} = K \tag{3-7}$$

Equation (3-7) is known as the *general gas law* and permits the calculation of simultaneous changes in the pressure, volume, and temperature of a gas from known conditions.

Example 3-4   The density of nitrogen at 0°C and 760 mmHg is 1.25 g/l. What is the density of nitrogen at 30°C and 730 mmHg?
solution   For 1 l of nitrogen at 0°C and 760 mmHg, Eq. (3-7) yields

$$K = \frac{PV}{T} = \frac{760\ \text{mmHg} \times 1\ \text{l}}{273\ \text{K}} = 2.78\ \frac{(\text{mmHg})(\text{l})}{\text{K}}$$

Note that the temperature must be expressed as the absolute temperature. At 30°C and 730 mmHg, the new volume will be

$$V = \frac{KT}{P} = \frac{(2.78\ \cancel{\text{mmHg}})(\text{l})}{\cancel{\text{K}}} \times \frac{303\ \cancel{\text{K}}}{730\ \cancel{\text{mmHg}}} = 1.15\ \text{l}$$

This calculation can be made more directly noting that

$$\frac{P_1 V_1}{T_1} = K = \frac{P_2 V_2}{T_2}$$

Thus

$$V_2 = \frac{P_1}{P_2} \times \frac{T_2}{T_1} \times V_1 = \frac{760}{730} \times \frac{303}{273} \times 1\ \text{l} = 1.15\ \text{l}$$

Over relatively small changes in pressure, volume, or temperature, the general gas law is quite accurate, and extremely useful. Over wide changes of pressure, volume, or temperature, however, many of the more common gases show considerable variation from the general gas law. Since the general gas law has been derived by theoretical methods, one of which involves the statistical mechanics of an "ideal" gas, the general gas law is frequently referred to as the *ideal gas law*. In order to treat actual gases more accurately, the ideal gas law is modified as follows:

$$\left(P + \frac{a}{V^2}\right)(V - b) = KT \tag{3-8}$$

This form of the gas law is known as *van der Waals' law*. The van der Waals constants [*a* and *b* in Eq. (3-8)] are given in Table 3-1 for a number of common gases. [Since Eq. (3-8) involves the concept of molecular weights, its use is illustrated later in this chapter.]

Standard Conditions

In using the gas laws in either of the forms indicated above, it is necessary to have a set of values for the gas under consideration for which the pressure, volume, and temperature are known in order to calculate what these parameters would be under another set of conditions. For convenience, it is desirable to have one specific set of conditions which can be used as a general reference. For work involving gases, this is most commonly a pressure of 1 atm (760 mmHg) and a temperature of 0°C (273 K). These conditions are known as *standard conditions* and are frequently abbreviated STP (i.e., standard temperature and pressure).

Example 3-5 A gas occupies a volume of 25 l at 700 mmHg and 35°C. What is its volume at STP?

solution $\dfrac{P_1 V_1}{T_1} = \dfrac{P_2 V_2}{T_2}$

$$V_2 = V_1 \times \frac{P_1}{P_2} \times \frac{T_2}{T_1}$$

$$V_2 = 25 \text{ l} \times \frac{700 \text{ mmHg}}{\text{mmHg}} \times \frac{273 \text{ K}}{308 \text{ K}} = 20.4 \text{ l}$$

Avogadro's Law

When measured at constant temperature and pressure, the volume of gases consumed or produced in a chemical reaction is in the ratio of small, whole numbers. This principle,

**TABLE 3-1   Van der Waals'
Constants for Common Gases**

Gas	a (l^2-atm)	b (l)
H_2	0.244	0.0266
He	0.034	0.0237
N_2	1.39	0.0391
O_2	1.36	0.0318
Cl_2	6.49	0.0562
NH_3	4.17	0.0371
CO	1.49	0.0399
CO_2	3.59	0.0427

known as *Gay-Lussac's law of combining volumes,* is of considerable importance in chemical work. An extension of this law, known as *Avogadro's law,* states that equal volumes of all gases, at the same temperature and pressure, contain the same number of molecules. Avogadro's law provides a means of determining the molecular weight of a gas. (The molecular weight of a substance is the combined relative weights of its atoms. This will be discussed later in this chapter.)

As a consequence of Avogadro's law, it has been found that 1 gram molecular weight of a gas (i.e., its molecular weight in grams) occupies 22.4 l at STP. Furthermore, from other measurements, it is possible to calculate the actual number of molecules present. This turns out to be 6.02×10^{23} molecules, and the number 6.02×10^{23} is known as Avogadro's number. As we shall see later, Avogadro's number is involved in a variety of chemical calculations.

Since a definite number of molecules of a gas at a particular temperature and pressure will occupy a particular volume, this volume has a particular weight of gas associated with it. The ideal gas law can be rewritten to reflect this situation:

$$PV = nRT \tag{3-9}$$

in which nR replaces the constant K in Eq. (3-7). In Eq. (3-9), n represents the number of molecular weights, or *mols,* of gas present and R is a constant known as the *universal gas constant.* R has values shown in Table 3-2 for various pressure, volume, temperature, and weight units.

Example 3-6   Calculate the volume occupied by 8.00 g of oxygen at 25°C and 790 mmHg. (The molecular weight of oxygen is 32.0.)

solution   Number of mols present $= m/M = n$, where m is the weight of gas and M is its molecular weight.

$$n = \frac{8.00}{32.0} = 0.25 \text{ mol}$$

From Eq. (3-9) and Table 3-2,

$$V = \frac{nRT}{P} = \frac{0.25 \text{ mol} \times 0.0821 \text{ l-atm}}{(790 \text{ mmHg})(\text{mol})(\text{K})} \times 298 \text{ K}$$

$$V = 0.0077 \frac{\text{l-atm}}{\text{mmHg}}$$

This obviously involves a mixed-pressure unit, which can be readily resolved, since 1 atm = 760 mmHg.

$$V = 0.0077 \frac{\text{l-atm}}{\text{mmHg}} \times \frac{760 \text{ mmHg}}{\text{atm}} = 5.9 \text{ l}$$

Notice also that 32.0 g of oxygen occupies 22.4 l at STP. An alternate, but equivalent, calculation would be

$$\text{Volume of 8.0 g of O}_2 \text{ at STP} = \frac{8.0}{32.0} \times 22.4 \text{ l} = 5.6 \text{ l}$$

TABLE 3-2   Values of the Universal Gas Constant

Units	R
cal/(g mol)(K)	1.987
J/(g mol)(K)	8.314
cm³-atm/(g mol)(K)	82.06
l-atm/(g mol)(K)	0.0826
l-mmHg/(g mol)(K)	62.36
l-bar/(g mol)(K)	0.08314
l-kg/(cm²)(g mol)(K)	0.08478
ft³-atm/(lb mol)(K)	1.314
ft³-mmHg/(lb mol)(K)	998.9
Btu/(lb mol)(°R)	1.987
hp-h/(lb mol)(°R)	0.0007805
kWh/(lb mol)(°R)	0.0005819
ft³-atm/(lb mol)(°R)	0.7302
ft³-inHg/(lb mol)(°R)	21.85
ft³-mmHg/(lb mol)(°R)	555.0
ft³-lb/(ft³)(lb mol)(°R)	10.73
ft-lb/(lb mol)(°R)	1,545.0

Correcting for temperature and pressure,

$$V = 5.6 \text{ l} \times \frac{760}{790} \times \frac{298}{273} = 5.9 \text{ l}$$

In a similar manner, van der Waals' law can be rewritten

$$\left(P + \frac{a}{V^2}\right)(V - b) = nRT \tag{3-10}$$

Example 3-7   What is the pressure of 2.0 mols of ammonia (NH_3) at 27°C contained in a steel cylinder whose volume is 1.5 l?

solution   Since the gas is obviously being contained at a high pressure, Eq. (3-10) is used [in preference to Eq. (3-9)]. From Table 3-1, the values for a and b are 4.17 l²-atm and 0.0371 l, respectively.

$$\left(P + \frac{a}{V^2}\right)(V - b) = nRT$$

$$\left(P + \frac{4.17}{1.5^2}\right)(1.5 - 0.0371) = 2(0.08207)(300)$$

Note that the value of R in atm/(mol)(K) is used, and that the temperature is therefore expressed in K.

$$(P + 1.85)(1.46) = 49.24$$
$$P + 1.85 = 33.7$$
$$P = 31.9 \text{ atm}$$

Partial Pressures

As an intuitive extension of Avogadro's principle, we expect that when two or more gases are confined in a single vessel, the total pressure, volume, and temperature will depend upon the total number of molecules present. In a way, therefore, each kind of molecule acts as if it occupied the total volume all by itself, and the total pressure of all the gases is therefore the sum of the individual pressures exerted by each of the gases occupying the given volume independently. This is shown mathematically:

$$P = p_1 + p_2 + p_3 \tag{3-11}$$

In Eq. (3-11), P is the total pressure, and the p's are the pressures due to each of the different kinds of gases present. Each individual p is known as a partial pressure, i.e., that pressure which would be exerted by that particular kind of gas if it were present in the vessel

TABLE 3-3   Vapor Pressure of Water at Various Temperatures

$t°C$	p(mm)	$t°C$	p(mm)	$t°C$	p(mm)
−15 (ice)	1.24	12	10.5	27	26.7
−10 (ice)	1.95	13	11.2	28	28.3
−5 (ice)	3.01	14	12.0	29	30.0
0	4.58	15	12.8	30	31.8
1	4.93	16	13.6	31	33.7
2	5.29	17	14.5	32	35.7
3	5.69	18	15.5	33	37.7
4	6.10	19	16.5	34	39.9
5	6.54	20	17.5	35	42.2
6	7.01	21	18.7	40	55.3
7	7.51	22	19.8	50	92.5
8	8.05	23	21.1	60	149.4
9	8.61	24	22.4	70	233.7
10	9.21	25	23.8	80	355.1
11	9.84	26	25.2	90	525.8
				100	760.0

From Conway Pierce and R. Nelson Smith, "General Chemistry Workbook," 3d ed., W. H. Freeman and Company, San Francisco, 1965.

all by itself. It follows as a natural consequence of Eq. (3-11) that if only one gas is present, all the p's except p_1 will be zero and P will equal p_1.

Example 3-8 What are the partial pressures of nitrogen and oxygen in air if the total pressure is 760 mmHg? (The volume composition of air is approximately 79 percent nitrogen and 21 percent oxygen.)

solution $p_{\text{nitrogen}} = 0.79 \times 760 = 600$ mm
$p_{\text{oxygen}} = 0.21 \times 760 = 160$ mm

In chemical work, partial pressures have particular significance in quantitatively relating wet and dry gases. At any temperature, water will exert a certain vapor pressure. That is, the partial pressure of water vapor in equilibrium with liquid water will have a particular value at a particular temperature. Table 3-3 gives the value of the vapor pressure of water at various temperatures. If a gas is in contact with water, it will pick up water vapor until the partial pressure of the water vapor assumes the value shown in Table 3-3. Whether a gas is wet or dry can have an important effect in gas problems, as illustrated by the following examples.

Example 3-9 A 500-cm^3 flask is filled over water with oxygen at a barometric pressure of 740 mm and a temperature of 25°C. What would the volume of the gas be if it were dry?
solution From Table 3-3, the vapor pressure of water at 25°C is 24 mm.

$$p_{oxygen} = P - p_{H_2O}$$
$$= 740 \text{ mm} - 24 \text{ mm} = 716 \text{ mm}$$
$$V = 500 \text{ cm}^3 \times \frac{716 \text{ mm}}{740 \text{ mm}} = 484 \text{ cm}^3$$

Example 3-10 What is the volume of 1 mol of a wet gas measured at 770 mm and 35°C?
solution At STP 1 mol of a dry gas has a volume of 22.4 l. If it is measured wet at 770 mm and 35°C, its partial pressure will be $770 - 42.2$ or 728 mm, since from Table 3-3, the vapor pressure of water at 35°C is 42.2 mm.

$$V = 22.4 \text{ l} \times \frac{333 \text{ K}}{298 \text{ K}} \times \frac{760 \text{ mm}}{728 \text{ mm}} = 26.1 \text{ l}$$

Partial Volumes

Analogous to partial pressures, the *partial volume* of a gas in a mixture of gases is the volume that gas would occupy by itself at the total pressure of the gas mixture. The total volume is thus the sum of the partial volumes of each of the gases present.

3-2 MATTER

Elements and Compounds

Pure substances are composed of a single element or compound. Where more than one element or compound or a combination of elements and compounds is present, the material is referred to as a *mixture*. An *element* is a material made up of only a single kind of atoms. When two or more different kinds of atoms combine, i.e., react, to form a new material, the resultant material is called a *compound*. Thus sodium (Na) and chlorine (Cl$_2$) are elements, even though chlorine normally appears as a chlorine molecule resulting from the combination of two chlorine atoms. When sodium and chlorine react, however, a new material, sodium chloride (NaCl), is formed. This new material, being composed of two different kinds of atoms, is a compound. When elements or compounds are mixed without a chemical reaction taking place, so that the original materials retain their identity, the result is simply a mixture.

An atom is the smallest particle of an element exhibiting all the chemical properties of that element. Its structure determines the chemical and physical properties of an element, and the structural differences give rise to the different elements. According to classical theory, an atom is made up of a nucleus surrounded by orbiting electrons, and the composition of the nucleus for the most part determines the mass of the atom and the number of electrons that surround it. Within the nucleus, there are protons and electrons. The proton is a fundamental particle having an extremely high mass and carrying a single positive electrical charge. The electron is a particle with an extremely small mass, but it carries an electrical charge that is equal and opposite to that of the proton, namely, a single negative charge. Some of the protons in the nucleus combine with electrons to form neutrons. These are therefore particles having no electrical charge, being made up of particles having one positive charge and one negative charge of equal value. Since the mass of the proton is much larger than that of the electron, the effect of the latter is practically negligible, and a neutron essentially has the mass of a proton. It so happens that in all atomic nuclei there is a surplus of protons so that the nucleus always carries a net positive charge. This positive charge is counterbalanced by the orbiting electrons, and there must therefore be an equal number of electrons in the outer orbits (or shells) for the atom as a whole to be electrically neutral. The electrons in the outer shells are not free to assume any position but are subject to various forces which tend to confine them to very definite orbital paths. For example, the first shell, i.e., the orbits closest to the nucleus, can be occupied by no more than two electrons. In a similar manner, additional shells are built up which can contain 8, 18, or 32 electrons. The specific electronic configurations for the first 20 elements are shown in Fig. 3-2. The configuration of the electrons in the outer shells, and in the outermost shell in particular, gives the atom its chemical properties.

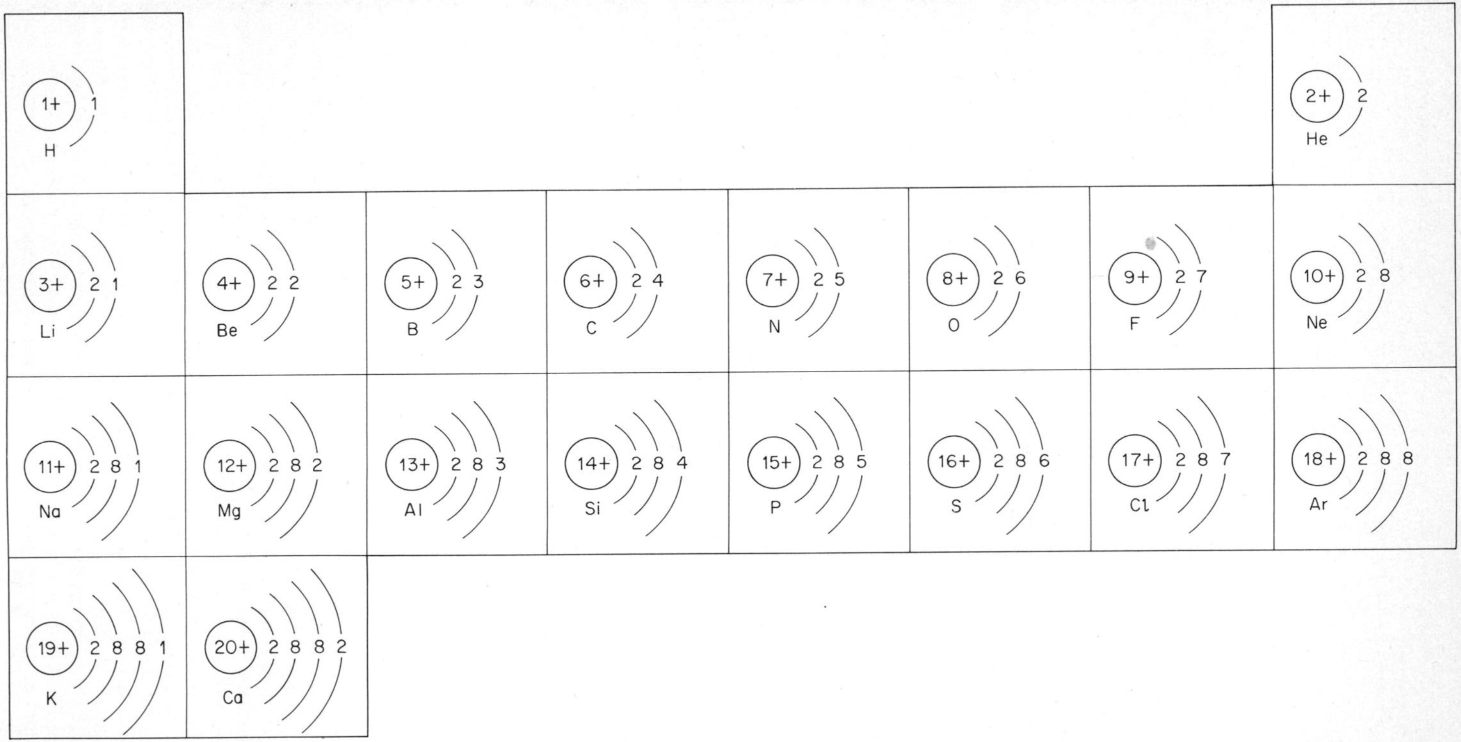

Fig. 3-2 Electronic configuration of the first 20 elements. The numbers shown outside the nucleus are the number of electrons in the indicated orbits. Note that the number of electrons in orbit around the nucleus is equal to the net positive charge in the nucleus.

Atomic and Molecular Weights

The simplest of the atoms is hydrogen, consisting of only 1 proton in the nucleus and 1 electron in orbit around it. The next element, helium, contains 2 protons and 2 neutrons in the nucleus and therefore has 2 electrons in orbits around the nucleus to balance the electrical charge of the protons. A helium atom has about 4 times the mass of a hydrogen atom since its nucleus contains 2 protons and 2 neutrons. Lithium, the next element, has 3 protons and 4 neutrons in the nucleus and has 3 orbiting electrons. An atom of lithium has 7 times the mass of a hydrogen atom. If oxygen, which has 8 protons and 8 neutrons in its nucleus, is assigned a weight value of 16, then the relative weights of all the other elements can be readily established, since these relative weights would be approximately the number of protons plus neutrons in the various nuclei. The fact that the atomic weight is not exactly a whole number is due in part to the fact that the electron, while it has an extremely low mass compared with that of the proton or neutron, nevertheless does contribute something to the mass of the atom. For another thing, it sometimes happens that the nucleus of an atom is deficient in neutrons or may contain a surplus of neutrons. Since the neutron does not affect the net electrical charge of the nucleus, this deficiency or surplus does not affect the electrical balance and therefore does not affect the number and positions of the orbiting electrons. An atom with a surplus or deficiency of neutrons, therefore, has exactly the same chemical properties as its normal counterpart; but it will have a different atomic weight. Atoms of the same species which have the same chemical properties but different atomic weights are called *isotopes*, and many elements contain a natural mixture of one or more isotopes. Thus the average value of the atomic weight for naturally occurring elements will reflect this mixture. Most tables giving atomic weights (such as Table 1-11) give the average atomic weight for naturally occurring elements; that is, they automatically take into account the natural isotope distribution for each element.

Since atomic weights can now all be made relative to the oxygen atom, which has been assigned a value of 16, atomic weights for all the elements can be established. The *atomic weight* of an element is its weight in grams relative to oxygen having an assigned atomic weight of 16 g. The atomic weight of chlorine is thus 35.5, that of tin 118.7 g, and so on. Since compounds are made up of definite combinations of atoms, called molecules, we can now also assign weights to molecules. Such weights are called *molecular weights*. The molecular weights of a large number of compounds are given in Table 1-11. Thus the molecular weight of sodium chloride, consisting of 1 atom of sodium and 1 atom of chlorine, is 58.5 g, since 1 atomic weight of sodium is 23.0 and 1 atomic weight of chlorine is 35.5. In this way, weights can be assigned for all elements and compounds.

Example 3-11   Using the atomic weights given in Table 1-11, calculate the molecular weights of the following compounds (and check with the molecular weights for these compounds, also given in Table 1-11):

a. Cadmium hydroxide $Cd(OH)_2$
b. Manganese dioxide MnO_2
c. Mercuric chloride $HgCl_2$
d. Hydrated sodium sulfate $Na_2SO_4 \cdot 10H_2O$
e. Ethyl alcohol C_2H_5OH
f. n-Propylamine $C_3H_7NH_2$

solution

a. Cadmium (Cd)   112.41
   O_2 32.00
   H_2 2.016

   $Cd(OH)_2 =$ 146.43

b. Manganese (Mn)   54.93
   O_2 32.00

   $MnO_2 =$ 86.93

c. Mercury (Hg)   200.61
   Cl_2 70.91

   $HgCl_2 =$ 271.52

d. Na_2SO_4 (2 × 22.997)   45.994
   S   32.06
   O_4 (2 × 32.00)   64.00
   H_2O (10 × 2.016)   20.16
   O_{10} (5 × 32.00)   160.00

   $Na_2SO_4 \cdot 10H_2O =$ 322.21

e. C_2 (2 × 12.01)   24.02
   H_5 (5 × 1.008)   5.040
   O   16.00
   H   1.008

   $C_2H_5OH =$ 46.07

f. C_3 (3 × 12.01)   36.03
   H_7 (7 × 1.008)   7.056
   N   14.01
   H_2 2.016

   n-Propylamine =   59.11

Note that Table 1-11 gives only molecular weights, and that many elements (O_2, Cl_2, N_2, H_2, etc.) form diatomic molecules. For such elements, the atomic weights are readily obtained by noting how many atoms make up the molecule and dividing the given molecular weight by that number.

It is extremely important to note that the atoms can be arranged according to the number of protons in their nuclei. Starting with hydrogen, having 1 proton, and progressing by units up to the presently known level of lawrencium, having 103 protons in its nucleus, each of the elements can be assigned a number which reflects the number of protons in its nucleus and therefore also the number of electrons orbiting the nucleus. This number is called the *atomic number* of the element, and each element has its own unique atomic number. Table 3-4 lists the elements in order of their atomic numbers, and also gives the electronic configuration of each element.

TABLE 3-4   Elements in Order of Atomic Number, and Electronic Configurations of the Elements

Element	Atomic number	Shell designations						
		K	L	M	N	O	P	Q
H	1	1						
He	2	2						
Li	3	2	1					
Be	4	2	2					
B	5	2	3					
C	6	2	4					
N	7	2	5					
O	8	2	6					
F	9	2	7					
Ne	10	2	8					
Na	11	2	8	1				
Mg	12	2	8	2				
Al	13	2	8	3				
Si	14	2	8	4				
P	15	2	8	5				
S	16	2	8	6				
Cl	17	2	8	7				
Ar	18	2	8	8				
K	19	2	8	8	1			
Ca	20	2	8	8	2			
Sc	21	2	8	9	2			
Ti	22	2	8	10	2			
V	23	2	8	11	2			
Cr	24	2	8	13	1			
Mn	25	2	8	13	2			
Fe	26	2	8	14	2			
Co	27	2	8	15	2			
Ni	28	2	8	16	2			
Cu	29	2	8	18	1			
Zn	30	2	8	18	2			
Ga	31	2	8	18	3			
Ge	32	2	8	18	4			
As	33	2	8	18	5			
Se	34	2	8	18	6			
Br	35	2	8	18	7			
Kr	36	2	8	18	8			
Rb	37	2	8	18	8	1		
Sr	38	2	8	18	8	2		
Y	39	2	8	18	9	2		
Zr	40	2	8	18	10	2		
Nb	41	2	8	18	12	1		
Mo	42	2	8	18	13	1		
Tc	43	2	8	18	14	1		
Ru	44	2	8	18	15	1		
Rh	45	2	8	18	16	1		
Pd	46	2	8	18	18			
Ag	47	2	8	18	18	1		
Cd	48	2	8	18	18	2		
In	49	2	8	18	18	3		
Sn	50	2	8	18	18	4		
Sb	51	2	8	18	18	5		
Te	52	2	8	18	18	6		

TABLE 3-4   Elements in Order of Atomic Number, and Electronic Configurations of the Elements —(Continued)

Element	Atomic number	K	L	M	N	O	P	Q
		Shell designations						
I	53	2	8	18	18	7		
Xe	54	2	8	18	18	8		
Cs	55	2	8	18	18	8	1	
Ba	56	2	8	18	18	8	2	
La	57	2	8	18	18	9	2	
Ce	58	2	8	18	20	8	2	
Pr	59	2	8	18	21	8	2	
Nd	60	2	8	18	22	8	2	
Pm	61	2	8	18	23	8	2	
Sm	62	2	8	18	24	8	2	
Eu	63	2	8	18	25	8	2	
Gd	64	2	8	18	25	9	2	
Tb	65	2	8	18	27	8	2	
Dy	66	2	8	18	28	8	2	
Ho	67	2	8	18	29	8	2	
Er	68	2	8	18	30	8	2	
Tm	69	2	8	18	31	8	2	
Yb	70	2	8	18	32	8	2	
Lu	71	2	8	18	32	9	2	
Hf	72	2	8	18	32	10	2	
Ta	73	2	8	18	32	11	2	
W	74	2	8	18	32	12	2	
Re	75	2	8	18	32	13	2	
Os	76	2	8	18	32	14	2	
Ir	77	2	8	18	32	17		
Pt	78	2	8	18	32	17	1	
Au	79	2	8	18	32	18	1	
Hg	80	2	8	18	32	18	2	
Tl	81	2	8	18	32	18	3	
Pb	82	2	8	18	32	18	4	
Bi	83	2	8	18	32	18	5	
Po	84	2	8	18	32	18	6	
At	85	2	8	18	32	18	7	
Rn	86	2	8	18	32	18	8	
Fr	87	2	8	18	32	18	8	1
Ra	88	2	8	18	32	18	8	2
Ac	89	2	8	18	32	18	9	2
Th	90	2	8	18	32	18	10	2
Pa	91	2	8	18	32	20	9	2
U	92	2	8	18	32	21	9	2
Np	93	2	8	18	32	22	9	2
Pu	94	2	8	18	32	24	8	2
Am	95	2	8	18	32	25	8	2
Cm	96	2	8	18	32	25	9	2
Bk	97	2	8	18	32	26	9	2
Cf	98	2	8	18	32	28	8	2
Es	99	2	8	18	32	29	8	2
Fm	100	2	8	18	32	30	8	2
Md	101	2	8	18	32	31	8	2
No	102	2	8	18	32	32	8	2
Lw	103	2	8	18	32	32	9	2

Valency and Chemical Bonds

The term valence is used to describe the units with which various elements generally react with each other. In any atomic configuration, there are fixed numbers of electrons which can occupy stable orbital positions. As indicated earlier, 2 electrons represent the stable configuration in the first shell, 8 in the second, and 8 or 18 in the third, etc. If the outermost shell has a deficiency of electrons, it has a tendency to attract electrons so as to complete a stable electronic configuration. On the other hand, if there is a surplus of electrons in the outermost shell, the atom will tend to give up this electron in order to produce a stable electronic configuration. The number of electrons which an atom will gain or lose in order to complete its outermost shell is called its *valence*. Frequently, the term *electrovalence* is used to indicate that there is a net electrical charge on the atom after it has undergone the loss or acquisition of electrons in the process of completing its outermost shell. Thus sodium, which tends to lose one electron and thereby is left with a single positive charge, is said to have an electrovalence, or frequently simply a valence, of +1. In a similar manner, chlorine, which tends to gain an electron to complete its outermost shell and therefore ends up with a surplus of negative charges, is said to have a valence of −1.

When one atom capable of giving up an electron comes into the vicinity of another atom capable of accepting an electron, an actual electron transfer occurs, thus satisfying the needs of both atoms. However, the resulting individual atoms now have an electrical charge, the atom donating the electron remaining with a positive charge, while the atom receiving the electron acquires a negative charge. Atoms which have acquired an electrical charge by virtue of the gain or loss of their valency electrons are called *ions*. Those ions carrying a positive charge are called *cations,* and those having a negative charge are called *anions*. Since the atoms involved in the electronic exchange are now electrically charged, and in fact are oppositely charged, they attract each other and tend to become closely associated, generally in a very stable combination called a *compound*. Since definite numbers of electrons are involved in filling definite numbers of vacancies, the number and arrangement of the atoms forming a particular molecule is clearly and reproducibly established. As an example, when sodium and chlorine react, each sodium atom gives up an electron to a chlorine atom, resulting in a compound consisting of one sodium ion and one chloride ion. Thus, the *formula,* or representation of the atomic ratios present in a molecule for sodium chloride, is NaCl, indicating that the sodium chloride molecule contains 1 atom of sodium and 1 atom of chlorine.

When an electron actually moves from one atom to another, the force holding the resulting ions together is electrical, and is referred to as a *polar bond*. This is illustrated in Fig. 3-3a, in which an atom of lithium reacts with an atom of chlorine to form lithium chloride.

In some instances, it is not actually necessary for an electron to leave one atom and join another in order to form a bond which will hold the atoms together. If donor atoms come in

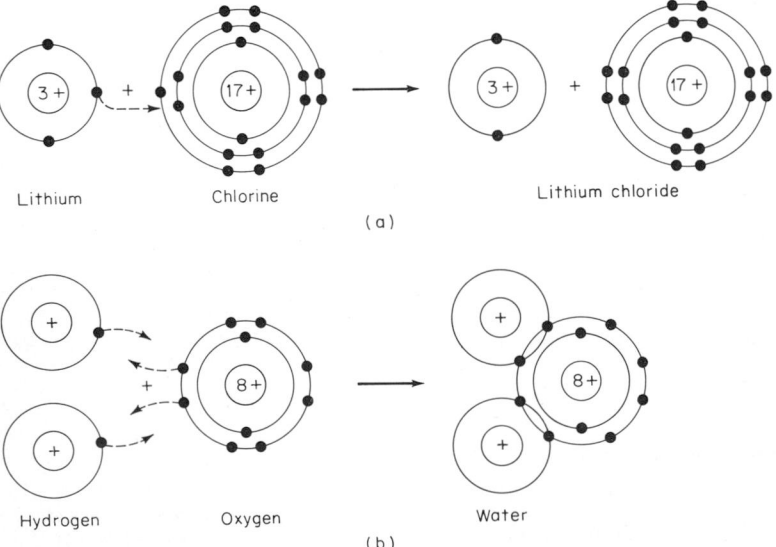

Lithium   Chlorine   Lithium chloride

(a)

Hydrogen   Oxygen   Water

(b)

Fig. 3-3 Formation of chemical bonds.

close proximity to acceptor atoms, it is possible for certain electrons to do a sort of double duty; that is, they can remain with their original atoms, but the orbits can be so overlapped as to complete the outer shells of both atoms by a "sharing" of electrons. This is illustrated in Fig. 3-3b, in which two hydrogen atoms, each with a single electron in its outermost shell, come in proximity to an oxygen atom with a deficiency of two electrons in its outermost shell. The two hydrogen atoms then share their electrons with the oxygen atom, thereby satisfying the requirements for the outermost shells of all three atoms. This type of bonding is called a *covalent bond*. In some covalent-type bonds, the shared electrons can be more closely associated with one atom or the other. Such a covalent bond actually takes on some of the characteristics of a polar bond, and not only can be very stable but can have some of the properties of both types of bonds. Water, the compound illustrated in Fig. 3-3b, typically exhibits such dual properties. In general, however, a polar bond is usually more stable than a covalent bond. Polar bonds are usually associated with inorganic compounds, and covalent bonds are generally characteristic of organic compounds.

Periodic Table

Even prior to a formalized atomic theory, it was noted that metals could be classified by the type of chloride they formed. For example, sodium, potassium, lithium, and silver formed monochlorides ($NaCl$, KCl, etc.). Other metals, such as lead, iron, tin, zinc, and calcium, formed dichlorides ($PbCl_2$, $FeCl_2$, etc.); while still others, such as aluminum, iron, and chromium, formed trichlorides ($AlCl_3$, $FeCl_3$, etc.). Such materials as carbon, silicone, and titanium, formed tetrachlorides (CCl_4, $SiCl_4$, etc.). It was also observed that certain elements, as for example, iron, formed more than one kind of chloride. In addition, it was found that metals that formed monochlorides also formed oxides in which the ratio of metal to oxygen atoms was 2:1 (such as lithium oxide, Li_2O, and potassium oxide, K_2O). Those metals which formed dichlorides formed oxides in which the atomic ratio of metal to oxygen was 1:1 (such as FeO and CaO). Those metals which formed trichlorides formed oxides in which the ratio of metal to oxygen atoms was 2:3 (such as Al_2O_3 and Fe_2O_3). Finally, those elements which formed tetrachlorides formed oxides in which the atomic ratio of metal to oxygen was 1:2 (such as SiO_2 and TiO_2).

This suggested that a number of metals have generally similar chemical combining properties; and with the advent of the atomic theory and the explanation of chemical properties in terms of the need to satisfy the electronic configuration of the outermost shell, it was possible to arrange the atoms in order of increasing atomic number such that elements having similar combining properties are placed under each other in columnar form. When this was done, it was found that the columns contained elements whose atomic numbers differed by eight units, and correlated with the electronic configuration of the atoms, shown in Table 3-4. In this way, it was possible to classify each atom in terms of its atomic number (period) and its valence (group), resulting in a chart in which the groups are generally vertical columns and the periods are horizontal rows. In this way, the chemical properties and combining weights of the various atoms could be correlated, and in fact, sufficiently well correlated to predict the existence of not yet discovered elements the chemical and physical properties for which were deduced from those of known related elements.

As the *periodic table*, as it was known, was refined, it appeared that there were significant deviations in the properties of certain elements which could not be entirely explained by the electronic configurations of its outermost orbits, or correlated with the element's position in the table. For one thing, as already pointed out, certain elements exhibited two or even more valences. As additional information concerning atomic configurations became known, it was also apparent that some of the chemical properties of an element are associated with orbits other than the outermost shell; and in fact, some reactions involve the completion of inner orbits as well as the outermost ones. Thus, the original eight periods were expanded to include subperiods. All elements belonging to the same subperiod have extremely similar chemical properties; whereas all elements belonging to the same period have generally similar chemical properties. A modern form of the periodic table is given in Table 3-5.

Ionization and Equivalent Weights

Under certain circumstances the ions of a molecule can be physically separated from each other and actually become individual, charged particles. One way of bringing this about is by dissolving the material in a solvent, such as water, in which case the impact of the water

TABLE 3-5 Periodic Chart of the elements

IA	IIA	IIIB	IVB	VB	VIB	VIIB	VIII	VIII	VIII	IB	IIB	IIIA	IVA	VA	VIA	VIIA	Inert gases
Light metals							**Transition metals**							**Nonmetals**			
1 **H** 1.0080																	2 **He** 4.003
3 **Li** 6.940	4 **Be** 9.013											5 **B** 10.82	6 **C** 12.011	7 **N** 14.008	8 **O** 16.000	9 **F** 19.00	10 **Ne** 20.183
11 **Na** 22.991	12 **Mg** 24.32											13 **Al** 26.98	14 **Si** 28.09	15 **P** 30.975	16 **S** 32.066	17 **Cl** 35.457	18 **Ar** 39.944
19 **K** 39.100	20 **Ca** 40.08	21 **Sc** 44.96	22 **Ti** 47.90	23 **V** 50.95	24 **Cr** 52.01	25 **Mn** 54.94	26 **Fe** 55.85	27 **Co** 58.94	28 **Ni** 58.71	29 **Cu** 63.54	30 **Zn** 65.38	31 **Ga** 69.72	32 **Ge** 72.60	33 **As** 74.91	34 **Se** 78.96	35 **Br** 79.916	36 **Kr** 83.80
37 **Rb** 85.48	38 **Sr** 87.63	39 **Y** 88.92	40 **Zr** 91.22	41 **Nb** 92.91	42 **Mo** 95.95	43 **Tc** [99]	44 **Ru** 101.1	45 **Rh** 102.91	46 **Pd** 106.4	47 **Ag** 107.880	48 **Cd** 112.41	49 **In** 114.82	50 **Sn** 118.70	51 **Sb** 121.76	52 **Te** 127.61	53 **I** 126.91	54 **Xe** 131.30
55 **Cs** 132.91	56 **Ba** 137.36	57–71 See Lanthanide series	72 **Hf** 178.50	73 **Ta** 180.95	74 **W** 183.86	75 **Re** 186.22	76 **Os** 190.2	77 **Ir** 192.2	78 **Pt** 195.09	79 **Au** 197.0	80 **Hg** 200.61	81 **Tl** 204.39	82 **Pb** 207.21	83 **Bi** 209.00	84 **Po** 210	85 **At** [210]	86 **Rn** 222
87 **Fr** [223]	88 **Ra** 226.05	89–101 See Actinide series															

Lanthanide series (rare-earth elements)

57 **La** 138.92	58 **Ce** 140.13	59 **Pr** 140.92	60 **Nd** 144.27	61 **Pm** [145]	62 **Sm** 150.35	63 **Eu** 152.0	64 **Gd** 157.26	65 **Tb** 158.93	66 **Dy** 162.51	67 **Ho** 164.94	68 **Er** 167.27	69 **Tm** 168.94	70 **Yb** 173.04	71 **Lu** 174.99

Actinide series (rare-earth elements)

89 **Ac** 227	90 **Th** 232.05	91 **Pa** 231	92 **U** 238.07	93 **Np** [237]	94 **Pu** [242]	95 **Am** [243]	96 **Cm** [247]	97 **Bk** [249]	98 **Cf** [251]	99 **Es** [254]	100 **Fm** [253]	101 **Md** [256]	102 **No** [254]	103 **Lw** [257]

From Conway Pierce and R. Nelson Smith, "General Chemistry Workbook," 3d ed., W. H. Freeman and Company, San Francisco, 1965.

molecules causes a disruption of the molecular structure of the dissolved compound, resulting in a release of free ions. The process whereby the ions of a molecule become free particles is called *ionization*. It generally occurs when a polar material is dissolved in a suitable solvent. While the ions in such a solution are mobile and free to move throughout the solution, the fact that they are still charged particles restricts their freedom of action. As an example, individual ions of opposite polarity coming in close proximity to each other will recombine to form the original compound, only to be ionized again at some later time by the continuing action of the solvent molecules. Thus ionization of a material in solution is a process of continuous ionization, recombination, and reionization. Ultimately, a dynamic equilibrium is established and the concentrations of ions and molecules assumes fixed values. Arrows pointing in both directions are used in writing ionization reactions, as in Eqs. (3-12a) to (3-12e):

$$NaCl \rightleftharpoons Na^+ + Cl^- \qquad (3\text{-}12a)$$

$$Na_2SO_4 \rightleftharpoons 2Na^+ + SO_4^{--} \qquad (3\text{-}12b)$$

$$H_2O \rightleftharpoons H^+ + OH^- \qquad (3\text{-}12c)$$

$$Al_2(SO_4)_3 \rightleftharpoons 2Al^{3+} + 3SO_4^{--} \qquad (3\text{-}12d)$$

$$NaHCO_3 \rightleftharpoons Na^+ + HCO_3^- \qquad (3\text{-}12e)$$
$$\Updownarrow$$
$$H^+ + CO_3^{--}$$

In ionic equations, the ions are always shown with their electrical charge. Groups of atoms (called radicals and discussed later in this chapter) retain their identity and yield appropriately charged ions as illustrated in Eqs. (3-12b) to (3-12e). In Eq. (3-12e), showing the ionization of sodium bicarbonate into sodium and bicarbonate ions, a secondary ionization of bicarbonate ions into hydrogen and carbonate ions is also shown. Thus, sodium bicarbonate ionizes, or *dissociates*, into sodium ions (Na^+), bicarbonate ions (HCO_3^-), hydrogen ions (H^+), and carbonate ions (CO_3^{--}).

In ionic equations, those substances which are insoluble, or ionize only weakly, are not written in ionic form, even though a very small fraction may ionize. Thus, calcium carbonate is written $CaCO_3$, and only very rarely as $Ca^{++} + CO_3^{--}$. Ammonium hydroxide is highly soluble but only weakly ionized and is usually written in molecular form, NH_4OH, rather than in its ionic components, $NH_4^+ + OH^-$. To show that ammonium hydroxide is only weakly dissociated, the arrows in the equilibrium equation are given different lengths:

$$NH_4OH \xleftarrow{\quad} NH_4^+ + OH^- \qquad (3\text{-}13)$$

In writing ionic equations, not only must all ions and compounds balance, but there must be electrical equivalence as well.

$$BaCO_3 + 2HC_2H_3O_2 \longrightarrow Ba^{++} + 2C_2H_3O_2^- + H_2O + CO_2 \qquad (3\text{-}14a)$$

$$BaCl_2 + 2H^+ + SO_4^{--} \longrightarrow BaSO_4\downarrow + 2H^+ + 2Cl^- \qquad (3\text{-}14b)$$

Since $2H^+$ appears on both sides of Eq. (3-14b), the latter can be written

$$BaCl_2 + SO_4^{--} \longrightarrow BaSO_4\downarrow + 2Cl^- \qquad (3\text{-}14c)$$

It should be noted that while Eq. (3-14b) is more complete, since it shows both sides as being electrically neutral, Eq. (3-14c) shows the essentials of the reaction, and both sides of the equation are electrically balanced.

The *equivalent weight* of an atom is its atomic weight divided by the charge of the ion it forms. Thus iron, which has an atomic weight of 55.8, will have an equivalent weight of 27.9 (i.e., $55.8 \div 2$) when its ions are divalent (Fe^{++}), or an equivalent weight of 18.6 (i.e., $55.8 \div 3$) if its ions are trivalent (Fe^{3+}). Equivalent weights of reacting elements will exactly combine with each other without a surplus or a deficiency of one material or the other at the end of the reaction. Thus, 27.9 g of iron will exactly react with 8.0 g of oxygen (to produce 43.9 g of iron oxide, FeO) since the equivalent weight of oxygen is $16.0 \div 2$. As a more general definition, the equivalent weight of an element is its atomic weight divided by its valence.

Example 3-12 What are the equivalent weights of the following elements or radicals: aluminum, sulfate, nickel?

solution From Table 1-11, the atomic weight of aluminum is 26.97, and since it forms trivalent ions (Al^{3+}), the equivalent weight of Al is $26.97/3 = 8.99$ g.

Similarly for SO_4^{--}, which has a valence of -2,

$$
\begin{array}{lr}
\text{S} & 32.06 \\
\text{O}_4 \ (4 \times 16.00) & 64.00 \\
\hline
\text{SO}_4 & 96.06
\end{array}
$$

Equivalent weight of $SO_4 = 96.06/2 = 48.03$ g. (Note that the *sign* of the valence is always taken as positive in calculating equivalent weights, regardless of whether a cation or anion is involved.)

Ni, with an atomic weight of 58.69, forms divalent (Ni^{++}) or trivalent (Ni^{3+}) ions. Therefore, its equivalent weight is $58.69/2 = 29.35$ g or $58.69/3 = 19.56$ g, depending on how the nickel reacts.

The concept of equivalent weights has been extended to include neutralization reactions, for which it is defined as the weight of a base needed to neutralize one molecular weight of hydrochloric acid (HCl), or the weight of an acid needed to neutralize one molecular weight of sodium hydroxide (NaOH), as the case may be. Thus, the molecular weight of sulfuric acid (as given in Table 1-11) is 98.0, but since 1 molecular weight of sulfuric acid is capable of neutralizing 2 molecular weights of sodium hydroxide, the equivalent weight of sulfuric acid is $98.0 \div 2$, or 49.0. The concept of equivalent weights is extremely important in stoichiometric calculations which deal with the quantitative manner in which elements and molecules react.

Example 3-13 How many equivalents are present in the following:
  a. 32 g of H_3PO_4
  b. 332 g of $Ca(OH)_2$

solution
$$H_3PO_4 + 3NaOH \longrightarrow Na_3PO_4 + 3H_2O$$

Thus, 1 molecular weight of H_3PO_4 will neutralize 3 molecular weights of NaOH. Since the molecular weight of H_3PO_4 (from Table 1-11) is 98.04, its equivalent weight is $98.04 \div 3 = 32.68$ g.

$$Ca(OH)_2 + 2HCl \longrightarrow CaCl_2 + 2H_2O$$

Thus 1 molecular weight of $Ca(OH)_2$ will neutralize 2 molecular weights of HCl. Since the molecular weight of $Ca(OH)_2$ is 74.10, its equivalent weight is $74.10 \div 2 = 37.05$ g.

3-3 INORGANICS

Generalized Reactions

The concept of valency also leads to the principle of *combining weights,* the basis of which is that chemical reactions always take place between whole atoms, whole molecules, or a specific arrangement of atoms acting as a recognizable entity called a *radical*. Usually, a radical does not change when it participates in a chemical reaction. However, if a radical does undergo a change during a chemical reaction, the types of atoms and their arrangement in the radical do not change, so that the original radical is readily recognizable in whatever form it appears in the products. Since they are only a part of a molecule, radicals carry an electrical charge equal to the net charge of its atoms. Thus the sulfate radical, which consists of 1 sulfur atom in a valence state of $+6$, and 4 oxygen atoms each having a valence of -2, has an effective valence of -2 (i.e., $+6 - 2 - 2 - 2 - 2 = -2$), and is written SO_4^{--}. When a sulfate radical combines with a metal having a valence of $+4$, such as lead in the tetravalent state, the formula of the resulting compound is $Pb(SO_4)_2$, indicating that two sulfate radicals are needed to satisfy electrically one tetravalent lead atom. In the case of a combination with a divalent metal, i.e., a metal having a valence of $+2$, such as calcium, the resulting compound would be calcium sulfate having the formula $CaSO_4$. The parentheses around the sulfate radical are omitted in the latter case, since it is understood that only a single sulfate group is involved. Similarly, the combination of a sulfate radical with a monovalent metal, i.e., one having a valence of $+1$, such as sodium, would result in sodium sulfate, the formula for which is Na_2SO_4. Radicals can therefore participate in chemical reactions in a manner similar to that of atoms. Table 3-6 lists the common radicals and their valences.

The fundamental types of reactions between atoms (or radicals) are based on the types of bonds produced, as illustrated in Fig. 3-4. In a *polar reaction*, the valency electrons transfer from one atom to another, creating stable electrical balances, and, therefore, stable com-

TABLE 3-6 Common Radicals with Their Valences and Combining Weights

Radical	Formula and charge	Molecular (combining) weight
Acetate	$C_2H_3O_2^-$	59.05
Ammonium	NH_4^+	18.05
Arsenate	AsO_4^{3-}	138.92
Bicarbonate	HCO_3^{--}	61.02
Bisulfate	HSO_4^-	97.07
Borate	$B_4O_7^{--}$	122.81
Bromate	BrO_3^-	127.91
Carbonate	CO_3^{--}	60.01
Chlorate	ClO_3^-	83.45
Chromate	CrO_4^{--}	116.00
Cyanate	OCN^-	42.02
Cyanide	CN^-	26.02
Dihydrogen phosphate	$H_2PO_4^-$	96.99
Hydrogen phosphate	HPO_4^{--}	95.98
Hypochlorite	OCl^-	51.45
Iodate	IO_3^-	174.90
Manganate	MnO_4^{--}	118.94
Nitrate	NO_3^-	62.01
Nitride	N^{3-}	14.01
Nitrite	NO_2^{--}	46.01
Perchlorate	ClO_4^-	99.45
Permanganate	MnO_4^-	118.94
Phosphate	PO_4^{3-}	94.97
Phosphide	P^{3-}	30.97
Silicate	SiO_3	76.09
Sulfate	SO_4^{--}	96.06
Sulfide	S	32.06
Sulfite	SO_3	80.06
Thiosulfate	S_2O_3	112.12

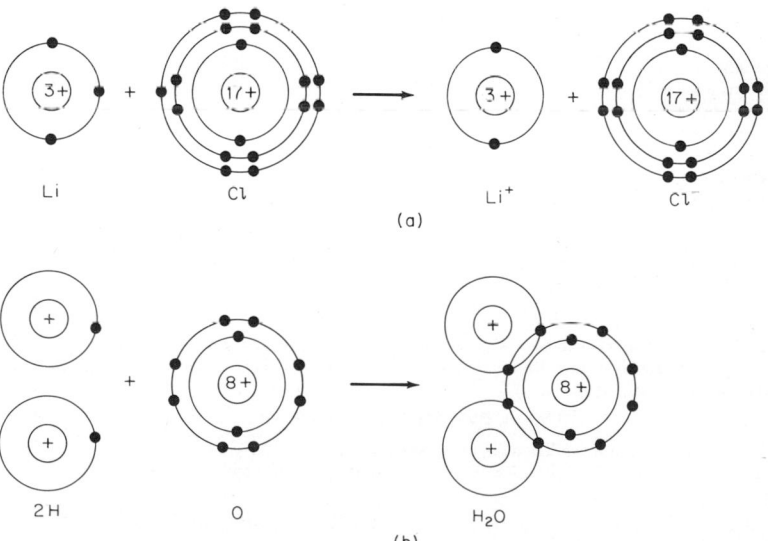

Fig. 3-4 Types of reactions. (*a*) Polar reaction: An electron moves from one atom to the other, the atoms being then held together by the electronic charges. (*b*) Covalent reaction: Electrons are shared between two atoms, the atoms then being held together to satisfy the spatial arrangement necessary for sharing.

pounds. In *covalent reactions*, the valency electrons are mutually shared by the reacting atoms, and depending on the specific nature of the reacting atoms and the resulting electronic arrangement, the resulting compound can be either stable or unstable. Reactions can also occur by a change in the arrangement of the atoms in reacting molecules. The formation of ammonia (NH_3) from nitrogen and hydrogen results from such a reaction:

$$3H_2 + N_2 \longrightarrow 2NH_3 \qquad (3\text{-}15)$$

Still other reactions can occur by a reassociation of groups of atoms. The precipitation of barium sulfate ($BaSO_4$) by mixing solutions of sodium sulfate and barium chloride is typical of such reactions:

$$Na_2SO_4 + BaCl_2 \longrightarrow 2NaCl + BaSO_4 \qquad (3\text{-}16)$$

Whether reacting individually or as radicals, the total number and types of atoms that enter into a chemical reaction must be exactly equal to the number and types of atoms that are in the products. Since each atom of a given type has a very definite weight, it must follow that the total weight of all the reactants must be equal to the total weight of all the products. Known as the *law of conservation of matter,* this forms the basis for calculations relating to the quantities of the various materials involved in chemical reactions. The fact that the total weight of materials that enter into a chemical reaction must be equal to the total weight of the products results in what is referred to as a *material balance,* and this provides the means for quantitatively determining the relationships between reactants and products.

Stoichiometry

Stoichiometry is that branch of chemistry which deals with the quantitative relationships between reactants and products. Thus, stoichiometry deals not only with the number and kinds of atoms involved in a chemical reaction but also with the balance between the total weight of reactants and products. The first step, therefore, in solving problems involving stoichiometric calculations is to balance the equation of the reaction in terms of the numbers and types of atoms involved.

Example 3-14   Balance the following skeleton equation for the reaction between iron and hydrochloric acid (an equation showing simply the reactants and products is called a skeleton equation):

$$Fe + HCl \longrightarrow FeCl_2 + H_2$$

solution   Inspection of the skeleton equation shows that the left-hand side of the equation (in the reactants) is deficient one hydrogen atom and one chlorine atom. This can be remedied by providing another molecule of HCl:

$$Fe + 2HCl \longrightarrow FeCl_2 + H_2 \qquad \text{(balanced)}$$

Example 3-15   Balance the following skeleton equation for the discharge reaction occurring in a lead-acid storage battery:

$$PbO_2 + Pb + H_2SO_4 \longrightarrow PbSO_4 + H_2O$$

solution   Inspection shows that we are deficient an atom of lead in the products. This can be corrected by adding one molecule of lead sulfate:

$$PbO_2 + Pb + H_2SO_4 \longrightarrow 2PbSO_4 + H_2O$$

(Note that the atom of lead can be added only as a constituent of one of the products, in this case lead sulfate.) However, now we are deficient in sulfate radicals in the reactants and must add one molecule of sulfuric acid:

$$PbO_2 + Pb + 2H_2SO_4 \longrightarrow 2PbSO_4 + H_2O$$

Now the lead and sulfate radicals are balanced, but we still have two extra hydrogen atoms and one extra oxygen atom in the reactants. This means that we should have two molecules of water in the products:

$$PbO_2 + Pb + 2H_2SO_4 \longrightarrow 2PbSO_4 + 2H_2O \qquad \text{(balanced)}$$

Example 3-16   Balance the following skeleton equation for the complete oxidation of glucose by oxygen:

$$C_6H_{12}O_6 + O_2 \longrightarrow CO_2 + H_2O$$

solution Balancing the carbons indicates that we get six molecules of carbon dioxide in the products:

$$C_6H_{12}O_6 + O_2 \longrightarrow 6CO_2 + H_2O$$

Balancing the hydrogens indicates that we also get six molecules of water in the products:

$$C_6H_{12}O_6 + O_2 \longrightarrow 6CO_2 + 6H_2O$$

To balance the equation, we need ten more atoms of oxygen in the reactants, and these can be supplied by five more molecules of oxygen:

$$C_6H_{12}O_6 + 6O_2 \longrightarrow 6CO_2 + 6H_2O \quad \text{(balanced)}$$

Example 3-17 Balance the following skeleton equation describing the liberation of chlorine by the action of hydrochloric acid on potassium permanganate ($KMnO_4$):

$$KMnO_4 + HCl \longrightarrow MnCl_2 + Cl_2 + H_2O + KCl$$

solution Balancing first the oxygen atoms yields

$$KMnO_4 + HCl \longrightarrow MnCl_2 + Cl_2 + 4H_2O + KCl$$

The hydrogens now need balancing:

$$KMnO_4 + 8HCl \longrightarrow MnCl_2 + Cl_2 + 4H_2O + KCl$$

This leaves only the chlorine out of balance, which we can bring into balance by increasing the potassium chloride on the product side:

$$KMnO_4 + 8HCl \longrightarrow MnCl_2 + Cl_2 + 4H_2O + 2KCl$$

However, this now throws the potassium out of balance, and further corrections are needed:

$$2KMnO_4 + 8HCl \longrightarrow MnCl_2 + Cl_2 + 4H_2O + 2KCl$$
$$2KMnO_4 + 8HCl \longrightarrow 2MnCl_2 + Cl_2 + 4H_2O + 2KCl$$
$$2KMnO_4 + 8HCl \longrightarrow 2MnCl_2 + Cl_2 + 8H_2O + 2KCl$$
$$2KMnO_4 + 16HCl \longrightarrow 2MnCl_2 + Cl_2 + 8H_2O + 2KCl$$
$$2KMnO_4 + 16HCl \longrightarrow 2MnCl_2 + 5Cl_2 + 8H_2O + 2KCl \quad \text{(balanced)}$$

Sometimes equations are written in ionic form, in which case the electrical charges must also be balanced, since each side of the equation must be electrically neutral.

Example 3-18 Balance the following equation describing the precipitation of silver chloride by calcium chloride:

$$Ag^+ + NO_3^- + Ca^{++} + Cl^- \longrightarrow AgCl + Ca^{++} + NO_3^-$$

solution This is balanced as far as the type of atoms is concerned, but the reactants side of the equation is not electrically neutral, nor are the charges balanced. Corrections are therefore needed:

$$Ag^+ + NO_3^- + Ca^{++} + 2Cl^- \longrightarrow AgCl + Ca^{++} + NO_3^-$$
$$Ag^+ + NO_3^- + Ca^{++} + 2Cl^- \longrightarrow 2AgCl + Ca^{++} + NO_3^-$$
$$2Ag^+ + NO_3^- + Ca^{++} + 2Cl^- \longrightarrow 2AgCl + Ca^{++} + NO_3^-$$
$$2Ag^+ + 2NO_3^- + Ca^{++} + 2Cl^- \longrightarrow 2AgCl + Ca^{++} + NO_3^-$$
$$2Ag^+ + 2NO_3^- + Ca^{++} + 2Cl^- \longrightarrow 2AgCl + Ca^{++} + 2NO_3^- \quad \text{(balanced)}$$

We have seen that each atom has a certain weight associated with it relative to the weight of the oxygen atom and that when the relative weight is expressed in grams, the quantity of the element present is a gram atomic weight, or simply an atomic weight. The number of atoms present in an atomic weight of any element is a fixed quantity (and equal to Avogadro's number). Thus, 16 g of oxygen contains the same number of atoms as does 32 g of sulfur, 1 g of hydrogen, or 108 g of silver, etc. In an analogous manner, when the relative weight is expressed in pounds, the quantity of the element present is a *pound atomic weight*. The number of atoms present in a pound atomic weight of any element is also a fixed quantity (and equal to 453.6 times the number present in a gram atomic weight). Thus, there are the same number of atoms in 16 lb of oxygen, 32 lb of sulfur, 1 lb of hydrogen, or 108 lb of silver. Since one atom of sulfur reacts with two atoms of oxygen to yield one molecule of sulfur dioxide:

$$S + O_2 \longrightarrow SO_2 \tag{3-17}$$

we can calculate the weights of the reactants that are exactly necessary to react fully. Thus, 32 g of sulfur will exactly react with 32 g of oxygen (i.e., 2×16) to yield 64 g, or one molecular weight, of sulfur dioxide. Similarly, 32 lb of sulfur will exactly react with 32 lb of oxygen to yield 64 lb of sulfur dioxide. Note that it is the ratios between the reacting substances and products that are important, and that these are established by the atomic and molecular weights of the materials involved in the reaction.

The number appearing in front of an atom or molecule in a chemical equation is called a *coefficient* and refers to the total number of the given atoms or molecules involved in the reaction. The number appearing slightly below an atom or radical is called a *subscript* and refers to the number of such atoms or radicals within the molecule:

2NaCl means two molecules of sodium chloride.

$4PbO_2$ means four molecules of lead dioxide.

Na_2SO_4 means one molecule of sodium sulfate consisting of two atoms of sodium, one of sulfur, and four of oxygen.

$2Cu(NO_3)_2$ means two molecules of cupric nitrate, each in turn containing two nitrate radicals for each copper atom. Note that each cupric nitrate molecule contains two nitrogen atoms (one in each nitrate radical) and six oxygen atoms (three in each nitrate radical).

$MgSO_4 \cdot 7H_2O$ means that one molecule of magnesium sulfate is associated with seven molecules of water (i.e., the hydrated form to be discussed later).

Certain atoms rarely appear as individual atoms but rather combine with their own species to appear as molecules. It is for this reason that hydrogen usually exists as a diatomic molecule whose formula is written H_2 (rather than 2H). H_2 indicates that two hydrogen atoms have combined to form an hydrogen molecule, whereas 2H would indicate that two atoms have not combined to form a molecule but rather are acting as two separate particles of matter. With these fundamental characteristics fixed, it is now possible to determine how much material has to react with other materials to produce known quantities of products. The procedures involved are illustrated by Examples 3-19 to 3-23.

Example 3-19 When heated together, zinc and sulfur react to form zinc sulfide, a white material frequently used as a pigment. How much sulfur is required to react with 100 g of zinc, and how much zinc sulfide is produced?

solution The first step in any stoichiometry problem is to write a balanced equation for the given reaction:

$$Zn + S \longrightarrow ZnS$$

In this case, the equation is balanced, and we can proceed. From Table 1-11, we find that the atomic weight of zinc is 65.4 g and the atomic weight of sulfur is 32.0 g. These will react to produce 97.4 g of zinc sulfide. These weight ratios can be written above each item in the equation, thus:

$$\begin{array}{ccc} 65.4 & 32.0 & 97.4 \\ Zn + & S \longrightarrow & ZnS \end{array}$$

(Note that the weight of the product is equal to the sum of the weights of the reactants.) Below the equation we can then write the actual weights in which we are interested:

$$\begin{array}{ccc} 65.4 & 32.0 & 97.4 \\ Zn + & S \longrightarrow & ZnS \\ 100\,g & X\,g \end{array}$$

The actual weights underneath the above equation must be in the same proportion as the relative weights above the equation, or

$$\frac{65.4}{32.0} = \frac{100}{X}$$

$$X = \frac{32.0}{65.4} \times 100 = 49.0 \text{ g}$$

In addition, the actual weight of the product (or products, as the case may be) must also be in proportion to the relative weights shown above the equation:

$$\begin{array}{ccc} 65.4 & 32.0 & 97.4 \\ Zn + & S \longrightarrow & ZnS \\ 100\,g & X\,g \end{array}$$

$$\frac{X}{100} = \frac{97.4}{65.4}$$

$$X = \frac{97.4}{65.4} \times 100 = 148.9 \text{ g}$$

or

$$\begin{array}{cccc} 65.4 & 32.0 & & 97.4 \\ \text{Zn} + & \text{S} & \longrightarrow & \text{ZnS} \\ & 49.0 \text{ g} & & X \text{ g} \end{array}$$

$$\frac{X}{49.0} = \frac{97.4}{32.0}$$

$$X = \frac{97.4}{32.0} \times 49.0 = 149.1 \text{ g}$$

check:

$$100 \text{ g} + 49.0 \text{ g} = 148.9 \text{ g}$$
$$149.0 \text{ g} = 148.9 \text{ g}$$

Example 3-20   If 50 g of hydrochloric acid is mixed with 50 g of sodium hydroxide, will the resultant mixture be acidic, alkaline, or neutral?
solution

$$\begin{array}{ccc} 36.5 & 40.0 & \\ \text{HCl} + & \text{NaOH} & \longrightarrow \text{NaCl} + \text{H}_2\text{O} \\ 50 & X & \end{array}$$

$$\frac{X}{50} = \frac{40.0}{36.5}$$

$$X = \frac{40.0}{36.5} \times 50 = 54.8 \text{ g}$$

It thus takes 54.8 g of NaOH to exactly neutralize 50 g of HCl. Since we have only 50 g of NaOH, there is a deficiency of NaOH, and the resultant mixture will therefore be acidic.

Example 3-21   How many pounds of 10 percent sulfuric acid solution are required to neutralize 80 lb of 20 percent sodium hydroxide, and what will be the strength of the resultant sodium sulfate solution?
solution   In 80 lb of 20% NaOH solution we have $80 \times 0.20 = 16$ lb of NaOH.

$$\begin{array}{ll} \text{NaOH} + \text{H}_2\text{SO}_4 \longrightarrow \text{Na}_2\text{SO}_4 + \text{H}_2\text{O} & \text{(skeleton equation)} \\ 2\text{NaOH} + \text{H}_2\text{SO}_2 \longrightarrow \text{Na}_2\text{SO}_4 + 2\text{H}_2\text{O} & \text{(balanced equation)} \end{array}$$

$$\begin{array}{cccc} 2 \times 40.0 & 98 & 142.0 & 2 \times 18.0 \\ 2\text{NaOH} + & \text{H}_2\text{SO}_4 \longrightarrow & \text{Na}_2\text{SO}_4 + & 2\text{H}_2\text{O} \end{array}$$

$$\begin{array}{cccc} 80.0 & 98.0 & 142.0 & 36.0 \\ 2\text{NaOH} + & \text{H}_2\text{SO}_4 \longrightarrow & \text{Na}_2\text{SO}_4 + & 2\text{H}_2\text{O} \\ 16 & X & & \end{array}$$

$$\frac{X}{16} = \frac{98.0}{80.0}$$

$$X = \frac{98.0}{80.0} \times 16 = 19.6 \text{ lb of H}_2\text{SO}_4$$

which is equivalent to

$$\frac{19.6}{0.10} = 196 \text{ lb of 10\% H}_2\text{SO}_4 \text{ solution}$$

Calculating now the weight of Na_2SO_4 formed:

$$\frac{X}{16} = \frac{142.0}{80.0}$$

$$X = \frac{142.0}{80.0} \times 16 = 28.4 \text{ lb}$$

and the weight of water formed:

$$\frac{X}{16} = \frac{36.0}{80.0}$$

$$X = \frac{36.0}{80.00} \times 16 = 7.2 \text{ lb}$$

The total water present at the end of the reaction is therefore the sum of the water in the original solutions plus the water formed by the neutralization reaction itself:

$$\text{Total H}_2\text{O} = 0.80(80) + 0.90(196) + 7.2 = 247.6 \text{ lb}$$

The final solution therefore weighs $28.4 + 247.6 = 276.0$ lb, and the strength of the resultant Na_2SO_4 solution is

$$\frac{28.4}{276.0} \times 100 = 10.3 \text{ percent}$$

Example 3-22   In a chemical plant 2 tons of nitrobenzene ($C_6H_5NO_2$) are converted into 1.3 tons of aniline ($C_6H_5NH_2$). What is the efficiency (yield) of this plant?
solution

$$\underset{\underset{2}{123.2}}{C_6H_5NO_2} + 3Fe + 6HCl \longrightarrow \underset{\underset{X}{93.1}}{C_6H_5NH_2} + 3FeCl2 + 2H_2O$$

$$\frac{X}{93.1} = \frac{2}{123.1}$$

$$X = \frac{2}{123.1} \times 93.1 = 1.5 \text{ tons}$$

If the plant were 100 percent efficient, it would produce 1.5 tons of $C_6H_5NH_2$. Its efficiency, or yield, is therefore

$$\frac{1.3}{1.5} \times 100 = 86.7 \text{ percent}$$

Example 3-23   Elemental phosphorus is made by heating phosphate rock [$Ca_3(PO_4)_2 + SiO_2$] in an electric furnace with coke. If the phosphate rock is 70% $Ca_3(PO_4)_2$, 25% SiO_2, and 5% inerts, while the coke is 85% C and 15% inerts, and the sand is 95% SiO_2 and 5% inerts, what should the composition of the furnace charge be, based on 1 ton of phosphate rock?

$$Ca_3(PO_4)_2 \text{ in phosphate rock} = 0.70(1) = 0.70 \text{ ton}$$
$$SiO_2 \text{ in phosphate rock} = 0.25 \text{ ton}$$

solution   Determining first the weight of SiO_2 required:

$$\underset{\underset{0.70}{310.2}}{Ca_3(PO_4)_2} + \underset{60.1}{5C} + \underset{\underset{X}{180.3}}{3SiO_2} \longrightarrow 3CaSiO_3 + 5CO + 2P$$

$$\frac{X}{180.3} = \frac{0.70}{310.2}$$

$$X = \frac{0.70}{310.2} \times 180.3 = 0.41 \text{ ton}$$

Of this, 0.25 ton is available from the phosphate rock, requiring that only 0.16 ton be made up in the form of sand. Since the sand is 95% SiO_2, the total weight of sand that will have to go into the charge is

$$\frac{0.16}{0.95} = 0.17 \text{ ton}$$

The carbon required for the reduction is similarly determined:

$$\frac{X}{60.1} = \frac{0.70}{310.2}$$

$$X = \frac{0.70}{310.2} \times 60.1 = 0.14 \text{ ton}$$

Since the carbon content of the coke is 85 percent, the total weight of coke that will have to go into the charge is

$$\frac{0.14}{0.85} = 0.17 \text{ ton}$$

The charge composition is therefore:

Phosphate rock	1 ton
Sand	0.14 ton (280 lb)
Coke	0.17 ton (340 lb)

Completion of Reactions

All the calculations above refer to reactions that go to completion, that is, reactions that come to a definite stopping point. A reaction will go to completion when at least one of the reactants is used up. In the majority of practical situations, a reaction may terminate with a surplus of one or more of the reactants still present. To terminate a reaction, not only must at least one of the reactants be used up, but something else must occur to prevent the products of the reaction from going back in the reverse direction, since this is always a theoretical possibility. Thus it is necessary to remove at least one of the reaction products from the vicinity of the reaction in order to prevent it from reversing. This is usually accomplished through the formation of a gas or an insoluble reaction product. In the case of a gas, one of the reaction products escapes from the vessel in which the reaction is taking place, so that at least one of the materials necessary to cause the reaction to go in the reverse direction is removed. In effect, the formation of an insoluble material accomplishes that same thing, since the insoluble material is removed from the solution in which the reaction is taking place. This is illustrated by Fig. 3-5. Those reactions for which no gas is formed, or which do not result in an insoluble reaction

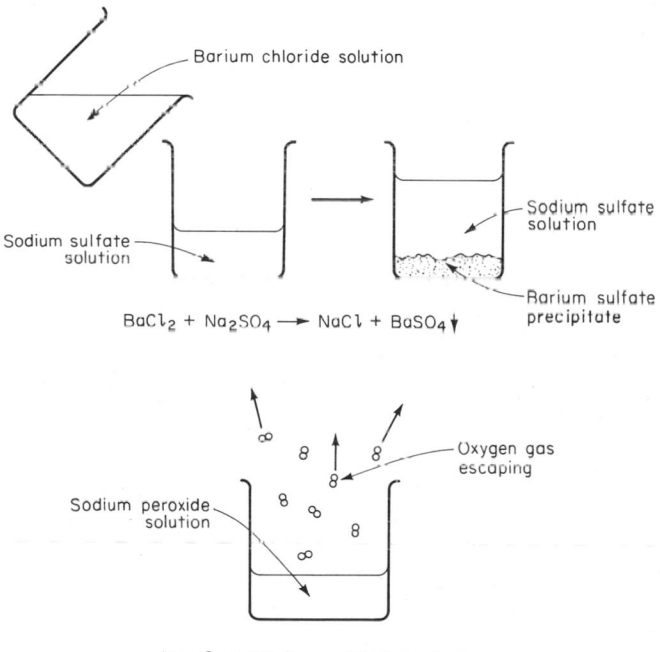

Fig. 3-5 Reactions going to completion. The downward pointing arrow in the first reaction indicates that $BaSO_4$ is insoluble and settles to the bottom. The upward pointing arrow of the second reaction indicates that O_2 is a gas and escapes into the atmosphere.

product, give rise to an *equilibrium* situation in which the reaction can indeed go in either direction, as shown in Fig. 3-6.

It is interesting to note that it is sometimes possible for an insoluble material actually to enter into a chemical reaction. In reality, there is no such thing as a truly insoluble material, since those things we generally refer to as insoluble do indeed have some slight solubility. A reaction can occur with an insoluble material if a more insoluble material results from the reaction. In this case, the more insoluble material is more thoroughly removed from the reaction vessel. Reactions of this type are illustrated in Fig. 3-7.

Equilibrium

As discussed earlier in this chapter, an equilibrium condition is one at which the concentrations of the various ions assume fixed values. It is possible to describe ionization equilibrium conditions quantitatively by means of *ionization constants*. In the case of a weak acid, such as acetic acid, only a small fraction of the molecules dissociate:

$$HC_2H_3O_2 \quad \Longrightarrow \quad H^+ + C_2H_3O_2^- \qquad (3\text{-}18)$$

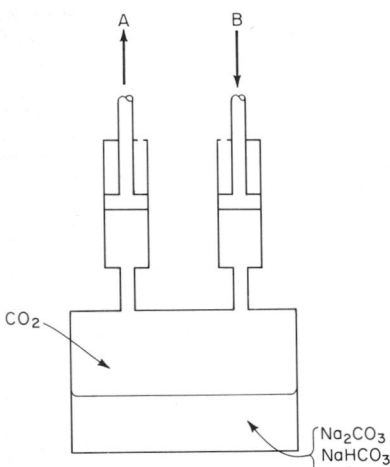

$$Na_2CO_3 + H_2O + CO_2 \rightleftharpoons 2NaHCO_3$$

Fig. 3-6 Illustration of an equilibrium reaction. Decreasing the CO_2 pressure by moving the left piston upward (as at A) tends to send the reaction to the left. Increasing the CO_2 pressure by moving the right piston down (as at B) tends to send the reaction to the right. Moving both cylinders simultaneously, and at the same rate, results in no net reaction; i.e., the reaction proceeds in both directions at the same time and at the same rates, and the system is said to be in dynamic equilibrium.

Since this reaction is at equilibrium, we can set up the following mathematical expression:

$$K_i = \frac{[\text{Conc. } H^+][\text{Conc. } C_2H_3O_2^-]}{[\text{Conc. } HC_2H_3O_2]} \qquad (3\text{-}19)$$

which defines the ionization constant K_i. The concentration of hydrogen ions in mols per liter, as written above, can be shortened to simply $[H^+]$. Since the other concentrations can be written similarly, Eq. (3-19) becomes

$$K_i = \frac{[H^+][C_2H_3O_2^-]}{[HC_2H_3O_2]} \qquad (3\text{-}20)$$

For convenience, concentrations are generally expressed in mols per liter. A solution containing 1 g mol of a material dissolved in 1 l of solution is called a *one molar* solution (abbreviated 1 M). (While molar concentrations are most often used in chemical work, occasionally concentrations are expressed in *molal* units. A *one molal* solution contains 1 g mol of the material dissolved in 1 l of water.)

Example 3-24 From acidity measurements, $[H^+]$ for a 0.10 M solution of acetic acid is found to be 1.32×10^{-3} mol/l. What is the ionization constant for acetic acid? (A 0.10 M solution of acetic acid contains 0.10 mol of acetic acid per liter.)

solution Since the ionization of acetic acid yields equal concentrations of H^+ and $C_2H_3O_2^-$ [see Eq. (3-15)],

$$[C_2H_3O_2^-] = [H^+] = 1.32 \times 10^{-3}$$

and this is also the number of mols that has dissociated from the total of 0.10 mol. This leaves a concentration of $100 \times 10^{-3} - 1.32 \times 10^{-3}$, or 98.68×10^{-3} mol of undissociated $HC_2H_3O_2$. Thus the ionization constant of acetic acid is

$$K_i = \frac{(1.32 \times 10^{-3})(1.32 \times 10^{-3})}{98.68 \times 10^{-3}} = 1.8 \times 10^{-5}$$

Ionization constants vary with temperature, usually in a fairly complex manner. Nevertheless, a great many measurements are carried out at a room temperature of 25°C, and tables of ionization constants, such as Table 3-7, usually refer to this temperature.

For the general case in which M_aN_b ionizes according to the equation

$$M_aN_b \rightleftharpoons aM^{+b} + bN^{-a} \qquad (3\text{-}21)$$

the ionization constant is given by the generalized equation

$$K_i = \frac{[M^{+b}]^a[N^{-a}]^b}{[M_aN_b]} \qquad (3\text{-}22)$$

In a similar manner, all equilibrium reactions, i.e., those which can proceed in either direction, can be described by an *equilibrium constant*. In terms of the generalized equation,

$$aA + bB + cC + \cdots \rightleftharpoons mM + nN + \cdots \qquad (3\text{-}23)$$

the equilibrium constant is defined by the following relation:

$$K_e = \frac{[M]^m [N]^n}{[A]^a [B]^b [C]^c} \tag{3-24}$$

As in the case of the ionization constant, each of the quantities in the brackets is the concentration of that material after equilibrium has been achieved. In gaseous reactions, the concentration of each gas is proportional to its partial pressure. Thus, for gaseous reactions, Eq. (3-24) can be written

$$K_e = \frac{p_M{}^m \cdot p_N{}^n}{p_A{}^a \cdot p_B{}^b \cdot p_C{}^c} \tag{3-25}$$

The specific value of K_e depends on the concentration units used, and these should be ascertained when using tables of equilibrium constants. The reason for this is that K_e can be a dimensionless number, in which case the concentration units used to calculate K_e will not affect its numerical value; or it can have dimensions of [concentration]^n, in

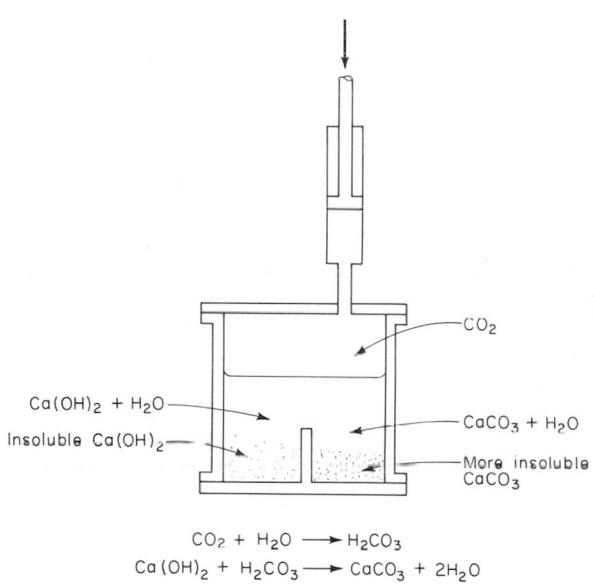

$$CO_2 + H_2O \longrightarrow H_2CO_3$$
$$Ca(OH)_2 + H_2CO_3 \longrightarrow CaCO_3 + 2H_2O$$

Fig. 3-7 Insoluble reactant forming a more insoluble product.

which case the value of K_e depends on the units in which the concentration is expressed. To illustrate this, if there are only two reactants in Eq. (3-23), and all the coefficients are equal,

$$aA + aB = aM + aN \tag{3-26}$$

the equilibrium constant will be

$$K_e = \frac{[M]^a [N]^a}{[A]^a [B]^a} \tag{3-27}$$

in which case all the concentration units will cancel out. In all other cases, however, there will be some uncanceled concentration units, so that the units of K_e depend on the values of the reaction coefficients a, b, c, etc. Table 3-8 gives the equilibrium constants for a number of gaseous reactions at certain specific temperatures. In this table, gas concentrations are expressed as partial pressures in atmospheres.

TABLE 3-7   Ionization Constants for Weak Acids and Weak Bases at 25°C

Substance	Formula	Constant
Weak acids:		
Acetic	$HC_2H_3O_2 = H^+ + C_2H_3O_2^-$	1.8×10^{-5}
Boric	$H_3BO_3 = H^+ + H_2BO_3^-$	5.8×10^{-10}
Carbonic	$H_2CO_3 = H^+ + HCO_3^-$	$K_1 = 4.5 \times 10^{-7}$
	$HCO_3^- = H^+ + CO_3^-$	$K_2 = 6 \times 10^{-11}$
Cyanic	$HCNO = H^+ + CNO^-$	2×10^{-4}
Formic	$HCHO_2 = H^+ + CHO_2^-$	2×10^{-4}
Hydrazoic	$HN_3 = H^+ + N_3^-$	1.9×10^{-5}
Hydrocyanic	$HCN = H^+ + CN^-$	4×10^{-10}
Hydrofluoric	$HF = H^+ + F^-$	7.2×10^{-4}
Hydrogen sulfide	$H_2S = H^+ + HS^-$	$K_1 = 1.0 \times 10^{-7}$
	$HS^- = H^+ + S^-$	$K_2 = 1.3 \times 10^{-13}$
Nitrous	$HNO_2 = H^+ + NO_2^-$	4.5×10^{-4}
Oxalic	$H_2C_2O_4 = H^+ + HC_2O_4^-$	$K_1 = 5.9 \times 10^{-2}$
	$HC_2O_4^- = H^+ + C_2O_4^-$	$K_2 = 6.4 \times 10^{-5}$
Phosphoric	$H_3PO_4 = H^+ + H_2PO_4^-$	$K_1 = 7.5 \times 10^{-3}$
	$H_2PO_4^- = H^+ + HPO_4^-$	$K_2 = 2 \times 10^{-7}$
	$HPO_4^- = H^+ + PO_4^-$	$K_3 = 1 \times 10^{-12}$
Phosphorus	$H_3PO_3 = H^+ + H_2PO_3^-$	$K_1 = 1.7 \times 10^{-2}$
Bisulfate ion	$HSO_4^- = H^+ + SO_4^-$	$K_2 = 1.2 \times 10^{-2}$
Sulfurous	$H_2SO_3 = H^+ + HSO_3^-$	$K_1 = 1.2 \times 10^{-2}$
	$HSO_3^- = H^+ + SO_3^-$	$K_2 = 1 \times 10^{-7}$
Weak bases:		
Ammonium hydroxide	$NH_4OH = NH_4^+ + OH^-$	1.8×10^{-5}
Methyl ammonium hydroxide	$CH_3NH_3OH = CH_3NH_3^+ + OH^-$	4.4×10^{-4}
Ethyl ammonium hydroxide	$C_2H_5NH_3OH = C_2H_5NH_3^+ + OH^-$	5.6×10^{-4}
Dimethyl ammonium hydroxide	$(CH_3)_2NH_2OH = (CH_3)_2NH_2^+ + OH^-$	7.5×10^{-4}
Trimethyl ammonium hydroxide	$(CH_3)_3NHOH = (CH_3)_3NH^+ + OH^-$	7.4×10^{-5}
Phenyl ammonium hydroxide	$C_6H_5NH_3OH = C_6H_5NH_3^+ + OH^-$	3.8×10^{-10}
Pyridinium hydroxide	$C_5H_5NHOH = C_5H_5NH^+ + OH^-$	1.4×10^{-9}
Water	$H_2O = H^+ + OH^-$	1×10^{-14}

From Conway Pierce and R. Nelson Smith, "General Chemistry Workbook," 3d ed., W. H. Freeman and Company, San Francisco, 1965.

TABLE 3-8   Equilibrium Constants for Typical Gaseous Reactions

Equilibrium	Temp (°C)	K_e
$H_2 \rightleftharpoons 2H$	1000	7.0×10^{-18}
	2000	3.1×10^{-6}
$Cl_2 \rightleftharpoons 2Cl$	1000	2.45×10^{-7}
	2000	0.570
$N_2O_4 \rightleftharpoons 2NO_2$	25	0.143
	45	0.671
$2H_2O \rightleftharpoons 2H_2 + O_2$	1000	6.9×10^{-15}
	1700	6.4×10^{-8}
$2H_2S \rightleftharpoons 2H_2 + S_2$	1130	0.0260
	1200	0.0507
$H_2 + Cl_2 \rightleftharpoons 2HCl$	1200	2.51×10^4
	1800	1.12×10^3
$SO_3 + \frac{1}{2}O_2 \rightleftharpoons SO_3$	900	6.55
	1000	1.86
$CO_2 + H_2 \rightleftharpoons CO + H_2O$	700	0.534
	1000	0.719

From Conway Pierce and R. Nelson Smith, "General Chemistry Workbook," 3d ed., W. H. Freeman and Company, San Francisco, 1965.

Example 3-25 Calculate the percentage of H_2 that dissociates into its constituent atoms when at equilibrium at a total pressure of 0.1 atm and 1000°C.

solution
$$H_2 \rightleftharpoons 2H$$

From Table 3-6,
$$K_e = 7.0 \times 10^{-18}$$

Thus
$$K_e = \frac{p_H^2}{p_{H_2}} = 7.0 \times 10^{-18}$$
$$p_H^2 = 7.0 \times 10^{-18}\,(p_{H_2})$$

Also
$$p_{H_2} + p_H = 0.1 \text{ atm}$$
$$p_H = 0.1 - p_{H_2}$$

Thus
$$(0.1 - p_{H_2})^2 - 7.0 \times 10^{-18}\,(p_{H_2})$$

This equation is solvable for p_{H_2}, but it may be tedious to do so. Consequently, the method of trial and retrial may be more convenient. Assume $p_H = 0.01$ atm; then $p_{H_2} = 0.09$ atm, and $K_e = 0.0001/0.09 g = 0.0011$ or 1.1×10^{-3}. This is much too high. Try $p_H \doteq 1.0 \times 10^{-9}$, in which case p_{H_2} will be for all intents and purposes equal to 0.1 atm. For these values,
$$K_e = \frac{1.0 \times 10^{-18}}{0.10} = 1.0 \times 10^{-17}$$

This is still too large, and p_H must be even less; try $p_H = 1.0 \times 10^{-10}$, from which
$$K_e = \frac{1.0 \times 10^{-20}}{0.10} = 1.0 \times 10^{-19}$$

This is too small; consequently p_H must lie between 1.0×10^{-9} and 1.0×10^{-10}. Try $p_H = 8.0 \times 10^{-10}$, for which
$$K_e = \frac{64.0 \times 10^{-20}}{0.10} = 64.0 \times 10^{-19} = 6.4 \times 10^{-18}$$

This is fairly close, and $p_H = 8.0 \times 10^{-10}$ atm is acceptable as an answer. If a more accurate answer is required, the process can be repeated to any degree of accuracy required. It should be noted that the problem requires an answer in terms of percentage H_2 dissociated. For a partial pressure of H equivalent to 8.0×10^{-10} atm, the partial pressure of H_2 from which it came must be 4.0×10^{-10} atm (since $H_2 \rightleftharpoons 2H$). Therefore, the percent dissociated will be
$$\frac{4.0 \times 10^{-10}}{0.10} \times 100 = 4.0 \times 10^{-7} \text{ percent}$$

Example 3-26 A 3:1 mixture, by volume, of H_2 and N_2 is charged into a pressure vessel and heated to 450°C, at which temperature the *equilibrium* pressure is found to be 55 atm. If the equilibrium composition of the gas is 9.3% NH_3, 22.6% N_2, and 68.1% H_2 by volume, what is the equilibrium constant?

solution From Sec. 3-1, the partial pressure of a gas in a mixture is proportional to the fraction of the volume of that gas in the mixture. Thus
$$p_{NH_3} = (0.093)(55 \text{ atm}) = 5.1 \text{ atm}$$
$$p_{N_2} = (0.226)(55 \text{ atm}) = 12.4 \text{ atm}$$
$$p_{H_2} = (0.681)(55 \text{ atm}) = \frac{37.5 \text{ atm}}{55.0 \text{ atm}}$$

The reaction involved is
$$N_2 + 3H_2 \rightleftharpoons 2NH_3$$

Therefore,
$$K_e = \frac{p_{NH_3}^2}{p_{N_2} \cdot p_{H_2}^3}$$
$$K_e = \frac{(5.0)^2}{12.4(37.5)^3} = 4.0 \times 10^{-5}$$

The basic equations describing equilibrium reactions indicate the manner in which some control over these reactions can be exercised. For the reaction of Eq. (3-23), increasing the concentration of N must cause a decrease in the concentration of M and a simultaneous increase in the concentrations of A, B, and C in order for the value of K_e in Eq. (3-24) to remain constant. In Example 3-26, increasing $[N_2]$ would cause the equilibrium to favor increased production of ammonia, since the basic relationship

$$K_e = \frac{p_{NH_3}^2}{p_{N_2} \cdot p_{H_2}^3}$$

must be satisfied. The practicality of this is important. Since hydrogen is much more expensive than nitrogen, and as complete a conversion of hydrogen to ammonia as possible is desired, the concentration of H_2 in the reaction vessel is made as high as possible. In a similar manner, the precipitation of barium sulfate, as shown in Eq. (3-16), can be made more complete by carrying out the precipitation in a high concentration of SO_4^{--} ion, a condition important in quantitative analyses. As a generalization, known as *Le Chatelier's principle*, whenever a system at equilibrium is disturbed, the concentrations of the materials present shift in such a manner as to counteract the disturbance and restore the equilibrium.

Solubility Products

In the case of a reaction involving the precipitation of an insoluble material, the equilibrium constant for the reaction has special significance, since the so-called "insoluble" material is soluble to some extent, and the extent of this solubility can determine the course of a reaction. When we have an insoluble material, such as barium sulfate ($BaSO_4$), the liquid is saturated with respect to $BaSO_4$ (the concentration of $BaSO_4$ in solution therefore having a fixed value). At the surface of a particle of $BaSO_4$ in contact with saturated $BaSO_4$ solution, Ba^{++} ions and SO_4^{--} ions are constantly going into solution and redepositing from the solution. Incidentally, this process results in a refinement and growth of the $BaSO_4$ crystals, since the smaller, or less perfect, crystals dissolve slightly more rapidly than the larger, or more perfect, crystals. The equilibrium reaction is

$$BaSO_4 \rightleftharpoons Ba^{++} + SO_4^{--} \tag{3-28}$$

and we can apply the mathematical relationship

$$K_e = \frac{[Ba^{++}] \cdot [SO_4^{--}]}{[BaSO_4(s)]} \tag{3-29}$$

The designation (s) denotes the presence of solid $BaSO_4$, indicating that we are dealing with a saturated solution. Equation (3-29) can be rewritten:

$$K_e \cdot [BaSO_4(s)] = [Ba^{++}] \cdot [SO_4^{--}] \tag{3-30}$$

Since $[BaSO_4(s)]$ has a fixed, i.e., constant value, $K_e \cdot [BaSO_4(s)]$ is also a constant and is called the *solubility product* K_{sp}:

$$K_{sp} = [Ba^{++}] \cdot [SO_4^{--}] \tag{3-31}$$

In more generalized form, for a salt of composition $M_a N_b$,

$$M_a N_b \rightleftharpoons aM^{+b} = bN^{-a} \tag{3-32}$$

and the solubility product is

$$K_{sp} = [M^{+b}]^a \cdot [N^{-a}]^b \tag{3-33}$$

The solubility product of a salt is dependent on temperature, since the solubility of any salt is affected by temperature. As in the case of other equilibrium constants, the value of the solubility product depends on the units in which the concentrations are expressed. Most frequently, as in the case of Table 3-9, which gives the solubility product for a number of insoluble compounds, the concentrations are usually in mols per liter.

Example 3-27 100 ml of a saturated solution of silver acetate ($AgC_2H_3O_2$) at 20°C is evaporated, leaving a dry residue weighing 0.73 g. What is the solubility product of silver acetate?
solution One liter of solution would contain $0.73 \times 10 = 7.30$ g or $7.30/166.9 = 0.0437$ mol of $AgC_2H_3O_2$; i.e., its concentration is 0.0437 mol/l.

$$AgC_2H_3O_2 \rightleftharpoons Ag^+ + C_2H_3O_2^-$$
$$K_{sp} = [Ag^+] \cdot [C_2H_3O_2^-]$$

Assuming all the dissolved material is ionized (an assumption which is very reasonable since the

TABLE 3-9   Solubility Products at 20°C

ACETATES		Fe(OH)$_2$	1×10^{-15}
		Fe(OH)$_3$	1×10^{-38}
AgOAc	2×10^{-3}	Mg(OH)$_2$	1×10^{-11}
CARBONATES		Mn(OH)$_2$	4×10^{-14}
Ag$_2$CO$_3$	8×10^{-12}	Pb(OH)$_2$	1×10^{-16}
BaCO$_3$	5×10^{-9}	Sn(OH)$_2$	1×10^{-26}
CaCO$_3$	4.8×10^{-9}	Zn(OH)$_2$	1×10^{-17}
CuCO$_3$	1×10^{-10}	OXALATES	
FeCO$_3$	2×10^{-11}		
MgCO$_3$	1×10^{-5}	CaC$_2$O$_4$	2×10^{-9}
MnCO$_3$	9×10^{-11}	MgC$_2$O$_4$	9×10^{-5}
PbCO$_3$	1×10^{-13}	BaC$_2$O$_4$	1×10^{-7}
SrCO$_3$	1×10^{-9}	SULFATES	
CHROMATES		Ag$_2$SO$_4$	1.2×10^{-5}
		BaSO$_4$	1×10^{-10}
Ag$_2$CrO$_4$	1×10^{-12}	CaSO$_4 \cdot 2H_2O$	2.4×10^{-5}
BaCrO$_4$	2×10^{-10}	Hg$_2$SO$_4$	6×10^{-7}
PbCrO$_4$	2×10^{-14}	PbSO$_4$	2×10^{-8}
SrCrO$_4$	3.6×10^{-5}	SrSO$_4$	2.8×10^{-7}
HALIDES		SULFIDES	
AgCl	1.6×10^{-10}		
AgBr	4×10^{-13}	Ag$_2$S	10^{-51}
AgI	1×10^{-16}	Bi$_2$S$_3$	10^{-72}
CaF$_2$	4×10^{-11}	CdS	10^{-28}
Hg$_2$Cl$_2$	1×10^{-18}	CoS	10^{-21}
PbCl$_2$	1.7×10^{-5}	CuS	10^{-40}
PbI$_2$	9×10^{-9}	FeS	10^{-22}
SrF$_2$	4×10^{-9}	HgS	10^{-54}
HYDROXIDES		MnS	10^{-16}
		NiS	10^{-21}
Al(OH)$_3$	1×10^{-33}	PbS	10^{-28}
Ca(OH)$_2$	8×10^{-6}	SnS	10^{-28}
Cd(OH)$_2$	1.2×10^{-14}	Tl$_2$S	10^{-22}
Cr(OH)$_3$	1×10^{-30}	ZnS	10^{-23}
Cu(OH)$_2$	6×10^{-20}		

From Lloyd E. Malm and Harper W. Frantz, "College Chemistry in the Laboratory, No. 2," W. H. Freeman and Company, San Francisco, 1954.

concentration of dissolved salt is so small), the dissolved Ag$_2$C$_2$H$_3$O$_2$ will give rise to 0.0437 mol/l of Ag$^+$, and also to 0.0437 mol/l of C$_2$H$_3$O$_2$$^{--}$. Thus

$$K_{sp} = 0.0437 \cdot 0.0437 = 0.00191 = 1.91 \times 10^{-3}$$

The solubility product is extremely useful in computing whether a precipitate will form when two solutions are mixed.

Example 3-28   Will a precipitate form if 100 ml of a solution containing 0.0001 g NaCl/ml is added to 200 ml of a solution containing 0.00308 g/l of silver nitrate (AgNO$_3$)?

solution
$$AgNO_3 + NaCl \longrightarrow AgCl \downarrow + NaNO_3$$
$$AgCl \rightleftharpoons Ag^+ + Cl^-$$

Since we have 300 ml of solution after mixing, the concentration of NaCl, and therefore of Cl$^-$ ion, is

$$[NaCl] = [Cl^-] = 100 \text{ ml} \times \frac{0.0001 \text{ g}}{\text{ml}} \times \frac{\text{mol}}{58.45 \text{ g}} \times \frac{1}{300 \text{ ml}} \times \frac{1,000 \text{ ml}}{1}$$

$$= 0.00057 \text{ mol/l}$$

In a similar manner, the concentration of silver nitrate in the final solution is

$$[AgNO_3] = [Ag^{\pm}] = 200 \text{ ml} \times \frac{0.00308 \text{ g}}{1} \times \frac{1}{1,000 \text{ ml}} \times \frac{1}{300 \text{ ml}} \times \frac{1,000 \text{ ml}}{1} \times \frac{\text{mol}}{169.89 \text{ g}}$$

$$= 1.20 \times 10^{-5} \text{ mols/l}$$
$$[Ag^+] \cdot [Cl^-] = (1.20 \times 10^{-5})(5.7 \times 10^{-4}) = 6.84 \times 10^{-9}$$

The solubility product of AgCl (from Table 3-7) $= 1.6 \times 10^{-10}$. Since $[Ag^+] \cdot [Cl^-]$ is larger than the solubility product for AgCl, the ionic concentrations are too great for the solution to tolerate and AgCl will precipitate until $[Ag^+] \cdot [Cl^-]$ equals 1.6×10^{-10}.

From the solubility of silver chloride, it is possible to calculate the concentrations of Ag^+ ion and Cl^- ion in a saturated solution of AgCl:

$$K_{sp} = [Ag^+] \cdot [Cl^-] = 1.6 \times 10^{-10}$$

Since $[Ag^+] = [Cl^-]$,

$$[Ag^+]^2 = 1.6 \times 10^{-10}$$
$$[Ag^+] = [Cl^-] = 1.27 \times 10^{-5} \text{ mols/l}$$

Now if we add additional chloride ion, as, for example, by adding sodium chloride, the amount of silver ion the solution can tolerate must be reduced, since the solubility-product relationship still governs how much Ag^+ and Cl^- can coexist in solution.

The reduction in the concentration of an ion by the addition of another ion, from some other source but common to other ions already in solution, is known as the *common-ion effect*. As can be seen, the common-ion effect can be an important means of reducing the solubility of particular ions where such is desirable. The common-ion effect is best illustrated by an actual example.

Example 3-29 A saturated solution of AgCl contains 1.27×10^{-5} mol/l of AgCl. If sodium chloride is added such that the Cl^- ion concentration is increased to 1.0×10^{-3} mol/l, what is the concentration of Ag^+ ion?

solution
$$[Ag^+] = \frac{K_{sp}}{[Cl^-]}$$

$$[Ag^+] = \frac{1.6 \times 10^{-10}}{1.0 \times 10^{-3}} = 1.6 \times 10^{-7} \text{ mols/l}$$

As can be seen, the concentration of Ag^+ ion is slightly more than $\frac{1}{100}$th after addition of NaCl as it was originally.

Frequently an insoluble material can be made to undergo a chemical reaction if the solubility of at least one of the products is less than that of the reactant. An indication of whether or not such a reaction will take place can be obtained by the relative magnitudes of the solubility products. Zinc hydroxide $[Zn(OH)_2]$ is an insoluble material having a solubility product of 1×10^{-17}. The addition of a soluble sulfide such as hydrogen sulfide (H_2S) will convert the $Zn(OH)_2$ to ZnS:

$$Zn(OH)_2 + H_2S \longrightarrow ZnS + 2H_2O \qquad (3\text{-}34)$$

Since the solubility product of ZnS is low, 1×10^{-23}, the concentration of zinc ions that a solution can tolerate in the presence of sulfide ions is much less than can be tolerated in the presence of hydroxyl ions.

Example 3-30 Will $1 M$ HCl dissolve copper sulfide (CuS)? ($1 M$ is the abbreviation for 1 molar, or 1 mol/l.)
solution From Table 3-7, K_{sp} for CuS is 1×10^{-40}. Since $CuCl_2$, the copper salt that would result from the reaction, is soluble, and since only a very small concentration of copper ions can be tolerated by a solution also containing sulfide ions, no appreciable amount of solubility will occur.

Example 3-31 If bismuth nitrate $[Bi(NO_3)_3]$ solution were added to copper sulfide, would the copper dissolve?

solution
$$CuS \rightleftharpoons Cu^{++} + S^{--}$$
$$Bi_2S_3 \rightleftharpoons 2Bi^{3+} + 3S^{--}$$

These reactions indicate that Cu^{++} ions and Bi^{3+} ions would compete for the available S^{--} ions. From the copper sulfide, since $[Cu^{++}] = [S^{--}]$, $[S^{--}]$ would be

$$[S^{--}]^2 = 1 \times 10^{-40}$$
$$[S^{--}] = 1 \times 10^{-20}$$

Similarly, from the Bi_2S_3 (for which $[Bi^{3+}] = \frac{2}{3}[S^{--}]$),

$$(\tfrac{2}{3}[S^{--}])^2([S^{--}])^3 = 1 \times 10^{-72}$$
$$\tfrac{4}{9}[S^{--}]^2[S^{--}]^3 = 1 \times 10^{-72}$$

$$0.44[S^{--}]^5 = 1 \times 10^{-72}$$
$$[S^{--}]^5 = 2.25 \times 10^{-72}$$
$$[S^{--}] = 1.2 \times 10^{-14.4}$$

Thus the concentration of sulfide ions that a solution can tolerate in the presence of Cu^{++} is much smaller than what it can tolerate in the presence of Bi^{3+}, and $Bi(NO_3)_3$ solution will not dissolve CuS. In this case, it is interesting to note that K_{sp} for Bi_2S_3 is much smaller than K_{sp} for CuS, and a simple inspection of solubility products could lead to the erroneous conclusion that bismuth nitrate dissolves copper sulfide.

Chelation

So far we have been dealing with salts that are easily ionized. It is also possible to have soluble salts that are so slightly ionized as to yield very few ions, even though the salt itself is in true solution. Thus it is possible to remove ions from a solution by tying them up in a soluble, but only slightly ionized, compound in much the way the formation of an insoluble salt removes ions from solution. The process of forming slightly ionized, but soluble, compounds is called *chelation*, and those materials which can do this efficiently are called *chelating agents*. Usually, chelating agents are organic compounds and are specific for a single, or a small group of ions. Thus, an iron-chelating agent can dissolve rust by forming an iron chelate:

$$Fe(OH)_3 + \text{iron-chelating agent} \longrightarrow \text{iron chelate} + 3\,OH \qquad (3\text{-}35)$$

The unique property of the iron chelate is that even though it is highly soluble, the concentration of Fe^{3+} ions it yields is less than the $[Fe^{3+}]$ from "insoluble" ferric hydroxide. [Ferric hydroxide can be considered as a "hydrated" rust, i.e., $Fe_2O_3 + 3H_2O \rightleftharpoons 2Fe(OH)_3$.]

Chapter 4

Review of Fundamentals (Part 2)

4-1 SOLUTIONS

Types of Solutions

A solution is a homogeneous mixture in which one or more materials are dispersed in another material. The materials being dispersed, or dissolved, are called *solutes*, and the material in which they are dispersed is called the *solvent*. There can be solutions of solids in a solid, such as glass, in which metallic oxides are dissolved in silicon dioxide (SiO_2), or solder, in which tin is dissolved in lead. Liquids can be dissolved in other liquids, such as alcohol dissolved in water. When two liquids can dissolve in each other in all proportions, such as alcohol and water, they are said to be *miscible*. Liquids which are not mutually soluble, such as water and gasoline, are *immiscible*. Gas-gas solutions are also common, air, for example, being a solution of oxygen in nitrogen. It is interesting to note that all gases are completely miscible with any other gas or gases. In the case of solid-solid, liquid-liquid, and gas-gas solutions, it is not always clear which component is the solute and which is the solvent. By convention, the minor component, in terms of amount present, is considered to be the solute, the major component being the solvent.

In many other cases, solutions can consist of combinations of solids, liquids, or gases. Sugar dissolved in water is an example of a solid dissolved in a liquid. Ammonium hydroxide (NH_4OH) is ammonia (NH_3, a gas) dissolved in water (a liquid), and so on. Frequently, a solid or liquid is dispersed in another liquid without truly dissolving. In such a dispersion, the two entities, or *phases*, can still be distinguished and will separate if allowed to stand. However, if the dispersed material is fine enough, it will remain suspended in the solvent. Such a dispersion is known as a *colloidal suspension*, or frequently (but incorrectly) as a colloidal solution. The formation of solutions is illustrated in Fig. 4-1.

In addition to the composition of the solute and solvent, solutions are defined in terms of how much solute and solvent are present, or their *concentration*. There are a great many ways of measuring solution concentrations, but basically these all resolve into a weight or volume of solute per unit weight or volume of solvent, or per unit weight or volume of total solution. The most common concentration terms are described below.

Concentrations expressed in *volume percent* refer to the volume of the solute expressed as a percent of the total volume of solution. Generally speaking, volume percents are used for liquid solutes in liquid solvents, and for gas mixtures. For solids dissolved in liquids, or solids dissolved in solids, *weight percent* is commonly used to express the concentration of the solution. Weight percent refers to the weight of solute as a percent of the total weight of solution. Frequently, it is convenient to use a mixed basis, such as *weight-volume percent*.

Weight-volume percent refers to the weight of solute expressed as a percent of the volume of solvent. In expressing concentrations in terms of mixed percents, the specific units of weight and volume used must be stipulated.

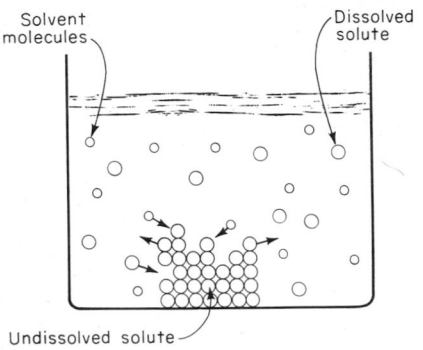

Solvent molecules

Dissolved solute

Undissolved solute

Fig. 4-1 Process of dissolution. Solvent molecules impacting undissolved solute to free solute particles to go into solution. Solute molecules also return to the undissolved solute. At equilibrium, the rate at which solute particles dissolve is equal to the rate at which they crystallize out of solution.

Example 4-1 A quantity of toluene having a volume of 78 cm^3 is dissolved in xylene. If the volume of the final solution is 0.5 l, what is the volume percent concentration of the solution?

solution

$$\frac{78}{500} \times 100 = 15.6 \text{ percent}$$

Example 4-2 5 gal of kerosene are used to dissolve 3 lb of paraffin. What is the concentration of the solution expressed as a weight-volume percent based on pounds and liters?

solution   From Table 1-5, 1 gal = 3.785 l. The volume of kerosene used is therefore 5 × 3.785 = 18.93 l, and the solution concentration in weight-volume percent is

$$\frac{3}{18.93} \times 100 = 15.8 \text{ percent}$$

In laboratory work, solution concentrations are generally expressed as a weight of solute per unit of solution, or per unit of solvent volume. One of the most common means for expressing concentrations is in mols per liter. In this case, the mol is used as the unit of weight. A *molar* solution (abbreviated 1 *M*) contains one molecular weight of the substance in grams, dissolved in sufficient solvent to give exactly 1 l of solution. A solution containing 1½ molecular weights of the substance in 1 l of solution would be a 1.5 *M* solution. A *molal* solution contains 1 molecular weight of the substance dissolved in 1 l of solvent. No abbreviation is used for the word molal, to avoid confusion with the more commonly used molar concentrations.

The concept of equivalent weights was developed in Chap. 3. An equivalent (i.e., an equivalent weight) is also a means of expressing weight, and can therefore also be used in expressing solution concentrations. A solution containing one equivalent weight of a substance in 1 l of solution is said to be a *normal* solution (abbreviated 1*N*). A 2*N* solution would be one with two equivalents per liter, and so on.

Example 4-3 How much silver nitrate (AgNO$_3$) is dissolved in 2.5 l in a 0.5 *M* solution?
solution The molecular weight of AgNO$_3$ is 169.9. There must therefore be 2.5 × 169.9 = 425 g of AgNO$_3$ in 2.5 l of a 1 *M* solution. For 2.5 l of 0.5 *M* solution, there would be 0.5 × 425 = 212.5 g of AgNO$_3$ present.

Example 4-4 10 g of AgNO$_3$ is dissolved in 1 l of solution. What is the molarity of this solution?
solution The number of molecular weights of AgNO$_3$ present is 10/169.9 = 0.06. The solution concentration is therefore 0.06 *M*.

Example 4-5 How much sulfuric acid must be dissolved to make 1 l of 1*N* solution?
solution The molecular weight of H$_2$SO$_4$ is 98.08, but since one molecular weight of sulfuric acid can neutralize two molecular weights of sodium hydroxide, its equivalent weight is half of its molecular weight, or 49.04 g, and this is therefore the amount of H$_2$SO$_4$ present in 1 l of a 1*N* solution.

Example 4-6 What is the normality of a solution of H$_2$SO$_4$ having a specific gravity of 1.066?
solution From Table 4-1, this would be a 10 percent solution. The weight of 1 l of this solution would be

$$1,000 \times 1.066 = 1,066 \text{ g}$$

of which 10 percent, or 106.6 g, is H_2SO_4. Since the equivalent weight of H_2SO_4 is 49.04, the number of equivalents present would be

$$\frac{106.6}{49.04} = 2.17$$

The solution is therefore 2.17 N.

The importance of expressing solution strengths in normality is that equal volumes of solutions of the same normality will exactly react with each other. This results directly from the fact that equal volumes of solutions of the same normality will contain the same number of equivalents, and we recall that one of the properties of an equivalent weight of a substance is that it will exactly react with an equivalent weight of another material. The concept of solution normality is of considerable importance in analytical work. In a procedure called *titration*, the amount of solution of known concentration, referred to as a *standard solution*, required to react with a given amount of solution of unknown concentration is determined. This is done, as illustrated in Fig. 4-2, by slowly admitting a standard solution from a buret into a flask containing the unknown solution until an *indicator*, such as litmus in an acid-base neutralization, shows that the reaction between the two solutions has been completed. The volume of standard solution used is then read directly from the buret, or volume-calibrated glass tube.

Example 4-7   What volume of 1.35N acid is needed to neutralize 39.4 cm^3 of a 1.16N base?

solution $V_{acid} \times N_{acid} = V_{base} \times N_{base}$
$$1.35\, V_{acid} = 39.4 \times 1.16$$
$$V_{acid} = 33.9\ cm^3$$

For a given solute dissolved in a given solvent, the specific gravity of the solution will depend upon its concentration. Specific gravity, because of the ease with which it can be measured (as described in Chap. 1) is commonly used to determine solution concentrations, and a great many tables, such as Table 4-1, relating the concentrations of particular solutions with their specific gravities, are available.

Example 4-8   The concentration of a brine (NaCl solution) is 3.5 percent. How much salt is present in 1,000 gal of this solution?

solution   From Table 4-1, a 2% NaCl solution has a specific gravity of 1.0125, and that of a 4 percent solution is 1.0268. By interpolation, the specific gravity of a 3.5 percent solution is

$$\frac{X - 1.0125}{1.0268 - 1.0125} = \frac{3.5 - 2}{4 - 2}$$

$$\frac{X}{0.0143} - \frac{1.0125}{0.0143} = \frac{1.5}{2}$$

$$\frac{X}{0.0143} = \frac{1.5}{2} + \frac{1.0125}{0.0143}$$

$$X = 0.0107 + 1.0125$$

sp. gr. of 3.5 percent solution = 1.0232

Since the density of water is 8.345 lb/gal, the density of the brine is 1.0232 (8.345) = 8.54

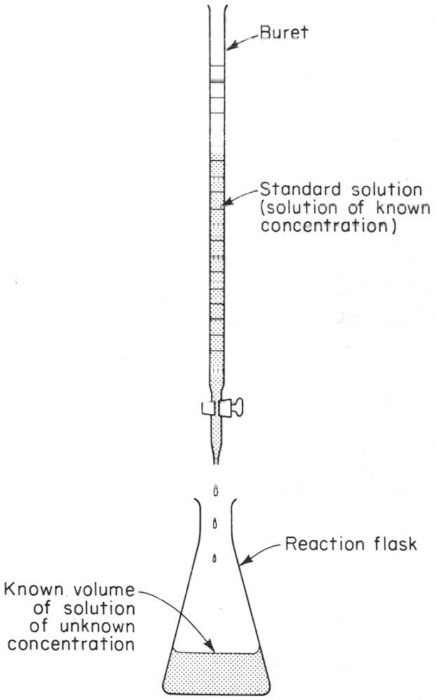

Fig. 4-2 Titration. Standard solution from the buret is carefully added to the known volume of solution of unknown concentration until an indicator shows that the standard solution has exactly reacted with the solution in the flask. Usually the indicator will undergo a change in color at the end point.

lb/gal. The weight of 1,000 gal of brine is therefore 1,000 (8.54) = 8,540 lb, and the amount of salt present is

$$8,540 \times 0.035 = 299\ lb$$

Since specific gravity was one of the earliest forms of control of chemical operations, certain industries developed their own systems for reporting solution concentrations, and

TABLE 4-1   Specific Gravities of Aqueous Solutions
(All values for room temperature unless otherwise indicated)

INORGANIC SOLUTIONS

Ammonium Chloride (NH₄Cl)

%	0°C.	10°C.	20°C.	30°C.	50°C.	80°C.	100°C.
1	1.0033	1.0029	1.0013	0.9987	0.910	0.9749	0.9617
2	1.0067	1.0062	1.0045	1.0018	.940	.9780	.9651
4	1.0135	1.0126	1.0107	1.0077	.999	.9842	.9718
8	1.0266	1.0251	1.0227	1.0195	1.0116	.9963	.9849
12	1.0391	1.0370	1.0344	1.0310	1.0231	1.0081	.9975
16	1.0510	1.0485	1.0457	1.0422	1.0343	1.0198	1.0096
20	1.0625	1.0596	1.0567	1.0532	1.0454	1.0312	1.0213
24	1.0736	1.0705	1.0674	1.0641	1.0564	1.0426	1.0327

Ammonium Chromate [(NH₄)₂CrO₄]

%	d	°C.
3.80	1.0219	20
10.52	1.0627	13.7
19.75	1.1189	19.6
28.04	1.1707	

Ammonium Nitrate (NH₄NO₃)

%	0°C.	10°C.	25°C.	40°C.	60°C.	80°C.
1.0	1.0043	1.0039	1.0011	0.9961	0.9870	0.9755
2.0	1.0088	1.0082	1.0051	1.0000	.9908	.9793
4.0	1.0178	1.0168	1.0132	1.0079	.9985	.9869
8.0	1.0358	1.0340	1.0297	1.0238	1.0142	1.0024
12.0	1.0539	1.0515	1.0464	1.0400	1.0301	1.0181
16.0	1.0721	1.0691	1.0633	1.0565	1.0462	1.0342
20.0	1.0905	1.0870	1.0806	1.0734	1.0627	1.0506
24.0	1.1090	1.1051	1.0982	1.0907	1.0796	1.0673
28.0	1.1277	1.1234	1.1161	1.1082	1.0968	1.0844
30.0	1.1371	1.1327	1.1252	1.1171	1.1055	1.0931
40.0	1.1862	1.1810	1.1727	1.1640	1.1515	1.1385
50.0	1.2380	1.2320	1.2229	1.2136	1.2006	1.1868

Ammonium Sulfate [(NH₄)₂SO₄]

%	0°C.	20°C.	40°C.	80°C.	100°C.
1	1.0061	1.0041	0.980	0.9777	0.9644
2	1.0124	1.0101	1.0039	.9836	.9705
4	1.0248	1.0220	1.0155	.9953	.9826
8	1.0495	1.0456	1.0387	1.0187	1.0066
12	1.0740	1.0691	1.0619	1.0421	1.0303
16	1.0980	1.0924	1.0849	1.0653	1.0539
20	1.1215	1.1154	1.1077	1.0883	1.0772
24	1.1448	1.1383	1.1304	1.1111	1.1003
28	1.1677	1.1609	1.1529	1.1338	1.1232
35	1.2072	1.2000	1.1919	1.1731	1.1629
40	1.2350	1.2277	1.2196	1.2011	1.1910
50	1.2899	1.2825	1.2745	1.2568	1.2466

Aluminum Sulfate [Al₂(SO₄)₃]

%	d	%	d
1	1.0093	16	1.1770
2	1.0195	20	1.2272
4	1.0404	24	1.2803
8	1.0837	26	1.3079
12	1.1293		

Ammonia (NH₃)

%	-15°C.	-10°C.	-5°C.	0°C.	5°C.	10°C.	20°C.	25°C.
1	....	0.9943	0.9954	0.9959	0.9958	0.9955	0.9939	0.993
2	....	.9906	.9915	.9919	.9917	.9913	.9895	.988
4	....	.9834	.9840	.9842	.9837	.9832	.9811	.980
8	0.970	.9701	.9701	.9695	.9686	.9677	.9651	.964
12	.958	.9576	.9571	.9561	.9548	.9534	.9501	.948
16	.947	.9461	.9450	.9435	.9420	.9402	.9362	.934
20	....	.9353	.9335	.9316	.9296	.9275	.9229	
24	....	.9249	.9226	.9202	.9179	.9155	.9101	
28	....	.9150	.9122	.9094	.9067	.9040	.8980	
30	....	.9101	.9070	.9040	.9012	.8983	.8920	

%	d
32	0.889
36	.877
40	.865
45	.849
50	.832
60	.796
70	.755
80	.711
90	.665
100	.618

Ammonium Acetate (CH₃COONH₄)

%	d
1	0.9992
2	1.0013
4	1.0055
8	1.0136
12	1.0216
16	1.0294
20	1.0368
24	1.0439
28	1.0507
30	1.0540
35	1.0618
40	1.0691
45	1.0760

Ammonium Bichromate [(NH₄)₂Cr₂O₇]

%	d
1	1.0051
2	1.0108
4	1.0223
8	1.0463
12	1.0715
16	1.0981
20	1.1263

Calcium Hypochlorite* (CaOCl₂)

% total salt	d
2	1.0169
4	1.0345
6	1.0520
8	1.0697
10	1.0876
12	1.1060

* $CaOCl_2 = 89.15\%$;
$CaCl_2 = 7.31\%$;
$Ca(ClO_3)_2 = 0.26\%$;
$Ca(OH)_2 = 2.92\%$.

Calcium Hydroxide [Ca(OH)₂]

%	d	d
0.05	0.9979	0.99773
.10	1.0044	.99833
.15	1.0110	.99904

Calcium Nitrate [Ca(NO₃)₂]

%	6°C.	18°C.	25°C.	30°C.
2*	1.0157	1.0137	1.0120	1.0105
4	1.0316	1.0291	1.0272	1.0256
8	1.0641	1.0608	1.0585	1.0565
12	1.0979	1.0937	1.091	1.0887
16	1.1350	1.1279	1.1250	1.1224
20	1.1694	1.1636	1.1602	1.1575
25	1.2168	1.2106	1.2065	1.2032
30	.....	1.260		
35	.....	1.311		
40	.....	1.365		
45	.....	1.422		

Chromic Acid (CrO₃)

%	d
1	1.006
2	1.014
6	1.045
10	1.076
16	1.27

%	d
20	1.163
26	1.220
30	1.260
40	1.371
50	1.505
60	1.663

Arsenic Acid (H₃AsO₄)

%	d
1	1.0057
2	1.0124
6	1.0398
10	1.0681
16	1.1128

%	d
20	1.1447
30	1.231
40	1.3370
50	1.4602
60	1.6070
70	1.7811

Barium Chloride (BaCl₂)

%	0°C.	20°C.	40°C.	60°C.	80°C.	100°C.
2	1.0181	1.0159	1.0096	1.0004	0.9890	0.9755
4	1.0368	1.0341	1.0275	1.0181	1.0066	.9931
8	1.0760	1.0721	1.0646	1.0551	1.0434	1.0299
12	1.1178	1.1128	1.1047	1.0948	1.0827	1.0692
16	1.1627	1.1564	1.1478	1.1373	1.1249	1.1113
20	1.2105	1.2031	1.1938	1.1828	1.1702	1.1563
24	.....	1.2531	1.2430	1.2316	1.2186	1.2045
26	.....	1.2793	1.2688	1.2571	1.2440	1.2298

Cadmium Nitrate [Cd(NO₃)₂]

%	d
2	1.0154
4	1.0326
8	1.0683
12	1.1061
16	1.1468

%	d
20	1.1904
25	1.2488
30	1.3124
40	1.4590
50	1.6556

Calcium Chloride (CaCl₂)

%	-5°C.	0°C.	20°C.	30°C.	40°C.	60°C.	80°C.	100°C.	120°C.	140°C.
2		1.0171	1.0148	1.0120	1.0084	0.9994	0.9881	0.9748	0.9596	0.9428
4		1.0346	1.0316	1.0286	1.0249	1.0158	1.0046	.9915	.9765	.9601
8	1.0708	1.0703	1.0659	1.0626	1.0585	1.0492	1.0382	1.0257	1.0111	.9954
12	1.1083	1.1072	1.1015	1.0978	1.0937	1.0840	1.0730	1.0610	1.0466	1.0317
16	1.1471	1.1454	1.1386	1.1345	1.1301	1.1202	1.1092	1.0973	1.0835	1.0691
20	1.1874	1.1853	1.1775	1.1730	1.1684	1.1581	1.1471	1.1352	1.1219	1.1080
25	.....	1.2376	1.2284	1.2236	1.2186	1.2079	1.1965	1.1846		
30	.....	1.2922	1.2816	1.2764	1.2709	1.2597	1.2478	1.2359		
35	.....	.....	1.3373	1.3316	1.3255	1.3137	1.3013	1.2893		
40	.....	.....	1.3957	1.3895	1.3826	1.3700	1.3571	1.3450		

TABLE 4-1   Specific Gravities of Aqueous Solutions—(Continued)

INORGANIC SOLUTIONS

Ferric Nitrate [Fe(NO₃)₃]

%	d (20°C)
1	1.0065
2	1.0144
4	1.0304
8	1.0636
12	1.0989
16	1.1359
20	1.1748
25	1.2281

Ferrous Sulfate (FeSO₄)

%	15°C	18°C	20°C
0.2	.....	1.00068	1.0002
0.4	.....	1.00275	1.0022
0.8	.....	1.00645	1.0062
1.0	1.0090	1.0085	1.0082
4.0	1.0380	1.0375	
8.0	1.0790	1.0785	
12.0	1.1235	1.1220	
16.0	1.1690	1.1675	
20.0	1.2150	1.2135	

Hydrogen Cyanide (HCN)

%	d
1	0.998
2	.996
4	.993
8	.984
12	.971
16	.956
82	.752
90	.724
100	.691

Ferric Sulfate [Fe₂(SO₄)₃]

%	d
1	1.0072
2	1.0157
4	1.0327
8	1.0670
12	1.1028
16	1.1409
20	1.1811
30	1.3073
40	1.4487
50	1.6127
60	1.7983

Hydrogen Bromide (HBr)

%	4°C	10°C	25°C
1.0	1.0073	1.0068	1.0041
2.0	1.0146	1.0139	1.0111
4.0	1.0295	1.0285	1.0255
6.0	1.0448	1.0435	1.0402
8.0	1.0604	1.0589	1.0552
10.0	1.0764	1.0747	1.0707
12.0	1.0928	1.0910	1.0867
14.0	1.1097	1.1078	1.1032
16.0	1.1272	1.1251	1.1202
18.0	1.1453	1.1430	1.1377
20.0	1.1640	1.1615	1.1557
22.0	1.1832	1.1806	1.1743
24.0	1.2030	1.2003	1.1935
26.0	1.2235	1.2206	1.2134
28.0	1.2446	1.2415	1.2340
30.0	1.2663	1.2630	1.2552
40.0	1.3877	1.3838	1.3736
50.0	1.5305	1.5257	1.5127
60.0	1.6950	1.6892	1.6731
65.0	1.7854	1.7792	1.7613

Chromium Chloride (CrCl₃)

%	Violet (d)	Green (d)	Equilibrium mixture of violet and green (d)
1	1.0076	1.0071	1.0075
2	1.0166	1.0157	1.0165
4	1.0349	1.0332	1.0347
8	1.0724	1.0691	1.0722
12	1.1114	1.1065	1.1111
14	1.1316		

Copper Nitrate [Cu(NO₃)₂]

%	d
12	1.107
16	1.147
20	1.189
25	1.248

Copper Sulfate (CuSO₄)

%	0°C	20°C	40°C
1	1.0104	1.0086	1.0024
4	1.0429	1.0401	1.0332
8	1.0887	1.084	1.0764
12	1.1379	1.1308	1.1222
16	.....	1.180	.....
18	.....	1.206	.....

Cuprous Chloride (Cu₂Cl₂)

%	0°C	20°C	40°C
1	1.0095	1.0072	1.002
4	1.0387	1.036	1.0305
8	1.0788	1.0754	1.0682
12	1.1208	1.1165	1.107
16	1.1653	1.1595	1.151
20	1.2121	1.2052	1.1953

Ferric Chloride (FeCl₃)

%	0°C	10°C	20°C	30°C
1	1.0086	1.0084	1.0068	1.0040
2	1.0174	1.0168	1.0152	1.0122
4	1.0347	1.0341	1.0324	1.0292
8	1.0703	1.0692	1.0669	1.0636
12	1.1088	1.107	1.1040	1.1006
16	1.1475	1.1449	1.1418	1.1386
20	1.1870	1.1847	1.1820	1.1786
25	1.2400	1.2380	1.2340	1.2290
30	1.2970	1.2950	1.2910	1.2850
35	1.3605	1.3580	1.3530	1.3475
40	1.4280	1.4235	1.4175	1.4115
45	.....	1.4920	1.4850	.....
50	.....	1.5610	1.5510	.....

Magnesium Chloride (MgCl₂)

%	0°C.	20°C.	40°C.	60°C.	80°C.	100°C.
2	1.0168	1.0146	1.0084	0.9995	0.9883	0.9753
4	1.0338	1.0311	1.0248	1.0159	1.0050	.9923
8	1.0683	1.0646	1.0580	1.0493	1.0388	1.0269
12	1.1035	1.0989	1.0921	1.0836	1.0735	1.0622
16	1.1395	1.1342	1.1272	1.1188	1.1092	1.0984
20	1.1764	1.1706	1.1635	1.1552	1.1460	1.1359
25	1.2246	1.2184	1.2111	1.2031	1.1942	1.1847
30	1.2754	1.2688	1.2614	1.2535	1.2451	1.2360

Magnesium Sulfate (MgSO₄)

%	0°C.	20°C.	30°C.	40°C.	50°C.	60°C.	80°C.
2	1.0210	1.0186	1.0158	1.0123	1.0081	1.0032	0.9916
4	1.0423	1.0392	1.0362	1.0326	1.0283	1.0234	1.0118
8	1.0858	1.0816	1.0782	1.0743	1.0700	1.0650	1.0534
12	1.1309	1.1256	1.1220	1.1179	1.1135	1.1083	1.0968
16	1.1777	1.1717	1.1679	1.1637	1.1592		
20	1.2264	1.2198	1.2159	1.2117	1.2072		
26	1.3052	1.2951	1.2922	1.2879	1.2836		

Nickel Sulfate (NiSO₄)

%	d (80°C.)
1	1.0091
2	1.0198
4	1.0415
8	1.0852
12	1.1325
16	1.1825
18	1.2090

Nickel Nitrate [Ni(NO₃)₂]

%	d
1	1.0065
2	1.0150
4	1.0325
8	1.0688
12	1.1070
16	1.1480
20	1.191
30	1.311
35	1.377

Nickel Chloride (NiCl₂)

%	d
1	1.0082
2	1.0179
4	1.0375
8	1.0785
12	1.1217
16	1.1674
20	1.2163
30	1.353

Hydrogen Chloride (HCl)

%	-5°C.	0°C.	10°C.	20°C.	40°C.	60°C.	80°C.	100°C.
1	1.0048	1.0048	1.0052	1.0032	0.9973	0.9861	0.9768	0.9636
2	1.0104	1.0106	1.0100	1.0082	1.0019	.9950	.9819	.9688
4	1.0213	1.0213	1.0202	1.0181	1.0116	1.0026	.9919	.9791
6	1.0321	1.0319	1.0303	1.0279	1.0211	1.0121	1.0016	.9892
8	1.0428	1.0423	1.0403	1.0376	1.0305	1.0215	1.0111	.9992
10	1.0536	1.0528	1.0504	1.0474	1.0400	1.0310	1.0206	1.0090
12	1.0645	1.0634	1.0607	1.0574	1.0497	1.0406	1.0302	1.0188
14	1.0754	1.0741	1.0711	1.0675	1.0594	1.0502	1.0398	1.0286
16	1.0864	1.0849	1.0815	1.0776	1.0692	1.0598	1.0494	1.0383
18	1.0975	1.0958	1.0920	1.0878	1.0790	1.0694	1.0590	1.0479
20	1.1087	1.1067	1.1025	1.0980	1.0888	1.0790	1.0685	1.0574
22	1.1200	1.1177	1.1131	1.1083	1.0986	1.0886	1.0780	1.0668
24	1.1314	1.1287	1.1238	1.1187	1.1085	1.0982	1.0874	1.0761
26	1.1426	1.1396	1.1344	1.1290	1.1183	1.1076	1.0967	1.0853
28	1.1537	1.1505	1.1449	1.1392	1.1280	1.1169	1.1058	1.0942
30	1.1648	1.1613	1.1553	1.1493	1.1376	1.1260	1.1149	1.1030
32	....	....	....	1.1593				
34				1.1691				
36				1.1789				
38				1.1885				
40				1.1980				

Hydrogen Peroxide (H₂O₂)

%	d	%	d
1	1.0022	26	1.0959
2	1.0053	28	1.1040
4	1.0131	30	1.1122
6	1.0204	35	1.1327
8	1.0277	40	1.1536
10	1.0351	45	1.1749
12	1.0425	50	1.1966
14	1.0499	55	1.2188
16	1.0574	60	1.2416
18	1.0649	70	1.2897
20	1.0725	80	1.3406
22	1.0802	90	1.3931
24	1.0880	100	1.4465

Hydrogen Fluoride (HF)

%	20°C.	0°C.
5	1.020	1.017
10	1.040	1.035
20	1.080	1.070
30	1.119	1.101
40	1.159	1.130
50	1.198	1.155
60	1.235	
70	1.258	
80	1.259	
90	1.178	
95	1.089	
100	1.0005	

Hydrofluosilic Acid (H₂SiF₆)

%	d	%	d
1	1.0080	16	1.1373
2	1.0161	20	1.1748
4	1.0324	25	1.2235
8	1.0661	30	1.2742
12	1.1011	34	1.3162

TABLE 4-1 Specific Gravities of Aqueous Solutions—(Continued)

INORGANIC SOLUTIONS

Nitric Acid (HNO₃)

%	0°C.	5°C.	10°C.	15°C.	20°C.	25°C.	30°C.	40°C.	50°C.	60°C.	80°C.	100°C.
1	1.0058	1.00572	1.00534	1.00464	1.00364	1.00241	1.0009	0.9973	0.9931	0.9882	0.9767	0.9632
2	1.0117	1.01149	1.01099	1.01018	1.00909	1.00778	1.0061	1.0025	.9982	.9932	.9816	.9681
3	1.0176	1.01730	1.01668	1.01576	1.01457	1.01318	1.0114	1.0077	1.0033	.9982	.9865	.9730
4	1.0236	1.02315	1.02240	1.02137	1.02008	1.01861	1.0168	1.0129	1.0084	1.0033	.9915	.9779
5	1.0296	1.02904	1.02816	1.02702	1.02563	1.02408	1.0222	1.0182	1.0136	1.0084	.9965	.9829
6	1.0357	1.03497	1.03397	1.03272	1.03122	1.02958	1.0277	1.0235	1.0188	1.0136	1.0015	.9879
7	1.0418	1.0410	1.0399	1.0385	1.0369	1.0352	1.0333	1.0289	1.0241	1.0188	1.0066	.9929
8	1.0480	1.0471	1.0458	1.0443	1.0427	1.0409	1.0389	1.0344	1.0295	1.0241	1.0117	.9980
9	1.0543	1.0532	1.0518	1.0502	1.0485	1.0466	1.0446	1.0399	1.0349	1.0294	1.0169	1.0032
10	1.0606	1.0594	1.0578	1.0561	1.0543	1.0523	1.0503	1.0455	1.0403	1.0347	1.0221	1.0083
11	1.0669	1.0656	1.0639	1.0621	1.0602	1.0581	1.0560	1.0511	1.0458	1.0401	1.0273	1.0134
12	1.0733	1.0718	1.0700	1.0681	1.0661	1.0640	1.0618	1.0567	1.0513	1.0455	1.0326	1.0186
13	1.0797	1.0781	1.0762	1.0742	1.0721	1.0699	1.0676	1.0624	1.0568	1.0509	1.0379	1.0238
14	1.0862	1.0845	1.0824	1.0803	1.0781	1.0758	1.0735	1.0681	1.0624	1.0564	1.0432	1.0289
15	1.0927	1.0909	1.0887	1.0865	1.0842	1.0818	1.0794	1.0739	1.0680	1.0619	1.0485	1.0341
16	1.0992	1.0973	1.0950	1.0927	1.0903	1.0879	1.0854	1.0797	1.0737	1.0675	1.0538	1.0393
17	1.1057	1.1038	1.1014	1.0989	1.0964	1.0940	1.0914	1.0855	1.0794	1.0731	1.0592	1.0444
18	1.1123	1.1103	1.1078	1.1052	1.1026	1.1001	1.0974	1.0913	1.0851	1.0787	1.0646	1.0496
19	1.1189	1.1168	1.1142	1.1115	1.1088	1.1062	1.1034	1.0972	1.0908	1.0843	1.0700	1.0547
20	1.1255	1.1234	1.1206	1.1178	1.1150	1.1123	1.1094	1.1031	1.0966	1.0899	1.0754	1.0598
21	1.1322	1.1300	1.1271	1.1242	1.1213	1.1185	1.1155	1.1090	1.1024	1.0956	1.0808	1.0650
22	1.1389	1.1366	1.1336	1.1306	1.1276	1.1247	1.1217	1.1150	1.1083	1.1013	1.0862	1.0701
23	1.1457	1.1433	1.1402	1.1371	1.1340	1.1310	1.1280	1.1210	1.1142	1.1070	1.0917	1.0753
24	1.1525	1.1501	1.1469	1.1437	1.1404	1.1374	1.1343	1.1271	1.1201	1.1127	1.0972	1.0805
25	1.1594	1.1569	1.1536	1.1503	1.1469	1.1438	1.1406	1.1332	1.1260	1.1185	1.1027	1.0857
26	1.1663	1.1638	1.1603	1.1569	1.1534	1.1502	1.1469	1.1394	1.1320	1.1244	1.1083	1.0910
27	1.1733	1.1707	1.1670	1.1635	1.1600	1.1566	1.1533	1.1456	1.1381	1.1303	1.1139	1.0963
28	1.1803	1.1777	1.1738	1.1702	1.1666	1.1631	1.1597	1.1519	1.1442	1.1362	1.1195	1.1016
29	1.1874	1.1847	1.1807	1.1770	1.1733	1.1697	1.1662	1.1582	1.1503	1.1422	1.1251	1.1069
30	1.1945	1.1917	1.1876	1.1838	1.1800	1.1763	1.1727	1.1645	1.1564	1.1482	1.1307	1.1122
31	1.2016	1.1988	1.1945	1.1906	1.1867	1.1829	1.1792	1.1708	1.1625	1.1542	1.1363	1.1175
32	1.2088	1.2059	1.2014	1.1974	1.1934	1.1896	1.1857	1.1772	1.1687	1.1602	1.1419	1.1228
33	1.2160	1.2131	1.2084	1.2043	1.2002	1.1963	1.1922	1.1836	1.1749	1.1662	1.1476	1.1281
34	1.2233	1.2203	1.2155	1.2113	1.2071	1.2030	1.1988	1.1901	1.1812	1.1723	1.1533	1.1335
35	1.2306	1.2275	1.2227	1.2183	1.2140	1.2098	1.2055	1.1966	1.1876	1.1784	1.1591	1.1390
36	1.2375	1.2344	1.2294	1.2249	1.2205	1.2163	1.2119	1.2028	1.1936	1.1842	1.1645	1.1440
37	1.2444	1.2412	1.2361	1.2315	1.2270	1.2227	1.2182	1.2089	1.1995	1.1899	1.1699	1.1490
38	1.2513	1.2479	1.2428	1.2381	1.2335	1.2291	1.2245	1.2150	1.2054	1.1956	1.1752	1.1540
39	1.2581	1.2546	1.2494	1.2446	1.2399	1.2354	1.2308	1.2210	1.2112	1.2013	1.1805	1.1589

	C1	C2	C3	C4	C5	C6	C7	C8	C9	C10	C11	C12
40	1.1638	1.1858	1.2069	1.2170	1.2270	1.2370	1.2427	1.2463	1.2511	1.2560	1.2613	1.2649
41	1.1687	1.1911	1.2126	1.2229	1.2330	1.2432	1.2480	1.2527	1.2576	1.2626	1.2680	1.2717
42	1.1735	1.1963	1.2182	1.2287	1.2390	1.2494	1.2543	1.2591	1.2641	1.2692	1.2747	1.2786
43	1.1783	1.2015	1.2238	1.2345	1.2450	1.2556	1.2606	1.2655	1.2706	1.2758	1.2814	1.2854
44	1.1831	1.2067	1.2294	1.2403	1.2510	1.2618	1.2669	1.2719	1.2771	1.2824	1.2880	1.2922
45	1.1879	1.2119	1.2350	1.2461	1.2570	1.2680	1.2752	1.2783	1.2836	1.2890	1.2947	1.2990
46	1.1927	1.2171	1.2406	1.2519	1.2630	1.2742	1.2795	1.2847	1.2901	1.2955	1.3014	1.3058
47	1.1976	1.2223	1.2462	1.2577	1.2690	1.2804	1.2858	1.2911	1.2966	1.3021	1.3080	1.3126
48	1.2024	1.2275	1.2518	1.2635	1.2750	1.2867	1.2921	1.2975	1.3031	1.3087	1.3147	1.3194
49	1.2073	1.2328	1.2575	1.2693	1.2811	1.2929	1.2984	1.3040	1.3096	1.3153	1.3214	1.3263
50	1.2118	1.2377	1.2628	1.2748	1.2867	1.2987	1.3043	1.3100	1.3157	1.3215	1.3277	1.3327
51	1.2163	1.2425	1.2680	1.2802	1.2923	1.3045	1.3102	1.3160	1.3218	1.3277	1.3339	1.3391
52	1.2208	1.2473	1.2731	1.2856	1.2978	1.3102	1.3160	1.3219	1.3278	1.3338	1.3401	1.3454
53	1.2252	1.2521	1.2782	1.2909	1.3035	1.3159	1.3218	1.3278	1.3338	1.3399	1.3462	1.3517
54	1.2296	1.2568	1.2833	1.2961	1.3087	1.3215	1.3275	1.3336	1.3397	1.3459	1.3523	1.3579
55	1.2339	1.2615	1.2883	1.3013	1.3141	1.3270	1.3331	1.3393	1.3455	1.3518	1.3583	1.3640
56	1.2382	1.2661	1.2932	1.3064	1.3194	1.3324	1.3386	1.3449	1.3512	1.3576	1.3642	1.3700
57	1.2424	1.2706	1.2981	1.3114	1.3246	1.3377	1.3441	1.3505	1.3569	1.3634	1.3700	1.3759
58	1.2466	1.2751	1.3029	1.3164	1.3298	1.3430	1.3495	1.3560	1.3625	1.3691	1.3757	1.3818
59	1.2507	1.2795	1.3077	1.3213	1.3348	1.3482	1.3548	1.3614	1.3680	1.3747	1.3813	1.3875
60	1.2547	1.2839	1.3124	1.3261	1.3398	1.3533	1.3600	1.3667	1.3734	1.3801	1.3868	1.3931
61	1.2587	1.2881	1.3169	1.3308	1.3447	1.3583	1.3651	1.3719	1.3787	1.3855	1.3922	1.3986
62	1.2625	1.2922	1.3213	1.3354	1.3494	1.3632	1.3700	1.3769	1.3838	1.3907	1.3975	1.4039
63	1.2661	1.2962	1.3255	1.3398	1.3540	1.3679	1.3748	1.3818	1.3888	1.3958	1.4027	1.4091
64						1.3725	1.3795	1.3866	1.3936	1.4007	1.4078	....
65						1.3770	1.3841	1.3913	1.3984	1.4055	1.4128	....
66						1.3814	1.3887	1.3959	1.4031	1.4103	1.4177	....
67						1.3857	1.3932	1.4004	1.4077	1.4150	1.4224	....
68						1.3900	1.3976	1.4048	1.4122	1.4196	1.4271	....
69						1.3942	1.4019	1.4091	1.4166	1.4241	1.4317	....
70						1.3983	1.4061	1.4134	1.4210	1.4285	1.4362	....
71						1.4023	1.4102	1.4176	1.4252	1.4328	1.4406	....
72						1.4063	1.4142	1.4218	1.4294	1.4371	1.4449	....
73						1.4103	1.4182	1.4258	1.4335	1.4413	1.4491	....
74						1.4142	1.4221	1.4298	1.4376	1.4454	1.4532	....
75						1.4180	1.4259	1.4337	1.4415	1.4494	1.4573	....
76						1.4217	1.4295	1.4375	1.4454	1.4533	1.4613	....
77						1.4253	1.4333	1.4413	1.4492	1.4572	1.4652	....
78						1.4288	1.4369	1.4450	1.4529	1.4610	1.4690	....
79						1.4323	1.4404	1.4486	1.4565	1.4647	1.4727	....
80						1.4357	1.4439	1.4521	1.4601	1.4683	1.4764	....
81						1.4391	1.4473	1.4555	1.4636	1.4718	1.4800	....
82						1.4424	1.4507	1.4589	1.4670	1.4753	1.4835	....
83						1.4456	1.4541	1.4622	1.4704	1.4787	1.4869	....
84						1.4487	1.4572	1.4655	1.4737	1.4820	1.4903	....
85						1.4518	1.4608	1.4686	1.4769	1.4852	1.4936	....
86						1.4548	1.4633	1.4716	1.4799	1.4883	1.4968	....
87						1.4577	1.4662	1.4745	1.4829	1.4913	1.4999	....
88						1.4605	1.4690	1.4773	1.4858	1.4942	1.5029	....
89						1.4631	1.4715	1.4800	1.4885	1.4970	1.5058	....

TABLE 4-1 Specific Gravities of Aqueous Solutions—(Continued)

INORGANIC SOLUTIONS

Nitric Acid (HNO₃)—(Concluded)

%	0°C.	5°C.	10°C.	15°C.	20°C.	25°C.	30°C.	40°C.	50°C.	60°C.	80°C.	100°C.
90	.....	1.5085	1.4997	1.4911	1.4826	1.4741	1.4656					
91	.....	1.5111	1.5023	1.4936	1.4850	1.4766	1.4681					
92	.....	1.5136	1.5048	1.4960	1.4873	1.4789	1.4704					
93	.....	1.5156	1.5068	1.4979	1.4892	1.4807	1.4722					
94	.....	1.5177	1.5088	1.4999	1.4912	1.4826	1.4741					
95		1.5198	1.5109	1.5019	1.4932	1.4846	1.4761					
96		1.5220	1.5130	1.5040	1.4952	1.4867	1.4781					
97		1.5244	1.5152	1.5062	1.4974	1.4889	1.4802					
98		1.5278	1.5187	1.5096	1.5008	1.4922	1.4835					
99		1.5327	1.5235	1.5144	1.5056	1.4969	1.4881					
100		1.5402	1.5310	1.5217	1.5129	1.5040	1.4952					

Potassium Bicarbonate (KHCO₃)

°C.	1%	2%	4%	6%	8%	10%
0	1.0066	1.0134	1.0270			
10	1.0064	1.0132	1.0268	1.0396	1.0534	1.0674
15	1.0058	1.0125	1.0260			
20	1.0049	1.0117	1.0252			
30	1.0024	1.0092	1.0228			
40	0.9990	1.0058	1.0195			
50	.9949	1.0017	1.0154			
60	.9901	.9969	1.0106			
80	.9786	.9855	0.9993			
100	.9653	.9722	.9860			

Potassium Bromide (KBr)

%	d
1	1.0054
2	1.0127
6	1.0426
12	1.0903
20	1.1601
30	1.2593
40	1.3746

Perchloric Acid (HClO₄)

%	15°C.	20°C.	50°C.	%	15°C.	20°C.	50°C.
1	1.0050	1.0035	0.9933	28	1.1900	1.1851	1.1645
2	1.0109	1.0090	0.9986	30	1.2067	1.2013	1.1800
4	1.0228	1.0199	1.0086	32	1.2239	1.2183	1.1960
6	1.0348	1.0309	1.0205	34	1.2418	1.2359	1.2130
8	1.0471	1.0422	1.0320	36	1.2603	1.2542	1.2310
10	1.0597	1.0536	1.0440	38	1.2794	1.2732	1.2490
12	1.0726	.....	1.0560	40	1.2991	1.2927	1.2680
14	1.0859	.....	1.0680	45	1.3521	1.3450	1.3180
16	1.0995	.....	1.0810	50	1.4103	1.4018	1.3730
18	1.1135	.....	1.0940	55	1.4733	1.4636	1.4320
20	1.1279	.....	1.1070	60	1.5389	1.5298	1.4950
22	1.1428	.....	1.1205	65	1.6059	1.5986	1.5620
24	1.1581	.....	1.1345	70	1.6736	1.6680	1.6290
26	1.1738	1.1697	1.1490				

Phosphoric Acid (H₃PO₄)

°C.	2%	6%	14%	20%	26%	35%	50%	75%	100%
0	1.0113	1.0339	1.0811	1.1192	1.1567	1.221	1.341	1.579	1.870
10	1.0109	1.0330	1.0792	1.1167	1.1529	1.216	1.335	1.572	1.862
20	1.0092	1.0309	1.0764	1.1134	1.1484	1.211	1.329		
30	1.0065	1.0279	1.0728	1.1094					
40	1.0029	1.0241	1.0685	1.1048					

Potassium Hydroxide (KOH)

%	d
1.0	1.0083
2.0	1.0175
4.0	1.0359
6.0	1.0544
8.0	1.0730
10.0	1.0918
15.0	1.1396
20.0	1.1884
25.0	1.2387
30.0	1.2905
35.0	1.3440
40.0	1.3991
45.0	1.4558
50.0	1.5143
51.7	1.5355 (sat'd. soln.)

Potassium Chrome Alum [$K_2Cr_2(SO_4)_4$]

%	d
1	1.007
2	1.016
6	1.052
10	1.089
14	1.129
20	1.193
30	1.315
40	1.456
50	1.615

Potassium Nitrate (KNO_3)

%	0°C	10°C	20°C	40°C	60°C	80°C	100°C
1	1.00554	1.00615	1.00447	0.99825	0.9890	0.9776	0.9641
2	1.01326	1.01262	1.01075	1.00430	.9949	.9834	.9699
4	1.02577	1.02366	1.02344	1.01652	1.0068	.9951	.9816
8	1.05419	1.05226	1.04940	1.04152	1.0313	1.0192	1.0056
12	1.08221	1.07963	1.07620	1.06740	1.0567	1.0442	1.0304
16	......	......	1.10392	1.09432	1.0831	1.0703	1.0562
20	......	......	1.13261	1.12240	1.1106	1.0974	1.0831
24	......	......	1.16233	1.15175	1.1391	1.1256	1.1110

Potassium Sulfate (K_2SO_4)

%	d
1	1.0063
2	1.0145
4	1.0310
6	1.0477
8	1.0646
10	1.0817

Potassium Dichromate ($K_2Cr_2O_7$)

%	d
1	1.0032
2	1.0122
4	1.0264
6	1.0408
8	1.0554
10	1.0703

Potassium Carbonate (K_2CO_3)

%	0°C	10°C	20°C	40°C	60°C	80°C	100°C
1	1.0094	1.0039	1.0072	1.0010	0.9919	0.9803	0.9670
2	1.0189	1.0132	1.0163	1.0098	1.0005	.9889	.9756
4	1.0381	1.0369	1.0345	1.0276	1.0183	1.0063	.9951
8	1.0768	1.0746	1.0715	1.0640	1.0538	1.0418	1.0291
12	1.1160	1.1131	1.1096	1.1013	1.0905	1.0786	1.0663
16	1.1562	1.1530	1.1490	1.1399	1.1290	1.1170	1.1049
20	1.1977	1.1941	1.1898	1.1801	1.1692	1.1570	1.1451
24	1.2405	1.2366	1.2320	1.2219	1.2105	1.1986	1.1869
28	1.2846	1.2804	1.2756	1.2652	1.2533	1.2418	1.2301
30	1.3071	1.3028	1.2979	1.2873	1.2753	1.2640	1.2522
35	1.3646	1.3600	1.3548	1.3440	1.3324	1.3206	1.3089
40	1.4244	1.4195	1.4141	1.4029	1.3913	1.3795	1.3678
45	1.4867	1.4815	1.4759	1.4644	1.4523	1.4408	1.4290
50	1.5517	1.5462	1.5404	1.5285	1.5169	1.5048	1.4928

Potassium Chromate (K_2CrO_4)

%	d
1	1.0066
2	1.0147
4	1.0311
8	1.0647
12	1.0999
16	1.1366
20	1.1748
24	1.2147
28	1.2566
30	1.2784

Potassium Chlorate ($KClO_3$)

°C.	1%	2%	3%	4%
0	1.0061	1.0124	1.0189	1.0256
10	1.0059	1.0122	1.0187	1.0254
20	1.0045	1.0109	1.0174	1.0241
30	1.0020	1.0085	1.0151	1.0218
40	0.9986	1.0051	1.0116	1.0183
60	.9895	.9959	1.0024	1.0091
80	.9781	.9845	.9910	0.9977
100	.9646	.9709	.9774	.9840

Potassium Chloride (KCl)

%	0°C.	20°C.	25°C.	40°C.	60°C.	80°C.	100°C.
1.0	1.00661	1.00462	1.00342	0.99847	0.9894	0.9780	0.9646
2.0	1.01335	1.01103	1.00977	1.00471	.9955	.9842	.9708
4.0	1.02690	1.02391	1.02255	1.01727	1.0083	.9966	.9834
8.0	1.05431	1.05003	1.04847	1.04278	1.0333	1.0219	1.0088
12.0	1.08222	1.07679	1.07506	1.06897	1.0592	1.0478	1.0350
16.0	1.11068	1.10434	1.10245	1.09600	1.0646	1.0746	1.0619
20.0	1.13973	1.13280	1.13072	1.12399	1.1133	1.1024	1.0897
24.0	......	1.16226	1.15995	1.15299	1.1425	1.1311	1.1185
28.0	......	......	......	1.18304	1.1723	1.1609	1.1483

%	110°C.	120°C.	130°C.	140°C.
3.79	0.9733	0.9663	0.9583	0.9502
7.45	.9978	.9899	.9827	.9745
13.62	1.0388	1.0313	1.0238	1.0159

TABLE 4-1 Specific Gravities of Aqueous Solutions—(Continued)

INORGANIC SOLUTIONS

Sodium Chlorate (NaClO₃)

%	d	%	d
1	1.0053	18	1.1288
2	.0121	20	.1449
4	.0258	22	.1614
6	.0397	24	.1782
8	.0538	26	.1953
10	.0681	28	.2128
12	.0827	30	.2307
14	.0977	32	.2491
16	.1131	34	.2680

Sodium Chloride (NaCl)

%	0°C.	10°C.	25°C.	40°C.	60°C.	80°C.	100°C.
1	1.00747	1.00707	1.00409	0.99908	0.9900	0.9785	0.9651
2	1.01509	1.01442	1.01112	1.00593	.9967	.9852	.9719
4	1.03038	1.02920	1.02530	1.01977	1.0103	.9988	.9855
8	1.06121	1.05907	1.05412	1.04798	1.0381	1.0264	1.0134
12	1.09244	1.08946	1.08365	1.07699	1.0667	1.0549	1.0420
16	1.12419	1.12056	1.11401	1.10688	1.0962	1.0842	1.0713
20	1.15663	1.15254	1.14533	1.13774	1.1268	1.1146	1.1017
24	1.18999	1.18557	1.17776	1.16971	1.1584	1.1463	1.1331
26	1.20709	1.20254	1.19443	1.18614	1.1747	1.1626	1.1492

Sodium Chromate (Na₂CrO₄)

%	d
1	1.0074
2	.0164
4	.0344
8	.0718
12	.1110
16	.1518
20	.1942
24	.2383
26	.2611

Potassium Sulfite (K₂SO₃)

%	d
1	1.0073
2	.0155
4	.0322
8	.0667
12	.1026
16	.1402
20	.1793
24	.2197
26	.2404

Sodium Acetate (NaC₂H₃O₂)

%	d
1	1.0033
2	.0084
4	.0186
8	.0392
12	.0598
18	.0807
20	.1021
26	.1351
28	.1462

Sodium Arsenate (Na₃AsO₄)

%	d
1	1.0097
2	.0207
4	.0431
8	.0892
10	.1130
12	.1373

Sodium Bichromate (Na₂Cr₂O₇)

%	d
1	1.006
2	.013
4	.027
8	.056
12	.084
16	.112
20	.140
24	.166
28	.193
30	.207
35	.244
40	.279
45	.312
50	.342

Sodium Bromide (NaBr)

%	d
1	1.0060
2	.0139
4	.0298
8	.0631
10	.0803
12	.0981
20	.1745
30	.2841
40	.4138

Sodium Formate (HCOONa)

%	d
1	1.003
2	.009
4	.022
8	.048
12	.074
16	.100
20	.127
24	.155
28	.184
30	.199
35	.236
40	.274

Sodium Carbonate (Na₂CO₃)

%	0°C.	10°C.	20°C.	30°C.	40°C.	60°C.	80°C.	100°C.
1	1.0109	1.0103	1.0086	1.0058	1.0022	0.9929	0.9814	0.9683
2	1.0219	1.0210	1.0190	1.0159	1.0122	1.0027	.9910	.9782
4	1.0439	1.0423	1.0398	1.0363	1.0323	1.0223	1.0105	.9980
8	1.0878	1.0850	1.0816	1.0775	1.0732	1.0625	1.0503	1.0380
12	1.1319	1.1284	1.1244	1.1200	1.1150	1.1039	1.0914	1.0787
14	1.1543	1.1506	1.1463	1.1417	1.1365	1.1251	1.1125	1.0996
16				1.1636				
18				1.1859				
20				1.2086				
24				1.2552				
28				1.3031				
30				1.3274				

Sodium Hydroxide (NaOH)

%	0°C	15°C	20°C	40°C	60°C	80°C	100°C
1	1.0124	1.01065	1.0095	1.0033	0.994	0.9824	0.9693
2	1.0244	1.02198	1.0207	1.0139	1.0045	.9929	.9797
4	1.0482	1.04441	1.0428	1.0352	1.0254	1.0139	1.0009
8	1.0943	1.08887	1.0869	1.0780	1.0676	1.0560	1.0432
12	1.1399	1.13327	1.1309	1.1210	1.1101	1.0983	1.0855
16	1.1849	1.17761	1.1751	1.1645	1.1531	1.1408	1.1277
20	1.2296	1.22183	1.2191	1.2079	1.1950	1.1833	1.1700
24	1.2741	1.26582	1.2629	1.2512	1.2388	1.2259	1.2124
28	1.3182	1.3094	1.3064	1.2942	1.2814	1.2682	1.2546
32	1.3614	1.3520	1.3490	1.3362	1.3232	1.3097	1.2960
36	1.4030	1.3933	1.3900	1.3768	1.3634	1.3498	1.3750
40	1.4435	1.4334	1.4300	1.4164	1.4027	1.3889	1.3750
44	1.4825	1.4720	1.4685	1.4545	1.4405	1.4266	1.4127
48	1.5210	1.5102	1.5065	1.4922	1.4731	1.4641	1.4503
50	1.5400	1.5290	1.5253	1.5109	1.4957	1.4827	1.4690

Sodium Nitrate (NaNO₃)

%	0°C	20°C	40°C	60°C	80°C	100°C
1	1.0071	1.0069	0.9986	0.9894	0.9779	0.9644
2	1.0144	1.0117	1.0050	.9956	.9840	.9704
4	1.0293	1.0254	1.0180	1.0082	.9964	.9826
8	1.0537	1.0552	1.0447	1.0340	1.0218	1.0078
12	1.0831	1.0819	1.0724	1.0609	1.0481	1.0340
16	1.1123	1.1118	1.1013	1.0892	1.0757	1.0614
20	1.1525	1.1429	1.1314	1.1187	1.1048	1.0901
24	1.1859	1.1722	1.1629	1.1496	1.1351	1.1200
28	1.2234	1.2005	1.1955	1.186	1.1667	1.1513
30	1.2330	1.2256	1.2122	1.1980	1.1830	1.1674
35	1.2834	1.2701	1.2560	1.2413	1.2258	1.2100
40	1.3345	1.3175	1.3027	1.2875	1.2715	1.2555
45	1.3663	1.3663	1.3528	1.3371	1.3206	1.3044

Sodium Nitrite (NaNO₂)

%	d
1	1.0058
2	1.0125
4	1.0260
8	1.0535
12	1.0816
16	1.1103
20	1.1394

Sodium Silicate

	Concentration, %												
	1	2	4	8	10	14	20	24	30	36	40	45	50
					Density								
Na₂O/3.9SiO₂	1.006	1.014	1.030	1.063	1.080	1.116	1.172	1.211	1.275	1.365	1.445	1.520	1.594
Na₂O/3.36SiO₂	1.006	1.014	1.030	1.065	1.083	1.120	1.179	1.222	1.290		1.450		
Na₂O/2.40SiO₂	1.007	1.016	1.034	1.071	1.090	1.130							
Na₂O/2.44SiO₂													
Na₂O/2.06SiO₂	1.007	1.016	1.035	1.073	1.093	1.134	1.200	1.247	1.309	1.387			
Na₂O/1.69SiO₂	1.007	1.017	1.036	1.077	1.098	1.141	1.210	1.259	1.32.	1.397	1.424		

Sodium Sulfate (Na₂SO₄)

%	0°C	20°C	30°C	40°C	60°C	80°C	100°C
1	1.0094	1.0073	1.0046	1.0010	0.9919	0.9805	0.9571
2	1.0189	1.0164	1.0135	1.0098	1.0007	.9892	.9758
4	1.0381	1.0348	1.0315	1.0276	1.0184	1.0068	.9934
8	1.0773	1.0724	1.0682	1.0639	1.0544	1.0425	1.0290
12	1.1174	1.1109	1.1062	1.1015	1.0915	1.0775	1.0561
16	1.1585	1.1506	1.1456	1.1406	1.1299	1.1175	1.1042
20	1.2008	1.1915	1.1865	1.1813	1.1696	1.1559	
24	1.2443	1.2336	1.2292	1.2257			

TABLE 4-1 Specific Gravities of Aqueous Solutions—(Continued)

INORGANIC SOLUTIONS

Sodium Sulfide (Na₂S)		Sodium Sulfite (Na₂SO₃)		Sodium Thiosulfate (Na₂S₂O₃)		Sodium Thiosulfate Pentahydrate (Na₂S₂O₃·5H₂O)		Stannic Chloride (SnCl₄)		Stannous Chloride (SnCl₂)	
%	d	%	d	%	d	%	d	%	d	%	d
1	1.0098	1	1.0078	1	1.0065	1	1.0052	1	1.007	1	1.0068
2	1.0211	2	1.0172	2	1.0148	2	1.0105	2	1.015	2	1.0146
4	1.0440	4	1.0363	4	1.0315	4	1.0211	4	1.031	4	1.0306
8	1.0907	8	1.0751	8	1.0654	8	1.0423	8	1.064	8	1.0638
12	1.1388	12	1.1146	12	1.1003	12	1.0639	12	1.099	12	1.0986
16	1.1885	16	1.1549	16	1.1365	16	1.0863	16	1.135	16	1.1353
18	1.2140	18	1.1755	20	1.1740	20	1.1087	20	1.173	20	1.1743
				24	1.2128	24	1.1322	24	1.212	24	1.2159
				28	1.2532	28	1.1558	28	1.255	28	1.2603
				30	1.2739	30	1.1676	30	1.278	30	1.2837
				35	1.3273	40	1.2297	35	1.337	35	1.3461
				40	1.3827	50	1.2954	40	1.403	40	1.4145
								45	1.475	45	1.4897
								50	1.555	50	1.5729
								55	1.644	55	1.6656
								60	1.742	60	1.7695
								65	1.851	65	1.8865
								70	1.971		

Sulfuric Acid (H₂SO₄)

%	0°C.	10°C.	15°C.	20°C.	25°C.	30°C.	40°C.	50°C.	60°C.	80°C.	100°C.
1	1.0074	1.0068	1.0060	1.0051	1.0038	1.0022	0.9986	0.9944	0.9895	0.9779	0.9645
2	1.0147	1.0138	1.0129	1.0118	1.0104	1.0087	1.0050	1.0006	.9956	.9839	.9705
3	1.0219	1.0206	1.0197	1.0184	1.0169	1.0152	1.0113	1.0067	1.0017	.9900	.9766
4	1.0291	1.0275	1.0264	1.0250	1.0234	1.0216	1.0176	1.0129	1.0078	.9961	.9827
5	1.0364	1.0344	1.0332	1.0317	1.0300	1.0281	1.0240	1.0192	1.0140	1.0022	.9888
6	1.0437	1.0414	1.0400	1.0385	1.0367	1.0347	1.0305	1.0256	1.0203	1.0084	.9950
7	1.0511	1.0485	1.0469	1.0453	1.0434	1.0414	1.0371	1.0321	1.0266	1.0146	1.0013
8	1.0585	1.0556	1.0539	1.0522	1.0502	1.0481	1.0437	1.0386	1.0330	1.0209	1.0076
9	1.0660	1.0628	1.0610	1.0591	1.0571	1.0549	1.0503	1.0451	1.0395	1.0273	1.0140
10	1.0735	1.0700	1.0681	1.0661	1.0640	1.0617	1.0570	1.0517	1.0460	1.0338	1.0204
11	1.0810	1.0773	1.0753	1.0731	1.0710	1.0686	1.0637	1.0584	1.0526	1.0403	1.0269
12	1.0886	1.0846	1.0825	1.0802	1.0780	1.0756	1.0705	1.0651	1.0593	1.0469	1.0335
13	1.0962	1.0920	1.0898	1.0874	1.0851	1.0826	1.0774	1.0719	1.0661	1.0536	1.0402
14	1.1039	1.0994	1.0971	1.0947	1.0922	1.0897	1.0844	1.0788	1.0729	1.0603	1.0469
15	1.1116	1.1069	1.1045	1.1020	1.0994	1.0968	1.0914	1.0857	1.0798	1.0671	1.0537
16	1.1194	1.1145	1.1120	1.1094	1.1067	1.1040	1.0985	1.0927	1.0868	1.0740	1.0605
17	1.1272	1.1221	1.1195	1.1168	1.1141	1.1113	1.1057	1.0998	1.0938	1.0809	1.0674
18	1.1351	1.1298	1.1271	1.1243	1.1215	1.1187	1.1129	1.1070	1.1009	1.0879	1.0744
19	1.1430	1.1375	1.1347	1.1318	1.1290	1.1261	1.1202	1.1142	1.1081	1.0950	1.0814

20	1.0885	1.1021	1.1153	1.1215	1.1275	1.1335	1.1365	1.1394	1.1424	1.1453	1.1510
21	1.0957	1.1093	1.1226	1.1288	1.1349	1.1410	1.1441	1.1471	1.1501	1.1531	1.1590
22	1.1029	1.1166	1.1299	1.1362	1.1424	1.1486	1.1517	1.1548	1.1579	1.1609	1.1670
23	1.1102	1.1239	1.1373	1.1437	1.1500	1.1563	1.1594	1.1626	1.1657	1.1688	1.1751
24	1.1176	1.1313	1.1448	1.1512	1.1576	1.1640	1.1672	1.1704	1.1736	1.1768	1.1832
25	1.1250	1.1388	1.1523	1.1588	1.1653	1.1718	1.1750	1.1783	1.1816	1.1848	1.1914
26	1.1325	1.1463	1.1599	1.1665	1.1730	1.1796	1.1829	1.1862	1.1896	1.1929	1.1996
27	1.1400	1.1539	1.1676	1.1742	1.1808	1.1875	1.1909	1.1942	1.1976	1.2010	1.2078
28	1.1476	1.1616	1.1753	1.1820	1.1887	1.1955	1.1989	1.2023	1.2057	1.2091	1.2160
29	1.1553	1.1693	1.1831	1.1898	1.1966	1.2035	1.2069	1.2104	1.2138	1.2173	1.2243
30	1.1630	1.1771	1.1909	1.1977	1.2046	1.2115	1.2150	1.2185	1.2220	1.2255	1.2326
31	1.1708	1.1849	1.1988	1.2057	1.2126	1.2196	1.2232	1.2267	1.2302	1.2338	1.2409
32	1.1787	1.1928	1.2068	1.2137	1.2207	1.2278	1.2314	1.2349	1.2385	1.2421	1.2493
33	1.1866	1.2008	1.2148	1.2218	1.2289	1.2360	1.2396	1.2432	1.2468	1.2504	1.2577
34	1.1946	1.2088	1.2229	1.2300	1.2371	1.2443	1.2479	1.2515	1.2552	1.2588	1.2661
35	1.2027	1.2169	1.2311	1.2383	1.2454	1.2526	1.2563	1.2599	1.2636	1.2672	1.2746
36	1.2109	1.2251	1.2394	1.2466	1.2538	1.2610	1.2647	1.2684	1.2720	1.2757	1.2831
37	1.2192	1.2334	1.2477	1.2550	1.2622	1.2695	1.2732	1.2769	1.2805	1.2843	1.2917
38	1.2276	1.2418	1.2561	1.2635	1.2707	1.2780	1.2818	1.2855	1.2891	1.2929	1.3004
39	1.2361	1.2503	1.2646	1.2720	1.2793	1.2866	1.2904	1.2941	1.2978	1.3016	1.3091
40	1.2446	1.2589	1.2732	1.2806	1.2880	1.2953	1.2991	1.3028	1.3065	1.3103	1.3179
41	1.2532	1.2675	1.2819	1.2893	1.2967	1.3041	1.3079	1.3116	1.3153	1.3191	1.3268
42	1.2619	1.2762	1.2907	1.2981	1.3055	1.3129	1.3167	1.3205	1.3242	1.3280	1.3357
43	1.2707	1.2850	1.2996	1.3070	1.3144	1.3218	1.3256	1.3294	1.3332	1.3370	1.3447
44	1.2796	1.2939	1.3086	1.3160	1.3234	1.3308	1.3346	1.3384	1.3423	1.3461	1.3538
45	1.2886	1.3029	1.3177	1.3251	1.3325	1.3399	1.3437	1.3476	1.3515	1.3553	1.3630
46	1.2976	1.3120	1.3269	1.3343	1.3417	1.3492	1.3530	1.3569	1.3608	1.3646	1.3724
47	1.3067	1.3212	1.3362	1.3435	1.3510	1.3586	1.3624	1.3663	1.3702	1.3740	1.3819
48	1.3159	1.3305	1.3455	1.3528	1.3604	1.3680	1.3719	1.3758	1.3797	1.3835	1.3915
49	1.3253	1.3399	1.3549	1.3623	1.3699	1.3775	1.3814	1.3854	1.3893	1.3931	1.4012
50	1.3348	1.3494	1.3644	1.3719	1.3795	1.3872	1.3911	1.3951	1.3990	1.4029	1.4110
51	1.3444	1.3590	1.3740	1.3816	1.3893	1.3970	1.4009	1.4049	1.4088	1.4128	1.4209
52	1.3540	1.3687	1.3837	1.3914	1.3991	1.4069	1.4109	1.4148	1.4188	1.4228	1.4310
53	1.3637	1.3785	1.3936	1.4013	1.4091	1.4169	1.4209	1.4248	1.4289	1.4329	1.4412
54	1.3735	1.3884	1.4036	1.4113	1.4191	1.4270	1.4310	1.4350	1.4391	1.4431	1.4515
55	1.3834	1.3984	1.4137	1.4214	1.4293	1.4372	1.4412	1.4453	1.4494	1.4535	1.4619
56	1.3934	1.4085	1.4239	1.4317	1.4396	1.4475	1.4516	1.4557	1.4598	1.4640	1.4724
57	1.4035	1.4187	1.4342	1.4420	1.4500	1.4580	1.4621	1.4662	1.4703	1.4746	1.4830
58	1.4137	1.4290	1.4446	1.4524	1.4604	1.4685	1.4726	1.4768	1.4809	1.4852	1.4937
59	1.4240	1.4393	1.4551	1.4629	1.4709	1.4791	1.4832	1.4875	1.4916	1.4959	1.5045
60	1.4344	1.4497	1.4656	1.4735	1.4816	1.4898	1.4940	1.4983	1.5024	1.5067	1.5154
61	1.4449	1.4602	1.4762	1.4842	1.4923	1.5006	1.5048	1.5091	1.5133	1.5177	1.5264
62	1.4554	1.4708	1.4869	1.4950	1.5031	1.5115	1.5157	1.5200	1.5243	1.5287	1.5375
63	1.4660	1.4815	1.4977	1.5058	1.5140	1.5225	1.5267	1.5310	1.5354	1.5398	1.5487
64	1.4766	1.4923	1.5086	1.5167	1.5250	1.5335	1.5378	1.5421	1.5465	1.5510	1.5600
65	1.4873	1.5031	1.5195	1.5277	1.5361	1.5446	1.5490	1.5533	1.5578	1.5623	1.5714
66	1.4981	1.5140	1.5305	1.5388	1.5472	1.5558	1.5602	1.5646	1.5691	1.5736	1.5828
67	1.5089	1.5249	1.5416	1.5499	1.5584	1.5671	1.5715	1.5760	1.5805	1.5850	1.5943
68	1.5198	1.5359	1.5528	1.5611	1.5697	1.5785	1.5829	1.5874	1.5920	1.5965	1.6059
69	1.5307	1.5470	1.5640	1.5724	1.5811	1.5899	1.5944	1.5989	1.6035	1.6081	1.6176

TABLE 4-1   Specific Gravities of Aqueous Solutions—(Continued)

INORGANIC SOLUTIONS

Sulfuric Acid (H_2SO_4)—(Concluded)

%	0°C.	10°C.	15°C.	20°C.	25°C.	30°C.	40°C.	50°C.	60°C.	80°C.	100°C.
70	1.6293	1.6198	1.6151	1.6105	1.6059	1.6014	1.5925	1.5838	1.5753	1.5582	1.5417
71	1.6411	1.6315	1.6268	1.6221	1.6175	1.6130	1.6040	1.5952	1.5867	1.5694	1.5527
72	1.6529	1.6433	1.6385	1.6338	1.6292	1.6246	1.6155	1.6067	1.5981	1.5806	1.5637
73	1.6648	1.6551	1.6503	1.6456	1.6409	1.6363	1.6271	1.6182	1.6095	1.5919	1.5747
74	1.6768	1.6670	1.6622	1.6574	1.6526	1.6480	1.6387	1.6297	1.6209	1.6031	1.5857
75	1.6888	1.6789	1.6740	1.6692	1.6644	1.6597	1.6503	1.6412	1.6322	1.6142	1.5966
76	1.7008	1.6908	1.6858	1.6810	1.6761	1.6713	1.6619	1.6526	1.6435	1.6252	1.6074
77	1.7128	1.7026	1.6976	1.6927	1.6878	1.6829	1.6734	1.6640	1.6547	1.6361	1.6181
78	1.7247	1.7144	1.7093	1.7043	1.6994	1.6944	1.6847	1.6751	1.6657	1.6469	1.6286
79	1.7365	1.7261	1.7209	1.7158	1.7108	1.7058	1.6959	1.6862	1.6766	1.6575	1.6390
80	1.7482	1.7376	1.7323	1.7272	1.7221	1.7170	1.7069	1.6971	1.6873	1.6680	1.6493
81	1.7597	1.7489	1.7435	1.7383	1.7331	1.7279	1.7177	1.7077	1.6978	1.6782	1.6594
82	1.7709	1.7599	1.7544	1.7491	1.7437	1.7385	1.7281	1.7180	1.7080	1.6882	1.6692
83	1.7815	1.7704	1.7649	1.7594	1.7540	1.7487	1.7382	1.7279	1.7179	1.6979	1.6787
84	1.7916	1.7804	1.7748	1.7693	1.7639	1.7585	1.7479	1.7375	1.7274	1.7072	1.6878
85	1.8009	1.7897	1.7841	1.7786	1.7732	1.7678	1.7571	1.7466	1.7364	1.7161	1.6966
86	1.8095	1.7983	1.7927	1.7872	1.7818	1.7763	1.7657	1.7552	1.7449	1.7245	1.7050
87	1.8173	1.8061	1.8006	1.7951	1.7897	1.7842	1.7736	1.7632	1.7529	1.7324	1.7129
88	1.8243	1.8132	1.8077	1.8022	1.7968	1.7914	1.7809	1.7705	1.7602	1.7397	1.7202
89	1.8306	1.8195	1.8141	1.8087	1.8033	1.7979	1.7874	1.7770	1.7669	1.7464	1.7269
90	1.8361	1.8252	1.8198	1.8144	1.8091	1.8038	1.7933	1.7829	1.7729	1.7525	1.7331
91	1.8410	1.8302	1.8248	1.8195	1.8142	1.8090	1.7986	1.7883	1.7783	1.7581	1.7388
92	1.8453	1.8346	1.8293	1.8240	1.8188	1.8136	1.8033	1.7932	1.7832	1.7633	1.7439
93	1.8490	1.8384	1.8331	1.8279	1.8227	1.8176	1.8074	1.7974	1.7876	1.7681	1.7485
94	1.8520	1.8415	1.8363	1.8312	1.8260	1.8210	1.8109	1.8011	1.7914		
95	1.8544	1.8439	1.8388	1.8337	1.8286	1.8236	1.8137	1.8040	1.7944		
96	1.8560	1.8457	1.8406	1.8355	1.8305	1.8255	1.8157	1.8060	1.7965		
97	1.8569	1.8466	1.8414	1.8361	1.8314	1.8264	1.8166	1.8071	1.7977		
98	1.8567	1.8463	1.8411	1.8361	1.8310	1.8261	1.8163	1.8068	1.7976		
99	1.8551	1.8445	1.8393	1.8342	1.8292	1.8242	1.8145	1.8050	1.7958		
100	1.8517	1.8409	1.8357	1.8305	1.8255	1.8205	1.8107	1.8013	1.7922		

Zinc Chloride (ZnCl₂)

%	0°C.	20°C.	40°C.	60°C.	80°C.	100°C.
2	1.0192	1.0167	1.0099	1.0003	0.9882	0.9739
4	1.0384	1.0350	1.0274	1.0172	1.0044	.9894
8	1.0769	1.0715	1.0624	1.0508	1.0369	1.0211
12	1.1159	1.1085	1.0980	1.0853	1.0704	1.0541
16	1.1558	1.1468	1.1350	1.1212	1.1055	1.0888
20	1.1970	1.1886	1.1736	1.1590	1.1428	1.1255
30	1.3062	1.2928	1.2778	1.2614	1.2458	1.2252
40	1.4329	1.4173	1.4003	1.3824	1.3637	1.3441
50	1.5860	1.5681	1.5495	1.5300	1.5097	1.4892
60	......	1.749				
70	......	1.962				

Zinc Sulfate (ZnSO₄)

%	20°C.
2	1.019
4	1.0403
6	1.0620
8	1.0842
10	1.1071
12	1.1308
14	1.1553
16	1.1806

Zinc Bromide (ZnBr₂)

%	0°C.	20°C.	40°C.	60°C.	80°C.	100°C.
2	1.0188	1.0167	1.0102	1.0008	0.9890	0.9751
4	1.0381	1.0354	1.0285	1.0187	1.0065	0.9921
8	1.0777	1.0738	1.0660	1.0554	1.0422	1.0270
12	1.1186	1.1135	1.1046	1.0932	1.0789	1.0629
16	1.1609	1.1544	1.1445	1.1320	1.1169	1.1000
20	1.2043	1.1965	1.1855	1.1720	1.1560	1.1382
30	1.3288	1.3170	1.3030	1.2868	1.2688	1.2489
40	1.477	1.462	1.445	1.427	1.406	1.385
50	1.661	1.643	1.623	1.602	1.579	1.555
60	1.891	1.869	1.845	1.822	1.797	1.771
65	2.026	2.002	1.976	1.951	1.924	1.898

Zinc Nitrate [Zn(NO₃)₂]

%	18°C.	%	18°C.
2	1.0154	18	1.1652
4	1.0322	20	1.1865
6	1.0496	25	1.2427
8	1.0675	30	1.3029
10	1.0859	35	1.3678
12	1.1048	40	1.4378
14	1.1244	45	1.5134
16	1.1445	50	1.5944

TABLE 4-1 Specific Gravities of Aqueous Solutions—(Continued)

ORGANIC SOLUTIONS

Formic Acid (HCOOH)

%	0°C.	15°C.	20°C.	30°C.
0	0.9999	0.9991	0.9982	0.9957
1	1.0028	1.0019	1.0019	0.9980
2	1.0059	1.0045	1.0044	1.0004
3	1.0090	1.0072	1.0070	1.0028
4	1.0120	1.0100	1.0093	1.0053
5	1.0150	1.0124	1.0115	1.0075
6	1.0179	1.0151	1.0141	1.0101
7	1.0207	1.0177	1.0170	1.0125
8	1.0237	1.0204	1.0196	1.0149
9	1.0266	1.0230	1.0221	1.0173
10	1.0295	1.0256	1.0246	1.0197
11	1.0324	1.0281	1.0271	1.0221
12	1.0351	1.0306	1.0296	1.0244
13	1.0379	1.0330	1.0321	1.0267
14	1.0407	1.0355	1.0345	1.0290
15	1.0435	1.0380	1.0370	1.0313
16	1.0463	1.0405	1.0393	1.0336
17	1.0491	1.0430	1.0417	1.0358
18	1.0518	1.0455	1.0441	1.0381
19	1.0545	1.0480	1.0464	1.0404
20	1.0571	1.0505	1.0488	1.0427
21	1.0598	1.0532	1.0512	1.0451
22	1.0625	1.0556	1.0537	1.0473
23	1.0652	1.0580	1.0561	1.0496
24	1.0679	1.0604	1.0585	1.0518
25	1.0706	1.0627	1.0609	1.0540
26	1.0733	1.0652	1.0633	1.0564
27	1.0760	1.0678	1.0656	1.0587
28	1.0787	1.0702	1.0681	1.0609
29	1.0813	1.0726	1.0705	1.0632
30	1.0839	1.0750	1.0729	1.0654
31	1.0866	1.0774	1.0753	1.0676
32	1.0891	1.0798	1.0777	1.0699
33	1.0916	1.0821	1.0800	1.0721
34	1.0941	1.0844	1.0823	1.0743
35	1.0966	1.0867	1.0847	1.0766
36	1.0993	1.0892	1.0871	1.0788
37	1.1018	1.0916	1.0895	1.0810
38	1.1043	1.0940	1.0919	1.0832
39	1.1069	1.0964	1.0940	1.0854
40	1.1095	1.0988	1.0963	1.0876
41	1.1122	1.1012	1.0990	1.0898
42	1.1148	1.1036	1.1015	1.0920
43	1.1174	1.1060	1.1038	1.0943
44	1.1199	1.1084	1.1062	1.0965
45	1.1224	1.1109	1.1085	1.0987
46	1.1249	1.1133	1.1108	1.1009
47	1.1274	1.1156	1.1130	1.1031
48	1.1299	1.1179	1.1157	1.1053
49	1.1324	1.1202	1.1185	1.1076
50	1.1349	1.1225	1.1207	1.1098
51	1.1374	1.1248	1.1223	1.1120
52	1.1399	1.1271	1.1244	1.1142
53	1.1424	1.1294	1.1269	1.1164
54	1.1448	1.1318	1.1295	1.1186
55	1.1472	1.1341	1.1320	1.1208
56	1.1497	1.1365	1.1342	1.1230
57	1.1523	1.1388	1.1361	1.1253
58	1.1548	1.1411	1.1381	1.1274
59	1.1573	1.1434	1.1401	1.1295
60	1.1597	1.1458	1.1424	1.1317
61	1.1621	1.1481	1.1448	1.1338
62	1.1645	1.1504	1.1473	1.1360
63	1.1669	1.1526	1.1493	1.1382
64	1.1694	1.1549	1.1517	1.1403
65	1.1718	1.1572	1.1543	1.1425
66	1.1742	1.1595	1.1565	1.1446
67	1.1766	1.1618	1.1584	1.1467
68	1.1790	1.1640	1.1604	1.1489
69	1.1813	1.1663	1.1628	1.1510
70	1.1835	1.1685	1.1655	1.1531
71	1.1858	1.1707	1.1677	1.1552
72	1.1882	1.1729	1.1702	1.1573
73	1.1906	1.1751	1.1728	1.1595
74	1.1929	1.1773	1.1752	1.1615
75	1.1953	1.1794	1.1769	1.1656
76	1.1976	1.1816	1.1785	1.1656
77	1.1999	1.1837	1.1801	1.1676
78	1.2021	1.1859	1.1818	1.1697
79	1.2043	1.1881	1.1837	1.1717
80	1.2065	1.1902	1.1806	1.1737
81	1.2088	1.1924	1.1876	1.1758
82	1.2110	1.1944	1.1896	1.1778
83	1.2132	1.1965	1.1914	1.1798
84	1.2154	1.1985	1.1929	1.1817
85	1.2176	1.2005	1.1953	1.1837
86	1.2196	1.2025	1.1976	1.1856
87	1.2217	1.2045	1.1994	1.1875
88	1.2237	1.2064	1.2012	1.1893
89	1.2258	1.2084	1.2028	1.1910
90	1.2278	1.2102	1.2044	1.1927
91	1.2297	1.2121	1.2059	1.1945
92	1.2316	1.2139	1.2078	1.1961
93	1.2335	1.2157	1.2099	1.1978
94	1.2354	1.2174	1.2117	1.1994
95	1.2372	1.2191	1.2140	1.2008
96	1.2390	1.2208	1.2158	1.2022
97	1.2408	1.2224	1.2170	1.2036
98	1.2425	1.2240	1.2183	1.2048
99	1.2441	1.2257	1.2202	1.2061
100	1.2456	1.2273	1.2212	1.2073

Acetic Acid (CH₃COOH)

%	0°C.	10°C.	15°C.	20°C.	25°C.	30°C.	40°C.
0	0.9999	0.9997	0.9991	0.9982	0.9971	0.9957	0.9922
1	1.0016	1.0013	1.0006	0.9996	0.9987	0.9971	0.9934
2	1.0033	1.0029	1.0021	1.0012	1.0000	0.9984	0.9946
3	1.0051	1.0044	1.0036	1.0025	1.0013	0.9997	0.9958
4	1.0070	1.0060	1.0051	1.0040	1.0027	1.0011	0.9970
5	1.0088	1.0076	1.0066	1.0055	1.0041	1.0024	0.9982
6	1.0106	1.0092	1.0081	1.0069	1.0055	1.0037	0.9994
7	1.0124	1.0108	1.0096	1.0083	1.0068	1.0050	1.0006
8	1.0142	1.0124	1.0111	1.0097	1.0081	1.0063	1.0018
9	1.0159	1.0140	1.0126	1.0111	1.0094	1.0076	1.0030
10	1.0177	1.0156	1.0141	1.0125	1.0107	1.0089	1.0042
11	1.0194	1.0171	1.0155	1.0139	1.0120	1.0102	1.0054
12	1.0211	1.0187	1.0170	1.0154	1.0133	1.0115	1.0065
13	1.0228	1.0202	1.0184	1.0168	1.0146	1.0127	1.0077
14	1.0245	1.0217	1.0199	1.0182	1.0159	1.0139	1.0088
15	1.0262	1.0232	1.0213	1.0195	1.0172	1.0151	1.0099
16	1.0278	1.0247	1.0227	1.0209	1.0185	1.0163	1.0110
17	1.0295	1.0262	1.0241	1.0223	1.0198	1.0175	1.0121
18	1.0311	1.0276	1.0255	1.0236	1.0210	1.0187	1.0132
19	1.0327	1.0291	1.0269	1.0250	1.0223	1.0198	1.0142

n							
20	1.0153	1.0210	1.0235	1.0263	1.0283	1.0305	1.0343
21	1.0164	1.0222	1.0248	1.0276	1.0297	1.0319	1.0354
22	1.0174	1.0235	1.0260	1.0288	1.0310	1.0333	1.0374
23	1.0185	1.0240	1.0272	1.0301	1.0323	1.0347	1.0389
24	1.0195	1.0252	1.0283	1.0313	1.0336	1.0361	1.0404
25	1.0205	1.0267	1.0295	1.0326	1.0349	1.0375	1.0419
26	1.0215	1.0275	1.0307	1.0338	1.0362	1.0388	1.0434
27	1.0225	1.0285	1.0318	1.0349	1.0374	1.0401	1.0449
28	1.0234	1.0299	1.0329	1.0361	1.0386	1.0414	1.0463
29	1.0244	1.0310	1.0340	1.0372	1.0399	1.0427	1.0477
30	1.0253	1.0320	1.0350	1.0384	1.0411	1.0440	1.0491
31	1.0262	1.0332	1.0361	1.0395	1.0423	1.0453	1.0505
32	1.0272	1.0344	1.0372	1.0406	1.0435	1.0465	1.0519
33	1.0281	1.0354	1.0382	1.0417	1.0446	1.0477	1.0532
34	1.0289	1.0364	1.0392	1.0428	1.0458	1.0489	1.0545
35	1.0298	1.0377	1.0402	1.0438	1.0469	1.0501	1.0558
36	1.0306	1.0383	1.0412	1.0449	1.0480	1.0513	1.0571
37	1.0314	1.0393	1.0422	1.0459	1.0491	1.0524	1.0584
38	1.0322	1.0399	1.0432	1.0469	1.0501	1.0535	1.0596
39	1.0330	1.0408	1.0441	1.0479	1.0512	1.0546	1.0608
40	1.0338	1.0416	1.0450	1.0488	1.0522	1.0557	1.0621
41	1.0346	1.0425	1.0460	1.0498	1.0532	1.0568	1.0633
42	1.0355	1.0433	1.0469	1.0507	1.0542	1.0578	1.0644
43	1.0361	1.0441	1.0477	1.0516	1.0551	1.0588	1.0656
44	1.0368	1.0449	1.0486	1.0525	1.0561	1.0598	1.0667
45	1.0375	1.0456	1.0495	1.0534	1.0570	1.0608	1.0679
46	1.0382	1.0464	1.0503	1.0542	1.0579	1.0618	1.0689
47	1.0389	1.0471	1.0511	1.0551	1.0588	1.0627	1.0699
48	1.0395	1.0478	1.0518	1.0559	1.0597	1.0636	1.0709
49	1.0402	1.0486	1.0526	1.0567	1.0605	1.0645	1.0720
50	1.0408	1.0492	1.0534	1.0575	1.0613	1.0654	1.0729
51	1.0414	1.0499	1.0542	1.0582	1.0622	1.0663	1.0738
52	1.0421	1.0506	1.0549	1.0590	1.0629	1.0671	1.0748
53	1.0427	1.0512	1.0555	1.0597	1.0637	1.0679	1.0757
54	1.0432	1.0518	1.0562	1.0604	1.0644	1.0687	1.0765
55	1.0438	1.0524	1.0568	1.0611	1.0651	1.0694	1.0774
56	1.0443	1.0531	1.0574	1.0618	1.0658	1.0701	1.0782
57	1.0448	1.0536	1.0580	1.0624	1.0665	1.0708	1.0790
58	1.0453	1.0542	1.0586	1.0631	1.0672	1.0715	1.0798
59	1.0458	1.0547	1.0592	1.0637	1.0678	1.0722	1.0805
60	1.0462	1.0552	1.0597	1.0642	1.0684	1.0728	1.0813
61	1.0466	1.0557	1.0602	1.0648	1.0690	1.0734	1.0320
62	1.0470	1.0562	1.0607	1.0653	1.0696	1.0740	1.0326
63	1.0473	1.0566	1.0612	1.0658	1.0701	1.0746	1.0333
64	1.0477	1.0571	1.0616	1.0662	1.0706	1.0752	1.0338
65	1.0480	1.0575	1.0621	1.0666	1.0711	1.0757	1.0344
66	1.0483	1.0578	1.0624	1.0671	1.0716	1.0762	1.0850
67	1.0486	1.0582	1.0628	1.0675	1.0720	1.0767	1.0856
68	1.0489	1.0585	1.0631	1.0678	1.0725	1.0771	1.0860
69	1.0491	1.0588	1.0634	1.0682	1.0729	1.0775	1.0865
70	1.0493	1.0590	1.0637	1.0685	1.0732	1.0779	1.0869
71	1.0495	1.0592	1.0640	1.0687	1.0736	1.0783	1.0874
72	1.0496	1.0594	1.0642	1.0690	1.0738	1.0786	1.0877
73	1.0497	1.0595	1.0644	1.0693	1.0741	1.0789	1.0881
74	1.0498	1.0596	1.0645	1.0694	1.0743	1.0792	1.0884
75	1.0499	1.0597	1.0647	1.0696	1.0745	1.0794	1.0887
76	1.0499	1.0598	1.0648	1.0698	1.0746	1.0796	1.0889
77	1.0499	1.0598	1.0648	1.0699	1.0747	1.0797	1.0891
78	1.0498	1.0598	1.0648	1.0700	1.0747	1.0795	1.0893
79	1.0497	1.0597	1.0648	1.0700	1.0747	1.0793	1.0894
80	1.0495	1.0596	1.0647	1.0700	1.0747	1.0798	1.0895
81	1.0493	1.0594	1.0646	1.0699	1.0745	1.0797	1.0895
82	1.0490	1.0592	1.0644	1.0698	1.0743	1.0796	1.0895
83	1.0487	1.0589	1.0642	1.0696	1.0741	1.0795	1.0893
84	1.0483	1.0585	1.0638	1.0693	1.0738	1.0793	1.0893
85	1.0479	1.0582	1.0635	1.0689	1.0735	1.0790	1.0891
86	1.0473	1.0576	1.0626	1.0685	1.0731	1.0787	1.0887
87	1.0467	1.0571	1.0626	1.0680	1.0726	1.0783	1.0883
88	1.0460	1.0564	1.0620	1.0675	1.0721	1.0778	1.0877
89	1.0453	1.0557	1.0613	1.0668	1.0715	1.0773	1.0872
90	1.0445	1.0549	1.0605	1.0661	1.0708	1.0766	1.0865
91	1.0436	1.0541	1.0597	1.0652	1.0700	1.0753	1.0857
92	1.0426	1.0530	1.0587	1.0643	1.0690	1.0749	1.0848
93	1.0414	1.0518	1.0577	1.0632	1.0680	1.0739	1.0838
94	1.0401	1.0506	1.0564	1.0619	1.0667	1.0727	1.0826
95	1.0386	1.0491	1.0551	1.0605	1.0652	1.0714	1.0813
96	1.0368	1.0473	1.0535	1.0588	1.0632	.....	1.0798
97	1.0348	1.0454	1.0516	1.0570	1.0611	.....	1.0780
98	1.0325	1.0431	1.0495	1.0549	1.0590	.....	1.0759
99	1.0299	1.0407	1.0468	1.0524	1.0567	.....	1.0730
100	1.0271	1.0380	1.0440	1.0498	1.0545		1.0697

TABLE 4-1 Specific Gravities of Aqueous Solutions—(Continued)

ORGANIC SOLUTIONS

Oxalic Acid (H₂C₂O₄)

$$Oxalic\ Acid\ (H_2C_2O_4)$$

%	d	%	d
1	1.0035	8	1.0280
2	1.0070	10	1.0350
4	1.0140	12	1.0420

Methyl Alcohol (CH₃OH)

$$Methyl\ Alcohol\ (CH_3OH)$$

%	0°C	10°C	15.56°C	20°C	15°C
0	0.9999	0.9997	0.9990	0.9982	0.99913
1	9981	9980	9973	9965	99727
2	9963	9962	9955	9948	99543
3	9946	9945	9938	9931	99370
4	9930	9929	9921	9914	99198
5	9914	9912	9904	9896	99029
6	9899	9896	9889	9880	98864
7	9884	9881	9872	9863	98701
8	9870	9865	9857	9847	98547
9	9856	9849	9841	9831	98394
10	9842	9834	9826	9815	98241
11	9829	9820	9811	9799	98093
12	9816	9805	9796	9784	97945
13	9804	9791	9781	9768	97802
14	9792	9778	9766	9754	97660
15	9780	9764	9752	9740	97518
16	9769	9751	9738	9725	97377
17	9758	9739	9723	9710	97237
18	9747	9726	9709	9696	97096
19	9736	9713	9695	9681	96955
20	9725	9700	9680	9666	96814
21	9714	9687	9666	9651	96673
22	9702	9673	9652	9636	96533
23	9690	9660	9638	9622	96392
24	9678	9646	9624	9607	96251
25	9666	9632	9609	9592	96108
26	9654	9618	9595	9576	95963
27	9642	9604	9580	9562	95817
28	9629	9590	9565	9546	95668
29	9616	9575	9550	9531	95518
30	9604	9560	9535	9515	95366
31	9590	9546	9521	9499	95213
32	9576	9531	9505	9483	95056
33	9563	9516	9489	9466	94896
34	9549	9500	9473	9450	94734
35	9534	9484	9456	9433	94570
36	9520	9469	9440	9416	94404
37	9505	9453	9422	9398	94237
38	9490	9437	9405	9381	94067
39	9475	9420	9387	9363	93894
40	9459	9403	9369	9345	93720
41	9443	9387	9351	9327	93543
42	9427	9370	9333	9309	93365
43	9411	9352	9315	9290	93185
44	9395	9334	9297	9272	93001
45	9377	9316	9279	9252	92815
46	9360	9298	9261	9234	92627
47	9342	9279	9242	9214	92436
48	9324	9260	9223	9196	92242
49	9306	9240	9204	9176	92048
50	9287	9221	9185	9156	91852
51	9269	9202	9166	9135	91653
52	9250	9182	9146	9114	91451
53	9230	9162	9126	9094	91248
54	9211	9142	9106	9073	91044
55	9191	9122	9086	9052	90839
56	9172	9101	9065	9032	90631
57	9151	9080	9045	9010	90421
58	9131	9060	9024	8988	90210
59	9111	9039	9002	8968	89996
60	9090	9018	8980	8946	89781
61	9068	8998	8958	8924	89563
62	9046	8977	8936	8902	89341
63	9024	8955	8913	8879	89117
64	9002	8933	8890	8856	88890
65	8980	8911	8867	8834	88662
66	8958	8888	8844	8811	88433
67	8935	8865	8820	8787	88203
68	8913	8842	8797	8763	87971
69	8891	8818	8771	8738	87739
70	0.8869	0.8794	0.8748	0.8715	0.87507
71	8847	8770	8726	8690	87271
72	8824	8747	8702	8665	87033
73	8801	8724	8678	8641	86792
74	8778	8699	8653	8616	86546
75	8754	8676	8629	8592	86300
76	8729	8651	8604	8567	86051
77	8705	8626	8579	8542	85801
78	8680	8602	8554	8518	85551
79	8657	8577	8529	8494	85300
80	8634	8551	8503	8469	85048
81	8610	8527	8478	8446	84794
82	8585	8501	8452	8420	84536
83	8560	8475	8426	8394	84274
84	8535	8449	8400	8366	84009
85	8510	8422	8374	8340	83742
86	8483	8394	8347	8314	83475
87	8456	8367	8320	8286	83207
88	8428	8340	8294	8258	82937
89	8400	8314	8267	8230	82667
90	8374	8287	8239	8202	82396
91	8347	8261	8212	8174	82124
92	8320	8234	8185	8146	81849
93	8293	8208	8157	8118	81568
94	8266	8180	8129	8090	81285
95	8240	8152	8101	8062	80999
96	8212	8124	8073	8034	80713
97	8186	8096	8045	8005	80428
98	8158	8068	8016	7976	80143
99	8130	8040	7987	7948	79859
100	8102	8009	;7959	7917	79577

Ethyl Alcohol (C₂H₅OH)

%	10°C	15°C	20°C	25°C	30°C	35°C	40°C
0	0.99973	0.99913	0.99823	0.99708	0.99568	0.99406	0.99225
1	.99785	.99725	.99636	.99520	.99379	.99217	.99034
2	.99602	.99542	.99453	.99336	.99194	.99031	.98846
3	.99426	.99365	.99275	.99157	.99014	.98849	.98663
4	.99258	.99195	.99103	.98984	.98839	.98672	.98485
5	.99098	.99032	.98938	.98817	.98670	.98501	.98311
6	.98946	.98877	.98780	.98656	.98507	.98335	.98142
7	.98801	.98729	.98627	.98500	.98347	.98172	.97975
8	.98660	.98584	.98478	.98346	.98189	.98009	.97808
9	.98524	.98442	.98331	.98193	.98031	.97846	.97641
10	.98393	.98304	.98187	.98043	.97875	.97635	.97475
11	.98267	.98171	.98047	.97897	.97723	.97527	.97312
12	.98145	.98041	.97910	.97753	.97573	.97371	.97150
13	.98026	.97914	.97775	.97611	.97424	.97216	.96989
14	.97911	.97790	.97643	.97472	.97278	.97063	.96829
15	.97800	.97669	.97514	.97334	.97133	.96911	.96670
16	.97692	.97552	.97387	.97199	.96990	.96760	.96512
17	.97583	.97435	.97259	.97062	.96844	.96607	.96352
18	.97473	.97313	.97129	.96923	.96697	.96452	.96189
19	.97363	.97191	.96997	.96782	.96547	.96294	.96023
20	.97252	.97068	.96864	.96639	.96395	.96134	.95856
21	.97139	.96944	.96729	.96495	.96242	.95973	.95687
22	.97024	.96818	.96592	.96348	.96087	.95809	.95515
23	.96907	.96689	.96453	.96199	.95929	.95643	.95343
24	.96787	.96558	.96312	.96048	.95769	.95476	.95168
25	.96665	.96424	.96168	.95895	.95607	.95306	.94991
26	.96539	.96287	.96020	.95738	.95442	.95133	.94810
27	.96406	.96144	.95867	.95576	.95272	.94955	.94625
28	.96268	.95996	.95710	.95410	.95098	.94774	.94438
29	.96125	.95844	.95548	.95241	.94922	.94590	.94248
30	.95977	.95686	.95382	.95067	.94741	.94403	.94055
31	.95823	.95524	.95212	.94890	.94557	.94214	.93860
32	.95665	.95357	.95038	.94709	.94370	.94021	.93662
33	.95502	.95186	.94860	.94525	.94180	.93825	.93461
34	.95334	.95011	.94679	.94337	.93986	.93626	.93257
35	.95162	.94832	.94494	.94146	.93790	.93425	.93051
36	.94986	.94650	.94306	.93952	.93591	.93221	.92843
37	.94805	.94464	.94114	.93756	.93390	.93016	.92634
38	.94620	.94273	.93919	.93556	.93186	.92808	.92422
39	.94431	.94079	.93720	.93353	.92979	.92597	.92208

%	10°C	15°C	20°C	25°C	30°C	35°C	40°C
40	.94236	.93882	.93518	.93148	.92770	.92385	.91992
41	.94042	.93682	.93314	.92934	.92553	.92170	.91774
42	.93842	.93478	.93107	.92729	.92344	.91952	.91554
43	.93639	.93271	.92897	.92516	.92128	.91733	.91332
44	.93433	.93062	.92685	.92301	.91910	.91513	.91108
45	.93226	.92852	.92472	.92085	.91692	.91291	.90884
46	.93017	.92640	.92257	.91868	.91472	.91069	.90660
47	.92806	.92426	.92041	.91649	.91250	.90845	.90434
48	.92595	.92211	.91823	.91429	.91023	.90621	.90207
49	.92377	.91995	.91604	.91208	.90805	.90396	.89979
50	.92125	.91776	.91384	.90985	.90580	.90168	.89750
51	.91943	.91555	.91160	.90750	.90353	.89940	.89519
52	.91723	.91333	.90936	.90534	.90125	.89710	.89288
53	.91502	.91110	.90711	.90307	.89896	.89479	.89056
54	.91279	.90885	.90485	.90079	.89667	.89248	.88823
55	.91055	.90659	.90258	.89850	.89437	.89016	.88589
56	.90831	.90433	.90031	.89621	.89206	.88784	.88356
57	.90607	.90207	.89803	.89392	.88975	.88552	.88122
58	.90381	.89980	.89574	.89159	.88744	.88319	.87888
59	.90154	.89752	.89344	.88931	.88512	.88085	.87653
60	.89927	.89523	.89113	.88699	.88278	.87851	.87417
61	.89698	.89293	.88882	.88446	.88044	.87615	.87180
62	.89468	.89062	.88650	.88233	.87809	.87379	.86943
63	.89237	.88830	.88417	.87998	.87574	.87142	.86705
64	.89006	.88597	.88183	.87753	.87337	.86905	.86466
65	.88774	.88364	.87948	.87527	.87100	.86667	.86227
66	.88541	.88130	.87713	.87288	.86863	.86429	.85987
67	.88308	.87895	.87477	.87054	.86625	.86190	.85747
68	.88074	.87660	.87241	.86817	.86387	.85950	.85507
69	.87839	.87424	.87004	.86579	.86148	.85710	.85266
70	.87602	.87187	.86766	.86340	.85908	.85470	.85025
71	.87365	.86949	.86527	.86100	.85667	.85228	.84783
72	.87127	.86710	.86297	.85839	.85426	.84986	.84540
73	.86885	.86470	.86047	.85618	.85184	.84743	.84297
74	.86648	.86229	.85806	.85376	.84941	.84500	.84053
75	.86408	.85988	.85564	.85134	.84698	.84257	.83809
76	.86165	.85747	.85322	.84891	.84455	.84013	.83564
77	.85927	.85505	.85079	.84647	.84211	.83768	.83319
78	.85685	.85262	.84835	.84403	.83966	.83523	.83074
79	.85442	.85018	.84590	.84158	.83720	.83277	.82827

TABLE 4-1 Specific Gravities of Aqueous Solutions—(Continued)

ORGANIC SOLUTIONS

Ethyl Alcohol (C₂H₅OH)—(Concluded)

%	10°C.	15°C.	20°C.	25°C.	30°C.	35°C.	40°C.
80	197	.84772	344	.83911	473	029	578
81	.84950	636	096	664	224	.82780	329
82	702	453	.83848	415	.82974	530	079
83	453	275	599	164	724	279	.81828
84	203	103	348	.82913	473	027	576
85	.83951	525	095	660	220	.81774	322
86	697	271	.82840	405	.81965	519	067
87	441	014	583	148	708	262	.80811
88	181	.82754	323	.81888	448	003	552
89	.82919	492	062	626	186	.80742	291
90	654	227	.81797	362	.80922	478	028
91	386	.81959	529	094	655	211	.79761
92	114	688	257	.80823	384	.79941	491
93	.81839	413	.80983	549	111	669	220
94	561	134	705	272	.79835	393	.78947
95	278	.80852	424	.79991	555	114	670
96	.80991	566	138	706	271	.78831	388
97	698	274	.79846	415	.78981	542	100
98	399	.79975	547	117	684	247	.77806
99	094	670	243	.78814	382	.77946	507
100	.79784	360	.78934	506	075	641	203

Mixtures of C₂H₅OH and H₂O at 20°C.

% alcohol by weight	Tenths of %									
	0	1	2	3	4	5	6	7	8	9
0	0.99823	804	785	766	748	729	710	692	673	655
1	636	618	599	581	562	544	525	507	489	471
2	453	435	417	399	381	363	345	327	310	292
3	275	257	240	222	205	188	171	154	137	120
4	103	087	070	053	037	020	003	*987	*971	*954
5	.98938	922	906	890	874	859	843	827	811	796
6	780	765	749	734	718	703	688	673	658	642
7	627	612	597	582	567	553	538	523	508	493
8	478	463	449	434	419	404	389	374	360	345
9	331	316	301	287	273	258	244	229	215	201
10	187	172	158	144	130	117	103	089	075	061
11	047	033	019	006	*992	*978	*964	*951	*937	*923
12	.97910	896	883	869	855	842	828	815	801	788
13	775	761	748	735	722	709	696	683	670	657
14	643	630	617	604	591	578	565	552	539	526
15	514	501	488	475	462	450	438	425	412	400
16	387	374	361	349	336	323	310	297	284	272
17	259	246	233	220	207	194	181	168	155	142
18	129	116	103	089	076	063	050	037	024	010
19	.96997	984	971	957	944	931	917	904	891	877
20	864	850	837	823	810	796	783	769	756	742
21	729	716	702	688	675	661	647	634	620	606
22	592	578	564	551	537	523	509	495	481	467
23	453	439	425	411	396	382	368	354	340	326
24	312	297	283	269	254	240	225	211	196	182
25	168	153	139	124	109	094	080	065	050	035
26	020	005	*990	*975	*959	*944	*929	*914	*898	*883
27	.95867	851	836	820	805	789	773	757	742	726
28	710	694	678	662	646	630	613	597	581	565
29	548	532	516	499	483	466	450	433	416	400

Table of function values (arguments 30–100). Entries give the first two decimal places only where they change; an asterisk (*) indicates a change in the first two decimal places.

Arg	0	1	2	3	4	5	6	7	8	9
30	.95382	365	349	332	315	298	281	264	247	230
31	.95212	195	178	161	143	126	108	091	074	056
32	.95038	020	003	*985	*967	*950	*932	*914	*896	*878
33	.94860	842	824	806	788	770	752	734	715	697
34	.94679	660	642	624	605	587	568	550	531	512
35	.94494	475	456	438	419	400	382	363	344	325
36	.94306	287	268	249	230	211	192	172	155	134
37	.94114	095	075	056	036	017	*997	*978	*958	*939
38	.93919	899	879	859	840	820	800	780	760	740
39	.93720	700	680	660	640	620	599	579	559	539
40	.93518	498	478	458	437	417	396	376	356	335
41	.93314	294	273	253	232	212	191	171	149	129
42	.93107	086	065	044	023	002	*981	*960	*939	*918
43	.92897	876	855	834	812	791	770	749	728	707
44	.92685	664	642	621	600	579	557	536	515	453
45	.92472	450	429	408	386	365	343	322	300	279
46	.92257	236	214	193	171	150	123	106	085	*037
47	.92041	019	*997	*976	*954	*932	*910	*889	*867	*845
48	.91823	801	780	758	736	714	692	670	648	626
49	.91604	582	560	538	516	494	472	450	428	466
50	0.91384	361	339	317	295	272	250	228	206	183
51	.91160	138	116	093	071	049	026	004	*981	*959
52	.90936	914	891	869	846	824	801	779	756	734
53	.90711	689	666	644	621	598	576	553	531	508
54	.90485	463	440	417	395	372	349	327	304	281
55	.90258	236	213	190	167	145	122	099	076	054
56	.90031	008	*985	*962	*939	*917	*894	*871	*848	*825
57	.89803	780	757	734	711	688	665	643	620	597
58	.89574	551	528	505	482	459	436	413	390	557
59	.89344	321	298	275	252	229	206	183	160	137
60	.89113	090	067	044	021	*998	*975	*951	*928	*905
61	.88882	859	836	812	789	766	743	720	696	673
62	.88650	626	603	580	557	533	510	487	463	440
63	.88417	393	370	347	323	300	277	253	230	206
64	.88183	160	136	113	089	066	042	019	*995	*972

Arg	0	1	2	3	4	5	6	7	8	9
65	.87948	925	901	878	854	831	807	784	760	737
66	.87713	689	666	642	619	595	572	548	524	501
67	.87477	454	440	405	383	359	336	312	288	265
68	.87241	218	194	170	147	123	099	075	052	028
69	.87004	*981	*957	*933	*909	*885	*862	*838	*814	*790
70	.86766	742	718	694	671	647	623	599	575	551
71	.86527	503	479	455	431	407	383	339	335	311
72	.86287	263	239	215	191	167	143	119	095	071
73	.86047	022	*998	*974	*950	*926	*902	*878	*854	*830
74	.85805	781	757	733	709	685	661	636	612	588
75	.85564	540	515	491	467	443	419	394	370	346
76	.85322	297	273	249	225	200	176	152	128	103
77	.85079	055	061	005	*982	*958	*933	*909	*884	*860
78	.84835	811	737	762	738	713	689	664	640	615
79	.84590	566	517	517	492	467	443	418	393	369
80	.84344	319	294	270	245	220	196	171	146	121
81	.84095	072	047	022	*997	*972	*947	*923	*898	*873
82	.83848	823	738	773	748	723	698	674	649	624
83	.83599	574	549	523	498	473	448	423	398	373
84	.83343	323	227	272	247	222	196	171	146	120
85	.83095	070	044	019	*994	*968	*943	*917	*892	*866
86	.82840	815	739	763	737	712	686	660	635	609
87	.82583	557	531	505	479	453	427	401	375	349
88	.82323	297	251	245	219	193	167	140	114	088
89	.82062	035	039	*983	*956	*930	*903	*877	*850	*824
90	.81797	770	714	717	690	664	637	610	583	556
91	.81529	502	475	443	421	394	366	339	312	285
92	.81257	230	233	175	148	120	093	066	038	010
93	.80983	955	998	900	872	844	817	789	761	733
94	.80705	677	649	621	593	565	537	509	480	452
95	.80424	395	357	333	310	281	253	224	195	166
96	.80153	109	090	054	022	*993	*963	*934	*905	*875
97	.79845	816	737	757	722	698	668	638	608	578
98	.79547	517	487	455	426	396	365	335	305	274
99	.79243	213	132	154	120	089	059	028	*997	*966
100	.78934									

* Indicates change in the first two decimal places.

TABLE 4-1 Specific Gravities of Aqueous Solutions—(Continued)

ORGANIC SOLUTIONS

Specific Gravity (60°/60°F.) (15.56°/15.56°C.) of Mixtures by (Volume) of C₂H₅OH and H₂O

% alcohol by volume at 60°F.	Tenths of %									
	0	1	2	3	4	5	6	7	8	9
0	1.00000	*985	*970	*955	*940	*925	*910	*895	*880	865
1	0.99850	835	820	806	791	776	761	747	732	717
2	703	688	674	659	645	630	616	602	587	573
3	559	545	531	516	502	488	474	460	446	432
4	419	405	391	378	364	350	336	323	309	296
5	282	269	255	242	228	215	202	189	176	163
6	150	137	124	111	098	085	073	060	047	035
7	022	009	*997	*984	*972	*960	*947	*935	*923	*911
8	.98899	*887	875	863	851	838	826	814	803	791
9	779	767	755	743	731	720	708	696	684	672
10	661	649	637	625	614	602	590	579	567	556
11	544	532	521	509	498	487	475	464	452	441
12	430	419	408	396	385	374	363	352	341	330
13	319	308	297	286	275	264	254	243	232	221
14	210	200	190	179	168	157	147	136	125	115
15	104	093	083	072	062	051	040	030	019	009
16	.97998	988	977	967	956	946	936	925	915	905
17	895	885	875	864	854	844	834	824	814	804
18	794	784	774	764	754	744	734	724	714	704
19	694	684	674	664	654	645	635	625	615	605
20	596	586	576	566	556	546	536	526	516	506
21	496	486	476	466	456	446	436	425	415	405
22	395	385	375	365	354	344	334	324	313	303
23	293	283	272	262	252	241	231	221	210	200
24	189	179	168	158	147	137	126	116	105	095
25	084	073	063	052	042	031	020	010	*999	*988
26	.96978	967	957	946	935	924	914	903	892	881
27	870	859	848	837	826	815	804	793	782	771
28	760	749	738	727	715	704	693	682	671	659
29	648	637	625	614	603	591	580	568	557	546
30	534	522	511	499	488	476	464	453	441	429
31	418	406	394	382	370	358	346	334	321	309
32	296	284	271	259	246	234	221	209	196	183
33	170	157	144	132	119	106	093	080	067	054
34	041	028	015	002	*988	*975	*962	*948	*935	*921
35	.95908	894	881	867	854	840	826	812	798	784
36	770	756	742	728	714	700	685	671	657	643
37	628	614	599	585	570	556	541	526	512	497
38	482	467	452	437	423	408	393	378	362	347
39	332	317	302	286	271	256	240	225	209	194
40	178	162	147	131	115	100	084	068	052	036
41	020	004	*988	*972	*956	*940	*923	*907	*891	*875
42	.94858	842	825	809	792	776	759	743	726	710
43	693	676	660	643	626	609	592	575	558	541
44	524	507	490	473	455	438	421	403	386	369
45	351	334	316	298	281	263	245	228	210	192
46	174	156	138	120	102	084	066	048	030	011
47	.93993	975	956	938	920	901	883	864	845	827
48	808	789	771	752	733	714	695	676	657	638
49	619	600	581	562	543	523	504	485	465	446
50	0.93426	407	387	368	348	328	309	289	270	250
51	230	210	190	171	151	131	111	091	071	051
52	031	011	*991	*971	*951	*931	*911	*890	*870	*850
53	.92830	810	789	769	749	728	708	688	667	647
54	626	605	585	564	544	523	502	482	461	440
55	419	398	377	357	336	315	294	273	252	231
56	210	189	168	147	126	105	084	062	041	020
57	.91999	978	956	935	914	892	871	849	827	806
58	784	762	741	719	697	675	653	631	610	588
59	565	543	521	499	477	455	433	410	388	366
60	344	322	299	277	255	232	210	188	165	143
61	120	097	075	052	030	007	*984	*962	*939	*916
62	.90893	870	847	825	802	779	756	733	710	687
63	664	641	618	595	572	549	526	503	480	457
64	434	411	388	365	341	318	295	272	249	225
65	202	179	155	132	108	085	061	038	014	*991
66	.89967	943	920	896	872	848	825	801	777	753
67	729	705	681	657	633	609	585	561	537	513
68	489	465	441	416	392	368	343	319	295	270
69	245	220	196	171	147	122	098	073	048	024

(Top table — continued, % 70–100)

%										
70	.88999	974	950	925	900	875	850	825	801	776
71	.751	725	700	675	650	625	600	574	549	524
72	.499	474	448	423	397	372	346	321	296	270
73	.244	218	193	167	141	116	090	064	039	013
74	.87987	961	935	910	884	858	832	806	780	754
75	.728	702	676	650	623	597	571	545	518	492
76	.465	439	412	386	359	332	306	279	252	226
77	.199	172	145	118	092	065	038	011	*984	*957
78	.86929	902	875	847	820	793	766	738	711	684
79	.656	629	601	574	546	518	491	463	435	408
80	.380	352	324	296	269	241	213	185	157	129
81	.100	072	044	015	*987	*959	*931	*902	*874	*846
82	.85817	789	760	732	703	674	646	617	588	560
83	.531	502	473	444	415	386	357	328	299	270
84	.240	211	181	152	122	093	063	033	004	*974
85	673	703	734	764	794	824	854	884	914	.8-944
86	398	428	459	490	520	551	581	612	612	642
87	056	088	119	150	181	212	243	274	305	336
88	*739	*771	*803	*835	*867	*899	*930	*962	*994	025
89	415	447	480	513	545	578	610	643	675	.83707
90	083	116	150	183	216	249	282	315	349	382
91	*741	*776	*810	*845	*879	*913	*947	*981	015	049
92	387	423	458	494	529	555	600	635	670	.8705
93	022	059	096	133	170	206	243	279	315	351
94	642	681	719	757	796	834	871	909	947	.8984
95	247	287	327	367	407	446	486	525	564	683
96	*834	*876	*918	*960	001	042	084	125	165	286
97	401	445	489	533	577	620	664	707	750	.8792
98	*937	*985	033	080	127	173	219	265	311	356
99	441	492	543	593	643	693	743	752	841	.75689
100										389

* Indicates change in first two decimal places.

n-Propyl Alcohol (C₃H₇OH)

%	0°C.	15°C.	30°C.	%	0°C.	15°C.	30°C.	%	0°C.	15°C.	30°C.	%	0°C.	15°C.	30°C.	%	0°C.	15°C.	30°C.
0	0.9999	0.9991	0.9957	20	.9789	.9723	0.9645	40	0.9430	0.9331	0.9226	60	0.9033	0.8922	0.8807	80	0.8634	0.8516	0.8394
1	.9982	.9974	.9940	21	.9776	.9705	.9622	41	.9411	.9310	.9205	61	.8994	.8902	.8736	81	.8614	.8496	.8373
2	.9967	.9960	.9924	22	.9763	.9688	.9602	42	.9391	.9290	.9184	62	.8974	.8882	.8745	82	.8594	.8475	.8352
3	.9952	.9944	.9908	23	.9748	.9670	.9585	43	.9371	.9269	.9164	63	.8954	.8861	.8724	83	.8574	.8454	.8352
4	.9939	.9929	.9893	24	.9733	.9651	.9563	44	.9352	.9243	.9143	64	.8954	.8841	.8724	84	.8554	.8434	.8311
5	.9926	.9915	.9877	25	.9717	.9633	.9543	45	.9332	.9223	.9122	65	.8934	.8820	.8703	85	.8534	.8413	.8290
6	.9914	.9902	.9862	26	.9700	.9614	.9522	46	.9311	.9207	.9100	66	.8913	.8800	.8632	86	.8513	.8393	.8269
7	.9904	.9890	.9848	27	.9682	.9594	.9501	47	.9291	.9186	.9079	67	.8894	.8779	.8652	87	.8492	.8372	.8248
8	.9894	.9877	.9834	28	.9664	.9576	.9481	48	.9272	.9165	.9057	68	.8874	.8759	.8641	88	.8471	.8351	.8227
9	.9883	.9864	.9819	29	.9646	.9556	.9460	49	.9252	.9145	.9036	69	.8854	.8739	.8620	89	.8450	.8330	.8206
10	.9874	.9852	.9804	30	.9627	.9535	.9439	50	.9232	.9124	.9015	70	.8835	.8719	.8600	90	.8429	.8308	.8185
11	.9865	.9840	.9790	31	.9608	.9516	.9418	51	.9213	.9104	.8994	71	.8815	.8700	.8530	91	.8408	.8287	.8164
12	.9857	.9828	.9775	32	.9589	.9495	.9396	52	.9192	.9084	.8973	72	.8795	.8680	.8539	92	.8387	.8266	.8142
13	.9849	.9817	.9760	33	.9570	.9474	.9375	53	.9173	.9064	.8952	73	.8776	.8659	.8518	93	.8364	.8244	.8120
14	.9841	.9806	.9746	34	.9550	.9454	.9354	54	.9153	.9044	.8931	74	.8756	.8639	.8518	94	.8342	.8221	.8098
15	.9833	.9793	.9730	35	.9530	.9434	.9333	55	.9132	.9023	.8911	75	.8736	.8618	.8497	95	.8320	.8199	.8077
16	.9825	.9780	.9714	36	.9511	.9413	.9312	56	.9112	.9005	.8890	76	.8716	.8598	.8477	96	.8296	.8176	.8054
17	.9817	.9768	.9698	37	.9491	.9392	.9289	57	.9093	.8983	.8869	77	.8695	.8577	.8456	97	.8272	.8153	.8031
18	.9808	.9752	.9680	38	.9471	.9372	.9269	58	.9073	.8963	.8849	78	.8675	.8556	.8435	98	.8248	.8128	.8008
19	.9800	.9739	.9661	39	.9450	.9351	.9247	59	.9053	.8942	.8828	79	.8655	.8536	.8414	99	.8222	.8104	.7984
																100	.8194	.8077	.7958

TABLE 4-1   Specific Gravities of Aqueous Solutions—(Continued)

ORGANIC SOLUTIONS

Isopropyl Alcohol (C$_3$H$_7$OH)

%	0°C	15°C	20°C	30°C
0	0.9999	0.99913	0.9982	0.9957
1	.9980	.9972	.9962	.9939
2	.9962	.9954	.9944	.9921
3	.9946	.9936	.9926	.9904
4	.9930	.9920	.9909	.9887
5	.9916	.9904	.9893	.9871
6	.9902	.9890	.9877	.9855
7	.9890	.9875	.9862	.9839
8	.9878	.9862	.9847	.9824
9	.9866	.9849	.9833	.9809
10	.9856	.98362	.9820	.9794
11	.9846	.9824	.9808	.9778
12	.9838	.9812	.9797	.9764
13	.9829	.9800	.9786	.9750
14	.9821	.9788	.9776	.9735
15	.9814	.9777	.9765	.9720
16	.9806	.9765	.9754	.9705
17	.9799	.9753	.9743	.9690
18	.9792	.9741	.9731	.9675
19	.9784	.9728	.9717	.9658
20	.9777	.97158	.9703	.9642
21	.9768	.9703	.9688	.9624
22	.9759	.9689	.9669	.9606
23	.9749	.9674	.9651	.9587
24	.9739	.9659	.9634	.9569
25	.9727	.9642	.9615	.9549
26	.9714	.9624	.9597	.9529
27	.9699	.9605	.9577	.9509
28	.9684	.9586	.9558	.9488
29	.9669	.9568	.9540	.9467
30	.9652	.95493	.9520	.9446
31	.9634	.9530	.9500	.9426
32	.9615	.9510	.9481	.9405
33	.9596	.9489	.9460	.9383
34	.9577	.9468	.9440	.9361

%	0°C	15°C	20°C	30°C
35	.9557	.9446	.9419	.9338
36	.9536	.9424	.9399	.9315
37	.9514	.9401	.9377	.9292
38	.9493	.9379	.9355	.9269
39	.9472	.9356	.9333	.9246
40	.9450	.93333	.9310	.9224
41	.9428	.9311	.9287	.9201
42	.9406	.9288	.9264	.9177
43	.9384	.9266	.9239	.9154
44	.9361	.9243	.9215	.9130
45	.9338	.9220	.9191	.9106
46	.9315	.9197	.9165	.9082
47	.9292	.9174	.9141	.9059
48	.9270	.9150	.9117	.9036
49	.9247	.9127	.9093	.9013
50	.9224	.91043	.9069	.8990
51	.9201	.9081	.9044	.8966
52	.9178	.9058	.9020	.8943
53	.9155	.9035	.8996	.8919
54	.9132	.9011	.8971	.8895
55	.9109	.8988	.8946	.8871
56	.9086	.8964	.8921	.8847
57	.9063	.8940	.8896	.8823
58	.9040	.8917	.8874	.8800
59	.9017	.8893	.8850	.8777
60	.8994	.88690	.8825	.8752
61	.8970	.8845	.8800	.8728
62	.8947	.8821	.8776	.8704
63	.8924	.8798	.8751	.8680
64	.8901	.8775	.8727	.8656
65	.8878	.8752	.8702	.8631
66	.8854	.8728	.8679	.8607
67	.8831	.8705	.8656	.8583
68	.8807	.8682	.8632	.8559
69	.8784	.8658	.8609	.8535

%	0°C	15°C	20°C	30°C
70	0.8761	0.86346	0.8584	0.8511
71	.8738	.8611	.8560	.8487
72	.8714	.8588	.8537	.8464
73	.8691	.8564	.8513	.8440
74	.8668	.8541	.8489	.8416
75	.8644	.8517	.8464	.8392
76	.8621	.8493	.8439	.8368
77	.8598	.8470	.8415	.8344
78	.8575	.8446	.8391	.8321
79	.8551	.8422	.8366	.8297
80	.8528	.83979	.8342	.8273
81	.8503	.8374	.8317	.8248
82	.8479	.8350	.8292	.8224
83	.8456	.8326	.8268	.8200
84	.8432	.8302	.8243	.8175
85	.8408	.8278	.8219	.8151
86	.8384	.8254	.8194	.8127
87	.8360	.8229	.8169	.8101
88	.8336	.8205	.8145	.8078
89	.8311	.8180	.8120	.8053
90	.8287	.81553	.8096	.8029
91	.8262	.8130	.8072	.8004
92	.8237	.8104	.8047	.7979
93	.8212	.8079	.8023	.7954
94	.8186	.8052	.7998	.7929
95	.8160	.8026	.7973	.7904
96	.8133	.7999	.7949	.7878
97	.8106	.7972	.7925	.7852
98	.8078	.7945	.7901	.7826
99	.8048	.7918	.7877	.7799
100	.8016	.78913	.7854	.7770

Glycerol

Glycerol, %	Density 15°C	15.5°C	20°C	25°C	30°C
100	1.26415	1.26381	1.26108	1.25802	1.25495
99	1.26160	1.26125	1.25850	1.25545	1.25235
98	1.25900	1.25865	1.25590	1.25290	1.24975
97	1.25645	1.25610	1.25335	1.25030	1.24710
96	1.25385	1.25350	1.25080	1.24770	1.24450
95	1.25130	1.25095	1.24825	1.24515	1.24190
94	1.24865	1.24830	1.24560	1.24250	1.23930
93	1.24600	1.24565	1.24300	1.23985	1.23670
92	1.24340	1.24305	1.24035	1.23725	1.23410
91	1.24075	1.24040	1.23770	1.23460	1.23150
90	1.23810	1.23775	1.23510	1.23200	1.22890
89	1.23545	1.23510	1.23245	1.22935	1.22625
88	1.23280	1.23245	1.22975	1.22665	1.22360
87	1.23015	1.22980	1.22710	1.22400	1.22095
86	1.22750	1.22710	1.22445	1.22135	1.21830
85	1.22485	1.22445	1.22180	1.21870	1.21565
84	1.22220	1.22180	1.21915	1.21605	1.21300
83	1.21955	1.21915	1.21650	1.21340	1.21035
82	1.21690	1.21650	1.21380	1.21075	1.20770
81	1.21425	1.21385	1.21115	1.20810	1.20505
80	1.21160	1.21120	1.20850	1.20545	1.20240
79	1.20885	1.20845	1.20575	1.20275	1.19970
78	1.20610	1.20570	1.20305	1.20005	1.19705
77	1.20335	1.20300	1.20030	1.19735	1.19435
76	1.20060	1.20025	1.19760	1.19465	1.19170
75	1.19785	1.19750	1.19485	1.19195	1.18900
74	1.19510	1.19480	1.19215	1.18925	1.18635
73	1.19235	1.19205	1.18940	1.18650	1.18365
72	1.18965	1.18930	1.18670	1.18380	1.18100
71	1.18690	1.18655	1.18395	1.18110	1.17830
70	1.18415	1.18385	1.18125	1.17840	1.17565
69	1.18135	1.18105	1.17850	1.17565	1.17290
68	1.17860	1.17830	1.17575	1.17295	1.17020
67	1.17585	1.17555	1.17300	1.17020	1.16745
66	1.17305	1.17275	1.17025	1.16745	1.16470

Glycerol, %	Density 15°C	15.5°C	20°C	25°C	30°C
65	1.17030	1.17000	1.16750	1.16475	1.16195
64	1.16755	1.16725	1.16475	1.16200	1.15925
63	1.16480	1.16445	1.16205	1.15925	1.15650
62	1.16200	1.16170	1.15930	1.15655	1.15375
61	1.15925	1.15895	1.15655	1.15380	1.15100
60	1.15650	1.15615	1.15380	1.15105	1.14830
59	1.15370	1.15340	1.15105	1.14835	1.14555
58	1.15095	1.15065	1.14830	1.14560	1.14285
57	1.14815	1.14785	1.14555	1.14285	1.14010
56	1.14535	1.14510	1.14280	1.14015	1.13740
55	1.14260	1.14230	1.14005	1.13740	1.13470
54	1.13980	1.13955	1.13730	1.13465	1.13195
53	1.13705	1.13680	1.13455	1.13195	1.12925
52	1.13425	1.13400	1.13180	1.12920	1.12650
51	1.13150	1.13125	1.12905	1.12650	1.12380
50	1.12870	1.12845	1.12630	1.12375	1.12110
49	1.12600	1.12575	1.12360	1.12110	1.11845
48	1.12325	1.12305	1.12090	1.11840	1.11580
47	1.12055	1.12030	1.11820	1.11575	1.11320
46	1.11780	1.11760	1.11550	1.11310	1.11055
45	1.11510	1.11490	1.11280	1.11040	1.10795
44	1.11235	1.11215	1.11010	1.10775	1.10530
43	1.10950	1.10945	1.10740	1.10510	1.10265
42	1.10690	1.10670	1.10470	1.10240	1.10005
41	1.10415	1.10400	1.10200	1.09975	1.09740
40	1.10145	1.10130	1.09930	1.09710	1.09475
39	1.09875	1.09860	1.09665	1.09445	1.09215
38	1.09605	1.09590	1.09400	1.09180	1.08955
37	1.09340	1.09320	1.09135	1.08915	1.08690
36	1.09070	1.09050	1.08865	1.08655	1.08430
35	1.08800	1.08780	1.08600	1.08390	1.08165
34	1.08530	1.08515	1.08335	1.08125	1.07905
33	1.08265	1.08245	1.08070	1.07860	1.07645
32	1.07995	1.07975	1.07800	1.07600	1.07380
31	1.07725	1.07705	1.07535	1.07335	1.07120

Glycerol, %	Density 15°C	15.5°C	20°C	25°C	30°C
30	1.07455	1.07435	1.07270	1.07070	1.06855
29	1.07195	1.07175	1.07010	1.06815	1.06605
28	1.06935	1.06915	1.06755	1.06560	1.06355
27	1.06670	1.06655	1.06495	1.06305	1.06105
26	1.06410	1.06390	1.06240	1.06055	1.05855
25	1.06150	1.06130	1.05980	1.05800	1.05605
24	1.05885	1.05870	1.05720	1.05545	1.05350
23	1.05625	1.05610	1.05465	1.05290	1.05100
22	1.05365	1.05350	1.05205	1.05035	1.04850
21	1.05100	1.05090	1.04950	1.04780	1.04600
20	1.04840	1.04825	1.04590	1.04525	1.04350
19	1.04590	1.04575	1.04440	1.04280	1.04105
18	1.04335	1.04325	1.04195	1.04035	1.03860
17	1.04085	1.04075	1.03945	1.03790	1.03615
16	1.03835	1.03825	1.03695	1.03545	1.03370
15	1.03580	1.03570	1.03450	1.03300	1.03130
14	1.03330	1.03320	1.03200	1.03055	1.02885
13	1.03080	1.03070	1.02955	1.02805	1.02640
12	1.02830	1.02820	1.02705	1.02560	1.02395
11	1.02575	1.02565	1.02455	1.02315	1.02150
10	1.02325	1.02315	1.02210	1.02070	1.01905
9	1.02085	1.02075	1.01970	1.01835	1.01670
8	1.01840	1.01835	1.01730	1.01600	1.01440
7	1.01600	1.01590	1.01495	1.01360	1.01205
6	1.01360	1.01350	1.01255	1.01125	1.00970
5	1.01120	1.01110	1.01015	1.00890	1.00735
4	1.00875	1.00870	1.00780	1.00655	1.00505
3	1.00635	1.00630	1.00540	1.00415	1.00270
2	1.00395	1.00385	1.00300	1.00180	1.00035
1	1.00155	1.00145	1.00060	0.99945	0.99800
0	0.99913	0.99905	0.99823	0.99708	0.99568

TABLE 4-1 Specific Gravities of Aqueous Solutions—(Continued)

ORGANIC SOLUTIONS

Hydrazine (N_2H_4)

%	d	%	d
1	1.0002	30	1.0305
2	1.0013	40	1.038
4	1.0034	50	1.044
8	1.0077	60	1.047
12	1.0121	70	1.046
16	1.0164	80	1.040
20	1.0207	90	1.030
24	1.0248	100	1.011
28	1.0286		

Densities of Aqueous Solutions of Miscellaneous Organic Compounds

d (resp., d_w, d_s) = density of the solution [resp., water; resp., the pure liquid solute] in g per ml. p_s (resp., p_w) = weight % of solute (resp., water) in the solution. "Range" = range of applicability of the equation.

Section A. $d = d_w + Ap_s + Bp_s^2 + Cp_s^3$

Name	Formula	t, °C	Range, p_s	A	B	C
Acetaldehyde	C_2H_4O	18	0–30	$+0.0_3255$	$-0._616$	
Acetamide	C_2H_5NO	15	0–6	$+0._4639$	$+0._4171$	
Acetone	C_3H_6O	15	0–100	$-0._8856$	$-0._4449$	$-0._5588$
		20	0–100	$-0._7648$	$-0._1193$	$+0._8272$
		25	0–100	$-0._1009$	$-0._9682$	$-0._8624$
Acetonitrile	C_2H_3N	15	0–16	$-0._1233$	$-0._3529$	$-0._5327$
Allyl alcohol	C_3H_6O	0	0–89	$-0._1171$	$-0._904$	$-0._656$
Benzenepentacarboxylic acid	$C_{11}H_6O_{10}$	25	0–0.6	$-0._3729$	$-0._2024$	$+0._72984$
Butyl alcohol (n-)	$C_4H_{10}O$	20	0–7.9	$-0._5615$	$-0._1232$	
Butyric acid (n-)	$C_4H_8O_2$	18	0–10	$-0._1651$	$-0._117$	$+0._611$
		25	0–62	$+0._4414$	$+0._285$	$-0._71291$
Chloral hydrate	$C_2H_3Cl_3O_2$	15	0–70	$-0._5135$	$+0._131$	$-0._4366$
		30	0–78	$-0._4489$	$+0._166$	$-0._6549$
Chloroacetic acid	$C_2H_3ClO_2$	20	0–90	$-0._4455$	$-0._2802$	$+0._622$
Citric acid (hydrate)	$C_6H_8O_7 + H_2O$	25	0–86	$-0._4401$	$-0._2198$	$+0._17$
Dichloroacetic acid	$C_2H_2Cl_2O_2$	20	0–50	$-0._3648$	$-0._1887$	$-0._7534$
		25	0–30	$-0._3602$	$+0._302$	$-0._7534$
Diethylamine hydrochloride	$C_4H_{12}ClN$	21	0–97	$-0._3824$	$+0._552$	
Ethylamine hydrochloride	C_2H_6ClN	21	0–36	$-0._4427$	$+0._1141$	
Ethylene glycol	$C_2H_6O_2$	0	0–65	$-0._4427$	$+0._537$	$-0._47$
		15	0–100	$+0._1193$	$+0._537$	$-0._5248$
Ethyl ether	$C_4H_{10}O$	20	0–6	$+0._1483$	$+0._676$	
		25	0–5	$-0._221$	$-0._307$	
tartrate	$C_8H_{14}O_6$	15	0–95	$+0._2367$	$+0._358$	$-0._6005$
Formaldehyde	CH_2O	15	0–40	$+0._2518$	$-0._6658$	$+0._6542$

Substance	Formula	t	Conc. range			
Formamide	CH_3NO	25	22–95	$+0.0_11217$	$+0.0_31199$	-0.0_72529
Furfural	$C_5H_4O_2$	20	0–8	$+0.0_21827$	$+0.0_2366$	
		25	0–3	$+0.0_21664$	$+0.0_221$	
Isoamyl alcohol	$C_5H_{12}O$	20	0–2.5	-0.0_2146	$+0.0_23$	
Isobutyl alcohol	$C_4H_{10}O$	15	0–3	-0.0_2169	$+0.0_238$	
		20				
Isobutyric acid	$C_4H_8O_2$	15	0–3	$+0.0_252$		
		18		$+0.0_245$		
		25	0–12	$+0.0_237$		
Isovaleric acid	$C_5H_{10}O_2$	25	0–3	-0.0_2253	-0.0_2282	
Lactic acid	C_3H_6O	25	0–3	$+0.0_2231$	$+0.0_6185$	
Maleic acid	$C_4H_4O_4$	25	0–40	$+0.0_234$	$+0.0_675$	
Malic acid	$C_4H_6O_5$	20	0–40	-0.0_23933	$+0.0_6957$	
		25	0–40	-0.0_23736	$+0.0_41066$	
Malonic acid	$C_4H_6O_4$	20	0–20	$+0.0_2389$	-0.0_674	
Methyl acetate	$C_3H_6O_2$	20		-0.0_240	$+0.0_6975$	
glucoside (α–)	$C_7H_{14}O_6$	0	26–51	$+0.0_23336$	$+0.0_6975$	$+0.01544$
		30	26–54	-0.0_23151	$+0.0_6454$	$+0.0_6978$
Nicotine	$C_{10}H_{14}N_2$	20	0–63	$+0.0_2642$	$+0.0_655$	-0.0_6687
Nitrophenol (p–)	$C_6H_5NO_3$	0	0–1.5	$+0.0_23216$	$+0.0_23185$	$+0.0_441$
Oxalic acid	$C_2H_2O_4$	15	0–4	-0.0_25898	$+0.0_68$	$+0.0_4254$
		17.5		$+0.0_2494$	$+0.0_68$	$+0.0_4208$
		20		$+0.0_2494$	-0.0_41996	
Phenol	C_6H_6O	15	0–5	-0.0_25264	-0.0_31607	
		80	0–65	$+0.0_25108$	-0.0_6283	
Phenylglycolic acid	$C_8H_8O_3$	25	0–11	$+0.0_2111$	$+0.0_686$	
Picoline (α–)	C_6H_7N	25	0–77	$+0.0_2462$	$+0.0_623$	
(β–)	C_6H_7N	25	0–63	$+0.0_2207$	-0.0_51405	-0.0_4167
Propionic acid	$C_3H_6O_2$	18	0–19	$+0.0_2386$	-0.0_613	
		25	0–40	-0.0_2683	$+0.0_4172$	
Pyridine	C_5H_5N	25	0–60	$+0.0_295$	-0.0_699	$+0.0_361$
Resorcinol	$C_6H_6O_2$	18	0–52	$+0.0_29245$	-0.0_6204	-0.0_628
Succinic acid	$C_4H_6O_4$	25	0–15	$+0.0_2229$	$+0.0_6519$	-0.0_619
			0–3.5	$+0.0_2304$		
Tartaric acid (d, l, or dl)	$C_4H_6O_6$	15	0–50	$+0.0_24482$	$+0.0_4185$	
		17.5	0–50	$+0.0_24455$	$+0.0_4185$	
		20	0–50	$+0.0_24432$	$+0.0_41837$	
		30	0–50	$+0.0_24335$	$+0.0_4185$	
		40	0–50	$+0.0_24265$	$+0.0_4185$	
		50	0–50	$+0.0_24205$	$+0.0_4185$	
		60	0–50	$+0.0_24155$	$+0.0_4185$	

4-29

TABLE 4-1 Specific Gravities of Aqueous Solutions—(Continued)

ORGANIC SOLUTIONS

Densities of Aqueous Solutions of Miscellaneous Organic Compounds—(Concluded)

Name	Formula	t, °C.	Range, p_a	A	B	C
Tetraethyl ammonium chloride	$C_8H_{20}ClN$	21	0–63	$+0.0_41884$	$+0.0_6$	$+0.0_7122$
Thiourea	CH_4N_2S	15	0–7	$+0.0_22995$	$+0.0_3374$	
		$\{12.5$	0–61	$+0.0_4499$	$+0.0_4153$	
Trichloroacetic acid	$C_2HCl_3O_2$	20	10–30	$+0.0_25053$	$+0.0_41387$	$+0.0_61038$
		25	0–94	$+0.0_25051$	$+0.0_6119$	-0.0_669
Triethylamine hydrochloride	$C_6H_{16}ClN$	21	0–54	$+0.0_6$	$+0.0_3558$	$+0.0_9957$
Trimethyl carbinol	$C_4H_{10}O$	20	0–100	-0.0_3117	-0.0_41908	$+0.0_9887$
		25	0–100	$+0.0_31286$	-0.0_4176	$+0.0_61216$
		$\{14.8$	0–12	$+0.0_33213$	-0.0_44802	-0.0_72573
Urea	CH_4N_2O	18	0–51	$+0.0_32718$	$+0.0_41552$	-0.0_72285
		20	0–35	$+0.0_22702$	$+0.0_33712$	-0.0_61379
		25	0–10	$+0.0_22728$	-0.0_41817	-0.0_73437
Urethane	$C_3H_7NO_2$	20	0–56	$+0.0_21278$	-0.0_3245	
Valeric acid (n-)	$C_5H_{10}O_2$	25	0–3	$+0.0_334$	-0.0_327	

Section B. $d = d_s + A p_w + B p_w^2 + C p_w^3$

Name	Formula	d_s	t, °C.	Range, p_w	A	B	C
Butyl alcohol (n-)	$C_4H_{10}O$	0.8097	20	0–20	$+0.0_22103$	-0.0_3113	
Butyric acid (n-)	$C_4H_8O_2$	0.9534	25	0–38	$+0.0_31854$	-0.0_22314	
Ethyl ether	$C_4H_{10}O$	0.7077	0	0–1.1	$+0.0_334$	$+0.0_336$	
Isobutyl alcohol	$C_4H_{10}O$	$\{0.8170$	15	0–14	$+0.0_22437$	$+0.0_3285$	
		0.8055		0–16	$+0.0_2224$	-0.0_4129	
Isobutyric acid	$C_4H_8O_2$	0.9425	26	0–80	$+0.0_31808$	-0.0_42358	$+0.0_41253$
Nicotine	$C_{10}H_{14}N_2$	1.0093	20	0–40	$+0.0_3199$	-0.0_3331	$+0.0_3315$
Picoline (α-)	C_6H_7N	0.9404	25	0–30	$+0.0_22715$	-0.0_3393	
(β-)	C_6H_7N	0.9515	25	0–40	$+0.0_31925$	-0.0_3352	$+0.0_625$
Pyridine	C_5H_5N	0.9776	25	0–40	$+0.0_31157$	-0.0_3536	-0.0_22
Trimethyl carbinol	$C_4H_{10}O$	0.7856	20	0–20	$+0.0_22287$	$+0.0_3275$	

Adapted from Robert H. Perry and Cecil H. Chilton (eds.), "Chemical Engineers' Handbook," 5th ed., McGraw-Hill Book Company, New York, 1973.

many such systems have survived to the present. In the alcoholic-beverage industry, for example, ethyl alcohol–water solutions are reported in *proof* units. Proof is the weight percent concentration of the alcohol multiplied by 2. Thus pure (i.e., 100%) alcohol is 200 proof. Hydrometers are available calibrated directly in proof units, zero proof corresponding to a specific gravity of 1.000, while 200 proof corresponds to a specific gravity of 0.794 (i.e., the specific gravity of pure ethyl alcohol). The most common of these specific-gravity scales used to express solution concentrations are summarized in Table 4-2, which also cites equivalent values to permit convenient conversion from one system to another.

4-2 OXIDATION-REDUCTION

Electronic Basis

When an atom loses one or more electrons, it is said to be *oxidized*. Conversely, when it gains one or more electrons it is said to be *reduced*. *Oxidation*, therefore, is characterized by an increase in the positive charge of an atom:

$$Fe^{++} - 1e \longrightarrow Fe^{3+}$$

Reduction is characterized by an increase in the negative charge of an atom:

$$S^0 + 2e \longrightarrow S^{--}$$

In S^0, the superscript indicates that the atom is electrically neutral, i.e., in its atomic state. The outstanding characteristic of such reactions is that while there is a gain or loss of electrons by the atoms involved, all chemical systems have to be electrically neutral, so that there must be a simultaneous and equivalent gain of electrons by one atom and loss by another, as illustrated in Fig. 4-3. Stating this another way, oxidation must always be accompanied by an equivalent reduction. To emphasize this, such reactions are called *oxidation-reduction* reactions, or sometimes simply redox (*reduction-oxidation*) reactions. The degree to which an atom is oxidized is called its *state of oxidation*, reduction being considered as negative oxidation, and *oxidation numbers* are used to describe the state of oxidation.

As indicated above, all elemental atoms are electrically neutral and are therefore assigned an oxidation number of zero. To determine the oxidation number of atoms in various compounds, the oxidation numbers of hydrogen and oxygen in water, $+1$ and -2, respectively, are used as fundamental references. Thus, in barium oxide (BaO), barium has an oxidation number of $+2$, since oxygen has a value of -2, and the compound as a whole

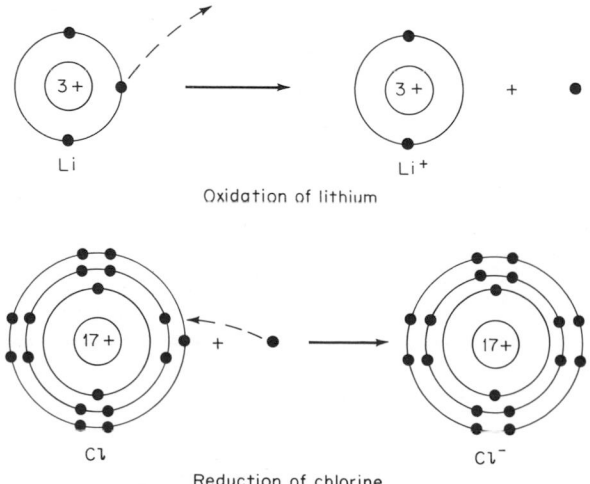

Oxidation of lithium

Reduction of chlorine

Fig. 4-3 Electronic changes in redox reactions.

TABLE 4-2 Liquid Density Scales Used to Express Solution Concentrations

$$\text{°Bé.} = 145 - \frac{145}{\text{sp. gr.}} \text{ (heavier than } H_2O\text{); } \quad \text{°Bé.} = \frac{140}{\text{sp. gr.}} - 130 \text{ (lighter than } H_2O\text{); } \quad \text{°Tw.} = \frac{\text{sp. gr. } 60°/60° \text{ F.} - 1}{0.005}; \quad \text{°A.P.I.} = \frac{141.5}{\text{sp. gr.}} - 131.5$$

Lighter than water

sp. gr. 60/60	°Bé.	°A.P.I.	Lb. per gal. at 60°F. wt. in air	Lb. per cu. ft. at 60°F. wt. in air
0.600	103.33	104.33	4.9929	37.350
0.605	101.40	102.38	5.0346	37.662
0.610	99.51	100.47	5.0763	37.973
0.615	97.64	98.58	5.1180	38.285
0.620	95.81	96.73	5.1597	38.597
0.625	94.00	94.90	5.2014	38.910
0.630	92.22	93.10	5.2431	39.222
0.635	90.47	91.33	5.2848	39.534
0.640	88.75	89.59	5.3265	39.845
0.645	87.05	87.88	5.3682	40.157
0.650	85.38	86.19	5.4098	40.468
0.655	83.74	84.53	5.4515	40.780
0.660	82.12	82.89	5.4932	41.092
0.665	80.53	81.28	5.5349	41.404
0.670	78.96	79.69	5.5766	41.716
0.675	77.41	78.13	5.6183	42.028
0.680	75.88	76.59	5.6600	42.340
0.685	74.38	75.07	5.7017	42.652
0.690	72.90	73.57	5.7434	42.963
0.695	71.44	72.10	5.7851	43.275
0.700	70.00	70.64	5.8268	43.587
0.705	68.58	69.21	5.8685	43.899
0.710	67.18	67.80	5.9101	44.211
0.715	65.80	66.40	5.9518	44.523
0.720	64.44	65.03	5.9935	44.834
0.725	63.10	63.67	6.0352	45.146
0.730	61.78	62.34	6.0769	45.458
0.735	60.48	61.02	6.1186	45.770
0.740	59.19	59.72	6.1603	46.082
0.745	57.92	58.43	6.2020	46.394
0.750	56.67	57.17	6.2437	46.706
0.755	55.43	55.92	6.2854	47.018
0.760	54.21	54.68	6.3271	47.330
0.765	53.01	53.47	6.3688	47.642
0.770	51.82	52.27	6.4104	47.953
0.775	50.65	51.08	6.4521	48.265
0.780	49.49	49.91	6.4938	48.577
0.785	48.34	48.75	6.5355	48.889
0.790	47.22	47.61	6.5772	49.201
0.795	46.10	46.49	6.6189	49.513
0.800	45.00	45.38	6.6606	49.825
0.805	43.91	44.28	6.7023	50.137
0.810	42.84	43.19	6.7440	50.448
0.815	41.78	42.12	6.7857	50.760
0.820	40.73	41.06	6.8274	51.072
0.825	39.70	40.02	6.8691	51.384
0.830	38.67	38.98	6.9108	51.696
0.835	37.66	37.96	6.9525	52.008
0.840	36.67	36.95	6.9941	52.320
0.845	35.68	35.96	7.0358	52.632
0.850	34.71	34.97	7.0775	52.943
0.855	33.74	34.00	7.1192	53.255
0.860	32.79	33.03	7.1609	53.567
0.865	31.85	32.08	7.2026	53.879
0.870	30.92	31.14	7.2443	54.191
0.875	30.00	30.21	7.2860	54.503
0.880	29.09	29.30	7.3277	54.815
0.885	28.19	28.39	7.3694	55.127
0.890	27.30	27.49	7.4111	55.438
0.895	26.42	26.60	7.4528	55.750
0.900	25.56	25.72	7.4944	56.062
0.905	24.70	24.85	7.5361	56.374
0.910	23.85	23.99	7.5777	56.685
0.915	23.01	23.14	7.6194	56.997
0.920	22.17	22.30	7.6612	57.310
0.925	21.35	21.47	7.7029	57.622
0.930	20.54	20.65	7.7446	57.934
0.935	19.73	19.84	7.7863	58.246
0.940	18.94	19.03	7.8280	58.557
0.945	18.15	18.24	7.8697	58.869
0.950	17.37	17.45	7.9114	59.181
0.955	16.60	16.67	7.9531	59.493
0.960	15.83	15.90	7.9947	59.805
0.965	15.08	15.13	8.0364	60.117
0.970	14.33	14.38	8.0780	60.428
0.975	13.59	13.63	8.1197	60.740
0.980	12.86	12.89	8.1615	61.052
0.985	12.13	12.15	8.2032	61.364
0.990	11.41	11.43	8.2449	61.676
0.995	10.70	10.71	8.2866	61.988
1.000	10.00	10.00	8.3283	62.300

Heavier than water

sp. gr. 60/60	°Bé.	°Tw.	Lb. per gal. at 60°F. wt. in air	Lb. per cu. ft. at 60°F. wt. in air
1.005	0.72	1	8.3700	62.612
1.010	1.44	2	8.4117	62.924
1.015	2.14	3	8.4534	63.236
1.020	2.84	4	8.4950	63.547
1.025	3.54	5	8.5367	63.859
1.030	4.22	6	8.5784	64.171
1.035	4.90	7	8.6201	64.483
1.040	5.58	8	8.6618	64.795
1.045	6.24	9	8.7035	65.107
1.050	6.91	10	8.7452	65.419
1.055	7.56	11	8.7869	65.731
1.060	8.21	12	8.8286	66.042
1.065	8.85	13	8.8703	66.354
1.070	9.49	14	8.9120	66.666
1.075	10.12	15	8.9537	66.978
1.080	10.74	16	8.9954	67.290
1.085	11.36	17	9.0371	67.602
1.090	11.97	18	9.0787	67.914
1.095	12.58	19	9.1204	68.226
1.100	13.18	20	9.1621	68.537
1.105	13.78	21	9.2038	68.849
1.110	14.37	22	9.2455	69.161
1.115	14.96	23	9.2872	69.473
1.120	15.54	24	9.3289	69.785
1.125	16.11	25	9.3706	70.097
1.130	16.68	26	9.4123	70.409
1.135	17.25	27	9.4540	70.721
1.140	17.81	28	9.4957	71.032
1.145	18.36	29	9.5374	71.344
1.150	18.91	30	9.5790	71.656
1.155	19.46	31	9.6207	71.968
1.160	20.00	32	9.6624	72.280
1.165	20.54	33	9.7041	72.592
1.170	21.07	34	9.7458	72.904
1.175	21.60	35	9.7875	73.216
1.180	22.12	36	9.8292	73.528
1.185	22.64	37	9.8709	73.840
1.190	23.15	38	9.9126	74.151
1.195	23.66	39	9.9543	74.463
1.200	24.17	40	9.9960	74.775

TABLE 4-2 Liquid Density Scales Used to Express Solution Concentrations—(Continued)

Sp. gr. 60/60°	°Bé.	°Tw.	Lb. per gal. at 60°F. wt. in air	Lb. per cu. ft. at 60°F. wt. in air
1.205	24.67	41	10.0377	75.087
1.210	25.17	42	10.0793	75.399
1.215	25.66	43	10.1210	75.711
1.220	26.15	44	10.1627	76.022
1.225	26.63	45	10.2044	76.334
1.230	27.11	46	10.2461	76.646
1.235	27.59	47	10.2878	76.958
1.240	28.06	48	10.3295	77.270
1.245	28.53	49	10.3712	77.582
1.250	29.00	50	10.4129	77.894
1.255	29.46	51	10.4546	78.206
1.260	29.92	52	10.4963	78.518
1.265	30.38	53	10.5380	78.830
1.270	30.83	54	10.5797	79.141
1.275	31.27	55	10.6214	79.453
1.280	31.72	56	10.6630	79.765
1.285	32.16	57	10.7047	80.077
1.290	32.60	58	10.7464	80.389
1.295	33.03	59	10.7881	80.701
1.300	33.46	60	10.8298	81.013
1.305	33.89	61	10.8715	81.325
1.310	34.31	62	10.9132	81.636
1.315	34.73	63	10.9549	81.948
1.320	35.15	64	10.9966	82.260
1.325	35.57	65	11.0383	82.572
1.330	35.98	66	11.0800	82.884
1.335	36.39	67	11.1217	83.196
1.340	36.79	68	11.1634	83.508
1.345	37.19	69	11.2051	83.820
1.350	37.59	70	11.2467	84.131
1.355	37.99	71	11.2884	84.443
1.360	38.38	72	11.3301	84.755
1.365	38.77	73	11.3718	85.067
1.370	39.16	74	11.4135	85.379
1.375	39.55	75	11.4552	85.691
1.380	39.93	76	11.4969	86.003
1.385	40.31	77	11.5386	86.315
1.390	40.68	78	11.5803	86.626
1.395	41.06	79	11.6220	86.938
1.400	41.43	80	11.6637	87.250
1.405	41.80	81	11.7054	87.562
1.410	42.16	82	11.7471	87.874
1.415	42.53	83	11.7888	88.186
1.420	42.89	84	11.8304	88.498
1.425	43.25	85	11.8721	88.810
1.430	43.60	86	11.9138	89.121
1.435	43.95	87	11.9555	89.433
1.440	44.31	88	11.9972	89.745
1.445	44.65	89	12.0389	90.057
1.450	45.00	90	12.0806	90.369
1.455	45.34	91	12.1223	90.681
1.460	45.68	92	12.1640	90.993
1.465	46.02	93	12.2057	91.305
1.470	46.36	94	12.2473	91.616
1.475	46.69	95	12.2890	91.928
1.480	47.03	96	12.3307	92.240
1.485	47.35	97	12.3724	92.552
1.490	47.68	98	12.4141	92.864
1.495	48.01	99	12.4558	93.176
1.500	48.33	100	12.4975	93.488
1.505	48.65	101	12.5392	93.800
1.510	48.97	102	12.5809	94.112
1.515	49.29	103	12.6226	94.424
1.520	49.61	104	12.6643	94.735
1.525	49.92	105	12.7060	95.047
1.530	50.23	106	12.7477	95.359
1.535	50.54	107	12.7894	95.671
1.540	50.84	108	12.8310	95.983
1.545	51.15	109	12.8727	96.295
1.550	51.45	110	12.9144	96.607
1.555	51.75	111	12.9561	96.918
1.560	52.05	112	12.9978	97.230
1.565	52.35	113	13.0395	97.542
1.570	52.64	114	13.0812	97.854
1.575	52.94	115	13.1229	98.166
1.580	53.23	116	13.1646	98.478
1.585	53.52	117	13.2063	98.790
1.590	53.81	118	13.2480	99.102
1.595	54.09	119	13.2897	99.414
1.600	54.38	120	13.3313	99.725
1.605	54.66	121	13.3730	100.037
1.610	54.94	122	13.4147	100.349
1.615	55.22	123	13.4564	100.661
1.620	55.49	124	13.4981	100.973
1.625	55.77	125	13.5398	101.285
1.630	56.04	126	13.5815	101.597
1.635	56.32	127	13.6232	101.909
1.640	56.59	128	13.6649	102.220
1.645	56.85	129	13.7066	102.532
1.650	57.12	130	13.7483	102.844
1.655	57.39	131	13.7900	103.156
1.660	57.65	132	13.8317	103.468
1.665	57.91	133	13.8734	103.780
1.670	58.17	134	13.9150	104.092
1.675	58.43	135	13.9567	104.404
1.680	58.69	136	13.9984	104.715
1.685	58.95	137	14.0401	105.027
1.690	59.20	138	14.0818	105.339
1.695	59.45	139	14.1235	105.651
1.700	59.71	140	14.1652	105.963
1.705	59.96	141	14.2069	106.275
1.710	60.20	142	14.2486	106.587
1.715	60.45	143	14.2903	106.899
1.720	60.70	144	14.3320	107.210
1.725	60.94	145	14.3737	107.522
1.730	61.18	146	14.4153	107.834
1.735	61.42	147	14.4570	108.146
1.740	61.67	148	14.4987	108.458
1.745	61.91	149	14.5404	108.770
1.750	62.14	150	14.5821	109.082
1.755	62.38	151	14.6238	109.394
1.760	62.61	152	14.6655	109.705
1.765	62.85	153	14.7072	110.017
1.770	63.08	154	14.7489	110.329
1.775	63.31	155	14.7906	110.641
1.780	63.54	156	14.8323	110.953
1.785	63.77	157	14.8740	111.265
1.790	64.00	158	14.9157	111.577
1.795	64.22	159	14.9574	111.889
1.800	64.44	160	14.9990	112.200
1.805	64.67	161	15.0407	112.512
1.810	64.89	162	15.0824	112.824
1.815	65.11	163	15.1241	113.136
1.820	65.33	164	15.1658	113.448
1.825	65.55	165	15.2075	113.760
1.830	65.77	166	15.2492	114.072
1.835	65.98	167	15.2909	114.384
1.840	66.20	168	15.3326	114.696
1.845	66.41	169	15.3743	115.007
1.850	66.62	170	15.4160	115.318
1.855	66.83	171	15.4577	115.630
1.860	67.04	172	15.4993	115.943
1.865	67.25	173	15.5410	116.255
1.870	67.46	174	15.5827	116.567
1.875	67.67	175	15.6244	116.879
1.880	67.87	176	15.6661	117.191
1.885	68.08	177	15.7078	117.503
1.890	68.28	178	15.7495	117.814
1.895	68.48	179	15.7912	118.126
1.900	68.68	180	15.8329	118.438
1.905	68.88	181	15.8746	118.740
1.910	69.08	182	15.9163	119.062
1.915	69.28	183	15.9580	119.374
1.920	69.48	184	15.9996	119.686
1.925	69.68	185	16.0413	119.998
1.930	69.87	186	16.0830	120.309
1.935	70.06	187	16.1247	120.621
1.940	70.26	188	16.1664	120.933
1.945	70.45	189	16.2081	121.245
1.950	70.64	190	16.2498	121.557
1.955	70.83	191	16.2915	121.869
1.960	71.02	192	16.3332	122.181
1.965	71.21	193	16.3749	122.493
1.970	71.40	194	16.4166	122.804
1.975	71.58	195	16.4583	123.116
1.980	71.77	196	16.5000	123.428
1.985	71.95	197	16.5417	123.740
1.990	72.14	198	16.5833	124.052
1.995	72.32	199	16.6250	124.364
2.000	72.50	200	16.6667	124.676

Adapted from Robert H. Perry and Cecil H. Chilton (eds.), "Chemical Engineers' Handbook," 5th ed., McGraw-Hill Book Company, New York, 1973.

must be electrically neutral. Frequently, a radical will contain an atom that can exist in more than one state of oxidation, and this permits the radical as a whole to participate in redox reactions. In chloric acid ($HClO_3$), for example, chlorine has an oxidation number of $+5$ [since $O_3 = 3(-2) = -6$ and $H = +1$], but in perchloric acid ($HClO_4$) its oxidation number is $+7$. Table 4-3 gives the oxidation numbers for several elements and radicals in their usual oxidation states.

Balancing the equation of redox reactions involves a balance of electrons as well as of atoms. To do this, it is frequently convenient to separate the oxidation and reduction reactions. This is illustrated by the following example.

Example 4-9   Balance the equation

$$PbO_2 + Pb + H_2SO_4 \longrightarrow PbSO_4 + H_2O$$

(*Note:* This is the reaction that occurs during discharge of a lead-acid battery.)

solution   Inspection of the equation indicates that lead goes from an oxidation state of $+4$ in PbO_2 to a state of $+2$ in $PbSO_4$. This is a reduction reaction:

$$Pb^{+4} + 2e \longrightarrow Pb^{+2}$$

(It should be remembered that an electron e carries a charge of -1, so that the equation is balanced

TABLE 4-3   Oxidation Numbers of the Common Elements

Element	Symbol	Oxidation number	Typical ion (name)
Aluminum	Al	3	Al^{3+}
Antimony	Sb	3	Sb^{3+} (antimonous)
		5	Sb^{5+} (antimonic)
Arsenic	As	3	As^{3+} (arsenous)
		5	As^{5+} (arsenic)
Barium	Ba	2	Ba^{++}
Beryllium	Be	2	Be^{++}
Bismuth	Bi	3	Bi^{++}
		5	BiO_3^- (bismuthate)
Boron	B	3	B^{3+}
Bromine	Br	1	Br^- (bromide)
		3	BrO_2^- (bromite)
		5	BrO_3^- (bromate)
		7	BrO_4^- (perbromate)
Cadmium	Cd	2	Cd^{++}
Calcium	Ca	2	Ca^{++}
Carbon	C	2	(monoxide)
		4	(dioxide)
Cerium	Ce	3	Ce^{3+} (cerous)
		4	Ce^{4+} (ceric)
Chlorine	Cl	1	Cl^- (chloride)
		3	ClO_2^- (chlorite)
		5	ClO_3^- (chlorate)
		7	ClO_4^- (perchlorate)
Chromium	Cr	2	Cr^{++} (chromous)
		3	Cr^{3+} (chromic)
		6	$Cr_2O_7^{--}$ (dichromate)
Cobalt	Co	2	Co^{++} (cobaltous)
		3	Co^{3+} (cobaltic)
Copper	Cu	1	Cu^+ (cuprous)
		2	Cu^{++} (cupric)
Fluorine	F	1	F^- (fluoride)
Gold	Au	1	Au^+ (aurous)
		3	Au^{3+} (auric)
Hydrogen	H	1	H^+
Iodine	I	1	I^- (iodide)
		3	IO_2^- (iodite)
		5	IO_3 (iodate)
		7	IO_4 (periodate)
Iron	Fe	2	Fe^{++} (ferrous)

TABLE 4-3   Oxidation Numbers of the Common Elements—(Continued)

Element	Symbol	Oxidation number	Typical ion (name)
		3	Fe^{3+} (ferric)
Lead	Pb	2	Pb^{++} (plumbous)
		4	Pb^{4+} (plumbic)
Lithium	Li	1	Li^+
Magnesium	Mg	2	Mg^{++}
Manganese	Mn	2	Mn^{++} (manganous)
		3	Mn^{3+} (manganic)
		4	Mn^{4+} (permanganic)
		6	MnO_4^{--} (manganate)
		7	MnO_4^- (permanganate)
Mercury	Hg	1	Hg^+ (mercurous)
		2	Hg^{++} (mercuric)
Nickel	Ni	2	Ni^{++} (nickelous)
		3	Ni^{3+} (nickelic)
Nitrogen	N	3	NO_2^- (nitrite)
		5	NO_3^- (nitrate)
Oxygen	O	2	(oxide)
Palladium	Pd	2	Pd^{++} (palladous)
		4	Pd^{4+} (palladic)
Phosphorus	P	3	PO_3^{3-} (phosphite)
		5	PO_4^{3-} (phosphate)
Platinum	Pt	2	Pt^{++} (platinous)
		4	Pt^{4+} (platinic)
Potassium	K	1	K^+
Radium	Ra	2	Ra^{++}
Silver	Ag	1	Ag^+
		2	Ag^{++}
Sodium	Na	1	Na^+
Strontium	Sr	2	Sr^{++}
Sulfur	S	2	S^{--} (sulfide)
		4	SO_3^{--} (sulfite)
		6	SO_4^{--} (sulfate)
Tin	Sn	2	Sn^{++} (stannous)
		4	Sn^{4+} (stannic)
Uranium	U	4	UO_3^{--} (uranyl)
		6	UO_5^{--} (peruranyl)
Vanadium	V	3	V^{3+}
		5	VO_3^- (vanadate)
Zinc	Zn	2	Zn^{++}

electrically.) Simultaneously, lead goes from an oxidation state of zero in Pb to $+2$ in $PbSO_4$. This is an oxidation reaction:

$$Pb^0 - 2e \longrightarrow Pb^{+2}$$

These equations can now be added, whereupon the electron terms cancel. Physically, this simply indicates that the Pb^0 loses electrons while an equal number are absorbed by the Pb^{+4}.

$$Pb^0 + Pb^{+4} \longrightarrow 2Pb^{+2}$$

However, this also indicates that $2\ PbSO_4$ are required on the right side of the original equation:

$$PbO_2 + Pb + H_2SO_4 \longrightarrow 2PbSO_4 + H_2O$$

and this now requires an additional H_2SO_4:

$$PbO_2 + Pb + 2H_2SO_4 \longrightarrow 2PbSO_4 + H_2O$$

Finally, to complete the material balance, an additional H_2O is produced:

$$PbO_2 + Pb + 2H_2SO_4 \longrightarrow 2PbSO_4 + 2H_2O$$

An *oxidizing agent* is one which can bring about the oxidation of a material, being itself reduced in the process, while a *reducing agent* is one which can bring about the reduction of a material, being itself oxidized in the process. In the above example, PbO_2 is an oxidizing agent and Pb a reducing agent. Usually, however, the terms oxidizing agent and reducing agent are reserved for materials that have particularly strong oxidizing or reducing properties. Table 4-4 lists common oxidizing and reducing agents. The strength of an oxidizing or reducing agent is expressed as its *oxidation* or *reduction potential*.

Forms of Redox Reactions

Oxidation-reduction can take place in many ways. In *direct oxidation*, the reaction can proceed simply by bringing the reactants together under proper circumstances, as in the burning of carbon in oxygen ($C + O_2 \longrightarrow CO_2$) or the oxidation of alcohol by silver oxide ($C_2H_5OH + AgO \longrightarrow CH_3COOH + 4Ag + H_2O$). In *oxidation in aqueous media,* the reactants are dissolved in water to ionize them and thereby facilitate the necessary electron migration. In this form of redox reaction, the reaction may involve both ions and molecules, as in the oxidation of chlorate ions to perchlorate ions ($ClO_3^- + H_2O \longrightarrow ClO_4^- + 2H^+ + e$). The aqueous medium tends to moderate the reaction, so that it is usually not as rapid (or as violent) as direct oxidation reactions. In *electrochemical* redox reactions, the individual oxidation and reduction reactions can be physically separated, with the electronic transfer occurring through an external wire. The fact that the electron transfer occurs through a wire permits electrochemical redox reactions to be controlled and measured to a high degree of precision.

4-3   ELECTROCHEMISTRY

Electrolytic Cells

When redox reactions take place by directly mixing the oxidizing agent and the reducing agent, the electron transfer occurs directly between the two atoms involved. It is possible, however, for an oxidizing agent to gain an electron from an electrode in an electrolytic cell, the electrode receiving the electrons from another electrode in the same cell to which it is connected by a wire. This other electrode in turn receives its electrons from a reducing

TABLE 4-4   Common Oxidizing and Reducing Agents

In order of approximate decreasing strength

OXIDIZING AGENTS
Sodium bismuthate
Potassium persulfate
Potassium permanganate
Manganese dioxide
Chlorine
Potassium dichromate
Bromine
Cerric sulfate
Hydrogen peroxide
Nitric acid
Ferric sulfate
Potassium ferricyanide
Oxygen
Iodine

REDUCING AGENTS
Zinc
Iron
Stannous chloride
Hydrogen sulfide
Sodium sulfite
Arsenious acid
Oxalic acid
Mercurous chloride
Potassium ferrocyanide
Sodium nitrite
Potassium bromide
Manganous sulfate

TABLE 4-5   Ionic Mobility of Various Ions

For dilute solutions at 18°C under a potential gradient of 1 V/cm)

Ion	Velocity (cm/s)
K^+	0.00066
NH_4^+	0.00066
Na^+	0.00045
Li^+	0.00036
Ag^+	0.00057
$Cr_2O_7^{--}$	0.00047
H^+	0.00320
Cl^-	0.00069
NO_3^-	0.00064
ClO_3^-	0.00057
OH^-	0.00181
Cu^{++}	0.00031

From F. H. Getman and F. Daniels, "Outlines of Theoretical Chemistry," 6th ed., John Wiley & Sons, Inc., New York, 1937.

agent in the cell. The individual oxidation and reduction reactions can thus be physically separated, i.e., take place in different places in the same cell, with the required electrons flowing from the source to the sink via an external wire. The system consisting of two dissimilar electrodes immersed in a conducting solution is referred to as an *electrolytic cell*, or simply a cell. Figure 4-4 illustrates the components and representations used in describing electrolytic cells. When an electric current is set up between two electrodes by virtue of the reaction of an oxidizing agent and a reducing agent at the electrode surfaces, the cell is called a *battery*.

When an electric current is supplied to the electrodes in an electrochemical cell, an oxidation reaction and a reduction reaction will occur at the electrode surfaces. In order for the cell to continue to function, unreacted ions must migrate to each electrode, and reacted ions must move away from the electrodes. Cations, being positively charged ions, are attracted by, and migrate to the negatively charged electrode, from which they acquire electrons and are consequently reduced. The electrode at which reduction takes place and to which cations are attracted is called the *cathode*. Conversely, anions, being negatively charged ions, are attracted by, and migrate to the positively charged electrode, where they give up electrons and are consequently oxidized. The electrode at which oxidation takes place and to which anions are attracted is called the *anode*.

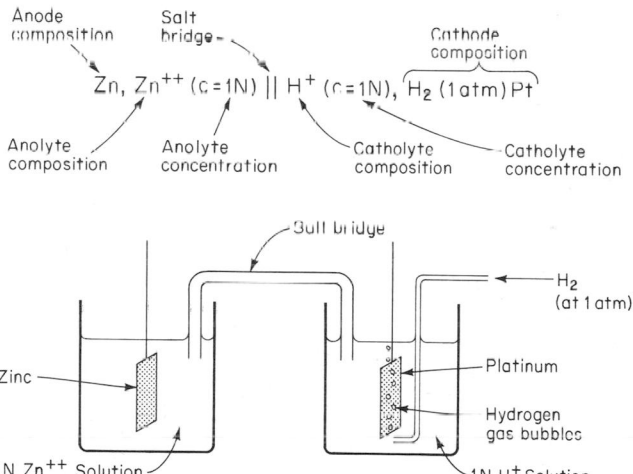

Fig. 4-4 Representation of electrolytic cells. (The actual cell system is shown below its conventional representation.)

The movement of anions and cations (being charged particles) through the solution in which the electrodes are immersed essentially constitutes an electric current, and the solution will conduct electricity. Such a solution, i.e., one containing mobile ions which can move under the influence of an electric potential and thereby carry an electric current, is called an *electrolyte*. The most common electrolytes consist of ionizible salts, acids, or bases dissolved in water. However, many nonaqueous liquids, such as molten salts, ionize or act as solvents for salts which do ionize. In either case, such liquids support ionic conduction and are also classified as electrolytes. Aluminum chloride ($AlCl_3$), sodium chloride (NaCl), and calcium chloride ($CaCl_2$) are typical salts that melt to form electrolytes. Pyridine [$N(CH)_5$], liquid ammonia (NH_3), and molten cryolite ($NaFAlF_3$) are typical nonaqueous solvents in which certain solutes will ionize to render the liquid conductive thereby forming an electrolyte.

Electrolytic Conduction

How well an electrolyte will conduct an electrolytic current, i.e., its *conductivity*, will depend on the number and types of ions present, and how easily these move through the electrolyte. The ease with which different ions move through an electrolyte under the influence of an electric potential is referred to as the *ion mobility*. Table 4-5 gives the ionic mobilities for various ions. In the electrolyte, the current is carried by negative ions moving

TABLE 4-6   Transference Numbers of Selected Cations

Cation	Source of cation	Temp (°C)	Concentration (mols/l)					
			0.005	0.01	0.02	0.05	0.10	0.20
Na^+	NaCl	0	0.387	0.387	0.387	0.386	0.385	
		18	0.396	0.396	0.396	0.395	0.393	0.390
		30	0.404	0.404	0.404	0.404	0.403	
	Na_2SO_4	18		0.392	0.390	0.383		
K^+	KCl	0	0.493	0.493	0.493	0.493	0.492	0.491
		18	0.496	0.496	0.496	0.496	0.495	0.494
		30	0.498	0.498	0.498	0.498	0.497	0.496
	K_2SO_4	18		0.494	0.492	0.490		
		25				0.496	0.494	0.493
Li^+	LiCl	18		0.332	0.328	0.320	0.313	0.304
NH_4^+	NH_4Cl	18		0.492	0.492	0.492		
Ag^+	$AgNO_3$	18		0.471	0.471	0.471	0.471	
		30	0.481	0.481	0.481	0.481	0.481	0.481
H^+	HCl	0	0.847	0.846	0.844	0.839	0.834	
		18	0.832	0.833	0.833	0.834	0.835	0.837
H^+	HNO_3	20	0.839	0.840	0.841	0.844		
H^+	H_2SO_4	20			0.822	0.822	0.822	0.821
Ba^{++}	$BaCl_2$	0	0.439	0.437	0.432			
		25				0.438	0.427	0.415
Ba^{++}	$Ba(NO_3)_2$	25				0.456	0.456	0.456
Mg^{++}	$MgSO_4$	18	0.388	0.385	0.381	0.373		
Cu^{++}	$CuSO_4$	18			0.375	0.375	0.373	0.361
Ca^{++}	$CaCl_2$	20	0.440	0.432	0.424	0.413	0.404	0.395

From F. H. Getman and F. Daniels, "Outlines of Theoretical Chemistry," 6th ed., John Wiley & Sons, Inc., New York, 1937.

toward the anode, and by positive ions moving toward the cathode. However, because different ions have different mobilities, it is not necessary that each ion carry the same fraction of the current. The ions that move faster will carry a larger fraction of the electricity through the solution in a given time. The fraction of the total electricity carried by a particular ion is its *transference number*. The transference numbers of common ions are given in Table 4-6. The measurement of transference numbers is based on changes in electrolyte concentrations in the vicinity of the electrodes. The apparatus used is shown in Fig. 4-5. To understand how transference numbers relate to electrolyte-concentration changes at the electrodes, consider the electrolysis of HCl as diagramed in Fig. 4-6. Table 4-6 indicates that the transference number of H^+ is about five times that of Cl^- in HCl; i.e., the H^+ ions will carry five times the electricity carried by the Cl^- ions. In Fig. 4-6, six H^+ ions are deposited on the cathode (i.e., $3H_2$ are liberated). These ions will be replaced by five H^+ ions from the middle compartment. At the anode, six Cl^- ions must be deposited (i.e., $3Cl_2$ are liberated) since equivalent numbers of electrons must be used at both electrodes. However, since Cl^- ions have a relatively low mobility, only one Cl^- ion will move from the middle compartment to the anode compartment. In other words, H^+ ions have carried five times the electricity carried by the Cl^- ions. The

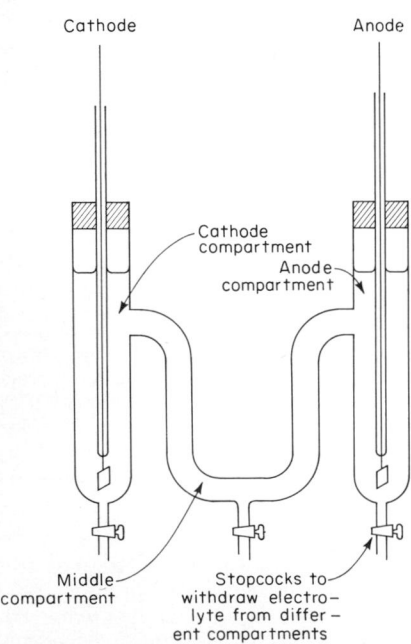

Cathode   Anode

Cathode compartment
Anode compartment
Middle compartment
Stopcocks to withdraw electrolyte from different compartments

Fig. 4-5 Transference-number apparatus.

net result, as shown in Fig. 4-6, is that the concentration of HCl in the cathode compartment is now greater than the concentration of HCl in the anode compartment. The concentration of HCl in both anode and cathode compartments, however, is less than in the middle compartment, which remains unchanged, and the net loss is equal to the total HCl decomposed at the electrode surfaces.

As in the case of electronic conduction in a wire, ionic conduction in an electrolyte follows Ohm's law. The *specific conductance* of an electrolyte is the reciprocal of the resistance between opposite faces of a 1-cm cube, and has units of reciprocal ohms (Ω^{-1} or mho). This is analogous to the conditions shown in Fig. 2-7. In the case of an electrolyte, it is frequently convenient to refer to its *equivalent conductance*. Consider two electrodes 1 cm apart and of such an area that the electrolyte between them contains one equivalent of solute. If the electrolyte is $1N$, 1,000 cm³ of solution will be required, and the electrodes must each have an area of 1,000 cm². If the solution is $0.5N$, 2,000 cm³ of solution will be required for one equivalent of solute to be between the electrodes, and the area of each electrode must therefore be 2,000 cm², as illustrated in Fig. 4-7. The conductance of the solution between such electrodes is its *equivalent conductance.* The uppercase Greek letter lambda (Λ) is the usual symbol for equivalent conductance.

The equivalent conductance of an electrolyte depends upon the concentration of the solute, increasing as the concentration decreases. To understand this, consider 10 cm³ of $1N$ copper sulfate ($CuSO_4$) solution placed between two copper electrodes 1 cm apart. As the conductance is numerically equal to the current flowing when a potential of 1 V is applied across the electrodes (since $I = EL$, where L is the conductance and is equal to $1/R$), the current flowing will be $\frac{1}{100}$ of the equivalent conductance because there is $\frac{1}{100}$ of an equivalent in the electrolyte between the electrodes. If 40 cm³ of water is stirred into the electrolyte between the electrodes, we still have one equivalent of solute in the electrolyte, but the current flowing will actually increase (in the case of the $CuSO_4$ solution, actually from 0.26 to 0.38 A). This increase is due not only to the use of a larger area of the electrodes but also to a greater degree of ionization of the $CuSO_4$ as the solution becomes more dilute. In other words, in the more dilute solution there are actually more current-carrying ions. Eventually, however, the solution becomes so dilute that there is no further increase in equivalent conductance. At this point, the solution has reached its *limiting equivalent conductance*. Conductance at zero concentration or conductance at infinite dilution are alternate terms for this condition. At the limiting conductance, the solution is so dilute that the solute is completely ionized, and further dilution does not increase the number of ions available for electrical conduction. Table 4-7 gives the equivalent conductance of various ions and electrolytes.

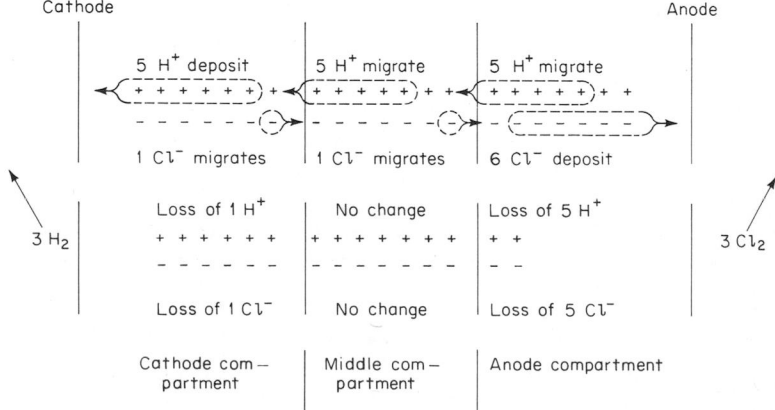

Fig. 4-6 Electrolyte-concentration changes due to transference phenomena. The illustration shows the electrolysis of HCl, with the + signs representing hydrogen ions and the − signs representing chloride ions. Hydrogen ions have a transference number about five times that of the chloride ions and, as shown, carry five times the electricity. Note that the electrolyte is electrically neutral at all points but that the loss in acidity is much greater in the anode compartment than in the cathode compartment.

Electrolysis

The process of inducing a redox reaction to occur by means of the application of an electric current to electrodes immersed in an electrolyte is called *electrolysis*. Electrolysis can take place in aqueous or fused-salt (nonaqueous) electrolytes, as shown in Fig. 4-8. The nineteenth century physicist Michael Faraday first formulated the laws of electrolysis which now bear his name.

Faraday's Laws

Faraday's laws of electrolysis state that (1) the weight of a substance liberated at an electrode is directly proportional to the amount of electricity that has passed through the

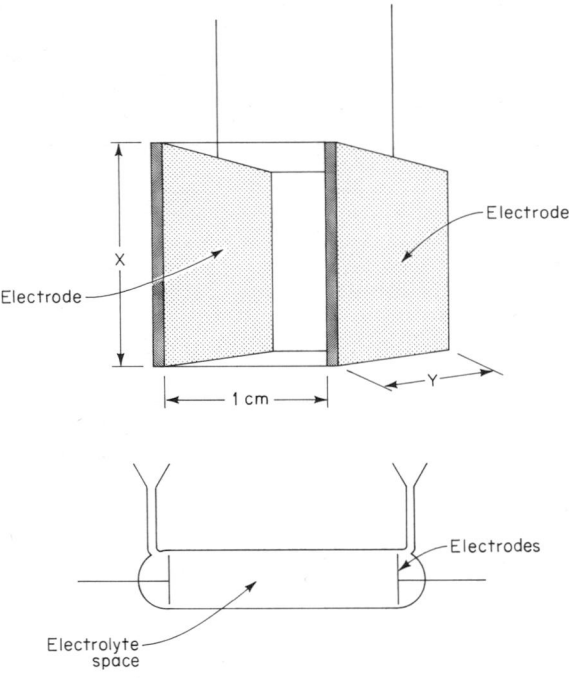

Fig. 4-7 Equivalent conductance and conductance cell. When the space between the electrodes, which has a volume of $1 \text{ cm} \times X \text{ cm} \times Y \text{ cm} = XY \text{ cm}^3$, is filled with electrolyte containing 1 equivalent weight of an acid, base, or salt, the conductivity of the electrolyte between the electrodes is its equivalent conductance at the particular concentration involved. The concentration is equal to $1/0.001XY$ expressed as normality (since there is one equivalent weight in $0.001XY$l). In the conductivity cell shown above, a "cell constant" is determined by filling the cell with electrolyte of known conductivity. This avoids the necessity of accurately controlling the geometry of the cell.

cell, and (2) the weights of different substances produced by the passage of a given amount of electricity through a cell are proportional to the equivalent weights of the substances.

Faraday's laws can be understood by reference to a typical electrolysis such as the decomposition of molten NaCl. At the cathode, one electron is required to reduce each sodium ion to metallic sodium:

$$\text{Na}^+ + e \longrightarrow \text{Na}^0$$

If Avogadro's number of electrons pass through the cell, one equivalent of metallic sodium (23.0 g) will be produced at the cathode. The corresponding quantity of electricity, 96,500 coulombs, is called a *faraday* (F). If 2F are passed through the cell, 46.0 g of sodium will

TABLE 4-7 Equivalent Conductance of Aqueous Solutions

ELECTROLYTES AT 25°C

Solute	Infinite dilution	0.0005	0.001	0.005	0.01	0.02	0.05	0.1
				Concentration (equivalents/1)				
$AgNO_3$	133.36	131.36	130.51	127.20	124.76	121.41	115.24	109.14
$BaCl_2$	139.98	135.96	134.34	128.02	123.94	119.09	111.48	105.19
$CaCl_2$	135.84	131.93	130.36	124.25	120.36	115.65	108.47	102.4
$Ca(OH)_2$	257.9			232.9	225.9	213.9		
$CuSO_4$	133.6	121.6	115.26	94.07	83.12	72.20	59.05	50.58
HCl	426.16	422.74	421.36	415.80	412.00	407.24	399.09	391.32
KBr	151.9			146.09	143.43	140.48	135.68	131.39
KCl	149.86	147.81	146.95	143.35	141.27	138.34	133.37	128.96
$KClO_4$	140.04	138.76	137.87	134.16	131.46	127.92	121.62	115.20
$K_3Fe(CN)_6$	174.5	166.4	163.1	150.7				
$K_4Fe(CN)_6$	184.5		167.24	146.09	134.83	122.82	107.70	97.87
$KHCO_3$	118.0	116.10	115.34	112.24	110.08	107.22		
KI	150.38			144.37	142.18	139.45	134.97	131.11
KIO_4	127.92	125.80	124.94	121.24	118.51	114.14	106.72	98.12
KNO_3	144.96	142.77	141.84	138.48	132.82	132.41	126.31	120.40
$LiCl$	115.03	113.15	112.40	109.40	107.40	104.65	100.11	95.86
$LiClO_4$	105.98	104.18	103.44	100.57	98.61	96.18	92.20	88.56
$MgCl_2$	129.40	125.61	124.11	118.31	114.55	110.04	103.08	97.10
NH_4Cl	149.7		146.8	143.5	141.28	138.33	133.29	128.75
$NaCl$	126.45	124.50	123.74	120.65	118.51	115.51	111.06	106.74
$NaClO_4$	117.48	115.64	114.87	111.75	109.59	106.96	102.40	98.43
NaI	126.94	125.36	124.25	121.25	119.24	116.70	112.79	108.78
$NaH_3C_2O_2$	91.0	89.2	88.5	85.72	83.76	81.24	76.92	72.80
$NaH_5C_3O_2$	85.9		83.5	80.9	79.1	76.6		
$NaH_7C_4O_2$	82.70	81.04	80.31	77.58	75.76	73.39	69.32	65.27
$NaOH$	247.8	245.6	244.7	240.8	238.0			
Na_2SO_4	129.9	125.74	124.15	117.15	112.44	106.78	97.75	89.98
$SrCl_2$	135.80	131.90	130.33	124.24	120.24	115.54	108.25	102.19
$ZnSO_4$	132.8	121.4	114.53	95.49	84.91	74.24	61.20	52.64

EQUIVALENT IONIC CONDUCTANCE AT INFINITE DILUTION

Ion	\multicolumn{6}{c}{Temp (°C)}					
	0	18	25	50	75	100
K^+	40.4	64.6	74.5	115	159	206
Na^+	26.0	43.5	50.9	82	116	155
NH_4^+	40.2	64.5	74.5	115	159	207
Ag^+	32.9	54.3	63.5	101	143	188
Ba^{++}	33	55	65	104	149	200
Ca^{++}	30	51	60	98	142	191
La^{3+}	35	61	72	119	173	235
Cl^-	41.1	65.5	75.5	116	160	207
NO_3^-	40.4	61.7	70.6	104	140	178
$C_2H_3O_2^-$	20.3	34.6	40.8	67	96	130
SO_4^{--}	41	68	79	125	177	234
$Fe(CN)_6^{4-}$	58	95	111	173	244	321
H^+	240	314	350	465	565	644
OH^-	105	172	192	284	360	439

From Robert C. Weast (ed.), "Handbook of Chemistry and Physics," The Chemical Rubber Co., Cleveland, 1964.

be produced, and so on. At the same time, one equivalent of chlorine (35.5 g) will be liberated at the anode for each faraday of electricity:

$$2\ Cl^- \longrightarrow Cl_2 + 2\ e$$

The passage of 2F will liberate 71.0 g of chlorine, and so on.

Example 4-10 In the electrolysis of copper sulfate ($CuSO_4$), what weight of copper is plated onto the cathode by a current of 0.85 A flowing for 30 min?

solution The number of faradays (recalling that 1 coulomb is 1 A flowing for 1 s) will be

$$F = 0.85 \, A \times 30 \, \text{min} \times \frac{60 \, s}{\text{min}} \times \frac{1 \, C}{A\text{-}s} \times \frac{F}{96{,}500 \, C} = 0.0159 \, F$$

The cathode reaction is $Cu^{++} + 2 \, e \longrightarrow Cu^0$, so that the equivalent weight of copper is its atomic weight divided by 2:

$$\frac{63.57}{2} = 31.79 \, g$$

and the weight of copper deposited will thus be

$$0.0159 \times 31.79 = 0.51 \, g$$

Example 4-11 While the copper is being deposited on the cathode in Example 4-10, what volume of oxygen is liberated (at STP) at the anode?
solution The anode reaction is

$$2H_2O \longrightarrow 4H^+ + O_2 + 4 \, e$$

indicating that 4 F is needed to produce 1 mol of oxygen, which would then occupy 22.4 l at STP. Since 0.0159 F, is used in Example 4-10, the volume of O_2 produced will be

$$22.4 \, l \times \frac{0.0159}{4} = 0.089 \, l \equiv 89 \, cm^3$$

Example 4-12 How much electrical energy is theoretically needed to produce 1 ton of aluminum by fused-salt electrolysis if the cell voltage is 4.9 V?
solution The cathode reaction is $Al^{3+} + 3 \, e \longrightarrow Al^0$, so that the equivalent weight of Al is one-third of its atomic weight:

$$\frac{26.97}{3} = 8.99 \, g \quad \text{or} \quad \frac{8.99}{453} = 0.0198 \, lb$$

Thus, to deposit 1 ton of aluminum will require

$$\frac{2{,}000}{0.0198} = 101{,}000 \, F$$

$$101{,}000 \, F \times \frac{96{,}500 \, C}{F} \times \frac{1 \, A\text{-}s}{C} \times \frac{1 \, h}{3{,}600 \, s} \times 4.9 \, V \times \frac{1 \, W}{VA} = 13{,}300{,}000 \, Wh \equiv 13{,}300 \, kWh$$

Faraday's laws thus relate the amount of electricity flowing through an electrolytic cell and the amount of reaction that takes place at each electrode. Frequently, it would appear that somewhat less reaction takes place than that required by Faraday's laws. As an example, the passage of 1 F should deposit 29.3 g of nickel from nickelous sulfate ($NiSO_4$) solution:

$$NI^{++} + 2 \, e \longrightarrow Ni$$

Usually, somewhat less nickel, perhaps only 20 g, will deposit. This results from the fact that there is a competing reaction also taking place at the cathode, namely, the reduction of

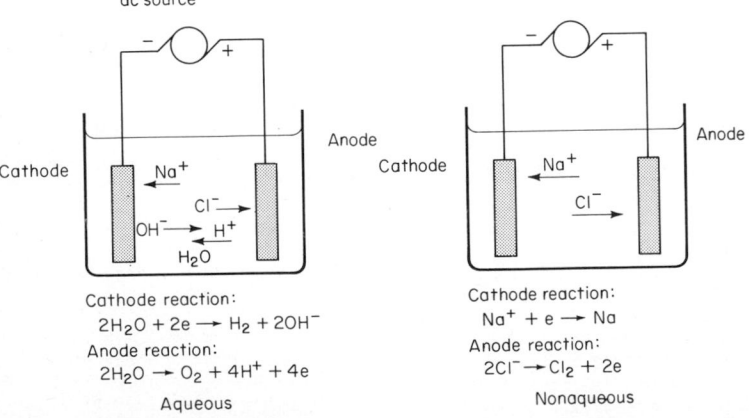

Fig. 4-8 Electrolysis in aqueous and nonaqueous electrolytes.

hydrogen ions (from the water of the electrolyte) to form gaseous hydrogen:

$$2H^+ + 2\,e \longrightarrow H_2$$

Some of the electricity goes into this reaction. The *faradic efficiency* is used to describe the proportion of the electricity that goes into the desired reaction:

$$\text{Faradic efficiency} = \frac{\text{weight of material liberated}}{\text{theoretical weight}} \times 100$$

Example 4-13 In chromium plating, chromic oxide (CrO_3) dissolved in sulfuric acid solution is used as the electrolyte. What is the faradic efficiency of the deposition if the passage of 8.60 A for 30 min deposits 0.15 g of chromium?
solution The cathode reaction is $Cr^{6+} + 6\,e \longrightarrow Cr^0$. Thus the equivalent weight of Cr is one-sixth of its atomic weight or 52.01/6 = 8.67 g. The number of faradays used is

$$8.60\ \cancel{A} \times 30\ \cancel{\text{min}} \times \frac{60\ \cancel{s}}{\cancel{\text{min}}} \times \frac{1\ \cancel{C}}{\cancel{A}\text{-}\cancel{s}} \times \frac{1\ F}{96{,}500\ \cancel{C}} = 0.16\ F$$

The weight of Cr that would be deposited at 100 percent efficiency is therefore 0.16 × 8.67 = 1.39 g. The faradic efficiency is therefore 0.15/1.39 × 100 or 10.8 percent.

Example 4-14 A battery using a zinc anode was found to produce 0.35 A for 28 h, during which time it consumed 12.35 g of zinc. How efficiently was the zinc used?
solution The anode reaction is $Zn^0 \longrightarrow Zn^{++} + 2\,e$, so that the equivalent weight of zinc is one-half of its atomic weight:

$$\frac{65.38}{2} = 32.69\ g$$

$$\text{Faradays generated} = 0.35\ \cancel{A} \times 28\ \cancel{h} \times \frac{3{,}600\ \cancel{s}}{\cancel{h}} \times \frac{F}{96{,}500\ \cancel{A}\cancel{s}} = 0.366\ F$$

This should have consumed 0.366 × 32.69 = 11.95 g of zinc. Instead, 12.35 g of zinc was consumed, so that the efficiency of zinc utilization was

$$\frac{11.95}{12.35} \times 100 = 96.8\ \text{percent}$$

Coulometers

A great many electrochemical reactions take place at virtually 100 percent faradic efficiency. Special cells used to measure how much electricity has passed through them by means of how much reaction has taken place are called *coulometers* (since they essentially measure electricity in terms of coulombs). Several coulometers have been developed for accuracy and convenience. Of these, the silver coulometer, illustrated in Fig. 4-9, is perhaps the most accurate and is widely accepted as a fundamental standard. Other coulometers use the deposition (or dissolution) of mercury or iodine, while others sense the oxidation or reduction of such metals as cadmium. The essential feature of any coulometer, however, is that the electrode reaction being measured occurs at 100 percent current efficiency.

Example 4-15 After a period of operation, 0.576 g of silver was deposited on the platinum cathode of a silver coulometer. How much electricity passed through the cell? If the cell was run for 30.0 min, what was the average current?
solution The cathode reaction is $Ag^+ + e \longrightarrow Ag^0$, so that the equivalent weight of silver is equal to its atomic weight, 107.88 g.

$$F = \frac{0.576}{107.88} = 0.00534 \equiv 0.00534 \times 96{,}500 = 515\ C$$

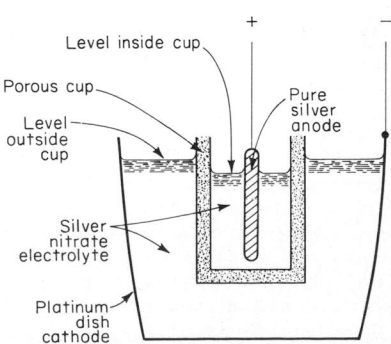

Fig. 4-9 Silver coulometer. The porous cup prevents any material which may flake off the anode from reaching the cathode. Keeping the catholyte level slightly higher than the anolyte level prevents any of the latter from moving toward the cathode. These measures keep the catholyte pure. The amount of silver that deposits is readily measured by weighing the platinum cathode before and after deposition of the silver.

Since the cell was run for 30.0 min, or 1,800 s, the average current was

$$\frac{515}{1,800} = 0.286 \text{ A}$$

Electrode Potentials

When a metal is immersed in a solution of its ions, two tendencies exist; namely, metallic ions in the solution tend to acquire electrons and be deposited on the electrode, and atoms on the electrode tend to lose electrons and go into solution as ions. In the case of a metal which tends to plate out of solution (such as copper), the metal ions receive electrons from the electrode, leaving the latter positively charged. This positive charge prevents other positively charged ions from reaching the electrode, so that the tendency for further reaction is quickly counteracted, and the electrode assumes a positive electrical charge with respect to the electrolyte in which it is immersed. In the case of other metals which tend to go into solution (such as zinc), the formation of metal ions leaves a surplus of electrons in the electrode, which thereby becomes negatively charged. This negative charge quickly prevents additional positively charged ions from leaving the electrode because of the attractive force set up by the negative charge. Such an electrode therefore assumes a negative charge with respect to the electrolyte in which it is immersed. The electrical charge thus acquired by an electrode is called its *single-electrode potential,* metals tending to dissolve being designated as *electronegative,* those tending to plate out being designated as *electropositive.*

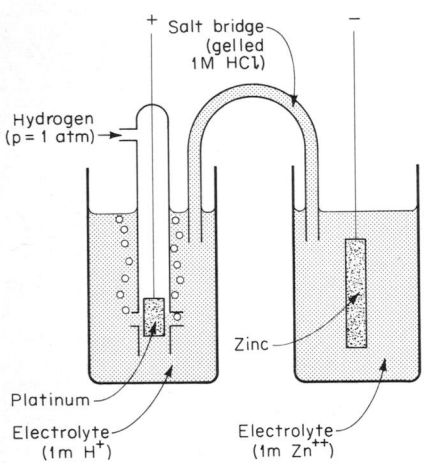

The electrode potential depends on the composition of the electrode, the concentration of the metal ions in solution, and the temperature of the system. The electrode potential of a metal immersed in a solution of one of its salts such that the concentration of the solution with respect to the metal is $1N$ is called the *normal-electrode potential* of the metal. At any given temperature, this is characteristic of the particular metal and is a very reproducible quantity.

One of the problems in using electrode potentials is that the absolute value of the potential of the electrolyte cannot be directly measured, since the very act of "attaching" a wire sets up an electrode-electrolyte system. Consequently, the potential of an electrode immersed in an electrolyte, or its *single-electrode po-*

Fig. 4-10 Standard hydrogen electrode and salt bridge. In this case, the standard electrode is shown being used against a standard zinc electrode. (The potential of this cell is 0.763 V.) This cell can be represented as follows: Zn, Zn^{++} (1 m Zn^{++})//H^+ (1 $m$$H^+$), H_2 (Pt).

tential, is measured against an arbitrary standard electrode whose potential is taken as a base point, or zero potential. Because it is a highly reproducible system, the hydrogen electrode is taken as the *primary standard electrode* and assigned a potential of zero. The fundamental construction of the hydrogen standard electrode is shown in Fig. 4-10, which also shows the use of a *salt bridge* to connect the standard electrode to the single-electrode system, or *half cell,* under examination. In this way, the potential of any half cell can be related to the standard hydrogen electrode. This has been done for many electrode systems, with the resultant values shown in Table 4-8 for standard electrode potentials and in Table 4-9 for other electrode systems. In standard electrodes, the concentration of the soluble reactant in the electrolyte is, for practical purposes, 1 m. The potential of any cell is the algebraic difference of the potentials of its half cells.

Example 4-16 What is the potential of a cell consisting of standard zinc and copper electrodes?
solution From Table 4-8, the potential of the zinc electrode ($Zn^0 \longrightarrow Zn^{++} + 2\,e$) is $+0.758$ V; that of the copper electrode ($Cu^0 \longrightarrow Cu^{++} + 2\,e$) is -0.344 V. The potential of the cell is therefore

$$0.758 - (-0.344) = 0.758 + 0.344 = 1.102 \text{ V}$$

TABLE 4-8 Standard Oxidation-Reduction Potentials

E_0 in volts referred to the hydrogen electrode

25°C

Reaction	E_0	Reaction	E_0
$Li = Li^+ + E^-$	$+2.957$	$Bi = Bi^{3+} + 3E^-$	-0.2
$Rb = Rb^+ + E^-$	$+2.924$	$2Ta + 5H_2O = Ta_2O_5 + 10H^+ + 10E^-$	ca. -0.2
$K = K^+ + E^-$	$+2.922$	$Pt + 4Cl^- = PtCl_4^- + 2E^-$	ca. -0.2
$Sr = Sr^{++} + 2E^-$	$+2.92$	$Ag + Cl^- = AgCl + E^-$	-0.223
$Ba = Ba^{++} + 2E^-$	$+2.90$	$As + 3H_2O = H_3AsO_3 + 3H^+ + 3E^-$	-0.24
$Ca = Ca^{++} + 2E^-$	$+2.87$	$Mo + 3H_2O = MoO_3 + 6H^+ + 6E^-$	-0.25
$Na = Na^+ + E^-$	$+2.712$	$2Hg + 2Cl^- = Hg_2Cl_2 + 2E^-$	-0.270
$Mg = Mg^{++} + 2E^-$	$+2.40$	$PbO + 2OH^- = PbO_2 + H_2O + 2E^-$	-0.3
$Al = Al^{3+} + 3E^-$	$+1.7$	$V + H_2O = VO^{++} + 2H^+ + 4E^-$	-0.3
$Be - Be^{++} + 2E^-$	$+1.69$	$Cu = Cu^{++} + 2E^-$	-0.344
$U = U^{+4} + 4E^-$	$+1.4$	$V^{3+} + H_2O = VO^{++} + 2H^+ + E^-$	-0.4
$Mn = Mn^{++} + 2E^-$	$+1.1$	$4OH^- = O_2 + 2H_2O + 4E^-$	-0.40
$CN^- + 2OH^- = CNO^- + H_2O + 2E^-$	$+0.97$	$PtCl_4^- + 2Cl^- = PtCl_6^- + 2E^-$	ca. -0.40
$Fe + 2OH^- = Fe(OH)_2 + 2E^-$	$+0.86$	$U^{+4} + 2H_2O = UO_2^{++} + 4H^+ + 2E^-$	-0.41
$\tfrac{1}{2}H_2 + OH^- = H_2O \mid E$	$+0.828$	$S + 3H_2O = H_2SO_3 + 4H^+ + 4E^-$	-0.47
$Tl + I^- = TlI + E^-$	$+0.77$	$Fe(CN)_6^{4-} = Fe(CN)_6^{3-} + E^-$	-0.49
$Hg + HS^- + OH^- = HgS + H_2O + 2E^-$	$+0.77$	$H_3AsO_3 + H_2O = H_3AsO_4 + 2H^+ + 2E^-$	-0.49
$Zn = Zn^{++} + 2E^-$	$+0.758$	$Ni(OH)_2 + 2OH^- = NiO_2 \cdot 2H_2O + 2E^-$	-0.49
$Zn + 3OH^- = HZnO_2^- + H_2O + 2E^-$	$+0.72$		
$H_2Te = Te + 2H^+ + 2E^-$	ca. $+0.7$	$2Ag + CO_3^- = Ag_2CO_3 + 2E^-$	-0.50
$Fe(OH)_2 + OH^- = Fe(OH)_3 + E$	$+0.65$	$MoO^{3+} + 2H_2O = MoO_4 + 4H^+ + E^-$	-0.5
$Cr = Cr^{++} + 2E$	$+0.6$	$Cu = Cu^+ + E^-$	-0.51
$Pb + 2OH^- = PbO + H_2O + 2E^-$	$+0.58$	$2I^- = I_2 + 2E^-$	-0.535
$S^- = S + 2E^-$	$+0.51$	$3I^- = I_3^- + 2E^-$	-0.54
$H_2Se = Se + 2H^+ + 2E^-$	ca. $+0.5$	$Hg_2Cl_2 + 2Cl^- = 2HgCl_2 + 2E^-$	-0.63
$Ga = Ga^{3+} + 3E^-$	$+0.5$	$MnO_4^= = MnO_4^- + E^-$	-0.66
$Ag + 2CN^- = Ag(CN)_2^- + E^-$	$+0.5$	$H_2O_2 = O_2 + 2H^+ + 2E^-$	-0.68
$Fe = Fe^{++} + 2E^-$	$+0.44$	$Ag + BrO_3^- = AgBrO_3 + E^-$	-0.68
$Cr^{++} = Cr^{3+} + E^-$	$+0.4$	$C_6H_4(OH)_2 = C_6H_4O_2 \text{ (quinone)}$	
$H_2 = 2H^+(10^{-7}M) + 2E^-$	$+0.414$	$\quad + 2H^+ + 2E^-$	-0.70
$Cd = Cd^{++} + 2E^-$	$+0.397$	$MnO_2 + 4OH^- = MnO_4^= + 2H_2O$	-0.71
$In = In^{3+} + 3E^-$	$+0.38$	$\quad + 2E^-$	
$Ti^{++} = Ti^{3+} + E^-$	$+0.37$	$Fe^{++} = Fe^{3+} + E^-$	-0.74
$2Cu + 2OH^- = Cu_2O + H_2O + 2E^-$	$+0.34$	$Se + 3H_2O = H_2SeO_3 + 4H^+ + 4E^-$	-0.74
$Tl = Tl^+ + E^-$	$+0.336$	$H_3SbO_3 + H_2O = H_3SbO_4 + 2H^+ + 2E^-$	-0.75
$Pb + SO_4^- = PbSO_4 + 2E$	$+0.31$	$2Hg = Hg_2^{++} + 2E^-$	-0.798
$P + 4H_2O = H_3PO_4 + 5H^+ + 5E^-$	$+0.3$	$Ag = Ag^+ + E$	-0.799
$Co(CN)_6^{4-} = Co(CN)_6^{3-} + E^-$	$+0.3$	$CuI = Cu^{++} + I^- + E^-$	-0.85
$Co = Co^{++} + 2E^-$	$+0.29$	$Hg = Hg^{++} + 2E^-$	-0.86
$Ni = Ni^{++} + 2E^-$	$+0.22$	$2H_2O + NH_4^+ = HNO_2 + 7H^+ + 6E^-$	-0.86
$V^{++} = V^{3+} + E^-$	$+0.2$	$3OH^- = HO_2^- + H_2O + 2E^-$	-0.87
$Cu + I^- = CuI + E^-$	$+0.17$	$CoO + 2OH^- = CoO_2 + H_2O + 2E^-$	-0.9
$Ag + I^- = AgI + E^-$	$+0.15$	$Hg_2^{++} = 2Hg^{++} + 2E^-$	-0.92
$Cu_2O + 2OH^- = 2CuO + H_2O + 2E^-$	$+0.15$	$Cl^- + 2OH^- = ClO^- + H_2O + 2E^-$	-0.94
$Sn = Sn^{++} + 2E^-$	$+0.13$	$NO + 2H_2O = NO_3^- + 4H^+ + 3E^-$	-0.94
$Pb = Pb^{++} + 2E^-$	$+0.12$	$HNO_2 + H_2O = NO_3^- + 3H^+ + 2E^-$	-0.95
$9OH^- + NH_3 = NO^- + 6H_2O + 8E^-$	$+0.12$	$NO + H_2O = HNO_2 + H^+ + E^-$	-0.98
$2Hg + 2I^- = Hg_2I_2 + 2E^-$	$+0.04$	$I^- + H_2O = HIO + H^+ + 2E^-$	-0.99
$2Ag + H_2S = Ag_2S + 2H^+ + 2E^-$	$+0.036$	$OsO_2Cl_4^- = OsO_4 + 4H^+$	
$Cu + H_2S = CuS + 2H^+ + 2E^-$	$+0.02$	$\quad + 4Cl^- + 2E^-$ ca.	ca. -1.0
$H_2 = 2H^+ + 2E^-$	0.000	$2Br^- = Br_2 + 2E^-$	-1.065
$2OH^- + NO_2^- = NO_3^- + H_2O + 2E^-$	0.0	$I^- + 3H_2O = IO_3^- + 6H^+ + 6E^-$	-1.09
$HCN + H_2O = HCNO + 2H^+ + 2E^-$	0.0	$VO^{++} + 2H_2O = HVO_3 + 3H^+ + E^-$	-1.1
$Sb + 3H_2O = H_3SbO_3 + 3H^+ + 3E^-$	ca. 0.0	$Tl^+ = Tl^{3+} + 2E^-$	-1.2
$W + 3H_2O = WO_3 + 6H^+ + 6E^-$	ca. 0.0	$H_2SeO_3 + H_2O = H_2SeO_4 + 2H^+ + 2E^-$	ca. -1.2
$WO^{3+} + 2H_2O = WO_3 + 4H^+ + E^-$	ca. 0.0	$2H_2O = O_2 + 4H^+ + 4E^-$	-1.23
$Ti^{3+} + H_2O = TiO^{++} + 2H^+ + E^-$	-0.04	$PdCl_4^- + 2Cl^- = PdCl_6^- + 2E^-$	-1.3
$Hg + 2OH^- = HgO + H_2O + 2E^-$	-0.099	$Cr^{3+} + 4H_2O = HCrO_4^- + 7H^+ + 3E^-$	-1.3
$Ag + Br^- = AgBr + E^-$	-0.10	$Br^- + H_2O = HBrO + H^+ + 2E^-$	-1.33
$2Hg + 2Br^- = Hg_2Br_2 + 2E^-$	-0.13	$Mn^{++} + 2H_2O = MnO_2 + 4H^+ + 2E^-$	-1.33
$Sn^{++} = Sn^{4+} + 2E^-$	-0.13	$Cl^- + 4H_2O = ClO_4^- + 8H^+ + 8E^-$	-1.35
$H_2O + H_2SO_3 = SO_4^- + 4H^+ + 2E^-$	-0.14	$2Cl^- = Cl_2 + 2E^-$	-1.359
$Cu^+ + Cu^{++} + E^-$	-0.17	$2Au + 3H_2O = Au_2O_3 + 6H^+ + 6E^-$	-1.362
$H_2S = S + 2H^+ + 2E^-$	-0.17		

TABLE 4-8   Standard Oxidation-Reduction Potentials—(Continued)

Reaction	E_0	Reaction	E_0
$I^- + 4H_2O = IO_4^- + 8H^+ + 8E^-$	-1.4	$Fe^{3+} + 4H_2O = FeO_4^- + 8H^+ + 3E^-$	ca. -1.7
$Br^- + 3H_2O = BrO_3^- + 6H^+ + 6E^-$	-1.42	$Bi^{3+} + 6H_2O = HBiO_3 + 5H^+ + 2E^-$	ca. -1.7
$Pb^{++} + 2H_2O = PbO_2 + 4H^+ + 2E^-$	-1.44	$PbSO_4 + 2H_2O = PbO_2 + 4H^+ + SO_4^-$	-1.7
$Cl^- + 3H_2O = ClO_3^- + 6H^+ + 6E^-$	-1.45	$\quad + 2E^-$	
$Cl^- + H_2O = HClO + H^+ + 2E^-$	-1.50	$2H_2O = H_2O_2 + 2H^+ + 2E^-$	-1.78
$Mn^{++} = Mn^{3+} + E^-$	ca. -1.5	$Co^{++} = Co^{3+} + F^-$	-1.8
$Au = Au^+ + E^-$	ca. -1.5	$Ni^{++} + 4H_2O = NiO_2 \cdot 2H_2O + 4H^+$	
$2SO_4^- + 2H^+ = H_2S_2O_8 + 2E^-$	ca. -1.5	$\quad + 2E^-$	-1.8
$Ce^{3+} + 2H_2O = CeO_2 + 4H^+ + E^-$	-1.5	$O_2 + H_2O = O_3 + 2H^+ + 2E^-$	-1.9
$Mn^{++} + 4H_2O = MnO_4^- + 8H^+ + 5E^-$	-1.52	$2F^- = F_2 + 2E^-$	-2.8
$MnO_2 + 2H_2O = MnO_4^- + 4H^+ + 3E^-$	-1.63		

From C. D. Hodgman (ed.), "Handbook of Chemistry and Physics," The Chemical Rubber Co., Cleveland, 1942.

Standard electrode potentials are usually measured in electrolyte which is 1 m with respect to the ion involved. However, many electrochemical reactions involve insoluble materials, in which case the concentration of the electrode material in solution is so small as to have little or no effect on the electrode potential. In fact, the various types of electrodes for which standard electrode potentials have been measured can be categorized as follows:

1. Metal, metal ion (example: copper in a solution of copper ions)
2. Inert electrode, nonmetal, nonmetal ion (example: platinum in contact with solid iodine in a solution of iodide ions)
3. Inert electrode, ions of one valence state, ions of another valence state (example: platinum immersed in a solution of ferric ions, but containing ferrous ions as well)
4. Inert electrode, gas, ions (example: platinum partially surrounded by hydrogen gas, with the rest immersed in an acid, i.e., a solution containing hydrogen ions)
5. Inert electrode, organic solute in one state of oxidation, organic solute in another state of oxidation (example: platinum immersed in an electrolyte containing nitrobenzene, but also containing some aniline)
6. Amalgam electrode, ion (example: a solution of sodium in mercury, i.e., sodium amalgam, immersed in a solution containing sodium ions)
7. Active electrode, insoluble salt or oxide, ions (example: metallic silver coated with silver chloride, immersed in a solution containing silver ions)

The above electrode systems are described according to a convention in which a representation of the cell is given such that the electrons move in the external circuit from left to right. The conventions for representing cells are illustrated in Fig. 4-10.

While the standard hydrogen electrode is the fundamental reference electrode, it is frequently troublesome to set up and use. Consequently, other *reference* electrodes, such as the calomel electrode, are generally used. As illustrated in Fig. 4-11, the *calomel electrode* is made up of mercury in contact with mercurous chloride (HgCl, i.e., calomel) and an electrolyte of potassium chloride (KCl) solution saturated with HgCl. If the concentration of the KCl solution is $1N$, the electrode is referred to as the *normal calomel electrode.* Other calomel electrodes in use are the saturated and decinormal, referring to the use of saturated KCl and $0.1N$ KCl as electrolytes, respectively. Not only is the calomel electrode convenient to use, but its potential is extremely stable and reproducible. Table 4-10 gives the potential of calomel electrodes with respect to the normal hydrogen electrode.

Frequently it is necessary to have a directly measurable source of a stable and reproducible potential. This is achieved by combining two suitable half cells to form a *standard cell.* The *Weston cell,* the essentials of which are shown in Fig. 4-12, is probably the most extensively used standard cell. The emf of the Weston cell is 1.0184 V at 18°C. The potential of all electrode systems is temperature-dependent, and one of the virtues of the Weston cell is its low temperature coefficient. The emf of the Weston cell at any temperature $t°C$ is

$$E_t = 1.0184 - 0.000041\,(t - 18) \tag{4-1}$$

TABLE 4-9 Potentials of Electrochemical Reactions at 25°C

Reaction	Potential (V)	Reaction	Potential (V)
$Ag(s) = Ag^+ + e$	+0.7996	$Cu(s) = Cu^{++} + 2e$	+0.3452
$Ag^+ = Ag^{++} + e$	+1.914	$Cu(\text{sat. amalgam}) = Cu^{++} + 2e$	+0.3495
$Ag(s) + Br^- = AgBr(s) + e$	+0.0713	$Cu^+ = Cu^{++} + e$	+0.17
$Ag(s) + BrO_3^- = AgBrO_3(s) + e$	+0.680	$Cu(s) + Cl^- = CuCl(s) + e$	+0.1287
$Ag(s) + Cl^- = AgCl(s) + e$	+0.2221	$Cu(s) + H_2S(g) = CuS(s) + 2H^+ + 2e$	-0.259
$Ag(s) + 2CN^- = Ag(CN)_2^- + e$	-0.5	$Cu(s) + I^- = CuI(s) + e$	-0.17
$2Ag(s) + CO_3^= = Ag_2CO_3(s) + 2e$	+0.500	$2Cu(s) + 2OH^- = Cu_2O(s) + H_2O(l) + 2e$	-0.344
$2Ag^+ + CrO_4^= = Ag_2CrO_4(s) + 2e$	+0.4463	$CuCl_2^- = Cu^{++} + 2Cl^- + e$	+0.455
$2Ag(s) + H_2S(g) = Ag_2S(s) + 2H^+ + 2e$	-0.0366	$Cu\text{-}Hg + (SO_4^= \text{ in } CuSO_4 \cdot 5H_2O$	
$Ag(s) + I^- = AgI(s) + e$	-0.1523	$\text{sat.}) = CuSO_4 \cdot 5H_2O(s) + 2e$	+0.2684
$Al(s) = Al^{3+} + 3e$	-1.7	$CuI(s) = Cu^{++} + I^- + e$	+0.85
$Al\text{-}Hg + 3OH^- = Al(OH)_3(s) + 3e$	-1.62	$Cu_2O(s) + 2OH^- = 2CuO(s, \text{ aged})$	
$As(s) + 2H_2O(l) = HAsO_2 + 3H^+ + 3e$	+0.2375	$\quad + H_2O(l) + 2e$	-0.154
$As(s) + 3H_2O = H_3AsO_3 + 3H^+ + 3e$	+0.24	$Cu_2O(s) + 2OH^- + H_2O(l) =$	
$H_3AsO_3 + H_2O = H_3AsO_4 + 2H^+ + 2e$	+0.49	$\quad 2Cu(OH)_2(s) + 2e$	-0.082
$Au(s) = Au^+ + e$	+1.5	$2F = F_2 + 2e$	+2.88
$Au(s) = Au^{3+} + 3e$	+1.36	$Fe(s) = Fe^{++} + 2e$	-0.441
$2Au(s) + 3H_2O(l) = Au_2O_3(s) + 6H^+ + 6e$	+1.363	$Fe^{++} = Fe^{3+} + e$	+0.782
$Ba(s) = Ba^{++} + 2e$	-2.90	$Fe^{3+} + 4H_2O = FeO_4^{--} + 8H^+ + 3e$	+1.7
$Ba\,Hg = Ba^{++} + 2e$	-1.5700	$Fe(s) + 2OH^- = Fe(OH)_2(s) + 2e$	-0.86
$Be(s) = Be^{++} + 2e$	-1.69	$Fe(CN)_6^- = Fe(CN)_6^- + e$	+0.36
$Bi(s) = Bi^{3+} + 3e$	+0.277	$Fe(OH)_2(s) + OH^- = Fe(OH)_3(s) + e$	-0.65
$Bi(s) + Cl^- + H_2O(l) = BiOCl(s)$		$K_4Fe(CN)_6 = K_3Fe(CN)_6 + K^+ + e$	+0.4866
$\quad + 2H^+ + 3e$	+0.1588	$Ga(s) = Ga^{3+} + 3e$	0.5
$Bi(s) + 4Cl^- = BiCl_4^- + 3e$	+0.1678	$\frac{1}{2}H_2(g) = H^+ + e$	0.0000
$Bi^{3+} + 6H_2O = HBiO_3 + 5H^+ + 2e$	+1.7	$H_2(g) = 2H^+(10^{-7}M) + 2e$	-0.4141
$Br^- = \frac{1}{2}Br_2(l) + e$	+1.0648	$\frac{1}{2}H_2(g) + OH^- = H_2O(l) + e$	-0.8295
$Br^- + 3H_2O = BrO_3^- + 6H^+ + 6e$	+1.42	$2H_2O = H_2O_2 + 2H^+ + 2e$	+1.78
$Br^- + H_2O = HBrO + H^+ + 2e$	+1.33	$2H_2O = O_2(g) + 4H^+ + 4e$	+1.23
$\frac{1}{2}Br_2(l) + 3H_2O = BrO_3^- + 6H^+ + 5e$	+1.491	$H_2O(l) = O(g) + 2H^+ + 2e$	+2.419
$C_6H_4(OH)_2 = C_6H_4O_2 + 2H^+ + 2e$	+0.6992	$H_2O_2 = O_2(g) + 2H^+ + 2e$	+0.68
$Ca(s) = Ca^{++} + 2e$	-2.763	$2Hg(l) = Hg_2^{++} + 2e$	+0.7986
Calomel electrode, sat. KCl	+0.2446	$Hg^{++} = \frac{1}{2}Hg_2^{++} + e$	+0.9011
Calomel electrode, normal KCl	+0.2809	$2Hg(l) + 2Br^- = Hg_2Br_2(s) + 2e$	+0.1385
Calomel electrode, molal KCl	+0.2816	$2Hg(l) + 2Cl^- = Hg_2Cl_2(s) + 2e$	+0.2676
Calomel electrode, decinormal KCl	+0.3334	$Hg(l) + HS^- + OH^- = HgS(s) +$	
$Cd(s) = Cd^{++} + 2e$	-0.4024	$\quad H_2O + 2e$	-0.77
$Cd\text{-}Hg = Cd^{++} + 2e$	-0.3519	$2Hg(l) + 2I^- = Hg_2I_2(s) + 2e$	-0.0416
$Cd\text{-}Hg + (2Br^- \text{ in } CdBr_2 \cdot 4H_2O$		$Hg_2Cl_2(s) + 2Cl^- = 2HgCl_2 + 2e$	+0.63
$\text{sat.}) = CdBr_2 \cdot 4H_2O + 2e$	-0.4182	$Hg(l) + 2OH = HgO(s, \text{ red}) +$	
$Cd\text{-}Hg + (2Cl^- \text{ in } CdCl_2 \text{ sat.}) =$		$\quad H_2O(l) + 2e$	+0.0969
$\quad CdCl_2(s) + 2e$	-0.4034	$Hg(l) + 2OH^- = HgO(s, \text{ yellow})$	
$Cd\text{-}Hg + (2I^- \text{ in } CdI_2 \text{ sat.}) =$		$\quad + H_2O(l) + 2e$	+0.0976
$\quad CdI_2(s) + 2e$	-0.4588	$2Hg(l) + SO_4^{--} = Hg_2SO_4(s) + 2e$	+0.6141
$Cd\text{-}Hg + 2OH^- = CdO(s) + H_2O(l) + 2e$	-0.726	$Hg(l), HgO(s), Ba(OH)_2(s), H_2O$	+0.1462
$Cd\text{-}Hg + 2OH^- = Cd(OH)_2(s) + 2e$	-0.761	$Hg(l), HgO(s), Ca(OH)_2(s), H_2O$	+0.192
$Cd\text{-}Hg + (SO_4^= \text{ in } CdSO_4 \cdot 8/3H_2O$		$I^- = \frac{1}{2}I_2(s) + e$	+0.5356
$\text{sat.}) = CdSO_4 \cdot 8/3H_2O + 2e$	-0.4346	$3I^- = I_3^- + 2e$	+0.54
$Ce^{3+} = Ce^{4+} + e(H_2SO_4 \text{ soln.})$	+1.44	$I^- + H_2O = HIO + H^+ + 2e$	+0.99
$Ce^{3+} = Ce^{4+} + e(HNO_3 \text{ soln.})$	+1.6095	$I^- + 3H_2O = IO_3^- + 6H^+ + 6e$	+1.09
$Ce^{3+} + 2H_2O = CeO_2 + 4H^+ + e$	+1.5	$I^- + 4H_2O = IO_4^- + 8H^+ + 8e$	+1.4
$Cl^- = \frac{1}{2}Cl_2 + e$	+1.3583	$\frac{1}{2}I_2(s) + 3H_2O = IO_3^- + 6H^+ + 5e$	+1.195
$Cl^- + H_2O = HClO + H^+ + 2e$	+1.50	$In(s) = In^{3+} + 3e$	-0.340
$Cl^- + 3H_2O = ClO_3^- + 6H^+ + 6e$	+1.45	$IrCl_6^{3-} = IrCl_6^{--} + e$	+0.97
$Cl^- + 4H_2O = ClO_4^- + 8H^+ + 8e$	+1.35	$K(s) = K^+ + e$	-2.9241
$Cl^- + 2OH^- = ClO^- + H_2O + 2e$	+0.94	$Li(s) = Li^+ + e$	-2.9595
$CN^- + 2OH^- = CNO^- + H_2O + 2e$	-0.97	$Mg(s) = Mg^{++} + 2e$	-2.40
$HCN + H_2O = HCNO + 2H^+ + 2e$	0.0	$Mn(s) = Mn^{++} + 2e$	-1.1
$Co(s) = Co^{++} + 2e$	-0.29	$Mn^{++} = Mn^{3+} + e$	+1.5
$Co^{++} = Co^{3+} + e$	+1.817	$Mn^{++} + 2H_2O = MnO_2(s) + 4H^+ + 2e$	+1.236
$Co(CN)_6^{4-} = Co(CN)_6^{3-} + e$	-0.3	$Mn^{++} + 4H_2O = MnO_4^- + 8H^+ + 5e$	+1.509
$CoO(s) + 2OH^- = CoO_2(s) + H_2O + 2e$	+0.9	$MnO_2(s) + 2H_2O = MnO_4^- + 4H^+ + 3e$	+1.691
$Cr(s) = Cr^{++} + 2e$	-0.557	$MnO_2(s) + 4OH^- = MnO_4^{--} + 2H_2O + 2e$	+0.71
$Cr^{++} = Cr^{3++} + e$	-0.400	$MnO_4^{--} = MnO_4^- + e$	+0.664
$Cr^{3+} + 4H_2O = HCrO_4^- + 7H^+ + 3e$	+1.3	$Mo(s) + 3H_2O = MoO_3(s) + 6H^+ + 6e$	+0.25
$Cu(s) = Cu^+ + e$	+0.51	$Mo(CN)_8^{4-} = Mo(CN)_8^{3-} + e$	+0.7260

TABLE 4-9   Potentials of Electrochemical Reactions at 25°C—(Continued)

Reaction	Potential (V)	Reaction	Potential (V)
$MoO^{3+} + 2H_2O = MoO_3(s) + 4H^+ + e$	+0.5	$H_2SO_3 + H_2O = SO_4^{--} + 4H^+ + 2e$	+0.14
$Na(s) = Na^+ + e$	-2.7146	$Sb(s) + H_2O(l) = SbO^+ + 2H^+ + 3e$	-0.212
$Na\text{-}Hg + (Cl^- \text{ in NaCl sat.}) = NaCl(s) + e$	-1.8378	$Sb(s) + 3H_2O = H_3SbO_3 + 3H^+ + 3e$	0.0
		$2Sb(s) + 3H_2O(l) = Sb_2O_3(s) + 6H^+ + 6e$	+0.1445
$NH_3 + 9OH^- = NO_3^- + 6H_2O + 8e$	-0.12	$H_3SbO_3 + H_2O = H_3SbO_4 + 2H^+ + 2e$	+0.75
$NH_4^+ + 2H_2O = HNO_2 + 7H^+ + 6e$	+0.86	$Se(s) + 3H_2O = H_2SeO_3 + 4H^+ + 4e$	+0.74
$NO + H_2O = HNO_2 + H^+ + e$	+0.98	$H_2Se = Se + 2H^+ + 2e$	-0.5
$NO + 2H_2O = NO_3^- + 4H^+ + 3e$	+0.94	$H_2SeO_3 + H_2O = H_2SeO_4 + 2H^+ + 2e$	+1.2
$NO_2^- + 2OH^- = NO_3^- + H_2O + 2e$	0.0	$Sn(s) = Sn^{++} + 2e$	-0.136
$HNO_2 + H_2O = NO_3^- + 3H^+ + 2e$	+0.95	$Sn^{++} = Sn^{4+} + 2e(0.1 \ M \ HCl)$	+0.070
$Ni(s) = Ni^{++} + 2e$	-0.227	$Sn^{++} = Sn^{4+} + 2e(0.53 \ M \ HCl)$	+0.144
$Ni^{++} + 4H_2O = NiO_2 \cdot 2H_2O + 4H^+ + 2e$	+1.8	$Sn^{++} = Sn^{4+} + 2e(2.0 \ M \ HCl)$	+0.133
$Ni(OH)_2(s) + 2OH^- = NiO_2 \cdot 2H_2O + 2e$	+0.49	$Sr\text{-}Hg = Sr^{++} + 2e$	-1.7932
$O_2(g) + H_2O = O_3(g) + 2H^+ + 2e$	+2.07	$2Ta(s) + 5H_2O = Ta_2O_5 + 10H^+ + 10e$	+0.2
$3OH^- = HO_2^- + H_2O + 2e$	+0.87	$Te(s) = Te^{4+} + 4e$	-0.5682
$4OH^- = O_2 + 2H_2O + 4e$	+0.40	$H_2Te = Te + 2H^+ + 2e$	-0.7
$OsO_2Cl_4^- + 2H_2O = OsO_4 + 4H^+ + 4Cl^- + 2e$	+1.0	$Ti^{++} = Ti^{3+} + e$	+0.37
$P(s) + 4H_2O = H_3PO_4 + 5H^+ + 5e$	-0.3	$Ti^{3+} + H_2O = TiO^{++} + 2H^+ + e$	+0.04
$Pb(s) = Pb^{++} + 2e$	-0.126	$Ti^{3+} + 2SO_4^{--} = Ti(SO_4)_2 + e$	+0.04
$Pb\text{-}Hg + 2Cl^- = PbCl_2(s) + 2e$	-0.2623	$Tl(s) = Tl^+ + e$	-0.336
$Pb\text{-}Hg + 2I^- = PbI_2(s) + 2e$	-0.3580	$Tl^+ = Tl^{3+} + 2e$	+1.2466
$Pb(s) + 2OH^- = PbO(s, red) + H_2O(l) + 2e$	-0.5786	$Tl\text{-}Hg = Tl^+ + e$	-0.3360
		$Tl(s) + I^- = TlI(s) + e$	-0.7715
$Pb(s) + 2OH^- = PbO \ (s, yellow) + H_2O(l) + 2e$	-0.575	$Tl\text{-}Hg + Cl^- = TlCl(s) + e$	-0.5545
		$Tl\text{-}Hg + Br^- = TlBr(s) + e$	-0.6058
$Pb(s) + 2OH^- = PbO(s) + H_2O(l) + 2e$	-0.576	$Tl\text{-}Hg + SO_4^{--} = Tl_2SO_4(s) + 2e$	-0.4360
$Pb\text{-}Hg + 2OH^- = Pb(OH)_2(s) + 2e$	-0.568	$U = U^{4+} + 4e$	-1.4
$Pb(s) + H_2S(g) = PbS(s) + 2H^+ + 2e$	+0.070	$U^{4+} + 2H_2O = UO_2^{++} + 4H^+ + 2e$	+0.41
$Pb(s) + SO_4^{--} = PbSO_4(s) + 2e$	-0.3447	$U(SO_4)_2 + 2H_2O(l) = UO_2SO_4 + 4H^+ + SO_4^{--} + 2e$	+0.358
$Pb\text{-}Hg + SO_4^{--} = PbSO_4(s) + 2e$	-0.3505	$V^{++} = V^{3+} + e$	-0.2
$Pb^{++} + 2H_2O = PbO_2(s) + 4H^+ + 2e$	+1.467	$V^{3+} + H_2O = VO^{++} + 2H^+ + e$	+0.4
$PbO(s) + 2OH^- = PbO_2(s) + H_2O(l) + 2e$	+0.27	$V(s) + H_2O = VO^{++} + 2H^+ + 4e$	+0.3
$PbSO_4(s) + 2H_2O(l) = PbO_2(s) + 4H^+ + SO_4^{--} + 2e$	+1.6797	$VO^{++} + 2H_2O = HVO_3 + 3H^+ + e$	+1.1
$PdCl_4^{--} + 2Cl^- = PdCl_6^- + 2e$	+1.3	$VOSO_4 + 2H_2O(l) = HVO_3 + H_2SO_4 + e$	+0.92
$Pt(s) + 4Cl^- = PtCl_4^{--} + 2e$	+0.2	$\frac{1}{2}(VO)_2SO_4 + \frac{1}{2}SO_4^{--} = VOSO_4 + e$	+0.30
$PtCl_4^{--} + 2Cl^- = PtCl_6^{--} + 2e$	+0.717	$VSO_4 + H_2O(l) = \frac{1}{2}(VO)_2SO_4 + 2H^+ + \frac{1}{2}SO_4^- + e$	-0.21
$[Pt(CN)_4]^{--} + 2Cl^- = [Pt(CN)_4Cl_2]^{--} + 2e$	+0.879	$W(s) + 3H_2O = WO_3(s) + 6H^+ + 6e$	0.0
Quinhydrone electrode, $H^+(a = 1)$	+0.6992	$W(CN)_8^{4-} = W(CN)_8^{3-} + e$	+0.485
$Rb(s) = Rb^+ + e$	-2.9259	$WO^{3+} = WO_3 + 4H^+ + e$	0.0
$S^{--} = S(rhombic) + 2e$	-0.51	$Zn(s) = Zn^{++} + 2e$	-0.7614
$S(rhombic) + 3H_2O = H_2SO_3 + 4H^+ + 4e$	+0.47	$Zn\text{-}Hg = Zn^{++} + 2e$	-0.7614
		$Zn(s) + 2OH^- = ZnO(s) + H_2O(l)$	-1.2483
$H_2S = S(rhombic) + 2H^+ + 2e$	+0.17	$Zn(s) + 3OH^- = HZnO_2^- + H_2O + 2e$	-0.72
$2SO_4^{--} + 2H^+ = H_2S_2O_8 + 2e$	+1.5	$Zn\text{-}Hg + (SO_4^{--} \text{ in } ZnSO_4 \cdot 7H_2O \text{ sat.}) = ZnSO_4 \cdot 7H_2O(s) + 2e$	-0.7993

From C. D. Hodgman (ed.), "Handbook of Chemistry and Physics," The Chemical Rubber Co., Cleveland, 1942.

Example 4-17   What is the potential of a Weston cell at 35°C?
solution   From Eq. (4-1),

$$E_t = 1.0184 - 0.000041 \ (35\text{-}18)$$
$$E_t = 1.0177 \text{ V}$$

As we have seen, the potential of a given half cell is affected by many factors, such as temperature, electrolyte concentration, and pressure (in the case of an electrode reaction involving a gas). The *Nernst equation* gives the emf of an electrode system when its standard electrode potential is known:

$$E = E_0 - \frac{RT}{nF} \ln Q \qquad (4\text{-}2)$$

By combining the constants (and converting from natural to base 10 logarithms), a more convenient form of the Nernst equation is obtained:

$$E = E_0 - \frac{0.591}{n} \log Q \tag{4-3}$$

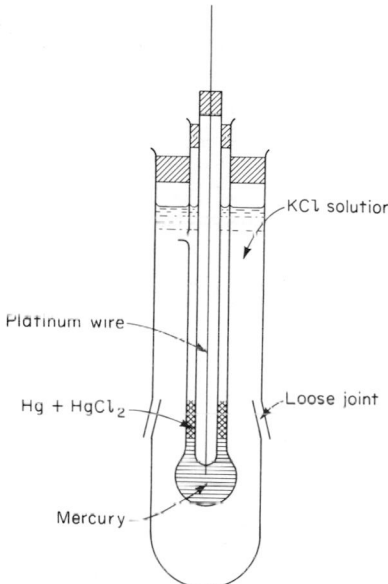

Fig. 4-11 Standard calomel electrode. (For the normal calomel electrode, the KCl solution would be 1N.)

TABLE 4-10 Calomel-Electrode Potentials with Respect to the Normal Hydrogen Electrode (at 18°C)

Type of calomel electrode	Potential (V)
Saturated	+0.2480
Normal	+0.2828
Decinormal	+0.3358

where E_0 is the standard potential, R the gas-law constant in electrical units (8.314 V − K), T the absolute temperature (K), n the electron change involved in the oxidation (or reduction) reaction, and F the faraday constant (96,500 C). Q, for want of a more precise definition, is the *activity quotient*. In the Nernst equation, Q can be expressed in many forms, depending on what is wanted. In one form, Q is the quotient of the concentration of the electrolyte divided by the concentration of 1N electrolyte:

$$Q = \frac{[\text{electrolyte}]}{[\text{standard electrolyte}]} \tag{4-4}$$

For gas electrodes under a given pressure, Q is given by

$$Q = p_{\text{gas}} \tag{4-5}$$

(recalling that the standard pressure is 1 atm). Since Q is a quotient, care must be taken to have both numerator and denominator in the same units.

Example 4-18 What is the half-cell potential for copper immersed in a solution that is $2 \times 10^{-16}\ m$ with respect to Cu^{++}?

solution From Table 4-8, E_0 for Cu immersed in a solution that is $1\ m$ with respect to Cu^{++} is -0.344 V. Since the reaction $Cu^0 \longrightarrow Cu^{++} + 2\,e$ involves a two-electron change, $n = 2$. Substituting the appropriate values in Eq. (4-3) yields

$$E = -0.344 - \frac{0.0591}{2} \log 2 \times 10^{-16}$$

$$= -0.344 - 0.0296\,(0.301 - 16)$$
$$= -0.344 + 0.465 = +0.121\ \text{V}$$

Example 4-19 What is the half-cell potential for a hydrogen electrode that is operating at a hydrogen pressure of 15 atm?

solution From Eq. (4-5), $Q = 15$.

Since for the reaction $H^0 \longrightarrow H^+ + e$, $n = 1$ and $E_0 = 0$, Eq. (4-3) yields

$$E = -0.0591 \log 15$$
$$= -0.0591 \ (1.1761) = -0.070 \ V$$

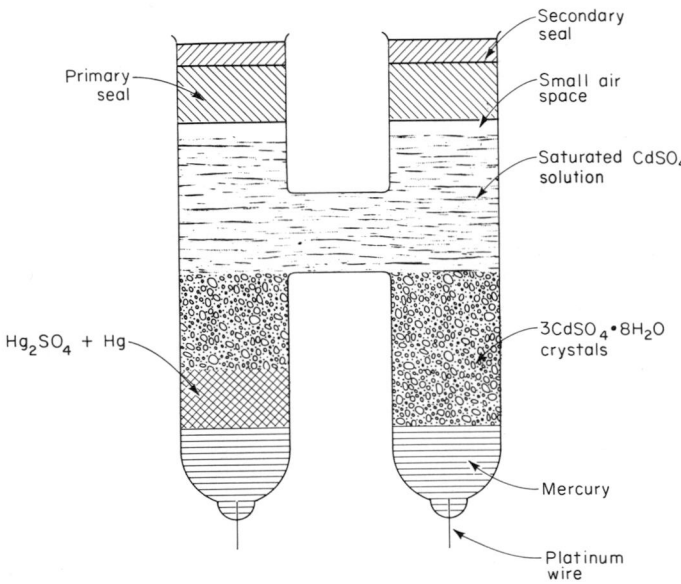

Primary seal

Secondary seal

Small air space

Saturated $CdSO_4$ solution

$3CdSO_4 \cdot 8H_2O$ crystals

$Hg_2SO_4 + Hg$

Mercury

Platinum wire

Fig. 4-12 Weston standard cell.

As will be discussed in connection with analytical procedures, the Nernst equation has made the potential of a half cell so precisely predictable that potentiometric measurements have become an important analytical and control tool.

Chapter 5

Thermochemistry and Kinetics

5-1 THERMOCHEMISTRY

Principles

Just as matter is neither created nor destroyed during a chemical reaction, energy is also neither created nor destroyed. Thus all energy brought to a reaction by the reactants emerges as energy associated with the products, or is given up to (or absorbed from) the surroundings. Known as the law of *conservation of energy*, this refers to all the energy involved in a reaction, in whatever form, i.e., heat, light, work, electricity, radiation, etc. In a great many chemical reactions the energy change associated with the reaction manifests itself as heat, and it is the heat *balance* between the reactants and products with which thermochemistry is concerned. In a great many reactions the energy content of the products is not as great as the energy content of the reactants. Such reactions give up the excess energy to their surroundings, most frequently in the form of heat. When heat is liberated in this manner, the reactants, the reaction products, and the containing vessel absorb the heat energy so released, thus causing their temperature to rise. Such reactions are said to be *exothermic*. In other reactions, however, energy must be absorbed from the reactants, reaction products, and/or the surroundings in order to satisfy the needs of the energy relationships of the reactions, and this manifests itself as a decrease in the temperature of the reaction materials and containing vessel. Such reactions are called *endothermic*. The amount of heat absorbed or liberated by a chemical reaction is measured in a calorimeter of the type described in Chap. 2.

Thermochemical calculations are based on two fundamental laws. The first of these, originally formulated by Lavoisier in 1780, states that the quantity of heat required to decompose a chemical compound is equal to that which was evolved in the formation of that compound. Thus, to decompose 1 mol of silver oxide (Ag_2O), it is necessary to supply 6.9 kcal, since that amount of heat was liberated in the original oxidation of the silver:

$$2Ag + \tfrac{1}{2}O_2 \longrightarrow Ag_2O \qquad \Delta H = -6.9 \text{ kcal} \tag{5-1}$$

As will be discussed later, ΔH is the heat change for the reaction. Lavoisier's law thus indicates that

$$Ag_2O \longrightarrow 2Ag + \tfrac{1}{2}O_2 \qquad \Delta H = 6.9 \text{ kcal} \tag{5-2}$$

The second law, which is the basis of many thermochemical calculations, is known as the *law of constant heat summation*, frequently referred to as Hess' law, after G. H. Hess, who established this principle experimentally in 1840. This law states that for any chemical reaction the change in heat content between reactants and products is a fixed value, whether the reaction occurs in one step or in several steps. Thus thermochemical equations can be treated as algebraic equations and can be added, subtracted, multiplied by a factor, etc., without destroying their equality. As an illustration, carbon dioxide can be formed by the direct combustion of carbon:

$$C + O_2 \longrightarrow CO_2 \qquad \Delta H = -94.4 \, \text{kcal} \qquad (5\text{-}3)$$

or by the combustion of carbon to carbon monoxide, and then by the combustion of CO to CO_2:

$$
\begin{array}{llr}
C + \frac{1}{2} O_2 \longrightarrow CO & \Delta H = -26.7 \, \text{kcal} & (5\text{-}4) \\
CO + \frac{1}{2} O_2 \longrightarrow CO_2 & \Delta H = -67.7 \, \text{kcal} & (5\text{-}5) \\
\hline
C + CO + O_2 \longrightarrow CO + CO_2 & \Delta H = -94.4 \, \text{kcal} & \\
C + O_2 \longrightarrow CO_2 & \Delta H = -94.4 \, \text{kcal} & (5\text{-}6)
\end{array}
$$

or

Heat Content and Enthalpy

In discussing energy changes during chemical reactions, it is convenient to think of a given substance as having a certain amount of heat associated with it when it is in a particular state and at a given temperature. This heat content, known as its *enthalpy,* is the heat energy stored by the material as sensible heat associated with its temperature, and as latent heat associated with its state. The symbol for enthalpy is H, and as indicated earlier, the change in the enthalpy associated with the reactants and products of a reaction is designated ΔH. For exothermic reactions ΔH is negative, meaning that the heat content, or enthalpy, of the reaction products is less than that of the reactants. Conversely ΔH is positive for endothermic reactions, since the enthalpy of the products is greater than that of the reactants.

Heat of Reaction

The ordinary equations used to indicate the nature of the reactants and products in a chemical reaction show only the specific materials that react, the specific materials that result, and their quantitative relationship. They do not, however, show the energy changes that take place, and a knowledge of the energy changes can be extremely important in understanding the reaction. Consequently, the heat given up during an exothermic reaction, or absorbed during an endothermic reaction, is shown by thermochemical equations:

$$
\begin{array}{lr}
CH_4(g) + 2O_2(g) \longrightarrow CO_2(g) + 2H_2O(l) + 212{,}800 \, \text{cal} & (5\text{-}7) \\
S(s) + O_2(g) \longrightarrow SO_2(g) + 69{,}300 \, \text{cal} & (5\text{-}8) \\
C(\text{graphite}) + O_2(g) \longrightarrow CO_2(g) + 94{,}400 \, \text{cal} & (5\text{-}9)
\end{array}
$$

In thermochemical equations, the states of the reactants and products must be indicated, since their enthalpies are dependent upon their state and temperature. The state of each material is indicated by a lowercase letter in parentheses immediately following the material, g standing for gas, s for solid, and l for liquid. When several forms exist in a particular state, the specific form is indicated. As an example, the graphitic form of carbon is shown as $C(\text{graphite})$. Furthermore, since enthalpy also depends on the temperature, by convention, the heat content of the reactants and products is referred to 25°C and 1 atm. These are the *standard conditions* for *heats of reactions,* and the heat of reaction at these conditions is known as the *standard heat of reaction,* designated $\Delta H°$. Frequently, however, the superscript is omitted, since practically all tabularized data of heats of reactions are given at standard conditions. In Eqs. (5-7) to (5-9) the heat liberated by each reaction is shown as a product; i.e., in addition to yielding CO_2 and H_2O, Eq. (5-7) shows that the reaction also yields 212,800 cal. Heats of reactions, however, are usually given in terms of $\Delta H°$, that is, the enthalpy change (at standard conditions) between the reactants and reaction products. Equations (5-7) to (5-9) are thus more generally written in the form

$$
\begin{array}{lr}
CH_4(g) + 2O_2(g) \longrightarrow CO_2(g) + 2H_2O(l) & \Delta H° = -212.8 \, \text{kcal} \quad (5\text{-}10) \\
S(s) + O_2(g) \longrightarrow SO_2(g) & \Delta H° = -69.3 \, \text{kcal} \quad (5\text{-}11) \\
C(\text{graphite}) + O_2(g) \longrightarrow CO_2(g) & \Delta H° = -94.4 \, \text{kcal} \quad (5\text{-}12)
\end{array}
$$

Note that Eqs. (5-10) to (5-12) give $\Delta H°$ in kilocalories. Knowing the specific heats of the reactants and products, and the heats associated with changes in state, such as the heat of fusion or the heat of vaporization, the heat content of the reactants and products can be readily calculated at conditions other than standard.

Example 5-1 The standard heat of reaction for the reaction Ca + S $\longrightarrow$ CaS is -114.3 kcal. However, this reaction is usually carried out at elevated temperatures. If the reaction is carried out at 225°C, what is the heat of reaction?

solution To heat the reactants from 25 to 225°C will require $40.0(0.163)(200) + 32.1(0.181)(95) + 32.1(13.2) + 32.1(0.220)(105) = 3.02$ kcal. The first term represents the heat necessary to raise the temperature of 1 mol (40.0 g) of Ca from 25 to 225°C. The second term refers to the heating of 1 mol (32.1 g) of S from 25°C to its melting point, 120°C. The third term is the heat of fusion of the sulfur, and the fourth term represents the heating of the liquid sulfur from its melting point to the reaction temperature. Similarly, the heat evolved in cooling the product to 25°C is $72.1(0.19)(200) = 2.74$ kcal. Since the standard heat of reaction is -114.3 kcal, the heat of reaction at 225°C will be

$$\Delta H_{225} = -3.02 - 114.3 + 2.74 = -114.6 \text{ kcal}$$

(The signs are obtained as follows: To cool the reactants from 225 to 25°C will evolve heat, hence the minus sign. The standard heat of reaction also evolves heat. However, in heating the products from 25 to 225°C, heat is absorbed, and hence the plus sign.)

In a combustion reaction, such as that illustrated in Eq. (5-7), the water will be produced as water vapor, since the reaction probably takes place at a high temperature. However, at standard conditions, the water produced will have condensed to liquid water, and in so doing will have given up some sensible heat and its heat of vaporization to the surroundings. The $\Delta H°$ for this reaction, 212.8 kcal, thus represents the gross heat liberated by the reaction, i.e., after the reaction products are brought to standard conditions, 25°C and 1 atm. In a like manner, if the reactants are not at standard conditions, corrections must be made to account for the heat changes necessary to bring the reactants to standard conditions, as will be discussed later in this chapter.

Example 5-2 When methane burns according to the reaction $CH_4 + O_2 \longrightarrow CO_2 + H_2O$, the heat of reaction is -212.8 kcal. If the combustion were to be used to heat a reactor, what maximum heat would be available from the burning of 1 mol of methane if (a) the reactor is at 25°C, and (b) the reactor is at 125°C? (Assume that the heat necessary to heat the reactants is about the same as that given up by the products in cooling to the specified temperatures.)

solution 1. Since the reactor is at 25°C, the water vapor formed in the combustion will condense on the reactor, thereby giving up its latent heat of vaporization to the reactor. Thus all the heat of reaction will be available to heat the reactor, i.e., 212.8 kcal.

2. In the case of the reactor being at 125°C, no water vapor will condense and the heat available to the reactor will be the heat of reaction less the heat of vaporization of the water formed. Since 1 mol of CH_4 forms 2 mols, or 36 g of water, the latent heat of vaporization is $36 (0.540) = 19.4$ kcal, and the heat available to the reactor is thus $212.8 - 19.4 = 193.4$ kcal.

Heat of Formation

Considering the large number of chemical reactions of all types, the task of measuring heats of reaction and tabularizing the results would be extremely time-consuming and cumbersome. In addition, it may not always be convenient, or even possible, to make the necessary measurements, as with reactions that take place only under specialized conditions (such as very high temperatures, or humidity-free atmospheres). Fortunately, the law of constant heat summation can be used in combination with the concept of heats of formation to simplify matters. The *heat of formation* of a compound, designated ΔH_f, is defined as the heat liberated (or absorbed) in forming that compound from its elements (the heat content of any element, in its standard state, being taken as zero). From the reaction,

$$C(s) + O_2(g) \longrightarrow CO_2(g) \qquad \Delta H° = -94.4 \text{ kcal} \qquad (5\text{-}13)$$

it follows that the heat of formation of carbon dioxide is -94.4 kcal. The heats of formation for a large number of compounds have been tabulated in Table 5-1. Since the heat content (enthalpy) of a material does not depend on the steps it took to get to any particular condition, the heat of reaction can be calculated from heats of formation by the following equation:

$$\Delta H = \Delta H_f \text{ (products)} - \Delta H_f \text{ (reactants)} \qquad (5\text{-}14)$$

Thus to calculate the heat of reaction for any reaction, it is only necessary to know the heats of formation of the reactants and products.

TABLE 5-1 Heats and Free Energies of Formation (Organic and Inorganic Compounds)

Compound	State*	Heat of formation ΔH at 25°C, kcal/mol	Free energy of formation ΔF at 25°C, kcal/mol
Aluminum:			
Al	c	0.00	0.00
$AlBr_3$	c	−123.4	
$AlBr_3$	aq	−209.5	−189.2
Al_4C_3	c	−30.8	−29.0
$AlCl_3$	c	−163.8	
$AlCl_3$	aq	−243.9	−209.5
AlF_3	c	−329	
AlF_3	aq	−360.8	−312.6
AlI_3	c	−72.8	
AlI_3	aq	−163.4	−152.5
AlN	c	−57.7	−50.4
$Al(NH_4)(SO_4)_2$	c	−561.19	−486.17
$Al(NH_4)(SO_4)_2 \cdot 12H_2O$	c	−1419.36	−1179.26
$Al(NO_3)_3 \cdot 6H_2O$	c	−680.89	−526.32
$Al(NO_3)_3 \cdot 9H_2O$	c	−897.59	
Al_2O_3	c, corundum	−399.09	−376.87
$Al(OH)_3$	c	−304.8	−272.9
$Al_2O_3 \cdot SiO_2$	c, sillimanite	−648.7	
$Al_2O_3 \cdot SiO_2$	c, disthene	−642.4	
$Al_2O_3 \cdot SiO_2$	c, andalusite	−642.0	
$3Al_2O_3 \cdot 2SiO_2$	c, mullite	−1874	
Al_2S_3	c	−121.6	
$Al_2(SO_4)_3$	aq	−820.99	
$Al_2(SO_4)_3$	c	−893.9	−739.53
$Al_2(SO_4)_3 \cdot 6H_2O$	c	−1268.15	−759.3
$Al_2(SO_4)_3 \cdot 18H_2O$	c	−2120	−1103.39
Antimony:			
Sb	c	0.00	0.00
$SbBr_3$	c	−59.9	
$SbCl_3$	c	−91.3	−77.8
$SbCl_5$	l	−104.8	
SbF_3	c	−216.6	
SbI_3	c	−22.8	
Sb_2O_3	c, I, orthorhombic	−165.4	−146.0
Sb_2O_3	c, II, octahedral	−166.6	
Sb_2O_4	c	−213.0	−186.6
Sb_2O_5	c	−230.0	−196.1
Sb_2S_3	c, black	−38.2	−36.9
Arsenic:			
As	c	0.00	0.00
$AsBr_3$	c	−45.9	
$AsCl_3$	l	−80.2	−70.5
AsF_3	l	−223.76	−212.27
AsH_3	g	43.6	37.7
AsI_3	c	−13.6	
As_2O_3	c	−154.1	−134.8
As_2O_5	c	−217.9	−183.9
As_2S_3	c	−20	
As_2S_5	amorphous	−34.76	−20
Barium:			
Ba	c	0.00	0.00
$BaBr_2$	c	−180.38	
$BaBr_2$	aq	−185.67	−183.0
$BaCl_2$	c	−205.25	−196.5
$BaCl_2$	aq	−207.92	
$Ba(ClO_3)_2$	aq	−176.6	−134.4
$Ba(ClO_4)_2$	c	−170.0	
$Ba(ClO_4)_2$	aq	−210.2	−155.3
$Ba(CN)_2$	aq	−48	
$Ba(CNO)_2$	c	−212.1	−180.7
$BaCN_2$	c	−63.6	
$BaCO_3$	c, witherite	−284.2	−271.4
$BaCrO_4$	c	−342.2	
BaF_2	c	−287.9	−265.3
BaF_2	aq	−284.6	
BaH_2	c	−40.8	−31.5
$Ba(HCO_2)_2$	aq	−459	−414.4
BaI_2	c	−144.6	
BaI_2	aq	−155.17	−158.52
$Ba(IO_3)_2$	c	−264.5	
$Ba(IO_3)_2$	aq	−237.50	−198.35
$BaMoO_4$	c	−370	
BaN_2	c	−90.7	
$Ba(NO_2)_2$	c	−184.5	
$Ba(NO_2)_2$	aq	−179.05	−150.75
$Ba(NO_3)_2$	c	−236.99	
$Ba(NO_3)_2$	aq	−227.74	−189.94
BaO	c	−133.0	
$Ba(OH)_2$	c	−225.9	
$Ba(OH)_2$	aq	−237.76	−209.02
$BaO \cdot SiO_2$	c	−363	
$Ba_3(PO_4)_2$	c	−992	
$BaPtCl_6$	c	−284.9	
BaS	c	−111.2	
$BaSO_3$	c	−282.5	
$BaSO_4$	c	−340.2	−313.4
$BaWO_4$	c	−402	
Beryllium:			
Be	c	0.00	0.00
$BeBr_2$	c	−79.4	
$BeBr_2$	aq	−142	−127.9
$BeCl_2$	c	−112.6	
$BeCl_2$	aq	−163.9	−141.4
BeI_2	c	−39.4	
BeI_2	aq	−112	−103.4
Be_3N_2	c	−134.5	−122.4
BeO	c	−145.3	−138.3

Heats and Free Energies of Formation (kcal/mol)

Left-hand table — columns: Formula | State | ΔH_f | ΔG_f

Formula	State	ΔH_f	ΔG_f
Beryllium (Cont.):			
Be(OH)$_2$	c	-215.6	-270.8
BeS	c	-56.1	
BeSO$_4$	c	-281	
	aq		-254.8
Bismuth:			
Bi	c	0.00	0.00
BiCl$_3$	c	-90.5	-70.7
	aq	-101.6	-76.4
BiI$_3$	c	-24	
BiO	aq	-27	
	c	-49.5	-43.2
Bi$_2$O$_3$	c	-137.1	-117.9
Bi(OH)$_3$	c	-171.1	
Bi$_2$S$_3$	c	-43.9	
Bi$_2$(SO$_4$)$_3$	c	-607.1	-39.1
Boron:			
B	c	0.00	0.00
	l	-52.7	
BBr$_3$	g	-44.6	-50.9
BCl$_3$	g	-94.5	-90.8
BF$_3$	g	-265.2	-261.0
B$_2$H$_6$	g	7.5	19.9
BN	c	-32.1	-27.2
B$_2$O$_3$	c	-302.0	-282.9
	gls	-297.6	-280.3
B(OH)$_3$	c	-260.0	-229.4
B$_2$S$_3$	c	-56.6	
Bromine:			
Br$_2$	l	0.00	0.00
	g	7.47	0.931
BrCl	g	3.06	-0.63
Cadmium:			
Cd	c	0.00	0.00
CdBr$_2$	c	-75.8	-70.7
	aq	-76.6	-67.6
CdCl$_2$	c	-92.149	-81.889
	aq	-96.44	-81.2
Cd(CN)$_2$	aq	36.2	
CdCO$_3$	c	-178.2	-163.2
CdI$_2$	c	-48.40	-43.22
Cd$_3$N$_2$	c	39.8	
Cd(NO$_3$)$_2$	aq	-115.67	-71.05
CdO	c	-62.35	-55.28
Cd(OH)$_2$	c	-135.0	-113.7
CdS	c	-34.5	-33.6
CdSO$_4$	c	-222.23	-194.65
	aq	-232.635	
Calcium:			
Ca	c	0.00	0.00
CaBr$_2$	c	-162.20	
	aq	-187.19	-181.86
CaC$_2$	c	-14.8	-16.0
CaCl$_2$	c	-190.6	-179.8
	aq	-209.15	-195.56
CaCN$_2$	c	-85	
Ca(CN)$_2$	c	-43.3	
	aq	-54.0	

Right-hand table — columns: Formula | State | ΔH_f | ΔG_f

Formula	State	ΔH_f	ΔG_f
Calcium (Cont.):			
CaCO$_3$	c, calcite	-239.5	-270.8
	c, aragonite	-239.54	-270.57
CaCO$_3$·MgCO$_3$	c	-538.8	
CaC$_2$O$_4$	c	-332.2	
Ca(C$_2$H$_3$O$_2$)$_2$	aq	-356.3	-311.3
CaF$_2$	c	-364.1	-264.1
	aq	-290.2	-35.7
CaH$_2$	c	-286.5	
Ca$_2$	c	-46	
CaN$_2$	c	-128.49	-157.37
Ca(NO$_3$)$_2$	c	-156.63	-88.2
	aq	-103.2	-177.38
Ca(NO$_3$)$_2$·2H$_2$O	c	-224.05	
Ca(NO$_3$)$_2$·3H$_2$O	c	-228.29	-293.57
Ca(NO$_3$)$_2$·4H$_2$O	c	-367.95	-351.58
CaO	c	-439.05	-409.32
Ca(OH)$_2$	c	-509.43	-144.3
	aq	-151.7	-213.9
CaO·SiO$_2$	c, II, wollastonite	-235.58	-207.9
	c, I, pseudowollastonite	-239.2	-357.5
		-377.9	
CaS	c	-376.6	-356.6
CaSO$_4$	c, insoluble form α	-338.73	-113.1
	c, soluble form α	-336.58	-311.9
	c, soluble form β	-335.52	-309.8
CaSO$_4$·½H$_2$O	c	-376.13	-308.8
CaSO$_4$·2H$_2$O	c	-479.33	
CaWO$_4$	c	-387	-425.47
Carbon:			
C	c, graphite	0.00	0.00
	c, diamond	0.453	0.685
CO	g	-26.416	-32.808
CO$_2$	g	-94.052	-94.260
CH$_4$ methane	g	-17.889	-12.140
C$_2$H$_6$ ethane	g	-20.236	-7.860
C$_3$H$_8$ propane	g	-24.820	-5.614
C$_4$H$_{10}$ n-butane	g	-29.812	-3.754
C$_4$H$_{10}$ isobutane	g	-31.452	-4.296
C$_5$H$_{12}$ n-pentane	g	-35.00	-1.96
C$_5$H$_{12}$ 2-methylbutane	l	-41.36	-2.21
C$_5$H$_{12}$ 2,2-dimethylpropane	g	-36.92	-3.50
C$_6$H$_{14}$ n-hexane	g	-42.85	-3.59
C$_6$H$_{14}$ 2-methylpentane	g	-39.67	-3.64
	l	-39.96	0.05
C$_7$H$_{16}$ 3-methylpentane	g	-47.52	-0.91
C$_6$H$_{14}$ 2,2-dimethylbutane	g	-41.66	-0.96
	l	-48.82	-1.73
C$_6$H$_{14}$ 2,5-dimethylbutane	g	-41.02	-0.29
	g	-48.28	-1.12
	g	-44.35	-2.35
C$_7$H$_{16}$ n-heptane	l	-51.00	-2.88
	g	-42.49	-0.73
	l	-49.48	-1.44
	g	-44.89	2.09
	l	-53.63	0.42

Adapted from Robert H. Perry and Cecil H. Chilton (eds.), "Chemical Engineers' Handbook," 5th ed., McGraw-Hill Book Company, New York, 1973.

* For aqueous solutions, heats of formation given for very dilute solutions; free energy of formation given for 1 m solution concentrations.

TABLE 5-1   Heats and Free Energies of Formation (Organic and Inorganic Compounds)—(Continued)

Compound	State	Heat of formation ΔH at 25°C, kcal/mol	Free energy of formation ΔF at 25°C, kcal/mol
Carbon (*Cont.*):			
C_7H_{16} 2-methylhexane	g	−46.60	0.98
	l	−54.93	−0.47
C_7H_{16} 3-methylhexane	g	−45.96	−1.10
	l	−54.35	−0.39
C_7H_{16} 3-ethylpentane	g	−45.34	2.59
	l	−53.77	1.06
C_7H_{16} 2,2-dimethylpentane	g	−49.29	0.09
	l	−57.05	−1.08
C_7H_{16} 2,3-dimethylpentane	g	−47.62	0.16
	l	−55.81	−1.27
C_7H_{16} 2,4-dimethylpentane	g	−48.30	−0.72
	l	−56.17	−0.49
C_7H_{16} 3,3-dimethylpentane	g	−48.17	0.63
	l	−56.07	−0.69
C_7H_{16} 2,2,3-trimethylbutane	g	−48.96	0.76
	l	−56.63	−0.43
C_8H_{18} n-octane	g	−49.82	4.14
	l	−59.74	1.77
C_8H_{18} 2-methylheptane	g	−51.50	3.06
	l	−60.98	0.92
C_8H_{18} 3-methylheptane	g	−50.82	3.29
	l	−60.34	1.12
C_8H_{18} 4-methylheptane	g	−50.69	4.00
	l	−60.17	1.86
C_8H_{18} 3-ethylhexane	g	−50.40	3.95
	l	−59.88	1.80
C_8H_{18} 2,2-dimethylhexane	g	−53.71	−2.56
	l	−62.63	−0.72
C_8H_{18} 2,3-dimethylhexane	g	−51.13	4.23
	l	−60.40	2.17
C_8H_{18} 2,4-dimethylhexane	g	−52.44	2.80
	l	−61.47	0.89
C_8H_{18} 2,5-dimethylhexane	g	−53.21	2.50
	l	−62.26	0.59
C_8H_{18} 3,3-dimethylhexane	g	−52.61	3.17
	l	−61.58	1.23
C_8H_{18} 3,4-dimethylhexane	g	−50.91	4.97
	l	−60.23	2.86
C_8H_{18} 2-methyl-3-ethylpentane	g	−50.48	5.08
	l	−59.69	3.03
C_8H_{18} 3-methyl-3-ethylpentane	g	−51.38	4.76
	l	−60.46	2.69
C_8H_{18} 2,2,3-trimethylpentane	g	−52.61	4.09
	l	−61.44	2.22
C_8H_{18} 2,2,4-trimethylpentane	g	−53.57	3.13
	l	−61.97	1.51
C_8H_{18} 2,3,3-trimethylpentane	g	−51.73	4.52
	l	−60.63	2.54
Carbon (*Cont.*):			
C_8H_{18} 2,3,4-trimethylpentane	g	−51.97	4.32
	l	−60.98	2.34
C_8H_{18} 2,2,3,3-tetramethyl-butane	g	−53.99	4.88
	c	−64.23	2.74
C_2H_4 ethylene	g	12.496	16.282
C_3H_6 propylene	g	4.879	14.964
C_4H_8 1-butene	g	0.280	17.217
C_4H_8 cis-2-butene	g	−1.362	16.007
C_4H_8 trans-2-butene	g	−2.405	15.323
C_4H_8 2-methyl-2-propene	g	−3.343	14.574
C_5H_{10} 1-pentene	g	−5.000	18.787
C_5H_{10} cis-2-pentene	g	−6.710	17.173
C_5H_{10} trans-2-pentene	g	−7.590	16.575
C_5H_{10} 2-methyl-1-butene	g	−8.680	15.509
C_5H_{10} 3-methyl-1-butene	g	−6.920	17.874
C_5H_{10} 2-methyl-2-butene	g	−10.170	14.267
C_2H_2 acetylene	g	54.194	50.000
C_3H_4 methylacetylene	g	44.319	46.313
C_4H_6 1-butyne	g	39.70	48.52
C_4H_6 2-butyne	g	35.374	44.725
C_5H_8 1-pentyne	g	34.50	50.17
C_5H_8 2-pentyne	g	30.80	46.41
C_5H_8 3-methyl-1-butyne	g	32.60	49.12
C_6H_6 benzene	g	19.820	30.989
	l	11.718	29.756
C_7H_8 toluene	g	11.950	29.228
	l	2.867	27.282
C_8H_{10} ethylbenzene	g	7.120	31.208
	l	−2.977	28.614
C_8H_{10} o-xylene	g	−4.540	29.177
	l	−5.841	26.370
C_8H_{10} m-xylene	g	4.120	28.405
	l	−6.075	25.730
C_8H_{10} p-xylene	g	4.290	28.952
	l	−5.838	26.310
C_9H_{12} n-propylbenzene	g	1.870	32.810
	l	−9.178	29.600
C_9H_{12} isopropylbenzene	g	0.940	32.738
	l	−9.848	29.708
C_9H_{12} 1-methyl-2-ethylben-zene	g	0.290	31.323
	l	−11.110	27.973
C_9H_{12} 1-methyl-3-ethylben-zene	g	−0.460	29.177
	l	−11.670	26.977
C_9H_{12} 1-methyl-4-ethylben-zene	g	−0.780	30.217
	l	−11.920	27.041
C_9H_{12} 1,2,3-trimethylbenzene	g	−2.290	29.319
	l	−14.013	25.679

Carbon (Cont.):

Formula / Name	State		
C₉H₁₂ 1,2,4-trimethylbenzene	g	-3.330	27.912
	l	-14.785	24.462
C₉H₁₂ 1,3,5-trimethylbenzene	g	-3.840	28.172
	l	-15.184	24.832
C₅H₁₀ cyclopentane	g	-18.46	9.23
	l	-25.31	8.70
C₆H₁₂ methylcyclopentane	g	-25.50	8.55
	l	-33.08	7.53
C₇H₁₄ ethylcyclopentane	g	-30.38	10.59
	l	-39.09	8.84
C₆H₁₂ cyclohexane	g	-29.43	7.59
	l	-37.34	6.39
C₇H₁₄ methylcyclohexane	g	-37.00	6.52
	l	-45.46	4.86
C₈H₁₆ ethylcyclohexane	g	-41.06	9.38
	l	-50.73	6.96
CH₄O methanol	g	-48.08	-38.62
	l	-57.04	-39.80
C₂H₆O ethanol	g	-52.23	-40.23
	l	-66.35	-41.76
C₃H₈O n-propanol	g	-61.17	-38.83
	l	-71.87	-38.84
C₃H₈O isopropanol	g	-62.41	-39.20
	l	-74.32	-38.83
C₄H₁₀O n-butanol	g	-67.81	-38.88
	l	-79.61	-40.37
C₄H₁₀O isobutanol	g	-69.05	-38.25
	l	-81.06	-39.36
C₂H₆O₂ ethylene glycol	g	-92.53	-71.26
	l	-107.91	-76.44
C₃H₈O₃ glycerol	g	-159.16	-113.65
	l	-21.71	-6.26
C₆H₆O phenol		-37.80	-11.02
C₇H₈O cresol		-16.1	-6.94
C₂H₄O ethylene oxide	g	-43.06	-26.06
C₂H₆O dimethyl ether	g	-51.3	
C₄H₁₀O diethyl ether	g	-65.2	-27.75
CH₂O formaldehyde	g	-28.29	-26.88
C₂H₄O acetaldehyde	g	-39.72	-31.46
C₃H₄O acrolein	l	-20.50	-15.57
		-27.97	-16.17
C₃H₆O propionaldehyde	g	-49.15	-33.96
C₄H₈O n-butyraldehyde	l	-52.40	-73.24
C₇H₆O benzaldehyde		-9.57	5.85
C₈H₈O p-toluic aldehyde		-21.23	2.24
		-17.78	4.09
C₂H₂O ketene	g	-14.78	14.30
C₃H₆O acetone		-18.78	-13.32
	l	-51.79	-36.45
C₅H₁₀O diethylketone		-59.32	-37.16
CH₂O₂ formic acid		-73.8	-80.24
		-86.67	-82.7
½(CH₂O₂)₂ bimolecular formic acid	g	-97.8	
		-93.85	-81.90

Carbon (Cont.):

Formula / Name	State		
C₂H₄O₂ acetic acid	g, l	-104.72	-91.24
		-116.2	-93.56
C₃H₆O₂ propionic acid	g,l	-108.75	-88.27
		-121.7	-91.65
C₂H₄O₃ hydroxyacetic acid	l	-155.33	-125.57
C₆H₁₀O₄ adipic acid	l	-216.19	-163.96
		-235.51	-177.17
C₂H₄O₂ methyl formate	g,l	-84.69	-71.37
		-95.26	-71.53
C₃H₆O₂ methyl acrylate	g,l	-70.10	-56.78
		-82.76	-58.13
C₄H₈O₂ ethyl acetate	g,l	-102.02	-74.93
		-110.72	-76.11
C₅H₁₀O₂ ethyl propionate	g,l	-112.36	-77.37
		-122.16	-79.16
C₄H₆O₃ acetic anhydride	g,l	-148.82	-119.29
		-155.16	-121.75
C₆H₁₀O₃ propionic anhydride	g,l	-147.32	-109.78
		-161.53	-113.66
CS₂ carbon disulfide	g,l	28.11	16.13
		-33.83	-40.85
COS carbonyl sulfide	g	73.82	71.02
C₂N₂ cyanogen	g	31.1	27.94
HCN hydrogen cyanide	g	25.2	29.0
	aq	19.81	26.8
CH₃N acetonitrile	g	-6.7	6.6
		-12.24	10.01
C₂H₅N methylamine	g	-16.45	14.38
C₂H₅N ethylamine	g	-15.60	19.55
C₃H₉N propylamine	g	-14.37	31.52
C₄H₁₁N butylamine	g	-24.90	28.84
C₆H₁₃N hexamethylene-imine	g	11.18	24.30
CH₂N₂ cyanamide	c	9.15	24.18
		33.34	61.43
C₆H₈N₂ adiponitrile	c	19.19	54.63
C₆H₁₆N₂ hexamethylenediamine	g,c	-30.57	28.91
CH₅N₃ guanidine	c	-27.48	7.34
CH₃N₃ melamine		-30.68	6.33
CH₃NO formamide	l	-19.33	40.80
C₂H₇NO ethanolamine		-44.64	-36.60
CH₄N₂O urea	c	-62.52	27.50
		-77.55	-46.45
		-79.634	-47.118

Cerium:

Ce	c	0.00	0.00
CeN		-78.2	-70.8

Cesium:

Cs	c	0.00	0.00
CsBr	c	-97.64	
CsC.	c	-91.39	-94.86
CsC.	aq	-106.31	
Cs₂CO₃	aq	-102.01	-101.61
CsF	c	-271.88	
		-131.67	
		-140.48	-135.98
CsH	c	-12	-7.30

TABLE 5-1 Heats and Free Energies of Formation (Organic and Inorganic Compounds)—(Continued)

Compound	State	Heat of formation ΔH at 25°C, kcal/mol	Free energy of formation ΔF at 25°C, kcal/mol
Cesium: (Cont.):			
$CsHCO_3$	c	-230.6	
	aq	-226.6	-210.56
CsI	c	-83.91	-82.61
	aq	-75.74	
$CsNH_2$	c	-28.2	
$CsNO_3$	c	-121.14	
	aq	-111.54	-96.53
Cs_2O	c	-82.1	
$CsOH$	c	-100.2	
	aq	-117.0	-107.87
Cs_2S	c	-87	
Cs_2SO_4	c	-344.86	
	aq	-340.12	-316.66
Chlorine:			
Cl_2	g	0.00	0.00
ClF	g	-25.7	
ClO	g	33	
ClO_2	g	24.7	29.5
ClO_3	g	37	
Cl_2O	g	18.20	22.40
Cl_2O_7	g	63	
Chromium:			
Cr	c	0.00	0.00
$CrBr_3$	aq	-122.7	
Cr_3C_2	c	-21.008	-21.20
Cr_7C_3	c	-16.378	-16.74
$CrCl_2$	c	-103.1	-93.8
	aq	-102.1	
CrF_2	aq	-152	
CrF_3	c	-231	
CrI_2	c	-63.7	-64.1
Cr_2O_3	c	-139.3	
$Cr_2(SO_4)_3$	aq	-268.8	-249.3
			-626.3
Cobalt:			
Co	c	0.00	0.00
$CoBr_2$	c	-55.0	
	aq	-73.61	-61.96
CoC	c	9.49	7.08
$CoCl_2$	c	-76.9	-66.6
	aq	-95.58	-75.46
$CoCO_3$	c	-172.39	-155.36
CoF_2	c	-172.98	-144.2
CoI_2	c	-24.2	
Co_3N	c	-43.15	-37.4
$Co(NO_3)_2$	aq	-102.8	
Co_2O	c	-114.9	-65.3
CoO	c	-57.5	
Co_3O_4	c	-196.5	
Cobalt: (Cont.):			
$Co(OH)_2$	c	-131.5	-108.9
$Co(OH)_3$	c	-177.0	-142.0
CoS	c	-22.3	-19.8
Co_2S_3	c	-40.0	
$CoSO_4$	aq	-216.6	-188.9
Columbium:			
Cb	c	0.00	0.00
Cb_2O_5	c	-462.96	
Copper:			
Cu	c	0.00	0.00
$CuBr$	c	-26.7	-23.8
$CuBr_2$	aq	-34.0	
$CuCl$	c	-42.4	-33.25
$CuCl_2$	c	-31.4	-24.13
$CuClO_4$	aq	-48.83	1.34
$Cu(ClO_3)_2$	aq	-64.7	15.4
$Cu(ClO_4)_2$	aq	-28.3	-5.5
CuI	c	-17.8	-16.66
	aq	-4.8	
		-11.9	
Cu_3N	c	17.78	
$Cu(NO_3)_2$	c	-73.1	-8.76
	aq	-83.6	-36.6
CuO	c	-38.5	-31.9
Cu_2O	c	-43.00	-38.13
$Cu(OH)_2$	c	-108.9	-85.5
CuS	c	-18.97	-11.69
Cu_2S	c	-11.6	-20.56
$CuSO_4$	c	-184.7	-158.3
	aq	-200.78	-160.19
		-179.6	-152.0
Erbium:			
Er	c	0.00	0.00
$Er(OH)_3$	c	-326.8	
Fluorine:			
F_2	g	0.00	0.00
F_2O	g	5.5	9.7
Gallium:			
Ga	c	0.00	0.00
$GaBr_3$	c	-92.4	
$GaCl_3$	c	-125.4	
GaN	c	-26.2	
Ga_2O	aq	-84.3	
Ga_2O_3	c	-259.9	
Germanium:			
Ge	c	0.00	0.00

Tables of heats of formation (ΔHf°, kcal/mol) continued.

Section 1

Formula	State	Value
Germanium: (Cont.):		
Ge₃N₄	c	-15.7
GeO₂	c	-128.6
Gold:		
Au	c	0.00
AuBr	c	-3.4
AuBr₃	c	-14.5
AuCl	aq	-11.0
AuCl₃	c	-8.3
	c	-28.3
	aq	-32.96
AuI	c	0.2
Au₂O₃	aq	11.0
Au(OH)₃	c	-100.6
Hafnium:		
Hf	c	0.00
HfO₂	c	-271.1
Hydrogen:		
H₃AsO₃	aq	-175.6
H₃AsO₄	c	-214.9
	c	-214.8
HBr	g	-8.66
	aq	-28.80
HBrO	aq	-25.4
HBrO₃	aq	-11.51
HCl	g	-22.063
	aq	-39.85
HCN	g	31.1
	aq	24.2
HClO	aq	-28.18
HClO₃	aq	-31.4
HClO₄	aq	-116.74
HC₂H₃O₂	aq	-196.7
H₂C₂O₄	c	-194.6
HCOOH	l	-97.8
	aq	-98.0
H₂CO₃	aq	-167.19
HF	g	-64.2
	aq	-75.75
HI	g	6.27
	aq	-13.47
HIO	aq	-38
HIO₃	c	-56.77
	aq	-54.8
HN₃	g	70.3
	l	-31.99
HNO₃	l	-41.35
	aq	-49.210
HNO₃·H₂O	l	-112.91
HNO₃·3H₂O	l	-252.15
H₂O	g	-57.7979
	l	-68.3174
H₂O₂	aq	-45.16
	c	-45.80
H₃PO₂	aq	-145.5
	aq	-145.6

Section 2 — Hydrogen (Cont.):

Formula	State	Value
H₃PO₃	o	0.00
H₃PO₄	c	24.47
H₂S	aq	4.21
	g	-0.76
H₂S₂	l	18.71
H₂SO₃	l	0.00
H₂SO₄	l	-258.2
H₂Se	g	-153.04
H₂SeO₃	aq	-183.93
H₂SeO₄	c	-12.72
H₂SiO₃	c	-24.58
H₂SiO₄	c	-19.90
H₂Te	g	5.00
H₂TeO₂	c	-22.778
H₂TeO₄	aq	-31.330
Indium:		
I₂Br₃	c	27.94
	c	26.55
I₂Cl₃	aq	-0.25
	c	-10.70
I₂	aq	-93.56
Iodine:		
I⁻	aq	96.8
IBr	c	-165.64
ICl	g	-82.7
ICl₃	g	-85.1
I₂O₅	g	-149.0
	c	-64.7
Iridium:		
I⁻	c	0.365
I-Cl	c	-12.35
I-Cl₂	c	-23.33
I-Cl₃	c	-32.25
I-Fe	l	78.50
I-O₂	c	-17.57
Iron:		
Fe	c,α	-19.05
FeBr₂	c	-78.36
FeBr₃	aq	-193.70
Fe(CO)₅	aq	-54.6351
	c	-56.6699
FeCl₂	c,siderite	-28.23
FeCl₃	aq	-31.47
	c	-120.0
	aq	

Section 3

Formula	State	Value 1	Value 2
Hydrogen (Cont.):			
H₃PO₆	o	-204.0	-232.2
	c		-232.2
H₃PO₄	aq	-270.0	-305.2
H₂S	g	-7.85	-309.32
	l		-4.77
H₂SO₃	l	-128.54	-9.38
			-3.6
H₂SO₄	g	17.0	-146.88
	aq	18.4	-193.69
H₂Se	c	-101.36	-212.03
H₂SeO₃	c	-247.9	20.5
H₂SeO₄	g	33.1	18.1
H₂TeO₂	c	-115.7	-126.5
H₂TeO₄	aq		-122.4
	aq		-130.23
			-143.4
Indium:		0.00	-267.8
I₂Br₃	c	-97.2	-340.6
I₂Cl₃	aq	-117.5	36.9
I₂	aq	-60.5	-145.0
			-145.0
Iodine:		0.00	-165.6
I⁻	c	4.63	0.00
IBr	g	-1.24	-97.2
ICl	g	-6.05	-112.9
ICl₃	g		-128.5
I-Fe	c		-145.6
I-O₂	c		-56.5
Iridium:		0.00	-67.2
	c	-16.9	-4.8
	c	-32.0	-222.47
	l	-46.5	
Iron:		0.00	0.00
Fe	c,α	-69.47	-57.15
FeBr₂	aq	-76.26	-78.7
FeBr₃	aq	4.24	-95.5
Fe(CO)₅	c		5.69
FeCl₂	c	-154.8	-187.6
FeCl₃	l	-72.6	-172.4
	l	-83.0	-81.9
			-100.0
	aq	-96.5	-96.4
			-128.5

TABLE 5-1 Heats and Free Energies of Formation (Organic and Inorganic Compounds)—(Continued)

Compound	State*	Heat of formation ΔH at 25°C, kcal/mol	Free energy of formation ΔF at 25°C, kcal/mol
Iron: (Cont.):			
FeF_2	aq	−177.2	−151.7
FeI_2	c	−24.2	
	aq	−47.7	−45
FeI_3	aq	−49.7	−39.5
FeN	c	−2.55	−0.862
$Fe(NO_3)_2$	aq	−118.9	−72.8
$Fe(NO_3)_3$	aq	−156.5	−81.3
FeO	c	−64.62	−59.38
Fe_2O_3	c	−198.5	−179.1
Fe_3O_4	c	−266.9	−242.3
$Fe(OH)_2$	c	−135.9	−115.7
$Fe(OH)_3$	c	−197.3	−166.3
$FeO.SiO_2$	c	−273.5	
FeP	c	−13	
$FeSi$	c	−19.0	
FeS	c	−22.64	−23.23
FeS_2	c, pyrites	−38.62	−35.93
	c, marcasite	−33.0	
$FeSO_4$	aq	−221.3	−195.5
	aq	−236.2	−196.4
$Fe_2(SO_4)_3$	aq	−653.3	−533.4
$FeTiO_3$	?, ilmenite	−295.51	−277.06
Lanthanum:			
La	c	0.00	0.00
$LaCl_3$	c	−253.1	
	aq	−284.7	
LaH_3	aq	−160	
LaN	c	−72.0	**−64.6**
La_2O_3	c	−539	
LaS	c	−148.3	
La_2S_3	c	−351.4	
$La_2(SO_4)_3$	aq	−972	
Lead:			
Pb	c	0.00	0.00
$PbBr_2$	c	−66.24	−62.06
	aq	−56.4	−54.97
$PbCO_3$	c, cerussite	−167.6	−150.0
$Pb(C_2H_3O_2)_2$	c	−232.6	
	aq	−234.2	−184.40
PbC_2O_4	c	−205.3	
$PbCl_2$	c	−85.68	−75.04
	aq	−82.5	−68.47
PbF_2	aq	−159.5	−148.1
PbI_2	c	−41.77	−41.47
$Pb(NO_3)_2$	aq	−106.88	−58.3
PbO	c, red	−52.40	−45.53
	c, yellow	−51.72	−43.88
PbO_2	c	−65.0	−52.0
Pb_3O_4	c	−172.4	−142.2
Lead: (Cont.):			
$Pb(OH)_2$	c	−123.0	−102.2
PbS	c	−22.38	−21.98
$PbSO_4$	c	−218.5	−192.9
Lithium:			
Li	c	0.00	0.00
$LiBr$	c	−83.75	
	aq	−95.40	−95.28
$LiBrO_3$	aq	−77.9	−65.70
Li_2C_2	c	−13.0	
$LiCN$	aq	−31.4	−31.35
$LiCNO$	aq	−101.2	−94.12
$LiC_2H_3O_2$	aq	−183.9	−160.00
Li_2CO_3	aq	−289.7	−269.8
	c	−293.1	−267.58
$LiCl$	c	−97.63	
	aq	−106.45	−102.03
$LiClO_3$	aq	−87.5	−70.95
$LiClO_4$	aq	−106.3	−81.4
LiF	c	−145.57	−136.40
	aq	−144.85	
LiH	c	−22.9	
$LiHCO_3$	aq	−231.1	−210.98
LiI	c	−65.07	
	aq	−80.09	−83.03
$LiIO_3$	aq	−121.3	−102.95
Li_3N	c	−47.45	−37.33
$LiNO_3$	c	−115.350	
	aq	−115.88	−96.95
Li_2O	c	−142.3	−138.0
Li_2O_2	c	−151.9	
$LiOH$	c	−116.58	−106.44
	aq	−121.47	−108.29
$LiOH.H_2O$	c	−188.92	
$Li_2O.SiO_2$	gls	−374	
Li_2Se	c	−84.9	
	aq	−95.5	−105.64
Li_2SO_4	c	−340.23	−314.66
	aq	−347.02	
$Li_2SO_4.H_2O$	c	−411.57	−375.07
Magnesium:			
Mg	c	0.00	0.00
$Mg(AsO_4)_2$	c	−731.3	
	aq	−749	−630.14
$MgBr_2$	c	−123.9	
	aq	−167.33	−156.94
$Mg(CN)_2$	aq	−39.7	−29.08
$MgCN_2$	c	−61	
$Mg(C_2H_3O_2)_2$	aq	−344.6	−286.38
$MgCO_3$	c	−261.7	−241.7

Heat of formation and free energy of formation (kg-cal per mole)

Left half

Compound	State	ΔH°	ΔF°
Magnesium (Cont.):			
MgCl₂	c	−153.220	−143.77
	aq	−189.76	
MgCl₂·H₂O	c	−230.970	−205.93
MgCl₂·2H₂O	c	−305.810	−267.20
MgCl₂·4H₂O	c	−453.820	−387.98
MgCl₂·6H₂O	c	−597.240	−505.45
MgF₂	c	−263.8	
MgI₂	c	−86.8	
	aq	−136.79	−132.45
MgMoO₄	c	−329.9	
Mg₃N₂	c	−115.2	−100.8
Mg(NO₃)₂	c	−188.770	−140.66
	aq	−209.927	−160.28
Mg(NO₃)₂·2H₂O	c	−336.625	
Mg(NO₃)₂·6H₂O	c	−624.48	−496.03
MgO	c	−143.84	−136.17
MgO·SiO₂	c	−347.5	−326.7
Mg(OH)₂	c, ppt.	−221.90	−200.17
	c, brucite	−223.9	−193.3
MgS	c	−84.2	
MgSO₄	c	−304.94	−277.7
	aq	−325.4	
MgTe	c	−25	
MgWO₄	c	−345.2	−283.88
Manganese:			
Mn	c, α	0.00	0.00
	γ	−91	
MnBr₂	c	−106	−97.8
Mn₃C	c	1.1	1.26
Mn(C₂H₃O₂)₂	aq	−270.3	−227.2
		−282.7	−192.5
MnCO₃	c	−211	
MnC₂O₄	c	−240.9	
MnCl₂	c	−112.0	−102.2
	aq	−128.9	
MnF₂	c	−206.1	−180.0
MnI₂	c	−49.8	
	aq	−76.2	−73.3
Mn₃N₂	aq	−57.77	−46.49
Mn(NO₃)₂	c	−134.9	
	aq	−148.0	−101.1
Mn(NO₃)₂·6H₂O	c	−557.07	−441.2
MnO	c	−92.04	−86.77
MnO₂	c	−124.58	−111.49
Mn₂O₃	c	−229.5	−209.9
Mn₃O₄	c	−331.65	−306.22
MnO·SiO₂	c	−301.3	−282.1
Mn(OH)₂	c	−163.4	−143.1
Mn(OH)₃	c	−221	−190
Mn₃(PO₄)₂	c	−756	
MnS	c, green	−26.3	−27.5
MnSe	aq	−47.0	−48.0
MnSO₄	c	−254.18	−228.41
	aq	−265.2	
Mn₂(SO₄)₃		−635	
		−657	
Mercury:			
Hg	l	0.00	0.00

Right half

Compound	State	ΔH°	ΔF°
Mercury (Cont.):			
HgBr	g	23	18.8
HgBr₂	c	−40.68	−38.8
	aq	−38.4	−9.74
Hg(C₂H₃O₂)₂	c	−196.3	−139.2
HgCl₂	c	−53.4	−42.2
	aq	−50.3	−23.25
Hg₂Cl₂	c	−63.13	−50.3
Hg(CN)₂	c	62.8	52.25
	aq	66.25	
Hg₂C₂O₄	c	−159.3	−24.0
HgH	g	19	14
HgI	g, red	57.1	23
HgI₂	c	−25.3	−26.53
	g	33	−13.09
Hg₂I₂	aq	−28.88	−15.65
Hg(NO₃)₂	aq	−56.8	
Hg₂(NO₃)₂	aq	−58.5	
HgO	c, red	−21.6	−13.94
	c, yellow ppt.	−20.8	
Hg₂O	c	−21.6	−12.80
HgS	c, red	−10.7	
	c, black		−8.80
HgSO₄	c	−166.6	−149.12
Hg₂SO₄	c	−177.34	
Molybdenum:			
Mo	c	0.00	0.00
Mo₂C	c	4.36	2.91
Mo₂N	c	−8.3	
MoO₂	c	−130	−118.0
MoO₃	c	−180.39	−162.01
MoS₂	c	−56.27	−54.19
MoS₃	c	−61.48	−57.38
Nickel:			
Ni	c	0.00	0.00
NiBr₂	c	−53.4	−60.7
Ni₃C	c	9.2	8.88
Ni(C₂H₃O₂)₂	aq	−249.6	−190.1
Ni(CN)₂	aq	230.9	66.3
NiCl₂	aq	−75.0	−74.19
		−94.34	
NiF₂	c	−157.5	−142.9
	aq	−171.6	
NiI₂	aq	−22.4	−36.2
		−42.0	
Ni(NO₃)₂	aq	−101.5	−64.0
		−113.5	
NiO	c	−58.4	−51.7
Ni(OH)₂	aq	−129.8	−105.6
Ni(OH)₃	c	−163.2	
NiS	aq	−20.4	
NiSO₄	aq	−216	−187.6
		−231.3	
Nitrogen:			
N₂	g	0.00	0.00
NF₃	g	−27	
NH₃	g	−10.96	−3.903
	aq	−19.27	

TABLE 5-1   Heats and Free Energies of Formation (Organic and Inorganic Compounds)—(Continued)

Compound	State*	Heat of formation ΔH at 25°C, kcal/mol	Free energy of formation ΔF at 25°C, kcal/mol
Nitrogen: (Cont.):			
NH_4Br	c	-64.57	-43.54
	aq	-60.27	
$NH_4C_2H_3O_2$	c	-148.1	
	aq	-148.58	-108.26
NH_4CN	c	-0.7	
	aq	3.6	20.4
NH_4CNS	c	-17.8	
	aq	-12.3	4.4
$(NH_4)_2CO_3$	aq	-223.4	-164.1
$(NH_4)_2C_2O_4$	aq	-266.3	
	c	-260.6	-196.2
NH_4Cl	c	-75.23	-48.59
	aq	-71.20	
NH_4ClO_4	c	-69.4	-21.1
	aq	-63.2	
$(NH_4)_2CrO_4$	c	-276.9	
	aq	-271.3	-209.3
NH_4F	c	-111.6	
	aq	-110.2	-84.7
NH_4I	c	-48.43	
	aq	-44.97	-31.3
NH_4NO_3	c	-87.40	
	aq	-80.89	
NH_4OH	aq.	-87.59	
$(NH_4)_2S$	aq	-55.21	-14.50
$(NH_4)_2SO_4$	aq	-281.74	-215.06
	c	-279.33	-214.02
N_2H_4	l	12.06	
$N_2H_4 \cdot H_2O$	l	-57.96	
$N_2H_4 \cdot H_2SO_4$	c	-232.2	
N_2O	g	19.55	24.82
NO	g	21.600	20.719
NO_2	g	7.96	12.26
N_2O_3	g	2.23	23.41
N_2O_4	l	-10.0	
	g	11.6	
$NOBr$	g	12.8	19.26
$NOCl$	g		16.1
Osmium:			
Os	c	0.00	0.00
OsO_4	c	-93.6	-70.9
	g	-80.1	-68.1
Oxygen:			
O_2	g	0.00	0.00
O_3	g	33.88	38.86
Palladium:			
Pd	c	0.00	0.00
PdO	c	-20.40	
Phosphorus:			
P	c, white ("yellow")	0.00	0.00
	c, red ("violet")	-4.22	-1.80

Compound	State	Heat of formation ΔH at 25°C, kcal/mol	Free energy of formation ΔF at 25°C, kcal/mol
Phosphorus: (Cont.):			
P	g	150.35	141.88
P_2	g	33.82	24.60
P_4	g	13.2	5.89
PBr_3	l	-45	
	g	-60.6	-65.2
PCl_3	l	-76.8	-63.3
	g	-70.0	-73.2
PCl_5	g	-91.0	-1.45
PH_3	g	2.21	
PI_3	c	-10.9	
P_2O_5	c	-360.0	
$POCl_3$	g	-138.4	-127.2
Platinum:			
Pt	c	0.00	0.00
$PtBr_4$	c	-40.6	
	aq	-50.7	
$PtCl_2$	c	-34	
$PtCl_4$	c	-62.6	
	aq	-82.3	
PtI_4	c	-18	
$Pt(OH)_2$	c	-87.5	-67.9
PtS	c	-20.18	-18.55
PtS_2	c	-26.64	-24.28
Potassium:			
K	c	0.00	0.00
K_3AsO_4	aq	-323.0	-355.7
K_2AsO_4	c	-390.3	
KH_2AsO_4	c	-271.2	-236.7
KBr	c	-94.06	-90.8
	aq	-89.19	-92.0
$KBrO_3$	c	-81.58	-60.30
	aq	-71.68	
$KC_2H_3O_2$	c	-173.80	-156.73
	aq	-177.38	-97.76
KCl	c	-104.348	-98.76
	aq	-100.164	-69.30
$KClO_3$	c	-93.5	-72.86
	aq	-81.34	
$KClO_4$	c	-103.8	
KCN	c	-101.14	-28.08
	aq	-28.1	
$KCNO$	c	-25.3	-90.85
	aq	-99.6	
$KCNS$	c	-94.5	-44.08
	aq	-47.0	
K_2CO_3	c	-274.01	-264.04
	aq	-280.90	
$K_2C_2O_4$	c	-319.9	
	aq	-315.5	-293.1

Potassium (Cont.):

Formula	State		
K₂CrO₄	c	−333.4	
	aq	−328.2	
K₂Cr₂O₇	c	−488.5	−306.3
	aq	−472.1	
KF	c	−134.50	−440.9
	aq	−138.36	
K₃Fe(CN)₆	c	−48.4	−133.13
	aq	−34.5	
K₄Fe(CN)₆	c	−131.8	
	aq	−119.9	
KH	c	−10	−5.3
KHCO₃	c	−229.8	
	aq	−224.85	
KI	c	−78.88	−207.71
	aq	−73.95	−77.37
KIO₃	c	−121.69	−79.76
	aq	−115.18	−101.87
KIO₄	c	−98.1	−99.68
	aq	−192.9	
KMnO₄	c	−182.5	−169.1
K₂MoO₄	aq	−364.2	−168.0
KNH₂	c	−28.25	−342.9
KNO₂	c	−86.0	
KNO₃	aq	−118.03	−75.9
	c	−109.79	−94.29
	aq	−86.2	−93.68
K₂O	c	−86.2	
K₂O.Al₂O₃.4H₂O	c, leucite	−1579.6	
	gls	−1368.2	
K₂O.Al₂O₃.6H₂O	c, adularia	−1810.7	
	c, microcline	−1784.5	
	gls	−1747	
KOH	c	−102.02	−105.0
	aq	−114.96	
K₃PO₄	aq	−397.5	−443.3
KH₂PO₄	c	−478.7	−326.1
	aq	−362.7	
K₂PtCl₄	c	−254.7	
K₂PtCl₆	aq	−242.6	−226.5
	c	−299.5	−263.6
K₂Se	aq	−286.1	
K₂SeO₄	c	−83.4	−99.10
K₂S	aq	−267.1	−240.0
	aq	−121.5	
K₂SO₃	aq	−110.75	−111.44
K₂SO₄	c	−267.7	
	aq	−269.7	−251.3
K₂SO₄.Al₂(SO₄)₃.2H₂O	c	−342.65	−314.62
K₂SO₄.Al₂(SO₄)₃.24H₂O	aq	−336.48	−310.96
K₂S₂O₆	c	−1178.38	−1068.48
	c	−2895.44	−2455.68
	c	−418.62	

Rhenium:

Formula	State		
Re	c	0.00	0.00
ReF₆	g	−274	

Rhodium:

Formula	State		
Rh	c	0.00	0.00
RhO	c	−21.7	

Rhodium (Cont.):

Formula	State			
Rh₂O	c	−22.7		
Rh₂O₃	c	−68.3		

Rubidium:

Formula	State			
Rb	c	0.00	0.00	
RbBr	c	−95.82	−52.50	
	c	−45.0	−93.38	
RbCN	g	−90.54		
Rb₂CO₃	aq	−25.9		
	aq	−273.22		
RbCl	aq	−282.61	−263.78	
	c	−105.06	−98.48	
	g	−53.6	−57.9	
RbF	aq	−101.06	−100.13	
	c	−133.23		
RbHCO₃	aq	−139.31	−134.5	
	aq	−230.01		
RbI	c	−225.59	−209.07	
	aq	−81.04		
RbNH₂	c	−31.2	−40.5	
RbNO₃	c	−74.57	−81.13	
	aq	−27.74		
Rb₂O	c	−119.22	−95.05	
Rb₂O₂	aq	−110.52		
RbOH	c	−82.9		
	c	−107		
	aq	−101.3	−106.39	
	aq	−115.8		

Ruthenium:

Formula	State			
Ru	c	0.00	0.00	
RuS₂	c	−46.99	−44.11	

Selenium:

Formula	State			
Se	c, I, hexagonal	0.00	0.00	
	c, II, red, monoclinic	0.2		
Se₂Cl₂	l	−22.06	−13.73	
SeF₆	g	−246	−222	
SeO₂	c	−56.33		

Silicon:

Formula	State			
Si	c	0.00	0.00	
SiB₄	c	−93.0		
SiC	c	−28	−27.4	
SiCl₄	l	−150.0	−133.9	
	g	−142.5	−133.0	
SiF₄	g	−370	−360	
SiH₄	g	−14.8	−9.4	
SiN₄	c	−29.8		
SiO₂	c, cristobalite, 1600° form	−179.25	−154.74	
	c, cristobalite, 1100° form	−202.62		
	c, quartz	−202.46	−190.4	
	c, tridymite			

Silver:

Formula	State			
Ag	c	0.00	0.00	
AgBr	c	−23.90	−23.02	
Ag₂C₂	c	84.5		
AgC₂H₃O₂	c	−95.9		
	aq	−91.7	−70.86	

TABLE 5-1 Heats and Free Energies of Formation (Organic and Inorganic Compounds)—(Continued)

Compound	State*	Heat of formation ΔH at 25°C., kcal/mol	Free energy of formation ΔF at 25°C., kcal/mol
Silver: (Cont.):			
$AgCN$	c	33.8	38.70
Ag_2CO_3	c	-119.5	-103.0
$Ag_2C_2O_4$	c	-158.7	
$AgCl$	c	-30.11	-25.98
AgF	c	-48.7	
	aq	-53.1	-47.26
AgI	c	-15.14	-16.17
$AgIO_3$	c	-42.02	-24.08
$AgNO_2$	c	-11.6	3.76
	aq	-2.9	9.99
$AgNO_3$	c	-29.4	-7.66
	aq	-24.02	-7.81
Ag_2O	c	-6.95	-2.23
Ag_2S	c	-5.5	-7.6
Ag_2SO_4	c	-170.1	-146.8
	aq	-165.8	-139.22
Sodium:			
Na	c	0.00	0.00
Na_3AsO_3	aq	-314.61	
Na_3AsO_4	c	-366	
	aq	-381.97	-341.17
$NaBr$	c	-86.72	
	aq	-86.33	-87.17
$NaBrO$	aq	-78.9	
$NaBrO_3$	c	-68.89	-57.59
$NaC_2H_3O_2$	c	-170.45	
	aq	-175.450	-152.31
$NaCN$	c	-22.47	
	aq	-22.29	-23.24
$NaCNO$	c	-96.3	
	aq	-91.7	-86.00
$NaCNS$	c	-39.94	
	aq	-38.23	-39.24
Na_2CO_3	c	-269.46	-249.55
	aq	-275.13	-251.36
$NaCO_2NH_2$	c	-142.17	
$Na_2C_2O_4$	c	-313.8	-283.42
	aq	-309.92	
$NaCl$	c	-98.321	-91.894
	aq	-97.324	-93.92
$NaClO_3$	c	-83.59	
	aq	-78.42	-62.84
$NaClO_4$	c	-101.12	
	aq	-97.66	-73.29
Na_2CrO_4	c	-319.8	-296.58
	aq	-323.0	
$Na_2Cr_2O_7$	c	-465.9	-431.18
NaF	c	-135.94	-129.0
	aq	-135.711	-128.29

Compound	State	Heat of formation ΔH at 25°C., kcal/mol	Free energy of formation ΔF at 25°C., kcal/mol
Sodium (Cont.):			
NaH	c	-14	-9.30
$NaHCO_3$	c	-226.0	-202.66
	aq	-222.1	-202.87
NaI	c	-69.28	
	aq	-71.10	-74.92
$NaIO_3$	c	-112.300	-94.84
Na_2MoO_4	c	-364	-333.18
	aq	-358.7	
$NaNO_2$	c	-86.6	
	aq	-83.1	-71.04
$NaNO_3$	c	-111.71	-87.62
	aq	-106.880	-88.84
Na_2O	c	-99.45	-90.06
Na_2O_2	c	-119.2	-105.0
$Na_2O.SiO_2$	c	-383.91	-361.49
$Na_2O.Al_2O_3.3SiO_2$	c	-1180	
$Na_2O.Al_2O_3.4SiO_2$	c, natrolite	-1366	
$NaOH$	c	-101.96	-90.60
	aq	-112.193	-100.18
Na_3PO_3	c	-389.1	
Na_3PO_4	c	-457	
	aq	-471.9	-428.74
Na_2PtCl_4	aq	-237.2	-216.78
Na_2PtCl_6	c	-272.1	
	aq	-280.9	
Na_2Se	c	-59.1	
	aq	-78.1	-89.42
Na_2SeO_4	c	-254	
	aq	-261.5	-230.30
Na_2S	c	-89.8	
	aq	-105.17	-101.76
Na_2SO_3	c	-261.2	-240.14
	aq	-264.1	-241.58
Na_2SO_4	c	-330.50	-302.38
	aq	-330.82	-301.28
$Na_2SO_4.10H_2O$	c	-1033.85	-870.52
Na_2WO_4	c	-391	-345.18
	aq	-381.5	
Strontium:			
Sr	c	0.00	0.00
$SrBr_2$	c	-171.0	
	aq	-187.24	-182.36
$Sr(C_2H_3O_2)_2$	c	-358.60	
	aq	-364.4	-311.80
$Sr(CN)_2$	aq	-59.5	-54.50
$SrCO_3$	c	-290.9	-271.9
$SrCl_2$	c	-197.84	
	aq	-209.20	-195.86
SrF_2	c	-289.0	
$Sr(HCO_3)_2$	aq	-459.1	-413.76

Formula	State		
Strontium: (Cont.):			
SrI₂	c	−136.1	
	aq	−156.70	−157.87
Sr₃N₂	c	−91.4	−76.5
Sr(NO₃)₂	c	−233.2	−185.70
	aq	−228.73	−133.7
SrO	c	−140.8	
SrO.SiO₂	gls	−364	
SrO₂	c	−153.3	−139.0
	c	−153.6	
Sr(OH)₂	c	−228.7	−208.27
	aq	−239.4	
Sr₃(PO₄)₂	c	−980	−881.54
SrS	aq	−113.1	−109.78
	aq	−120.4	
SrSO₄	c	−345.3	−309.30
	aq	−345.0	
SrWO₄	aq	−393	
Sulfur:			
S	c, rhombic	0.00	0.00
	c, monoclinic	−0.071	−0.023
	l, λ	0.257	0.071
	g	53.25	43.57
S₂	g	31.02	19.36
S₆	g	27.78	13.97
S₈	g	27.090	12.770
S₂Br₂	l	−4	
SCl₄	l	−13.7	−5.90
S₂Cl₂	l	−14.2	
S₂Cl₄	l	−24.1	
SF₆	g	−262	−237
SO	g	−19.02	−12.75
SO₂	g	−70.94	−71.68
SO₃	c, α	−103.03	−88.59
	c, β	−105.09	−88.28
	c, γ	−105.92	−88.22
	l	−109.34	−88.34
		−82.04	−88.98
		−89.80	−74.06
SO₂Cl₂	l		−75.06
		0.00	0.00
Tantalum:			
Ta	c	0.00	0.00
TaN	c	−51.2	−45.11
Ta₂O₅	c	−486.0	−453.7
Tellurium:			
Te	c	0.00	0.00
TeBr₄	c	−49.3	−57.4
TeCl₄	c	−77.4	
TeF₆	g	−315	−292
TeO₂	c	−77.56	−64.66
Thallium:			
Tl	c	0.00	0.00
TlBr	c	−41.5	−39.43
	aq	−28.0	−32.34
TlCl	c	−49.37	−44.46
	aq	−38.4	−39.09

Formula	State		
Thallium: (Cont.):			
TlCl₃	c	−82.4	−44.25
TlF	aq	−91.0	
	c	−77.6	−73.46
TlI	c	−31.1	−31.3
	aq	−12.7	−20.09
TlNO₃	c	−58.4	−36.32
	aq	−48.4	−34.01
Tl₂O	c	−43.18	
Tl₂O₃	c	−120	
TlOH	c	−57.44	−45.54
	aq	−53.9	−45.35
Tl₂S	c	−22	
Tl₂SO₄	c	−222.8	−197.79
	aq	−214.1	−191.62
Thorium:			
Th	c	0.00	0.00
ThBr₄	c	−281.5	−295.31
ThC₂	c	−352.0	
ThCl₄	c	−45.1	−322.32
ThI₄	aq	−335	−246.33
Th₃N₄	aq	−392	
ThO₂	c, "soluble"	−292.0	−282.3
Th(OH)₄	c	−309.0	−280.1
	aq	−291.6	
Th(SO₄)₂	c	−336.1	
		−632	−549.2
		−668.1	
Tin:			
Sn	c, II tetragonal	0.00	0.00
	c, III "gray," cubic	0.6	1.1
SnF₂	aq	−61.4	−55.43
SnF₄	aq	−60.0	
SnCl₂	aq	−94.8	−97.66
SnCl₄	l	−110.6	−68.94
		−83.6	−110.4
		−81.7	−124.67
		−127.3	−30.95
SnI₂	aq	−157.6	−60.75
SnO	c	−38.9	−123.6
Sn(OH)₂	c	−33.3	−115.95
Sn(OH)₄	c	−67.7	−226.00
		−138.1	
SnS		−136.2	
		−268.9	
		−18.61	
Titanium:			
Ti	c	0.00	0.00
TiC	c	−110.1	−109.2
TiCl₄	l	−181.4	−165.5
TiN	g	−80.0	−73.17
TiO₂	c, III, rutil	−225.0	−211.9
	c, amorphous	−214.1	−201.4
Tungsten:			
W	c	0.00	0.00
WO₂	c	−130.5	−118.3
WO₃	aq	−195.7	−177.3
WS₂	aq	−84	

Compound	State*	Heat of formation ΔH at 25°C, kcal/mol	Free energy of formation ΔF at 25°C, kcal/mol
Uranium:			
U.........	c	0.00	0.00
UCl2.....	c	-29	
UCl3.....	c	-213	
UCl4.....	c	-251	-249.6
UN4......	c	-274	-242.2
UO2......	c	-256.6	
UO2(NO3)2.6H2O..	c	-756.8	-617.8
UO3......	c	-291.6	
U3O8.....	c	-845.1	
Vanadium:			
V.........	c	0.00	0.00
VCl2.....	l	-147	
VCl3.....	l	-187	
VCl4.....	l	-165	
VN.......	c	-41.43	-35.08
V2O2.....	c	-195	
V2O3.....	c	-296	-277
V2O4.....	c	-342	-316
V2O5.....	c	-373	-342
Zinc:			
Zn........	c	0.00	0.00
ZnSb.....	c	-3.6	-3.88
ZnBr2....	c	-77.0	-72.9
	aq	-93.6	

Compound	State	Heat of formation ΔH at 25°C, kcal/mol	Free energy of formation ΔF at 25°C, kcal/mol
Zinc: (Cont.):			
Zn(C2H3O2)2..	c	-259.4	
	aq	-269.4	-214.4
Zn(CN)2..	c	17.06	-173.5
ZnCO3....	c	-192.9	-88.8
ZnCl2.....	c	-99.9	
	aq	-115.44	-166.6
ZnF2.....	aq	-192.9	
ZnI2......	c	-50.50	-49.93
	aq	-61.6	
Zn(NO3)2..	aq	-134.9	-87.7
ZnO.......	c, hexagonal	-83.36	-76.19
ZnO.SiO2..	c, rhombic	-282.6	
ZnS.......	c, wurtzite	-153.66	-44.2
ZnSO4....	c	-45.3	
	c	-233.4	
	aq	-252.12	-211.28
Zirconium:			
Zr........	c	0.00	0.00
ZrC.......	c	-29.8	-34.6
ZrCl4.....	c	-268.9	
ZrN.......	c	-82.5	-75.9
ZrO2......	c, monoclinic	-258.5	-244.6
Zr(OH)4...	c	-411.0	-307.6
ZrO(OH)2..	c	-337	

Example 5-3 Calculate the heat of reaction for the reaction $AlCl_3(s) + 3Na(s) \longrightarrow$ $3NaCl(s) + Al(s)$.

solution From Table 5-1, the heats of formation of $AlCl_3(s)$ and $NaCl(s)$ are -163.8 and -98.3 kcal, respectively. By convention, the heats of formation of $Na(s)$ and $Al(s)$ are zero. Thus,

$$AlCl_3(s) + 3Na(s) \longrightarrow 3NaCl(s) + Al(s)$$
$$-163.8 \qquad 0 \qquad\qquad 3(-98.3) \qquad 0$$
$$\Delta H = -294.9 - (-163.8) = -131.1 \text{ kcal}$$

Example 5-4 Calculate the heat of reaction for the burning of methane.

solution From Table 5-1, the heats of formation of $CH_4(g)$, $CO_2(g)$, and $H_2O(l)$ are -17.9, -94.1, and -68.3, respectively. Thus

$$CH_4(g) + 2O_2(g) \longrightarrow CO_2(g) + 2H_2O(l)$$
$$-17.9 \qquad 0 \qquad\qquad -94.1 \qquad 2(-68.3)$$
$$\Delta H = -94.1 - 136.6 - (-17.9) = -212.8 \text{ kcal}$$

Heats of Solution and Dilution

Heat effects also occur when a material dissolves or when it comes out of solution (as, for example, in crystallization). In many instances, the heat effects of dissolution can be very small, as when sodium chloride dissolves in water. On the other hand, many other materials produce a good deal of heat when they dissolve, as when sodium hydroxide dissolves in water; while for other materials, heat is absorbed (and cooling occurs) when they dissolve, as when sugar dissolves in water. The heat liberated (or absorbed) when 1 mol of a solute dissolves is known as its heat of solution ΔH_s. The heat of solution varies with the concentration of the solution that is formed, the total heat being greater as the solution produced becomes more dilute. This is sometimes confusing, since the temperature effects are diminished as the solution becomes more dilute (simply because there is more material to heat up or cool down). This can be understood by reference to Fig. 5-1, which shows the total heat liberated as a given number of mols of a particular solute dissolves in 1,000 g of solvent. In this figure, the final-solution concentration is expressed in mols per 1,000 g of solvent, or molality, and the corresponding heat of solution is referred to as the *molal heat of solution*.

Since more heat is liberated for a 2-molal solution of this solute than for a 1-molal solution, by an amount equal to $B - A$, we would expect a 2-molal solution to get hotter while the solute is dissolving than a 1-molal solution. However, B, the total amount of heat liberated in forming the 2-molal solution,

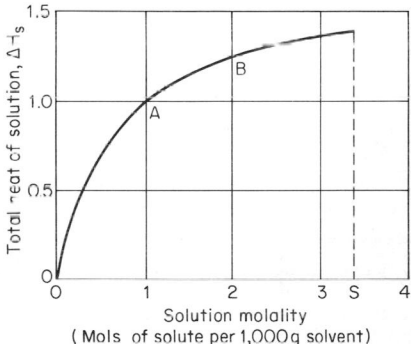

Fig. 5-1 Heat of solution as a function of final-solution concentration. When a 2 molal solution is diluted to a 1 molal solution, heat is liberated. The heat liberated in forming the original solution is given by point B (1.25 kcal). When this is diluted, we have to add an additional 1,000 g of solvent. This is the same as dissolving 2 mols of solute in 2,000 g of solvent, a process that would have liberated twice the heat shown at point A ($2 \times 1.0 = 2.0$ kcal). Thus, in diluting the above solution from 2 to 1 molal, $2.0 - 1.25 = 0.75$ kcal would be liberated. Point S represents the limit of solubility of the solute. (*From F. H. Getman and F. Daniels, "Outlines of Theoretical Chemistry," 6th ed., John Wiley & Sons, Ltd., New York, 1937.*)

is not as great as would be the case if the *same amount of solute* were used to produce a 1-molal solution (for which we would have had to use 2,000 g of solvent). In this case, the amount of heat liberated would be $2A$, since we have 2,000 g of solvent, and $2A$ is greater than B. However, we have twice as much solution to heat up, and consequently the temperature effects will not be as pronounced. Since the heat of solution depends on how much solute and how much solvent are present, tabulated data of heats of solution, such as Table 5-2, must give the specific formula on which the molecular weight is based, and the amount of solvent in which 1 mol of solute is dissolved.

Since the heat of solution depends on the concentration of the final solution, there must also be heat effects when a concentrated solution is diluted. Heat that is liberated (or

TABLE 5-2 Heats of Solution

INORGANIC COMPOUNDS DISSOLVED IN WATER

Substance	Dilution*	Formula	Heat of solution kcal/mol
Aluminum bromide	aq	AlBr₃	+85.3
chloride	600	AlCl₃	+77.9
	600	AlCl₃.6H₂O	+13.2
fluoride	aq	AlF₃	+31
	aq	AlF₃.3½H₂O	+19.0
	aq	AlF₃.3½H₂O	−1.7
iodide	aq	AlI₃	+89.0
sulfate	aq	Al₂(SO₄)₃	+126.
	aq	Al₂(SO₄)₃.6H₂O	+56.2
	aq	Al₂(SO₄)₃.18H₂O	+6.7
Ammonium bromide	aq	NH₄Br	−4.45
chloride	∞	NH₄Cl	−3.82
chromate	aq	(NH₄)₂CrO₄	−5.82
dichromate	600	(NH₄)₂Cr₂O₇	−12.9
iodide	aq	NH₄I	−3.56
nitrate	∞	NH₄NO₃	−6.47
perborate	aq	NH₄BO₃.H₂O	−9.0
sulfate	∞	(NH₄)₂SO₄	−2.75
sulfate, acid	800	NH₄HSO₄	+0.56
sulfite	aq	(NH₄)₂SO₃	−1.2
	aq	(NH₄)₂SO₃.H₂O	−4.13
Antimony fluoride	aq	SbF₃	−1.7
iodide	aq	SbI₃	−0.8
Arsenic acid	aq	H₃AsO₄	−0.4
Barium bromate	∞	Ba(BrO₃)₂.H₂O	−15.9
bromide	∞	BaBr₂	+5.3
	∞	BaBr₂.H₂O	−0.8
	∞	BaBr₂.2H₂O	−3.87
chlorate	aq	Ba(ClO₃)₂	−6.7
	aq	Ba(ClO₃)₂.H₂O	−10.6
chloride	aq	BaCl₂	+2.4
	aq	BaCl₂.H₂O	−2.17
	aq	BaCl₂.2H₂O	−4.5
cyanide	aq	Ba(CN)₂	+1.5
	aq	Ba(CN)₂.H₂O	−2.4
	aq	Ba(CN)₂.2H₂O	−4.9
iodate	∞	Ba(IO₃)₂	−9.1
	∞	Ba(IO₃)₂.H₂O	−11.3
iodide	aq	BaI₂	+10.5
	aq	BaI₂.H₂O	+2.7
	aq	BaI₂.2H₂O	+0.14
	aq	BaI₂.2½H₂O	−0.58
	aq	BaI₂.7H₂O	−6.61
nitrate	aq	Ba(NO₃)₂	−10.2
perchlorate	aq	Ba(ClO₄)₂	−2.8
	aq	Ba(ClO₄)₂.3H₂O	−10.5
sulfide	∞	BaS	+62.6
Beryllium bromide	aq	BeBr₂	+51.1
chloride	aq	BeCl₂	+72.6
iodide	aq	BeI₂	+13.5
sulfate	aq	BeSO₄	+7.9
	aq	BeSO₄.2H₂O	
	aq	BeSO₄.4H₂O	+1.1

Substance	Dilution*	Formula	Heat of solution kcal/mol
Bismuth iodide	aq	BiI₃	+3
Boric acid	aq	H₃BO₃	−5.4
Cadmium bromide	400	CdBr₂	+0.4
	400	CdBr₂.4H₂O	−7.3
chloride	400	CdCl₂	+3.1
	400	CdCl₂.H₂O	+0.6
	400	CdCl₂.2½H₂O	−3.00
nitrate	400	Cd(NO₃)₂.H₂O	+4.17
	400	Cd(NO₃)₂.4H₂O	−5.08
sulfate	400	CdSO₄	+10.69
	400	CdSO₄.H₂O	+6.05
	400	CdSO₄.2⅔H₂O	+2.51
Calcium acetate	∞	Ca(C₂H₃O₂)₂	+7.6
	∞	Ca(C₂H₃O₂)₂.H₂O	+6.5
bromide	∞	CaBr₂	−24.86
	∞	CaBr₂.6H₂O	−0.9
chloride	∞	CaCl₂	−4.9
	∞	CaCl₂.H₂O	+12.3
	∞	CaCl₂.2H₂O	+12.5
	∞	CaCl₂.4H₂O	+2.4
	∞	CaCl₂.6H₂O	−4.11
formate	400	Ca(CHO₂)₂	+0.7
iodide	∞	CaI₂	−28.0
	∞	CaI₂.8H₂O	+1.8
nitrate	∞	Ca(NO₃)₂	+4.1
	∞	Ca(NO₃)₂.H₂O	+0.7
	∞	Ca(NO₃)₂.2H₂O	−3.2
	∞	Ca(NO₃)₂.3H₂O	−4.2
	∞	Ca(NO₃)₂.4H₂O	−7.99
phosphate, mono-	aq	CaH₄(PO₄)₂.H₂O	−0.6
dibasic	aq	CaHPO₄.2H₂O	−1
sulfate	aq	CaSO₄	+5.1
	aq	CaSO₄.½H₂O	+3.6
	∞	CaSO₄.2H₂O	−0.18
Chromous chloride	aq	CrCl₂	+18.6
	aq	CrCl₂.3H₂O	+5.3
	aq	CrCl₂.4H₂O	+2.0
iodide	aq	CrI₂	+18.4
Cobaltous bromide	aq	CoBr₂	−1.25
	aq	CoBr₂.6H₂O	+18.5
chloride	400	CoCl₂	+9.8
	400	CoCl₂.2H₂O	−2.9
	400	CoCl₂.6H₂O	+18.8
iodide	aq	CoI₂	+15.0
sulfate	400	CoSO₄	−1.4
	400	CoSO₄.6H₂O	−3.6
	400	CoSO₄.7H₂O	+2.4
Cupric acetate	aq	Cu(C₂H₃O₂)₂	+0.5
formate	aq	Cu(CHO₂)₂	+0.3
nitrate	200	Cu(NO₃)₂	−2.6
	200	Cu(NO₃)₂.3H₂O	−10.7
	200	Cu(NO₃)₂.6H₂O	+15.9
sulfate	800	CuSO₄	

*mols of water used to dissolve 1 mol of the substance. ∞ means infinite dilution; aq means unspecified dilution.

Substance	Formula	mols H₂O*	ΔH
Cupric acetate (Cont.):			
	$CuSO_4 \cdot H_2O$	aq	+9.3
	$CuSO_4 \cdot 3H_2O$	1000	+3.65
	$CuSO_4 \cdot 5H_2O$	1000	-2.85
Cuprous sulfate	Cu_2SO_4	1000	+11.6
Ferric chloride	$FeCl_3$	aq	+31.7
	$FeCl_3 \cdot 2\tfrac12 H_2O$	1000	+21.0
	$FeCl_3 \cdot 6H_2O$	1000	+5.6
nitrate	$Fe(NO_3)_3 \cdot 9H_2O$	800	-9.1
Ferrous bromide	$FeBr_2$	aq	+18.0
chloride	$FeCl_2$	400	+17.9
	$FeCl_2 \cdot 2H_2O$	400	+8.7
	$FeCl_2 \cdot 4H_2O$	400	+2.7
iodide	FeI_2	aq	+23.3
sulfate	$FeSO_4$	400	+14.7
	$FeSO_4 \cdot H_2O$	400	+7.35
	$FeSO_4 \cdot 4H_2O$	400	+1.4
	$FeSO_4 \cdot 7H_2O$	400	-4.4
Lead acetate	$Pb(C_2H_3O_2)_2$	400	+1.4
	$Pb(C_2H_3O_2)_2 \cdot 3H_2O$	400	-5.9
bromide	$PbBr_2$	aq	-10.1
chloride	$PbCl_2$	aq	-3.4
formate	$Pb(CHO_2)_2$	aq	-6.9
nitrate	$Pb(NO_3)_2$	400	-7.61
Lithium bromide	$LiBr$	∞	+7.61
	$LiBr \cdot H_2O$	∞	+5.30
	$LiBr \cdot 2H_2O$	∞	+2.05
	$LiBr \cdot 3H_2O$	∞	-1.59
chloride	$LiCl$	∞	+8.66
	$LiCl \cdot H_2O$	∞	+4.45
	$LiCl \cdot 2H_2O$	∞	+1.07
	$LiCl \cdot 3H_2O$	∞	-1.98
fluoride	LiF	∞	-0.74
hydroxide	$LiOH$	∞	+4.74
	$LiOH \cdot \tfrac12 H_2O$	∞	+4.39
	$LiOH \cdot H_2O$	∞	+9.6
iodide	LiI	∞	+14.92
	$LiI \cdot \tfrac12 H_2O$	∞	+10.08
	$LiI \cdot H_2O$	∞	+6.93
	$LiI \cdot 2H_2O$	∞	+3.43
	$LiI \cdot 3H_2O$	∞	-0.17
nitrate	$LiNO_3$	∞	+0.466
	$LiNO_3 \cdot 3H_2O$	∞	-7.87
sulfate	Li_2SO_4	∞	+6.71
	$Li_2SO_4 \cdot H_2O$	∞	+3.77
Magnesium bromide	$MgBr_2$	∞	+43.7
	$MgBr_2 \cdot H_2O$	∞	+35.9
	$MgBr_2 \cdot 6H_2O$	∞	+19.8
chloride	$MgCl_2$	∞	+36.3
	$MgCl_2 \cdot 2H_2O$	∞	+20.8
	$MgCl_2 \cdot 4H_2O$	∞	+10.5
	$MgCl_2 \cdot 6H_2O$	∞	+3.4
iodide	MgI_2	∞	+50.2
nitrate	$Mg(NO_3)_2 \cdot 6H_2O$	∞	-3.7
phosphate	$Mg_3(PO_4)_2$	aq	+10.2
sulfate	$MgSO_4$	∞	+21.1
	$MgSO_4 \cdot H_2O$	∞	+14.0
Magnesium bromide (Cont.):			
	$MgSO_4 \cdot 2H_2O$	∞	+11.7
	$MgSO_4 \cdot 4H_2O$	∞	+4.9
	$MgSO_4 \cdot 6H_2O$	∞	+0.55
	$MgSO_4 \cdot 7H_2O$	∞	-3.18
sulfide	MgS	aq	+25.8
Manganic nitrate	$Mn(NO_3)_2$	400	+12.9
	$Mn(NO_3)_2 \cdot 3H_2O$	400	-3.9
	$Mn(NO_3)_2 \cdot 6H_2O$	400	-6.2
sulfate	$Mn_2(SO_4)_3$	aq	+22
Manganous acetate	$Mn(C_2H_3O_2)_2$	aq	+12.2
	$Mn(C_2H_3O_2)_2 \cdot 4H_2O$	aq	+1.6
bromide	$MnBr_2$	aq	+15
	$MnBr_2 \cdot H_2O$	aq	+14.4
	$MnBr_2 \cdot 4H_2O$	aq	+16.1
chloride	$MnCl_2$	400	+16.0
	$MnCl_2 \cdot 2H_2O$	400	+8.2
	$MnCl_2 \cdot 4H_2O$	400	+1.5
formate	$Mn(CHO_2)_2$	aq	+4.3
	$Mn(CHO_2)_2 \cdot 2H_2O$	aq	-2.9
iodide	MnI_2	aq	+26.2
	$MnI_2 \cdot H_2O$	aq	+24.1
	$MnI_2 \cdot 2H_2O$	aq	+22.7
	$MnI_2 \cdot 4H_2O$	aq	+19.9
	$MnI_2 \cdot 6H_2O$	aq	+21.2
sulfate	$MnSO_4$	400	+13.8
	$MnSO_4 \cdot H_2O$	400	+11.9
	$MnSO_4 \cdot 7H_2O$	400	-1.7
Mercuric acetate	$Hg(C_2H_3O_2)_2$	aq	-4.0
bromide	$HgBr_2$	aq	-2.4
chloride	$HgCl_2$	aq	-3.3
nitrate	$Hg(NO_3)_2 \cdot \tfrac12 H_2O$	aq	-0.7
Mercurous nitrate	$Hg_2(NO_3)_2 \cdot 2H_2O$	aq	-11.5
Nickel bromide	$NiBr_2$	aq	+19.0
	$NiBr_2 \cdot 3H_2O$	aq	+0.2
Nickel chloride	$NiCl_2$	800	+19.23
	$NiCl_2 \cdot 2H_2O$	800	+10.4
	$NiCl_2 \cdot 4H_2O$	800	+4.2
	$NiCl_2 \cdot 6H_2O$	800	-1.15
iodide	NiI_2	aq	+19.4
nitrate	$Ni(NO_3)_2$	200	+11.8
	$Ni(NO_3)_2 \cdot 6H_2O$	200	-7.5
sulfate	$NiSO_4$	200	+15.1
	$NiSO_4 \cdot 7H_2O$	200	-4.2
Phosphoric acid, ortho	H_3PO_4	400	+2.79
	$H_3PO_4 \cdot \tfrac12 H_2O$	400	-0.1
pyro-	$H_4P_2O_7$	aq	+25.9
	$H_4P_2O_7 \cdot 1\tfrac12 H_2O$	aq	+4.65
Potassium acetate	$KC_2H_3O_2$	∞	+3.55
aluminum sulfate	$KAl(SO_4)_2$	600	+48.5
	$KAl(SO_4)_2 \cdot 3H_2O$	600	+26.6
	$KAl(SO_4)_2 \cdot 12H_2O$		-10.1
bicarbonate	$KHCO_3$	2000	-5.1
bromate	$KBrO_3$	∞	-10.13
bromide	KBr	∞	-5.13
carbonate	K_2CO_3	∞	+6.58
	$K_2CO_3 \cdot \tfrac12 H_2O$	∞	+4.25

Adapted from Robert H. Perry and Cecil H. Chilton (eds.), "Chemical Engineers' Handbook," 5th ed., McGraw-Hill Book Company, New York, 1973.

TABLE 5-2 Heats of Solution—(Continued)

INORGANIC COMPOUNDS DISSOLVED IN WATER—(Continued)

Substance	Dilution*	Formula	Heat of solution kcal/mol
Potassium acetate (Cont.):			
	∞	$KC_2H_3O_2 \cdot 1\tfrac{1}{2}H_2O$	−0.43
chlorate	∞	$KClO_3$	−10.31
chloride	2185	KCl	−4.404
chromate	600	K_2CrO_4	−4.9
chrome sulfate		$KCr(SO_4)_2$	+55
		$KCr(SO_4)_2 \cdot H_2O$	+42
		$KCr(SO_4)_2 \cdot 2H_2O$	+33
		$KCr(SO_4)_2 \cdot 6H_2O$	+7
		$KCr(SO_4)_2 \cdot 12H_2O$	−9.5
cyanide	200	KCN	−3.0
dichromate	1600	$K_2Cr_2O_7$	−17.8
fluoride	∞	KF	+3.96
	∞	$KF \cdot 2H_2O$	−1.85
	∞	$KF \cdot 4H_2O$	−6.05
hydrosulfide	∞	KHS	+1.21
hydroxide	∞	KOH	+12.91
	∞	$KOH \cdot \tfrac{3}{4}H_2O$	+4.27
	∞	$KOH \cdot H_2O$	+3.48
	∞	$KOH \cdot 2H_2O$	+0.86
iodate	∞	KIO_3	−6.93
iodide	∞	KI	−5.23
nitrate	400	KNO_3	−8.633
oxalate	∞	$K_2C_2O_4$	−4.6
	400	$K_2C_2O_4 \cdot H_2O$	−7.5
perchlorate	400	$KClO_4$	−12.94
permanganate	aq	$KMnO_4$	−10.4
phosphate, dihydrogen	aq	KH_2PO_4	+4.7
pyrosulfite	aq	$K_2S_2O_5$	−11.0
	aq	$K_2S_2O_5 \cdot \tfrac{1}{2}H_2O$	−10.22
sulfate	∞	K_2SO_4	−6.32
sulfate, acid	800	$KHSO_4$	−3.10
sulfide	aq	K_2S	−11.0
sulfite	aq	K_2SO_3	+1.8
	aq	$K_2SO_3 \cdot H_2O$	+1.37
thiocyanate	∞	$KCNS$	−6.08
thionate, di-	∞	$K_2S_2O_6$	−13.0
thiosulfate	aq	$K_2S_2O_3$	−4.5
Silver acetate	aq	$AgC_2H_3O_2$	−5.4
nitrate	200	$AgNO_3$	−4.4
Sodium acetate	∞	$NaC_2H_3O_2$	+4.085
		$NaC_2H_3O_2 \cdot 3H_2O$	−4.665
arsenate	500	Na_3AsO_4	+15.6
	500	$Na_3AsO_4 \cdot 12H_2O$	−12.61
bicarbonate	1800	$NaHCO_3$	−4.1
borate, tetra-	900	$Na_2B_4O_7$	+10.0
	900	$Na_2B_4O_7 \cdot 10H_2O$	−16.8
bromide	∞	$NaBr$	−0.58
	∞	$NaBr \cdot 2H_2O$	−4.57
carbonate	∞	Na_2CO_3	+5.57
		$Na_2CO_3 \cdot H_2O$	+2.19

Substance	Dilution*	Formula	Heat of solution kcal/mol
Sodium carbonate (Cont.):			
	∞	$Na_2CO_3 \cdot 7H_2O$	−10.81
	∞	$Na_2CO_3 \cdot 10H_2O$	−16.22
chlorate	∞	$NaClO_3$	−5.37
chloride	∞	$NaCl$	−1.164
chromate	800	Na_2CrO_4	+2.50
	800	$Na_2CrO_4 \cdot 4H_2O$	−7.52
	800	$Na_2CrO_4 \cdot 10H_2O$	−16.0
cyanide	200	$NaCN$	−0.37
	200	$NaCN \cdot \tfrac{1}{3}H_2O$	−0.92
	200	$NaCN \cdot 2H_2O$	−4.41
fluoride	∞	NaF	−0.27
hydrosulfide	∞	$NaHS$	+4.62
	∞	$NaHS \cdot 2H_2O$	−1.49
Sodium hydroxide	∞	$NaOH$	+10.18
	∞	$NaOH \cdot \tfrac{1}{3}H_2O$	+8.17
	∞	$NaOH \cdot 2\tfrac{1}{3}H_2O$	+7.08
	∞	$NaOH \cdot 3\tfrac{1}{2}H_2O$	+6.48
	∞	$NaOH \cdot H_2O$	+5.17
iodide	∞	NaI	+1.57
	∞	$NaI \cdot 2H_2O$	+3.89
metaphosphate	600	$NaPO_3$	+3.97
nitrate	600	$NaNO_3$	−5.05
nitrite	aq	$NaNO_2$	−3.6
perchlorate	aq	$NaClO_4$	−4.15
phosphate, di-	1600	Na_2HPO_4	+5.21
tri-	1600	Na_3PO_4	+13
	1600	$Na_3PO_4 \cdot 12H_2O$	−15.3
phosphate		$Na_2HPO_4 \cdot 2H_2O$	−0.82
di-	1600	$Na_2HPO_4 \cdot 7H_2O$	−12.04
	1600	$Na_2HPO_4 \cdot 12H_2O$	−23.18
phosphite, mono-	600	NaH_2PO_3	−0.90
	600	$NaH_2PO_3 \cdot 2\tfrac{1}{2}H_2O$	−5.29
di-	800	$Na_2HPO_3 \cdot 5H_2O$	+9.30
pyrophosphate	800	$Na_4P_2O_7$	−4.54
	1600	$Na_4P_2O_7 \cdot 10H_2O$	+11.9
di-	1600	$Na_2H_2P_2O_7$	−11.7
	1200	$Na_2H_2P_2O_7 \cdot 6H_2O$	−2.2
	1200		−14.0
sulfate	∞	Na_2SO_4	+0.28
	∞	$Na_2SO_4 \cdot 10H_2O$	−18.74
sulfate, acid	800	$NaHSO_4$	+1.74
	800	$NaHSO_4 \cdot H_2O$	+0.15
sulfide	∞	Na_2S	+15.2
	∞	$Na_2S \cdot 4\tfrac{1}{2}H_2O$	+0.09
	∞	$Na_2S \cdot 5H_2O$	−6.54
	∞	$Na_2S \cdot 9H_2O$	−16.65
sulfite	aq	Na_2SO_3	+2.8
	aq	$Na_2SO_3 \cdot 7H_2O$	−11.1
thiocyanate	∞	$NaCNS$	−1.83
thionate, di-	aq	$Na_2S_2O_6$	−5.80
	aq	$Na_2S_2O_6 \cdot 2H_2O$	−11.86

Substance	Formula		Value
Sodium thiosulfate.........	Na₂S₂O₃	aq	+2.0
	Na₂S₂O₃.5H₂O	aq	-11.30
Stannic bromide.........	SnBr₄	aq	+15.5
Stannous bromide.........	SnBr₂	aq	-1.6
iodide.........	SnI₂	aq	-5.8
Strontium acetate.........	Sr(C₂H₃O₂)₂	aq	+6.2
	Sr(C₂H₃O₂)₂.½H₂O	∞	+5.9
bromide.........	SrBr₂	∞	+16.4
	SrBr₂.H₂O	∞	+9.25
	SrBr₂.2H₂O	∞	+6.5
	SrBr₂.4H₂O	∞	+0.4
	SrBr₂.6H₂O	∞	-6.1
chloride.........	SrCl₂	∞	+11.54
	SrCl₂.H₂O	∞	+6.4
	SrCl₂.2H₂O	∞	+2.95
	SrCl₂.6H₂O	∞	-7.1
iodide.........	SrI₂	∞	+20.7
	SrI₂.H₂O	∞	+12.65

Substance	Formula		Value
Strontium acetate (Cont.):			
	SrI₂.2H₂O	∞	+10.4
	SrI₂.6H₂O	∞	-4.5
nitrate.........	Sr(NO₃)₂	∞	-4.8
	Sr(NO₃)₂.4H₂O	∞	-12.4
sulfate.........	SrSO₄	∞	+0.5
Sulfuric acid, pyro-.........	H₂S₂O₇	∞	-18.08
Zinc acetate.........	Zn(C₂H₃O₂)₂	400	+9.8
	Zn(C₂H₃O₂)₂.H₂O	400	+7.0
	Zn(C₂H₃O₂)₂.2H₂O	400	+3.9
bromide.........	ZnBr₂	400	+15.0
chloride.........	ZnCl₂	400	+15.72
iodide.........	ZnI₂	aq	+11.6
nitrate.........	Zn(NO₃)₂.3H₂O	400	-5
	Zn(NO₃)₂.6H₂O	400	-6.0
sulfate.........	ZnSO₄	400	+18.5
	ZnSO₄.6H₂O	400	+10.0
	ZnSO₄.7H₂O	400	-0.8
		400	-4.3

TABLE 5-2 Heats of Solution—(Continued)

ORGANIC COMPOUNDS DISSOLVED IN WATER
(AT INFINITE DILUTION)

Solute	Heat of solution, cal/mol solute*
Acetic acid (solid), $C_2H_4O_2$	−2,251
Acetylacetone, $C_5H_8O_2$	−641
Acetylurea, $C_3H_6N_2O_2$	−6,812
Aconitic acid, $C_6H_6O_6$	−4,206
Ammonium benzoate, $C_7H_9NO_2$	−2,700
picrate	−8,700
succinate (n-)	−3,489
Aniline, hydrochloride, C_6H_8ClN	−2,732
Barium picrate	−4,708
Benzoic acid, $C_7H_6O_2$	−6,501
Camphoric acid, $C_{10}H_{16}O_4$	−502
Citric acid, $C_6H_8O_7$	−5,401
Dextrin, $C_{12}H_{20}O_{10}$	268
Fumaric acid, $C_4H_4O_4$	−5,903
Hexamethylenetetramine, $C_6H_{12}N_4$	−4,780
Hydroxybenzamide (m-), $C_7H_7NO_2$	−4,161
(HCl)	−7,003
(o-), $C_7H_7NO_2$	−4,340
(p-)	−5,392
Hydroxybenzoic acid (o-), $C_7H_6O_3$	−6,350
(p-), $C_7H_6O_3$	−5,781
Hydroxybenzyl alcohol (o-), $C_7H_8O_2$	−3,203
Inulin, $C_{36}H_{62}O_{31}$	−96
Isosuccinic acid, $C_4H_6O_4$	−3,420
Itaconic acid, $C_5H_6O_4$	−5,922
Lactose, $C_{12}H_{22}O_{11}.H_2O$	−3,705
Lead picrate	−7,098
(2H₂O)	−13,193
Magnesium picrate (8H₂O)	14,699
	−15,894
Maleic acid, $C_4H_4O_4$	−4,441
Malic acid, $C_4H_6O_5$	−3,150
Malonic acid, $C_3H_4O_4$	−4,493
Mandelic acid, $C_8H_2O_4$	−3,090
Mannitol, $C_6H_{14}O_6$	−5,260
Menthol, $C_{10}H_{20}O$	0
Nicotine dihydrochloride, $C_{10}H_{16}Cl_2N_2$	6,561
Nitrobenzoic acid (m-), $C_7H_5NO_4$	−5,593
(o-), $C_7H_5NO_4$	−5,306
(p-), $C_7H_5NO_4$	−8,891
Nitrophenol (m-), $C_6H_5NO_3$	−5,210
(o-), $C_6H_5NO_3$	−6,310
(p-), $C_6H_5NO_3$	−4,493

Solute	Heat of solution, cal/mol solute*
Oxalic acid, $C_2H_2O_4$	−2,290
(2H₂O)	−8,485
Phenol (solid), C_6H_6O	−2,605
Phthalic acid, $C_8H_6O_4$	−4,871
Picric acid, $C_6H_3N_3O_7$	−7,098
Piperic acid, $C_{12}H_{10}O_4$	−10,492
Piperonylic acid, $C_8H_6O_4$	−9,106
Potassium benzoate	−1,506
citrate	−2,820
tartrate (n-) (0.5 H₂O)	−5,562
Pyrogallol, $C_6H_6O_3$	−3,705
Pyrotartaric acid	−5,019
Quinone	−3,991
Raffinose, $C_{18}H_{32}O_{16}$ (5H₂O)	−9,703
Resorcinol, $C_6H_6O_2$	−3,960
Silver malonate (n-)	−9,799
Sodium citrate (tri-)	−5,270
picrate	−6,441
potassium tartrate	−1,817
(4H₂O)	−12,342
succinate (n-)	−2,390
(6H₂O)	−10,994
tartrate (n-) (2H₂O)	−1,121
Strontium picrate	−5,882
(6H₂O)	−7,887
Succinic acid, $C_4H_6O_4$	−14,412
Succinimide, $C_4H_5NO_2$	−6,405
Sucrose, $C_{12}H_{22}O_{11}$	−4,302
	−1,319
Tartaric acid (d-)	−3,451
Thiourea, CH_4N_2S	−5,330
Urea, CH_4N_2O	−3,609
acetate	−8,795
formate	−7,194
nitrate	−10,803
oxalate	−17,806
Vanillic acid	−5,160
Vanillin	−5,210
Zinc picrate	−11,496
(8H₂O)	−15,894

* + denotes heat evolved, and − denotes heat absorbed. All values are positive unless otherwise noted.

absorbed) when a solution goes from a higher concentration to a lower concentration is called its *heat of dilution*. The heat of dilution is the difference in the heats of solution between the original and final concentrations.

Example 5-5 If the solute of Fig. 5-1 has a molecular weight of 75, how much heat is liberated when 100 g of solute is dissolved in 2,000 g of water?
solution Mols of solute present = $100/75 = 1.33$. Molality of final solution: $1.33 \times 1,000/2,000 = 0.67$. Following the abscissa 0.67 upward in Fig. 5-1, the curve is intercepted at 0.85 kcal. However, this is for a solution containing 1,000 g of solvent. Since we have a total solution containing 2,000 g of solvent, the total heat liberated is $0.85 \times 2 = 1.70$ kcal.

Example 5-6 If the solution of Example 5-5 has a specific heat of 1.05 and was originally at 25°C, what is its final temperature (assuming no heat is lost to its surroundings)?
solution The total weight of solution is $2,000 + 100 = 2,100$ g. Thus

$$t_f - t_i = \frac{1,700 \text{ kcal}}{2,100 \text{ g}} \times \frac{g - °C}{1.05 \text{ kcal}} = 0.77$$

$$t_f = t_i + 0.77 = 25 + 0.77 = 25.8°C$$

Example 5-7 If the solution of Example 5-5 is diluted with an additional 2,000 g of solvent, how much heat is liberated?
solution If 4,000 g of solvent is used, solution molality will be 0.33. From Fig. 5-1, the heat liberated (per 1,000 g of solvent) is 0.60 kcal, or $0.60 \times 4 = 2.40$ kcal for the total solution (i.e., containing 100 g of solute in 4,000 g of solvent). From Example 5-5, the heat liberated in dissolving 100 g of solute in 2,000 g of solvent is 1.70 kcal. Therefore, the heat liberated in diluting this latter solution with an additional 2,000 g of solvent is $2.40 - 1.70 = 0.70$ kcal.

The heats of dilution of the common inorganic acids HNO_3, HCl, and H_2SO_4 are particularly pronounced and frequently of considerable importance. The heats of dilution of these acids are given in Table 5-3. This table presents data that are particularly convenient in calculating the heat effects between two acid concentrations.

Although usually of only minor importance (except for large-scale operations), when concentrating a solution, as, for example, by evaporation of its solvent, an amount of heat equal to the heat of dilution must be supplied or removed apart from the heat necessary to evaporate the solvent.

Heat of Neutralization

When dilute solutions of equivalent quantities of strong acids and strong bases are mixed, resulting in each solution neutralizing the other, practically the same amount of heat, 13.8 kcal, is evolved, regardless of which acid or base is used. Since the neutralization of a dilute strong acid by a dilute strong base essentially involves the combination of

TABLE 5-3   Heats of Dilution of Common Inorganic Acids

Mols H_2O added to	ΔH (kcal)		
	1 mol HNO_3 + 1 mol H_2O	1 mol HCl + 1 mol H_2O	1 mol H_2SO_4 + 1 mol H_2O
0.5			−2.11
1	−1.51	−5.97	−3.30
2	−2.44	−7.98	−5.02
4	−3.50	−9.59	
5			−7.70
8			−8.95
9	−4.04	−10.78	
19			−10.43
24	−4.23	−11.48	
49	−4.18	−11.70	−10.68
99	−4.16	−11.90	−10.90
199	−4.16	−11.95	−11.08
399	−4.16	−12.00	−11.40
799			−11.82
1,599			−12.30
	−4.16	−12.07	−14.09

From F. H. Getman and F. Daniels, "Outlines of Theoretical Chemistry," 6th ed., John Wiley & Sons, Inc., New York, 1937.

hydrogen and hydroxyl ions to form undissociated water, the reaction can be represented by the equations

$$H^+ + OH^- \longrightarrow H_2O \qquad \Delta H = -13.8 \text{ kcal} \tag{5-15}$$

or $\qquad H^+ + A^- + M^+ + OH^- \longrightarrow M^+ + A^- + H_2O \qquad \Delta H = -13.8 \text{ kcal} \tag{5-16}$

It should be noted that the heat of formation of liquid water, -68.4 kcal, is quite different from the heat of neutralization, and the two should not be confused. The heat of neutralization in other than dilute solutions will be different for different acid-base combinations and concentrations because of the degree of ionization of the reactants and products, and the degree of hydration of the hydrogen and hydroxyl ions of the reactants.

TABLE 5-4 Heats of Formation of Various Ions

Ion	Heat of formation (kcal)
METALS	
Hydrogen	0.0
Potassium	−60.3
Sodium	−57.6
Lithium	−66.4
Ammonium	−31.7
Magnesium	−110.0
Calcium	−129.5
Aluminum	−126.4
Manganese	−48.0
Iron (ous)	−20.8
Iron (ic)	−9.6
Cobalt	−16.5
Nickel	−15.3
Zinc	−36.6
Cadmium	−17.4
Copper (ous)	+16.0
Copper (ic)	+16.5
Mercury (ous)	+20.1
Silver	+24.9
Lead	−0.5
Tin (ous)	−2.4
NONMETALS	
Chloride	−39.6
Bromide	−28.6
Iodide	−13.3
Sulfate	−211.3
Sulfite	−146.8
Nitrite	−25.6
Nitrate	−49.8
Carbonate	−161.0
Hydroxyl	−54.6

From F. H. Getman and F. Daniels, "Outlines of Theoretical Chemistry," 6th ed., John Wiley & Sons, Inc., New York, 1937.

For different combinations of weak acids, weak bases, strong acids, and strong bases, the heat of neutralization will differ from -13.8 kcal, sometimes by a fairly substantial amount, as illustrated by the following neutralization reactions:

$$\text{HCOOH } (aq) + \text{NaOH } (aq) \longrightarrow \text{HCOONA } (aq) + \text{H}_2\text{O } (l)$$
$$\Delta H = -13.40 \text{ kcal} \tag{5-17}$$

$$\text{HCOOH } (aq) + \text{NH}_4\text{OH } (aq) \longrightarrow \text{HCOOHN}_4 \, (aq) + \text{H}_2\text{O } (l)$$
$$\Delta H = -11.90 \text{ kcal} \tag{5-18}$$

$$\text{HCN } (aq) + \text{NaOH } (aq) \longrightarrow \text{NaCN } (aq) + \text{H}_2\text{O } (l) \qquad \Delta H = -2.90 \text{ kcal} \tag{5-19}$$

Heat of Ionization

Since 13.8 kcal is liberated when water is formed from its ions, and since the heat of formation of water (liquid) is -68.4 kcal, it follows that the heat of formation of one equivalent of hydrogen and hydroxyl ions is $-68.4 + 13.8$, or -54.6 kcal. By convention, the heat of formation of H^+ is assigned a value of zero, so that the heat of formation of OH^- is -54.6 kcal. Having thus established relative heats of formation for two ions, it is an easy matter to assign values to other ions. For example, since the heat of formation of KOH (aq) is -114.9 kcal, the heat of formation of potassium ions must be $-114.9 + 54.6$, or -60.3 kcal. The heats of formation for a number of important ions are given in Table 5-4.

The convention that $\Delta H = 0$ for the reaction $H_2(g) \longrightarrow 2H^+ + 2e$ is a convenience but has no foundation in fact. It is known, for example, that the energy required to ionize hydrogen gas is indeed very large. However, even though the convention that the heat of formation of H^+ is zero is not correct, the usefulness of the arbitrary values cited in Table 5-4 is not lost, since these values can be used for calculating heats of reaction of ionic reactions:

$$Fe + Cu^{++} \longrightarrow Fe^{++} + Cu \tag{5-20}$$
$$\Delta H = -20.8 + 0 - (0 + 16.5) = -37.3 \text{ kcal}$$

(recalling that the heat of formation of an element is zero).

Example 5-8 What is the heat of reaction for the reaction between sodium metal and water?
solution $2Na(s) + 2H_2O(l) \longrightarrow H_2(g) + 2Na^+ + 2OH^-$. The heats of formation of Na^+ and OH^- are given in Table 5-4. Thus the heat of reaction is $2(-57.6) + 2(-54.6) - 2(-68.4) = -115.2 - 109.2 + 136.8 = -87.6$ kcal.

Example 5-9 What is the heat of reaction for the precipitation of silver bromide from a solution of silver nitrate by a solution of sodium bromide?
solution $(Na^+ + Br^-) + (Ag^+ + NO_3^-) \longrightarrow AgBr + (Na^+ + NO_3^-)$. From Table 5-4 the heats of formation of Na^+, Br^-, Ag^+, and NO_3^- are -57.6, -28.6, $+24.9$, and -49.8 kcal, respectively, and the heat of formation of AgBr, from Table 5-1, is -23.8 kcal. Thus $\Delta H = -23.8 + (-57.6) + (-49.8) - (-57.6) - (-28.6) - (+24.9) - (-49.8)$

$$\Delta H = -23.8 - (-28.6 + 24.9) = -20.1 \text{ kcal}$$

Effect of Temperature on Heat of Reaction

We have already indicated that the heats of reaction are given for a particular set of conditions, i.e., standard conditions, so that the enthalpy of the reactants and products can have known values. The calculation of heats of reaction for temperatures other than standard is illustrated in Fig. 5-2. If a chemical reaction is allowed to take place at T_1, and then is repeated at T_2, the amount of heat evolved for the two cases may be significantly different. If the initial state of the system is represented by a and the final state by c, we have the option of heating the reactants to T_2 and then allowing the reaction to proceed at T_2, or allowing the reaction to proceed at T_1 and then heating the products to T_2. If the path adc is taken, and the reaction takes place at T_1, the heat change will be ΔH_{T_1}. The heat absorbed by the products on being heated to T_2 will be the specific heat of the products multiplied by the temperature change: $C_{\text{products}} (T_2 - T_1)$. Similarly, for path abc, the heat absorbed by the reactants on being heated to T_2 will be $C_{\text{reactants}} (T_2 - T_1)$, and at T_2 the heat of reaction will be ΔH_{T_2}. Thus,

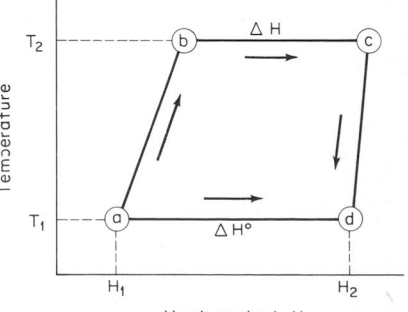

Fig. 5-2 Diagram illustrating relation between ΔH and $\Delta H°$. The net heat content of the products is the same for routes $abcd$ and ad.

$$\Delta H_{T_1} + C_{\text{products}} (T_2 - T_1) = \Delta H_{T_2} + C_{\text{reactants}} (T_2 - T_1) \tag{5-21}$$

If in going from T_1 to T_2 the reactants or products undergo a change in state, as, for example, one or more of the materials going from a solid to a liquid, or a liquid to a vapor, the heat absorbed in the change of state must be included. If $T = 25°C$, ΔH_T will be, by definition, ΔH_0.

Example 5-10   What is the heat of formation of water at $1000°C$?

solution   From Table 5-1, $\Delta H_f = -68.4$ kcal at $25°C$. To obtain the heat of formation at $1000°C$, it is necessary to subtract from ΔH_{25} the heat necessary to first heat the water formed at 25 to $100°C$, then the heat necessary to vaporize the water, and finally the heat necessary to raise the temperature of the water vapor from 100 to $1000°C$.

$$\text{Heat to raise 1 mol of } H_2O \text{ from 25 to } 100°C = 18 \text{ g} \times \frac{1 \text{ cal}}{°C} \times 75°C = 1{,}350 \text{ cal} = 1.35 \text{ kcal}$$

$$\text{Heat to vaporize 1 mol of } H_2O = 18 \text{ g} \times \frac{539 \text{ cal}}{g} = 9{,}702 \text{ cal} = 9.70 \text{ kcal}$$

$$\text{Heat to raise 1 mol of water vapor from 100 to } 1000°C = 18 \text{ g} \times \frac{0.48 \text{ cal}}{g - °C} \times 900°C$$
$$= 7{,}776 \text{ cal} = 7.78 \text{ kcal}$$

(*Note:* An average value for the specific heat of H_2O vapor over the temperature range of 100 to $1000°C$, 0.48 cal/g $-$ °C, is used in the above calculation.) The heat of formation of water at $1000°C$ is thus $-68.4 + 1.35 + 9.70 + 7.78 = -49.6$ kcal.

Heat of Combustion

Because of the frequency with which it occurs, the heat of reaction for the complete oxidation of a mol of a substance is referred to as the *heat of combustion*. (Because heats of combustion are usually large, occasionally the values are given in terms of heats of combustion per gram.) Oxidations take place rapidly and completely, and heats of combustion are relatively easy to measure with a high degree of accuracy. Since water is a product of most combustion reactions, it must be kept in mind that heats of combustion, such as those given in Table 5-5, refer to the maximum amount of heat liberated by the combustion. (In Example 5-10, the combustion of 1 mol of H_2 at $1000°C$ liberated only 49.6 kcal, even though the heat-of-combustion data indicated that the gross heat that could have been liberated is 68.4 kcal.)

Because they can be measured accurately, heats of combustion can be used to determine the heats of reaction in transforming one allotropic (physical) form of an element into another one. As an example, when equal weights of the three common allotropic forms of carbon are burned in an excess of oxygen, different amounts of heat are evolved:

$$C(s, \text{ diamond }) + O_2(g) \longrightarrow CO_2(g) \qquad \Delta H = -94.3 \text{ kcal} \qquad (5\text{-}22)$$
$$C(s, \text{ graphite}) + O_2(g) \longrightarrow CO_2(g) \qquad \Delta H = -94.8 \text{ kcal} \qquad (5\text{-}23)$$
$$C(s, \text{ amorphous}) + O_2(g) \longrightarrow CO_2(g) \qquad \Delta H = -97.7 \text{ kcal} \qquad (5\text{-}24)$$

It can be readily seen that amorphous carbon has the greatest heat content of the three forms of carbon (since it can liberate more heat on being burned to carbon dioxide). Thus, if this form of carbon were to be transformed to, say, graphite, the reaction would evolve 2.9 kcal:

$$C(s, \text{ amorphous}) \longrightarrow C(s, \text{ graphite}) \qquad \Delta H_0 = -2.9 \text{ kcal} \qquad (5\text{-}25)$$

(For calculating heats of reaction, amorphous carbon is considered to be the elemental form of carbon and is assigned a heat-of-formation value of zero.)

Example 5-11   The heat of combustion of white phosphorus to phosphorus pentoxide (P_2O_5) is found to be -365.8 kcal. What is the heat of combustion of red phosphorus to phosphorus pentoxide given that

$$P(s, \text{ white}) \longrightarrow P(s, \text{ red}) \qquad \Delta H = -4.3 \text{ kcal}$$

solution

$$2P(s, \text{ white}) + 2\tfrac{1}{2} O_2(g) \longrightarrow P_2O_5(s) \qquad \Delta H = -365.8 \text{ kcal}$$
$$2P(s, \text{ red}) \longrightarrow 2P(s, \text{ white}) \qquad \Delta H = +8.6 \text{ kcal}$$

$$\overline{2P(s, \text{ white}) + 2P(s, \text{ red}) + 2\tfrac{1}{2} O_2 \longrightarrow P_2O_5(s) + 2P(s, \text{ white})}$$
$$\Delta H = -357.2 \text{ kcal}$$

Note that the heat of reaction of 2 mols of red phosphorus is used since 2 mols of phosphorus are required to make 1 mol of P_2O_5, and also that the conversion of red to white phosphorus involves a change in the sign of the heat of reaction, since the latter reaction absorbs heat. Canceling $2P(s, \text{white})$, since it appears on both sides of the equation, yields

$$2P(s, \text{red}) + 2\tfrac{1}{2} O_2 \longrightarrow P_2O_5(s) \qquad \Delta H = -357.2 \text{ kcal}$$

The heats of combustion of naturally occurring materials are of considerable importance, particularly where the materials are to be used as fuels or food. In such cases, it is important to know how much heat energy will be released by the oxidation of the particular material. (It should be recalled that the heat released will be the same whether it occurs by direct combustion or by a series of reactions such as occur in metabolic processes.) The heats of combustion of liquid and gaseous fuels, coals, and a variety of natural materials are given in Tables 5-6 to 5-10. The heating values of common foodstuffs are given in Table 5-11. In a combustion of a fuel in which it is intended that another material be heated, the usual products of the combustion (CO_2 and H_2O) are used in their heated and gaseous form. As indicated earlier, the heat of combustion would be greater if the water vapor produced by the combustion were condensed to liquid water, thereby giving up some sensible heat as well as its latent heat of vaporization. Since the latter is accomplished by cooling the combustion gases, they give up more heat but do so at a lower temperature. When the heat of combustion is used essentially for heating to temperatures above the boiling point of water, the heat generated is referred to as the *net heat of combustion*, meaning that it is the heat available as burned, i.e., *without* condensation of moisture. Where the heating occurs in such a way as to permit the condensation of moisture from the combustion gases, the heat available is referred to as the *gross heat of combustion*. In using heat-of-combustion tables, care should be exercised as to whether net or gross heating values are presented.

5-2 FREE ENERGY AND ENTROPY

Free Energy

At this point, it is necessary to differentiate between heat and work, even though they are both forms of energy. Heat, of course, refers to energy that manifests itself as a rise in temperature, or by evaporation, freezing, or some other phase change. Work, on the other hand, manifests itself by the raising of a body, operation of an engine, generation of electricity, operation of an electric motor, or other useful electromechanical activity. If the energy can be made to do useful work, it is said to be *available energy*, whereas if it cannot be made to do useful work, it is said to be *unavailable energy*. Heat, as such, is isothermally unavailable energy, since it will not flow, and therefore can do no work unless there is a temperature difference to make it flow. (The word isothermal means at a uniform, given temperature.) On the other hand, chemical, electrical, and certain other types of energy can be converted to useful work under isothermal conditions. That part of the energy of a system which can be used isothermally to produce work, i.e., the isothermally available energy, is called the *free energy*, usually denoted by the symbol F.

Entropy

Since isothermally unavailable energy is associated with a particular (absolute) temperature T, the amount of unavailable energy at that temperature is $T \times S$, where S is the unavailable energy per unit temperature. While S is a somewhat abstract quantity, it does have tremendous mathematical impact on thermochemical calculations and is called *entropy*. Entropy has units of cal/K or Btu/$°$R.

Internal and External Energy

Another important concept is that of the internal and external energy of a system. The *internal energy* of a system is simply the total energy existing within the system, and is usually designated by the symbol E. Part of the internal energy of a system can be made to do isothermal work, and part cannot. Frequently, that part of the energy isothermally available to do work is called the *work content*, designated by the symbol A. The *external energy* is that part of the total energy of the system which can be utilized for the production of work, either directly (as in a gas driving a piston) or indirectly (as through the generation of electricity). No special symbol has been generally adopted for external energy.

TABLE 5-5 Heats of Combustion

Compound	Formula	State	Heat of combustion at 25°C, and constant pressure, to form					
			H₂O (liq.) and CO₂ (gas)			H₂O (gas) and CO₂ (gas)		
			kcal/mol	cal/g	Btu/lb	kcal/mol	cal/g	Btu/lb
Hydrogen........	H₂	gas	68.3174	33,887.6	60,957.7	57.7979	28,669.6	51,571.4
Carbon..........	C	solid, graph.	94.0518	7,831.1	14,086.8			
Carbon monoxide..	CO	gas	67.6361	2,414.7	4,343.6			
Paraffins								
Methane........	CH₄	gas	212.798	13,265.1	23,861	191.759	11,953.6	21,502
Ethane.........	C₂H₆	gas	372.820	12,399.2	22,304	341.261	11,349.6	20,416
Propane........	C₃H₈	gas	530.605	12,033.5	21,646	488.527	11,079.2	19,929
Propane........	C₃H₈	liq.	526.782	11,946.8	21,490	484.704	10,992.5	19,774
n-Butane.......	C₄H₁₀	gas	687.982	11,837.3	21,293	635.384	10,932.3	19,665
n-Butane.......	C₄H₁₀	liq.	682.844	11,748.9	21,134	630.246	10,843.9	19,506
2-Methylpropane (Isobutane)..	C₄H₁₀	gas	686.342	11,809.1	21,242	633.744	10,904.1	19,614
2-Methylpropane (Isobutane)..	C₄H₁₀	liq.	681.625	11,727.9	21,096	629.027	10,822.9	19,468
n-Pentane......	C₅H₁₂	gas	845.16	11,714.6	21,072	782.04	10,839.7	19,499
n-Pentane......	C₅H₁₂	liq.	838.80	11,626.4	20,914	775.68	10,751.5	19,340
2-Methylbutane (Isopentane)...	C₅H₁₂	gas	843.24	11,688.0	21,025	780.12	10,813.1	19,451
2-Methylbutane (Isopentane)...	C₅H₁₂	liq.	837.31	11,605.8	20,877	774.19	10,730.9	19,303
2,2-Dimethylpropane (Neopentane)..	C₅H₁₂	gas	840.49	11,649.8	20,956	777.37	10,775.0	19,382
2,2-Dimethylpropane (Neopentane)..	C₅H₁₂	liq.	835.18	11,576.2	20,824	772.06	10,701.4	19,250
n-Hexane.......	C₆H₁₄	gas	1,002.57	11,634.5	20,928	928.93	10,780.0	19,391
n-Hexane.......	C₆H₁₄	liq.	995.01	11,546.8	20,771	921.37	10,692.2	19,233
2-Methylpentane..	C₆H₁₄	gas	1,000.87	11,614.8	20,893	927.23	10,760.2	19,356
2-Methylpentane..	C₆H₁₄	liq.	993.71	11,531.7	20,743	920.07	10,677.1	19,206
3-Methylpentane..	C₆H₁₄	gas	1,001.51	11,622.2	20,906	927.87	10,767.6	19,369
3-Methylpentane..	C₆H₁₄	liq.	994.25	11,538.0	20,755	920.61	10,683.4	19,218
2,2-Dimethylbutane.	C₆H₁₄	gas	998.17	11,583.5	20,837	924.53	10,728.9	19,299
2,2-Dimethylbutane.	C₆H₁₄	liq.	991.52	11,506.2	20,698	917.88	10,651.7	19,161
2,3-Dimethylbutane.	C₆H₁₄	gas	1,000.04	11,605.2	20,876	926.40	10,750.6	19,338
2,3-Dimethylbutane.	C₆H₁₄	liq.	993.05	11,524.0	20,730	919.41	10,669.5	19,192
n-Heptane......	C₇H₁₆	gas	1,160.01	11,577.2	20,825	1,075.85	10,737.2	19,314
n-Heptane......	C₇H₁₆	liq.	1,151.27	11,489.9	20,668	1,067.11	10,650.0	19,157
2-Methylhexane...	C₇H₁₆	gas	1,158.30	11,560.1	20,795	1,074.14	10,720.2	19,284
2-Methylhexane...	C₇H₁₆	liq.	1,149.97	11,477.0	20,645	1,065.81	10,637.0	19,134
3-Methylhexane...	C₇H₁₆	gas	1,158.94	11,566.5	20,806	1,074.78	10,726.6	19,295
3-Methylhexane...	C₇H₁₆	liq.	1,150.55	11,482.8	20,655	1,066.39	10,642.8	19,145
3-Ethylpentane...	C₇H₁₆	gas	1,159.56	11,572.7	20,817	1,075.40	10,732.7	19,306
3-Ethylpentane...	C₇H₁₆	liq.	1,151.13	11,488.6	20,666	1,066.97	10,648.6	19,155
2,2-Dimethylpentane..	C₇H₁₆	gas	1,155.61	11,533.3	20,746	1,071.45	10,693.9	19,235
2,2-Dimethylpentane..	C₇H₁₆	liq.	1,147.85	11,455.8	20,607	1,063.69	10,615.9	19,096
2,3-Dimethylpentane..	C₇H₁₆	gas	1,157.28	11,549.9	20,776	1,073.12	10,710.0	19,265
2,3-Dimethylpentane..	C₇H₁₆	liq.	1,149.09	11,468.2	20,629	1,064.93	10,628.3	19,118
2,4-Dimethylpentane..	C₇H₁₆	gas	1,156.60	11,543.1	20,764	1,072.44	10,703.2	19,253
2,4-Dimethylpentane..	C₇H₁₆	liq.	1,148.73	11,464.6	20,623	1,064.57	10,624.7	19,112
3,3-Dimethylpentane..	C₇H₁₆	gas	1,156.72	11,544.4	20,766	1,072.57	10,704.5	19,255
3,3-Dimethylpentane..	C₇H₁₆	liq.	1,148.83	11,465.6	20,625	1,064.67	10,625.7	19,114
2,2,3-Trimethylbutane.	C₇H₁₆	gas	1,155.94	11,536.6	20,752	1,071.78	10,696.6	19,241
2,2,3-Trimethylbutane.	C₇H₁₆	liq.	1,148.27	11,460.0	20,614	1,064.11	10,620.1	19,104
n-Octane.......	C₈H₁₈	gas	1,317.45	11,533.9	20,747	1,222.77	10,705.0	19,256
n-Octane.......	C₈H₁₈	liq.	1,307.53	11,447.1	20,591	1,212.85	10,618.2	19,100
2-Methylheptane..	C₈H₁₈	gas	1,315.76	11,519.1	20,721	1,221.08	10,690.2	19,230
2-Methylheptane..	C₈H₁₈	liq.	1,306.28	11,436.1	20,572	1,211.60	10,607.2	19,080

Compound	Formula	State						
3-Methylheptane	C_8H_{18}	gas	1,316.44	11,525.1	20,732	1,221.76	10,696.2	19,240
3-Methylheptane	C_8H_{18}	liq.	1,306.92	11,441.7	20,582	1,212.24	10,612.8	19,091
4-Methylheptane	C_8H_{18}	gas	1,316.57	11,526.2	20,734	1,221.89	10,697.3	19,243
4-Methylheptane	C_8H_{18}	liq.	1,307.09	11,443.2	20,584	1,212.41	10,614.3	19,093
3-Ethylhexane	C_8H_{18}	gas	1,316.87	11,528.8	20,738	1,222.19	10,699.9	19,247
3-Ethylhexane	C_8H_{18}	liq.	1,307.59	11,445.8	20,589	1,212.71	10,616.9	19,098
2,2-Dimethylhexane	C_8H_{18}	gas	1,313.56	11,499.9	20,686	1,218.88	10,671.0	19,195
2,2-Dimethylhexane	C_8H_{18}	liq.	1,304.64	11,421.8	20,546	1,209.96	10,592.9	19,055
2,3-Dimethylhexane	C_8H_{18}	gas	1,316.13	11,522.4	20,727	1,221.45	10,693.5	19,236
2,3-Dimethylhexane	C_8H_{18}	liq.	1,306.86	11,441.2	20,581	1,212.18	10,612.3	19,090
2,4-Dimethylhexane	C_8H_{18}	gas	1,314.83	11,511.0	20,706	1,220.15	10,682.1	19,215
2,4-Dimethylhexane	C_8H_{18}	liq.	1,305.80	11,431.9	20,564	1,211.12	10,603.0	19,073
2,5-Dimethylhexane	C_8H_{18}	gas	1,314.05	11,504.2	20,694	1,219.37	10,675.3	19,203
2,5-Dimethylhexane	C_8H_{18}	liq.	1,305.00	11,424.9	20,551	1,210.32	10,596.0	19,060
3,3-Dimethylhexane	C_8H_{18}	gas	1,314.65	11,509.4	20,703	1,219.97	10,680.5	19,212
3,3-Dimethylhexane	C_8H_{18}	liq.	1,305.68	11,430.9	20,562	1,211.00	10,602.0	19,071
3,4-Dimethylhexane	C_8H_{18}	gas	1,316.36	11,524.4	20,730	1,221.68	10,695.5	19,239
3,4-Dimethylhexane	C_8H_{18}	liq.	1,307.04	11,442.8	20,583	1,212.36	10,613.9	19,092
2-Methyl-3-ethylpentane	C_8H_{18}	gas	1,316.79	11,523.1	20,737	1,222.11	10,699.2	19,246
2-Methyl-3-ethylpentane	C_8H_{18}	liq.	1,307.58	11,447.5	20,592	1,212.90	10,618.6	19,101
3-Methyl-3-ethylpentane	C_8H_{18}	gas	1,315.88	11,520.2	20,723	1,221.20	10,691.3	19,232
3-Methyl-3-ethylpentane	C_8H_{18}	liq.	1,306.80	11,440.7	20,580	1,212.12	10,611.8	19,089
2,2,3-Trimethylpentane	C_8H_{18}	gas	1,314.66	11,509.5	20,703	1,219.98	10,680.6	19,212
2,2,3-Trimethylpentane	C_8H_{18}	liq.	1,305.83	11,432.2	20,564	1,211.15	10,603.3	19,073
2,2,4-Trimethylpentane	C_8H_{18}	gas	1,313.69	11,501.0	20,688	1,219.01	10,672.1	19,197
2,2,4-Trimethylpentane	C_8H_{18}	liq.	1,305.29	11,427.5	20,556	1,210.61	10,598.6	19,065
2,3,3-Trimethylpentane	C_8H_{18}	gas	1,315.54	11,517.2	20,717	1,220.86	10,688.3	19,226
2,3,3-Trimethylpentane	C_8H_{18}	liq.	1,306.64	11,439.3	20,577	1,211.96	10,610.4	19,086
2,3,4-Trimethylpentane	C_8H_{18}	gas	1,315.29	11,515.0	20,713	1,220.61	10,686.1	19,222
2,3,4-Trimethylpentane	C_8H_{18}	liq.	1,306.28	11,436.1	20,572	1,211.60	10,607.2	19,080
2,2,3,3-Tetramethylbutane	C_8H_{18}	gas	1,313.27	11,497.3	20,682	1,218.59	10,668.4	19,191
2,2,3,3-Tetramethylbutane	C_8H_{18}	solid	1,303.03	11,407.7	20,520	1,208.35	10,578.8	19,029
n-Nonane	C_9H_{20}	gas	1,474.90	11,500.2	20,687	1,369.70	10,680.0	19,211
n-Nonane	C_9H_{20}	liq.	1,463.80	11,413.6	20,531	1,358.60	10,593.4	19,056
n-Decane	$C_{10}H_{22}$	gas	1,632.34	11,473.0	20,638	1,516.63	10,659.7	19,175
n-Decane	$C_{10}H_{22}$	liq.	1,620.06	11,386.7	20,483	1,504.35	10,573.4	19,020
n-Undecane	$C_{11}H_{24}$	gas	1,789.78	11,450.8	20,598	1,663.55	10,643.2	19,145
n-Undecane	$C_{11}H_{24}$	liq.	1,776.32	11,364.7	20,443	1,650.09	10,557.0	18,990
n-Dodecane	$C_{12}H_{26}$	gas	1,947.23	11,432.2	20,564	1,810.48	10,629.4	19,120
n-Dodecane	$C_{12}H_{26}$	liq.	1,932.59	11,346.5	20,410	1,795.84	10,543.4	18,966
n-Tridecane	$C_{13}H_{28}$	gas	2,104.67	11,416.5	20,536	1,957.40	10,617.6	19,099
n-Tridecane	$C_{13}H_{28}$	liq.	2,088.85	11,330.6	20,382	1,941.58	10,531.8	18,945
n-Tetradecane	$C_{14}H_{30}$	gas	2,262.11	11,402.9	20,512	2,104.32	10,607.5	19,081
n-Tetradecane	$C_{14}H_{30}$	liq.	2,245.11	11,317.2	20,358	2,087.30	10,521.8	18,927
n-Pentadecane	$C_{15}H_{32}$	gas	2,419.55	11,391.2	20,491	2,251.24	10,598.7	19,065
n-Pentadecane	$C_{15}H_{32}$	liq.	2,401.37	11,305.6	20,337	2,233.06	10,513.2	18,911
n-Hexadecane	$C_{16}H_{34}$	gas	2,577.00	11,380.9	20,472	2,398.17	10,591.1	19,052
n-Hexadecane	$C_{16}H_{34}$	liq.	2,557.64	11,295.4	20,318	2,378.81	10,505.6	18,898
n-Heptadecane	$C_{17}H_{36}$	gas	2,734.44	11,371.8	20,456	2,545.09	10,584.2	19,039
n-Heptadecane	$C_{17}H_{36}$	liq.	2,713.90	11,286.4	20,302	2,524.55	10,498.8	18,886
n-Octadecane	$C_{18}H_{38}$	gas	2,891.88	11,363.7	20,441	2,692.01	10,578.3	19,028
n-Octadecane	$C_{18}H_{38}$	liq.	2,870.16	11,278.4	20,288	2,670.29	10,493.0	18,875
n-Nonadecane	$C_{19}H_{40}$	gas	3,049.33	11,356.5	20,428	2,838.94	10,572.9	19,019
n-Nonadecane	$C_{19}H_{40}$	liq.	3,026.43	11,271.2	20,275	2,816.04	10,487.7	18,865
n-Eicosane	$C_{20}H_{42}$	gas	3,206.77	11,350.0	20,416	2,985.86	10,568.1	19,010
n-Eicosane	$C_{20}H_{42}$	liq.	3,182.69	11,264.7	20,263	2,961.78	10,482.8	18,857

TABLE 5-5 Heats of Combustion—(Continued)

Compound	Formula	State	Heat of combustion at 25°C and constant pressure, to form					
			H_2O (liq.) and CO_2 (gas)			H_2O (gas) and CO_2 (gas)		
			kcal/mol	cal/g	Btu/lb	kcal/mol	cal/g	Btu/lb
Alkyl benzenes								
Benzene	C_6H_6	gas	789.08	10,102.4	18,172	757.52	9,698.4	17,446
Benzene	C_6H_6	liq.	780.98	9,998.7	17,986	749.42	9,594.7	17,259
Methylbenzene (toluene)	C_7H_8	gas	943.58	10,241.4	18,422	901.50	9,784.7	17,601
Methylbenzene (toluene)	C_7H_8	liq.	934.50	10,142.8	18,245	892.42	9,686.1	17,424
Ethylbenzene	C_8H_{10}	gas	1,101.13	10,372.4	18,658	1,048.53	9,876.9	17,767
Ethylbenzene	C_8H_{10}	liq.	1,091.03	10,277.2	18,487	1,038.43	9,781.7	17,596
1,2-Dimethylbenzene (o-xylene)	C_8H_{10}	gas	1,098.54	10,348.0	18,614	1,045.94	9,852.5	17,723
1,2-Dimethylbenzene (o-xylene)	C_8H_{10}	liq.	1,088.16	10,250.2	18,438	1,035.56	9,754.7	17,547
1,3-Dimethylbenzene (m-xylene)	C_8H_{10}	gas	1,098.12	10,344.0	18,607	1,045.52	9,848.5	17,716
1,3-Dimethylbenzene (m-xylene)	C_8H_{10}	liq.	1,087.92	10,247.9	18,434	1,035.32	9,752.4	17,543
1,4-Dimethylbenzene (p-xylene)	C_8H_{10}	gas	1,098.29	10,345.6	18,610	1,045.69	9,850.1	17,719
1,4-Dimethylbenzene (p-xylene)	C_8H_{10}	liq.	1,088.16	10,250.2	18,438	1,035.56	9,754.7	17,547
n-Propylbenzene	C_9H_{12}	gas	1,258.24	10,469.1	18,832	1,195.12	9,943.9	17,887
n-Propylbenzene	C_9H_{12}	liq.	1,247.19	10,377.2	18,667	1,184.07	9,852.0	17,722
isopropylbenzene (cumene)	C_9H_{12}	gas	1,257.31	10,461.4	18,818	1,194.19	9,936.2	17,873
isopropylbenzene (cumene)	C_9H_{12}	liq.	1,246.52	10,371.6	18,657	1,183.40	9,846.4	17,712
1-Methyl-2-ethylbenzene	C_9H_{12}	gas	1,256.66	10,456.0	18,808	1,193.54	9,950.8	17,864
1-Methyl-2-ethylbenzene	C_9H_{12}	liq.	1,245.26	10,361.1	18,638	1,182.14	9,835.9	17,693
1-Methyl-3-ethylbenzene	C_9H_{12}	gas	1,255.92	10,449.8	18,797	1,192.80	9,924.6	17,853
1-Methyl-3-ethylbenzene	C_9H_{12}	liq.	1,244.71	10,356.5	18,630	1,181.59	9,831.3	17,685
1-Methyl-4-ethylbenzene	C_9H_{12}	gas	1,255.59	10,447.1	18,792	1,192.47	9,921.9	17,848
1-Methyl-4-ethylbenzene	C_9H_{12}	liq.	1,244.45	10,354.4	18,626	1,181.33	9,829.2	17,681
1,2,3-Trimethylbenzene (hemimellitene)	C_9H_{12}	gas	1,254.08	10,434.5	18,770	1,190.96	9,909.3	17,825
1,2,3-Trimethylbenzene (hemimellitene)	C_9H_{12}	liq.	1,242.36	10,337.0	18,594	1,179.24	9,811.8	17,650
1,2,4-Trimethylbenzene (pseudocumene)	C_9H_{12}	gas	1,253.04	10,425.8	18,754	1,189.92	9,900.7	17,809
1,2,4-Trimethylbenzene (pseudocumene)	C_9H_{12}	liq.	1,241.58	10,330.5	18,583	1,178.46	9,805.3	17,638
1,3,5-Trimethylbenzene (mesitylene)	C_9H_{12}	gas	1,252.53	10,421.6	18,747	1,189.41	9,896.4	17,802
1,3,5-Trimethylbenzene (mesitylene)	C_9H_{12}	liq.	1,241.19	10,327.2	18,577	1,178.07	9,802.1	17,632
n-Butylbenzene	$C_{10}H_{14}$	gas	1,415.44	10,546.3	18,971	1,341.80	9,997.6	17,984
n-Butylbenzene	$C_{10}H_{14}$	liq.	1,403.46	10,457.0	18,810	1,329.82	9,908.4	17,823
Alkyl cyclopentanes								
Cyclopentane	C_5H_{10}	gas	793.39	11,313.1	20,350	740.79	10,563.1	19,001
Cyclopentane	C_5H_{10}	liq.	786.54	11,215.5	20,175	733.94	10,465.4	18,825
Methylcyclopentane	C_6H_{12}	gas	948.72	11,273.4	20,279	885.60	10,523.3	18,950
Methylcyclopentane	C_6H_{12}	liq.	941.14	11,183.3	20,117	878.02	10,433.2	18,768
Ethylcyclopentane	C_7H_{14}	gas	1,106.21	11,266.9	20,267	1,032.57	10,516.9	18,918
Ethylcyclopentane	C_7H_{14}	liq.	1,097.50	11,178.2	20,108	1,023.86	10,428.2	18,758
n-Propylcyclopentane	C_8H_{16}	gas	1,263.56	11,260.9	20,256	1,179.40	10,510.8	18,907
n-Propylcyclopentane	C_8H_{16}	liq.	1,253.74	11,173.4	20,099	1,169.58	10,423.3	18,750
n-Butylcyclopentane	C_9H_{18}	gas	1,421.10	11,257.7	20,250	1,326.42	10,507.6	18,901
n-Butylcyclopentane	C_9H_{18}	liq.	1,410.10	11,170.5	20,094	1,315.42	10,420.5	18,745
Alkyl cyclohexanes								
Cyclohexane	C_6H_{12}	gas	944.79	11,226.7	20,195	881.67	10,476.7	18,846
Cyclohexane	C_6H_{12}	liq.	936.88	11,132.7	20,026	873.76	10,382.7	18,676
Methylcyclohexane	C_7H_{14}	gas	1,099.59	11,199.5	20,146	1,025.95	10,449.5	18,797
Methylcyclohexane	C_7H_{14}	liq.	1,091.13	11,113.3	19,991	1,017.49	10,363.3	18,642
Ethylcyclohexane	C_8H_{16}	gas	1,257.90	11,210.4	20,166	1,173.74	10,460.4	18,816
Ethylcyclohexane	C_8H_{16}	liq.	1,248.23	11,124.3	20,011	1,164.07	10,374.3	18,661
n-Propylcyclohexane	C_9H_{18}	gas	1,415.12	11,210.3	20,165	1,320.44	10,460.3	18,816

n-Propylcyclohexane	C_9H_{18}	liq.	1,404.34	11,124.9	20,012	1,309.66	10,374.9	18,663
n-Butylcyclohexane	$C_{10}H_{20}$	gas	1,572.74	11,213.0	20,170	1,467.54	10,463.0	18,821
n-Butylcyclohexane	$C_{10}H_{20}$	liq.	1,560.78	11,127.8	20,017	1,455.58	10,377.8	18,668

Monoolefins

Ethene (ethylene)	C_2H_4	gas	337.234	12,031.7	21,625	316.195	11,271.7	20,276
Propene (propylene)	C_3H_6	gas	491.987	11,692.3	21,032	460.428	10,942.3	19,683
1-Butene	C_4H_8	gas	649.757	11,581.3	20,833	607.679	10,831.3	19,484
cis-2-Butene	C_4H_8	gas	648.115	11,552.0	20,780	606.037	10,802.0	19,431
trans-2-Butene	C_4H_8	gas	647.072	11,533.4	20,747	604.994	10,783.4	19,397
2-Methylpropene (isobutene)	C_4H_8	gas	646.154	11,516.7	20,716	604.056	10,766.7	19,367
1-Pentene	C_5H_{10}	gas	806.85	11,505.1	20,696	754.25	10,755.1	19,346
cis-2-Pentene	C_5H_{10}	gas	805.34	11,433.5	20,657	752.74	10,733.5	19,308
trans-2-Pentene	C_5H_{10}	gas	804.26	11,468.1	20,629	751.66	10,718.1	19,280
2-Methyl-1-butene	C_5H_{10}	gas	803.17	11,432.6	20,601	750.57	10,702.6	19,252
3-Methyl-1-butene	C_5H_{10}	gas	804.93	11,477.7	20,646	752.33	10,727.7	19,297
2-Methyl-2-butene	C_5H_{10}	gas	801.68	11,431.3	20,563	749.08	10,681.3	19,214

Acetylenes

Ethyne (acetylene)	C_2H_2	gas	310.615	11,930.2	21,460	300.096	11,526.2	20,734
Propyne (methylacetylene)	C_3H_4	gas	463.109	11,559.8	20,794	442.070	11,034.6	19,849
1-Butyne (ethylacetylene)	C_4H_6	gas	620.86	11,478.7	20,648	589.302	10,895.2	19,599
2-Butyne (dimethylacetylene)	C_4H_6	gas	616.553	11,398.7	20,504	584.974	10,815.2	19,455
1-Pentyne	C_5H_8	gas	778.05	11,422.5	20,547	735.95	10,804.7	19,436
2-Pentyne	C_5H_8	gas	774.35	11,368.2	20,449	732.25	10,750.4	19,338
3-Methyl-1-butyne	C_5H_8	gas	776.15	11,394.6	20,497	734.05	10,776.8	19,386

TABLE 5-6   Heat of Combustion of Liquid Fuels

Fuel	cal/g	Btu/lb
Alcohol (denatured)	6.46	11,600
Crude oil (average)	10.7	19,200
Gas oil	10.67	19,200
Gasoline	11.53	20,800
Fuel oil (average)	10.50	18,900
Furnace oil	10.57	19,000
Kerosene	11.01	19,800

From C. D. Hodgman (ed.), "Handbook of Chemistry and Physics," The Chemical Rubber Co., Cleveland, 1942.

TABLE 5-7   Heat of Combustion of Gases

Gas	Formula	Molecular weight	Gross heating value (Btu/mol)	Net heating value (Btu/mol)
Acetylene	C_2H_2	26.02	562,000	543,000
Benzene	C_6H_6	78.05	1,413,000	1,356,000
Butane	C_4H_{10}	58.08	1,237,000	1,142,000
Butylene	C_4H_8	56.06	1,171,000	1,095,000
Carbon monoxide	CO	28.00	122,400	122,400
Ethane	C_2H_6	30.05	668,300	611,300
Ethylene	C_2H_4	28.03	622,400	584,400
Hydrogen	H_2	2.02	123,100	104,100
Methane	CH_4	16.03	384,000	346,000
Naphthalene	$C_{10}H_8$	128.06	2,219,000	2,143,000
Propane	C_3H_8	44.06	952,000	876,000
Propylene	C_3H_6	42.05	893,000	836,000
Toluene	C_7H_8	92.06	1,685,000	1,609,000
Xylene	C_8H_{10}	106.08	1,955,000	1,860,000

From C. D. Hodgman (ed.), "Handbook of Chemistry and Physics," The Chemical Rubber Co., Cleveland, 1942.

TABLE 5-8   Heat of Combustion of Gaseous Fuels

Fuel	Composition, %									Heating value (Btu/ft³)	
	CO_2	CO	C_2H_6	C_2H_4	H_2	CH_4	N_2	O_2	C_3H_8	Gross	Net
Blast-furnace gas	13.0	26.2			3.2		57.6			93	92
Blue-water gas	3.5	43.4			51.8		1.3			310	285
Carbureted water gas	1.5	33.9		12.8	35.2	14.8	1.8			578	529
Coal gas	1.1	9.9		6.6	47.0	34.0	2.3			634	560
Coke-oven gas	2.0	5.6	47.7	4.1	52.7	31.2	4.0	0.6	21.9	568	507
Natural gas*	0.1		47.7		29.0		1.4		21.9	1,655	1,494
Oil gas	2.8	10.6		2.7	53.5	27.0	3.4			516	261
Producer gas	5.7	22.0		0.4	10.5	2.6	58.8			136	128

From C. D. Hodgman (ed.), "Handbook of Chemistry and Physics," The Chemical Rubber Co., Cleveland, 1942.
* Average values from various sources.

TABLE 5-9 Heating Value of Coals

Location	Type	Btu/lb	Location	Type	Btu/lb
Alabama:			Ohio:		
Bibb, Belle Ellen	Bituminous	14,140	Columbia, New	Bituminous	12,730
Jefferson, Bessemer	Bituminous	14,620	Salisbury		
Shelby, Aldrich	Bituminous	13,650	Jefferson, Yellow	Bituminous	12,720
Alaska:			Creek		
Moose Creek	Bituminous	12,150	Oklahoma:		
Arkansas:			Coal, Lehigh	Bituminous	11,260
Hartford, Central	Bituminous	13,270	Latimer, Degnan	Bituminous	13,630
No. 10			Pittsburg, Ridgway	Bituminous	13,280
Huntington, No. 6	Semibituminous	13,700	Oregon:		
Central			Coos, Beaverhill	Subbituminous	9,030
Colorado:			Pennsylvania:		
Gunnison, Somerset	Bituminous	12,630	Armstrong, Mont-	Cannel	10,460
Weld, Erie	Subbituminous	9,520	gomeryville		
Illinois:			Armstrong, W. Kit-	Bituminous	13,040
Christian, Pana	Bituminous	10,860	tanning		
Franklin, Orient	Bituminous	12,160	Bedford, Hopewell	Bituminous	13,810
Williamson, Herrin	Bituminous	11,860	Cambria, Bakerton	Semibituminous	14,460
Indiana:			Cambria, Nanty Glo	Bituminous	14,380
Green, Jasonville	Bituminous	11,540	Cambria, Windber	Bituminous	14,620
Knox, South Bruce-	Bituminous	11,540	Jefferson, Punxsu-	Bituminous	13,860
ville			tawney		
Sullivan, Vandalia	Bituminous	11,420	Somerset, Seanor	Semibituminous	13,740
Iowa:			Rhode Island:		
Lucas, Chariton	Bituminous	10,240	Providence, Cranston	Anthracite	11,620
Polk, Altoona	Bituminous	10,240	Texas:		
Kansas:			Webb, Dolores	Cannel	11,070
Cherokee, Stone	Bituminous	13,080	Virginia:		
City			Montgomery,	Semibituminous	12,740
Crawford, Edison	Bituminous	12,500	Blacksburg		
Kentucky:			Pulaski, Guntan	Semianthracite	10,960
Christian, Manning-	Bituminous	11,680	Park		
ton			Tazewell, Pocahon-	Semibituminous	14,610
Webster	Bituminous	12,500	tas		
Maryland:			Wise, Josephine	Bituminous	13,270
Allegany, Frostburg	Semibituminous	13,430	Washington:		
Allegany, Ocean	Semibituminous	14,190	Kittitas, Ellensburg	Bituminous	11,010
Montana:			Thurston, Tono	Subbituminous	8,700
Carbon, Washoe	Subbituminous	10,550	West Virginia:		
Musselshell,	Bituminous	10,690	Brook, Collier	Bituminous	12,940
Roundup			Grant, Bismarck	Semibituminous	13,590
New Mexico:			Mineral, Emoryville	Bituminous	12,600
San Juan, Farming-	Bituminous	11,630	Ohio, Elm Grove	Bituminous	13,200
ton			Wyoming:		
North Dakota:			Lincoln, Elkol	Subbituminous	10,080
Ward, Burlington	Lignite	6,010	Lincoln, Green	Bituminous	13,310
Williams, Wheelock	Lignite	5,990	River		

From C. D. Hodgman (ed.), "Handbook of Chemistry and Physics," The Chemical Rubber Co., Cleveland, 1942.

At this point, a number of energy relations can be summarized as follows:

$$\text{Total energy} = \text{internal energy} + \text{external energy} \qquad (5\text{-}26)$$
$$H \quad = \quad E \quad + \quad PV$$
$$\text{Total energy} = \text{free energy} + \text{isothermally unavailable energy} \qquad (5\text{-}27)$$
$$H \quad = \quad F \quad + \quad TS$$
$$\text{Internal energy} = \text{work content} + \text{isothermally unavailable energy} \qquad (5\text{-}28)$$
$$E \quad = \quad A \quad + \quad TS$$

In Eq. (5-26), the product of pressure and volume PV is used as a typical work term. These equations can be combined and recombined in a number of ways, but a most important equation derives from Eq. (5-27) showing the changes in the pertinent quantities for any given reaction:

$$\Delta F = \Delta H - T \,\Delta S \qquad (5\text{-}29)$$

Free Energy of Reaction

The *free energy of reaction*, or more correctly, the free-energy change of a reaction, can be calculated in a manner analogous to the heat of reaction. For example, as in the case of heat of reaction, the free-energy change of a given reaction is the sum of the free-energy

TABLE 5-10   Heat of Combustion
of Various Substances

Substance	kcal/g of substance
Asphalt	9.532
Butter	9.200
Carbon, crystal to CO_2	7.859
Casein	5.860
Charcoal to CO_2	8.080
	8.137
Dynamite, 75%	1.290
Egg white	5.700
Egg yolk	8.100
Fats, animal, mean	9.500
Graphite	7.901
Gunpowder	0.720–0.750
Hemoglobin	5.900
Hydrogen, to liquid	33.900
Hydrogen, to liquid	34.500
Hydrogen, to gas	29.150
Oil, cottonseed	9.500
Oil, lard	9.200–9.400
Oil, olive	9.328–9.442
Oil, paraffin	9.800
Oil, rape	9.489
Oil, sperm	10.000
Paraffin	10.340
Pitch	8.400
Wood, beech	4.774
Wood, birch	4.771
Wood, oak	4.620
Wood, pine	5.085

From C. D. Hodgman (ed.), "Handbook of Chemistry and Physics," The Chemical Rubber Co., Cleveland, 1942.

changes of all the component reactions, the overall value being the same regardless of the path taken by intermediate reactions. Continuing the analogy to heats of reaction, there are *free-energies-of-formation* tables, such as Table 5-1, which are used in the same way as, and usually combined with, heat-of-formation tables. It should be noted that the values usually given by free-energy tables are in terms of energy $\Delta F°$, since the reactants and products are referred to their natural state at 25°C and 1 atm. Frequently, however, the superscript is omitted.

Example 5-12   Find the free-energy change for the reaction $H_2O(g) + Cl_2(g) \longrightarrow 2HCl(g) + \frac{1}{2}O_2(g)$.

solution   From Table 5-12, the free energies of formation of $H_2O(g)$ and $HCl(g)$ are -54.5 and -22.7 kcal, respectively.

$$
\begin{array}{lll}
a & H_2(g) + \frac{1}{2}O_2(g) \longrightarrow H_2O(g) & \Delta F = -54.5 \text{ kcal} \\
b & \frac{1}{2}H_2(g) + \frac{1}{2}Cl_2(g) \longrightarrow HCl(g) & \Delta F = -22.7 \\
c = 2b & H_2(g) + Cl_2(g) \longrightarrow 2HCl(g) & \Delta F = -45.4 \\
d = c - a & Cl_2(g) - \frac{1}{2}O_2(g) \longrightarrow 2HCl(g) - H_2O(g) & \Delta F = 9.1
\end{array}
$$

Transposing negative terms:

$$H_2O(g) + Cl_2(g) \longrightarrow 2HCl(g) + \frac{1}{2}O_2(g) \qquad \Delta F = 9.1 \text{ kcal}$$

TABLE 5-11   Heat of Combustion of Common Foodstuffs

Food	Fuel value (cal/100 g)	Food	Fuel value (cal/100 g)
ANIMAL FOODS		ANIMAL FOODS (continued)	
Cheese:		Meats: (continued)	
American, pale	453	Veal:	
Camembert	282	Chops	174
Cheddar	473	Liver	127
Cottage	112	Roast, leg	148
Cream, full	430	Milk:	
Neufchâtel	337	Buttermilk, plain	36.4
Roquefort	375	Condensed, sweetened	335
Swiss	443	unsweetened	172
Eggs:		Cream, 18%	201
Plain	159	40%	381
Scrambled	210	Human	62
Whites, boiled	55.1	Ice cream, plain	208
Yolks, boiled	376	Skimmed	37.5
Fish:		Whole	71.6
Caviar	300	Poultry:	
Clams	47.4	Chicken, broiled	111
Codfish, salt, cooked	90.4	Liver	141
Crab, canned	81.6	Roast	200
Haddock, fresh, raw	73.9	Duck	243
Halibut, fresh	125	Goose	403
Smoked	225	Quail	154
Herring, fresh	146	Squab	274
Smoked	299	Turkey, liver	144
Lobster, canned	86.0	Roast	285
Mackerel, fresh	80.5	VEGETABLE FOODS	
Oysters, raw	50.7	Bread:	
Salmon, canned	202	Biscuits	381
Fresh	132	Brown	231
Sardines, canned	278	Corn	266
Shrimp, canned	115	Gluten	256
Trout, fresh	169	Graham	267
Tuna, canned, oil	287	Griddle cakes, plain	200
Fresh	95	Muffins, cornmeal	273
Whitefish	154	Graham	252
Gelatin	376	Rye	260
Meats:		Rye and wheat	262
Beef:		Toast, plain	313
Chipped, dried	185	Waffles, plain	385
Corned, cooked	308	Wheat, cracked	250
Kidneys	115	White	266
Liver	133	Whole	251
Miscellaneous cuts, fat-free	119	Zwieback	434
Roast, loin	255	Cake:	
Rump, lean	213	Angel food	300
Steak, round	184	Cream puffs	397
Sirloin	249	Devil's food	371
Lamb:		Doughnuts	523.8
Chops, broiled	367	Fruit	388
Leg, roast	198	Gingerbread	368
Miscellaneous:		Sponge	396
Bologna	241	Cereals:	
Frankfurters	258	Barley, pearl	364
Mutton:		Bran, unwashed	359
Chop	270	Corn flakes	383
Leg, roast	313	Cornmeal	381
Pork:		Cream of wheat, raw	366
Bacon, smoked	646	Farina, light	371
Chop, loin, lean	260	Grapenuts	371
Ham, smoked, lean	274	Macaroni, cooked	91.5
Liver	136	Noodles	367
Roast	235	Oats, rolled, cooked	62.8
		Popcorn	413

TABLE 5-11   Heat of Combustion of Common Foodstuffs—(Continued)

Food	Fuel value (cal/ 100 g)	Food	Fuel value (cal/ 100 g)
VEGETABLE FOODS (continued)		VEGETABLE FOODS (continued)	
Cereals: (continued)		Fruits: (continued)	
Puffed rice	370	Grape juice	100
Puffed wheat	374	Grapes, Concord	99.2
Rice, cooked	112	Malaga	96
Spaghetti	366	Lemon juice	39.7
Tapioca	364	Lemons	45.2
Wheatena, raw	342	Loganberries	53
Wheat-germ meal	374	Muskmelon	40.8
Wheat, shredded	365	Nectarines	67.2
Chocolate, sweet	505	Orange juice	43
Unsweetened	631	Oranges	52.9
Cocoa	511	Peaches, canned	48.5
Cookies:		Fresh	41.9
Gingersnaps	418	Pears, canned	78.3
Hermits	400	Fresh	65.0
Macaroons	435	Pineapple, canned	158
Molasses	421	Fresh	44.1
Oatmeal	500	Plums, fresh	87.1
Sugar	423	Prunes, stewed	94.8
Vanilla wafers	451	Raisins	354
Crackers:		Raspberries, fresh	68.3
Boston	416	Rhubarb	23.1
Butter	427	Cooked (with sugar)	272
Educator	484	Strawberries, cooked	101
Graham	431	Raw	39.7
Oatmeal	434	Tangerines	125
Oyster	433	Watermelon	30.9
Pretzels	375	Nuts:	
Saltines	442	Almonds	679
Soda	424	Brazil	720
Flour:		Butternuts	698
Arrowroot	400	Chestnuts, dried	413
Barley	362	Coconut, prepared	689
Buckwheat	357	Filberts	725
Corn flour	363	Hickor	737
Starch	369	Peanuts, shelled	564
Gluten (high)	367	Pecans	757
Graham	368	Pine	627
Rice	370	Pistachios	660
Rye	359	Walnuts, black	685
Wheat, patent	362	English	728
Whole	369	Pie:	
Fruits:		Apple	280
Apple, baked	125	Custard	183
Raw	63.9	Lemon meringue	262
Sauce	161	Mince	308
Apricots, canned	75.0	Pumpkin	200
Fresh	59.5	Squash	185
Bananas	101	Pudding:	
Blackberries, canned	254	Apple tapioca	122
Blueberries, canned	60.6	Bread	156
Cantaloupe	40	Cottage	329
Cherries, canned	91.5	Rice	182
Fresh	80.5	Tapioca, cream	159
Cranberries, raw	47.4	Vegetables:	
Cranberry sauce	250	Artichoke, French	62.8
Currants, fresh	58.4	Asparagus, canned	18.7
Dates	356	Fresh	26.2
Figs, dried	325	Beans, kidney, canned	106
Fresh	83.8	Lima, dried	358
Grapefruit	47	Fresh	131

TABLE 5-11   Heat of Combustion of Common Foodstuffs—(Continued)

Food	Fuel value (cal/ 100 g)	Food	Fuel value (cal/ 100 g)
VEGETABLE FOODS (continued)		MISCELLANEOUS (continued)	
Vegetables: (continued)		Candy: (continued)	
Navy, dried	354	Gumdrops	350
String, fresh	43.0	Marshmallows	346
Beet greens, cooked	54.0	Mints, after-dinner	350
Beets, cooked	40.8	Molasses candy	447
Brussel sprouts	57.7	Nougat, chocolate	505
Cabbage	31.9	Nut bar	437
Carrots	46.3	Peanut brittle	537
Cauliflower	31.0	Fats:	
Celery	18.7	Butter	795
Chard, Swiss	25.0	Cod-liver oil	900
Corn, canned	100	Corn oil	900
Fresh on cob	104	Crisco	900
Cucumbers	17.6	Lard	930
Eggplant	28.7	Oleomargarine	777
Lentils, dried	357	Olive oil	900
Lettuce, green-leaf	19.8	Peanut oil	900
Mushrooms	46.3	Jelly, cherry	921
Okra	38.6	Salad dressing:	
Olives, green	309	Cream	360
Ripe	266	French	548
Onions	49.6	Mayonnaise	752
Parsnips	66.1	Thousand Island	413
Peas, canned	56.2	Soup:	
Dried	365	Asparagus, cream of	62.8
Green	103	Bouillon	11.0
Potatoes, boiled	97.0	Celery, cream of	55.1
Sweet	126	Chicken	60.6
Pumpkin	26.5	Chowder, clam	43.0
Radishes	29.8	Corn	91
Rutabagas	41.9	Consomme	12.1
Sauerkraut	27.6	Oyster stew	105
Spinach, cooked	57.3	Potato	85
Squash	47.4	Tomato	40.8
Tomatoes, canned	23.1	Vegetable	14.3
Fresh	23.1	Sugar:	
Turnips	40.8	Brown	389
Watercress	22.7	Granulated	410
Yeast, compressed	138	Maple	340
MISCELLANEOUS		Syrup:	
Candy:		Corn	340
Butterscotch	789	Honey	335
Chocolate creams	350	Maple	293
Fudge	400	Molasses	284
Sweet milk	500		

From C. D. Hodgman (ed.), "Handbook of Chemistry and Physics," The Chemical Rubber Co., Cleveland, 1942.

Example 5-13   What is the free energy of vaporization of water?
solution   From Table 5-12, the free energies of formation of $H_2O(g)$ and $H_2O(l)$ are -54.5 and -56.6 kcal, respectively.

$$a \qquad \tfrac{1}{2}H_2(g) + \tfrac{1}{2}O_2(g) \longrightarrow H_2O(g) \qquad \Delta F = -54.5 \text{ kcal}$$

$$b \qquad \tfrac{1}{2}H_2(g) + \tfrac{1}{2}O_2(g) \longrightarrow H_2O(l) \qquad \Delta F = -56.6$$

$$c = a - b \qquad 0 \longrightarrow H_2O(g) - H_2O(l) \qquad \Delta F = 2.1$$

Transposing $H_2O(l)$:

$$H_2O(l) \longrightarrow H_2O(g) \qquad \Delta F = 2.1 \text{ kcal}$$

Equilibrium

One of the most important applications of the concept of free energy is to establish whether or not a reaction can proceed spontaneously. As a general rule, those reactions for which ΔF is negative can proceed spontaneously. Among those reactions for which ΔF is negative, the larger the numerical value of ΔF, the more readily will the reaction proceed. The fundamental relationship between the equilibrium constant of a reaction K_e and the free-energy change of the reaction is given by the equation

$$\Delta F^\circ = -RT \ln K_e \tag{5-30}$$

where R is the universal gas constant and T is the absolute temperature. To convert this equation from natural logarithms to common logarithms (to the base 10), we simply multiply by 2.303:

$$\Delta F^\circ = -2.303 \, RT \log K \tag{5-30a}$$

Example 5-14 What is the equilibrium constant at 25° for the reaction between chlorine and hydrogen bromide $Cl_2 + 2HBr \longrightarrow Br_2 + 2HCl$?
solution From the data given in Table 5-1, ΔF for the reaction can be calculated:

$$\Delta F = \Delta F_{\text{products}} - \Delta F_{\text{reactants}}$$

$$\Delta F = [0 + 2(-22.8)] - [0 + 2(-12.7)] = -20.2 \text{ kcal}$$

$$\Delta F = -2.303 \, RT \log K_e$$

$$-20.2 \text{ kcal} = -2.303 \times \frac{0.001987 \text{ kcal}}{K} \times 298 \text{ K} \times \log K_e$$

$$\log K_e = 14.813 = 0.813 + 14$$
$$K_e = 6.50 \times 10^{14}$$

Thus, for the given reaction,

$$K_e = \frac{p_{Br_2} \times p^2_{HCl}}{p_{Cl_2} \times p^2_{HBr}} = 6.50 \times 10^{14}$$

If the available energy of a reaction is obtained as electricity, as in an electrochemical reaction, we would expect the free energy of the reaction to be related to the electromotive force, or the potential for doing work. Such is actually the case, and the relationship is given by the equation

$$\Delta G = -nFE \tag{5-31}$$

where n is the valence change involved in the reaction, F the faraday, or 96,500 C, and E the electromotive force of the cell in volts. In this equation, the free-energy change is designated as ΔG to avoid confusion with the use of F as the faraday. (As a matter of fact, G is displacing F as the symbol for free energy, after Gibbs, an early thermodynamicist.) It should be noted that the product of F and E is in joules, in which case it is necessary to divide by 4.183 to convert into calories. More conveniently a value of 23.1 kcal/V can be used as the value of F.

Since electrochemical measurements can be made with considerable accuracy, many thermodynamic values, free energies of reaction in particular, are derived from electrochemical measurements.

Example 5-15 What is the free energy of formation of silver chloride, given the single-electrode potentials of the silver electrode and chlorine electrode in chloride electrolyte?

$$Ag(s) + Cl^- \longrightarrow AgCl(s) \qquad E^0 = -0.225 \text{ V}$$
$$\tfrac{1}{2}Cl_2(g) + e \longrightarrow Cl^- \qquad E^0 = 1.3587 \text{ V}$$

(The superscript ^{0} is used to indicate that the value of E is given for the reactants and products in their standard states.)
solution The sum of the two half cells above yields

$$Ag(s) + \tfrac{1}{2}Cl_2(g) \longrightarrow AgCl(s) \qquad E- = 1.1362 \text{ V}$$
$$\Delta G^0 = -nFE$$

Since the formation of AgCl involves the passage of 1 faraday, $n = 1$.

$$\Delta G^\circ = -(1)\left(23.1 \, \frac{\text{kcal}}{\text{V}}\right)(1.1326 \text{ V}) = -26.2 \text{ kcal}$$

5-3   KINETICS

Rates of Reaction

While free energies of reaction and equilibrium constants indicate whether or not a reaction should take place, they provide practically no information as to the *rate* at which the reaction will take place, or how the concentration of a reactant or product changes with time.   When a reaction has a negative free-energy change, it can be assumed that the reaction will take place spontaneously, but the rate may be so slow that for all intents and purposes there is no reaction.   In other cases, the reaction may proceed with such rapidity as to produce explosive violence.   It is the province of *chemical kinetics* to analyze how rapidly reactions occur, and to try to predict and control reaction rates.   The factors that influence the rate at which a reaction will take place are (1) the nature of the reactants, (2) the temperature at which the reaction is proceeding, (3) the concentrations of the reactants, and (4) the presence of catalysts or inhibitors.

Specific materials can differ widely in the rate at which they undergo chemical change.   Sodium, for example, reacts vigorously with water at room temperature with the evolution of considerable heat:

$$2Na + 2H_2O \longrightarrow 2NaOH + H_2 \tag{5-32}$$

but iron reacts only very slowly:

$$Fe + 2H_2O \longrightarrow Fe(OH)_2 + H_2 \tag{5-33}$$

The fact that Eq. (5-32) evolves a good deal of heat raises the temperature of the reaction so that the reaction rate has a tendency to increase as the reaction proceeds.   Materials that readily enter into rapid chemical reactions are referred to as *reactive*, while those which do not readily enter into chemical reactions, or do so at generally negligible rates, are called *inert*.   It is interesting to note that the periodic chart of the elements (Table 3-5) also reveals something about the reactivity of the elements.   Reactivity increases for elements from the center to both the left and right.   Thus, Group IV elements are generally less reactive than those of Group III, which in turn are generally less reactive than those of Group II, and so on.   In the other direction, Group IV elements are generally less reactive than those of Group V, which in turn are generally less reactive than those in Group VI, and so on.   Group VIII is an exception in that it consists of completely inert elements.   In addition, within any one period, the higher (lighter) elements tend to be more reactive than the lower ones.   Thus fluorine is more reactive than chlorine, which in turn is more reactive than bromine.

The effect of temperature on reaction rate is very pronounced.   As a broad rule, the reaction rate for any given reaction will double for each rise of $10°C$ in the reaction temperature.   It is for this reason that the reaction of Eq. (5-32) tends to become violent.   Since most chemical reactions are brought about by molecular impacts, anything that increases the number and severity of the impacts tends to increase the reaction rate.   In the case of a gas, for which the distribution of molecular velocities is well known, the mechanism which results in higher reaction rates for higher reaction temperatures is illustrated in Fig. 5-3.

The concentration of the reactants can also be expected to have an influence on reaction rates, since the number and severity of the molecular impacts increase as the concentration of the reactants increases.   Again, as a broad rule, the reaction rate doubles each time the concentration of the reactants doubles.   It is for this reason that commercial practice is to speed up reactions by carrying them out under pressure or in concentrated solutions.   In the other direction, reactions that tend to be violent are moderated by carrying them out in dilute solutions.

Catalysts and Activation Energy

In recent years, the function of catalysts and inhibitors has been better understood.   A *catalyst* is a material which accelerates a chemical reaction but does not alter the reactants or products, nor is it itself consumed by the reaction.   Catalysts may act in different ways.   They may, for example, enter directly into intermediate reactions, to be returned to their original composition as a result of subsequent reactions that result in the desired products.   By whichever route they function, catalysts tend to reduce the activation energy required to initiate a reaction.   The *activation energy* is the minimum energy the reacting substances must have for the reaction to occur.   This can be illustrated by a ball rolling down an incline

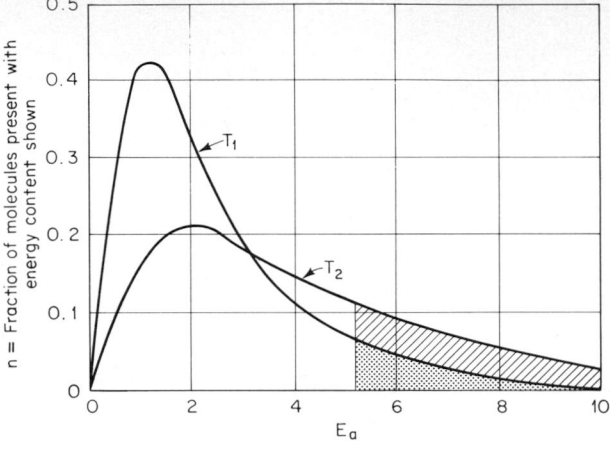

Fig. 5-3 Influence of temperature on reaction rate. Given that T_2 is higher than T_1, the fraction of molecules at T_2 having an energy greater than that required to initiate a chemical reaction, E_a, is considerably larger than at T_1. The total of both shaded areas represents the number of molecules having an energy E_a or greater at temperature T_2, whereas only the lower shaded area represents the number of such molecules at temperature T_1.

as in Fig. 5-4 in which the ball will not spontaneously roll down the hill until it receives sufficient energy to overcome the depression. It is ultimately the function of a catalyst to reduce the activation energy required to initiate a reaction, or to raise temporarily the energy content of the reactants until the activation energy is achieved, as illustrated in Fig. 5-5. *Inhibitors* are substances that slow down or moderate a reaction and function in a manner directly opposite that of a catalyst. Inhibitors are thus occasionally referred to as negative catalysts. Frequently, the activation barrier can be overcome without catalysts

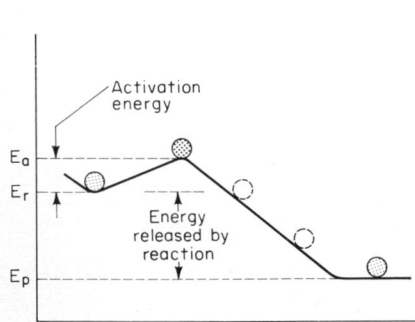

Fig. 5-4 Activation energy. Although the energy content of the products E_p is less than that of the reactants E_r, the reaction does not occur until enough energy is supplied to the reactants to bring it to E_a, the activated state, after which the reaction can proceed spontaneously in a manner analogous to the ball rolling downhill in the diagram.

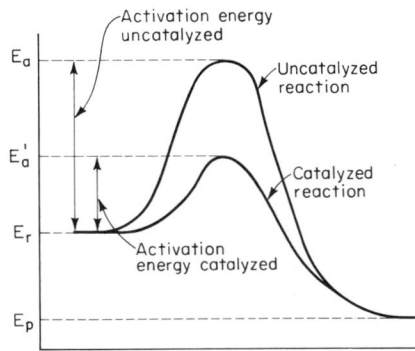

Fig. 5-5 Mechanism of catalyst action. The energy required to raise the reactants to an activated state without a catalyst $E_a - E_r$ is greater than when a catalyst is present, $E_a{}' - E_r$. Stating this another way, the activation energy required to bring about the reaction is less when a catalyst is present. Thus more molecules will have sufficient energy to enter into the reaction, and the reaction will proceed faster than if the reaction were not catalyzed.

through alternate paths involving *sequential* or chain reactions, and the formation of *intermediate* compounds:

$$A + B \longrightarrow AB \tag{5-34}$$

High activation energy, AB formed slowly

$$A + C \longrightarrow AC \tag{5-35}$$

Low activation energy, AC formed rapidly

$$AC + B \longrightarrow AB + C \tag{5-36}$$

Low activation energy, AB formed rapidly

Notice that while C is used up in Eq. (5-35), it is recovered as a product in Eq. (5-36). If Eqs. (5-35) and (5-36) occurred simultaneously in the same vessel, C would be a catalyst. If Eqs. (5-35) and (5-36) are sequential reactions, AC is referred to as an intermediate compound.

Rate Constants and Reaction Order

It has been found experimentally that reaction rates are proportional to the concentrations of the reactants raised to an appropriate power, or expressing this mathematically for the reaction $mA + nB \longrightarrow pC$,

$$\text{Rate} = k(\text{Conc. } A)^m (\text{Conc. } B)^n \tag{5-37}$$

In Eq. (5-37), k is a proportionality constant and is known as the *rate constant* for the particular reaction. The powers m and n, to which the concentrations of the reactants are raised, determine the *order* of the reaction. Thus if m in Eq. (5-37) is 1, the reaction is said to be a first-order reaction with respect to A. If n is 3, the reaction is said to be a third-order reaction with respect to B, and so on. The sum of m and n determines the overall reaction order. Thus, if $m = 1$ and $n = 3$, the reaction is a fourth-order reaction.

Example 5-16 What is the order of each of the following reactions:

$$Br_2 \longrightarrow 2Br$$
$$H_2 + I_2 \longrightarrow 2HI$$
$$2NO + O_2 \longrightarrow 2NO_2$$

solution For $Br_2 \longrightarrow 2Br$,

$$\text{Rate} = k[Br_2]$$

and m in Eq. (5-37) equals 1. This is therefore a first-order reaction.
For $H_2 + I_2 \longrightarrow 2HI$,

$$\text{Rate} = k[H_2][I_2]$$

and m and n in Eq. (5-37) both equal 1, and the reaction is a second-order reaction.
For $2NO + O_2 \longrightarrow 2NO_2$,

$$\text{Rate} = k[NO]^2[O_2]$$

Here $m = 2$ and $n = 1$ in Eq. (5-37), and the reaction is a third-order reaction.

A *first-order reaction* is one in which the rate of reaction is directly proportional to the concentration of one of the reacting substances:

$$\frac{\Delta c}{\Delta t} = -kc \tag{5-38}$$

in which c is the concentration of the reacting substance at time t, and k is the reaction-rate constant. (The rate constant is a number divided by time, i.e., has units such as 1/s). The minus sign indicates that the concentration of the reacting substance decreases with time. Another convenient equation for first-order reactions is

$$k = \frac{2.303}{t} \log \frac{c_0}{c} \tag{5-39}$$

in which c_0 is the original concentration of the reacting substance and c the concentration at time t.

Example 5-17 The decomposition of nitrogen pentoxide proceeds according to the reaction $N_2O_5 \longrightarrow 2NO_2 + \frac{1}{2}O_2$. If the reaction is 90 percent complete at the end of 12.0 min, what is its reaction-rate constant?

solution Since the reaction rate $= k[N_2O_5]$, this is a first-order reaction. At the start of the decomposition $[N_2O_5] = 100$, and 12.0 min later $[N_2O_5] = 10$. Thus, from Eq. (5-39),

$$k = \frac{2.303}{12.0 \text{ min}} \times \log \frac{100}{10}$$

$$= 0.192 \log 10/\text{min}$$

Since $\log 10 = 1$, $k = 0.192/\text{min}$.

Frequently, it is convenient to give reaction rates in terms of the *period of half-life*, or the time required for half of a given amount of material to have reacted. The relationship between half-life and reaction rate derives directly from Eq. (5-39):

$$k = \frac{2.303}{t_{1/2}} \log \frac{1}{1/2} \tag{5-40}$$

from which

$$t_{1/2} = \frac{0.693}{k} \tag{5-41}$$

Example 5-18 What is the half-life of nitrogen pentoxide?

solution From Example 5-17, $k = 0.192/\text{min}$ for the decomposition of N_2O_5. Thus, from Eq. (5-41),

$$t_{1/2} = \frac{0.693}{0.192} \text{ min} = 3.6 \text{ min}$$

The concept of half-life is widely used to describe the rate of decay of radioactive materials. Since radioactive decay can be measured with great accuracy, the half-lives of many radioactive materials are known to a high degree of precision. Table 5-12 gives the half-lives for a number of radioactive materials.

For *second-order reactions,* equations similar to Eqs. (5-38) and (5-39) can be used. Thus for a reaction of the type $A + B \longrightarrow AB$:

$$\frac{\Delta c_A}{\Delta t} = \frac{\Delta c_B}{\Delta t} = -kc_A c_B \tag{5-42}$$

TABLE 5-12   Half-Life Series of the Radioactive Elements

Element	Half-life
Thorium	1.39×10^{10} years
Uranium I	4.5×10^9
Uranium II	2.5×10^5
Ionium	8.0×10^4
Radium	$1,590$
Radium D	22
Mesothorium I	6.7
Radiothorium	1.9
Polonium	140 days
Uranium X_1	24.5
Radium E	5.0
Radon	3.82
Thorium X	3.64
Thorium B	10.16 hours
Mesothorium II	6.13
Thorium C	60.5 minutes
Radium B	26.8
Radium C	19.7
Radium A	3.05
Uranium X_2	68 seconds
Thoron	54.5
Thorium A	0.16
Radium C^1	1.5×10^{-4}
Thorium C^1	3×10^{-7}

where c_A and c_B are the concentrations of A and B, respectively, t is time, and k is the reaction-rate constant. Also:

$$k = \frac{2.303}{t(a - b)} \log \frac{b(a - x)}{a(b - x)} \qquad (5\text{-}43)$$

for which a and b are the initial concentrations of A and B, and x denotes the amount of A or B reacting in the interval of time t. (It should be noted that x is the same for both A and B, since the generalized equation calls for equimolar quantities of A and B to react.)

Example 5-19 The hydrolysis of ethyl acetate proceeds in aqueous solution according to the reactions

$$CH_3COOC_2H_5 + NaOH \longrightarrow CH_3COONa + C_2H_5OH$$

If the initial concentrations of NaOH and $CH_3COOC_2H_5$ were 0.00980 and 0.00486 mol/l, respectively, and the concentration of each reaction product is 0.00116 mol/l in 320 s, what is the reaction rate constant?

solution For use in Eq. (5-43), $a = 0.00980 - 0.00116 = 0.00864$; $b = 0.00486 - 0.00116 = 0.00370$. $\log \frac{b}{a} \frac{a - x}{b - x} = \log \frac{0.00486}{0.00980} \times \frac{0.00864}{0.00370} = \log 1.158 = 0.0640$. Substituting these values, and $t = 320$ s, in Eq. (5-43) yields

$$k = \frac{2.303}{320(0.00980 - 0.00486)} \times 0.0640$$

$$k = \frac{2.303}{320(0.00494)} \times 0.0640$$

$$= \frac{2.303}{1.58} \times 0.0640$$

$$= 1.46 \times 0.0640 = 0.093$$

For second-order reactions, k has dimensions of liters/mol-s.

Third-order reactions, of the general type $A + B + C \longrightarrow ABC$ are derived from the basic relationship:

$$\frac{\Delta C_A}{\Delta t} = \frac{\Delta c_B}{t} = \frac{\Delta c_C}{t} = -k c_A c_B c_C \qquad (5\text{-}44)$$

When A, B, and C are present in equivalent amounts,

$$\frac{\Delta x}{\Delta t} = k(a - x)^3 = k(b - x)^3 = k(c - x)^3 \qquad (5\text{-}45)$$

which results in the relationship

$$k = \frac{1}{t} \frac{x(2a - x)}{2a^2(a - x)^2} = \frac{1}{t} \frac{x(2b - x)}{2b^2(b - x)^2} = \frac{1}{t} \frac{x(2c - x)}{2c^2(c - x)^2} \qquad (5\text{-}46)$$

[in which the notations correspond to those used in Eq. (5-43)].

It should be noted that the third-order equations can be used for reactions of the type $2A + B \longrightarrow A_2B$ simply by making $C = A$ in the above general third-order reaction. Rate equations for higher than third-order reactions are handled in the same manner but tend to become unwieldy and are usually solved by computerized programs.

It should also be noted that there are many reactions whose rate is unaffected by concentration or temperature but is affected by some other limiting factor such as the absorption of radiation or the availability of reactive surfaces. The rate of such reactions, called *zero-order* reactions, is given by equations

$$\frac{\Delta c_A}{\Delta t} = k \qquad (5\text{-}47)$$

and

$$k = \frac{a - x}{t} \qquad (5\text{-}48)$$

Chapter 6

Organic Chemistry

6-1 CLASSIFICATION OF COMPOUNDS

Definition of Organic Chemistry

Defined in terms of its chief characteristic, organic chemistry is the chemistry of compounds of carbon. The term *organic* was introduced some 150 years ago to denote the fact that it was the chemistry of substances formed by living organisms. The great triumphs of synthetic chemistry, however, have shown that compounds of carbon can be produced from inorganic "nonliving" matter. The classic discovery of Wohler in 1828 that urea—an organic compound—could be made by heating ammonium cyanate—an inorganic compound—was the first recognition of the true nature of what we call organic chemistry today.

Another very important feature of organic chemistry is the ability of carbon atoms to form stable chains and rings by bonding to each other. Although atoms of several other elements show similar tendencies, none can match the vast number of combinations that exist in the family of carbon compounds.

Classification of Organic Compounds

The basis for the classification of *organic compounds* is the carbon skeletal structure to which other atoms and groups are attached. Hydrogen is the element most commonly associated with carbon in this manner, oxygen and nitrogen being other important elements in such compounds.

Each carbon atom has four electrons available for bonding with other atoms. When one atom of carbon shares an electron with another atom which, in turn, shares one of its electrons with the first atom, a *single bond* exists between the two atoms. Graphically, this is often represented as shown in Fig. 6-1a. When each atom contributes two electrons to the bond, we have a *double bond*, as shown in Fig. 6-1b. A *triple bond*, shown in Fig. 6-1c, occurs when three electrons are contributed by each atom.

Large numbers of compound combinations are possible with these three types of carbon-to-carbon bonds, especially when, in addition to hydrogen, elements such as oxygen, nitrogen, chlorine, sulfur, phosphorus, and bromine are considered. The single bond is referred to as a *saturated bond*. A double or triple bond is called an *unsaturated bond*. Saturation is often associated with lack of reactivity, while the unsaturated bond is often found to be highly reactive or ready to enter into other combinations.

Perhaps the most common way to classify organic compounds is by the general differences in structure and properties which are related to the carbon skeleton being either open carbon chains or closed rings, as illustrated in Fig. 6-2. Chains may be short, consisting of only several carbon atoms, or very long, consisting of hundreds or even thousands of carbon atoms. Compounds based on a carbon-chain structure are referred to as *aliphatic* compounds. Individual carbon rings, on the other hand, are limited to only several atoms each, although many rings may be combined in various ways to form molecules that also contain hundreds or thousands of carbon atoms. Among ring compounds, it is important to distinguish between the *cyclic* compounds, in which predominantly single bonds are involved, and those of the benzene or *aromatic* type in which we have a *resonant* bond, or a bond that oscillates between a single and double bond with its neighbors. These are illustrated in Fig. 6-3.

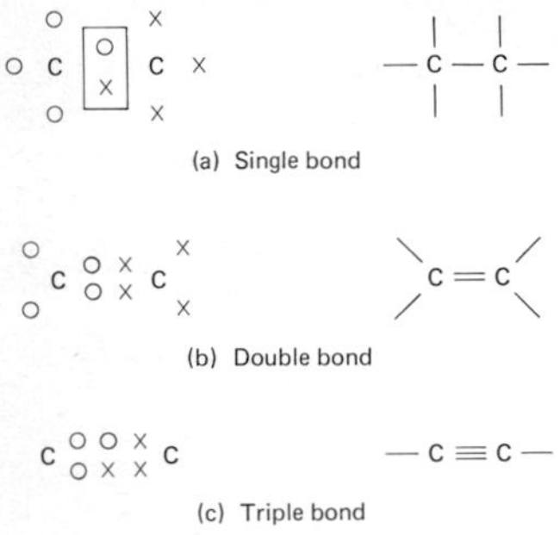

(a) Single bond

(b) Double bond

(c) Triple bond

Fig. 6-1 Representations of single, double, and triple bonds.

Classification of the vast number of organic compounds is further simplified by the fact that relatively few *functional groups* of atoms or reactive centers exist in the molecules. The carboxyl group, —COOH, the alcohol group, —OH, or the primary amine group, —NH$_2$, are typical of such reactive centers. For example, the presence of the carboxyl group helps classify the compound as an acid, the —OH group as an alcohol, etc.

Another helpful feature in the classification of organic compounds is the existence of substituent groups, sometimes called *radicals*, such as methyl, ethyl, and phenyl. In general, these are referred to as *alkyl* groups when they are derived from aliphatic hydrocarbons (i.e., methane), or *aryl* groups when they are derived from aromatic hydrocarbons (i.e., benzene). The following sections and the corresponding tables contain numerous examples of these

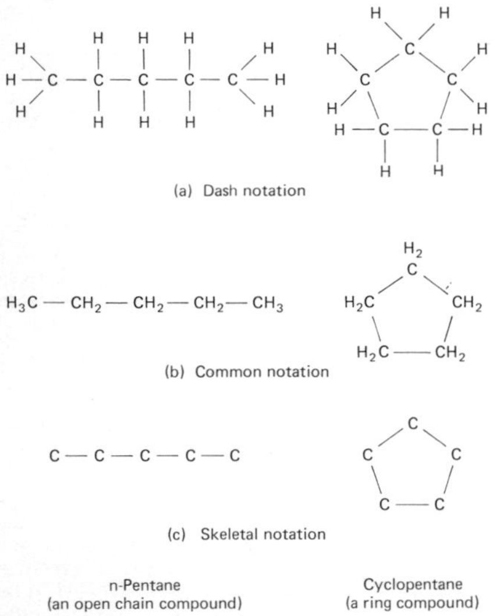

(a) Dash notation

(b) Common notation

(c) Skeletal notation

n-Pentane
(an open chain compound)

Cyclopentane
(a ring compound)

Benzene
(an aromatic ring compound)

Cyclohexane
(a cycloalkane)

Fig. 6-2 Representation of open carbon chains and carbon rings.

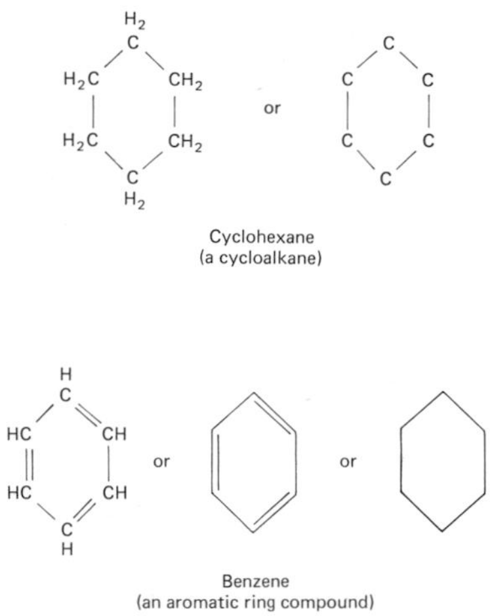

Fig. 6-3 Cyclic and aromatic rings.

simplifying classification features of organic compounds. However, as with all simplifications, important exceptions may exist.

Names of Organic Compounds

As in other branches of chemistry, common or "historic" names of organic compounds, such as wood alcohol for methanol or grain alcohol for ethanol, are often used. Such names are hard to use in a systematic way and therefore have only limited use. In general, the systematic classification of organic compounds may follow either a common-name method or the International Union of Pure and Applied Chemistry (IUPAC) convention, frequently referred to as the Geneva system. Figure 6-4 illustrates the use of these two systems.

It will be seen that it is necessary to know the common names of the hydrocarbons and the substituent groups for either system. The main features of the Geneva system are the use of the longest continuous chain of carbon atoms to establish the name of the compound, using the name of the corresponding hydrocarbon as the basic name of the compound. In both compounds shown in Fig. 6-4, the three-carbon propane chain is the basis for the names in the Geneva system, even though the common name of one of these compounds reflects the fact that its molecule contains four carbon atoms. Table 6-1 lists the first eight straight-chain saturated hydrocarbons (alkanes), which are used as the basis for the naming systems. The most common substituent groups are listed in Table 6-2.

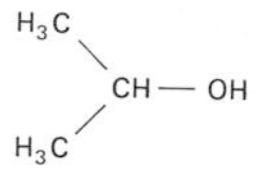

Isomerism

It will be seen from the material presented up to this point, including Tables 6-1 and 6-2, that a molecular formula such as C_4H_{10} may not be good enough to identify a compound because the atoms in a compound represented by this formula may be bonded to each other in different combinations, each such combination leading to distinctly different chemical and physical properties. This important characteristic of organic compounds is known as *isomerism,* and compounds which have the same kind and number of atoms, but in different structural arrangements, are called *isomers.* Thus, in the alkane series, we have no isomerism in methane, ethane, and propane, but in butane two isomers are possible, *n*-butane and isobutane. (Note that only the normal-chain hydrocarbons are shown in Table 6-1.) Isomerism in radicals can start with the propyl radical. Isomerism in compounds and radicals is illustrated in Fig. 6-5. It will be appreciated that the number of isomers of a compound may be very large. For instance, 18 octanes are theoretically possible, all of which have been prepared.

(a) Common name: Isopropyl alcohol
Geneva system: 2-Propanol

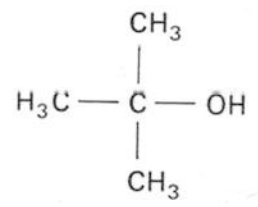

(b) Common name: tert(iary)-Butyl alcohol
Geneva system: 2-Methyl-2 propanol

Fig. 6-4 Comparison of nomenclature systems for organic compounds.

n-Butane

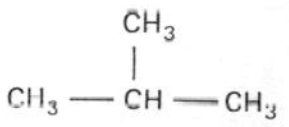

iso-Butane

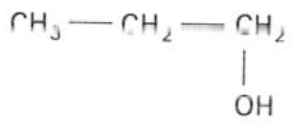

n-Propyl alcohol

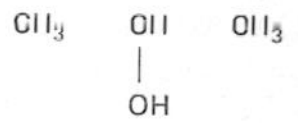

iso-Propyl alcohol

Fig. 6-5 Isomerism in compounds and radicals.

6-2 MAJOR CLASSES OF ORGANIC COMPOUNDS

Alkanes (Paraffins)

Also known as *paraffins, alkanes* are hydrocarbons in which the carbon atoms are connected only by single bonds. Table 6-3 lists the representative members of this class of compounds. Note that the class of alkanes includes gases, liquids, and solids, depending on the molecular weight of the compound. The alkane series represents a good example of the concept of *homology,* meaning that the melting points and boiling points of the compounds in the series increase in a regular manner as the molecular weight increases by the addition of $-CH_2$ units, as shown in Table 6-3. The generalized formula for the alkanes is C_nH_{2n+2}, indicating that each member of the series differs from its neighbors by one $-CH_2$ unit (having a molecular weight of 14).

Many of the alkanes occur naturally in petroleum oil. Isolation of the various members of the series by refining crude oil through distillation is an important method for the preparation of these substances. Thus crude oil, which is a complex mixture of hydrocarbons with carbon-chain lengths up to 35 to 40 carbon atoms, can yield gaseous compounds, such as ethane, propane, and butane, which are of great value as fuels, as well as liquids, such as gasoline and kerosene. Mineral oils, lubricants, asphalts, and tars are also obtained, being still longer carbon-chain members of the alkanes series. *Cracking* is an important industrial method for preparation of the lower members of the series from the higher ones. This process breaks down the heavier hydrocarbons by passing them through tubes containing suitable catalysts at temperatures of about 500 to 700°C.

TABLE 6-1   Carbon-Chain Names and Formulas

Formula	Structure	Simplified structure	Abbreviated structure	Name
CH_4	H—C—H (with H above and H below)	CH_4	C	Methane
C_2H_6	H—C—C—H	CH_3CH_3	C—C	Ethane
C_3H_8	H—C—C—C—H	$CH_3CH_2CH_3$	C—C—C	Propane
C_4H_{10}	H—C—C—C—C—H	$CH_3CH_2CH_2CH_3$	C—C—C—C	Butane
C_5H_{12}	H—C—C—C—C—C—H	$CH_3CH_2CH_2CH_2CH_3$	C—C—C—C—C	Pentane
C_6H_{14}	H—C—C—C—C—C—C—H	$CH_3CH_2CH_2CH_2CH_2CH_3$	C—C—C—C—C—C	Hexane
C_7H_{16}	H—C—C—C—C—C—C—C—H	$CH_3CH_2CH_2CH_2CH_2CH_2CH_3$	C—C—C—C—C—C—C	Septane
C_8H_{18}	H—C—C—C—C—C—C—C—C—H	$CH_3CH_2CH_2CH_2CH_2CH_2CH_2CH_3$	C—C—C—C—C—C—C—C	Octane

TABLE 6-2   Substituent Groups Used in Naming Compounds

CH_3—	Methyl
CH_3CH_2—	Ethyl
$CH_3CH_2CH_2$—	n-Propyl (normal propyl)
CH_3 CH— CH_3	Isopropyl
$CH_3CH_2CH_2CH_2$—	n-Butyl (normal butyl)
CH_3 CH—CH_2— CH_3	Isobutyl
CH_3 CH_3—C— CH_3	tert-Butyl (tertiary butyl)

TABLE 6-3   Alkane Compounds

Name	Formula, C_nH_{2n+2}	mp (°C)	bp (°C)	Sp. gr.
		GASES		
Methane	CH_4	−182.6	−161.7	0.4240*
Ethane	C_2H_6	−172.0	−88.6	0.5462*
Propane	C_3H_8	−187.1	−42.2	0.5824*
n-Butane	C_4H_{10}	−135.0	−0.5	0.5788*
		LIQUIDS		
n Pentane	C_5H_{12}	−129.7	36.1	0.6264
n-Hexane	C_6H_{14}	−94.0	68.7	0.6594
n-Heptane	C_7H_{16}	−90.5	98.4	0.6837
n-Octane	C_8H_{18}	−56.8	125.6	0.7028
n-Nonane	C_9H_{20}	−53.7	150.7	0.7179
n-Decane	$C_{10}H_{22}$	−29.7	174.0	0.7298
n-Undecane	$C_{11}H_{24}$	−25.6	195.8	0.7404
n-Dodecane	$C_{12}H_{26}$	−9.6	216.3	0.7493
n Tridecane	$C_{13}H_{28}$	−6	230	0.7568
n-Tetradecane	$C_{14}H_{30}$	5.5	251	0.7638
n-Pentadecane	$C_{15}H_{32}$	10	268	0.7688
n-Hexadecane	$C_{16}H_{34}$	18.1	280	0.7749
n Heptadecane	$C_{17}H_{36}$	22.0	303	0.7767
		SOLIDS		
n-Octadecane	$C_{18}H_{38}$	28.0	308	0.7767
n-Nonadecane	$C_{19}H_{40}$	32	330	0.7776
n-Eicosane	$C_{20}H_{42}$	36.4		0.7777
n-Heneicosane	$C_{21}H_{44}$	40.4		0.7782
n-Docosane	$C_{22}H_{46}$	44.4		0.7778
n-Tricosane	$C_{23}H_{48}$	47.4		0.7797
n-Tetracosane	$C_{24}H_{50}$	51.1		0.7786
n-Pentacosane	$C_{25}H_{52}$	53.3		
n-Triacontane	$C_{30}H_{62}$	66		
n-Pentatriacontane	$C_{35}H_{72}$	74.6		0.7814
n-Tetracontane	$C_{40}H_{82}$	81		
n-Pentacontane	$C_{50}H_{102}$	92		0.7940
n-Hexacontane	$C_{60}H_{122}$	99		
n-Dohexacontane	$C_{62}H_{126}$	101		
n-Tetrahexacontane	$C_{64}H_{130}$	102		
n-Heptacontane	$C_{70}H_{142}$	105		

* For liquid at boiling point.

Synthetic methods are also available for producing alkanes. The *Fischer-Tropsch process* is an important method for preparing mixtures of the lower homologs for use as fuels. Using cobalt or nickel catalysts at elevated temperatures, such as 200°C, carbon monoxide and hydrogen react to produce hydrocarbons. The generalized equation for this process is

$$nCO + (2n + 1)H_2 \longrightarrow C_nH_{2n+2} + nH_2O \tag{6-1}$$

Thus, to show the formation of heptane, for which $n = 7$, the simplified equation would be

$$7CO + 15H_2 \longrightarrow C_7H_{16} + 7H_2O \tag{6-2}$$

Another synthetic method for preparing alkanes involves a reduction of alkyl halides with hydrogen according to the simplified equation

$$R\text{-}Cl + H_2 \xrightarrow{\text{catalyst}} RH + HCl \tag{6-3}$$

where R stands for an alkyl group such as $-CH_3$ or $-C_2H_5$. Other methods are available, such as the *Wurtz synthesis:*

$$2R\text{-}I + 2Na \longrightarrow R\text{-}R + 2NaI \tag{6-4}$$

alkyl
iodide

alkane

the *Grignard reaction:*

$$R—I \xrightarrow{Mg} R—MgI \xrightarrow{H_2O} RH + Mg\underset{\diagdown OH}{\overset{I \diagup}{}} \tag{6-5}$$

Grignard
reagent

the decarboxylation reaction:

$$RCOOH + 2NaOH \xrightarrow{fuse} RH + Na_2CO_3 + H_2O \tag{6-6}$$

acid

and the *Kolbe synthesis,* shown here for the case of ethane from sodium acetate:

$$2CH_3COONa + H_2O \xrightarrow{electrolysis} \underbrace{C_2H_6 + 2CO_2}_{} + \underbrace{2NaOH + H_2}_{} \tag{6-7}$$

sodium   anodic   cathodic
acetate   products   products

Alkanes are relatively stable compounds, and their usefulness as raw materials for the production of other chemicals is rather limited in contrast to their enormous importance as fuels.

In organic reactions, one or more side reactions may take place simultaneously. This has the effect of reducing the amount of desired product from a particular reaction. The *yield* is defined as

$$\text{Yield (percent)} = \frac{\text{weight of product obtained}}{\text{weight of product theoretically possible}} \times 100 \tag{6-8}$$

Example 6-1   A decarboxylization of 100 g of acetic acid (CH_3COOH) was found to have produced 21.8 g of methane (CH_4). What yield was obtained?
solution   From Eq. (6-6).

$$\overset{60}{CH_3COOH} + 2NaOH \longrightarrow \overset{16}{CH_4} + Na_2CO_3 + H_2O$$

Thus, 100 g of CH_3COOH should produce

$$\frac{100}{60} \times 16 = 26.7 \text{ g of } CH_4$$

$$\text{Yield} = \frac{21.8}{26.7} \times 100 = 81.6 \text{ percent}$$

Oxidation of hydrocarbons is their most important reaction, as witnessed by the widespread use of gasoline, kerosene, and other fuels in the generation of power. A generalized equation for the combustion of alkanes is

$$C_nH_{2n+2} + \left[\frac{3n+1}{2}\right]O_2 \longrightarrow nCO_2 + (n+1)H_2O \tag{6-9}$$

Example 6-2   What is the equation representing the combustion of octane (C_8H_{18})?
solution   From Eq. (6-9) for $n = 8$,

$$C_8H_{18} + 12\tfrac{1}{2}\, O_2 \longrightarrow 8CO_2 + 9H_2O$$

Since whole integers should be used, both sides of this equation should be multiplied by 2:

$$2C_8H_{18} + 25O_2 \longrightarrow 16CO_2 + 18H_2O$$

Table 6-4 lists a group of compounds known as *cycloalkanes* or *cycloparaffins.* The properties of these substances are, in general, similar to those of the corresponding straight-chain alkanes. They differ mainly in the slightly higher boiling points and densities than their open-chain counterparts. The importance of cycloalkanes in the chemistry of many naturally occurring compounds has been brought to light in recent years by studies relating to their structural geometry.

TABLE 6-4   Cycloalkanes

Name	Formula C_nH_{2n}	Structure	mp (°C)	bp (°C)	Sp. gr.
Cyclopropane	C_3H_6	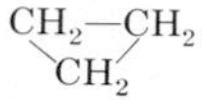 $\begin{array}{c}CH_2-CH_2\\ \diagdown CH_2 \diagup\end{array}$	−127	−32.9	0.688*
Cyclobutane	C_4H_8	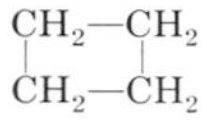 $\begin{array}{c}CH_2-CH_2\\ \mid\qquad\mid\\ CH_2-CH_2\end{array}$	−80	11	0.7038*
Cyclopentane	C_5H_{10}	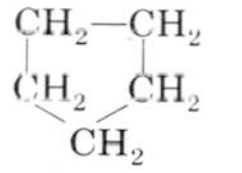 $\begin{array}{c}CH_2-CH_2\\ CH_2\quad CH_2\\ CH_2\end{array}$	−94	49.5	0.7460
Cyclohexane	C_6H_{12}	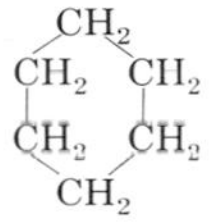 $\begin{array}{c}CH_2\\ CH_2\quad CH_2\\ CH_2\quad CH_2\\ CH_2\end{array}$	6.4	80.8	0.7781
Cycloheptane	C_7H_{14}	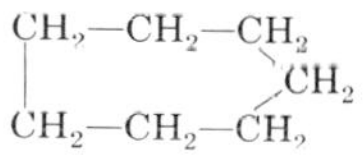 $\begin{array}{c}CH_2-CH_2-CH_2\\ \qquad\qquad CH_2\\ CH_2-CH_2-CH_2\end{array}$	−13	117	0.8100
Cyclooctane	C_8H_{16}	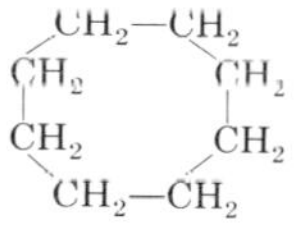 $\begin{array}{c}CH_2-CH_2\\ CH_2\qquad CH_2\\ CH_2\qquad CH_2\\ CH_2-CH_2\end{array}$	14	147	0.8304

*For liquid at boiling point.

Alkenes (Olefins)

Hydrocarbons containing at least one carbon-carbon double bond in their structure are called *alkenes* or *olefins*. Chain compounds of this type containing one double bond have a general formula C_nH_{2n}, where n equals the number of carbons in the chain. Table 6-5 presents typical alkenes and some of their physical properties. The alkenes do not differ greatly in physical properties from the corresponding alkanes, but their chemical reactivity is much greater.

In addition to their occurrence in petroleum and in products resulting from the cracking of petroleum, alkenes can be prepared synthetically by several routes:

TABLE 6-5   Alkenes

Name	Formula C_nH_{2n}	Carbon structure	mp (°C)	bp (°C)	Sp. gr.
Ethylene	C_2H_4	C=C	−169.4	−102.4	0.6100*
Propylene	C_3H_6	C·C=C	−185	−47.7	0.6104*
Butene-1	C_4H_8	C·C·C=C		−6.5	0.6255*
Butene-2 (*cis*[1])	C_4H_8	C·C=C·C	−139.3	−3.7	
Isobutylene	C_4H_8	$\begin{array}{c}C\cdot C{=}C\\ \mid\\ C\end{array}$	−140.7	−6.6	0.6266*
Pentene-1	C_5H_{10}	C·C·C·C=C		30.1	0.6429
2-Methylbutene-1	C_5H_{10}	$\begin{array}{c}C\cdot C\cdot C{=}C\\ \mid\\ C\end{array}$		31	0.6501
3-Methylbutene-1	C_5H_{10}	$\begin{array}{c}C\cdot C\cdot C{=}C\\ \mid\\ C\end{array}$		20.1	0.6340
Hexene-1	C_6H_{12}	C·C·C·C·C=C	−138	63.5	0.6747
Heptene-1	C_7H_{14}	C·C·C·C·C·C=C	−119	93.1	0.6976
Octene-1	C_8H_{16}	C·C·C·C·C·C·C=C	(−104)	122.5	0.7159

*For liquid at boiling point.

Dehydration of alcohols at high temperatures using sulfuric acid

$$\underset{\substack{\text{ethyl}\\\text{alcohol}}}{C_2H_5OH} \xrightarrow[\text{heat}]{H_2SO_4} \underset{\text{ethene}}{H_2C{=}CH_2} + H_2O \tag{6-10}$$

Catalytic processes involving dehydrogenation of alkanes

$$\underset{\text{ethane}}{H_3C{-}CH_3} \xrightarrow[\text{heat}]{\text{catalyst}} H_2C{=}CH_2 + H_2 \tag{6-11}$$

Removal of a hydrogen halide, such as hydrogen bromide, from an alkyl bromide

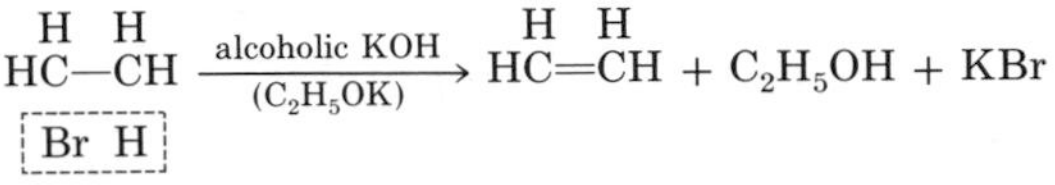

$$\underset{\substack{\text{ethyl}\\\text{bromide}}}{\overset{\text{H H}}{\underset{\boxed{\text{Br H}}}{HC{-}CH}}} \xrightarrow[(C_2H_5OK)]{\text{alcoholic KOH}} \overset{\text{H H}}{HC{=}CH} + C_2H_5OH + KBr \tag{6-12}$$

The most important reactions of alkenes are those in which additions to the double bond take place. This is in contrast to the alkanes, in which substitution-type reactions predominate. Typical reactions are:

Halogenation

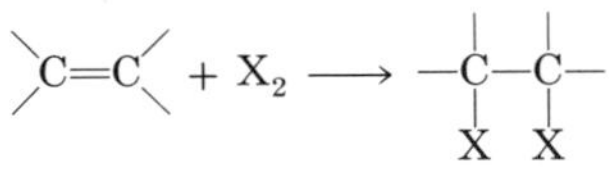

$$\overset{}{\underset{\text{alkene}}{>C{=}C<}} + X_2 \longrightarrow \underset{\substack{\text{dihalide}\\\text{ethane}}}{-\overset{|}{\underset{X}{C}}{-}\overset{|}{\underset{X}{C}}-} \tag{6-13}$$

Hydrogen halide additions

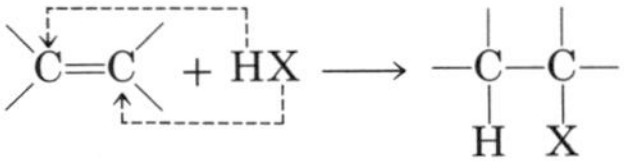

$$>C{=}C< + HX \longrightarrow -\overset{|}{\underset{H}{C}}{-}\overset{|}{\underset{X}{C}}- \tag{6-14}$$

Hydrogenation

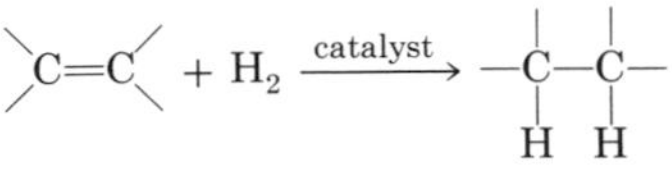

$$>C{=}C< + H_2 \xrightarrow{\text{catalyst}} -\overset{|}{\underset{H}{C}}{-}\overset{|}{\underset{H}{C}}- \tag{6-15}$$

A great variety of products may result from the many possible oxidation paths for alkenes. Ethylene glycol, for example, may be made by the controlled, partial oxidation of ethylene.

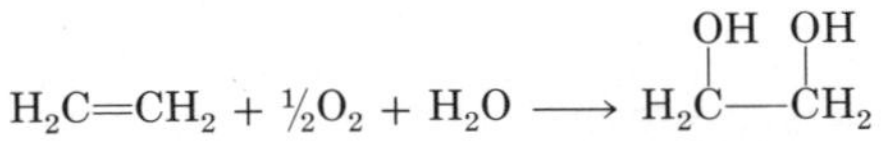

$$\underset{\text{ethylene}}{H_2C{=}CH_2} + \tfrac{1}{2}O_2 + H_2O \longrightarrow \underset{\substack{\text{ethylene}\\\text{glycol}}}{H_2\overset{\overset{\text{OH}}{|}}{C}{-}CH_2\overset{\overset{\text{OH}}{|}}{}} \tag{6-16}$$

  The ability of the double bond to "open up" is of great importance in polymerization processes. Some of the most important synthetic materials of commerce depend on this property of the olefinic bond. Polymeric materials, such as polybutadiene, polystyrene, polyisoprene (synthetic "rubbers"), and epoxy resins, are good examples.
  In respect to their commercial and scientific significance, ethylene and propylene, two alkenes, must be considered among the most useful building blocks of chemistry.

Alkynes

The *alkynes*, or *acetylenic hydrocarbons*, are characterized by the presence of a triple bond in their structure and have the general formula C_nH_{2n-2}. Acetylene gas is the simplest

TABLE 6-6 Alkynes

Name	Formula C_nH_{2n-2}	Carbon skeleton	mp (°C)	bp (°C)	Sp. gr.
Acetylene	C_2H_2	C≡C	−81.8	−83.4	0.6179*
Methylacetylene	C_3H_4	C·C≡C	−101.5	−23.3	0.6714*
Ethylacetylene	C_4H_6	C·C·C≡C	−122.5	8.6	0.6682*
Dimethylacetylene	C_4H_6	C·C≡C·C	−28	27.2	0.6937
Pentyne-1	C_5H_8	C·C·C·C≡C	−98	39.7	0.695
Pentyne-2	C_5H_8	C·C·C≡C·C	−101	55.5	0.7127
3-Methylbutyne-1	C_5H_8	C·C·C≡C C		28	0.665
Hexyne-1	C_6H_{10}	C·C·C·C·C≡C	−124	71	0.7195
Hexyne-2	C_6H_{10}	C·C·C·C≡C·C	−92	84	0.7305
Hexyne-3	C_6H_{10}	C·C·C≡C·C·C C	−51	82	0.7255
3,3-Dimethylbutyne-1	C_6H_{10}	C·C·C≡C C	−81	38	0.6686
Octadecyne-1	$C_{18}H_{34}$	C·(C)$_{15}$·C≡C	28	180 at 15 mm	0.8025

*For liquid at boiling point.

member of this class. Table 6-6 lists the lower members of this series with their more important physical properties. In general, the triple-bond compounds boil at temperatures somewhat higher than the corresponding olefins.

Acetylene is obtained by reacting calcium carbide with water. Calcium carbide is prepared by fusion of coke with lime.

$$CaO + 3C \xrightarrow[\text{furnace}]{\text{electric}} CaC_2 + CO$$

lime   coke   calcium
carbide
+
$$H_2O$$
↓
$$HC≡CH + Ca(OH)_2 \qquad (6\text{-}17)$$
acetylene

Other alkynes may be prepared by bromination of the corresponding alkenes, followed by reaction with alcoholic potassium hydroxide

$$\overset{H\ \ H}{HC{=}CH} + Br_2 \longrightarrow \overset{H\ \ H}{\underset{Br\ Br}{HC{-}CH}} \xrightarrow{C_2H_5OK} -C≡C- \qquad (6\text{-}18)$$

Of all the alkynes, acetylene is the most important member of the class in view of the fact that it is a key starting substance in many synthesis processes. As in the case of alkenes, addition-type reactions are important in the alkyne class. Products which may be industrially derived from the alkynes include acetaldehyde, acetic acid, vinyl acetylene, neoprene rubber, and other related polymers.

Alcohols

In general, alcohols are compounds which may be thought of as resulting from the replacement of a hydrogen atom in a hydrocarbon by a hydroxyl group. The formula R·OH is often used to represent this generalization. If the carbon atom to which the hydroxyl group is attached is connected with one alkyl group, the compound is classified as *primary alcohol*. If two alkyls are attached to this carbon, the alcohol is referred to as a *secondary alcohol*, or a *tertiary alcohol* when three alkyls are attached to the OH-bearing carbon. Table 6-7 gives examples of the three types for monohydric alcohols, or alcohols containing a single hydroxyl group. Polyhydric alcohols are those in which more than one hydroxyl group is present in a molecule. Table 6-8 lists the more important monohydric alcohols. Table 6-9 lists several typical polyhydric alcohols.

TABLE 6-7 Classification of an Alcohol (as Primary, Secondary, or Tertiary)

Butyl alcohols

Formula	Type	Common name	Carbinol name
$CH_3CH_2CH_2CH_2OH$	Primary	n-Butyl alcohol	n-Propylcarbinol
CH_3 $\quad$ CHCH$_2$OH CH_3	Primary	Isobutyl alcohol	Isopropylcarbinol
$CH_3CH_2CHCH_3$ $\qquad$ OH	Secondary	sec-Butyl alcohol	Methylethylcarbinol
CH_3 CH_3—COH CH_3	Tertiary	$tert$-Butyl alcohol	Trimethylcarbinol

TABLE 6-8 Monohydric Alcohols

Name	Formula	mp (°C)	bp (°C)	Sp. gr.
Methyl alcohol	CH_3OH	−97	64.7	0.792
Ethyl alcohol	CH_3CH_2OH	−114	78.3	0.789
n-Propyl alcohol	n-C_3H_7OH	−126	97.2	0.804
Isopropyl alcohol	i-C_3H_7OH	−88.5	82.3	0.786
Allyl alcohol	$CH_2\!=\!CHCH_2OH$	−129	97.0	0.855
n-Butyl alcohol	n-C_4H_9OH	−90	117.7	0.810
Isobutyl alcohol	$(CH_3)_2CHCH_2OH$	−108	107.9	0.802
sec-Butyl alcohol	$CH_3CH_2CH(OH)CH_3$		99.5	0.808
$tert$-Butyl alcohol	$(CH_3)_3COH$	25	82.5	0.789
n-Amyl alcohol	n-$C_5H_{11}OH$	−78.5	138.0	0.817
Isoamyl alcohol	$(CH_3)_2CHCH_2CH_2OH$	−117	131.5	0.812
$tert$-Amyl alcohol	$CH_3CH_2C(OH)(CH_3)_2$	−12	101.8	0.809
n-Hexyl alcohol	n-$C_6H_{13}OH$	−52	155.8	0.820
Cyclohexanol	$C_6H_{11}OH$	−24	161.5	0.962
n-Octyl alcohol	n-$C_8H_{17}OH$	−16	194.0	0.827
Capryl alcohol (octanol-2)	n-$C_6H_{13}CH(OH)CH_3$	−39	179.0	0.819
n-Decyl alcohol	n-$C_{10}H_{21}OH$	6	232.9	0.829
Lauryl alcohol	n-$C_{12}H_{25}OH$	24	259	0.831
Myristyl alcohol	n-$C_{14}H_{29}OH$	38	167/15 mm	0.824
Cetyl alcohol	n-$C_{16}H_{33}OH$	49	189/15 mm	0.798
Stearyl alcohol	n-$C_{18}H_{37}OH$	58.5	210.5/15 mm	0.812
Benzyl alcohol	$C_6H_5CH_2OH$	−15.3	205.4	1.046

TABLE 6-9 Polyhydric Alcohols

Name	Formula	mp (°C)	bp (°C)	Sp. gr.
Ethylene glycol	$HOCH_2CH_2OH$	−13	197	1.116
Diethylene glycol	$HOC_2H_4OC_2H_4OH$	−8	245	1.118
Triethylene glycol	$HOC_2H_4OC_2H_4OC_2H_4OH$	−7	287	1.125
Propylene glycol	$CH_3CHOHCH_2OH$		188	1.038
Dipropylene glycol	$(CH_3CHOHCH_2)_2O$		232	1.025
1,3-Butylene glycol	$C_4H_8(OH)_2$		208	1.006
Ethylhexanediol	$C_3H_7CH(OH)CH(C_2H_5)CH_2OH$	−40	244	0.942
Mannitol	$CH_2OH(CHOH)_4CH_2OH$	166		
Sorbitol	$CH_2OH(CHOH)_4CH_2OH$	110		

Note that mannitol and sorbitol can also be classified as carbohydrates. Furthermore, they have the same formula and are therefore isomers. However, in this case, each atom is attached to the same corresponding atoms in the molecule, but they are not arranged in the same planes in both molecules. Such isomers are called *stereo isomers*.

It is useful to consider the alcohols as the first oxidation product of the hydrocarbons as illustrated in Fig. 6-6.

The names of the alcohols are derived from their hydrocarbon radical; thus CH_3OH is methyl alcohol, C_2H_5OH is ethyl alcohol, etc.; but often the names are derived by adding the suffix "ol" to the name of the hydrocarbon; CH_3OH is thus called methanol, C_2H_5OH is ethanol, etc. The term carbinol is sometimes used in naming secondary and tertiary alcohols as derivatives of carbinol—a commonly used name for methanol. This system is illustrated in Fig. 6-7.

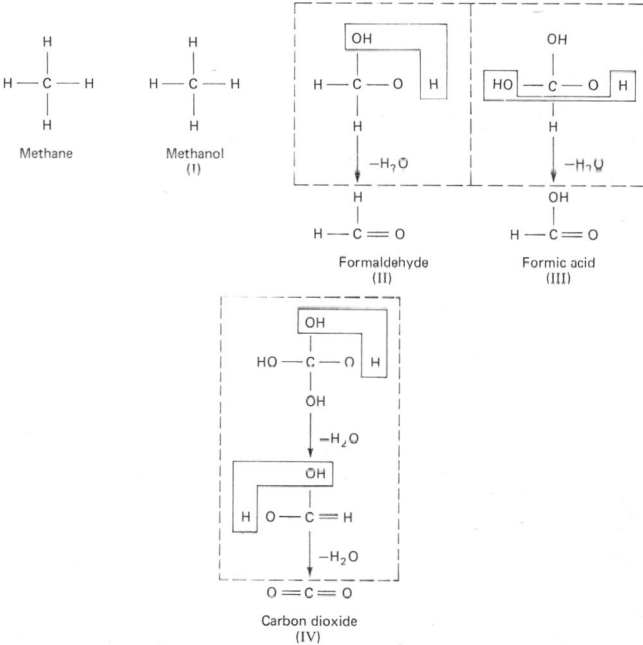

Fig. 6-6 Oxidation sequence of a hydrocarbon. The roman numerals under the compounds indicate the degree of oxidation of the methane. The carbon compounds enclosed by the dashed lines do not exist, and their hypothetical transformation into existing compounds is indicated by a loss of water molecules.

Of the lower alcohols, methanol and ethanol are by far the most important. Until about 50 years ago, methanol was obtained almost exclusively by the destructive distillation of wood (hence its old name wood alcohol). Modern production is by hydrogenation of carbon monoxide at high pressure and temperature in presence of catalysts:

$$CO + 2H_2 \xrightarrow[\text{pressure}]{\text{catalyst}} \underset{\text{methanol}}{CH_3OH} \tag{6-19}$$

Ethanol has been prepared by the time-honored method of fermentation of starch or sugar, but the synthetic method of producing ethanol from ethylene by the sulfuric acid process is also important. This method is illustrated in Fig. 6-8.

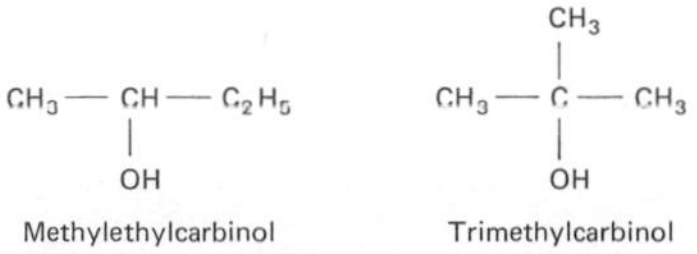

CH₃ — CH — C₂H₅
         |
         OH
Methylethylcarbinol

         CH₃
          |
CH₃ — C — CH₃
          |
         OH
Trimethylcarbinol

Fig. 6-7 Naming of secondary and tertiary alcohols as carbinol derivatives.

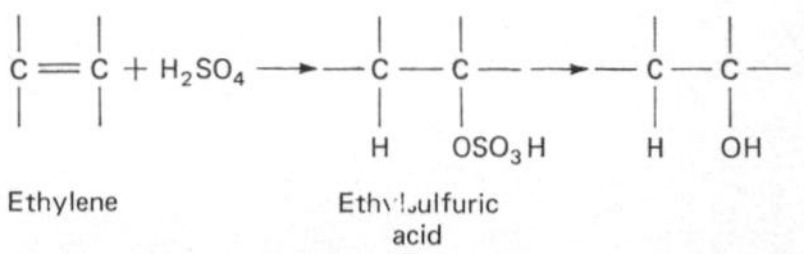

Ethylene   Ethylsulfuric acid

Fig. 6-8 Production of ethanol from ethylene.

The ethanol obtained by both processes is in dilute form and must be concentrated by distillation, as shown in Fig. 6-9, to produce the 95 percent commercial grade. The general laboratory techniques for preparation of alcohols include hydrolysis of alkyl halides:

$$RX + NaOH \xrightarrow{Ag_2O} ROH + NaX \tag{6-20}$$

and the much more useful Grignard synthesis:

$$\underset{\substack{\text{carbonyl} \\ \text{compound}}}{\text{>C=O}} + RMgX \longrightarrow \text{>C—OMgX} \xrightarrow{H_2O} \underset{\text{alcohol}}{\text{>C—OH}} + HoMgX \tag{6-21}$$

The carbonyl compound in the Grignard synthesis may be an aldehyde, ketone, or ester. Of the many reactions possible with alcohols, the more important ones are esterification, or the reaction of an alcohol with an organic acid, and oxidation. A general example of esterification is as follows:

$$\underset{\text{alcohol}}{ROH} + \underset{\text{acid}}{HOC—R'} \longrightarrow \underset{\text{ester}}{RO—C—R'} + H_2O \tag{6-22}$$

Note that in esterification H_2O is formed from an H that comes from the acid and an OH group that comes from the alcohol. The organic compound formed is called an *ester*. Oxidation of alcohols leads to aldehydes, ketones, and acids.

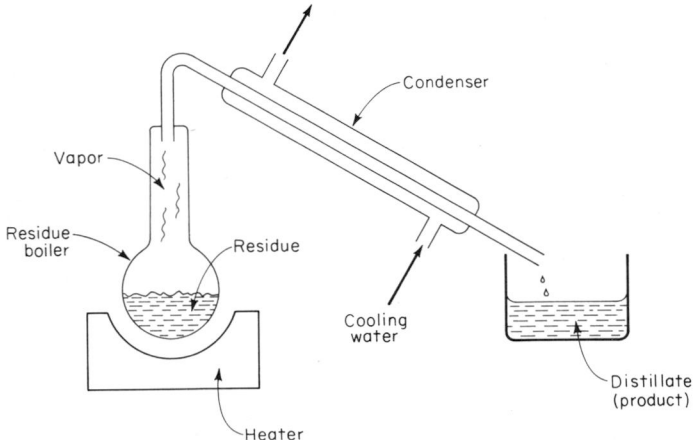

Fig. 6-9 Distillation as a purification process. Solids and nonvolatile components of the original feed remain in the boiler as a residue. The volatile components vaporize and subsequently condense. The distillate thus contains the volatile fractions of the original feed. By suitable temperature and pressure controls, a separation of the volatile components can also be made on the basis of their relative volatility.

In addition to the monohydric alcohols mentioned so far, two polyhydric alcohols are of importance:

$$\underset{\substack{\text{ethylene} \\ \text{glycol}}}{\begin{array}{c} CH_2OH \\ | \\ CH_2OH \end{array}} \qquad \underset{\text{glycerol}}{\begin{array}{c} CH_2OH \\ | \\ CHOH \\ | \\ CH_2OH \end{array}}$$

Ethylene glycol is produced from ethylene, and glycol can be made from propylene.

Aldehydes and Ketones

Owing to their ability to enter into a great variety of reactions, aldehydes and ketones are at the center of the advances made in organic synthesis. An aldehyde (*alcohol dehydro*genated) is a compound having a general formula

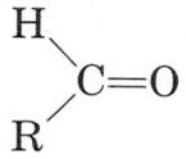

$$\begin{array}{c} H \\ \diagdown \\ \quad C{=}O \\ \diagup \\ R \end{array}$$

(i.e., a carbonyl [$>$C$=$O] compound with one alkyl group attached to the carbon, the remaining position filled by a hydrogen). A *ketone* (a name derived from acetone—the first member of the series) is a carbonyl compound in which two alkyl groups are attached to the carbonyl carbon

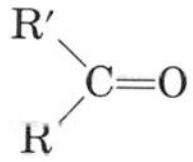

$$\begin{array}{c} R' \\ \diagdown \\ \quad C{=}O \\ \diagup \\ R \end{array}$$

Table 6-10 lists the properties of a selected group of aldehydes, and Table 6-11 shows similar data for ketones.

A number of more complex aldehydes and ketones occur in nature as components of perfume substances. Special methods are available for preparation of the simpler compounds of this class such as formaldehyde, acetaldehyde, and acetone:

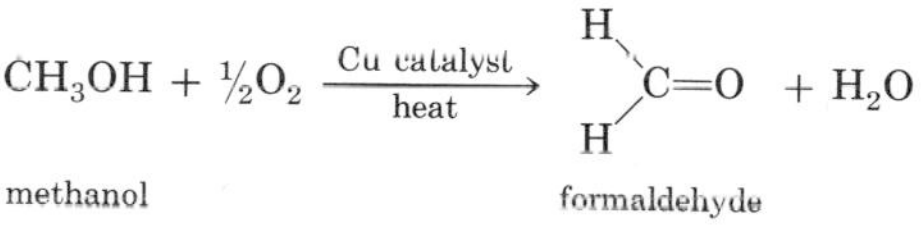

$$CH_3OH + \tfrac{1}{2}O_2 \xrightarrow[\text{heat}]{\text{Cu catalyst}} \begin{array}{c} H \\ \diagdown \\ \quad C{=}O \\ \diagup \\ H \end{array} + H_2O \qquad (6\text{-}23)$$

methanol   formaldehyde

$$HC{\equiv}CH + H_2O \xrightarrow[\text{catalyst}]{\text{Hg salt}} \begin{array}{c} O \\ \| \\ CH_3C{=}O \end{array} \qquad (6\text{-}24)$$

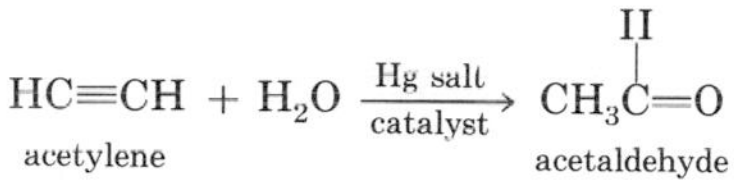

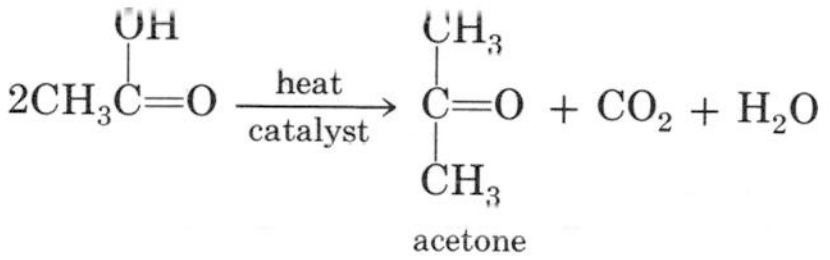

acetylene   acetaldehyde

$$2CH_3\overset{\ddot{O}H}{\underset{}{C}}{=}O \xrightarrow[\text{catalyst}]{\text{heat}} \begin{array}{c} CH_3 \\ | \\ C{=}O \\ | \\ CH_3 \end{array} + CO_2 + H_2O \qquad (6\text{-}25)$$

acetone

TABLE 6-10   Aldehydes

Name	Formula	mp (°C)	bp (°C)	Sp. gr.
Formaldehyde	CH_2O	-92	-21	0.815*
Acetaldehyde	CH_3CHO	-123	20.8	0.781
Chloral	CCl_3CHO	-57.5	97.8	1.512
Glyoxal	$O{=}HC \cdot CH{=}O$	15	50.4	1.14
Propionaldehyde	CH_3CH_2CHO	-81	48.8	0.807
n-Butyraldehyde	$CH_3CH_2CH_2CHO$	-97	74.7	0.817
Isobutyraldehyde	$(CH_3)_2CHCHO$	-66	61	0.794
n-Valeraldehyde	$CH_3CH_2CH_2CH_2CHO$	-92	103.7	0.819
Isovaleraldehyde	$(CH_3)_2CHCH_2CHO$	-51	92.5	0.803
n-Caproaldehyde	$CH_3(CH_2)_4CHO$		129	0.834
n-Heptaldehyde (oenanthol)	$CH_3(CH_2)_5CHO$	-45	155	0.850
Stearaldehyde	$CH_3(CH_2)_{16}CHO$	38		
Acrolein	$CH_2{=}CHCHO$	-88	52.5	0.841
Crotonaldehyde	$CH_3CH{=}CHCHO$	-76.5	104	0.859
Benzaldehyde	C_6H_5CHO	-56	179	1.046
Furfural	$\begin{array}{c} CH{-}CH \\ \| \quad \| \\ CH \quad C \\ \diagdown O \diagup \quad CHO \end{array}$	-31	162	1.156

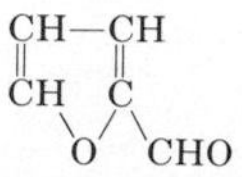

*For liquid at boiling point.

Major general techniques for preparation of aldehydes and ketones include:

Oxidation of primary alcohols to produce aldehydes

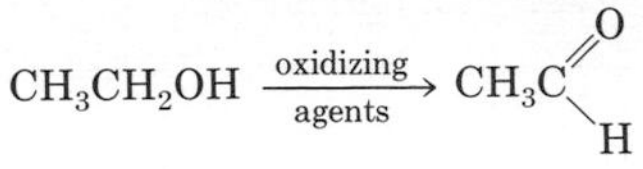

$$CH_3CH_2OH \xrightarrow[\text{agents}]{\text{oxidizing}} CH_3C{\overset{\displaystyle O}{\underset{\displaystyle H}{}}} \qquad (6\text{-}26)$$

aldehyde

Oxidation of secondary alcohols to produce ketones

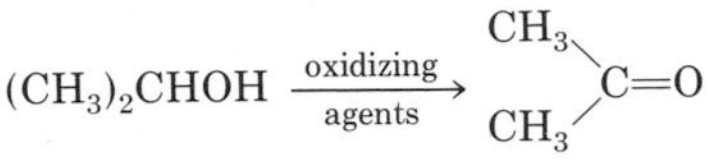

$$(CH_3)_2CHOH \xrightarrow[\text{agents}]{\text{oxidizing}} {CH_3 \atop CH_3}{>}C{=}O \qquad (6\text{-}27)$$

ketone

Pyrolysis of calcium salts of carboxylic acids

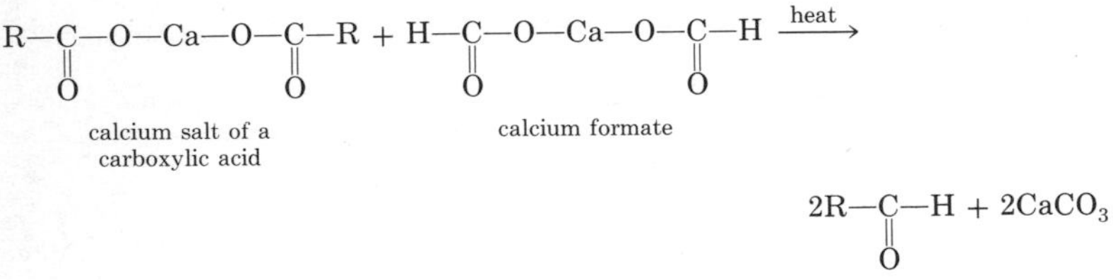

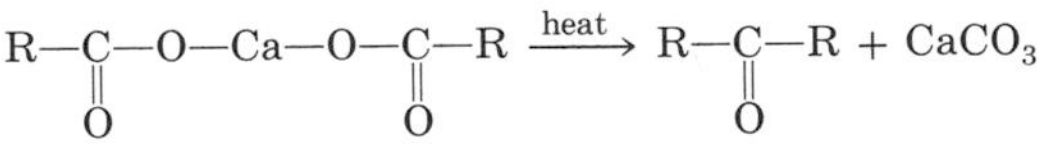

$$R{-}\underset{\displaystyle O}{C}{-}O{-}Ca{-}O{-}\underset{\displaystyle O}{C}{-}R + H{-}\underset{\displaystyle O}{C}{-}O{-}Ca{-}O{-}\underset{\displaystyle O}{C}{-}H \xrightarrow{\text{heat}}$$

calcium salt of a   calcium formate
carboxylic acid

$$2R{-}\underset{\displaystyle O}{C}{-}H + 2CaCO_3 \qquad (6\text{-}28)$$

aldehyde

$$R{-}\underset{\displaystyle O}{C}{-}O{-}Ca{-}O{-}\underset{\displaystyle O}{C}{-}R \xrightarrow{\text{heat}} R{-}\underset{\displaystyle O}{C}{-}R + CaCO_3 \qquad (6\text{-}29)$$

ketone

The versatility of aldehydes and ketones is best illustrated by the wide range of reactions that these compounds can undergo. Oxidation reactions lead to acids and other carboxylic

TABLE 6-11   Ketones

Name	Formula	mp (°C)	bp (°C)	Sp. gr.
Acetone	CH_3COCH_3	−95	56.1	0.7915
Methyl ethyl ketone	$CH_3COCH_2CH_3$	−86	79.6	0.805
Methyl n-propyl ketone	$CH_3CH_2CH_2COCH_3$	−77.8	102.1	0.812
Diethyl ketone	$CH_3CH_2COCH_2CH_3$	−42.0	101.7	0.814
Hexanone-2	$CH_3CH_2CH_2CH_2COCH_3$	−56.9	127.2	0.830
Hexanone-3	$CH_3CH_2CH_2COCH_2CH_3$		124	0.818
Methyl t-butyl ketone (pinacolone)	$CH_3COC(CH_3)_3$	−52.5	106.3	0.811
Di-n-propyl ketone	$(CH_3CH_2CH_2)_2CO$	−34	144.2	0.821
Diisopropyl ketone	$[(CH_3)_2CH]_2CO$		125	0.806
Diisobutyl ketone	$[(CH_3)_2CHCH_2]_2CO$		166	0.833
Di-n-amyl ketone	$(n\text{-}C_5H_{11})_2CO$	14.6	228	0.826
Stearone	$(n\text{-}C_{17}H_{35})_2CO$	88.5	345^{12mm}	0.793
Chloroacetone	CH_3COCH_2Cl	−44.5	119	1.162
s-Dichloroacetone	$ClCH_2COCH_2Cl$	45	173.4	1.383
Diacetyl	$CH_3COCOCH_3$		89	0.975
Acetylacetone	$CH_3COCH_2COCH_3$	−23.2	137	0.976
Mesityl oxide	$(CH_3)_2C{=}CHCOCH_3$	−59.0	131	0.863
Phorone	$[(CH_3)_2C{=}CH]_2CO$	28	198.2	0.885
Cyclohexanone	$CH_2{<}{\overset{CH_2CH_2}{\underset{CH_2CH_2}{>}}}CO$		156.7	0.949
Benzophenone	$C_6H_5COC_6H_5$	48	305.4	1.083

compounds. Reduction to alcohols is another important way in which these substances can be used. Aldehydes, because of the reactive nature of their carboxyl group, undergo many addition reactions of interest in synthetic preparations. Sodium hydrogen sulfide, hydrogen cyanide, derivatives of hydrazine, and hydroxylamine are examples of compounds which add readily to the carbonyl group. The *aldol condensation* is an interesting reaction illustrating the *dimerization* characteristic of an aldehyde, or the ability to react with itself.

$$CH_3\overset{H}{\underset{}{C}}=O + H\overset{H}{\underset{}{C}}H_2C=O \xrightarrow{OH^-} CH_3\overset{}{\underset{OH}{C}}HCH_2\overset{H}{\underset{}{C}}=O \qquad (6\text{-}30)$$

aldol

Acetal formation presents another example of the versatility of aldehydes:

$$R\overset{H}{\underset{}{C}}=O + 2C_2H_5OH \xrightarrow{H^+} R-\overset{OC_2H_5}{\underset{OC_2H_5}{CH}} + H_2O \qquad (6\text{-}31)$$

acetal

Ethers

Ethers may be regarded as derivatives of alcohols or water. Note the similarity in structure:

$$\underset{\text{alcohol}}{R-O-H} \qquad \underset{\text{water}}{H-O-H} \qquad \underset{\text{ether}}{R-O-R}$$

Properties of selected ethers are summarized in Table 6-12. Diethyl ether, commonly referred to simply as ether, is a very good solvent but is most widely known for its anesthetic properties. It is immiscible with water and hence is also a good medium for extracting a number of organic compounds from aqueous systems, as shown in Fig. 6-10. Diethyl ether may be prepared from ethanol:

$$C_2H_5OH \xrightarrow{H_2SO_4} \underset{\text{ether}}{C_2H_5-O-C_2H_5} + H_2O \qquad (6\text{-}32)$$

A general method for preparation of ethers is known as *Williamson's synthesis*:

$$\underset{\substack{\text{sodium} \\ \text{alkoxide}}}{RONa} + \underset{\substack{\text{alkyl} \\ \text{halide}}}{R'X} \longrightarrow \underset{\text{ether}}{R-O-R'} + NaX \qquad (6\text{-}33)$$

TABLE 6-12   Ethers

Name	Formula	mp (°C)	bp (°C)	Sp. gr.
Dimethyl ether	CH_3OCH_3	−140	−24.9	0.661*
Methyl ethyl ether	$CH_3OCH_2CH_3$		7.9	0.697*
Diethyl ether	$CH_3CH_2OCH_2CH_3$	−116	34.6	0.714
Di-n-propyl ether	$(CH_3CH_2CH_2)_2O$	−122	90.5	0.736
Diisopropyl ether	$(CH_3)_2CHOCH(CH_3)_2$		68	0.735
Methyl n-butyl ether	$CH_3OCH_2CH_2CH_2CH_3$	−116	70.3	0.744
Ethyl n-butyl ether	$CH_3CH_2OCH_3CH_2CH_2CH_3$		92	0.752
Di-n-butyl ether	$(CH_3CH_2CH_2CH_2)_2O$		141	0.769
Di-n-amyl ether	$(n\text{-}C_5H_{11})_2O$	−69	187.5	0.774
Diisoamyl ether	$[(CH_3)_2CHCH_2CH_2]_2O$		172.2	0.777
Di-n-hexyl ether	$(n\text{-}C_6H_{13})_2O$		208.8	
s-Di-(chloromethyl) ether	$ClCH_2OCH_2Cl$		106	1.315
α,β-Dichloroethyl ethyl ether	$CH_3CH_2OCHClCH_2Cl$		145	1.174
Di-(β-chloroethyl) ether	$CH_2ClCH_2OCH_2CH_2Cl$		178	1.213
Ethylene glycol dimethyl ether	$CH_3OCH_2CH_2OCH_3$		83	0.863
Divinyl ether	$CH_2=CHOCH=CH_2$		35	
Diallyl ether	$(CH_2=CHCH_2)_2O$		94	0.826
Diphenyl ether	$C_6H_5OC_6H_5$	26.9	259	1.072
Anisole	$C_6H_5OCH_3$	−37.3	154	0.994

*For liquid at boiling point.

Diethyl ether is used in the preparation of the Grignard reagent used extensively in organic synthesis work. Ether forms a complex with an alkyl magnesium halide which is known as the Grignard reagent dietherate:

$$\underset{\substack{\text{alkyl magnesium}\\\text{halide}}}{\text{RMX}} \quad + 2C_2H_5{-}O{-}C_2H_5 \longrightarrow \underset{\substack{\text{Grignard reagent}\\\text{dietherate}}}{\text{RMX}\cdot 2(C_2H_5{-}O{-}C_2H_5)} \qquad (6\text{-}34)$$
$$\underset{\substack{\text{diethyl}\\\text{ether}}}{\phantom{+ 2C_2H_5{-}O{-}C_2H_5}}$$

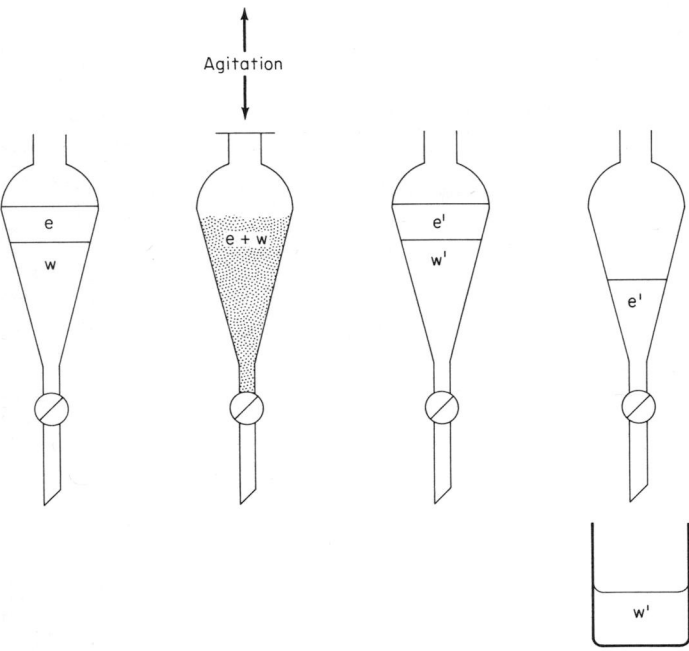

Fig. 6-10 Use of ether in extraction. e = pure ether; w = water containing solute to be extracted; e′ = ether containing solute; w′ = water with greatly reduced amount of solute.

Carboxylic Acids and Derivatives

Carboxylic compounds are those containing a carboxylic group, the name being derived from the *carb*onyl and hyd*roxyl* groups present on the same carbon atom.

$$\underset{\underset{O}{\|}}{-C}{-}OH \qquad \text{or} \qquad -COOH$$

When the remaining bond on the carbon is used by an alkyl, we have a carboxylic acid:

$$R{-}\underset{\underset{O}{\|}}{C}{-}OH \qquad \text{or} \qquad RCOOH$$

When the hydrogen is replaced by another alkyl, the resultant compound is called an *ester:*

$$R{-}\underset{\underset{O}{\|}}{C}{-}OR' \qquad \text{or} \qquad RCOOR'$$

Anhydrides may be thought of as compounds resulting from the removal of water from two molecules of a carboxylic acid:

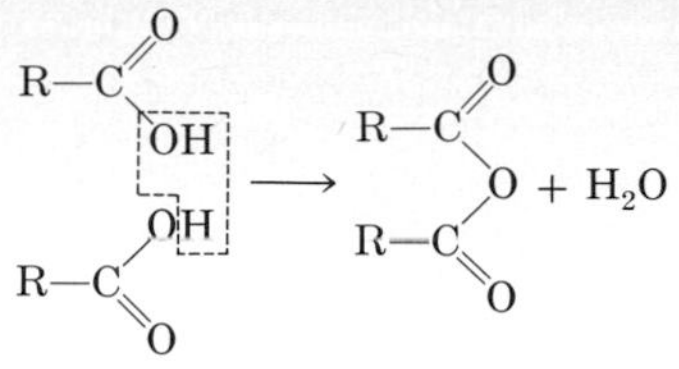

$$R-C\overset{O}{\underset{\substack{OH \\ OH \\ R-C\diagdown O}}{\diagup}} \longrightarrow \quad \begin{matrix} R-C\diagup^O \\ O \\ R-C\diagdown_O \end{matrix} + H_2O \qquad (6\text{-}35)$$

acid   anhydride

When the hydroxyl group of an acid is replaced by a halogen, we have an *acyl halide:*

$$R-\underset{O}{\overset{|}{C}}-OH \qquad R-\underset{O}{\overset{|}{C}}-Cl$$

acid   acyl chloride

When, instead of the halogen, an amine group is present, we have an *amide:*

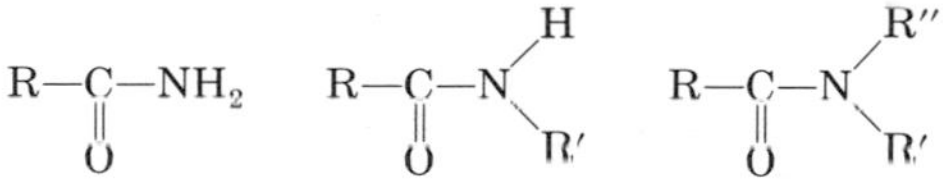

$$R-\underset{O}{\overset{|}{C}}-NH_2 \qquad R-\underset{O}{\overset{|}{C}}-N\diagup^H_{R'} \qquad R-\underset{O}{\overset{|}{C}}-N\diagup^{R''}_{R'}$$

Tables 6-13 to 6-17 list the typical carboxylic acids, esters, anhydrides, acyl halides, and amides.

Carboxylic compounds are widely found in nature as esters. Free acids also occur, but much less frequently. The first acid in the series, formic acid, is prepared industrially from sodium hydroxide and carbon monoxide:

$$NaOH + CO \xrightarrow[\text{pressure}]{\text{high temp}} H\underset{O}{\overset{|}{C}}ONa \xrightarrow{H_2SO_4} H\underset{O}{\overset{|}{C}}OH \qquad (6\text{-}36)$$

formic acid

Acetic acid used to be prepared from alcohol by oxidation in the presence of certain bacteria, but current methods involve oxidation of ethanol over a catalyst, or oxidation of acetaldehyde produced from acetylene. In general, acids are prepared by oxidation of primary alcohols or aldehydes, by hydrolysis of alkyl cyanides, and by a Grignard synthesis. The alkyl cyanide process may be illustrated as follows:

$$RX + KCN \longrightarrow RC{\equiv}N + KX$$

alkyl cyanide

$$+$$

$$H_2O$$

$$\Big\downarrow \begin{matrix} \text{acid} \\ \text{or} \\ \text{base} \end{matrix} \qquad (6\text{-}37)$$

$$RC{-}OH + NH_3$$
$$\overset{}{\underset{O}{\|}}$$

Carboxylic acids are generally weak compared with inorganic acids such as HCl. The ionization constants of many carboxylic acids are low and have values of the order of 10^{-5}. Reactions of acids with alcohols result in the formation of esters:

$$R-\underset{O}{\overset{|}{C}}-OH + HOR' \longrightarrow R-\underset{O}{\overset{|}{C}}-OR' + H_2O \qquad (6\text{-}38)$$

acid   alcohol   ester

An example of this general reaction is seen in the equation showing the formation of a *fat, glyceryl stearate:*

$$3C_{17}H_{35}COOH + C_3H_5(OH)_3 \longrightarrow C_3H_5(C_{17}H_{35}COO)_3 + 3H_2O \qquad (6\text{-}39)$$

stearic acid   glycerol   glyceryl stearate
(a glyceride or fat)

TABLE 6-13   Acids

MONOBASIC ALKYL ACIDS				
Acid	Formula	mp (°C)	bp (°C)	Sp. gr.
Formic	HCOOH	8.4	100.5	1.220
Acetic	CH_3COOH	16.6	118	1.049
Propionic	CH_3CH_2COOH	−22	141	0.992
n-Butyric	$CH_3CH_2CH_2COOH$	−4.7	162.5	0.959
Isobutyric	$(CH_3)_2CHCOOH$	−47	154.4	0.949
n-Valeric	$CH_3(CH_2)_3COOH$	−34.5	187.0	0.939
Trimethylacetic	$(CH_3)_3CCOOH$	35.5	163.8	0.905
Caproic	$CH_3(CH_2)_4COOH$	−1.5	205	0.929
n-Heptylic	$CH_3(CH_2)_5COOH$	−11	223.5	0.922
Caprylic	$CH_3(CH_2)_6COOH$	16	237	0.910
Pelargonic	$CH_3(CH_2)_7COOH$	12.5	254	0.907
Chloroacetic	$CH_2ClCOOH$	63	189.5	1.37
Bromoacetic	$CH_2BrCOOH$	50	208	1.934
Iodoacetic	CH_2ICOOH	82		
Dichloroacetic	$CHCl_2COOH$	10	193.5	1.563
Trichloroacetic	CCl_3COOH	58	196	1.617
α-Chloropropionic	$CH_3CHClCOOH$		186	1.306
β-Chloropropionic	CH_2ClCH_2COOH	39	204	
Glycolic	$HOCH_2COOH$	79		
Lactic	$CH_3CHOHCOOH$	18	122^{15mm}	1.249
Methoxyacetic	CH_3OCH_2COOH		204	1.777
Thioglycolic	$HSCH_2COOH$	−16.5	123^{29mm}	1.325
Cyanoacetic	$N{\equiv}CCH_2COOH$	66	dec.	
Glyoxylic	$O{=}CHCH_2COOH$			

MONOBASIC ARYL ACIDS			
Acid	Formula	mp (°C)	bp (°C)
Benzoic	$C_6H_5CO_2H$	121.7	249
o-Toluic	$o\text{-}CH_3C_6H_4CO_2H$	104	259
m-Toluic	$m\text{-}CH_3C_6H_4CO_2H$	111	263
p-Toluic	$p\text{-}CH_3C_6H_4CO_2H$	180	
o-Chlorobenzoic	$o\text{-}ClC_6H_4CO_2H$	141	
m-Chlorobenzoic	$m\text{-}ClC_6H_4CO_2H$	158	
p-Chlorobenzoic	$p\text{-}ClC_6H_4CO_2H$	243	
o-Bromobenzoic	$o\text{-}BrC_6H_4CO_2H$	149	
m-Bromobenzoic	$m\text{-}BrC_6H_4CO_2H$	155	
p-Bromobenzoic	$p\text{-}BrC_6H_4CO_2H$	252	
o-Nitrobenzoic	$o\text{-}NO_2 \cdot C_6H_4CO_2H$	148	
m-Nitrobenzoic	$m\text{-}NO_2 \cdot C_6H_4CO_2H$	142	
p-Nitrobenzoic	$p\text{-}NO_2 \cdot C_6H_4CO_2H$	240	
3,5-Dinitrobenzoic	$3,5\text{-}(NO_2)_2 \cdot C_6H_3CO_2H$	205	
Salicylic	$o\text{-}HOC_6H_4CO_2H$	159	
m-Hydroxybenzoic	$m\text{-}HOC_6H_4CO_2H$	200	
p-Hydroxybenzoic	$p\text{-}HOC_6H_4CO_2H$	215	
Anisic	$p\text{-}CH_3OC_6H_4CO_2H$	184	277
Gallic	$3,4,5\text{-}(HO)_3C_6H_2CO_2H$	253 dec.	
Syringic	$4\text{-}(HO)\text{-}3,5(CH_3O)_2C_6H_2CO_2H$	205	
Anthranilic	$o\text{-}NH_2 \cdot C_6H_4CO_2H$	145	
m-Aminobenzoic	$m\text{-}NH_2 \cdot C_6H_4CO_2H$	174	
p-Aminobenzoic	$p\text{-}NH_2 \cdot C_6H_4CO_2H$	187	

TABLE 6-13 Acids—(Continued)

DIBASIC ACIDS

Acid	Formula	mp (°C)	Solubility (g/100 g H₂O at 18°C)
Oxalic	HOOC · COOH	187	10.2
Malonic	HOOCCH₂COOH	135	138
Succinic	HOOCCH₂CH₂COOH	185	6.8
Glutaric	HOOCCH₂CH₂CH₂COOH	97.5	63.9
Adipic	HOOC(CH₂)₄COOH	151	1.4
Pimelic	HOOC(CH₂)₅COOH	105	2.5
Suberic	HOOC(CH₂)₆COOH	142	0.14
Azelaic	HOOC(CH₂)₇COOH	106	0.2
Sebacic	HOOC(CH₂)₈COOH	134	0.1

SATURATED FATTY ACIDS

Acid	Formula	mp (°C)	bp (°C)
Butyric	CH₃(CH₂)₂COOH	−4.7	163
Isovaleric	(CH₃)₂CHCH₂COOH	−51	174
Caproic	CH₃(CH₂)₄COOH	−1.5	205
Caprylic	CH₃(CH₂)₆COOH	16.5	237
Capric	CH₃(CH₂)₈COOH	31.3	269
Lauric	CH₃(CH₂)₁₀COOH	43.6	102/1 mm
Myristic	CH₃(CH₂)₁₂COOH	58.0	122/1 mm
Palmitic	CH₃(CH₂)₁₄COOH	62.9	139/1 mm
Stearic	CH₃(CH₂)₁₆COOH	69.9	160/1 mm
Arachidic	CH₃(CH₂)₁₈COOH	75.2	205/1 mm
Behenic	CH₃(CH₂)₂₀COOH	80.2	
Lignoceric	CH₃(CH₂)₂₂COOH	84.2	
Cerotic	CH₃(CH₂)₂₄COOH	87.7	

UNSATURATED FATTY ACIDS

Acid	Formula	mp (°C)
Decylenic	CH₂=CH(CH₂)₇COOH	
Dodecylenic	CH₃CH₂CH=CH(CH₂)₇COOH	
Palmitoleic	CH₃(CH₂)₅CH=CH(CH₂)₇COOH	
Oleic	CH₃(CH₂)₇CH=CH(CH₂)₇COOH	13, 16
Ricinoleic	CH₃(CH₂)₅CH(OH)CH₂CH=CH(CH₂)₇COOH	50
Petroselinic	CH₃(CH₂)₁₀CH=CH(CH₂)₄COOH	30
Vaccenic	CH₃(CH₂)₅CH=CH(CH₂)₉COOH	39
Linoleic	CH₃(CH₂)₄CH=CHCH₂CH=CH(CH₂)₇COOH	
Linolenic	CH₃CH₂CH=CHCH₂CH=CHCH₂CH=CH(CH₂)₇COOH	
Eleostearic	CH₃(CH₂)₃(CH=CH)₃(CH₂)₇COOH	49
Licanic	CH₃(CH₂)₃(CH=CH)₃(CH₂)₄CO(CH₂)₂COOH	75
Parinaric	CH₃CH₂(CH=CH)₄(CH₂)₇COOH	86
Tariric	CH₃(CH₂)₇C≡C(CH₂)₄COOH	
Gadoleic	CH₃(CH₂)₉CH=CH(CH₂)₇COOH	
Arachidonic	CH₃(CH₂)₄(CH=CHCH₂)₄(CH₂)₂COOH	
Cetoleic	CH₃(CH₂)₉CH=CH(CH₂)₉COOH	
Erucic	CH₃(CH₂)₇CH=CH(CH₂)₁₁COOH	33.5
Selacholeic or nervonic	CH₃(CH₂)₇CH=CH(CH₂)₁₃COOH	39

TABLE 6-14   Esters

Name	Formula	bp (°C)	Sp. gr.
Methyl formate	HCO_2CH_3	32	0.974
Ethyl formate	$HCO_2C_2H_5$	54	0.906
Methyl acetate	$CH_3CO_2CH_3$	57	0.924
Ethyl acetate	$CH_3CO_2C_2H_5$	77.1	0.901
n-Propyl acetate	$CH_3CO_2C_3H_7(n)$	101.7	0.886
n-Butyl acetate	$CH_3CO_2C_4H_9(n)$	126.5	0.882
n-Amyl acetate	$CH_3CO_2C_5H_{11}(n)$	147.6	0.879
Isobutyl acetate	$CH_3CO_2C_4H_9(i)$	118	0.871
t-Butyl acetate	$CH_3CO_2C(CH_3)_3$	97	0.896
Isoamyl acetate	$CH_3CO_2C_5H_{11}(i)$	142	0.876
n-Octyl acetate	$CH_3CO_2C_8H_{17}(n)$	210	0.885
Methyl propionate	$C_2H_5CO_2CH_3$	79.7	0.915
Ethyl propionate	$C_2H_5CO_2C_2H_5$	99.1	0.891
Methyl n-butyrate	$n\text{-}C_3H_7CO_2CH_3$	102.3	0.898
Ethyl n-butyrate	$n\text{-}C_3H_7CO_2C_2H_5$	121	0.879
Isoamyl n-butyrate	$n\text{-}C_3H_7CO_2C_5H_{11}(i)$	178.6	0.866
Methyl n-valerate	$n\text{-}C_4H_9CO_2CH_3$	127.3	0.910
Ethyl n-valerate	$n\text{-}C_4H_9CO_2C_2H_5$	145.5	0.877
Methyl isovalerate	$i\text{-}C_4H_9CO_2CH_3$	117	0.881
Isoamyl isovalerate	$i\text{-}C_4H_9CO_2C_5H_{11}(i)$	194	0.858
Ethyl n-heptylate	$n\text{-}C_6H_{13}CO_2C_2H_6$	187	0.872
Ethyl pelargonate	$n\text{-}C_8H_{17}CO_2C_2H_5$	228	0.866

TABLE 6-15   Anhydrides

Name	Formula	mp (°C)	bp (°C)	Sp. gr.
Acetic anhydride	$(CH_3CO)_2O$	−73	139.6	1.082
Propionic anhydride	$(C_2H_5CO)_2O$	−45	168	1.012
n-Butyric anhydride	$(n\text{-}C_3H_7CO)_2O$	−75	198	0.969
n-Valeric anhydride	$(n\text{-}C_4H_9CO)_2O$		218	0.929
Stearic anhydride	$(n\text{-}C_{17}H_{35}CO)_2O$	72		
Succinic anhydride	$\begin{array}{c}CH_2CO\\ \quad\quad\; O\\ CH_2CO\end{array}$	119.6	261	1.104
Benzoic anhydride	$(C_6H_5CO)_2O$	42	360	1.199
Phthalic anhydride	(structure)	132	284.5	1.527

TABLE 6-16   Acyl Halides

Name	Formula	bp (°C)	Sp. gr.
Acetyl fluoride	CH_3COF	20.5	0.993
Acetyl chloride	CH_3COCl	52	1.104
Acetyl bromide	CH_3COBr	76.7	1.52
Acetyl iodide	CH_3COI	108	1.98
Chloroacetyl chloride	$CH_2ClCOCl$	105	1.495
Bromoacetyl bromide	$CH_2BrCOBr$	150	2.317
Propionyl chloride	CH_3CH_2COCl	80	1.065
n-Butyryl chloride	$CH_3CH_2CH_2COCl$	102	1.028
Isobutyryl chloride	$(CH_3)_2CHCOCl$	92	1.017
n-Valeryl chloride	$CH_3CH_2CH_2CH_2COCl$	128	1.016
Isovaleryl chloride	$(CH_3)_2CHCH_2COCl$	113	
n-Caproyl chloride	$n\text{-}C_5H_{11}COCl$	153	
Capryl chloride	$CH_3(CH_2)_6COCl$	196	0.975
Stearoyl chloride	$CH_3(CH_2)_{16}COCl$	215[16mm]	
Benzoyl chloride	C_6H_5COCl	197.2	1.212

A reaction known as *saponification* converts a glyceride to a *soap*. A soap is a salt of a long-chain organic acid, called a *fatty acid*, and an inorganic alkali, such as sodium hydroxide.

$$C_3H_5(C_{17}H_{35}COO)_3 + 3NaOH \longrightarrow 3\underset{\text{soap}}{C_{17}H_{35}COONa} + \underset{\text{glycerol}}{C_3H_5(OH)_3} \qquad (6\text{-}40)$$

Anhydrides can be prepared by dehydration of the acid, but yields of this reaction are poor:

$$2R\underset{\underset{O}{\|}}{C}OH \longrightarrow R-\underset{\underset{O}{\|}}{C}-O-\underset{\underset{O}{\|}}{C}-R + H_2O \qquad (6\text{-}41)$$

Much better results are obtained by reacting a sodium salt of an acid with an acyl chloride:

$$R-\underset{\underset{O}{\|}}{C}ONa + Cl-\underset{\underset{O}{\|}}{C}-R \longrightarrow R-\underset{\underset{O}{\|}}{C}-O-\underset{\underset{O}{\|}}{C}-R + NaCl \qquad (6\text{-}42)$$

Acyl halides (chlorides) may be formed by reacting acids with a chloride such as PCl_3:

$$3R-\underset{\underset{O}{\|}}{C}-OH + PCl_3 \longrightarrow R-\underset{\underset{O}{\|}}{C}-Cl + H_3PO_4 \qquad (6\text{-}43)$$

TABLE 6-17   Amides

Name	Formula	mp (°C)	bp (°C)	Sp. gr.
Formamide	$HCONH_2$	2	193	1.139
Acetamide	CH_3CONH_2	82	222	1.159
Propionamide	$C_2H_5CONH_2$	80	213	1.042
n-Butyramide	$n\text{-}C_3H_7CONH_2$	116	216	1.032
n-Valeramide	$n\text{-}C_4H_9CONH_2$	106		1.023
n-Caproamide	$n\text{-}C_5H_{11}CONH_2$	101		0.999
Stearamide	$n\text{-}C_{17}H_{35}CONH_2$	109	251^{12mm}	

Amides can be formed from acids by reacting their esters, acyl chlorides, or anhydrides with ammonia. A technical method for the production of acetamide involves a pyrolysis of ammonium acetate:

$$\underset{\substack{\text{ammonium} \\ \text{acetate}}}{CH_3\underset{\underset{O}{\|}}{C}ONH_4} \longrightarrow \underset{\text{acetamide}}{CH_3\underset{\underset{O}{\|}}{C}NH_2} + H_2O \qquad (6\text{-}44)$$

An amide which deserves special mention is that of carbamic acid or *urea*:

$$H_2N-\underset{\underset{O}{\|}}{C}-NH_2$$

Carbamic acid $NH_2\underset{\underset{O}{\|}}{C}-OH$ has not been isolated but forms stable salts, esters, and amides.

Halogen Compounds

Organic compounds containing halogens are widely used as *intermediates* (that is, compounds used to prepare other compounds) or, if they are liquids, as solvents. In general, halogen compounds are insoluble in water but soluble in organic solvents. The liquid chlorides tend to be lighter than water, while the bromides and iodides are heavier. The general preparative methods include:

Preparation from alcohols and inorganic chlorides

$$3ROH + PCl_3 \longrightarrow 3RCl + H_3PO_4 \qquad (6\text{-}45)$$
$$ROH + SOCl_2 \longrightarrow RCl + HCl + SO_2 \qquad (6\text{-}46)$$
$$ROH + PCl_5 \longrightarrow RCl + POCl_3 + HCl \qquad (6\text{-}47)$$

Preparation from olefins, by halogenation and addition of hydrogen halides

$$\underset{\text{H}}{\overset{\text{H}}{H_3C-\overset{|}{C}=CH_2}} + Cl_2 \longrightarrow H_3C-\overset{\overset{\displaystyle H}{|}}{\underset{\underset{\displaystyle Cl}{|}}{C}}-\underset{\underset{\displaystyle Cl}{|}}{CH_2} \qquad (6\text{-}48)$$

$$H_3C-\overset{\overset{\displaystyle H}{|}}{C}=CH_2 + HX \longrightarrow H_3C-\overset{\overset{\displaystyle H}{|}}{\underset{\underset{\displaystyle X}{|}}{C}}-CH_3 \qquad (6\text{-}49)$$

In addition to the great value of halogen compounds as intermediates, a number of such compounds have important properties in themselves. *Chloroform* ($CHCl_3$), *ethylene dichloride* (CH_2Cl-CH_2Cl), and *tetrachloroethane* ($CHCl_2-CHCl_2$) are very useful as solvents. Carbon tetrachloride (CCl_4) occupies a special place in the cleaning industry since it is not only an excellent solvent but is nonflammable. The compound is prepared industrially by a reaction of chlorine with carbon disulfide:

$$\underset{\substack{\text{carbon} \\ \text{disulfide}}}{CS_2} + 3Cl_2 \xrightarrow{\text{catalyst}} \underset{\substack{\text{carbon} \\ \text{tetrachloride}}}{CCl_4} + S_2Cl_2 \qquad (6\text{-}50)$$

Vinyl chloride is a major halogen derivative of ethylene. It is the building block for the important vinyl plastics.

$$CH_2{=}CH_2 + Cl_2 \xrightarrow{\text{heat}} \underset{\underset{\text{vinyl chloride}}{Cl}}{CH_2{=}\overset{|}{CH}} + HCl \qquad (6\text{-}51)$$

Freon refrigerants are mixtures of *fluorochloromethanes*, such as $CFCl_3$ and CF_2Cl_2. Another fluorinated hydrocarbon, tetrafluoroethylene, is the monomer for the important Teflon family of plastics.

$$n(CF_2{=}CF_2) \longrightarrow \underset{\text{Teflon}}{-(CF_2-CF_2)_n-} \qquad (6\text{-}52)$$

Properties of typical halides are listed in Table 6-18.

TABLE 6-18   Alkyl and Aryl Halides

ALKYL MONOHALIDES				
Name	Formula	mp (°C)	bp (°C)	Sp. gr.
Methyl chloride	CH_3Cl	−97	−23.7	0.920*
Methyl bromide	CH_3Br	−93	4.6	1.732*
Methyl iodide	CH_3I	−64	42.3	2.279
Ethyl chloride	C_2H_5Cl	−139	13.1	0.910*
Ethyl bromide	C_2H_5Br	−119	38.4	1.430
Ethyl iodide	C_2H_5I	−111	72.3	1.933
n-Propyl chloride	$CH_3CH_2CH_2Cl$	−123	46.4	0.890
n-Propyl bromide	$CH_3CH_2CH_2Br$	−110	71	1.353
n-Propyl iodide	$CH_3CH_2CH_2I$	−101	102	1.747
Isopropyl chloride	$(CH_3)_2CHCl$	−117	36.5	0.860
Isopropyl bromide	$(CH_3)_2CHBr$	−89	59.5	1.310
Isopropyl iodide	$(CH_3)_2CHI$	−91	89.4	1.703
n-Butyl chloride	$CH_3(CH_2)_3Cl$	−123	78.1	0.884
n-Butyl bromide	$CH_3(CH_2)_3Br$	−112	101.6	1.275
Isobutyl chloride	$(CH_3)_2CHCH_2Cl$	−131	68.9	0.866
Isobutyl bromide	$(CH_3)_2CHCH_2Br$	−120	91.3	1.250
sec-Butyl chloride	$CH_3CH_2CHClCH_3$		68	0.871
sec-Butyl bromide	$CH_3CH_2CHBrCH_3$		91.3	1.251
tert-Butyl chloride	$(CH_3)_3CCl$	−28.5	51.0	0.851
tert-Butyl bromide	$(CH_3)_3CBr$	−20	73.3	1.222
n-Amyl bromide	$CH_3(CH_2)_4Br$	−95	129.7	1.223
Isoamyl bromide	$(CH_3)_2CHCH_2CH_2Br$	−112	120.7	1.215

*For liquid at boiling point.

TABLE 6-18   Alkyl and Aryl Halides—(Continued)

ALKYL MONOHALIDES—(Continued)

Name	Formula	mp (°C)	bp (°C)	Sp. gr.
tert-Amyl bromide	$(CH_3)_2CBrCH_2CH_3$		109.2	1.190
n-Hexyl bromide	$CH_3(CH_2)_5Br$		156	1.173
n-Octadecyl bromide	$CH_3(CH_2)_{17}Br$	34	170 at 0.5 mm	

ALKYL POLYHALIDES

Name	Formula	mp (°C)	bp (°C)	Sp. gr.
Methylene chloride	CH_2Cl_2	−96	40.8	1.336
Methylene bromide	CH_2Br_2	−53	98.2	2.46
Methylene iodide	CH_2I_2	5	180 dec.	3.322
Chloroform	$CHCl_3$	−63.5	61.2	1.489
Bromoform	$CHBr_3$	7.8	149.6	2.865
Iodoform	CHI_3	119		4.1
Carbon tetrafluoride	CF_4		−128	
Carbon tetrachloride	CCl_4	−23	76.8	1.575
Carbon tetrabromide	CBr_4	92.5	189.5	3.42
Dichlorodifluoromethane	CCl_2F_2	−155	−29.8	1.4
Ethylene dichloride (ethylene chloride)	$ClCH_2CH_2Cl$	−35.5	83.8	1.238
Ethylene dibromide	$BrCH_2CH_2Br$	10	131.7	2.182
Ethylidene chloride	CH_3CHCl_2	−97	57.3	1.174
Ethylidene bromide	CH_3CHBr_2		110	2.056
Tetrachloroethane (acetylene tetrachloride)	$CHCl_2CHCl_2$	−43	146.3	1.600
Hexachloroethane	CCl_3CCl_3	189	185	2.091
Dichloroethylene (acetylene dichloride)	$ClCH=CHCl$	50	48.4	1.259
Trichloroethylene	$CHCl=CCl_2$	−86	87	1.477
Trimethylene bromide	$Br(CH_2)_3Br$	−36	167	1.979
Tetramethylene bromide	$Br(CH_2)_4Br$	−21	198 dec.	1.79
Pentamethylene bromide	$Br(CH_2)_5Br$	−40	221	1.706
Hexamethylene bromide	$Br(CH_2)_6Br$		240 dec.	1.599

ARYL HALIDES

Name	Formula	mp (°C)	bp (°C)
Fluorobenzene	C_6H_5F	−45	85
Chlorobenzene	C_6H_5Cl	−45	132
Bromobenzene	C_6H_5Br	−30.6	155.5
Iodobenzene	C_6H_5I	−29	188.5
o-Chlorotoluene	o-$CH_3C_6H_4Cl$	−36	159
m-Chlorotoluene	m-$CH_3C_6H_4Cl$	−48	162
p-Chlorotoluene	p-$CH_3C_6H_4Cl$	7	162
o-Bromotoluene	o-$CH_3C_6H_4Br$	−26	182
m-Bromotoluene	m-$CH_3C_6H_4Br$	40	184
p-Bromotoluene	p-$CH_3C_6H_4Br$	28	184
o-Bromoanisole	o-$CH_3OC_6H_4Br$	liq.	222
p-Bromoanisole	p-$CH_3OC_6H_4Br$	11	223
p-Bromodimethylaniline	p-$(CH_3)_2NC_6H_4Br$	55	264
o-Dichlorobenzene	$C_6H_4Cl_2(1,2)$	liq.	179
p-Dichlorobenzene	$C_6H_4Cl_2(1,4)$	53	173
1,2,4-Trichlorobenzene	$C_6H_3Cl_3(1,2,4)$	17	213
1,2,3,4-Tetrachlorobenzene	$C_6H_2Cl_4(1,2,3,4)$	46	254
1,2,4,5-Tetrachlorobenzene	$C_6H_2Cl_4(1,2,4,5)$	138	245
Hexachlorobenzene	C_6Cl_6	228	332
p-Dibromobenzene	p-$C_6H_4Br_2$	89	218
o-Bromochlorobenzene	$C_6H_4ClBr(1,2)$	liq.	199
p-Bromochlorobenzene	$C_6H_4ClBr(1,4)$	67	196
o-Bromoiodobenzene	$C_6H_4BrI(1,2)$	liq.	257
p-Bromoiodobenzene	$C_6H_4BrI(1,4)$	92	252
o-Chloroiodobenzene	$C_6H_4ClI(1,2)$	liq.	235
p-Chloroiodobenzene	$C_6H_4ClI(1,4)$	57	227

Nitrogen-Containing Compounds

Compounds containing nitrogen give rise to such major classes of organic compounds as amines, amino acids, nitro compounds, amides, nitriles, azo, and diazo compounds. *Amines* may be thought of as compounds resulting from the substitution of alkyl groups for hydrogen atoms in the ammonia molecule, amines being classified as primary, secondary, or tertiary, to indicate whether one, two, or all three of the hydrogen atoms in the base ammonia molecule have been replaced by alkyl groups, as illustrated in Fig. 6-11. Properties of typical amines are listed in Table 6-19. Amines are of great importance in biochemistry, but a number of them find use in industrial chemical processes. They are organic bases (in general, stronger than ammonia), and their reactions are often derived from this property. The lower members of the group have a characteristic fishy ammonia odor. Many methods are available for preparation of amines, one of the most general being the *Hoffman alkylation* of ammonia.

$$NH_3 + RX \longrightarrow [RNH_2 \cdot HX] \xrightarrow{\text{NaOH}} \underset{\text{amine}}{RNH_2} \tag{6-53}$$

Usually mixtures of different amines are obtained by this method, since the reaction is difficult to control. Amines may also be prepared from alcohols, e.g.,

$$CH_3OH + NH_3 \xrightarrow{\text{heat}} \underset{\text{methylamine}}{CH_3NH_2} + H_2O \tag{6-54}$$

Degradation of acid compounds is another method for obtaining amines:

$$\underset{\underset{\text{amide}}{O}}{R-\overset{\|}{C}-NH_2} + NaOBr \xrightarrow{\text{NaOH}} RNH_2 + NaBr + CO_2 \tag{6-55}$$

Note that in this procedure the resultant amine has a shorter carbon chain, hence the term "degradation." Amines (primary and secondary) react with carboxylic acids to form *amides:*

$$\underset{O}{R-\overset{\|}{C}-OH} + R'NH_2 \longrightarrow \underset{\underset{\text{amide}}{O}}{R-\overset{\|}{C}-N\overset{R'}{\underset{H}{\diagup}}} + H_2O \tag{6-56}$$

A reaction with nitrous acid is specific for different types of amines. With primary amines, a corresponding alcohol is formed and nitrogen is evolved:

$$\underset{\underset{\text{primary amine}}{}}{R-N\underset{H}{\overset{H}{\diagup}}} + \underset{\underset{\text{nitrous acid}}{}}{HO-N=O} \longrightarrow ROH + N_2 \tag{6-57}$$

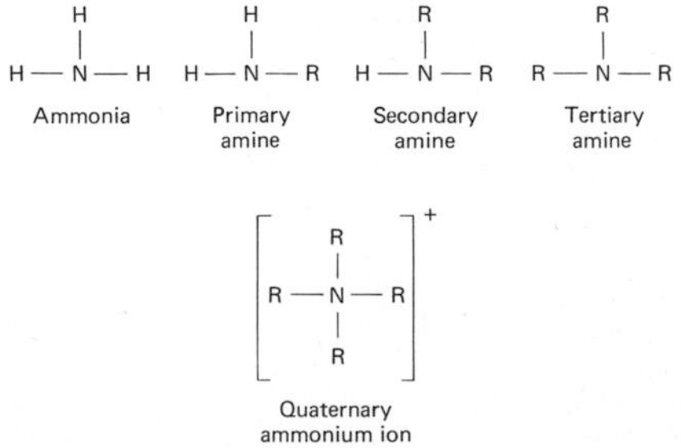

Fig. 6-11 Amines as substituted ammonias.

TABLE 6-19   Amines

ALKYL AMINES

Name	Formula	mp (°C)	bp (°C)	Sp. gr.
Methylamine	CH_3NH_2	−92.5	−6.5	0.699
Dimethylamine	$(CH_3)_2NH$	−96.0	7.4	0.680
Trimethylamine	$(CH_3)_3N$	−124.0	3.5	0.662
Ethylamine	$CH_3CH_2NH_2$	−80.6	16.6	0.689
Diethylamine	$(CH_3CH_2)_2NH$	−38.9	56.0	0.711
Triethylamine	$(CH_3CH_2)_3N$	−114.8	89.5	0.728
n-Propylamine	$CH_3CH_2CH_2NH_2$	−83.0	48.7	0.719
Di-n-propylamine	$(CH_3CH_2CH_2)_2NH$	−39.6	110.7	0.738
Tri-n-propylamine	$(CH_3CH_2CH_2)_3N$	−93.5	156	0.757
n-Butylamine	$CH_3CH_2CH_2CH_2NH_2$	−50.5	76	0.740
n-Amylamine	$CH_3CH_2CH_2CH_2CH_2NH_2$	−55.0	104	0.766
n-Hexylamine	$CH_3(CH_2)_5NH_2$	−19	130	
Laurylamine	$CH_3(CH_2)_{11}NH_2$	28	135^{16mm}	
Ethylenediamine	$H_2NCH_2CH_2NH_2$	8.5	117	0.892
Trimethylenediamine	$H_2NCH_2CH_2CH_2NH_2$		135.5	0.884
Tetramethylenediamine	$H_2NCH_2CH_2CH_2CH_2NH_2$	27	158	
Pentamethylenediamine	$H_2NCH_2CH_2CH_2CH_2CH_2NH_2$	9	178	0.855
Hexamethylenediamine	$H_2N(CH_2)_6NH_2$	39	196	
Ethanolamine	$HOCH_2CH_2NH_2$		171	1.022
Diethanolamine	$(HOCH_2CH_2)_2NH$	28	270	1.097
Triethanolamine	$(HOCH_2CH_2)_3N$	21	279^{150mm}	1.124
Allylamine	$CH_2=CHCH_2NH_2$		53.2	0.761
Aniline	$C_6H_5NH_2$	−6	184	1.022

ARYL AMINES

Name	Formula	mp (°C)	bp (°C)
Aniline	$C_6H_5NH_2$	−6	184
Methylaniline	$C_6H_5NHCH_3$	liq.	194
Dimethylaniline	$C_6H_5N(CH_3)_2$	2	193
Diethylaniline	$C_6H_5N(C_2H_5)_2$	−39	215
o-Toluidine	$CH_3C_6H_4NH_2(1,2)$	−15.5	197
m-Toluidine	$CH_3C_6H_4NH_2(1,3)$	liq.	203
p-Toluidine	$CH_3C_6H_4NH_2(1,4)$	44	200
o-Nitroaniline	$H_2NC_6H_4NO_2(1,2)$	71.5	
m-Nitroaniline	$H_2NC_6H_4NO_2(1,3)$	114	
p-Nitroaniline	$H_2NC_6H_4NO_2(1,4)$	146	
2,4-Dinitroaniline	$H_2NC_6H_3(NO_2)_2(1,2,4)$	187	
o-Phenylenediamine	$C_6H_4(NH_2)_2(1,2)$	103	257
m-Phenylenediamine	$C_6H_4(NH_2)_2(1,3)$	63	284
p-Phenylenediamine	$C_6H_4(NH_2)_2(1,4)$	140	267
o-Anisidine	$H_2NC_6H_4OCH_3(1,2)$	5.2	225
p-Anisidine	$H_2NC_6H_4OCH_3(1,4)$	57	244
p-Phenetidine	$H_2NC_6H_4OC_2H_5(1,4)$	2	254
o-Chloroaniline	$H_2NC_6H_4Cl(1,2)$	liq.	209
m-Chloroaniline	$H_2NC_6H_4Cl(1,3)$	liq.	236
p-Chloroaniline	$H_2NC_6H_4Cl(1,4)$	70	231
p-Bromoaniline	$H_2NC_6H_4Br(1,4)$	66	
2,4,6-Trichloroaniline	$H_2NC_6H_2Cl_3(1,2,4,6)$	78	262
2,4,6-Tribromoaniline	$H_2NC_6H_2Br_3(1,2,4,6)$	118	300
Diphenylamine	$C_6H_5NHC_6H_5$	54	302
Triphenylamine	$(C_6H_5)_3N$	126	348
Benzidine	$(4)H_2NC_6H_4—C_6H_4NH_2(4')$	127	401
o-Tolidine	$[—C_6H_3(CH_3)NH_2]_2(3,3',4,4')$	129	
o-Dianisidine	$[—C_6H_3(OCH_3)NH_2]_2(3,3',4,4')$	131	

TABLE 6-20   Amino Acids

NEUTRAL AMINO ACIDS	
Name	Formula
1. Glycine	$CH_2(NH_2)COOH$
2. Alanine	$CH_3CH(NH_2)COOH$
3. Serine	$HOCH_2CH(NH_2)COOH$
4. Cysteine	$HSCH_2CH(NH_2)COOH$
5. Cystine	$[-SCH_2CH(NH_2)COOH]_2$
6. Djenkolic acid	$H_2C[SCH_2CH(NH_2)COOH]_2$
7. Aminobutyric acid	$CH_3CH_2CH(NH_2)COOH$
8. Threonine	$CH_3CH(OH)CH(NH_2)COOH$
9. Valine	$(CH_3)_2CHCH(NH_2)COOH$
10. Norvaline	$CH_3(CH_2)_2CH(NH_2)COOH$
11. Methionine	$CH_3S(CH_2)_2CH(NH_2)COOH$
12. Leucine	$(CH_3)_2CHCH_2CH(NH_2)COOH$
13. Norleucine	$CH_3(CH_2)_3CH(NH_2)COOH$
14. Isoleucine	$CH_3CH_2CH(CH_3)CH(NH_2)COOH$
15. Citrulline	$H_2NCONH(CH_2)_3CH(NH_2)COOH$
16. Phenylalanine	
17. Tyrosine	
18. Diiodotyrosine (iodogorgoic acid)	
19. Thyroxine	
20. Dihydroxy-phenylalanine	
21. Dibromotyrosine	
22. Proline	
23. Hydroxyproline	
24. Tryptophane	

ACIDIC AMINO ACIDS	
1. Aspartic acid	$HOOCCH_2CH(NH_2)COOH$
2. Glutamic acid	$HOOC(CH_2)_2CH(NH)_2COOH$

TABLE 6-20   Amino Acids—(Continued)

Name	Formula
BASIC AMINO ACIDS	
1. Arginine	$NH_2C(=NH)NH(CH_2)_3CH(NH_2)COOH$
2. Lysine	$NH_2(CH_2)_4CH(NH_2)COOH$
3. Hydroxylysine	$NH_2CH_2CH(OH)(CH_2)_2CH(NH_2)COOH$
4. Canavanine	$NH_2C(=NH)NHO(CH_2)_2CH(NH_2)COOH$
5. Histidine	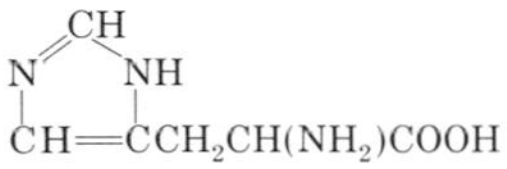 (structure)
6. Thiolhistidine	 (structure)

With secondary amines, nitrosoamines (typically, oily yellow liquids) are formed:

$$\underset{\substack{\text{secondary}\\\text{amine}}}{\overset{R}{\underset{R'}{\diagdown}}N-H} + HO-N=O \longrightarrow \underset{\text{nitrosoamine}}{\overset{R}{\underset{R'}{\diagdown}}N-N-O} + H_2O \qquad (6\text{-}58)$$

There is no reaction between the tertiary amines and nitrous acid. Thus the nitrous acid reaction is useful for analytical purpose as it helps to identify the types of amines present.

Amino acids are carboxylic acids in which the amino function is present on the carbon atom adjacent to the carbonyl atom. The amino group can occur on atoms further removed from the carboxyl carbon, but this is less frequent. The greatest importance of these compounds lies in that they are the building blocks of proteins. Table 6-20 lists typical amino acids. All amino acids are solids, and the presence of acidic and basic groups in the same molecule gives them their *amphoteric character*, or the ability to react as an acid or as a base.

$$\underset{\text{glycine}}{NH_2CH_2COOH} \rightleftharpoons \underset{\substack{\text{dipolar ion}\\\text{of glycine}}}{\overset{+}{N}H_3CH_2\overset{-}{C}O_2} \qquad (6\text{-}59)$$

The type of bonding found in the proteins is called a *peptide bond*, that is, one resulting from a loss of units of water when the hydroxyl is supplied from the acid group of one molecule and the hydrogen from the amino group of a neighboring acid molecule, as illustrated in Fig. 6-12.

Nitro compounds are characterized by the presence of the $-NO_2$ group in their structure. They are unstable and hence are often used as explosives. TNT (2,4,6-trinitrotoluene) is such a compound.

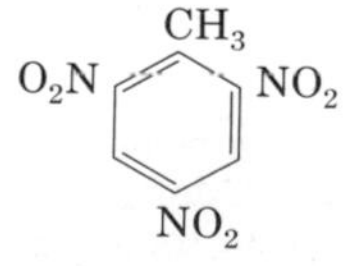

(structure of TNT)

This reactivity is also useful in a number of synthetic processes. *Nitroalkanes* can be prepared by direct reaction with nitric acid, but mixtures of nitro compounds are obtained and yields are generally low. The aromatic nitro compounds are, by far, more important than the aliphatic ones and are widely found in nature. A selected group of nitro compounds is listed in Table 6-21. The *amides* have been previously discussed as derivatives of carboxylic acid.

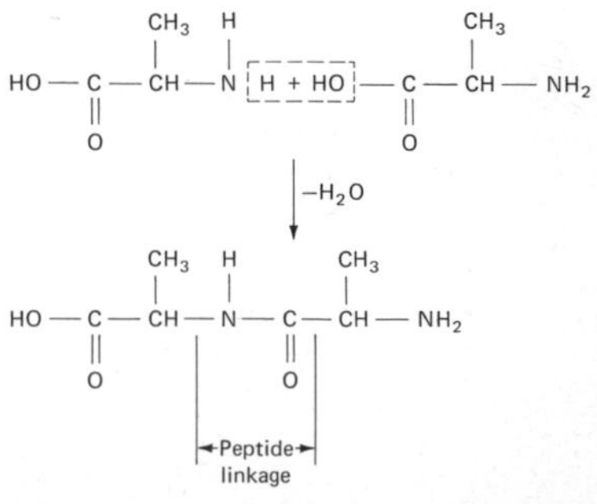

Fig. 6-12 Formation of peptide linkage.

Nitriles, or organic cyanides, are compounds with at least one carbon-nitrogen triple bond $-C \equiv N$. Reference to nitriles in connection with the preparation of acids and other reactions has already been made.

Azo compounds are substances having the $-N=N-$ group in their structures. They decompose readily to form *free radicals* and nitrogen.

$$R-N=N-R \longrightarrow 2R \cdot + N_2 \qquad (6-60)$$

azo compound   "free"
radical

The azo compounds are of great importance in the dye industry.

Diazo compounds are characterized by the presence of the $=N=N$ grouping in their molecules. Diazomethane $CH_2=N=N$ is frequently used as a methylating agent.

TABLE 6-21   Nitro Compounds

Name	Formula	mp (°C)	bp (°C)
Nitrobenzene	$C_6H_5NO_2$	5.7	210
o-Dinitrobenzene	$C_6H_4(NO_2)_2(1,2)$	118	319
m-Dinitrobenzene	$C_6H_4(NO_2)_2(1,3)$	89.8	303
p-Dinitrobenzene	$C_6H_4(NO_2)_2(1,4)$	174	299
1,3,5-Trinitrobenzene	$C_6H_3(NO_2)_3(1,3,5)$	122	
o-Nitrotoluene	$CH_3C_6H_4NO_2(1,2)$	$\left\{ \begin{array}{l} -9.5\alpha \\ -4\beta \end{array} \right\}$	222
m-Nitrotoluene	$CH_3C_6H_4NO_2(1,3)$	16	231
p-Nitrotoluene	$CH_3C_6H_4NO_2(1,4)$	52	238
2,4-Dinitrotoluene	$CH_3C_6H_3(NO_2)_2(1,2,4)$	70	
2,4,6-Trinitrotoluene	$CH_3C_6H_2(NO_2)_3(1,2,4,6)$	80.6	
2,4,6-Trinitro-*m*-xylene	$(CH_3)_2C_6H(NO_2)_3(1,3,2,4,6)$	182	
Picric acid	$HOC_6H_2(NO_2)_3(1,2,4,6)$	122.5	
2,4,6-Trinitroresorcinol	$(HO)_2C_6H(NO_2)_3(1,3,2,4,6)$	176	
Tetryl	$CH_3(NO_2)NC_6H_2(NO_2)_3(1,2,4,6)$	129	
o-Nitrochlorobenzene	$ClC_6H_4NO_2(1,2)$	32.5	245
m-Nitrochlorobenzene	$ClC_6H_4NO_2(1,3)$	47.9	236
p-Nitrochlorobenzene	$ClC_6H_4NO_2(1,4)$	83	239
2,4-Dinitrochlorobenzene	$ClC_6H_3(NO_2)_2(1,2,4)$	53	
Picryl chloride	$ClC_6H_2(NO_2)_3(1,2,4,6)$	83	
o-Nitrodiphenyl	$C_6H_6 \cdot C_6H_4NO_2(1,2)$	37	320
p-Nitrodiphenyl	$C_6H_5 \cdot C_6H_4NO_2(1,4)$	114	340

Sulfur Compounds

Of the many organic sulfur compounds, the thiols and the sulfonic acids are the most prominent. The *thiols* ("sulfur alcohols" or mercaptans) are the structural analogs of the alcohols

$$R-OH \qquad R-SH$$

alcohol   thiol

A number of thiols are found in petroleum oil. They have a strong characteristic, unpleasant odor. Although the chemistry of thiols is somewhat similar to that of the alcohols, the important differences are in the different susceptibility of the thiols to oxidizing agents; i.e., the sulfur compounds are more easily oxidized.

Sulfonic acids are important commercially as components of detergents. The sodium salts are generally used for this purpose.

$$RSO_3Na$$

sodium sulfonate

The fact that magnesium and calcium sulfonates are much more soluble in water than the fatty acid salts of these metals makes the sulfonate detergents more attractive than ordinary soaps for use with hard water. Sulfonic acids are strong acids, in fact, similar to sulfuric acid, and are widely used in the preparation of synthetics.

Arenes

Arenes, also known as the aromatic hydrocarbons, are characterized by the presence of the benzene ring in their structure. In general, the arene hydrocarbons are considerably more reactive than the alkanes, but the stability of the benzene ring is remarkably great. Thus most reactions of the arenes involve the substituent groups rather than the ring. The great stability of the benzene ring is ascribed to the phenomenon of *resonance*, or the rapid oscillation of its electrons from one position to another between the "fixed" atoms in the ring. The formula of benzene, C_6H_6, is represented by any one of the several notations shown in Fig. 6-13. A number of arenes are obtainable from coal tar. Table 6-22 lists typical

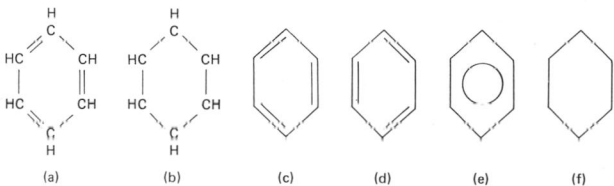

Fig. 6-13 Representations of the benzene structure. Structures *b* and *f* are the most commonly used forms, it being understood that the benzene ring has three resonating double bonds.

members of this group, including selected nitrogen- and oxygen-containing arenes. When two substituent groups are present on the benzene ring, the convention illustrated in Fig. 6-14 is used to name the various possible isomers.

When two or more benzene rings are "fused," i.e., have common ortho positions, the

TABLE 6-22   Selected Arenes

Name	Formula	mp (°C)	bp (°C)	Sp. gr.
Benzene	C_6H_6	5.6	80.1	0.874
Toluene	$C_6H_5CH_3$	−95.0	110.6	0.862
o-Xylene	$1,2\text{-}C_6H_4(CH_3)_2$	−27.1	142.7	0.875
m-Xylene	$1,3\text{-}C_6H_4(CH_3)_2$	−47.4	139.3	0.864
p-Xylene	$1,4\text{-}C_6H_4(CH_3)_2$	13.4	138.4	0.862
Ethylbenzene	$C_6H_5C_2H_5$	−94.4	136.1	0.867
n-Propylbenzene	$C_6H_5CH_2C_2H_5$	−99.2	159.4	0.858
Diphenyl	$C_6H_5\text{—}C_6H_5$	69.0	254.9	1.041
Diphenylmethane	$(C_6H_5)_2CH_2$	27.0	262.0	1.006
Triphenylmethane	$(C_6H_5)_3CH$	92.5	359.2	1.014
Naphthalene	$C_4H_4C_6H_4$	80.0	218.0	1.145
Anthracene	$C_4H_4C_6H_2C_4H_4$	216.0	340.0	1.250
Nitrobenzene	$C_6H_5NO_2$	5.7	210.9	1.199
m-Dinitrobenzene	$1,3\text{-}C_6H_4(NO_2)_2$	89.7	decomp.	1.571
p-Dinitrobenzene	$1,4\text{-}C_6H_4(NO_2)_2$	173.0	decomp.	1.625
o-Nitrotoluene	$1,2\text{-}C_6H_4CH_3NO_2$	−10.6	222.3	1.163
m-Nitrotoluene	$1,3\text{-}C_6H_4CH_3NO_2$	15.5	231.0	1.164
p-Nitrotoluene	$1,4\text{-}C_6H_4CH_3NO_2$	51.4	238.0	1.286
2,4-Dinitrotoluene	$1,2,4\text{-}C_6H_3CH_3(NO_2)_2$	69.6	decomp.	1.521
2,4,6-Trinitrotoluene	$1,2,4,6\text{-}C_6H_2CH_3(NO_2)_2$	80.7	decomp.	1.654
Aniline	$C_6H_5NH_2$	−6.2	184.3	1.027
o-Toluidine	$1,2\text{-}C_6H_4CH_3NH_2$	−21.0	200.0	1.003
m-Toluidine	$1,3\text{-}C_6H_4CH_3NH_2$	−31.5	203.0	0.996
p-Toluidine	$1,4\text{-}C_6H_4CH_3NH_2$	45.0	200.5	1.046
Methylaniline	$C_6H_5CH_2NH_2$	−57.0	196.0	0.991
Dimethylaniline	$C_6H_5N(CH_3)_2$	2.5	193.1	0.962
Diphenylamine	$(C_6H_5)_2NH$	53.0	302.0	1.159
Triphenylamine	$(C_6H_5)_3N$	127.0	348.0	0.774
Phenol	C_6H_5OH	41.0	182.6	1.082
o-Cresol	$1,2\text{-}C_6H_4CH_3OH$	30.1	190.8	1.047
m-Cresol	$1,3\text{-}C_6H_4CH_3OH$	10.0	201.0	1.034
p-Cresol	$1,4\text{-}C_6H_4CH_3OH$	36.0	201.1	1.035

resulting compounds are classified as *polynuclear aromatic hydrocarbons*. The examples shown in Fig. 6-15 have the carbon positions numbered in the conventional way to indicate positions of substituents when present.

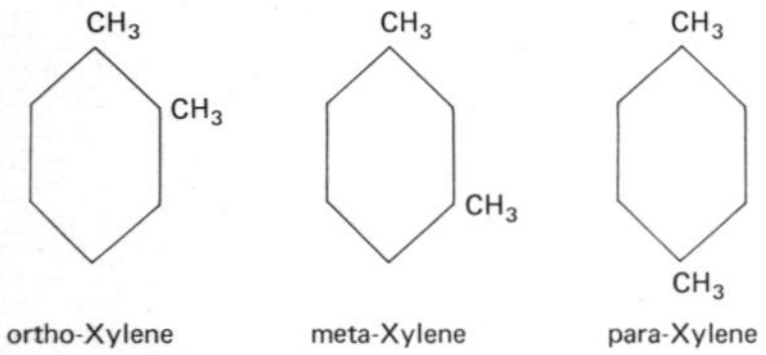

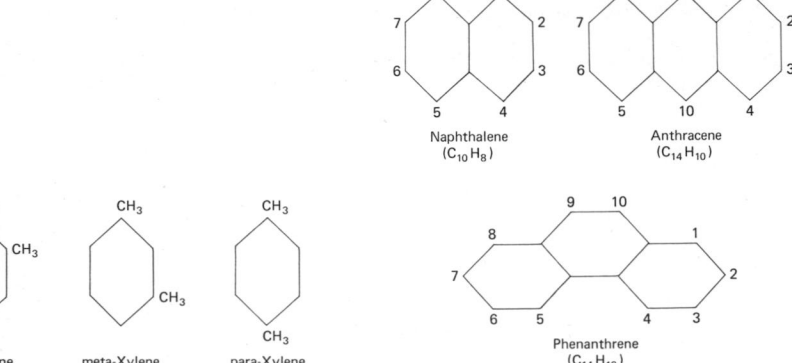

Fig. 6-14   Isomers of substituted benzene.

ortho-Xylene   meta-Xylene   para-Xylene

Naphthalene
($C_{10}H_8$)

Anthracene
($C_{14}H_{10}$)

Phenanthrene
($C_{14}H_{10}$)

Fig. 6-15   Representation of polynuclear aromatic hydrocarbons. The position of each carbon is identified by the numbers as shown. Positions 1 and 2 in naphthalene are also called the α and β positions.

Major types of reactions of aromatic rings are:
Substitution

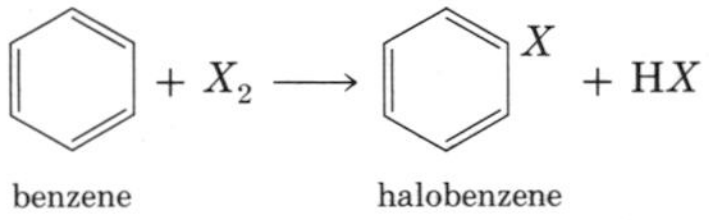

$$\text{benzene} + X_2 \longrightarrow \text{halobenzene}\ X + HX \qquad (6\text{-}61)$$

$$\text{benzene} + HNO_3 \longrightarrow \text{nitrobenzene}\ NO_2 + H_2O \qquad (6\text{-}62)$$

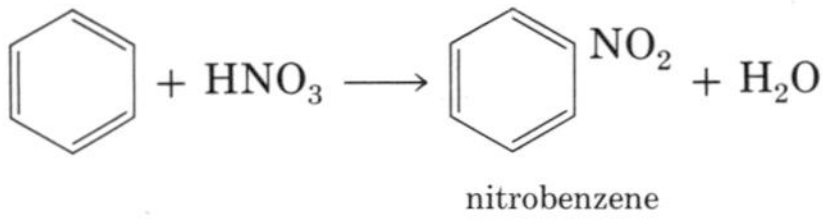

Friedel-Crafts reaction

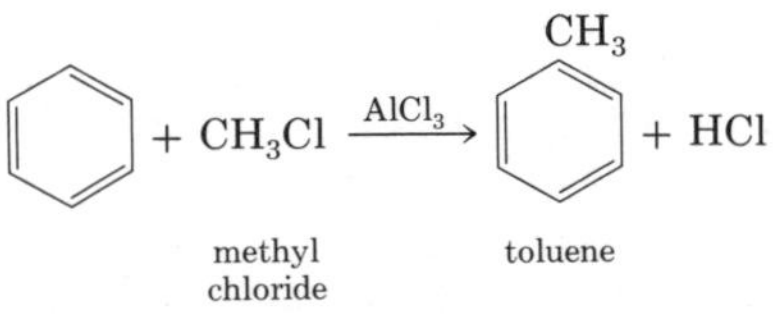

$$\text{benzene} + CH_3Cl \xrightarrow{\ AlCl_3\ } \text{toluene}\ CH_3 + HCl \qquad (6\text{-}63)$$

methyl chloride

Sulfonation

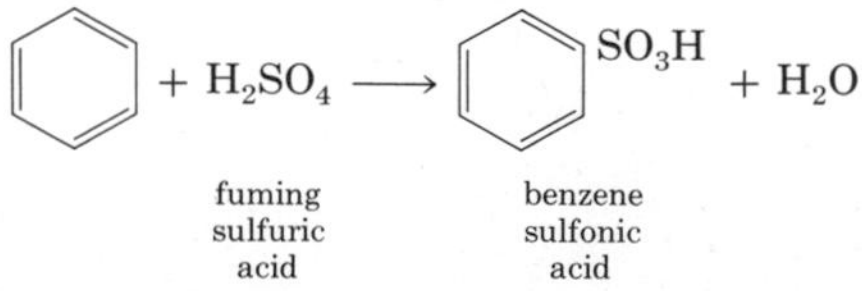

$$\text{benzene} + H_2SO_4 \longrightarrow \text{benzene sulfonic acid}\ SO_3H + H_2O \qquad (6\text{-}64)$$

fuming sulfuric acid

Complications arise when more than one substituent is being introduced. Other reactions of the arenes involve reactions of substituent groups that may be present. Many of these are similar to those already discussed in earlier sections under alcohols, amines, halogen compounds, etc.

Heterocyclic Compounds

Heterocyclic compounds are ring compounds in which not all atoms in the ring are of the same element. Examples of these compounds are given in Fig. 6-16. Heterocyclic compounds are important components of many natural products, as well as synthetic dyes and pharmaceuticals.

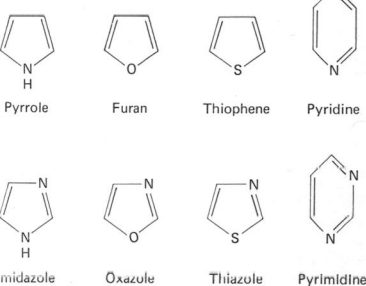

Pyrrole   Furan   Thiophene   Pyridine

Imidazole   Oxazole   Thiazole   Pyrimidine

Fig. 6-16 Heterocyclic compounds.

Carbohydrates

Carbohydrates occur in great abundance in plant and animal life. Sugars, starches, and celluloses are the important members of this group. The name carbohydrate is derived from the observation that hydrogen and oxygen are present in the carbohydrate molecule in the same ratio as in water, and that many carbohydrates can react with, or lose, water to form other carbohydrates:

$$C_{12}H_{22}O_{11} + H_2O \longrightarrow C_6H_{12}O_6 + C_6H_{12}O_6 \qquad (6\text{-}65)$$

sucrose   glucose   fructose

Empirical formulas, such as are used in Eq. (6-65), are inadequate to distinguish between two isomers, such as glucose and fructose. To do so requires information about their molecular structures as illustrated in Fig. 6-17.

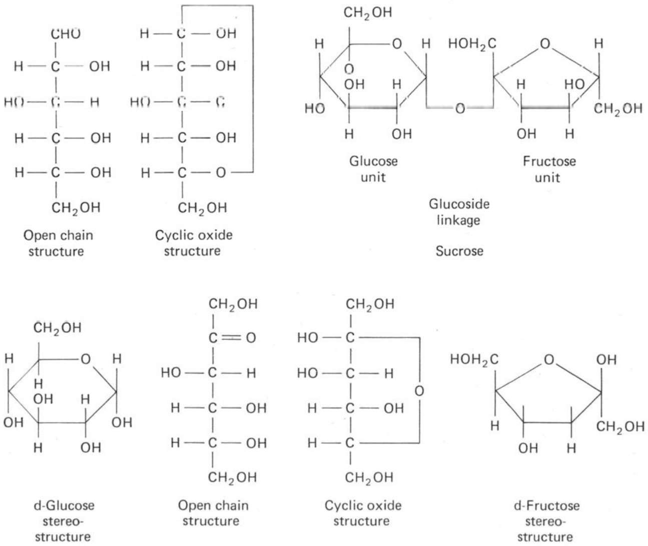

Open chain structure   Cyclic oxide structure   Glucose unit   Fructose unit   Glucoside linkage   Sucrose

d-Glucose stereo-structure   Open chain structure   Cyclic oxide structure   d-Fructose stereo-structure

Fig. 6-17 Structures of carbohydrates. Glucose and fructose are monosaccharides. Sucrose, a disaccharide, is composed of one glucose unit and one fructose unit combined through a glucoside linkage. The stereo structure indicates the spatial arrangement of the side chains with respect to each other and the ring structure.

Chapter 7

Metals and Alloys

7-1 PHYSICAL METALLURGY

Structure of Metals

In broad chemical terms, a metal is an element which loses electrons during a chemical reaction and thereby carries a positive charge in the compounds it forms. In the chemical sense, sulfur would be a metal in a compound such as sodium sulfate (Na_2SO_4) in which it exhibits a valence of $+6$, but would be a nonmetal in a compound such as lead sulfide in which it has a valence of -2. A great many elements show a dual nature, entering into reactions as either a metal or a nonmetal. All told, nearly four out of five elements can enter into chemical reactions as a metal. However, in common usage, the term metal is usually reserved for those materials which have a high electrical conductivity, a high thermal conductivity, and a light-reflective surface ("metallic" luster), and which can be deformed under a stress without breaking. There are other properties generally character-istic of metals, such as hardness and elasticity, but these are hard to relate quantitatively because of the extremes to which they can vary from one metal to another. If there is any universal characteristic of metallic solids, it is that they are all crystalline.

In a *crystalline material,* the atoms that go to make up the material are arranged in a fixed spatial pattern. Although there are a few exceptions, metal crystals fall into three fundamental crystal structures, face-centered cubic (fcc), hexagonal close-packed (hcp), and body-centered cubic (bcc). In the bcc structure, an atom is positioned at each corner of a cube, with an additional atom positioned at the center of the cube. In the hcp structure, an atom is positioned at each corner of a regular hexagon, two additional atoms in the centers of the top and bottom faces, and three more atoms in the body of the hexagon. The latter three are arranged in such a way as to form a triangle in a plane halfway between the top and bottom faces. In the fcc structure, an atom is positioned at each corner of a cube, with six additional atoms located at the center of each face. These unit structures are illustrated in Fig. 7-1. Metal bodies consist of large numbers of these structures interlocking in all directions through a sharing of atoms located at the corners and on the faces of adjoining unit crystals, as shown in Fig. 7-2. For the bcc crystal shown in the illustration, each of the corner atoms is shared by the four unit cubes that have a common corner at that point. Thus, only one-fourth of each corner atom can be assigned to any one unit cube. Since there are eight corner atoms and one body-centered atom (which is not shared with adjacent cubes) in the bcc arrangement, this type of structure is said to have three atoms per unit crystal.

Example 7-1 How many atoms per unit crystal are there in the fcc system?
solution Each of the eight corner atoms is shared by four adjacent unit cubes, while each of the six face atoms is shared by two adjacent unit cubes. Thus the atoms per unit crystal for the fcc system is

$$(\tfrac{1}{4} \times 8) + (\tfrac{1}{2} \times 6) = 5$$

It is interesting to note that, as might be expected, there is a relationship between the atomic structure of an atom and the system into which it crystallizes. This is brought out in Table 7-1, which gives the crystal structure of metals arranged in accordance with their position in the periodic table.

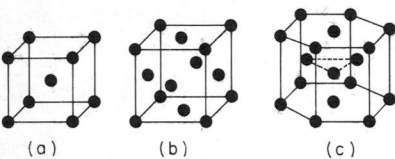

Fig. 7-1 Position of atoms in unit crystal structures. (a) bcc; (b) fcc; (c) hcp.

There are many instances in which the property of a metal is influenced by the various planes in which the atoms of its crystals are located. Consequently, a system in which three coordinate axes are set up parallel to the edges of the unit crystal, as illustrated in Fig. 7-3, has been devised to describe such planes. Units of length are set up along these axes equal to the lengths of the corresponding sides of the unit cell. Any plane will then intersect one, two, or all three of these axes, and the reciprocals of the intercepts are called the *Miller indices* of the plane. Since any actual crystal consists of a large number of unit crystals, the orientation of a given crystallographic plane is more important than its actual location, and the Miller indices are therefore given in terms of a parallel plane having the smallest whole-number indices. Figure 7-4 illustrates how Miller indices are obtained for a plane intersecting three coordinate axes. In a cube, each face intersects only one axis, the intercepts for each face being $(1,\alpha,\alpha)$, $(\alpha,1,\alpha)$, and $(\alpha,\alpha,1)$. The reciprocals of these numbers are $(1,0,0)$, $(0,1,0)$, and $(0,0,1)$, and the corresponding Miller indices are (100), (010), and (001), written as shown. It should be noted that the (200) or (300) plane is parallel to the (100) plane. The latter Miller indices for these planes, and all such parallel planes, are (100).

Example 7-2 What are the Miller indices for a plane that intersects X, Y, and Z coordinate axes at 3, 6, and 4, respectively?

solution The reciprocals of the intercepts are $\frac{1}{3}$, $\frac{1}{6}$, and $\frac{1}{4}$. Multiplying each of these by 12 (the least common denominator) gives the Miller indices (423). Note that a parallel plane having intercepts of 6, 12, and 8 would still have Miller indices of (423).

In an actual metal, crystals will tend to grow from a melt around a *nucleus* which is usually a unit crystal that forms spontaneously as the melt cools. The nucleus then forms a structure around which individual atoms can attach themselves as the melt continues to cool. The single unit crystal thus grows until it comes in contact with other growing crystals, thus forming *grains*. Since grains can grow to substantial size, they can be readily seen under a microscope, as in Fig. 7-5, in which the grain structure of pure metal, or a metal of uniform grain composition, is illustrated. When a melt is cooled quickly, nucleation, i.e., the formation of nuclei, occurs rapidly, producing many nuclei and resulting in a fine crystal structure. Slow cooling permits fewer nuclei to grow to larger sizes. In a microscopic examination (such as that illustrated in Fig. 7-5), the sample is sectioned so that the plane of observation cuts across different crystals in a random manner. Thus the actual grain-size distribution is not apparent from simple observation but must be deduced from the largest grain and the randomness of the crystal orientations, as illustrated in Fig. 7-6. The microscopic examination of alloys can be even more revealing, as will be discussed more fully later in this chapter.

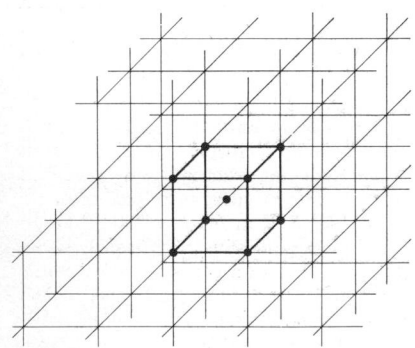

Fig. 7-2 Sharing of atoms by adjacent unit crystals. In the central unit bcc crystal shown in heavy outline, the corner atoms are shared with each of the unit cubes that come together at each corner. The body-centered atom, however, belongs exclusively to the unit crystal shown.

Alloys

An alloy is a metal formed by the mixture of two or more metals. The most common way of producing an alloy is to melt the metals to-

TABLE 7-1   Crystal Structure of the Metals

Li bcc hcp	Be hcp												
Na bcc	Mg hcp												Al fcc
K bcc	Ca fcc hcp bcc	Sc hcp fcc	Ti hcp bcc	V bcc	Cr bcc fcc	Mn	Fe bcc fcc	Co hcp fcc	Ni fcc	Cu fcc	Zn hcp	Ga	Ge
Rb bcc	Sr fcc hcp bcc	Y hcp	Zr hcp bcc	Nb bcc	Mn bcc	Tc hcp	Ru hcp	Rh fcc	Pd fcc	Ag fcc	Cd hcp	In	Sn
Cs bcc	Ba bcc	La hcp fcc	Hf hcp bcc	Ta bcc	W bcc	Re hcp	Os hcp	Ir fcc	Pt fcc	Au fcc	Hg	Tl hcp bcc	Pb fcc

Ce fcc	Pr hcp	Nd hcp	Pm	Sm	Eu bcc	Gd hcp	Tb hcp	Dy hcp	Ho hcp	Er hcp	Tm hcp	Yb fcc	Lu hcp

From Charles E. Mortimer, "Chemistry—A Conceptual Approach," Van Nostrand Reinhold, New York, 1967.

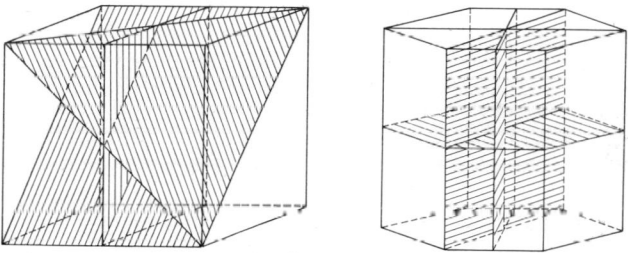

Fig. 7-3 Principal atomic planes in cubic and hexagonal unit crystals.

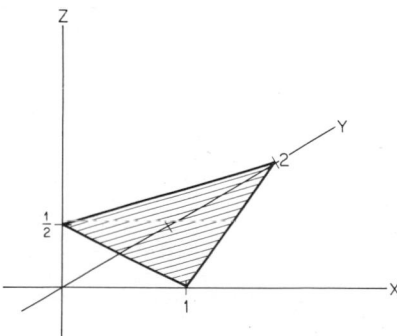

Fig. 7-4 Miller indices. The plane shown intersects the X axis at 1 scale unit, the Y axis at 2 on the same scale, and the Z axis at $\frac{1}{2}$. The reciprocals of the intercepts are 1, $\frac{1}{2}$, and 2. Multiplying each of these by 2 so as to avoid the use of a fraction gives the Miller indices for the plane 2, 1, 4. All planes parallel to the one shown will have the same Miller indices when the reciprocals of their intercepts are reduced to the lowest whole numbers.

gether to form a homogeneous liquid and allow the melt to cool and solidify. Other means of preparing alloys are by electrolytic codeposition of the metals, or by heat-treating a composite of the metals, such as would be obtained by pressing a mixture of metal powders. In the melt, the metals of an alloy may dissolve completely in each other and form a homogeneous solid solution on cooling; or they may crystallize separately and be present in the alloy as a uniform mixture of heterogeneous crystals; or they may combine chemically to form homogeneous intermetallic compounds. Alloys which form as a result of replacing an atom in a metal crystal with an atom of another metal without significantly altering the crystal system are called *substitutional alloys*. Those alloys in which the atoms of the alloying metal take up positions between unit crystals of the host metal are called *interstitial alloys*.

The properties of an alloy depend on the metals forming the alloy, the manner in which they combine, and the resulting crystalline composition. Usually, the properties of an alloy differ considerably from those of the constituent metals. Although there are no fixed rules, in a general way, alloys are usually harder, tougher, and more corrosion-resistant, and have lower melting points than their component metals. Because of the wide range of properties that alloying can develop, most metals used commercially are alloys. They are classified as *nonferrous*, containing no iron, and *ferrous*, containing iron. Steels are ferrous alloys containing carbon as an alloying element. Table 7-2 gives the compositions of the important ferrous and nonferrous alloys.

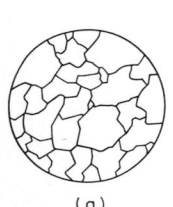 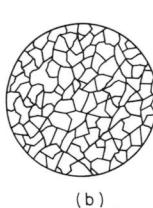

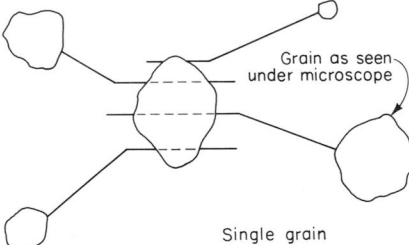

Fig. 7-5 Grain structure of a metal with uniform grain composition. (*a*) Slow cooling from melt; (*b*) fast cooling from melt.

Fig. 7-6 Influence of sectioning on microscopic appearance of a grain. In an alloy of uniform grain size, the largest grain size seen under the microscope is indicative of the actual grain size.

Phase Diagrams

When two or more metals are present in a melt that is freezing, the solid that forms usually has a composition and crystallographic structure that changes as the solidification process progresses, owing to changes in the solubility of the various components in each other as the temperature of the melt changes. This ultimately gives rise to the specific structure of the alloy, and thereby its specific physical and chemical properties. A quantity of matter that has a homogeneous composition and physical form is called a *phase*, and a *phase diagram* shows the relationships that exist among the various phases of an alloy, and the transformations to which they are subject.

If a melt of a pure metal is cooled, a cooling curve such as that shown in Fig. 7-7 will be obtained, with the onset of crystallization occurring at *A*. From *A* to *B*, additional crystals will form, and those already formed will grow, essentially at a constant temperature, the temperature being maintained by the release of the heat of fusion of the metal. At *B*, no more liquid remains and the solid simply then cools to room temperature. A somewhat different situation occurs if two metals are present. Figure 7-8 shows the cooling curves for a number of compositions of a *binary alloy*, i.e., one consisting of two metals *A* and *B*. For 100 percent *A* or 100 percent *B*, the cooling curves are similar to Fig. 7-7. However, for compositions containing both metals, the onset of crystallization at *A* is followed by a period of solidification at a falling temperature. This is because the crystals formed at *A* do not have the same composition as the liquid from which they were formed. As crystallization proceeds, a point is reached at *B* where the crystals that are forming have the same composition as the liquid from which they are being crystallized. At this point, the

TABLE 7-2 Composition of the Common Commercial Alloys

Name	Composition, %	Uses
Admiralty metal	Cu 70, Zn 29, Sn 1	Condenser tubes for use with salt water. Marine fittings
Alpax	Al 92–89, Si 4–11	General castings. Has lower shrinkage than Al-Cu alloy
Aluminum alloy—No. 12 S.A.E. No. 30	Al 92, Cu 8	This alloy with slight variations in composition is the standard aluminum casting alloy used in the United States
Aluminum brass	Cu 70–68, Zn 27–31, Al 1–3	Propeller blades, rudder frames, sea valves
Babbitt metal	Sn 70–90, Sb 7–24, Cu 2–22	Bearings and antifriction lining for bronze or steel bushings
"Genuine"	Sn 88.9, Sb 7.4, Cu 3.5	
Bell metal	Cu 80–75, Sn 20–25 (sometimes Ag, Ni, or other metals)	Bells, gongs, etc.
Brass:		
Gilding metal	Cu 99–80, Zn 1–20	Cheap jewelry, gold paint
Dutch metal	Cu 80–76, Zn 20–24	Thin sheets as substitute for gold leaf
Standard	Cu 73–66, Zn 27–34	Brass for cold working. Sheets, tubes, cartridges
White	Cu, less than 45	Ornamental castings not requiring strength
Brazing metal	Cu 85, Zn 15	
Britannia metal	Sn 95–90, Sb 5–10, Cu 1–3	Cheap tableware
Bronze:	Cu 90, Al 10	Hard, noncorrodible. Used in parts exposed to tanning, sulfite, and similar corrosive liquors
Aluminum		
Bearing	Cu 70–90, Sn 1–10, Pb 0–15, Zn 0–27	Bearings of various sorts
Gear	Cu 89, Sn 11	Used for heavy gears usually against steel
Gun	Cu 88, Sn 10–8, Zn 2, Pb 2	Strong valves and fittings
Phosphor	Cu 80–77, Sn 8–10, Pb 0–15, P 0–1.0	Bearing metal, wire, rods, steam fittings
Plastic	Cu 70–50, Pb 30–50, Ni trace	Bearing metal
Silicon	Cu 96, Si 1–4	Telegraph wires, electrical work, rivets, range boilers, fans
Chromel (Nichrome)	Ni 70–85, Cr 15–30 (approx.)	Resistance wire for heating units, crucibles, triangles, tongs (patented)
Constantan	Cu 60, Ni 40	Used with copper or iron to make thermocouples
Copper-beryllium Beryllium bronze	Cu 97–98, Be 2–3	Heat-treated to give 200,000 lb/in^2 tensile strength. Used for pump parts, springs, and nonsparking tools
Cupronickel	Cu 98–52, Ni 2–48	Projectile driving bands, rifle-bullet caps, electrical resistances, condenser tubes
Delta metal	Cu 60, Zn 40, Mn 0.5–2	See Sterro metal
Dow metal	Mg 85–92, Al 8–15	Standard magnesium alloys. Used as castings or forgings
Duralumin	Al 95.5, Cu 3.0, Mn 1.0, Mg 0.5 (approx.)	Strongest and best of aluminum alloys. Used in airplane and automobile parts
Fusible metals:		
Lipowitz	Bi 50, Pb 27, Sn 13, Cd 10	These and other ternary and quaternary alloys are used for fuse plugs for automatic sprinklers

TABLE 7-2 Composition of the Common Commercial Alloys—(Continued)

Name	Composition, %	Uses
Fusible metals: (continued)		
Wood's	Bi 38, Pb 31, Sn 15, Cd 16	
German silver		See Nickel silver
Gun metal	Cu 92–88, Sn 8–12	Gears, heavy hydraulic castings
Hercules metal	Aluminum brass with Fe	Same as aluminum brass with added toughness
Invar	Fe 64, Ni 36	Low coefficient of expansion. Used in clocks, precision instruments
Magnalium	Al 90–94, Mg 10–6	Scientific instruments, balance beams
Magnolia metal (lead-base babbitt)	Pb 78, Sb 16, Sn 6	Antifriction, bearing alloy
Manganese bronze	Cu 56, Zn 41, Sn 0.5, Fe 1, Mn 0.5, Al 1	Propeller blades. Noncorrodible and good wearing qualities
Manganin	Cu 82, Mn 15, Ni 2.3, Fe 0.6	High electrical resistance and low temperature coefficient
Monel metal	Ni 68, Cu 27, Fe, Mn, Si 5	Almost noncorrodible. Used for propeller blades, wire, sheets, valves, pumps, etc.
Muntz metal	Cu 60, Zn 40	Sheathing for ships, bolts, nuts, condenser tubes
Naval brass	Cu 60, Zn 39, Sn 1	Properties like Muntz metal. Less easily corroded by sea water
Nickelin	Cu 74.5, Ni 25, Fe 0.5	Resistance wire
Nickel silver (German silver)	Ni 18–25, Zn 20–30, Cu (remainder)	Tableware, cheap jewelry, base for silver plating
Palau	Pd, Au	Substitute for platinum in chemical crucibles, dishes, etc. (patented)
Pewter	Sn 85–90, Sb 15–10	Platters, bowls, cups, etc.
Platnite	Fe 54, Ni 46, C 0.15	Same coefficient of expansion as glass. Used as substitute for platinum in equipping incandescent lamps
Platinoid	Cu 60, Zn 24, Ni 14, W 1–2	High-resistance wire but not suitable for heating coils
Platinum-iridium	Pt 90, Ir 10	Standard meter and other standards. Thermocouple with platinum
Rheotan	Cu 52, Zn 18, Ni 25, Fe 5	High resistance but not suitable for heating coils
Shot metal	Pb 99, As 1	Casting bullets and small shot
Solder:		
Soft	Pb 67, Sn 33	Plumber's solder
Medium	Pb 50, Sn 50	
Hard	Pb 33, Sn 67	
Silver	Ag 10–80, Cu 16–52, Zn 4–38	High-melting solder
Speculum metal	Cu 70–65, Sn 30–35	Takes a high polish. Formerly used in reflectors for telescopes
Steel:		
Plain carbon	C 0.05–0.15	Boiler plate, rivets, sheet steel, casehardening stock
	C 0.15–0.25	Structural work, bridges, shafting
	C 0.25–0.40	Axles, connecting rods, piston rods
	C 0.4–0.75	Rails, steel castings
	C 0.6–0.8	Cutlery, woodworking tools, drills
	C 0.8–1.0	Springs, lathe tools, drills

TABLE 7-2   Composition of the Common Commercial Alloys—(Continued)

Name	Composition, %	Uses
Steel: (continued)		
	C 1.0–1.2	Large lathe tools, axes, knives
	C 1.2–1.5	Saws, files, balls for bearings, razors
Chromium	Cr less than 3	Projectiles, files
Chromium-tungsten	C 0.65–0.75, W 13.5–18.5, Cr 3.5–4.5, V 0.75–2.0	High-speed tools. May be run at 500–600°C without losing their edge
Chromium-vanadium	C 0.25–1, Cr 0.8–1.1, V 0.15	Gears, springs, general automobile parts
Manganese	Mn 12–14	Used on sharp railroad curves, frogs, switches, etc., where wear is hard
Nickel	Ni 3–4	Drive shafts, crankshafts, gears, and other automobile parts
Nickel-chromium	Ni 1–4, Cr 0.45–2	Armor plate, automobile parts subject to heavy stresses
Silicon	Si less than 5	Has high permeability and low hysteresis. Used in dynamo construction
Stainless	Cr 13–20	Used in cutlery, containers for corrosive liquids, etc.
	Cr 8–25, Ni 8–25	General-purpose corrosion-resistant metal
Sterro metal (Aich's metal, Delta metal)	Cu 60, Zn 38, Fe 2	Strong as mild steel and not easily corroded. Used in hydraulic cylinders, sea-water valves
Type metal	Pb 60–85, Sb 8–20, Sn 5–35	
Y alloy	Al 92.5, Cu 4, Ni 2, Mg 1.5	Used for general castings

From Robert S. Williams and Victor O. Homerberg, "Principles of Metallography," 4th ed., McGraw-Hill Book Company, New York, 1939.

compositions of the crystals being formed and the melt do not change with time, and the crystallization from B to C proceeds at a constant temperature as if a pure substance were crystallizing. At C no more liquid remains, and the solid proceeds to cool to room temperature.

The composition illustrated as 40 percent A is a special one in which the composition of the solid is the same as that of the melt throughout the entire freezing process. This special composition is known as a *eutectic alloy*, and the temperature at which a eutectic alloy solidifies is known as the *eutectic temperature*. The *eutectic* point is the point defined by the eutectic composition and the eutectic temperature.

When the various temperatures at which crystallization first begins (A points in Fig. 7-8) are plotted against composition, a *freezing-point diagram*, frequently referred to as an *equilibrium diagram*, is obtained, such as Fig. 7-9. If we were to cool a melt consisting of 75 percent A and 25 percent B starting at temperatures T_1 (point X_0), nothing much would happen until T_2 was reached, at which time crystals of A would begin to form. This leaves the remaining melt richer in B. Thus the freezing point of the melt would be lower, corresponding to a melt richer in B. As the mix continues to cool, more A freezes out of the melt, and this process continues with melt composition and temperature following line PQ. At point Q, further cooling simply solidifies the remaining melt, Q being the eutectic point. While the composition of the melt changes as the solidification process proceeds, the composition of the melt *plus* solid does not change, and the cooling follows line X_0-X_1. At temperature T_3, the mixture consists of solid A in a liquid of A and B. At T_4, the eutectic temperature, the last of the liquid phase freezes, and as the mixture cools still further to T_5, the mix consists of two crystals, pure A and eutectic of A plus B. Above line PQ all points represent temperature-composition combinations for which the system is liquid, and the line

PQ is called the *liquidus*. Below the line SQ, all the material is solid, and the line SQ is called the *solidus*. In the area defined by PQ, QS, and SP, both liquid and solid phases are present. Because the freezing-point diagram also reveals the phase composition of a mix at a given temperature, it is also referred to as a *phase diagram*.

In an analogous manner, if we were to cool a melt consisting of 20 percent A and 80 percent B starting at T_1 (point Y_0), the first solid would appear at T_6 and would be pure B. Further cooling would result in a mixture of B and liquid until T_4 is reached, at which temperature the entire mix would be solid and consist of pure B and eutectic. As before, QR is a liquidus and QT a solidus, with the area bounded by QR, RT, and TQ being a region in which both liquid and solid are present. The cooling of this latter melt would follow line Y_0-Y_1, since the overall composition does not change regardless of the manner in which A and B distribute themselves.

A composition of 40 percent A and 60 percent B would cool along line Z_0-Z_1. In this case, the first crystallization would take place at T_4 and the solid would consist of 40 percent A and 60 percent B, or the eutectic alloy. Freezing would take place at constant temperature, with the liquid and solid having the same composition. Below T_4, the mixture would be solid and consist entirely of eutectic alloy.

The phase diagram shown in Fig. 7-9 is for a simple system in which A and B are mutually

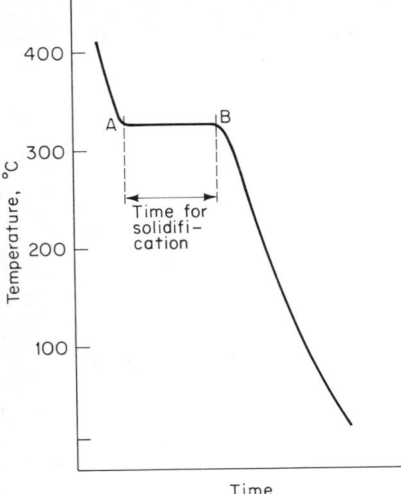

Fig. 7-7 Cooling curve of a pure metal. The metal illustrated is cadmium. Up to point A the liquid is cooling. At point A solidification begins, and continues until point B, at which point all the liquid has solidified. The curve beyond point B represents the cooling of the solid metal. The shape and duration of the cooling and solidification portions of the curve depend upon the amount of metal present and the rate at which heat is lost from the pot.

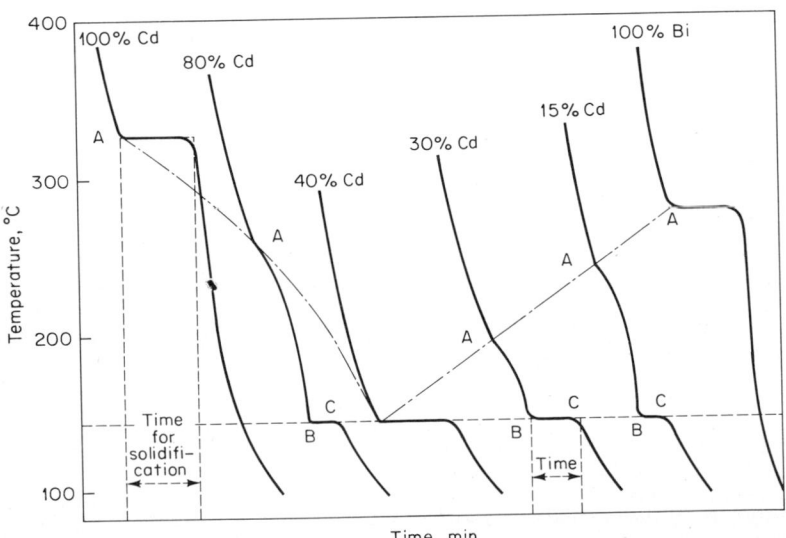

Fig. 7-8 Cooling curves of a binary alloy (cadmium-bismuth). (*From Robert S. Williams and Victor O. Homerberg, "Principles of Metallography," 4th ed., McGraw-Hill Book Company, New York, 1939.*)

soluble and there are no crystalline transformations in the solid phases. More frequently, however, the first freezing will not result in pure A, but rather in an alloy rich in A but containing some B. Furthermore, once formed, crystals may undergo changes such as, for example, from bcc to fcc. These changes can also be incorporated into a phase diagram, and while these may result in a complicated diagram, each section is generally similar to the binary system of Fig. 7-9.

The phase diagram is also useful in giving the actual composition of individual phases of the system at any temperature. In Fig. 7-9, for example, the composition of the system at point b is obtained by the *lever rule*:

$$\text{Percent solid} = \frac{\text{length of line } bc}{\text{length of line } ac} \times 100$$

$$\text{Percent liquid} = \frac{\text{length of line } ab}{\text{length of line } ac} \times 100$$

in which a, b, and c are defined in the figure.

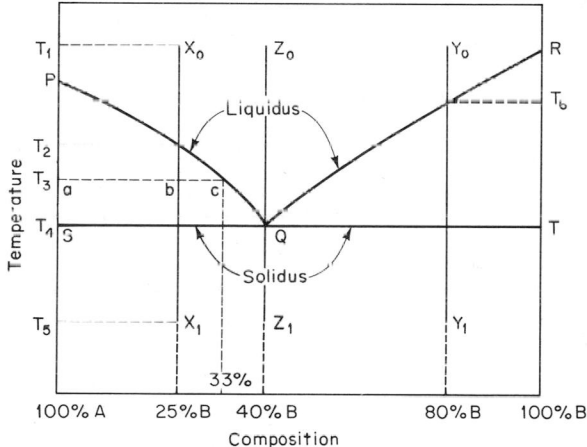

Fig. 7-9 Equilibrium diagram (also referred to as a freezing-point or phase diagram).

Example 7-3 In Fig. 7-9, what are the percentages and composition of the liquid and solid for the conditions represented by point b?

solution By the lever rule, the percent solid is

$$\frac{\overline{bc}}{\overline{ac}} \times 100 = \frac{33 - 25}{33} \times 100 = \frac{8}{33} \times 100 = 24.2 \text{ percent}$$

and the percent liquid is

$$\frac{\overline{ab}}{\overline{ac}} \times 100 = \frac{25}{33} \times 100 = 75.8 \text{ percent}$$

The composition of the solid is pure A (the composition at a) and the composition of the liquid is 33 percent B (the composition at c). Note that in, say, 100 g of metal there is 24.2 g of A in the solid and 0.67 (75.8) = 50.9 g of A in the liquid, for a total of 75.1 g of A, and (100 − 75.1) = 24.9 g of B, i.e., the composition corresponding to point b.

It is important to realize that phase diagrams refer to equilibrium conditions, meaning that they represent systems that have been brought to a given state slowly enough for all the required changes to take place. In actual practice, this is rarely the situation, particularly in solid-to-solid transformations, which may take place only very slowly, particularly at low temperatures. Even at elevated temperatures, sufficient time to allow crystals to form and grow properly is usually not available. As a result, an alloy structure may be very different from that which would be expected from its phase diagram. In the *heat treatment* of alloys, the phase composition and structure are altered by subjecting the alloy to a programmed temperature cycle. When an alloy at an elevated temperature is cooled very rapidly, a process called *quenching,* its phase composition and structure are "frozen" at that of the elevated temperature. The structure developed (or retained) by quenching depends on the temperature to which the alloy was heated and the subsequent rate of cooling. As its name suggests, quenching is frequently accomplished by plunging the heated metal into a liquid, water or oil being the most commonly used. Water has more quenching power (is capable of cooling more quickly) than oil. Air, blown rapidly over the specimen, is also a commonly used quenching agent, although its quenching power is considerably less than that of liquids. The quenching of steels is a common practice that greatly increases its hardness.

Frequently, the crystal structure of a metal may have been altered by severe mechanical working, or distorted by quenching, and it is desired to restore the metal to a more normal, stable structure. This is done by heating the metal to a temperature sufficiently high (but below the melting point) to permit recrystallization to occur, and maintaining this temperature for a time sufficient to allow the desired changes to take place, after which it is

TABLE 7-3 Recrystallization and Annealing Temperatures of the Common Commercial Metals

Metal	Temp (°C)
RECRYSTALLIZATION TEMPERATURES	
Iron	450
Nickel	600
Gold	200
Silver	200
Copper	200
Aluminum	150
Magnesium	150
Tungsten	1200
Molybdenum	890
Zinc	Room temp
Lead	Below room temp
Tin	Below room temp
ANNEALING TEMPERATURES	
Mild steel	550–650
Nickel	700–900
Copper	350–650
Brass	450–700
Aluminum and aluminum alloys	350–425

From P. G. Ormandy, "An Introduction to Metallurgical Laboratory Techniques," Pergamon Press, London, 1968.

allowed to cool slowly. This procedure is known as *annealing.* The recrystallization temperatures for common metals are given in Table 7-3, which also gives commonly used annealing temperatures. It should be noted that annealing temperatures are usually higher than recrystallization temperatures in order to accelerate the desired changes.

Frequently, hardened metals are further heat-treated to permit a relaxation of stresses in their lattice structure. This treatment, usually carried out for several hours at moderate temperatures (300 to 650°C), is known as tempering, and results in a tough metal with only a small sacrifice in hardness.

In the heat treatment of metals, control of the temperature during the heating period is very critical. Usually, *thermocouples* and optical *pyrometers* are used for temperature measurements. The method for calculating the temperature—emf characteristics of a given thermocouple is discussed in Chap. 2. For convenience, however, such information has been calculated and tabularized for the more commonly used thermocouple combinations. The pertinent values are given in Tables 7-4 to 7-8. While a wide range of temperatures can be covered by using different thermocouple combinations, temperature measurements in the

TABLE 7-4   Temperature-EMF Values, Platinum (90%), Rhodium (10%) vs. Platinum Thermocouple

EMF values in millivolts

°C	0°	10°	20°	30°	40°	50°	60°	70°	80°	90°
0	0	0.06	0.11	0.17	0.24	0.30	0.36	0.43	0.50	0.57
100	0.64	0.72	0.79	0.87	0.95	1.02	1.10	1.18	1.26	1.35
200	1.43	1.52	1.60	1.69	1.78	1.86	1.95	2.04	2.13	2.22
300	2.31	2.40	2.50	2.59	2.68	2.77	2.87	2.96	3.05	3.15
400	3.24	3.34	3.44	3.53	3.63	3.73	3.82	3.92	4.02	4.12
500	4.22	4.31	4.41	4.51	4.61	4.71	4.82	4.92	5.02	5.12
600	5.22	5.32	5.43	5.53	5.63	5.74	5.84	5.94	6.05	6.16
700	6.26	6.37	6.47	6.58	6.68	6.79	6.89	7.01	7.11	7.22
800	7.33	7.44	7.55	7.66	7.77	7.88	7.99	8.10	8.21	8.32
900	8.43	8.54	8.66	8.77	8.89	9.00	9.11	9.22	9.34	9.46
1000	9.57	9.68	9.80	9.92	10.03	10.15	10.27	10.38	10.50	10.62
1100	10.74	10.86	10.98	11.10	11.21	11.33	11.45	11.57	11.69	11.81
1200	11.93	12.05	12.17	12.29	12.41	12.53	12.65	12.77	12.89	13.01
1300	13.13	13.25	13.37	13.49	13.61	13.73	13.85	13.97	14.09	14.21
1400	14.33	14.45	14.58	14.70	14.82	14.94	15.06	15.19	15.31	15.43
1500	15.55	15.67	15.79	15.91	16.03	16.15	16.27	16.39	16.51	16.63
1600	16.75	16.87	16.99	17.11	17.23	17.35	17.47	17.59	17.71	17.83
1700	17.95	18.07	18.19	18.31	18.43	18.55				

From C. D. Hodgman (ed.), "Handbook of Chemistry and Physics," The Chemical Rubber Co., Cleveland, 1942.

TABLE 7-5   Temperature-EMF Values, Platinum (87%), Rhodium (13%) vs. Platinum Thermocouple

EMF values in millivolts

°C	0°	10°	20°	30°	40°	50°	60°	70°	80°	90°
0	0	0.06	0.12	0.18	0.25	0.31	0.38	0.45	0.52	0.60
100	0.67	0.75	0.83	0.90	0.99	1.07	1.15	1.23	1.32	1.40
200	1.49	1.58	1.67	1.76	1.85	1.94	2.03	2.12	2.21	2.30
300	2.40	2.49	2.59	2.68	2.77	2.87	2.98	3.08	3.19	3.29
400	3.40	3.51	3.61	3.72	3.82	3.93	4.04	4.15	4.25	4.36
500	4.47	4.58	4.69	4.81	4.92	5.03	5.14	5.26	5.37	5.49
600	5.60	5.72	5.83	5.95	6.06	6.18	6.30	6.42	6.53	6.65
700	6.77	6.89	7.01	7.13	7.25	7.37	7.49	7.62	7.74	7.87
800	7.99	8.12	8.24	8.37	8.49	8.62	8.75	8.88	9.00	9.13
900	9.26	9.39	9.52	9.66	9.79	9.92	10.05	10.18	10.32	10.45
1000	10.58	10.72	10.85	10.99	11.12	11.26	11.40	11.54	11.67	11.81
1100	11.95	12.09	12.23	12.38	12.52	12.66	12.80	12.94	13.09	13.23
1200	13.37	13.52	13.66	13.81	13.95	14.10	14.25	14.40	14.54	14.69
1300	14.84	14.99	15.14	15.30	15.45	15.60	15.75	15.90	16.06	16.21
1400	16.36	16.52	16.67	16.83	16.98	17.14	17.30	17.46	17.61	17.77
1500	17.93	18.09	18.25	18.42	18.58	18.74	18.90	19.06	19.23	19.39
1600	19.55	19.71	19.88	20.04	20.21	20.37				

From C. D. Hodgman (ed.), "Handbook of Chemistry and Physics," The Chemical Rubber Co., Cleveland, 1942.

TABLE 7-6   Temperature-EMF Values, Chromel P–Alumel Thermocouple

EMF values in millivolts

Temp (°C)	0	10	20	30	40	50	60	70	80	90
−200	−5.75									
−100	−3.49	−3.78	−4.05	−4.32	−4.57	−4.81	−5.03	−5.24	−5.43	−5.60
0	0.00	−0.39	−0.77	−1.14	−1.50	−1.86	−2.21	−2.55	−2.87	−3.19
0	0.00	0.40	0.80	1.20	1.161	2.02	2.43	2.85	3.26	3.68
100	4.10	4.51	4.92	5.33	5.73	6.13	6.53	6.93	7.33	7.73
200	8.13	8.53	8.93	9.34	9.74	10.15	10.56	10.97	11.38	11.80
300	12.21	12.62	13.04	13.45	13.87	14.29	14.71	15.13	15.55	15.97
400	16.39	16.82	17.24	17.66	18.08	18.50	18.93	19.36	19.78	20.21
500	20.64	21.07	21.49	21.92	22.34	22.77	23.20	23.62	24.05	24.48
600	24.90	25.33	25.75	26.18	26.60	27.03	27.45	27.87	28.29	28.72
700	29.14	29.56	29.98	30.40	30.82	31.23	31.65	32.07	32.48	32.90
800	33.31	33.71	34.12	34.53	34.94	35.35	35.75	36.16	36.56	36.96
900	37.36	37.76	38.16	38.56	38.96	39.35	39.75	40.14	40.53	40.92
1000	41.31	41.70	42.08	42.47	42.86	43.24	43.62	44.00	44.38	44.76
1100	45.14	45.52	45.89	46.27	46.64	47.01	47.38	47.75	48.12	48.48
1200	48.85	49.21	49.57	49.94	50.29	50.65	51.00	51.36	51.71	52.06
1300	52.41	52.75	53.10	53.45	53.79	54.13	54.47	54.81	55.15	55.48
1400	55.81									

From C. D. Hodgman (ed.), "Handbook of Chemistry and Physics," The Chemical Rubber Co., Cleveland, 1942.

TABLE 7-7   Temperature-EMF Values, Copper-Constantan Thermocouple

EMF values in millivolts

°C	0°	10°	20°	30°	40°	50°	60°	70°	80°	90°
−200	−5.54	−5.69								
−100	−3.35	−3.62	−3.89	−4.14	−4.38	−4.60	−4.82	−5.02	−5.20	−5.38
0	0	−0.38	−0.75	−1.11	−1.47	−1.81	−2.14	−2.46	−2.77	−3.06
0	0	0.40	0.80	1.20	1.61	2.03	2.47	2.91	3.36	3.81
100	4.28	4.70	5.23	5.71	6.20	6.70	7.21	7.72	8.23	8.76
200	9.29	9.82	10.34	10.90	11.46	12.01	12.57	13.14	13.71	14.28
300	14.86	15.44	16.03	16.62	17.22	17.82	18.42	19.02	19.63	

From C. D. Hodgman (ed.), "Handbook of Chemistry and Physics," The Chemical Rubber Co., Cleveland, 1942.

TABLE 7-8   Temperature-EMF Values, Iron-Constantan Thermocouple

EMF values in millivolts

°C	emf	°C	emf	°C	emf
0	0	500	27.41	1000	58.17
50	2.61	550	30.24	1050	61.33
100	5.28	600	33.13	1100	64.50
150	8.01	650	36.11	1150	67.67
200	10.77	700	39.19	1200	70.84
250	13.54	750	42.33		
300	16.30	800	45.49		
350	19.06	850	48.66		
400	21.83	900	51.83		
450	24.61	950	35.00		

From C. D. Hodgman (ed.), "Handbook of Chemistry and Physics," The Chemical Rubber Co., Cleveland, 1942.

higher temperature ranges, where it is not physically convenient to use thermocouples, are made by means of a pyrometer. There are various forms of pyrometers, but all work on the principle of focusing the radiation coming from a hot body onto a temperature-measuring device, such as a thermocouple, as shown in Fig. 7-10. In an optical pyrometer, the radiation is passed through a red filter and focused on a small screen, against which the filament of a bulb is displayed. The temperature of the filament is varied by a variable resistor between it and its power source (usually a calibrated dry cell) until the radiation from the filament matches that focused on the screen, at which point the filament seems to disappear, as illustrated in Fig. 7-11. The voltage across the bulb at this point is thus a measure of the temperature of the hot body. The theoretical equations relating the temperature of the hot body and the corresponding bulb voltage are difficult to handle. Consequently, most optical pyrometers are empirically calibrated (usually against thermocouples).

For many metallurgical purposes, where high accuracy is not necessary, *temperature-indicating crayons* or *melting cones* can be used. The crayons are usually used in a series of graded melting points. The surface to be heated is marked by these crayons, leaving chalklike lines. At the appropriate temperature, each chalk line melts to a liquid smear, a change that is easily detected visually. Temperature-indicating crayons can be used from about 60 to 850°C. From about 600 to 2000°C, *pyrometric cones* can be used. These are small tetrahedral pyramids that soften and bend as a result of being heated to a particular temperature for a given time. As such, they are useful in indicating the rate of heating as well as the approximate temperature, as illustrated in Fig. 7-12.

Microscopic Examination

As might be anticipated, the crystalline structure of a metal reflects not only the phase relationships of its components but its thermal and mechanical history as well. Figure 7.13 illustrates this for an alloy consisting of metals A and B. For the composition X percent of B at temperature T_1, the alloy will consist only of crystals of α and will have a microstructure illustrated by a. If this metal is cooled slowly to T_2, crystals of β will appear as illustrated by b, with the amounts of α and β given by the lever rule previously discussed. If the alloy is quenched to temperature T_2, the structure shown by a will be preserved even though it represents an unstable form. After aging, crystals of β will appear as at c as the metal structure tends to become more stable. Of course, the rate at which the stable structure forms at room or low temperatures depends on the system. Recrystallization can occur within a few hours, or (for all practical purposes) never at all. The grain-size distribution depends largely on the rate at which the crystals were formed from the melt, slow crystallization favoring the growth of large crystals as in a, fast crystallization producing small crystals as in d.

To study the microstructure of metals and alloys requires careful preparation of the sample, and specific techniques have been developed for different metals, or to bring out various specific features. The preparation of the sample for microscopic examination must be carried out in such a way as to avoid disturbing the microstructure, and

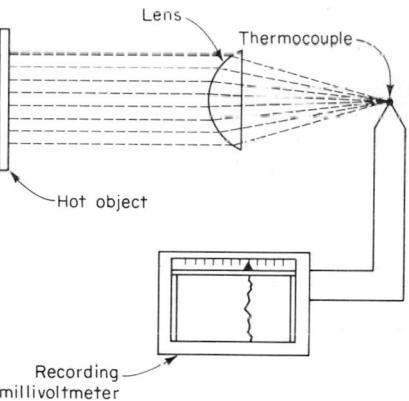

Fig. 7-10 Radiation pyrometer with chart recorder.

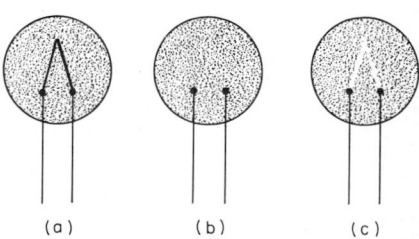

Fig. 7-11 Disappearing filament in an optical pyrometer. (*a*) Filament temperature lower than temperature being measured; (*b*) filament at same temperature as that being measured; (*c*) filament temperature higher than temperature being measured.

this means avoiding physical deformations and local temperature effects which can indeed alter the structure even if present for only short periods of time. The usual specimen for microscopic examination is about $\frac{3}{4}$ in in diameter and about $\frac{1}{2}$ in thick. Specimens below these proportions, such as thin metal sections, become difficult to handle and usually require some special mounting. A metallurgical cutoff machine using a carborundum cutoff disk provides a means of obtaining a sample while producing a fairly uniform surface. In the use of a cutoff machine, a sufficient amount of coolant fluid must be used to prevent localized heating. While the surface to be examined can be fairly flat when obtained with a cutoff tool, the imperfections are still very large compared with the grain structure to be examined, and it is necessary to prepare the surface further to remove scratches and other imperfections. This can usually be conveniently done by hand, using abrasive papers or

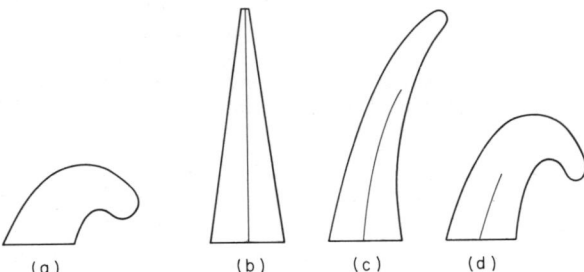

Fig. 7-12 Pyrometric cones. The "cones" are actually tetrahedrons with their axes tilted slightly (about 8°) from the normal so that it will "wilt" in a predictable direction. (*a*) Temperature too high, cone practically completely sagged; (*b*) temperature too low, cone has not melted to any degree; (*c*) proper temperature reached rapidly; (*d*) proper temperature reached more slowly.

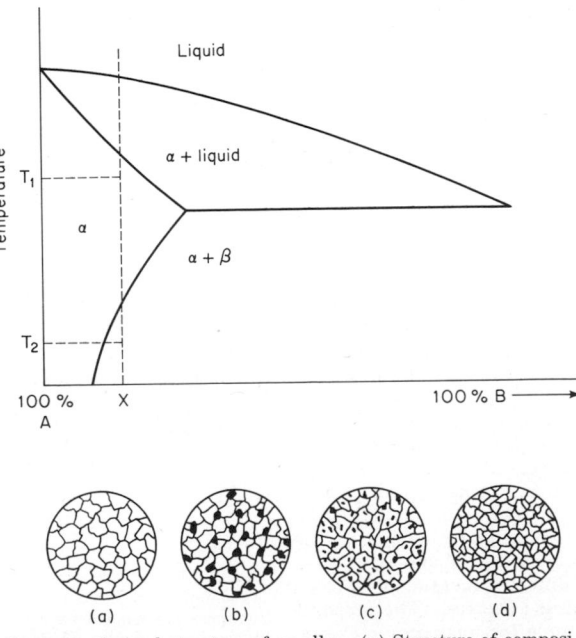

Fig. 7-13 Crystal structure of an alloy. (*a*) Structure of composition *X* at temperature T_1; (*b*) composition *X* cooled slowly to temperature T_2; (*c*) composition *X* quenched to temperature T_2; (*d*) fast cooling of composition *X* to temperature T_1.

fabrics in which the abrasive, usually carborundum or some other hard ceramic having carefully sized particles, does the actual abrading. The abrasive paper is specified by the largest mesh size through which all the particles will pass. Starting with a coarse grit, the scratch marks are rubbed out through a sequence of finer and finer grits. Again, to avoid surface heating, some lubricant such as water or oil is used to prevent heat buildup at the surface of the specimen as it is being abraded. Usually, for a well-cut metallurgical sample, only about four grits are needed to prepare the sample for final polishing. Usually, a grit size of 120 mesh down to a final grit of 300 to 400 mesh will be satisfactory for final polishing. The sequential abrasion of the specimen surface can usually be accomplished in a short time.

The final preparation of the surface involves polishing, and this is usually accomplished on a polishing wheel which is rotated at fairly high speeds by an electric motor. The wheel is covered with a soft fabric, usually a flannel material, and a water suspension of a fine abrading agent is used to saturate the polishing cloth. A number of polishing agents can be used, the most common being alumina prepared by calcining precipitated aluminum hydroxide. This material stays in suspension well, has a high abrasive power, and has a mesh size sufficiently small actually to produce a highly polished surface. Occasionally, where harder metals are involved, the abrasive can be rouge, a calcined precipitated iron oxide, or even diamond dust. Because the polishing wheel is rotated at high speed, centrifugal forces tend to remove the water in which the abrasive is suspended, and frequent water additions are required. Only a light pressure on the specimen is needed as it is kept in a continuous circular motion in a direction counter to the rotation of the wheel. This ensures a completely random polishing action. After the specimen has been brought to a high shine, care must be exercised to prevent scratches, dust, and other surface defects from developing. Usually, the sample is washed carefully in running water, using light finger pressure to dislodge any abrasives. The wash is completed in distilled water, and drying is accomplished by rinsing in alcohol and air drying in a stream of a gently heated air. Usually there is not too much difficulty in polishing a sample, but with some metals, corrosion can occur from the water in which the abrasive is suspended, and/or the wash waters, in which case special lubricants must be used. The prepared specimen is mounted on a glass slide by pressing it into soft clay, or a material similar to clay called plasticine, in a special press, illustrated in Fig. 7-14, which ensures that the surface of the specimen will be parallel to the surface of the glass slide and that the surface will not be injured during mounting. The sample is now ready for examination in the "as polished" condition.

Frequently, the size or shape of the sample is not convenient for conventional grinding and polishing. In this case, the specimen is positioned in a cylindrical cavity (usually about 1 in in diameter) and plastic molding powder is used to surround the specimen and fill the cylinder. A number of commercial molding powders are now available, but all follow the same general characteristics and preparation technique. The cylinder containing the plastic powder and specimen is then transferred to a hydraulic press in which a ram can exert pressure on the powder while it is being heated. Usually, pressures of 2,000 to 5,000 lb/in^2 are used at temperatures around 250 to 350°C, depending upon the specific molding powder used. When the pressure is removed and the sample is allowed to cool, it is rigidly held in a solid plastic mount, making it very convenient to perform subsequent preparation operations, such as grinding and polishing.

Where the sample is likely to be disturbed by the heat and pressure, as, for example, a porous material, a number of liquid polymers are now available to which various curing agents are added just before use. In this case, the sample is mounted in a cylinder as before, and the cylinder is filled with liquid plastic to which a curing agent has been added. After a period of time, usually several hours if heat is to be avoided, the plastic cures to a hard material and the sample can be readily removed from the cylinder.

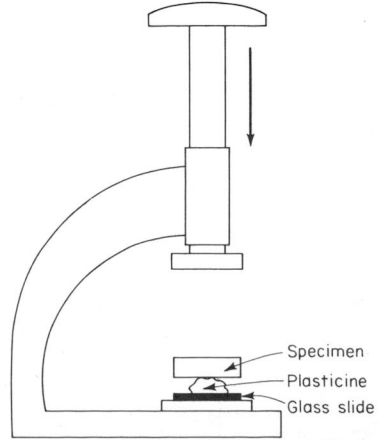

Fig. 7-14 Metallurgical-specimen-mounting press.

The highly reflective surface of the polished specimen makes it difficult to see the crystal structure. In addition, the polishing operation can sometimes cause some of the metal to flow and thus distort the crystal structure. To get down below the abnormalities introduced by the polishing, and to be able to see the crystal structure more readily, the samples are frequently put through a process called *etching*. In the etching operation, a controlled chemical corrosion takes place to remove the surface of the sample and expose the crystal structure beneath. Since a chemical reaction is occurring in the etching operation, and since the composition or orientation of the various crystals are different, there is a differential etching that further facilitates observation of the crystal structure. The specific etching solution used depends upon the nature of the metal. Table 7-9 lists the most commonly used etching reagents.

Determining the time the sample should be exposed to the etchant usually requires some experience with the particular metal under observation. If insufficient etching occurs, the

TABLE 7-9 Metal-etching Reagents

Metal	Reagent	
Common, quenched, and tempered steels	Nital, solution of nitric acid in alcohol, 2–4% in strength	
High-carbon steels	Picral, picric acid in alcohol, 5% in strength	
Stainless steels	Vilella etch, 3 parts concentrated hydrochloric acid, 1 part concentrated nitric acid, 6–8 parts glycerine	
	NOTE: First heat the specimen in hot water. The etchant must not be allowed to stand for long periods of time	
High-alloy steels in general	Hydrochloric acid	5 ml
	Picric acid	1 g
	Alcohol	100 ml
Chromium steels and nickel-chromium steels	Nitric acid	10 ml
	Hydrochloric acid	20 ml
	Glycerol	20–30 ml
High-speed steels and tungsten steels	Hydrochloric acid	10 ml
	Nitric acid	3 ml
	Methyl alcohol	100 ml
Pure aluminum	a. Tucker's etch, 180 ml hydrochloric acid, 60 ml nitric acid, 60 ml hydrofluoric acid, 100 ml water	
	NOTE: Reveals the macrostructure for macroscopic examination	
	b. 2% solution of hydrofluoric acid.	
	NOTE: Reveals the microstructure for microscopic examination	
Brasses and copper alloys	2% solution of ferric chloride	
Pure copper	880 ammonia, hydrogen peroxide (a few drops), Ammonium persulfate	
	NOTE: Make up fresh for each application	
Aluminum and its alloys	a. 10% solution sodium hydroxide in water	
	b. Sulfuric acid	20 ml
	Water	80 ml
	c. Nitric acid	20 ml
	Water	80 ml
Aluminum bronze	a. Nitric acid	20 ml
	Hydrofluoric acid	10 ml
	Water	150 ml
Pure lead	a. Nitric acid	
	b. Solution of perchloric acid in water	600 g/l
	c. Acetic acid	30 ml
	Hydrogen peroxide	(30 vol)
Tin	a. 2% nitric acid in alcohol	10 ml
	b. Acetic acid	50 ml
	Water	50 ml
	Hydrogen peroxide (30 vol)	1 drop
	c. Ammonium persulfate	5 ml
	Water	95 ml

From P. G. Ormandy, "An Introduction to Metallurgical Laboratory Techniques," Pergamon Press, London, 1968.

microstructure will not be seen clearly, and the specimen will exhibit a "fogginess" or "ghosting" of the structure. In this case, the specimen can be returned to the etching reagent for further treatment. If the microstructure is very dark, and possibly stained, the specimen has probably been overetched. In this case, the only corrective action is to return the specimen to the polishing wheel and start all over again. After etching, the specimens have to be thoroughly washed and dried, and this is accomplished by rinsing in running water, then in distilled water, and finally in alcohol, followed quickly by drying in a warmed airstream.

A special procedure known as electrolytic polishing and etching has been developed for certain metals, particularly hard metals, which are difficult to handle by mechanical

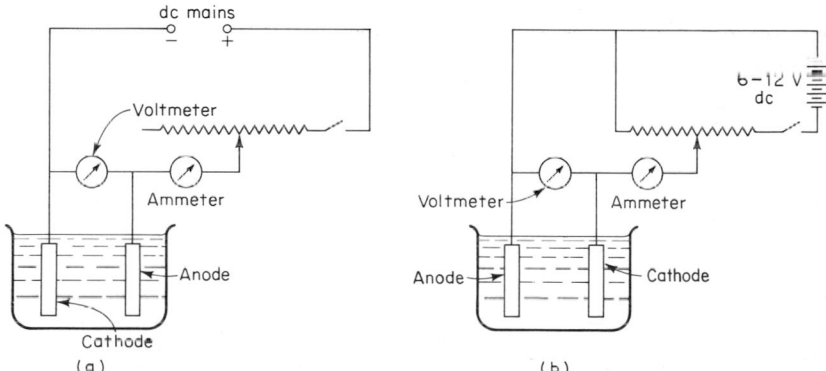

Fig. 7-15 Electrolytic etching and polishing circuits. (a) High current density circuit. (The bath will normally contain a cooling device, thermometer, and stirrer.) (b) Low current density circuit. (This circuit usually does not require cooling or temperature control of the bath. However, stirring is still required.)

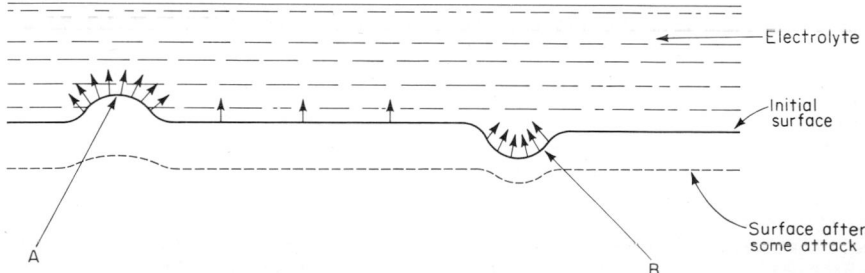

Fig. 7-16 Smoothing effect of electropolishing. At a protuberance, such as at A, the surface area over which attack can occur is large, and dissolved metal can easily diffuse away from the surface. At a pit, such as at B, the reverse is true. Thus the rate of attack is greater at a protuberance than at a pit, eventually smoothing the surface.

means. In this procedure, the specimen is made the anode in an electrolytic bath, as illustrated in Fig. 7-15. Specific electrolytes have been developed for certain alloys as given in Table 7-10. When a high current density is used, a smoothing effect produced by the chemical dissolution occurs, as illustrated in Fig. 7-16. When the desired degree of polishing has occurred, the current density is reduced, and the electrolytic corrosion is permitted to proceed evenly, thus bringing out the crystal structure of the metal.

TABLE 7-10   Electrolytic Polishing and Etching Electrolytes

Metal	Electrolyte	
Iron, alloy steels, carbon steels	Perchloric acid (density 1.61)	18.5 ml
	Acetic anhydride	76.5 ml
	Distilled water	5.0 ml
	NOTE: The solution should stand for 24 h before use	
Austenitic steels	Perchloric acid (density 1.61)	
	Acetic anhydride	
	NOTE: The solution should stand for 24 h before use	
Steels and general reagent for many metals	*a.* Perchloric acid (density 1.61)	20 ml
	Ethyl alcohol containing	
	3% ether	80 ml
	b. Perchloric acid (density 1.61)	10 ml
	Glacial acetic acid	100 ml
	c. Nitric acid	10 ml
	Methyl alcohol containing	
	5% ether	90 ml
Stainless steels (18% Cr, 8% Ni)	*a.* Orthophosphoric acid	37 ml
	Glycerol	56 ml
	Water	7 ml
	b. Orthophosphoric acid	67 ml
	Sulfuric acid	20 ml
	Chromic acid	2 g
	Water	11 ml
	NOTE: This is a slow-acting reagent	
	c. Sulfuric acid	15 ml
	Acetic acid	55 g
	Water	30 ml

From P. G. Ormandy, "An Introduction to Metallurgical Laboratory Techniques," Pergamon Press, London, 1968.

7-2   TESTING OF METALS

Tension

A number of standard tests have been developed to characterize the physical strength and properties of metals and alloys. Many of these are specific adaptations of tests discussed in Chap. 1 relating to the properties of matter in general. In the tension testing of metals, a test specimen is rigidly clamped into the jaws of a device in which the jaws can be pulled apart by a hydraulic or a mechanical system. The test machine has means of measuring the force being used to pull the jaws apart and the distance the jaws have moved. The test specimens are standardized as illustrated in Fig. 7-17 and are designed to develop a uniform stress in that section of the specimen which is actually under observation. As the force pulling the jaws apart is increased, the specimen elongates in a manner which is directly proportional to the applied force (as discussed under Young's modulus in chap. 1). At a particular point, the elongation of the specimen increases at a rate which is greater than the rate at which the force is applied. This point is called the *yield strength* of the material. As the force on the specimen is further increased, the specimen continues to elongate until the force reaches a value which is the maximum for the particular material. This point is known as the *tensile strength* of the material. After the tensile strength of the material is reached, the specimen will continue to elongate without an increase in the applied force. As a matter of fact, after the tensile strength is achieved, the specimen can continue to elongate with an actual decrease in the applied force, and it will do so until the specimen actually ruptures. The force at which the specimen ruptures is known as the *fracture strength* of the material. These relationships are illustrated in Fig. 7-18.

As the test specimen elongates, its cross-sectional area reduces. However, in most tests, the cross section of the test specimen is made large enough so that the reduction in cross section is of only minor consequence. For purposes of most tabularized data, the cross-sectional area of the specimen is taken as the original area. The elongation, on the other hand, is usually quite substantial, particularly if the testing is taken to the fracture point, and many tables of physical properties also give the percent elongation of the specimen as defined by the following equation:

$$\text{Percent elongation} = \frac{L_f - L_0}{L_0} \times 100 \tag{7-1}$$

in which L_f is the length of the specimen immediately prior to fracture and L_0 is the original length of the specimen. To some degree the percent elongation depends on the original length of the specimen, since failure usually occurs with fracture of a short, necked-down section of the specimen. It is therefore necessary to associate the percent elongation with a particular specimen length, usually 2 in in the case of metals.

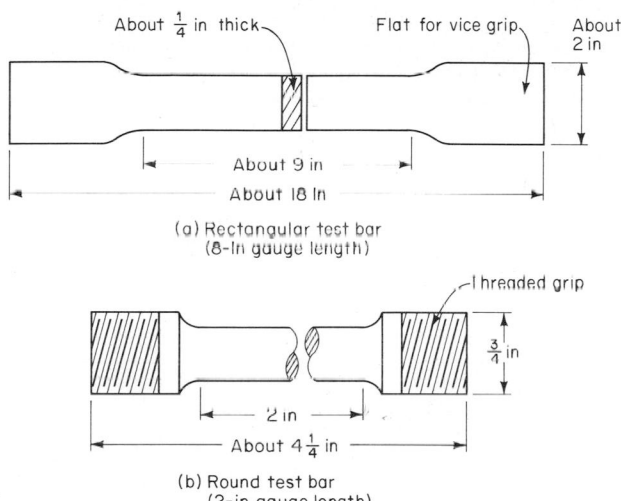

Fig. 7-17 Tension-test specimens. The rectangular specimen is usually used for materials of low tensile strength. The round specimen is usually used for materials of high tensile strength.

Example 7-4 At rupture, the gage marks on a 316 stainless steel specimen were measured at 2.84 in. If the gage marks were originally 2.00 in apart, what is the percent elongation for this particular steel?
solution

$$\text{Percent elongation} = \frac{2.84 - 2.00}{2.00} =$$

41 percent for 2-in specimen

Hardness

In metallurgical work, the term hardness usually refers to the ability of the metal to resist permanent deformation. A widely accepted and highly standardized hardness test was developed by J. A. Brinell about 1900. The Brinell test measures the indentation brought about by pressing a small, hardened steel ball into the surface of a specimen under the application of a given load. For most metals, the indenting ball is 10 mm in diameter and the load is 3,000 kg. The load is applied for a standard time, usually 30 s, through a damping mechanism which prevents a sudden impact type of in-

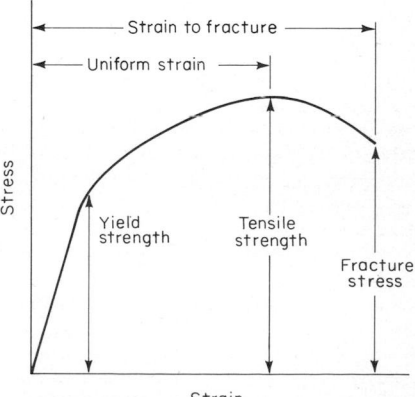

Fig. 7-18 Stress-strain curve showing yield strength, tensile strength, and fracture stress.

dentation. For soft metals, the load is reduced to 500 kg to avoid an excessively deep indentation, and for very hard metals, the ball may be tungsten carbide to avoid distortion of the indenter. After removal of the load and indenter, the diameter of the indentation is measured with a low-powered microscope. Usually two readings of the diameter of the impression are taken at right angles to each other, the average being used to calculate the Brinell hardness (BHN) by the following formula:

$$BHN = \frac{P}{(\pi D/2)(D - \sqrt{D^2 - d^2})} \tag{7-2}$$

where P = applied load (kg), D = diameter of the indenting ball (mm), and d = diameter of the indentation (mm).

Example 7-5 What is the Brinell hardness of a metal indented with a 5-mm ball under a load of 3,000 kg if the indent has a diameter of 1.5 mm?
solution

$$BHN = \frac{3,000}{(5\pi/2)(5 - \sqrt{(5)^2 - (1.5)^2})} = \frac{3,000}{(7.85)(5 - 4.77)} = 1,660$$

Some judgment is frequently needed in measuring the diameter of the indentation because of ridging or sinking, conditions illustrated by Fig. 7-19. Actually, these conditions are usually not too severe and do not present serious difficulties in measuring hardness. Where they do occur, however, some practice is necessary to develop skill in compensating for ridging or sinking. Frequently, for special purposes, other types of indentation devices are used. For example, in the Vickers hardness test, a square-based diamond pyramid is

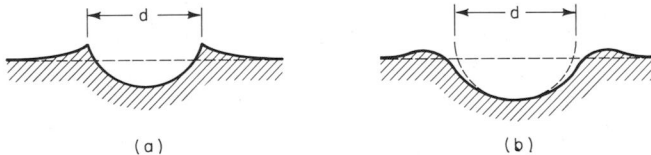

(a)   (b)

Fig. 7-19 Ridging (*a*) and sinking (*b*) from Brinell indentations.

used as the indenter. In another variation, known as the Rockwell hardness, the depth of indentation under a constant load is used as the measure of hardness. This has the tendency to eliminate personal error and does have the ability to distinguish small hardness differences. In the Rockwell hardness, a "minor" load of 10 kg is first applied to seat the specimen. This eliminates the need for careful surface preparation and reduces the tendency for ridging or sinking. A "major" load is then applied, and the depth of indentation is automatically indicated on a dial gage calibrated in arbitrary hardness numbers. Unlike the Brinell or Vickers hardness, which have units of kilograms per square millimeter, the Rockwell hardness is a purely arbitrary number and therefore requires that the testing device be carefully standardized. It has been found that a 120° diamond cone with a slightly rounded point is a useful indenter for hard metals and that $\frac{1}{15}$- and $\frac{1}{5}$-in-diameter steel balls are suitable for softer metals. The major loads for which the Rockwell hardness testers are standardized are 60, 100, and 150 kg. Thus, in expressing Rockwell hardness it is necessary to indicate the indenter and load used. The various standard combinations are given in Table 7-11.

TABLE 7-11   Rockwell Hardness Scales

Scale designation	Indenter	Minor load (kg)	Major load (kg)
A	$\frac{1}{8}$-in steel ball	10	60
B	$\frac{1}{16}$-in steel ball	10	100
C	120° diamond cone with slightly rounded tip	10	150

The A scale is used for soft materials, the B scale for intermediate hardnesses, and the C scale for hard materials.

The yield strength, tensile strength, elongation, and Brinell hardness of metals and alloys are given in Table 7-12.

Fatigue

When a metal is subjected to a repetitive stress, failure will occur at a stress which is much lower than that required to cause failure under a single application. Fatigue failures are those which occur under the application of a repetitive or cyclical stress. There are a great many ways of testing the fatigue properties of a metal, but these usually reflect the conditions under which the metal is intended to be used, and there has been no universally accepted testing procedure. Consequently, there are also few tabulated data on the fatigue characteristics of metals. About the most widely used testing procedure is known as a *Moore rotating-beam machine* in which a rotating shaft is operated at various speeds and stressed by loads applied to the center of the beam while it is being rotated. This produces

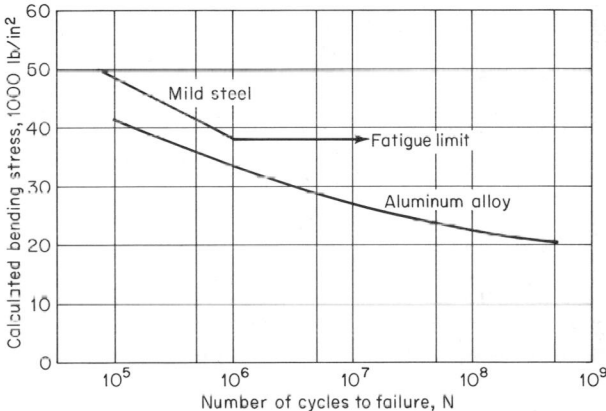

Fig. 7-20 Typical fatigue (*S-N*) curves.

a sinusoidal type of stress cycle. Frequently, if the stress is not large enough, there is no failure regardless of the number of cycles of stress involved. The stress below which no fatigue failure occurs is known as the *fatigue limit* of the metal.

The basic means of presenting engineering fatigue data is by an *S-N* curve, which shows the number of cycles *N* to failure for a given applied stress *S*. Typical *S-N* curves are shown in Fig. 7-20. It should be noted that the mild steel illustrated in this *S-N* curve shows a fatigue limit of about 38,000 lb/in^2, while the aluminum alloy shows no particular clearly defined fatigue limit but is not likely to fatigue at stresses below 20,000 lb/in^2 (for the range of cycles covered). Because fatigue failures are so closely related to the manner in which metal is used and are subject to a good deal of variation owing to such factors as scratches (which cause very high local stresses), it is common to fatigue-test metals with several duplicate samples using a test which is as closely related to the actual use conditions as possible. Metals that are subject to cyclical stresses are usually prepared and handled very carefully so as to avoid any mechanical condition which would tend to concentrate stresses, since even small imperfections can cause very high localized stress concentrations and represent the points at which fatigue failure is likely to occur. Fatigue failures also have a tendency to occur when the metal is subjected to cyclical stresses under conditions in which chemical attack or corrosion is also present. In this case, the fatigue failure occurs even more rapidly than when the corrosive conditions are not present. This condition is frequently referred to as *corrosion fatigue*. Small surface imperfections, such as can cause a localized high stress, can also cause a localized high corrosion, and this further increases the effect of the original surface imperfection, thereby causing unusually rapid fatigue failure.

TABLE 7-12 Properties of Metals and Alloys

Material	Nominal composition (essential elements), %	Form and condition	Yield strength (0.2% offset), 1000 lb./sq. in.	Tensile strength, 1000 lb./sq. in.	Elongation in 2 in., %	Hardness, Brinell
Low-alloy Irons and Steels						
Carbon steel A.I.S.I.-S.A.E. 1020	Fe bal., Mn 0.45, Si 0.25, C 0.20	Annealed / Hot-rolled / Hardened (water quench 1000°F. temper)	38 / 42 / 62	65 / 68 / 90	30 / 32 / 25	130 / 135 / 179
300-M	C 0.43, Mn 0.80, Si 1.60, Ni 1.85, Cr 0.85, Mo 0.38, V 0.08	Hardened (oil quench, 600°F. temper)	240	290	10	535
Wrought iron	Fe bal., Slag 2.5	Hot-rolled	30	48	30 (in 8 in.)	100
Ingot iron	Fe 99.9 plus	Hot-rolled / Annealed	29 / 19	45 / 38	26 / 45	90 / 67
Cast gray iron	C 3.4, Si 1.8, Mn 0.5, Fe bal.	Cast (as cast)	.....	25 min.	0.5 max.	180
Malleable iron	C 2.5, Si 1, Mn 0.55 max	Cast (annealed)	33	52	12	130
Ni-Tensyliron	C 2.7, Si 1.8, Mn 0.8, Ni 2.3, Cr 0.3, Fe bal., Mo 0.4	Cast (as cast) / Cast (heat-treat)	30 / 40	60 / 80	..... /	260 / 390
Ductile iron (Mg-containing)	C 3.4, Si 2.5, Mn 0.40, P 0.1 max., Ni 0-1, Mg 0.06, Fe bal.	Cast (as cast) / Cast (as cast) / Cast (quench, temper)	53 / 68 / 108	70 / 90 / 135	18 / 7 / 5	170 / 235 / 310
Ductile iron (Mg-containing) (heat-resistant)	C 3.3, Si 4.3, Mn 0.4, P 0.1 max., Ni 0-1, Mg 0.06, Fe bal.	Cast (annealed, as cast)	60	80	10	220
Ni-Resist ductile iron (Mg-containing)	C 2.8, Si 2.5, Mn 1, P 0.2 max., Ni 20, Cr 2, Mg 0.1, Fe bal.	Cast (as cast)	35	60	10	175
Ni-Hard type 2	C 2.7, Si 0.6, Mn 0.5, Ni 4.5, Cr 2.0, Fe bal.	Sand-cast / Chill-cast (temper)	..... /	55 / 75	..... /	550 / 625
Ni-Hard type 1	C 3.5, Si 0.6, Mn 0.5, Ni 4.5, Cr 2.0, Fe bal.	Sand-cast / Chill-cast (temper)	..... /	40 / 50	..... /	600 / 700
Ni-Resist type 1	C 2.8, Si 2.0, Mn 1.2, Ni 15.5, Cr 2.5, Cu 6.5, Fe bal.	Cast (as cast)	.....	27	2	150
Ni-Resist type 2	C 2.8, Si 2.0, Mn 1.0, Ni 20.0, Cr 2.5, Fe bal.	Cast (as cast)	.....	27	2	140
Ni-Resist type 3	C 2.6 max., Si 1.5, Mn 0.6, Ni 30.0, Cr 3.5, Fe bal.	Cast	.....	30	2	140
Ni-Resist type 4	C 2.6 max., Si 5.5, Mn 0.6, Ni 30.0, Cr 5.5, Fe bal.	Cast	.....	30	2	180

Material	Composition	Condition				
Ni-Resist type D-2	C 2.90 max., Si 2.5, Mn 1.0, P 0.2 max., Ni 20.0, Cr 2.5	Cast	34	62	14	160
Ni-Resist type D-3	C 2.60 max., Si 2.5, Mn 0.6, P 0.2 max., Ni 30.0, Cr 3.5	Cast	35	61	12	150
Ni-Resist type D-4	C 2.60 max., Si 5.5, Mn 0.6, P 0.2 max., Ni 30.0, Cr 5.5	Cast	41	66	25	190
Wrought Stainless Steels						
Stainless steel type 201	C 0.15 max., Mn 5.5–7.5, Cr 16.0–18.0, Ni 3.5–5.5, N 0.25 max.	Mill-annealed strip	50	115	60	194
Stainless steel type 202	C 0.15 max., Mn 7.5–10.0, Cr 17.0–19.0, Ni 4.0–6.0, N 0.25 max.	Mill-annealed strip	50	100	60	184
Stainless steel type 301	Fe bal., Cr 17, Ni 7, C 0.08–0.20	Annealed	30	100	72	160
		Cold-rolled	up to 165	up to 200	15[e]	385
Stainless steel type 302	Fe bal., Cr 18, Ni 8, C 0.08–0.20	Annealed	30	90	60	160
		Cold-rolled	up to 165	up to 190	8[e]	up to 400
Stainless steel type 304	Fe bal., Cr 19, Ni 9.0, C 0.08 max.	Annealed	30	85	62	160
		Cold-rolled	up to 160	up to 185	8[e]	up to 400
Stainless steel type 304L	Fe bal., Cr 19, Ni 10, C 0.03 max.	Annealed	30	80	60	150
		Cold-drawn	95	125	25	277
Stainless steel type 309	Fe bal., Cr 23, Ni 13, C 0.20 max.	Annealed	30	82	50	165
		Cold-rolled	up to 120	up to 150	4[e]	275
Stainless steel type 310	Fe bal., Cr 25, Ni 20, C 0.25 max.	Annealed	40	100	50	165
Stainless steel type 316	Fe bal., Cr 18, Ni 11, Mo 2.5, C 0.10 max.	Annealed	30	90	50	165
		Cold-rolled	up to 120	up to 150	8[e]	275
Stainless steel type 316L	Fe bal., Cr 17, Ni 12, C 0.03 max., Mo 2	Annealed	30	80	60	150
		Cold-drawn	60	90	45	190
Stainless steel types 321 and 347 (321 has Ti) (347 has Cb)	Fe bal., Cr 18, Ni 10, C 0.10 max., Ti 4 × carbon min. or Cb 3 × carbon min.	Annealed	30	85	50	160
		Cold-rolled	up to 120	up to 150	5[d]	300
Stainless steel type 330	Fe bal., Ni 35, Cr 16	Hot-rolled	55	100	35	200
		Cold-drawn (anneal)	.....	.....	.....	.....
		Cold-drawn (heat-treated)	115	150	15	300
Stainless steel AM 350	Fe bal., Cr 17, Ni 4, Mo 3, C 0.10	Annealed	45	156	21	205
		Hardened	153	195	12	382
Stainless steel type 410	Fe bal., Cr 12.5, C 0.15 max.	Annealed	40	75	30	150
		Heat-treated	115	150	15	300
Stainless steel type 414	Fe bal., Cr 12.5, Ni 2.5	Annealed	80	100	22	217
		Heat-treated	150	200	17	387
Stainless steel type 420	Fe bal., Cr 13, C 0.35	Annealed	60	98	28	180
		Heat-treated	200	250	8	480
Stainless steel type 430	Fe bal., Cr 16, C 0.12 max.	Annealed	40	70	35	165
Stainless steel type 431	Fe bal., Cr 16, Ni 2	Cold-rolled	95	110	10	225
		Annealed	85	120	25	250
		Heat-treated	150	195	20	400
Stainless steel type 446	Fe bal., Cr 25, C 0.35 max.	Annealed	50	80	30	165
Stainless steel 17-4 FH	Fe bal., Cr 17, Ni 4, Cu 4, Co 0.35, C 0.07	Annealed	110	150	12	363
		Hardened	180	195	13	404
Stainless steel 17-7 FH	Fe bal., Cr 17, Ni 7, Al 1, C 0.09	Annealed	40	130	30	165
		Hardened	185	200	9	404
Stainless steel HNM	Fe bal., Cr 18, Mn 3.5, Ni 0.5, C 0.30	Annealed	56	116	57	192
		Hardened	124	168	19	352

Adapted from Robert H. Perry and Cecil H. Chilton (eds.), "Chemical Engineers' Handbook," 5th ed., McGraw-Hill Book Company, New York, 1973.

TABLE 7-12 Properties of Metals and Alloys—(Continued)

Material	Nominal composition (essential elements), %	Form and condition	Typical mechanical properties			
			Yield strength (0.2% offset), 1000 lb./sq. in.	Tensile strength, 1000 lb./sq. in.	Elongation in 2 in., %	Hardness, Brinell
Stainless steel, stainless W	Fe bal., Cr 17, Ni 7, Ti 0.7, Al 0.2, C 0.07	Annealed	75–115	120–150	8–15	255
		Hardened	150–185	170–210	8–16	365
Carpenter stainless No. 20	C 0.07 max., Mn 0.75, Si 1.00, Cr 20.00, Ni 29.00, Mo 2.00 min., Cu 3.00 min.	Annealed	35	85	50	160
Cast Stainless Steels						
Cast 12 Cr Alloy (CA-15) (C means corrosion-resistant casting, Alloy Casting Institute designations)	C 0.15 max., Mn 1.00 max., Si 1.50 max., Cr 11.5–14, Ni 1.00 max., Fe bal.	Air-cooled from 1800°F. Tempered at 600°F.	150	200	7	390
		Air-cooled from 1800°F. Tempered at 1400°F.	75	100	30	185
Cast 12 Cr Alloy (CA-40)	C 0.20–0.40, Mn 1.00 max., Si 1.50 max., Cr 11.5–14, Ni 1.00 max., Fe bal.	Air-cooled from 1800°F. Tempered at 600°F.	165	220	1	470
		Air-cooled from 1800°F. Tempered at 1400°F.	67	110	18	212
Cast 20 Cr Alloy (CB-30)	C 0.30 max., Mn 1.00 max., Si 1.00 max., Cr 18–22, Ni 2 max.	Annealed	60	95	15	195
Cast 28-4 Alloy (CC-50)	C 0.50 max., Mn 1.00 max., Si 1.00 max., Cr 26–30, Ni 4.00 max., Fe bal.	As-cast	65	70	2	212
		Air-cooled from 1900°F.	65	97	18	210
Cast 29-9 Alloy (CE-30)	C 0.30 max., Mn 1.50 max., Si 2.00 max., Cr 26–30, Ni 8–11, Fe bal.	As-cast	60	95	15	170
Cast 20-10 Alloy (CF-8)	C 0.08 max., Mn 1.50 max., Si 2.00 max., Cr 18–21, Ni 8–11, Fe bal.	Water-quenched (1950°–2050°F.)	37	77	55	140
Cast 20-10 Alloy (CF-20)	C 0.20 max., Mn 1.50 max., Si 2.00 max., Cr 18–21, Ni 8–11, Fe bal.	Water-quenched (above 2000°F.)	36	77	50	163
Cast 20-10-2.5 Alloy (CF-8M)	C 0.08 max., Mn 1.50 max., Si 1.50 max., Cr 18–21, Ni 9–12, Mo 2.5, Fe bal.	Water-quenched (1950°–2050°F.)	42	80	50	156–170

Alloy	Composition	Condition				
Cast 18-38 Alloy (HU) (H means heat-resistant)	C 0.35-0.75, Mn 2.00 max, Si 2.50 max, Cr 17-21, Ni 37-41, Fe bal.	As-cast	40	70	9	170
		Aged	43	73	5	190
Cast 12-60 Alloy (HW)	C 0.35-0.75, Mn 2.00 max, Si 2.50 max, Cr 10-14, Ni 58-62, Fe bal.	As-cast	36	68	4	185
		Aged	52	84	4	205
Cast 15-65 Alloy (HX)	C 0.35-0.75, Mn 2.00 max, Si 2.50 max, Cr 15-19, Ni 64-68, Fe bal.	As-cast	36	65	9	176
		Aged	44	73	9	185
Cast 9 Cr Alloy (HA)	C 0.20 max, Mn 0.35-0.65 max, Si 1.00 max, Cr 8-10, Fe bal.	Annealed	65	95	23	180
		Normalized	8.	107	21	220
Cast 28-4 Alloy (HC)	C 0.50 max, Mn 1.00 max, Si 2.00 max, Cr 26-30, Ni 4 max, Fe bal.	As-cast	65	70	2	190
		Aged	80	115	18	
Cast 28-7 Alloy (HD)	C 0.50 max, Mn 1.50 max, Si 2.00 max, Cr 26-30, Ni 4-7, Fe bal.	As-cast	48	85	16	190
Cast 29-9 Alloy (HE)	C 0.20-0.50, Mn 2.00 max, Si 2.00 max, Cr 26-30, Ni 8-11, Fe bal.	As-cast	45	95	20	200
		Aged	55	90	10	270
Cast 21-10 Alloy (HF)	C 0.20-0.40, Mn 2.00 max, Si 2.00 max, Cr 19-23, Ni 9-12, Fe bal.	As-cast	45	85	35	165
		Aged	50	100	25	190
Cast 25-12 Alloy (HH)	C 0.20-0.50, Mn 2.00 max, Si 2.00 max, Cr 24-28, Ni 11-14, Fe bal.	As-cast—Type 1	50	80	25	185
		Type 2	40	85	15	180
		Aged—Type 1	55	86	11	200
		Type 2	45	92	8	200
Cast 28-15 Alloy (HI)	C 0.20-0.50, Mn 2.00 max, Si 2.00 max, Cr 26-30, Ni 14-18, Fe bal.	As-cast	45	80	12	180
		Aged	65	90	6	200
Cast 25-20 Alloy (HK)	C 0.20-0.60, Mn 2.00 max, Si 2.00 max, Cr 24-28, Ni 18-22, Fe bal.	As-cast	50	75	17	170
		Aged	50	85	10	190
Cast 30-20 Alloy (HL)	C 0.20-0.60, Mn 2.00 max, Si 2.00 max, Cr 28-32, Ni 18-22, Fe bal.	As-cast	52	82	19	192
Cast 20-25 Alloy (HN)	C 0.20-0.50, Mn 2.00 max, Si 2.00 max, Cr 19-23, Ni 23-27, Fe bal.	As-cast	38	68	17	160
Cast 15-35 Alloy (HT)	C 0.35-0.75, Mn 2.00 max, Si 2.50 max, Cr 13-17, Ni 33-37, Fe bal.	As-cast	40	70	10	180
		Aged	45	75	5	200
Cast 20-10-2.5 Alloy (CF-12M)	C 0.12 max, Mn 1.50 max, Si 1.50 max, Cr 18-21, Ni 9-12, Mo 2.5, Fe bal.	Water-quenched (from above 2000°F.)	42	80	50	156–170
Cast 20-10 Cb Alloy (CF-8C)	C 0.08 max, Mn 1.50 max, Si 1.50 max, Cr 18-21, Ni 9-12, Cb or Cb-Ta, Fe bal.	Water-quenched (1950°-2050°F.)	38	77	39	149
Cast 20-10 Alloy (CF-16F)	C 0.16 max, Mn 1.50 max, Si 2.00 max, Cr 18-21, Ni 9-12, Mo 1.5 max, Fe bal.	Water-quenched (from above 2000°F.)	40	77	52	150

TABLE 7-12 Properties of Metals and Alloys—(Continued)

Material	Nominal composition (essential elements), %	Form and condition	Yield strength (0.2% offset), 1000 lb./sq. in.	Tensile strength, 1000 lb./sq. in.	Elongation in 2 in., %	Hardness, Brinell
Cast 25-12 Alloy (CH-20)	C 0.20 max., Mn 1.50 max., Si 2.00 max., Cr 22–26, Ni 12–15, Fe bal.	Water-quenched (from above 2000°F.)	50	88	38	190
Cast 25-20 Alloy (CK-20)	C 0.20 max., Mn 1.50 max., Si 2.00 max., Cr 23–27, Ni 19–22, Fe bal.	Water-quenched (from 2100°F.)	38	76	37	144
Cast 25-20 Alloy (CN-7M)	C 0.07 max., Mn 1.50 max., Si 100, Cr 18–22, Ni 21–31, Fe bal.	As-cast	30	65	30–45	130–150
Other Cast Alloys						
Durimet 20	C 0.07, Mn 1.5, Si 1.5, Cr 20.0, Ni 29.0, Mo 2.0, Cu 3.0	Cast, annealed	30	65	48	130
Iron-silicon Alloy (Duriron)	Si 14.50, C 0.85, Mn 0.65, Mo nil, Fe bal.	Cast only	.....	16	Nil	520
Fe-Si-Mo Alloy (Durichlor 51)	Si 14.50, C 0.85, Mn 0.65, Cr 4.5, Fe bal.	Cast only	.....	16	Nil	520
Ni-Mo Alloy (Chlorimet 2)	Ni 62.00, Mo 32.00 Fe 6.00 max., Si 1.00, C 0.10	Cast	55	80	5	230
Ni-Cr-Mo Alloy (Chlorimet 3)	Ni 60.00, Cr 18.00, Mo 18.00, Fe 7.5 max., Si 1.00, C 0.07	Cast only	50	75	10	220
Illium G	Ni bal., Cr 22.5, Fe 6.5, Mo 6.4, Cu 6.5, Mn, Si	As-cast	50	68	32	200
Illium 98	Ni bal., Cr 28.00, Cu 5.5, Mn 1.25, Fe 1.00	Cast	.....	54	18	155
Nickel Alloys						
Nickel (pure)	Ni 99.99	Annealed	8.5	46	30	.....
Nickel (cast)	Ni 95.6, Cu 0.5, Fe 0.5, Mn 0.8, Si 1.5, C 0.8	As-cast	25	57	22	110
Nickel 200	Ni(+Co) 99.40, C 0.06, Mn 0.25, Fe 0.15, S 0.005, Si 0.05, Cu 0.05	Annealed	20	70	40	100
		Hot-rolled	25	75	40	110
		Cold-drawn	70	95	25	170
		Cold-rolled	95	105	5	210

Lower table

Alloy	Composition	Condition				
Low-carbon nickel 201	Ni(+Co) 99.50, C 0.02, Mn 0.20, Fe 0.15, S 0.005, Si 0.05, Cu 0.05	Annealed	15	60	50	90
		Hot-rolled	25	60	45	105
		Cold-rolled	65	95	15	150
Nickel 212	Ni(+Co) 97.85, C 0.05, Mn 1.95, Fe 0.05, S 0.005, Si 0.04, Cu 0.03	Annealed	35	75	40	140
		Hot-rolled	50	90	35	150
		Cold-drawn	80	100	25	190
Nickel 211	Ni(+Co) 95.00, C 0.10, Mn 4.75, Fe 0.05, S 0.005, Si 0.05, Cu 0.02	Annealed	35	75	40	140
		Hot-rolled	50	90	35	150
		Cold-drawn	80	100	25	190
Duranickel	Ni(+Co) 93.90, C 0.15, Mn 0.25, Fe 0.15, S 0.005, Si 0.55, Cu 0.05, Al 4.50, Ti 0.45	Annealed	45	100	40	160
		Annealed age-hardened	125	170	25	330
		Spring	.....	175	5	320
		Spring, age-hardened	.....	205	10	370
Permanickel	Ni(+Co) 98.65, C 0.25, Mn 0.10, Fe 0.10, S 0.005, Si 0.06, Cu 0.02, Ti 0.45, Mg 0.35	Annealed	45	105	45	160
		Annealed, age-hardened	125	175	25	325
		Spring	.....	180	5	
		Spring, age-hardened	195	210	10	
Nickel 205 Electronic grade	Ni(+Co) 99.55, C 0.09, Mn 0.20, Fe 0.05, S 0.005, Si 0.05, Cu 0.02	Annealed	20	70	40	100
Nickel 220 A.S.T.M. B239, grade 11	Ni(+Co) 99.65, C 0.06, Mn 0.10, Fe 0.05, S 0.005, Si 0.05, Cu 0.02, Mg 0.04	Annealed	20	70	40	100
Nickel 225	Ni(+Co) 99.50, C 0.07, Mn 0.10, S 0.005, Si 0.20, Cu 0.02, Fe 0.05	Annealed	20	70	40	100
Nickel 213	Ni(+Co) 94.2, Cu 0.5, Fe 0.5, Mn 0.8, Si 1.5, C 1.5	As-cast	30	55	20	105
Nickel 305	Ni(+Co) 91.5, Cu 0.5, Fe 0.5, Mn 0.8, Si 6.0, C 0.8	As-cast	62	85	2	220
		Annealed, aged	65	90	2	240

Upper table

Alloy	Composition	Condition				
Hastelloy Alloy B	Ni bal, Mo 28, Fe 5, Mn, Si	Sand-cast (anneal)	50	80	8	199
		Rolled (anneal)	56	120	50	215
Hastelloy Alloy C	Ni bal, Mo 13, Cr 16, Fe 5, W 4, Mn Si	Investment cast	54	85	14	209
		Sand-cast (anneal)	50	78	5	199
	......	Rolled (anneal)	71	130	45	204
		Investment cast	50	80	10	215
Hastelloy Alloy D	Ni bal, Si 10, Cu 3, Mn	Sand-cast (anneal)	118	118	0–2	321
Hastelloy Alloy G	Ni 44, Cr 22, Fe 20, Mo 6.5, Cb + Ta 2.1, Cu 2.0, C 0.05 max, W 1 max.	Sheet	46.2	102	61	B-84 Rockwell
		Plate	45.0	99.6	62	
		Sand-cast	38.8	87.5	30	
Hastelloy Alloy X	Co .5 max, Fe 18.5, Cr 22.0, Mo 9.0, W 0.3, C 0.15 max. 0.20 max. (cast), N bal. (wrought), C	Wrought sheet	52	113.2	41.0	194
		Mill-annealed	.....	67.0	17.0	172
		As investment cast	46.5			
Incoloy Alloy 800	Ni(—Co) 31.50, C 0.04, Mn 0.75, Fe 46 0.5, S 0.07, Si 0.55, Cu 0.27, Cr 20.30	Annealed	40	90	40	150
Incoloy Alloy 801	Ni(—Co) 32.50, C 0.04, Mn 0.85, Fe 44.60, S 0.07, Si 0.55, Cu 0.15, Cr 20.30, Ti 1.00	Annealed	50	90	40	160
Incoloy Alloy 804	Ni(—Co) 42.0, C 0.05, Mn 0.85, Fe 25.60, S 0.07, Si 0.60, Cu 0.40, Cr 29.70, Al 0.25, Ti 0.40	Annealed	45	100	40	175

TABLE 7-12 Properties of Metals and Alloys—(Continued)

Material	Nominal composition (essential elements), %	Form and condition	Yield strength (0.2% offset), 1000 lb./sq. in.	Tensile strength, 1000 lb./sq. in.	Elongation in 2 in., %	Hardness, Brinell
Inconel (wrought) Alloy 600	Ni(+Co) 76.40, C 0.04, Mn 0.20, Fe 7.20, S 0.007, Si 0.20, Cu 0.10, Cr 15.85	Annealed Cold-drawn	35 100	90 130	45 20	150 200
Inconel (cast) Alloy 610	Ni(+Co) 71.5, Cu 0.5, Fe 8.0, Mn 1.0, Si 2.0, Cr 16.0, C 0.20	As-cast	38	80	15	175
Inconel Alloy 625	Ni(+Co) 62.59, C 0.05, Mn 0.55, Fe 6.85, S 0.007, Si 0.35, Cu 0.05, Cr 20, Al 0.15, Ti 0.30, Cb(+Ta) 3.95	Annealed	75	125	50	176
Inconel Alloy 700	Ni 45.0, C 0.16, Mn 0.10, Fe 7.0, S 0.008, Si 0.25, Cr 15.0, Al 3.0, Ti 2.20, Mo 3.0, Co 28.0	Hot-rolled Heat-treated	106	168	......	321
Inconel Alloy 702	Ni(+Co) 78.00, C 0.02, Mn 0.05, Fe 0.30, S 0.007, Si 0.15, Cu 0.05, Cr 15.85, Al 3.00, Ti 0.60	Annealed	45	105	55	150
Incoloy Alloy 825	Ni 40, Cr 21, Fe 31, Mo 3.0, Cu 1.75, Mn 0.60, Si 0.40, C 0.05	Annealed Cold-drawn	45	95 155	40 10	185 255
Incoloy Alloy 901	Ni(+Co) 42.65, C 0.05, Mn 0.45, Fe 33.90, S 0.010, Si 0.40, Cu 0.10, Cr 13.45, Al 0.25, Ti 2.50, Mo 6.20	Annealed Annealed, aged	45 105	110 165	45 26	160 310
Nimonic 75	Ni(+Co) 77.40, C 0.10, Mn 0.45, Fe 0.50, S 0.007, Si 0.45, Cu 0.05, Cr 20.50, Al 0.15, Ti 0.35	Annealed	55	115	40	168
Nimonic 80A	N(+Co) 74.45, C 0.05, Mn 0.55, Fe 0.55, S 0.007, Si 0.20, Cu 0.05, Cr 20.45, Al 1.25, Ti 2.40	Annealed	60	115	60	185
Nimonic 90	Ni(+Co) 57.00, C 0.05, Mn 0.50, Fe 0.45, S 0.007, Si 0.20, Cu 0.05, Cr 20.55, Al 1.65, Ti 2.60, Co 16.90	Annealed	90	155	......	260

Table — Properties of Nickel Alloys, Monel Alloys, and Aluminum Alloys (no column headers visible)

Alloy	Composition	Condition				
Inconel Alloy 705 weldable	Ni(+Co) 68.5, Cu 0.5, Fe 9.0, Mn 1.0, Si 1.6, Cr 15.5, C 0.20, Cb added	As-cast	38	80	15	175
Inconel Alloy 705	Ni(+Co) 68.0, Cu 0.5, Fe 8.0, Mn 1.0, Si 5.5, Cr 15.5, C 0.20	As-cast	90	95	2	340
Inconel Alloy 713	Ni(+Co) bal., Cr 13.0, C 0.13, Mo 4.5, Cb 2.0, Al 6.0, Ti 06	Annealed and aged	95	100	2	340
		Investment-cast	102	120	6	.....
Inconel Alloy 721	Ni(+Co) 71.40, C 0.03, Mn 2.20, Fe 6.70, S 0.007, Si 0.10, Cu 0.04, Cr 16.40, Al 0.05, Ti 3.05	Annealed	45	110	50	160
		Annealed, age-hardened	100	145	7	300
Inconel Alloy 722	Ni(+Co) 74.35, C 0.05, Mn 0.60, Fe 6.50, S 0.007, Si 0.20, Cu 0.05, Cr 15.20, Al 0.60, Ti 2.40	Annealed	45	110	50	160
		Annealed, age-hardened	95	155	30	275
Inconel Alloy X-750	Ni(+Co) 72.85, C 0.04, Mn 0.65, Fe 6.80, S 0.007, Si 0.30, Cu 0.05, Cr 15.15, Al 0.75, Ti 2.50, Cb(+Ta) 0.85	Annealed	50	115	50	150
		Annealed, age-hardened	115	175	25	300
Monel Alloy 400	Ni(+Co) 66.15, C 0.12, Mn 0.90, Fe 1.35, S 0.005, Si 0.15, Cu 31.30	Annealed	35	75	40	125
		Hot-rolled	50	90	35	150
		Cold-drawn	80	110	25	190
		Cold-rolled	100	110	5	240
Monel Alloy 401	Ni(+Co) 44.0, Cu bal., Fe 1.0, Mn 0.8, C 0.10	Annealed	20	75	51	140
Monel Alloy R-405	Ni(+Co) 66.35, C 0.18, Mn 0.90, Fe 1.35, S 0.050, Si 0.15, Cu 31.00	Hot-rolled	45	85	35	145
		Cold-drawn	75	100	25	200
Monel Alloy 402 (obsolete)	Ni(+Co) 58.0, C 0.12, Mn 0.50, Fe 1.20, S 0.005, Si 0.10, Cu 39.55	Hot-rolled	65	90	35	175
		Cold-drawn	85	95	25	215
Monel Alloy 404	Ni(+Co) 55.85, C 0.12, Mn 0.10, Fe 0.15, S 0.006, Si 0.10, Cu 44	Cold-drawn	24	65	50	200
Monel Alloy 411 (A loy 19) weldable / Centrifugal castings only	Ni(+Co) 62.0, Cu 31.5, Fe 2.0, Mn 0.8, Si 1.5, C 0.20, Cb added	As-cast	35	75	35	140
Monel Alloy K-500	Ni(+Co) 65.25, C 0.15, Mn 0.30, Fe 1.00, S 0.005, Si 0.15, Cu 29.60, Al 2.75, Ti 0.45	Annealed	45	100	40	155
		Annealed, age-hardened	100	155	25	270
		Spring	140	150	5	300
		Spring, age-hardened	160	185	10	335
Monel Alloy 506 (A loy 16), centrifugal castings only	Ni(+Co) 63.0, Cu 30.5, Fe 1.5, Mn 0.8, Si 3.2, C 0.10	As-cast	70	115	10	265
Monel Alloy 505 (A loy 17), centrifugal castings only	Ni(+Co) 63.0, Cu 29.5, Fe 2.0, Mn 0.8, Si 4.0, C 0.08	Annealed	75	110	8	220
Monel Alloy 502	Ni(+Co) 64.75, C 0.25, Mn 0.50, Fe 1.00, S 0.005, Si 0.15, Cu 29.85, Al 2.85, Ti 0.45	As-cast, or annealed and aged	110	135	2	340
		Hot-finished, annealed	35	95	48	160
		Hot-finished, annealed, age-hardened	90	145	25	275
Aluminum and Alloys						
Aluminum Alloy No. 1100	Al 99 plus	Annealed-0	5	13	45	23
		Cold-rolled-H14	17	18	20	32
		Cold-rolled-H18	22	24	15	44
3003	Al bal., Mn 1.2	Annealed-0	6	16	40	28
		Cold-rolled-H14	21	22	16	40
		Cold-rolled-H18	27	29	10	55

TABLE 7-12 Properties of Metals and Alloys—(Continued)

Material	Nominal composition (essential elements), %	Form and condition	Yield strength (0.2% offset), 1000 lb./sq. in.	Tensile strength, 1000 lb./sq. in.	Elongation in 2 in., %	Hardness, Brinell
5052	Al bal., Mg 2.5, Cr 0.25	Annealed-0	13	28	30	47
		Cold-rolled and stabilized-H34	31	38	14	68
		Cold-rolled and stabilized-H38	37	42	8	77
5086	Al bal., Mn 0.5, Mg 4.0, C 0.15	Annealed	17	38	22	.....
		H34	37	47	10	
6063	Al bal., Si 0.4, Mg 0.7	Annealed-0	7	13	30	25
		Artificially aged-T5	21	27	12	60
		Heat-treated and artificially aged-T6	31	35	12	73
7075	Al bal., Zn 5.6, Cu 1.6, Mg 2.5, Cr 0.3	Annealed-0	15	33	17	60
		Heat-treated and artificially aged-T6	73	83	11	150
380	Al bal., Cu 3.5, Si 9.0	Die-cast-F	26	43	2.0	.....
43	Al bal., Si 5.0	Sand-cast-F	8	19	8	40
		Permanent mold-cast-F	9	23	10	45
		Die-cast-F	16	30	7	
195	Al bal., Cu 4.5, Si 0.8	Sand-cast; heat-treated-4	16	32	8.5	60
		Sand-cast; heat-treated and artificially aged-T6	24	36	5	75

Copper and Alloys

Material	Nominal composition (essential elements), %	Form and condition	Yield strength (0.2% offset), 1000 lb./sq. in.	Tensile strength, 1000 lb./sq. in.	Elongation in 2 in., %	Hardness, Brinell
Nickel silver 18% (wrought) 752 65-18	Cu 65, Zn 17, Ni 18	Annealed	25	58	40	70
		Cold-rolled (HT)	70	85	4	170
		Cold-drawn wire (HT)	.....	105		
Nickel silver 10% (wrought) 740 65-10	Cu 65, Zn 25, Ni 10	Annealed	20	55	45	60
		Cold-rolled (HT)	70	88	5	180
		Cold-drawn wire (HT)	.....	110		
Nickel silver 20% (cast) (11A)	Cu bal., Ni 20, Zn 6, Pb 5, Sn 4	Cast (as cast)	25	40	15	85
Cupronickel 10% 706	Cu 88.35, Ni 10, Fe 1.25, Mn 0.4	Annealed	22	44	45	.....
		Cold-drawn tube	57	60	15	
Cupronickel 30% 715	Cu 68.90, Ni 30, Mn 0.60, Fe 0.50	Annealed	22	55	45	70
		Cold-drawn	60	75	20	150
		Cold-rolled	70	77	5	155
Cupronickel 55-45 (Constantan)	Cu 55, Ni 45	Annealed	30	60	45	.....
		Cold-drawn	50	65	30	
		Cold-rolled	65	85	20	
Copper 102	Cu 99.9 plus	Annealed	10	32	45	42
		Cold-drawn	40	45	15	90
		Cold-rolled (HT)	40	46	5	100
Red brass (wrought) 230	Cu 85, Zn 15	Annealed	15	40	50	50
		Cold-drawn	55	70	15	120
		Cold-rolled	60	75	7	135
Red brass (cast)	Cu 85, Zn 5, Pb 5, Sn 5	Cast (as cast)	17	35	25	60
Gilding metal 210	Cu 95.0, Zn 5.0	Cold-rolled	50	56	5	114

Copper alloys — mechanical properties

Material	Composition	Condition				Hardness
Commercial bronze 220	Cu 90.0, Zn 10.0	Cold-rolled	54	61	5	125
Cartridge 70-30 brass 260	Cu 70.0, Zn 30.0	Cold-rolled	63	76	8	155
Architectural bronze 385	Cu 57.0, Zn 40.0, Pb 3.0	Annealed	20	60	30	95
Phosphor bronze 10% 524	Cu 90, Sn 10, P 0.25	Spring temper	.....	122	4	241
Phosphor bronze 5% 510	Cu 94.75, Sn 5, P 0.25	Annealed	20	50	50	60
		Cold-drawn wire (HT)	.....	130	2	.....
		Cold-rolled (HT)	65	80	8	160
Aluminum brass	Cu 76.0, Zn 22.0, Al 2.0, As trace	Annealed	27	60	55	82
Yellow brass (high brass 268)	Cu 65, Zn 35	Annealed	18	48	60	55
		Cold-drawn	55	70	15	115
		Cold-rolled (HT)	60	74	10	180
Naval brass 464	Cu 60, Zn 39.25, Sn 0.75	Annealed	22	56	40	90
		Cold-drawn	40	65	35	150
Admiralty brass 443 (inhibited)	Cu 71, Zn 28, Sn 1, As, Sb, or P present	Annealed	20	53	65	60
Muntz metal 280	Cu 60, Zn 40	Annealed	20	54	45	80
Manganese bronze 675	Cu 58.5, Zn 39.2, Fe 1, Sn 1, Mn 0.3	Annealed	30	60	30	95
		Cold-drawn	50	80	20	180
High-silicon bronze A 655	Cu 96, Si 3, Mn, Zn, or Fe	Annealed	22	58	60	70
		Cold-drawn	60	90	20	180
		Cold-rolled	60	95	-	190
Low-silicon bronze B 651	Cu 96, Si 0.8–2.0, Mn 0.7 max, Fe 0.8 max.	Annealed	15	40	50	F55
		Hardened	55	70	15	B80
Aluminum bronze 612	Cu 92, Al 8	Annealed	25	70	60	80
		Hard	65	105	-	210
Ni-Vee bronze type A	Cu 88, Ni 5, Sn 5, Zn 2	As-cast	22	50	40	85
		Tempered	40	65	10	130
		Heat-treated	55	85	10	180
Ni-Vee bronze type B	Cu 87, Ni 5, Sn 5, Pb 1, Zn 2	As-cast	20	45	30	80
		Tempered	30	60	8	120

Material	Composition	Condition				Rockwell
Copper beryllium 172	Be 1.9, Co 0.25, Cu bal.	Annealed (SA)	.....	70	45	B60
		Annealed (SA, HT)		175	6	C38
		Cold-rolled (HT)		110	5	B99
		Cold-rolled (HT, HT)		200	2	C42
Copper beryllium 170	Be 1.7, Co 0.25, Cu bal.	Annealed (SA)	.....	.....	.....	.....
		Annealed (SA, HT)		70	45	B60
		Cold-rolled (HT)		165	6	C35
		Cold-rolled (HT, HT)		.....	.....	.....
Copper beryllium 177	Be 0.55, Co 2.5, Cu bal.	Annealed (SA)	.....	45	27	B30
		Annealed (SA, HT)		110	9	B96
		Cold-rolled (HT)		78	6	B75
		Cold-rolled (HT, HT)		120	8	B98
Copper beryllium 176	Be 0.38, Co 1.55, Ag 2.0, Cu bal.	Annealed (SA)	.....	45	27	B30
		Annealed (SA, HT)		110	15	B96
		Cold-rolled (HT)		70	13	B73
		Cold-rolled (HT, HT)		120	14	B99

Lead and Alloys

Material	Composition	Condition				
Chemical lead	Pb 99.9, Cu 0.06, Bi 0.005 max.	Rolled	1.9	2.5	50	5
Antimonial lead	Pb 94, Sb 6	Cast	.....	6.8	22	12
		Rolled	.....	4.1	47	9
Tellurium lead	Pb 99.85, Te 0.04, Cu 0.06	Rolled	2.2	3	45	6
Soft solder 50-50	Sn 50, Pb 50	Cast	.....	6.8	50	14
Soft solder 60-40	Sn 60, Pb 40	Cast	.....	7.1	45	15

TABLE 7-12 Properties of Metals and Alloys—(Continued)

Material	Nominal composition (essential elements), %	Form and condition	Yield strength (0.2% offset), 1000 lb./sq. in.	Tensile strength, 1000 lb./sq. in.	Elongation in 2 in., %	Hardness, Brinell
Magnesium Alloys						
Magnesium alloy AZ92A	Mg bal., Al 9.0, Zn 2.0, Mn 0.10 min.	Sand-cast (as cast)	14	24	6	50
		Sand-cast (solution heat-treated)	14	40	12	55
		Sand-cast (solution heat-treated and aged)	19	40	5	83
		Sand-cast (age-hardened)	16	30	4	66
Magnesium alloy AZ31B	Mg bal., Al 3.0, An 1.0, Mn 0.20 min.	Rolled-plate (strain hardened then partially annealed)	24	37	18	.....
		Rolled-sheet (strain hardened then partially annealed)	32	42	15	73
		Annealed	22	37	21	56
		Extruded	28	38	14	.....
Magnesium alloy AZ80A	Mg bal., Al 8.5, Zn 0.5, Mn 0.15 min.	Extruded	36	49	11	60
		Extruded (age-hardened)	39	53	6	82
		Forged (age-hardened)	34	50	6	72
Magnesium alloy AZ91A and AZ91B	Mg bal., Al 9.0 Zn 0.6, Mn 0.13 min.	Die-cast (as cast)	22	33	3	67
Magnesium alloy AZ91C	Mg bal., Al 8.7, Zn 0.7, Mn 0.13 min.	Sand-cast (as cast)	14	24	2	52
		Sand-cast (solution heat-treated)	14	40	11	55
		Sand-cast (solution heat-treated and aged)	19	40	5	73
Magnesium alloy EZ33A	Mg bal., Zn 2.6, Zr 0.7, other elements 3.0	Sand-cast (age-hardened)	15	23	3	50
Magnesium alloy HK31A	Mg bal., Th 3.0, Zr 0.7	Sand-cast (solution heat-treated and aged)	15	30	8	55
		Rolled-sheet (strain-hardened then partially annealed)	29	37	8	57
Magnesium alloy HZ32A	Mg bal., Th 3.0, Zn 2.1, Zr 0.7	Sand-cast (age-hardened)	14	29	7	57
Magnesium alloy ZK60A	Mg bal., Zn 5.7, Zr 0.55	Extruded	37	49	14	75
		Extruded (age-hardened)	43	52	12	82
		Forged	38	49	13	

Titanium and Alloys

Material	Composition	Condition	(1)	(2)	(3)	(4)
Titanium (commercially pure)	Ti bal., Fe 0.2 max., N_2 0.05 max., C 0.08 max., H_2 0.015 max., O_2 0.4	Annealed	75	85	23	200
Titanium-chromium-iron-molybdenum alloy (Ti 5Al-2.5Sn)	Ti bal., Al 5, Sn 2.5, Fe 0.5 max., C 0.05 max., H_2 0.015 max., N_2 0.05 max.	Annealed	120	127	12	34 (Rockwell C)
Titanium-aluminum-vanadium alloy (Ti-6 Al-4V)	Ti bal., Al 6.0, V 4.0, Fe 0.25 max., C 0.08 max., H_2 0.0125 max., N_2 0.05 max.	Annealed	130	140	14	32 (Rockwell C)
		Heat-treated	145	155	12	37 (Rockwell C)

Material class	Sub-material	Composition	Condition	(A)	(B)	(C)	(D)
Gold	18K white gold	Au 75, Ni 18.5, Zn 5.25, Cu 1.25	Hard	.....	30	2	48
			Annealed	93	17.5	40	23
	Haynes Stellite Alloy 21	C 0.25, Cr 28, Ni 2.5, Mo 5.5, Co bal.	As investment cast	82.0	103	8.0	313 max.
	Hayes Stellite Alloy 31 (X-40 Cast)	C 0.50, Cr 25.5, Ni 11, W 7.5, Co bal.	As investment cast	80.0	113	8.0	313 max.
	Haynes Stellite Alloy 25	C 0.15 max., Cr 20.0, Ni 10.0, W 15.0, Mn 1.5, Co bal.	Wrought sheet, Mill annealed	63	140	60.0	244
	Haynes Stellite Alloy 36	C 0.43, Cr 19, Ni 10, W 15.0, Mn 1.5, Co bal.	As investment cast	90	103	5.0	298 max.
Iridium		Ir 100	Annealed	.....	36	.....	175
Molybdenum		Mo 99.9 plus	As-rolled	75	100	30	250
			Stress-relieved	75	100	30	240
			Recrystallized	50	70	45	190
0.5% Titanium molybdenum alloy		Mo bal., Ti 0.5	As-rolled	90	120	30	290
			Stress-relieved	90	120	30	280
			Recrystallized	60	80	40	200
Multimet N-155 Alloy		Ni 19.0–21.0, Co 18.5–21.0, Cr 20.0–22.5, Mo 2.5–3.5, W 2.0–3.0, Fe bal., C 0.08–0.16, N 0.10–0.20, Cb + Ta 0.75–1.25	Mill annealed sheet	58	118	49	194
			Mill annealed bar	54	111		189
			Sand-cast	54	98		180
			An investment cast	58	101		180 max.
Platinum		Pt (commercial)	Hard	.....	65	2	101
			Annealed	.....	27	28	65
Platinum-iridium		Pt 90, Ir 10	Hard	34	80	2	169
			Annealed	.....	53	23	104
Platinum-rhodium		Pt 90, Rh 10	Hard	.....	93	3	139
			Annealed	18.3	50	36	79
Platinum-ruthenium		Pt 90, Ru 10	Hard	.....	145	2	210
			Annealed	47.6	91	28	156

Other Non-ferrous Alloys

Material	Composition	Condition	(1)	(2)	(3)	(4)
Antimony	Sb 100	As-cast	.....	.....	.....	42
Bi-Pb-In-Cd-Sn alloy (Cerrolow-117)	Bi 44.70, Pb 22.60, Sn 8.30, Cd 5.30, In 19.10	Cast	.....	1.56	1.5	12
Bi-Pb-Sn-Cd alloy (Cerrobend)	Bi 50.00, Pb 26.70, Sn 13.30, Cd 10.00	Cast	.....	5.99	200	9.2
Bi-Pb alloy (Cerrobase)	Bi 55.50, Pb 44.50	Cast	.....	6.4	60–70	10.2
Bi-Sn alloy (Cerrotru)	Bi 58.00, Sn 42.00	Cast	.....	8.0	200	22
Columbium	.....	.....	35	40	.....	.....
Columbium 1Zr	Zr 0.8–1.2, Cb bal.	.....	135	48	.....	.....
Columbium F48	W 13.5–16.5, Mn 4.5–5.5, Zr 0.85–1.15, Cb bal.	.....	110	120	.....	.....
Columbium FS82	Ta 33, Zr 0.8, Cb bal.	.....	50	68	.....	.....
Columbium D31	Ti 10, Mo 10, Cb bal.	.....	90	100	.....	.....

TABLE 7-12 Properties of Metals and Alloys—(Continued)

Material	Nominal composition (essential elements), %	Form and condition	Yield strength (0.2% offset), 1000 lb./sq. in.	Tensile strength, 1000 lb./sq. in.	Elongation in 2 in., %	Hardness, Brinell
Palladium	Pd (commercial)	Hard	....	55	....	91
		Annealed	7.6	30	30	47
Palladium-ruthenium	Pd 95.5, Ru 4.5	Hard	51	132	3	184
		Annealed	85	85	26	120
Palladium-silver	Pd 60, Ag 40	Hard	94	100	....	176
		Annealed	15	47	40	87
Palladium alloy 934	Pd 35, Pt 10, Au 10, Ag 30, Cu 15	Annealed	61	96	24	180
		Heat-treated	125	146	10	Aged 280
Rhodium	Rh 100	Annealed	....	80	....	119
Silver (pure)	Ag 99.9 plus	Annealed	12	23	45	30
		Cold-rolled	38	43	6	90
Sterling silver	Ag 92.5, Cu bal.	Hard	50	64	4	125
		Annealed	20	41	26	65
Silver, coin	Ag 90, Cu bal.	Hard	53	65	4	125
		Annealed	23	42	26	70
Tantalum	Ta 99.9 plus	Annealed sheet	45	60	37	55
		Unannealed sheet	100	110	3	123
Tantalum 10W	W 10, Ta bal.	Annealed	158	160	....	....
Tin	Sn 100	As-cast	....	2.1	70	3.9
Tungsten	W	Hard (sheet)	360	400	....	....
		Annealed	....	....	....	290
		Hard (wire)	540	600	0-8	....
Zinc	Zn bal, Pb 0.08	Hot-rolled (long.)	....	19.5	65	38
		Hot-rolled (transv.)	....	23	50	....
		Cold-rolled (long.)	....	21	50	....
		Cold-rolled (transv.)	....	27	40	....
Zilloy-15	Zn bal, Mg 0.010, Cu 1.00	Hot-rolled (long.)	....	29	20	61
		Hot-rolled (transv.)	....	40	10	....
		Cold-rolled (long.)	....	36	25	80
		Cold-rolled (transv.)	....	46	10	....
Zilloy-40	Zn bal, Cu 1.00	Hot-rolled (long.)	....	24	50	52
		Hot-rolled (transv.)	....	30	35	....
		Cold-rolled (long.)	....	31	40	60
		Cold-rolled (transv.)	....	40	30	....
Zinc-aluminum alloy	Zn (99.99% pure remainder), Al 3.5-4.3, Mg 0.03-0.08, Cu 0.25 max.	Die-cast	....	41	10	82

	Composition	Condition				
Zinc-aluminum-copper alloy	Zn (99.99% pure remainder), Al 3.5–4.3, Mg 0.03–0.08, Cu 0.75–1.25	Die-cast	.....	47.6	7	91
Zirconium, commercial	O₂ 0.07, C 0.15; Hf 1.90, Zr bal.	Annealed	40	65	27	B80 (Rockwell)
Zircaloy 2	Hf 0.02, Sn 1.46, Fe 0.12, Ni 0.05, Zr bal., other 0.25	Annealed	50	75	22	B90 (Rockwell)
Zircaloy 3	Hf 0.02, Sn 0.25, Fe 0.25, Ni 0.05, Zr bal., other 0.20	Annealed	45	70	25	B85 (Rockwell)

Creep

The progressive deformation of a metal under a constant stress is called *creep*. For most metals, creep is not a particularly important factor at room temperature, but it can become significant at elevated temperatures. As in the case of fatigue, no widely accepted test procedure or tabularized data exist for the creep resistance of metals, and where this property becomes important, special tests are usually devised. The most common method of presenting creep data is by a plot of the logarithm of the stress against creep rate, as illustrated in Fig. 7-21.

Impact

In connection with fatigue and creep, we have seen that the manner in which stress is applied to a metal has a profound effect on the ability of the metal to withstand the stress. Perhaps the most devastating method of applying stress is by impact, that is, a stress which is applied suddenly or over a short period of time. Since in many uses metals are subjected to this kind of stress, impact testing has been well defined in terms of equipment and procedures for carrying out such tests. In impact testing, a sample is mounted in such a manner as to provide a very high stress over a very small area. This is accomplished by notching the test specimen. Perhaps the most widely used impact-test device is the *Izod tester*, in which the sample is mounted vertically with impact from a heavy swinging pendulum occurring at a particular point on the specimen. In another procedure, known as the *Charpy test*, the sample is mounted horizontally and impacted by a falling weight which strikes the sample directly above a V-shaped notch. The impact is usually measured by the energy absorbed in fracturing the metal. This is usually expressed in foot-pounds and is easily calculated from the weight of the pendulum and the angle through which it swings (for the Izod test), or the weight and the height from which it falls (in the Charpy test). As in the case of creep and fatigue, impact testing is usually carried out in a manner which is directly related to the manner in which the metal is ultimately going to be used. For this reason many of the tests are highly empirical, and tabulated data of impact resistance have not been extensively developed.

7-3   METALWORKING

Metal-shaping Processes

One of the important reasons that metals are so widely used in modern technology is the ease with which they can be shaped into useful, and sometimes intricate, forms. While a large number of procedures have been developed for shaping metals, all such operations fall

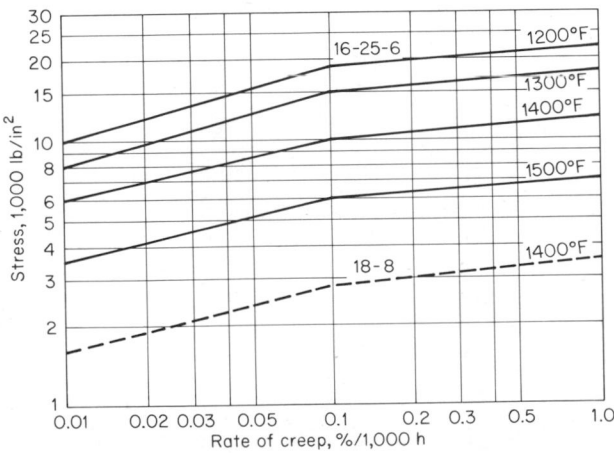

Fig. 7-21 Typical (high-strength-steel) stress-creep curves.

into relatively few categories, and these are based on the nature of the forces applied to the work during the shaping operation. These categories are:

1. Direct-compression processes
2. Indirect-compression processes
3. Tension processes
4. Bending processes
5. Shearing processes

In *direct-compression processes,* the metal flows at right angles to a compressive force applied to it. Forging and rolling are examples of direct-compression processes as illustrated in Fig. 7-22a and b. In *indirect-compression processes,* the primary force applied to the workpiece is usually a tension force, but by reaction of the workpiece against a die, compressive forces are indirectly developed. Such processes as wire and tube drawing, extrusion, and deep drawing are typical of this type of process and are illustrated in Fig. 7-22c, d, and e. In *tension processes,* the metal is stretched and shaped under a tensile force, as, for example, when a section of sheet metal is wrapped around a contour under the application of a tensile force, as illustrated in Fig. 7-22f. *Bending processes* occur when a compressive force is applied in such a manner as also to provide a moment which causes deformation of the metal, as illustrated in Fig. 7-22g. In *shearing processes,* compressive forces are applied to the metal under conditions which normally would cause bending, but in a shearing operation, the bending arm is sufficiently short to cause rupture of the metal, as illustrated in Fig. 7-22h.

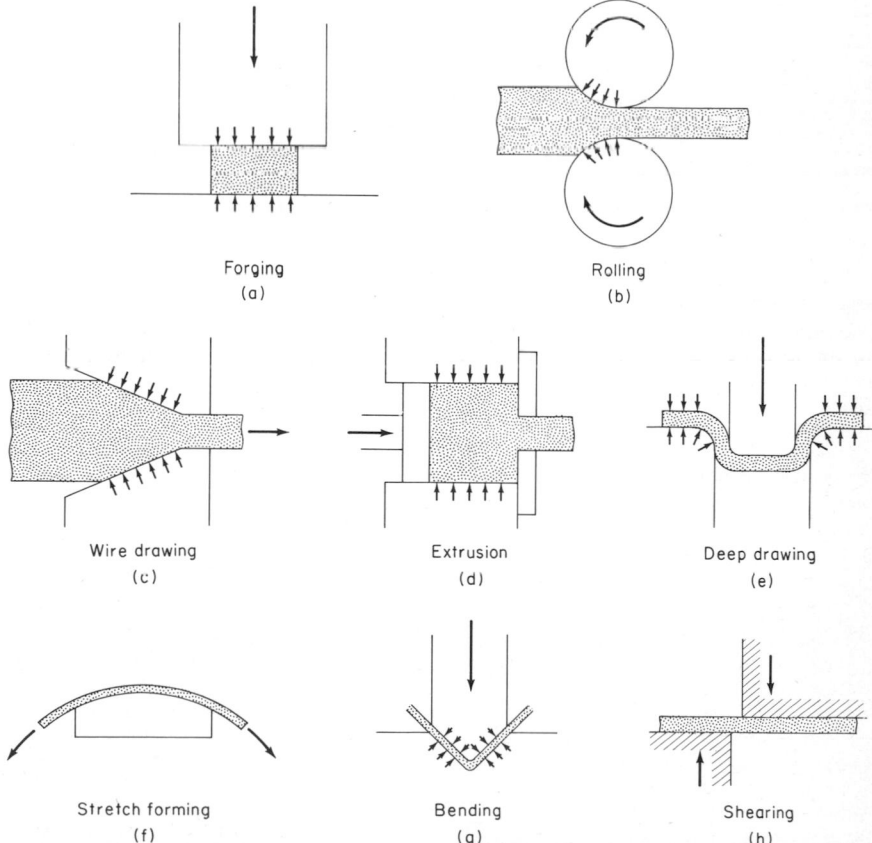

Forging
(a)

Rolling
(b)

Wire drawing
(c)

Extrusion
(d)

Deep drawing
(e)

Stretch forming
(f)

Bending
(g)

Shearing
(h)

Fig. 7-22 Metal shaping operations. (a) and (b) illustrate direct compression processes; (c), (d), and (e) illustrate indirect compression processes; (f) illustrates a tension process; (g) illustrates a bending process; (h) illustrates a shearing process.

Temperature usually plays an important part in metal-shaping operations. Because the plasticity of the metal increases with temperature, many operations such as forging, rolling, drawing, and extrusion are carried out with the metal at an elevated temperature. Such a procedure is referred to as *hot working*. Because the metal flows more easily at elevated temperatures, the forces necessary to cause the deformation can be smaller (thereby reducing the cost of the equipment, wear, and maintenance) and the operation can usually be carried out more rapidly. In other instances, the metal-shaping operation is carried out at room temperatures, such procedures being referred to as *cold working*. As previously discussed, cold working can cause some significant changes in the strength and ductility of the piece, the strength usually increasing and the ductility usually decreasing as the amount of cold working increases. The physical properties of a cold-worked piece can be altered by annealing. Frequently, where a cold-working operation is carried out in several sequential steps, the metal may work harden to the point where the latter operations are difficult to perform. In this case, the piece may be annealed after a given number of operations to permit the full sequence of operations to be applied without failure of the metal.

The speed of *deformation* of the forming processes is another important factor. Frequently, a brittle metal that would shatter when subjected to a high-speed impact can be successfully shaped if the force is applied more slowly. In other processes, such as deep drawing, the force necessary to cause the metal to deform decreases after the metal has started to flow, and the shaping operation can take place at reduced forces if the metal flow is not interrupted. Table 7-13 gives typical velocity ranges for different forming operations.

TABLE 7-13 Velocity Ranges for Typical Metal-forming Operations

Operation	Velocity (ft/s)
Tension test	$2 \times 10^{-6} – 2 \times 10^{-2}$
Hydraulic extrusion press	0.01–10
Mechanical press	0.5–5
Charpy impact test	10–20
Forging hammer	10–30
Explosive forming	100–400

Joining of Metals

There are basically four different ways to join metals: mechanical, welding, brazing or soldering, and through the use of nonmetallic adhesives. The *mechanical joining* of metals is largely accomplished by some form of nut and bolt, the various types in common usage being illustrated in Fig. 7-23. In the case of rivets, a similar variety of types exists, largely reflecting the shape of the rivet head.

Welding involves the joining of similar metals by melting the pieces together in the vicinity of the joint, thus essentially making one continuous piece of metal. Frequently, a filler or joining metal of the same composition as the pieces to be joined is used to add to the volume of metal in the melt area. In welding, a flame of oxygen and acetylene gas (or occasionally oxygen and hydrogen) is used as a source of heat. Such flames can provide temperatures up to 3000°C. This is a sufficiently high temperature to melt the metals in the restrictive area involved in the weld. However, because the thermal conductivity of the metal is usually high, the tendency is for the heat to be conducted quickly away from the joint, and this permits the welding operation to proceed from point to point, only a small area being molten at any one time. Welding is widely used for steels, aluminum, and other common metals. However, these metals have a tendency to oxidize when heated in air, and frequently the melt develops

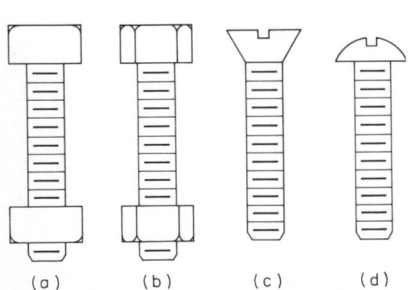

(a)   (b)   (c)   (d)

Fig. 7-23 Common nut and bolt metal fasteners. (*a*) Square head with square nut; (*b*) hex head with hex nut; (*c*) flat head or coach bolt; (*d*) round head or stove bolt.

an oxide coating that prevents the fusion of the molten metals; or the oxide may remain occluded in the weld, thereby reducing the strength of the joint. To prevent oxidation of the molten metal, a *flux* is used for all types of welding operations. Fluxes are usually reducing protective coatings that melt and protect the heated metal from exposure to air.

Other types of welding relate to the manner in which the heat is produced or the manner in which the metals are caused to flow together. In *electric-arc welding* an electric arc is struck between two carbon or tungsten electrodes, or between the metal to be welded and either a carbon or tungsten welding electrode. Frequently, a rod of filler metal, suitably coated by a flux, is used as one of the electrodes, with the metals to be joined forming the other electrode. In electric-arc welding, the highly localized heat generated by the arc is used to melt and join the metals. In *pressure welding* the two pieces of metal to be joined are heated to a high temperature, but just below the actual fusion point of the metals. The pieces to be joined are then held together firmly, and pressure is applied to them in a suitable press or in a hammering device. In pressure welding, recrystallization of the metals occurs at the surfaces and the crystals of the metal grow across the interface, thus producing a continuous metallic structure such as is characteristic of a welding operation. In *spot welding* a small area of molten metal is produced in the pieces to be joined by causing a large electric current to flow in the areas to be joined, while at the same time applying a high mechanical pressure to the area. This is accomplished by bringing two copper electrodes directly across the points to be welded, and applying a very high current for a very short (and accurately controlled) period. The electrical resistance of the system, including the interfacial contact resistances, causes the metals to be heated up and melt directly between the two electrodes.

In *brazing or soldering* the two metals are joined through the agency of a third metal. The characteristic required of the third metal, i.e., the solder or brazing metal, is that it "wet" the surface of the metals to be joined under conditions such that the solder or brazing metal is molten, while the metals to be joined are not. Because it wets the surfaces of the metals to be joined, the solder or brazing material actually forms a chemical bond with the two metals, which are then held together by the strength of the solder or the brazing metal. As in the case of welding, fluxes must be used to protect the metals and the solder or brazing alloy from air oxidation, even though the temperatures at which soldering and brazing are carried out are much lower than those required for welding. *Soft solders* are usually very low-melting alloys based on various compositions of lead and tin. Soft soldering usually requires the application of only low heat, say up to about 200°C. *Hard soldering* is more of a brazing operation using low-melting silver alloys. However, the temperature at which silver soldering or hard soldering is carried out is between 600 and 800°C. Above 800°C, special brazing alloys are used. Table 7-14 gives the composition and melting range of various soft solders, silver solders, and brazing alloys.

Powder Metallurgy

Powder metallurgy involves the shaping of metal pieces by pressing powders in a suitable mold. Although powder metallurgy is a fairly recently developed metal-shaping technique, it has already gained tremendous importance in that it is capable of forming small metal

TABLE 7-14   Soft Solders, Silver Solders, and Brazing Alloys

Composition (%)						Melting range (°C)	Application
Pb	Sn	Cu	Zn	Ag	Cd		
50	50					180–190	Lead
40	60					170–180	Zinc
30	70					160–170	Copper, brass
		16	4	80		740–800	Copper, brass, stainless steel
		20	15	65		640–730	Copper, brass, stainless steel
		15	17	50	18	620–640	Copper, brass, stainless steel
		50	50			870–880	Ferrous metals
	5	50	45			750–760	Ferrous metals

From P. G. Ormandy, "An Introduction to Metallurgical Laboratory Techniques," Pergamon Press, London, 1968.

parts into intricate shapes, and to be able to do so at high production rates. Powder metallurgy can also be used for the production of porous-metal objects. Modern practice now permits the development of a wide variety of physical characteristics in the powder-metal piece.

The preparation of the metal powder is of prime importance in determining how it will react to the pressing operation. The manner in which the powder flows into the mold and compacts depends upon the physical characteristics of the powder. Several methods have been developed to provide powders of different physical properties.

In *comminution processes,* the metal powder is prepared by a mechanical grinding operation. Such a process may be carried out in a ball mill or similar device and is usually applicable to hard metals like iron and steel alloys but not to ductile metals like aluminum or lead. *Comminution by machining* is accomplished by placing the alloy in a lathe chuck and applying a cutting tool to produce small chips of the metal. While comminution by machining is relatively expensive, it has found application for metals like magnesium which are not readily broken down by other means. Usually, to avoid surface oxidation, the machining operation is carried out in an inert atmosphere. Many metals such as aluminum, lead, tin, zinc, and copper (i.e., ductile metals) can be made into fine powders by *atomization,* a process in which the molten metal is forced through a nozzle against a jet of air, water, or inert gas in such a manner that the molten-metal stream is quickly broken up into fine droplets which then solidify. Still another process for producing fine metal powders is by *oxide reduction.* In this particular process, a powdered metallic oxide is reduced to the metal by passing hydrogen or carbon monoxide over the metal powder at an elevated temperature, but well below the melting point of the metal. Powders of metals which are difficult to melt, tungsten, molybdenum, or cobalt, for example, are frequently made by this process. By *electrolytic deposition* under special conditions, particularly fine metal powders can be produced to controlled particle sizes. Normally, the electrodeposition of a metal results in a continuous metallic body. However, if the electrodeposition is carried out at very high current densities and high temperatures, the deposited metal does not adhere to the cathode or to itself and can be readily scraped from the cathode as a fine powder. Electrolytic deposition can be used to produce powdered iron, copper, zinc, cadmium, nickel, tin, and lead.

The particle shape is largely governed by how the powder is prepared. The particle-size distribution of the powder is also important, particularly where high-density compacts are required. As shown in Fig. 7-24, the particle-size distribution has an important influence on the void volume of the powder. If only large spheres were present, the void volume would be larger than for the particle-size distribution shown. Even though the void volume is greatly altered by the pressing operation, the particle-size distribution retains its fundamental importance. Metal powders produced by any of the foregoing methods are classified

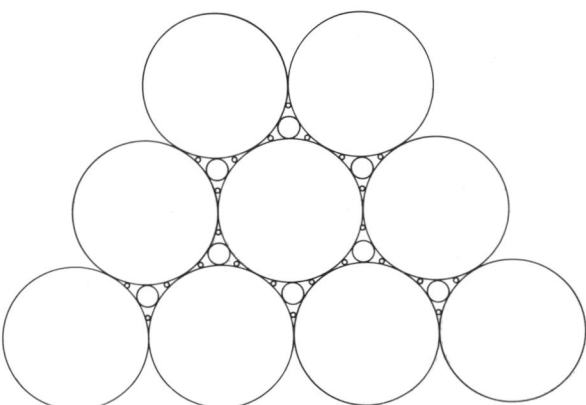

Fig. 7-24 Effect of particle size distribution on density of a pow-dered metal (prior to compaction). The fine particles fill the voids between the coarse particles.

by sieving through standard-mesh sieves. Table 7-15 gives the standard sieve numbers and mesh sizes used for classifying metal powders.

The actual preparation of metal pieces by powder-metallurgical techniques involves filling a cavity with the metal powder, then compacting the metal powder under a sufficient force from a punch to produce a piece that has enough strength to be taken from the mold and handled in subsequent processing. The initial strength of the piece is called the *green strength*. The two basic types of compacting presses, single-acting and double-acting, are illustrated in Fig. 7-25. This figure also indicates that single- and double-acting presses produce metals of somewhat different compaction properties. A third type of press currently coming into widespread use is known as an *isostatic press*. In this particular equipment, a metal compact is placed in a suitable mold, usually made of rubber, and the entire form is submerged in an oil the pressure of which can be raised to the desired point. The isostatic pressure (i.e., uniform in all directions) produced by the hydraulic system tends to give a piece of very uniform density.

Frequently, the initial compact does not have sufficient green strength to permit subsequent handling. In such cases, *binders* may be used to improve this characteristic. A *fugitive binder* is a wax or other volatile material that is used to give the piece a high green strength but is thoroughly removed (for example, by vaporization) during the subsequent processing. In a *permanent binder,* the binder itself actually forms part of the metal system and remains in the final piece. Frequently, permanent binders are themselves metals which alloy with the main metal during subsequent operations. The most important subsequent treatment for powder-metal pieces is *sintering*, in which the temperature of the metal is

TABLE 7-15 Sieve Numbers and Mesh Sizes

Sieve No.	Sieve opening		Mesh	
	(mm)	(in)	(per cm)	(per in)
2½	8.00	0.315	1	2.6
3	6.72	0.265	1.2	3.0
3½	5.66	0.223	1.4	3.6
4	4.76	0.187	1.7	4.2
5	4.00	0.157	2	5.0
6	3.36	0.132	2.3	5.8
7	2.83	0.111	2.7	6.8
8	2.38	0.094	3	7.9
10	2.00	0.079	3.5	9.2
12	1.68	0.066	4	10.8
14	1.41	0.0557	5	12.5
16	1.19	0.0468	6	14.7
18	1.00	0.0394	7	17.2
20	0.84	0.0331	8	20.2
25	0.71	0.0278	9	23.6
30	0.59	0.0234	11	27.5
35	0.50	0.0197	13	32.3
40	0.42	0.0166	15	37.9
45	0.35	0.0139	18	44.7
50	0.30	0.0117	20	52.4
60	0.25	0.0098	24	61.7
70	0.21	0.0083	29	72.5
80	0.177	0.0070	34	85.5
100	0.149	0.0059	40	101
120	0.125	0.0049	47	120
140	0.105	0.0041	56	143
170	0.088	0.0035	66	167
200	0.074	0.0029	79	200
230	0.062	0.0025	93	233
270	0.053	0.0021	106	270
325	0.044	0.0017	125	323

From C. D. Hodgman (ed.), "Handbook of Chemistry and Physics," The Chemical Rubber Co., Cleveland, 1942.

raised and the piece is held at the elevated temperature until diffusion has actually joined the particles of metal. By the process of sintering, powder-metal pieces can be brought to a very high strength, in many instances approaching the strength of the solid metal. Table 7-16 indicates conventional temperatures and times commonly used for sintering powder-metal pieces. Frequently, where the metal would be attacked by heating in air, a controlled reducing or inert atmosphere is used for sintering. Table 7-17 indicates the atmospheres commonly used for various metal-sintering operations.

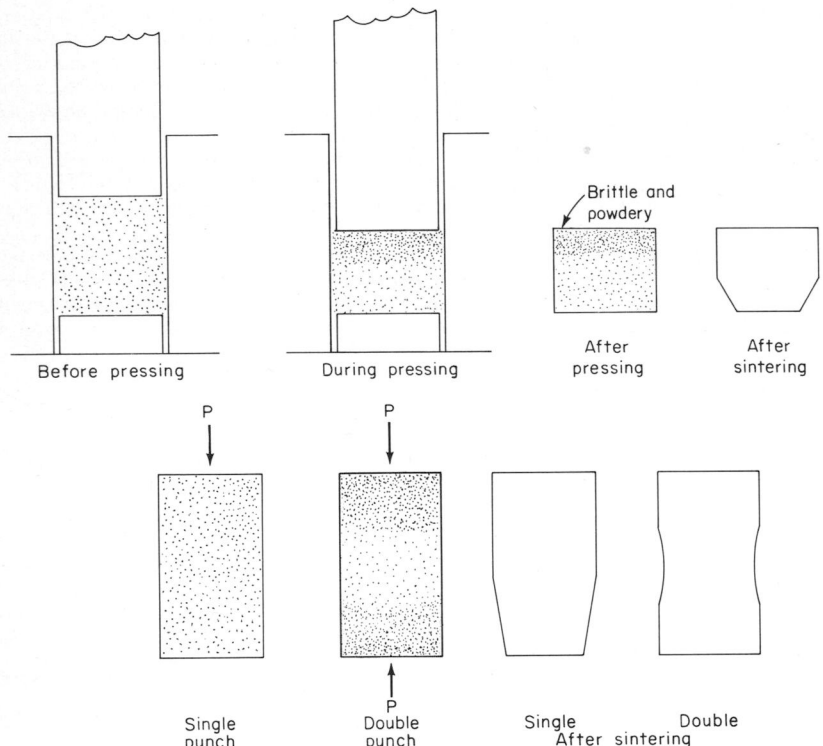

Fig. 7–25 Types of compacting presses. Note that the manner in which a powder-metal part shrinks during sintering is directly affected by the type of press used to prepare the compact.

TABLE 7-16 Typical Time-Temperature Values for Metal Sintering

Metal	Temp (°C)	Time
Bronze	700–800	1–2 h
Copper	850–950	1 h
Iron	900–1100	45 min
Magnet alloys	1200–1300	2 h
Refractory metals	2000–2500	30 min

From P. G. Ormandy, "An Introduction to Metallurgical Laboratory Techniques," Pergamon Press, London, 1968.

TABLE 7-17 Atmospheres for Metal Sintering

Metal	Atmosphere
Tungsten	Hydrogen
Molybdenum	Hydrogen
Cobalt/tungsten carbide	Hydrogen
Iron	Hydrogen
Nickel	Hydrogen
Copper	Hydrogen or vacuum
Chromium	Vacuum
Aluminum	Vacuum
Tantalum	Vacuum
Vanadium	Vacuum
Platinum	Air
Gold	Air
Silver	Air or hydrogen

From P. G. Ormandy, "An Introduction to Metallurgical Laboratory Techniques," Pergamon Press, London, 1968.

Chapter 8

Fluid Mechanics

8-1 FLUID FLOW

Flow Conditions

In describing the flow of a fluid, a line which gives the direction of flow of a given point at a given time is called a *streamline*. The flow is said to be *laminar* if the streamlines remain distinct and do not mingle with each other. Laminar flow is also called streamline or viscous flow, and usually occurs when the fluid is moving at low velocities. In contrast, when the streamlines are irregular and entwined, the flow is described as *turbulent*. Turbulent flow usually occurs when the fluid is moving at high velocities. As the velocity of the fluid increases, the transition from laminar to turbulent flow occurs very suddenly. Similarly, the transition from turbulent to laminar flow occurs suddenly as the velocity of the fluid decreases. The transition from laminar to turbulent flow usually occurs at a higher velocity than the reverse transition from turbulent to laminar flow, the velocities between the transition velocities lying in an unstable range, as illustrated in Fig. 8-1.

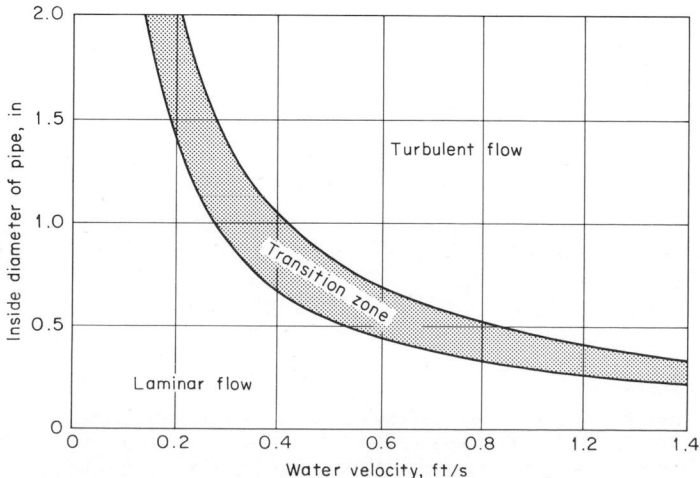

Fig. 8-1 Transition between laminar and turbulent flow. Conditions shown are for water flowing through round pipes at room temperature.

Example 8-1   Would a flow of 0.03 gal of water per minute through a 1-in-ID pipe be laminar or turbulent?

solution   From Table 1-5, 1 gal = 231 in^3.

$$\text{Cross-sectional area of pipe} = \frac{\pi(1)^2}{4} = 0.785 \text{ in}^2$$

$$\text{Velocity of water} = \frac{0.03 \text{ gal}}{\text{min}} \times \frac{231 \text{ in}^3}{\text{gal}} \times \frac{1}{0.785 \text{ in}^2}$$

$$= 8.8 \text{ in/min} \equiv 0.74 \text{ ft/min}$$

From Fig. 8-1, this velocity results in turbulent flow.

Frictional Losses

The energy required to make a fluid flow through a conduit is dependent on the nature of the fluid (liquid or gas, density, viscosity, etc.) and the nature of the conduit (square, round, wall roughness, etc.). The frictional resistance to flow results from the relative movement between fluid particles and the walls of the pipe or duct. Thus the friction, or frictional drag, is greatest nearest the conduit walls and least at some distance from the walls. Because the frictional drag varies from point to point in a conduit, the fluid velocity also varies, being greatest near the center of the conduit and least near the walls. A *velocity profile* shows the distribution of fluid velocities across a given cross section, as illustrated in Fig. 8-2. The *mean linear velocity* of a flowing fluid is defined by the equation

$$V = \frac{Q}{A} \tag{8-1}$$

where V = velocity (length/time), Q = quantity of fluid flowing past a given cross section per unit time (length^3/time), and A = cross-sectional area (length^2). Thus, if Q is in cubic feet per minute and A in square feet, V will have units of feet per minute. The mean linear velocity is used for most flow equations.

Laminar   Turbulent

Fig. 8-2   Velocity profiles for laminar and turbulent flow. In both cases, the velocity of the stream is greatest at the center of the pipe and falls off to a very low value near the wall. In turbulent flow the profile is blunter.

Example 8-2   What is the mean linear velocity of gasoline flowing at a rate of 50 gal/min through a 3-in-ID pipe?

solution

$$Q = \frac{50 \text{ gal}}{\text{min}} \times \frac{231 \text{ in}^3}{\text{gal}} = 11,550 \text{ in}^3/\text{min}$$

$$A = \frac{\pi(3 \text{ in})^2}{4} = 7.07 \text{ in}^2$$

$$V = \frac{Q}{A} = 11,550 \text{ in}^3 \times \frac{1}{7.07 \text{ in}^2} \times \frac{\text{ft}}{12 \text{ in}}$$

$$= 136 \text{ ft/min}$$

Reynolds Number

The wide variety of pipes and other conduits that must be considered, along with an equally wide variety of fluid properties, makes it difficult to describe flow conditions in a generalized manner. However, the ratio of inertial forces to viscous forces for a flowing fluid has been found to be characteristic of the flow conditions. This ratio, known as the Reynolds number N_{Re}, is defined by the equation

$$N_{\text{Re}} = \frac{DV\rho}{\mu} = 4\frac{\bar{R}_H G}{\mu} \qquad (8\text{-}2)$$

where D = diameter of pipe or round conduit (ft), R_H = a characteristic dimension of the flow channel known as the *hydraulic radius* and defined as the cross-sectional area divided by the wetted perimeter of the conduit (ft), V = mean linear velocity (ft/s), ρ = fluid density (lb/ft³), G = mass velocity (lb/s-ft²), and μ = viscosity of fluid (lb/ft-s). It should be noted that the Reynolds number, being the ratio of two forces, is dimensionless. Table 8-1 gives the hydraulic radii for a variety of conduits. The *critical Reynolds number* describes conditions when the transition from turbulent to laminar flow occurs as the velocity of the flowing fluid is reduced. The value of the critical Reynolds number is 2,400, although it may range from 2,000 to 3,000 depending on the specific nature of the fluid and the pipe or conduit through which it is flowing.

Example 8-3 What is the hydraulic radius of a pipe whose internal diameter is 3 in?
solution

$$R_H = \frac{\text{cross-sectional area of flow}}{\text{perimeter of pipe wetted by fluid}}$$

$$R_H = \frac{\pi D^2}{4} \times \frac{1}{\pi D} = \frac{D}{4}$$

TABLE 8-1   Hydraulic Radii for a Variety of Conduits

$$R_H = \frac{\text{area of stream cross section}}{\text{wetted perimeter}}; \text{ hydraulic diameter} = 4R_H$$

Shape of Cross Section	R_H
Pipes and ducts, running full:	
Circle, diam. $= D$	$\dfrac{D}{4}$
Annulus, inner diam. $= d$, outer diam. $= D$ . .	$\dfrac{D-d}{4}$
Square, side $= D$	$\dfrac{D}{4}$
Rectangle, sides a, b	$\dfrac{ab}{2(a+b)}$
Ellipse, major axis $= 2a$, minor axis $= 2b$ . . .	$\dfrac{ab}{K(a+b)}$ See Note
Open channels or partly filled ducts:	
Rectangle, depth $= y$, width $= b$	$\dfrac{by}{b+2y}$
Semicircle, free surface on a diam. D	$\dfrac{D}{4}$
Wide shallow stream on flat plate, depth $= y$.	y
Triangular trough, $\angle = 90$ deg., bisector vertical, depth $= y$, slant depth $= d$	$\dfrac{d}{4} = \dfrac{y}{2\sqrt{2}}$
Trapezoid (depth $= y$, bottom width $= b$): Side slope 60 deg. from horizontal	$y\left(\dfrac{b+y/\sqrt{3}}{b+4y/\sqrt{3}}\right)$
Side slope 45 deg	$\dfrac{yb+y^2}{b+2\sqrt{2}y}$
Film (thickness $= t$) on wall of vertical wetted wall tower of diameter $= D$	$t - t^2/D = t$ (approx.)

Adapted from Robert H. Perry and Cecil H. Chilton (eds.), "Chemical Engineers' Handbook," 5th ed., McGraw-Hill Book Company, New York, 1973.

For $(a - b)/(a + b) = $ 0.2   0.3   0.4   0.5   0.6   0.7   0.8   0.9   1.0
$K = $ 1.01   1.02   1.04   1.06   1.09   1.13   1.17   1.22   1.27

See also Table 8-1:

$$R_H = \frac{3 \text{ in}}{4} = 0.75 \text{ in}$$

Example 8-4 If the pipe of Example 8-3 runs only half full of liquid, what is the hydraulic radius of the flow?
solution

$$\text{Cross-sectional area of flow} = \frac{1}{2} \times \frac{\pi D^2}{4} = 3.53 \text{ in}^2$$

$$\text{Perimeter of pipe wetted by fluid} = \frac{1}{2}\pi D = 4.71 \text{ in}$$

$$R_H = \frac{3.53 \text{ in}^2}{4.71 \text{ in}} = 0.75 \text{ in}$$

Example 8-5 Calculate the Reynolds number for water flowing through a 0.6-in-ID pipe at 8 ft/s at room temperature. If the fluid were an oil with specific gravity of 0.85 and a viscosity of 0.009 lb/ft-s, what would the Reynolds number be?
solution From Table 1-16, the viscosity of water is 1 cP and the conversion factor to lb/ft-s is 0.672×10^{-3}, i.e., $\mu = 0.672 \times 10^{-3}$ lb/ft-s.

$$\rho = 62.4 \text{ lb/ft}^3$$

$$D = 0.6 \text{ in} \times \frac{\text{ft}}{12 \text{ in}} = 0.05 \text{ ft}$$

$$N_{Re} = \frac{DV\rho}{\mu} = 0.05 \text{ ft} \times \frac{8 \text{ ft}}{\text{s}} \times \frac{62.4 \text{ lb}}{\text{ft}^3}$$

$$\frac{\text{ft-s}}{0.672 \times 10^{-3} \text{ lb}} = 37,100$$

For the oil,

$$N_{Re} = 0.05 \text{ ft} \times \frac{8 \text{ ft}}{\text{s}} \times \frac{(0.85 \times 62.4) \text{ lb}}{\text{ft}^3} \times \frac{\text{ft-s}}{0.009 \text{ lb}} = 2,400$$

Example 8-6 What is the Reynolds number for water flowing in an open channel at a rate of 160 gal/min? The channel is 8 in wide and 6 in high and, at this flow, is half full.
solution

$$R_H = \frac{8 \text{ in} \times 3 \text{ in}}{(3 + 8 + 3) \text{ in}} = \frac{24 \text{ in}^2}{14 \text{ in}} = 1.71 \text{ in} \equiv 0.143 \text{ ft}$$

$$G = \frac{160 \text{ gal}}{\text{min}} \times \frac{8.35 \text{ lb}}{\text{gal}} \times \frac{\text{min}}{60 \text{ s}} \times \frac{1}{24 \text{ in}^2} \times \frac{144 \text{ in}^2}{\text{ft}^2}$$

$$= 133.6 \text{ lb/s-ft}^2$$

$$\mu = 0.672 \times 10^{-3} \text{ lb/ft-s}$$

$$N_{Re} = 4 \times 0.143 \text{ ft} \times \frac{133.6 \text{ lb}}{\text{s-ft}^2} \times \frac{1}{0.672 \times 10^{-3}}$$

$$= 114,000$$

Bernoulli's Principle

Essentially, all problems relating to fluid flow involve the pressure differential between two points, the rate of discharge (or the quantity of fluid flowing between the two points), and the size of the pipe or conduit. The relationship among these quantities can be derived from a total *energy balance* for the fluid flowing between the two points; i.e., the total energy of the fluid as it enters a given section is equal to its total energy as it leaves another given section downstream, plus any energy gains or losses between the two sections. This is known as *Bernoulli's principle*. Of course, the form of the energy associated with the fluid (kinetic, potential, thermal) may change as it flows between the two sections, but it must do so within the confines of the total energy balance. This is illustrated in Fig. 8-3. In mathematical terms, Bernoulli's principle can be expressed as

$$Z_1 + p_1v_1 + \frac{V_1^2}{2g} + J(H_1 + Q) + W = Z_2 + p_2v_2 + \frac{V_2^2}{2g} + JH_2 \qquad (8\text{-}3)$$

where Z = height above a given datum plant (ft), p = pressure (lb/ft^2), v = specific volume of the fluid (ft^3/lb), V = average linear velocity (ft/s), J = mechanical equivalent of heat (778 ft-lb/Btu), H = heat content of the fluid (Btu/lb), Q = heat interchange of the fluid with the surroundings (Btu/lb), and W = work interchange of the fluid with the surroundings (ft-lb/lb). This equation and the equation of continuity are used to solve a variety of fluid-flow problems. The *equation of continuity* mathematically expresses the condition in which no fluid enters or leaves the system between the two sections under consideration:

$$Q_1 = Q_2 = \frac{V_1 S_1}{v_1} = \frac{V_2 S_2}{v_2} \qquad (8\text{-}4)$$

in which Q = quantity of fluid flowing (lb/s), V = average linear velocity (ft/s), S = area of the cross section (ft^2), and v = specific volume of the fluid (ft^3/lb).

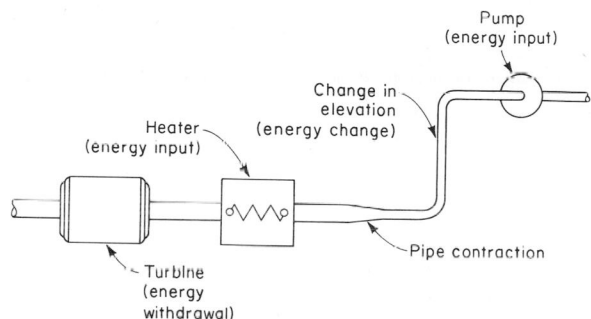

Fig. 8-3 Energy changes in a flowing fluid.

Example 8-7 Neglecting frictional losses, what is the downstream pressure for a water flow of 35 gal/min through a 0.6-in-ID pipe that expands to 1.1 in ID if the pressure in the smaller section is 18 lb/in^2?

solution In Eq. (8-3), $Z_1 = Z_2$, $H_1 = H_2$, and $Q = 0$. Thus

$$p_1 v_1 + \frac{V_1^2}{2g} = p_2 v_2 + \frac{V_2^2}{2g}$$

$$\text{Flow} = \frac{35 \text{ gal}}{\text{min}} \times \frac{\text{ft}^3}{7.48 \text{ gal}} \times \frac{\text{min}}{60 \text{ s}} = 0.078 \text{ ft}^3/\text{s}$$

$$V_1 = \frac{0.078 \text{ ft}^3}{\text{s}} \times \frac{1}{\pi(0.6)^2 \text{ in}^2} \times \frac{144 \text{ in}^2}{\text{ft}^2} = 39.7 \text{ ft/s}$$

$$V_2 = \frac{0.078 \text{ ft}^3}{\text{s}} \times \frac{4}{\pi(1.1)^2 \text{ in}^2} \times \frac{144 \text{ in}^2}{\text{ft}^2} = 11.8 \text{ ft/s}$$

$$\frac{V_1^2}{2g} = \frac{(39.7)^2 \text{ ft}^2}{\text{s}^2} \times \frac{\text{s}^2}{2 \times 32 \text{ ft}} = 24.63 \text{ ft}$$

$$\frac{V_2^2}{2g} = \frac{(11.8)^2 \text{ ft}^2}{\text{s}} \times \frac{\text{s}^2}{2 \times 32 \text{ ft}} = 2.18 \text{ ft}$$

$$p_2 v_2 - p_1 v_1 = (24.63 - 2.18) \text{ ft} = 22.45 \text{ ft}$$

For water, $v_1 = v_2 = 0.0160$ ft^3/lb:

$$(p_2 - p_1)v_1 = 22.45 \text{ ft}$$

$$p_2 - p_1 = 22.45 \text{ ft} \times \frac{\text{lb}}{0.016 \text{ ft}^3}$$

$$p_2 - p_1 = 1,403 \text{ lb/ft}^2 \equiv 9.7 \text{ lb/in}^2$$

$$p_1 = 18 \text{ lb/in}^2$$

$$p_2 = 9.7 + 18 = 27.7 \text{ lb/in}^2$$

Example 8-8 What is the downstream pressure under the conditions of Example 8-7 if the plane of the large-diameter pipe is 22 ft below that of the small diameter?

solution In this case, $Z_1 \neq Z_2$, and if Z_1 is assigned a value of 0 (i.e., is considered as the datum plane), then $Z_2 = -20$ ft.

$$Z_1 + p_1 v_1 + \frac{V_1^2}{2g} = Z_2 + p_2 v_2 + \frac{V_2^2}{2g}$$

Since $p_1 = \dfrac{18 \text{ lb}}{\text{in}^2} \equiv \dfrac{2{,}592 \text{ lb}}{\text{ft}^2}$

$$\frac{2{,}592 \text{ lb}}{\text{ft}^2} \times \frac{0.016 \text{ ft}^3}{\text{lb}} + 24.63 \text{ ft} = -20 \text{ ft} + p_2 \times \frac{0.016 \text{ ft}^3}{\text{lb}} + 2.18 \text{ ft}$$

$$41.5 + 24.63 = -20 + 0.016\, p_2 + 2.18$$

$$0.016\, p_2 = 83.95$$

$$p_2 = 5{,}247 \text{ lb/ft}^2 \equiv 36.4 \text{ lb/in}^2$$

Example 8-9   If the system described in Example 8-8 contains a 2-hp pump between the upstream and downstream points, what is the downstream pressure?
solution   From Table 2-10,

$$2 \text{ hp} \times \frac{550 \text{ ft-lb}}{\text{s-hp}} = 1{,}110 \text{ ft-lb/s}$$

From Example 8-7, the water flow is 0.078 ft³/s. The pressure produced by the pump is therefore

$$\frac{1{,}100 \text{ ft-lb}}{\text{s}} \times \frac{\text{s}}{0.078 \text{ ft}^3} = 14{,}100 \text{ lb/ft}^2$$

$$\equiv 97.9 \text{ lb/in}^2$$

The total pressure at the downstream section is

$$36.4 + 97.9 = 134.3 \text{ lb/in}^2$$

In Eq. (8-3), the units for each term reduce to a simple linear dimension (ft). Each such term is called a *head*, Z being the head due to elevation, pv the pressure head, $V^2/2g$ the velocity head, and so on. There is no special virtue in expressing energy in terms of a related linear dimension, but the practice is widespread among engineers.

Friction, as a fluid flows through a pipe or conduit, manifests itself as a rise in the temperature of the fluid. If Q in Eq. (8-3) is zero, as would be the case, for example, for a liquid flowing through an insulated pipe, $J(H_2 - H_1)$ would represent the heating of the fluid due to friction as it flows from section 1 to section 2, and Eq. (8-3) reduces to

$$Z_1 + p_1 v_1 + \frac{V_1^2}{2g} = Z_2 + p_2 v_2 + \frac{V_2^2}{2g} + F \tag{8-5}$$

where F = heat generated by friction (ft-lb/lb), or simply the frictional (head) loss. For a fluid flowing in a circular pipe, F is given by the *Fanning equation*:

$$F = \frac{4fL}{D} \times \frac{V^2}{2g} \tag{8-6}$$

where f = a dimensionless number, called the *Fanning friction factor, D* = pipe diameter (ft), L = length of pipe between the two sections under consideration (ft), V = average fluid velocity (ft/s), and g = gravitational constant (32.2 ft/sec²). The friction factor is influenced by other conditions, such as the diameter of the pipe and the roughness of its internal surface, the density, and the viscosity of the fluid, but it has been found that it correlates well with the Reynolds number, as shown in Fig. 8-4. This figure was developed over many years from a large number of test observations involving a large variety of fluids flowing through all types of conduits. At the critical Reynolds number, there is a break in the curve as the flow shifts from laminar to turbulent or vice versa. However, in the vicinity of the critical Reynolds number, the flow may be metastable; that is, it is possible to have laminar flow at Reynolds numbers higher than 2,400 if the condition is reached carefully (and usually slowly). Such a condition can convert suddenly to its stable form from a simple disturbance, as, for example, from a hammer tap. Since there can be some overrun in each direction in the vicinity of the critical Reynolds number, there can be some uncertainty as to the value of f in this area. However, the correlation of the friction factor and flow conditions via the Reynolds number provides a means of using the Bernoulli equation in solving practical flow problems.

Example 8-10   What is the friction factor for glycerin flowing through a 0.5-in-ID pipe at 16 gal/min at 20°C? What is the friction factor if the glycerin is at 100°C, at which temperature its viscosity is 13.5 cP?

solution

$$D = 0.5 \text{ in} \times \frac{\text{ft}}{12 \text{ in}} = 0.0417 \text{ ft}$$

$$\rho = 1.26 \times \frac{62.4 \text{ lb}}{\text{ft}^3} = 78.6 \text{ lb/ft}^3$$

$$V = \frac{16 \text{ gal}}{\text{min}} \times \frac{\text{ft}^3}{7.48 \text{ gal}} \times \frac{\text{min}}{60 \text{ s}} \times \frac{4}{\pi(0.5)^2 \text{ in}^2} \times \frac{144 \text{ in}^2}{\text{ft}^2}$$

$$\mu = 833 \text{ cP} \times \frac{0.672 \times 10^{-3} \text{ lb}}{(\text{cp})(\text{ft})(\text{s})}$$

$$= 0.560 \text{ lb/ft-s}$$

$$N_{\text{Re}} = \frac{DV\rho}{\mu} = 0.0417 \text{ ft} \times \frac{26.1 \text{ ft}}{\text{s}} \times \frac{78.6 \text{ lb}}{\text{ft}^3} \times \frac{\text{ft-s}}{0.560 \text{ lb}} = 153$$

From Fig. 8-4,

$$f = 0.11$$

At 100°C

$$\mu = 13.5(0.672 \times 10^{-3}) = 9.07 \times 10^{-3} \text{ lb/ft-s}$$

$$N_{\text{Re}} = \frac{0.0417 \times 26.1 \times 78.6}{9.07 \times 10^{-3}} = 9,400$$

From Fig. 8-4,

$$f = 0.0088.$$

(Note the tremendous reduction in the friction factor as a result of raising the temperature of the liquid to reduce its viscosity.)

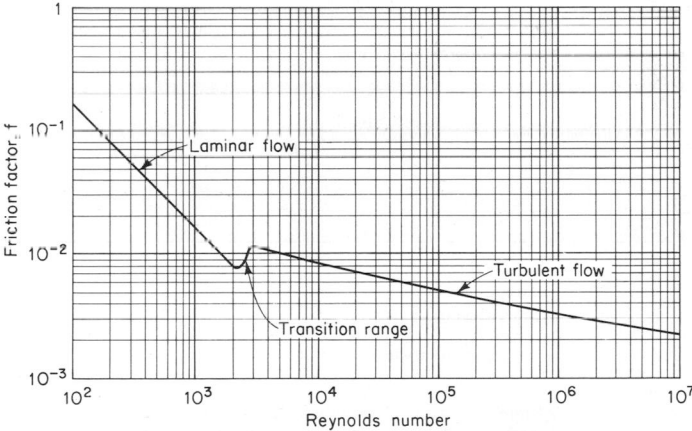

Fig. 8-4   Friction factors for straight, round pipes.

Example 8-11   What are the frictional head loss and pressure drop for the flow conditions described in Example 8-10 if the pipe is 80 ft long?
solution   For 20°C and a friction factor of 0.11,

$$F = \frac{4 \times 0.11 \times 80 \text{ ft}}{0.0417 \text{ ft}} \times \frac{(26.1)^2 \text{ ft}^2\text{-s}^2}{2 \times 32.1 \text{ ft-s}^2}$$

$$= 8,960 \text{ ft}$$

$$\Delta p = \frac{8,960 \text{ ft} \times 78.6 \text{ lb}}{\text{ft}^3} \times \frac{\text{ft}^2}{144 \text{ in}^2}$$

$$= 4,890 \text{ lb/in}^2$$

This would be an extremely high pressure drop, meaning that very heavy piping would be required and that it would be somewhat impractical to pump such a viscous liquid under the conditions given.

At 100°C and a friction factor of 0.0088,

$$F = \frac{4 \times 0.0088 \times 80}{0.0417} \times \frac{(26.1)^2}{2 \times 32.1} = 716 \text{ ft}$$

$$\Delta p = \frac{716 \times 78.6}{144} = 391 \text{ lb/in}^2$$

Thus, by heating the glycerin, it is possible to pump it under reasonable conditions. (In practice, a larger-diameter pipe would be used to reduce the pressure drop still further.)

Since the nature of the pipe or conduit affects the amount of friction, fittings such as elbows, tees, and valves make an important contribution to frictional losses, not only as a result of the turbulence they themselves introduce but also because of burrs and other projections that may intrude into the flow. For most work, the friction introduced by a fitting is considered to be equal to a given length of pipe of the diameter the fitting is intended to fit. This equivalent length L_e is added to the actual length of straight pipe to give the length of straight pipe that would have the same frictional losses as the actual pipe plus its fittings. The equivalent lengths of common fittings for turbulent-flow conditions are given in Table 8-2.

TABLE 8-2 Equivalent Lengths of Common Fittings

L_e/D ratios for standard fittings (turbulent flow)

90° elbows:	
Less than 1 in ID	20
1–2½ in	30
3–6 in	40
7–10 in	50
90° curves:	
Radius of centerline of	
curve = ID of pipe	20
Radius of centerline of	
curve = 2 to 8 times the	
ID of the pipe	10
Cross (all sizes)	50
Globe valves:	
Less than 1 in	30
1–2½ in	45
3–6 in	60
7–10 in	75
Tees:	
Less than 1 in	50
1–4 in	60
Larger than 4 in	70

Example 8-12 If the system described in Examples 8-10 and 8-11 also had four 90° elbows and one globe valve in the line, what additional frictional head losses would these introduce at each of the two temperature conditions?

solution From Table 8-2, L_e for the elbows is $20 \times D = 20 \times 0.0417 = 0.83$ ft. Since there are four such elbows, their total equivalent length is $4(0.83) = 3.34$ ft. For the globe valve, $L_e = 30 \times (0.0417) = 1.25$ ft.

For the conditions at 20°C,

$$F = \frac{4 \times 0.11 \times (3.34 + 1.25)}{0.0417} \times \frac{(26.1)^2}{2 \times 32.1}$$

$$= 514 \text{ ft}$$

At 100°C,

$$F = 514 \times \frac{0.0088}{0.11} = 41 \text{ ft}$$

An alternate procedure, which is more accurate since it takes the flow conditions into account to some extent, gives the frictional losses induced by the fittings in terms of the velocity of the fluid and, as such, is perhaps also more easily used. In this method, a factor K is defined as the frictional loss in terms of an equivalent number of velocity heads. Values of K for various fittings for turbulent-flow conditions are given in Table 8-3.

Example 8-13 What is the frictional head loss for water flowing at a rate of 16 gal/min through a 0.5-in-ID pipe? The pipe is 80 ft long and has four 90° elbows and one globe valve.

TABLE 8-3 Frictional Losses in Fittings Expressed as Equivalent Velocity Heads

Type of fitting or valve	Additional friction loss, equivalent No. of velocity heads, K
45° ell, standard	0.35
45° ell, long radius	0.2
90° ell, standard	0.75
Long radius	0.45
Square or miter	1.3
180° bend, close return	1.5
Tee standard, along run, branch blanked off	0.4
Used as ell, entering run	1.3
Used as ell, entering branch	1.5
Branching flow	1
Coupling	0.04
Union	0.04
Gate valve, open	0.17
$\frac{3}{4}$ open	0.9
$\frac{1}{2}$ open	4.5
$\frac{1}{4}$ open	24.0
Diaphragm valve, open	2.3
$\frac{3}{4}$ open	2.6
$\frac{1}{2}$ open	4.3
$\frac{1}{4}$ open	21.0
Globe valve, bevel seat, open	6.4
$\frac{1}{2}$ open	9.5
Composition seat, open	6.0
$\frac{1}{2}$ open	8.5
Plug disk, open	9.0
$\frac{3}{4}$ open	13.0
$\frac{1}{2}$ open	36.0
$\frac{1}{4}$ open	112.0
Angle valve, open	3.0
Y or blowoff valve, open	3.0
Plug cock $\theta = $ 5°	0.05
10°	0.29
20°	1.56
40°	17.3
60°	206.0
Butterfly valve $\theta = $ 5°	0.24
10°	0.52
20°	1.54
40°	10.8
60°	118.0
Check valve, swing	2.0
Disk	10.0
Ball	70.0
Foot valve	15.0
Water meter, disk	7.0
Piston	15.0
Rotary (star-shaped disk)	10.0
Turbine-wheel	6.0

Adapted from Robert H. Perry and Cecil H. Chilton (eds.), "Chemical Engineers' Handbook," 5th ed., McGraw-Hill Book Company, New York, 1973.

solution

$$D = 0.5 \text{ in} \times \frac{\text{ft}}{12 \text{ in}} = 0.0417 \text{ ft}$$

$$\rho = 62.4 \text{ lb/ft}^3$$

$$V = \frac{16 \text{ gal}}{\text{min}} \times \frac{\text{ft}^3}{7.48 \text{ gal}} \times \frac{\text{min}}{60 \text{ s}} \times \frac{4}{\pi(0.5)^2 \text{ in}^2} \times \frac{144 \text{ in}^2}{\text{ft}^2}$$

$$= 26.1 \text{ ft/s}$$

$$\mu = 0.672 \times 10^{-3} \text{ lb/ft} - \text{s}$$

$$N_{\text{Re}} = \frac{DV\rho}{\mu} = 0.0417 \text{ ft} \times \frac{26.1 \text{ ft}}{\text{s}} \times \frac{62.4 \text{ lb}}{\text{ft}^3} \times \frac{\text{ft} - \text{s}}{0.672 \times 10^{-3} \text{ lb}} = 101{,}000$$

$$f = 0.0054$$

For the straight pipe, the frictional head loss is

$$F = 4f \frac{L}{D} \times \frac{V^2}{2g} = 4 \times 0.0054 \times \frac{80 \text{ ft}}{0.0417 \text{ ft}} \times \frac{(26.1)^2 \text{ ft}^2\text{-s}}{2(32.1) \text{ ft-s}^2} = 440 \text{ ft}$$

From Table 8-3, K is 0.75 for a 90° elbow and 6.4 for an open globe valve. Since there are four elbows,

$$K = 4 \times 0.75 + 6.4 = 9.4$$

The friction introduced by the fittings is thus

$$F = K \frac{V^2}{2g} = 9.4 \times \frac{(26.1)^2}{2 \times 32.1} = 100 \text{ ft}$$

The total frictional head loss is therefore

$$F = 440 + 100 = 540 \text{ ft}$$

Since the fluid is water, this is equal to a pressure loss of

$$\Delta p = 540 \text{ ft} \times \frac{62.4 \text{ lb}}{\text{ft}^3} \times \frac{\text{ft}^2}{144 \text{ in}^2}$$

$$= 234 \text{ lb/in}^2$$

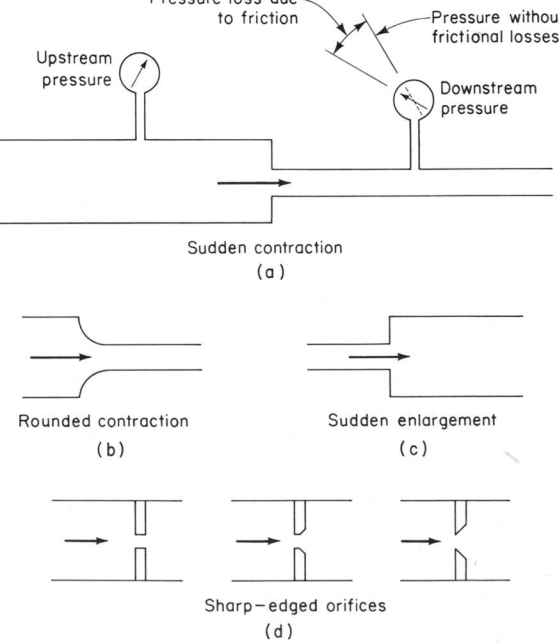

Fig. 8-5 Typical conditions causing frictional losses in pipes.

In addition to fittings, various other conduit conditions can induce frictional losses. Where a fluid enters a pipe from a larger vessel or there is a sudden reduction in the cross-sectional area of the pipe, as illustrated in Fig. 8-5a, the frictional-loss factor K can also be used. Values of K for various sudden contractions are given in Table 8-4 for turbulent flow. For rounded entrances or contractions, illustrated in Fig. 8-5b, the frictional-loss factor is about 0.04 for all turbulent-flow conditions. For laminar flow, entrance or contraction losses for sudden or rounded contractions are negligible.

Example 8-14   Water is draining from a large tank at a rate of 6 gal/min through a 0.5-in-ID pipe. What is the frictional head loss as the water enters the pipe through a simple entrance? What is the frictional head loss if the water enters through a rounded entrance?

solution   Since A_1, the area of the upstream cross-sectional flow area, is large compared with A_2, the downstream cross-sectional flow area, A_2/A_1 can be considered to be 0. Therefore, from Table 8-4, $K = 0.5$.

$$V_2 = \frac{6\ \text{gal}}{\text{min}} \times \frac{\text{ft}^3}{7.48\ \text{gal}} \times \frac{\text{min}}{60\ \text{s}} \times \frac{4}{\pi(0.5)^2\ \text{in}^2} \times \frac{144\ \text{in}^2}{\text{ft}^2} = 9.80\ \text{ft/s}$$

$$F = K\frac{V_2^2}{2g} = 0.5 \times \frac{(9.80)^2\ \text{ft}^2}{\text{s}^2} \times \frac{\text{s}^2}{2 \times 32.1\ \text{ft}}$$

$$= 0.75\ \text{ft}$$

For a rounded entrance, from Table 8-4, $K = 0.04$:

$$F = 0.75\ \text{ft} \times \frac{0.04}{0.5} = 0.060\ \text{ft}$$

TABLE 8-4   Friction-Loss Coefficient for Sudden Contraction in Turbulent Flow

A_2/A_1	0	0.2	0.4	0.6	0.8	1.0
K	0.5	0.45	0.36	0.21	0.07	0

K for rounded contractions = 0.04.

For a sudden enlargement of a pipe, as illustrated in Fig. 8-5c, the frictional-loss factor is given by the equation

$$K = \left(1 - \frac{A_1}{A_2}\right)^2 \tag{8-7}$$

where A_1 = cross-sectional area of the smaller (upstream) pipe and A_2 = cross-sectional area of the larger (downstream) pipe. By convention, K of Eq. (8-7) is applied to the upstream velocity to obtain the frictional head loss. For laminar flow frictional enlargement losses are usually negligible.

Example 8-15   Water flowing at a rate of 6 gal/min through a 0.5-in-ID pipe enters a short section of 0.3-in-ID pipe and then continues through a 0.5-in-ID pipe. What is the frictional head loss in passing through the restriction? The sudden contraction and sudden expansion are first treated as separate situations and the results added together. Note that the subscripts refer to the following conditions:

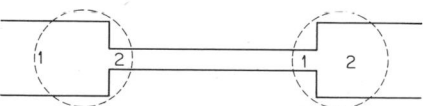

and that the friction coefficients are applied in either expansion or contraction to the velocity head in the smaller-diameter pipe.

solution

$$A_1 = \frac{\pi(0.5)^2\ \text{in}^2}{4} = 0.196\ \text{in}^2$$

$$A_2 = \frac{\pi(0.3)^2\ \text{in}^2}{4} = 0.0707\ \text{in}^2$$

$$\frac{A_2}{A_1} = 0.361$$

From Table 8-4, $K = 0.38$ (by interpolation between values for K at $A_2/A_1 = 0.2$ and $A_2/A_1 = 0.4$). Thus, for the sudden decrease in pipe diameter,

$$V_2 = \frac{6 \text{ gal}}{\text{min}} \times \frac{\text{ft}^3}{7.48 \text{ gal}} \times \frac{\text{min}}{60 \text{ s}} \times \frac{4}{\pi(0.3)^2 \text{ in}^2} \times \frac{144 \text{ in}^2}{\text{ft}^2} = 27.2 \text{ ft/s}$$

$$F = K\frac{V_2^2}{2g} = 0.38 \times \frac{(27.2)^2 \text{ ft}^2}{\text{s}^2} \times \frac{\text{s}^2}{2(32.1) \text{ ft}} = 4.37 \text{ ft}$$

For the sudden enlargement from Eq. (8-7),

$$K = \left(1 - \frac{A_1}{A_2}\right)^2$$

However, note that the subscripts 1 and 2 still refer to the upstream and downstream conditions, respectively. Thus $A_1 = 0.0707 \text{ in}^2$ and $A_2 = 0.196 \text{ in}^2$ (i.e., the reverse of the first part of this problem).

$$K = \left(1 - \frac{0.0707}{0.139}\right)^2 = (1 - 0.51)^2 = 0.240$$

This value of K is now applied to the *upstream* velocity head $V_1^2/2g$:

$$F = K\frac{V_1^2}{2g} = 0.240 \times \frac{(27.2)^2}{64.2}$$

$$= 2.77 \text{ ft}$$

Total frictional head loss $= 4.37 + 2.77 = 7.14$ ft

For a perforated plate or sharp-edged orifice placed in the stream, as in Fig. 8-5d, the frictional loss is given by the equation

$$L = \frac{\alpha^2}{64}(1 - \beta^4)D_2 \tag{8-8}$$

where L = equivalent length of straight pipe (ft) of diameter D_1 (ft), $\alpha = 6.38 + 2.33\,\beta^2$ (for $\beta \leqslant 0.8$), $\beta = D_2/D_1$, and D_2 = diameter of the orifice opening or equivalent diameter of the perforations for a perforated plate (ft). For orifices with rounded edges and turbulent flow, the frictional losses are generally negligible. The above are estimates that have broad utility but are of limited precision. Because of the complexities in describing fluid-flow conditions, where greater precision is required, actual measurements on the system itself, or a suitably scaled model, are generally made.

Example 8-16 What is the friction developed by a 0.1-in-diameter orifice opening placed in a 0.5-in-ID pipe, expressed as equivalent pipe lengths?
solution

$$\beta = \frac{D_2}{D_1} = \frac{0.1}{0.5} = 0.2$$

$$\beta^4 = 0.0016$$

Since $0.2 < 0.8$, Eq. (8-8) is applicable:

$$\alpha = 6.38 + 2.33\,\beta^2 = 6.38 + 0.093 = 6.47$$

$$L = \frac{\alpha^2}{64}(1 - \beta^4)\,D_2 = \frac{41.86}{64}(0.998)\frac{(0.1)}{12} \text{ ft}$$

$$= 0.005 \text{ ft}$$

8-2 FLOW MEASUREMENT

Static Pressure

A variety of flow-measuring devices depend upon determining the pressure differential between two points, or upon the conversion of a velocity head to a pressure head. In either case, a knowledge of the static pressure, i.e., the pressure of the system on a surface parallel to the flow (so that no kinetic forces act on it), is needed. The static pressure of a system can be readily measured by carefully tapping into the wall of the conduit so that no burr or other interference with the flow results. The static pressure can then be measured by any of the means discussed in Chap. 2. Where the flow conditions may not be exactly parallel to the walls of the pipe, or where slight anomalies in the pressure-sensing opening are to be minimized, an average static pressure at a particular cross section can be obtained directly from an averaging device called a *piezometer ring*, illustrated in Fig. 8-6. A *Pitot tube* is a device for converting the velocity head of a flowing fluid, that is, its kinetic energy, into an equivalent pressure head. This is accomplished by having the opening of the Pitot tube face directly into the stream so that the fluid "impacts" the opening. The pressure produced by this impact must be measured against the static pressure of the fluid, and consequently, most Pitot tubes are designed to measure the differential between the total pressure (static plus kinetic) and the static pressure directly, as shown in Fig. 8-7. As can be seen in Fig. 8-7, a Pitot tube measures the fluid velocity at the particular point where the tip is located. At this point, the velocity of the fluid is given by

$$V = \sqrt{\frac{2g\,(p_1 - p_0)}{\rho}} \qquad (8\text{-}9)$$

where V = velocity at the tip of the Pitot tube (ft/s), g = gravitational constant (32.2 ft/s^2), p_1 and p_0 = pressure at the tip of the Pitot tube and the static pressure, respectively (lb/ft^2), and ρ = density of the fluid (lb/ft^3). Since the fluid velocity can and usually will vary from point to point, a Pitot tube can be used to traverse the stream to obtain a velocity profile such as that in Fig. 8-2. For greater accuracy, or to determine the average flow from a single Pitot tube measurement, Pitot tubes can be calibrated against other flow measurements to determine the Pitot-tube coefficient C, which acts as an empirical correction for Eq. (8-9).

$$V = C \sqrt{\frac{2g\,(p_2 - p_0)}{\rho}} \qquad (8\text{-}10)$$

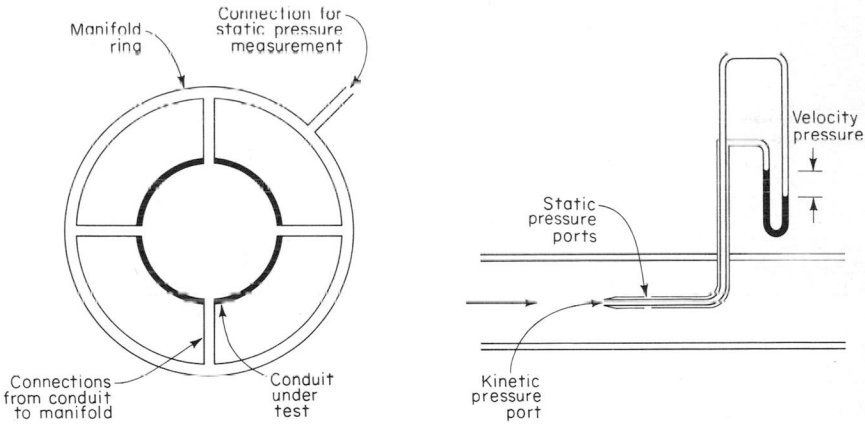

Fig. 8-6 Piezometer ring.   **Fig. 8-7** Pitot tube.

Example 8-17 Given a static pressure of 35 lb/in^2 and a velocity pressure reading on a Pitot tube of 37 lb/in^2, what is the velocity of the liquid if its specific gravity is 1.25?
solution

$$p_1 - p_0 = 37 - 35 = 2 \text{ lb/in}^2$$
$$\equiv 288 \text{ lb/ft}^2$$
$$\rho = 1.25 \times 62.4 = 78.0 \text{ lb/ft}^3$$
$$V = \sqrt{\frac{2g\,(p_1 - p_0)}{\rho}}$$
$$= \sqrt{\frac{2 \times 32.1 \text{ ft}}{\text{s}^2} \times \frac{288 \text{ lb}}{\text{ft}^2} \times \frac{\text{ft}^3}{78.0 \text{ lb}}}$$
$$= 15.4 \text{ ft/s}$$

Example 8-18 A liquid having a specific gravity of 0.94 is flowing through a pipe at a point where the velocity is 18.4 ft/s. What is the pressure due to this flow on the tip of a Pitot tube, expressed as inches of mercury?

solution From Eq. (8-9),

$$p_1 - p_0 = \frac{V^2 \rho}{2g}$$

$$= \frac{(18.4)^2 \text{ ft}^2}{\text{s}^2} \times \frac{\text{s}^2}{64.2 \text{ ft}} \times \frac{62.4 \text{ lb}}{\text{ft}^3} \times 0.94 = 309 \text{ lb/ft}^2$$

$$\equiv 2.14 \text{ lb/in}^2$$

From Table 1-11, $\rho_{Hg} = 13.55$

$$h = \frac{p}{\rho} = \frac{2.14 \text{ lb}}{\text{in}^2} \times \frac{144 \text{ in}^2}{\text{ft}^2} \times \frac{\text{ft}^3}{13.55 \ (62.4) \text{ lb}} = 0.36 \text{ ft}$$

$$\equiv 4.37 \text{ in}$$

Example 8-19 From other measurements, it is known that water is flowing through a 6-in-ID main at a rate of 350 gal/min. A Pitot meter inserted into the main shows a pressure differential of 3.5 lb/in². What is the Pitot-tube coefficient?
solution

$$V_{av} = \frac{350 \text{ gal}}{\text{min}} \times \frac{\text{ft}^3}{7.48 \text{ gal}} \times \frac{4}{\pi \ (0.5)^2 \text{ ft}^2} \times \frac{\text{min}}{60 \text{ s}}$$

$$= 3.97 \text{ ft/s}$$

$$V = C\sqrt{\frac{2g \, \Delta p}{\rho}} = C\sqrt{2x \frac{32.1 \text{ ft}}{\text{s}^2} \times \frac{3.5 \text{ lb}}{\text{in}^2} \times \frac{144 \text{ in}^2}{\text{ft}^2} \times \frac{\text{ft}^3}{62.4 \text{ lb}}}$$

$$3.97 = 22.8 \, C$$

$$C = 0.174$$

For smooth, round pipes, the maximum velocity will occur at the axis of the pipe, and Pitot tubes are therefore usually placed at the axis so as to give a maximum reading. The average velocity can be related to the maximum velocity through the Reynolds number, as shown in Fig. 8-8.

Example 8-20 For a water flow of 350 gal/min through a 4.6-in-ID pipe, what is the water velocity at the axis of the pipe?
solution

$$\text{Area of flow} = \frac{\pi (4.6)^2 \text{ in}^2}{4} \times \frac{\text{ft}^2}{144 \text{ in}^2} = 0.115 \text{ ft}^2$$

$$V_{av} = \frac{350 \text{ gal}}{\text{min}} \times \frac{\text{ft}^3}{7.48 \text{ gal}} \times \frac{\text{min}}{60 \text{ s}} \times \frac{1}{0.115 \text{ ft}^2} = 6.78 \text{ ft/s}$$

$$N_{Re} = \frac{DV\rho}{\mu} = 4.6 \text{ in} \times \frac{\text{ft}}{12 \text{ in}} \times \frac{6.78 \text{ ft}}{\text{s}} \times \frac{62.4 \text{ lb}}{\text{ft}^3} \times \frac{\text{ft} - \text{s}}{0.672 \times 10^{-3} \text{ lb}}$$

$$= 241,000$$

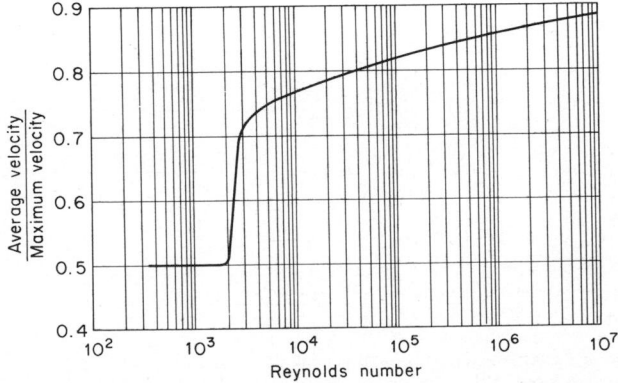

Fig. 8-8 Relation of average velocity and maximum velocity.

From Fig. 8-8,

$$\frac{V_{av}}{V_{max}} = 0.835$$

$$V_{max} = \frac{6.78}{0.835} = 8.12 \text{ ft/s}$$

Example 8-21 For water flowing through the system shown in Fig. 8-7, the Pitot-tube reading is 5.8 inHg. If the pipe has an ID of 4.6 in and the Pitot tube is axially located as shown, what is the flow rate of the water?

solution

$$\Delta p = 5.8 \text{ inHg} \times \frac{\text{ft}}{12 \text{ in}} \times \frac{13.55 \ (62.4) \text{ lb}}{\text{ft}^3}$$

$$= 408.7 \text{ lb/ft}^2$$

$$V^2_{max} = \frac{2g \, \Delta p}{\rho} = \frac{2 \times 32.1 \text{ ft}}{\text{s}^2} \times \frac{408.7 \text{ lb}}{\text{ft}^2} \times \frac{\text{ft}^3}{13.55 \ (62.4) \text{ lb}}$$

$$= 31.0 \text{ ft}^2/\text{s}^2$$

$$V_{max} = 5.57 \text{ ft/s}$$

Assume that $V_{av}/V_{max} = 0.8$, then $V_{av} = 0.8 \times 5.57 = 4.46$ ft/s. For this value of V_{av},

$$N_{Re} = 4.6 \text{ in} \times \frac{\text{ft}}{12 \text{ in}} \times \frac{4.46 \text{ ft}}{\text{s}} \times \frac{62.4 \text{ lb}}{\text{ft}^3} \times \frac{\text{ft-s}}{0.672 \times 10^{-3} \text{ lb}}$$

$$= 159,000$$

From Fig. 8-8, for $N_{Re} = 159,000$, $V_{av}/V_{max} = 0.83$, which is close to but not what was assumed. Consequently, the calculations will be repeated for an assumed value of $V_{av}/V_{max} = 0.83$.

$$V_{av} = 0.83 \ (5.57) = 4.62 \text{ ft/s}$$

$$N_{Re} = 159,000 \times \frac{4.62}{4.46} = 165,000$$

(The above is a shortcut in calculating N_{Re}, recognizing that for the same set of conditions, N_{Re} will be directly proportional to the average fluid velocities.)
From Fig. 8-8, for $N_{Re} = 165,000$, $V_{av}/V_{max} = 0.83$, which is consistent with the value for which the calculations were made. Occasionally, it may be necessary to repeat the calculations if the value of V_{av}/V_{max} does not come out close enough to the assumed value.

For $V_{av} = 4.62$ ft/s and $A = \pi \ (4.6)^2/4 \times \frac{1}{144} = 0.115$ ft²:

$$Q = \frac{4.62 \text{ ft}}{\text{s}} \times \frac{60 \text{ s}}{\text{min}} \times 0.115 \text{ ft}^2 \times \frac{7.48 \text{ gal}}{\text{ft}^3} = 238 \text{ gal/min}$$

Since the fluid velocity can and usually will vary from point to point, a Pitot tube can be used to traverse the stream to obtain a velocity profile such as that in Fig. 8-2.

Flowmeters

There are a great many flow-measuring devices that convert a velocity head into an equivalent pressure head by means of a constriction placed in the fluid stream. The most common forms of such devices are Venturi meters, nozzle meters, and orifice meters. These are illustrated in Fig. 8-9. In each of these devices, the velocity of the fluid increases as it moves through the constriction, causing a drop in the static pressure as required by Bernoulli's principle. This easily measurable drop in pressure is directly related to the velocity of the fluid. While mathematical equations exist for various meter conditions, these tend to be difficult to use, and the

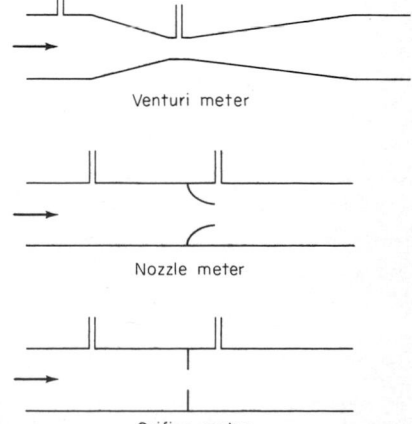

Venturi meter

Nozzle meter

Orifice meter

Fig. 8-9 Flowmeters showing location of pressure taps.

general practice is to calibrate each meter installation directly against known flows. It should be noted that the turbulence introduced by such meters develops friction which results in a permanent pressure loss, as illustrated by the pressure characteristics of an orifice meter in Fig. 8-10.

Example 8-22 Water is flowing through a 2-in-ID pipe at 65 gal/min. This flow is monitored by a Venturi meter whose throat is 0.78 in. What is the pressure differential between the entrance and the throat of the Venturi meter?

solution For this case, the Bernoulli equation [Eq. (8-3)] reduces to

$$p_1 v_1 + \frac{V_1^2}{2g} = p_2 v_2 + \frac{V_2^2}{2g}$$

$$V_1 = \frac{65 \text{ gal}}{\text{min}} \times \frac{\text{ft}^3}{7.48 \text{ gal}} \times \frac{\text{min}}{60 \text{ s}} \times \frac{4}{\pi(2)^2 \text{ in}^2} \times \frac{144 \text{ in}^2}{\text{ft}^2} = 6.64 \text{ ft/s}$$

$$V_2 = 65 \times \frac{1}{7.48} \times \frac{1}{60} \times \frac{4}{\pi(0.78)^2} \times 144$$

$$= 43.65 \text{ ft/s}$$

Since $v_1 = v_2$,

$$(p_1 - p_2)v_1 = \frac{V_2^2 - V_1^2}{2g}$$

$$(p_1 - p_2) \times \frac{\text{ft}^3}{62.4 \text{ lb}} = \frac{[(43.65)^2 - (6.64)^2] \text{ ft}^2}{\text{s}^2} \times \frac{\text{s}^2}{64.2 \text{ ft}}$$

$$= p_1 - p_2 = \frac{62.4 \times (1,905 - 44)}{64.2}$$

$$= 1,809 \text{ lb/ft}^2 \equiv 12.6 \text{ lb/in}^2$$

Another group of devices, usually used to measure the flow of gases and known as *anemometers,* utilize the kinetic energy of the stream to cool a wire to which a known amount of heat is being applied by an electric current, or to rotate vanes that integrate the amount of gas flowing past and give a readout as a velocity as shown in Fig. 8-11. Usually, anemometers are calibrated directly under the circumstances under which they are intended to be used.

Other forms of flowmeters, known as *quantity meters* because they measure the total quantity of fluid passing through, are widely used. Typical of these are wet-test gas meters, rotary-disk water meters, and rotary-piston meters, illustrated in Fig. 8-12. A wet-test meter is used for relatively small gas flows, which it can measure with considerable accuracy. It consists of a series of buckets partially submerged in water. The gas flow to be measured passes through the water into one of the buckets, changing its buoyancy and causing a rotational movement which, by means of suitable gearing, indicates the volume of gas directly on dials on the face of the meter. The buckets are vented into the atmosphere or back into the gas stream at an appropriate point, so that the motion of the buckets is continuous. A wet-test meter integrates the total quantity of gas passing through the instrument over the time it took to pass through. Thus a wet-test meter gives the average flow for that interval of time. In addition, it is important to note that the gas is saturated with water vapor as it passes through the water on its way to the buckets. As such, a wet-test meter measures the volume of gas plus the volume of water vapor necessary to saturate the gas.

A *rotary-disk meter,* perhaps the most commonly used liquid flowmeter, is the type usually used in the home. In a rotary-disk meter, a flat round plate, free to pivot at its center, is caused to flipflop within a specially devised chamber by the passage of a liquid through this chamber. This toggle motion is converted to a rotary movement and transmitted to direct-reading dials on the face of the meter by a suitable mechanical linkage. A rotary-

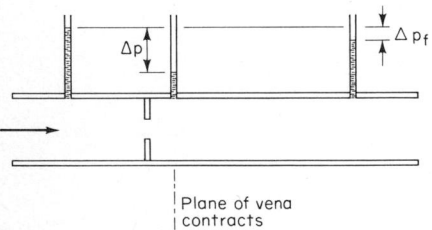

Fig. 8-10 Pressure characteristics of an orifice meter. The vena contracta is the section at which the pressure is a minimum, and therefore provides the greatest Δp reading. The plane of the vena contracta occurs about one orifice diameter downstream from the orifice. Δp = pressure drop dependent on flow rate; Δp_f = pressure loss due to friction.

disk meter is also an integrating device yielding the total flow for any given time interval.

Rotary-piston meters come in a variety of designs and are used to measure both gases (usually under high pressures) and liquids. In a rotary-piston meter, the fluid causes a piston to revolve, and this motion is translated to direct-reading dials on the face of the meter. Rotary-piston meters have particular utility in measuring dry gases, since no water vapor is introduced into the gas stream as with a wet-test meter.

Still another flow-measuring device uses the kinetic energy of the stream to suspend a weight against the force of gravity. Such meters, illustrated in Fig. 8-13, consist of a conical tube and a float. As the fluid flows from bottom to top, its velocity past the float decreases, so that the float assumes an equilibrium position determined by its weight and shape and the velocity of the fluid flowing past it at a particular height in the tube. To prevent sticking to the sides and to stabilize its attitude, small vanes are cut into the float so that they rotate, hence the name *rotameter*. Rotameters are best calibrated against known flows for the conditions under which they are to be used. Through the use of different tube and float designs, rotameters have found widespread use for measuring both liquid and gas flows.

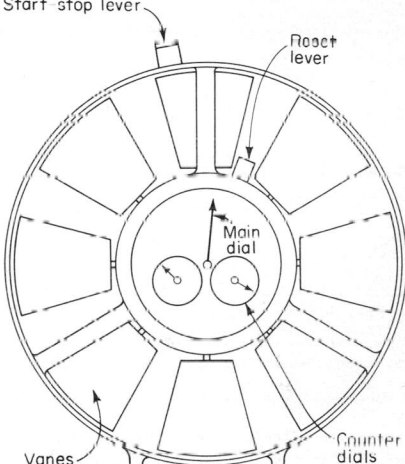

Weirs

To this point, we have considered conduits that are completely filled with the flowing fluid. A liquid flowing in an open channel or through any other partially filled conduit can be metered by means of a notched dam called a *weir*. The shape of the notch is used to describe the weir, the common ones being rectangular or triangular, as illustrated in Fig. 8-14. Flow through a rectangular weir is given the empirical formula

$$Q_v = 0.42(L - 0.2)h\sqrt{2gh} \quad (8\text{-}11)$$

where Q_v = volumetric rate of flow (ft³/s), L = width of the opening (ft), h = height of the crest (ft), and g = gravitational constant (32.1 ft/s²). The height of the crest is measured from the edge of the weir to the surface of the liquid far enough back from the weir so as not to be influenced by the curvature of

Fig. 8-11 Vane anemometer. The main dial and counter dials indicate how many linear feet of gas pass through the anemometer. Dividing this by the time it took for this gas to pass through, as measured by a stopwatch, gives the velocity of the gas.

the surface as the liquid approaches the weir. The surface of the liquid is usually taken at a distance at least $3h$ back from the edge of the weir. For a triangular weir, the flow is given by the empirical formula

$$Q_v = 0.312 \left(\tan \frac{\alpha}{2} \right) h^2 \sqrt{2gh} \tag{8-12}$$

where Q_v = volumetric rate of flow (ft³/s), α = angle of the notch, and h = height of the crest (ft).

Example 8-23 If the width of the rectangular weir shown in Fig. 8-14 is 9 in, what is the flow over the weir when the crest height is 2 in and when it is 4 in?
solution

$$L = 9 \text{ in} \times \frac{\text{ft}}{12 \text{ in}} = 0.75 \text{ ft}$$

For $h = 2$ in,
$$h = 2 \text{ in} \times \frac{\text{ft}}{12 \text{ in}} = 0.167 \text{ ft}$$

For $h = 4$ in,
$$h = \frac{4}{12} = 0.333 \text{ ft}$$

Applying Eq. (8-11) for $h = 0.167$ ft,

$$Q_v = 0.42(L - 0.2)h \sqrt{2gh}$$

$$= 0.42(0.75 - 0.2) \text{ ft} \times 0.167 \text{ ft} \times \sqrt{2 \times \frac{32.1 \text{ ft}}{\text{s}^2} \times 0.167 \text{ ft}}$$

$$= 0.42(0.55)(0.167)(3.27) \text{ ft}^3/\text{s}$$
$$= 0.126 \text{ ft}^3/\text{s}$$
$$= \frac{0.126 \text{ ft}^3}{\text{s}} \times \frac{7.48 \text{ gal}}{\text{ft}^3} \times \frac{60 \text{ s}}{\text{min}} = 56.5 \text{ gal/min}$$

For $h = 0.333$ ft,

$$Q_v = 0.42(0.55)(0.333) \sqrt{64.2 \ (0.333)}$$
$$= 0.356 \text{ ft}^3/\text{s} = 160 \text{ gal/min}$$

Example 8-24   For the triangular weir shown in Fig. 8-14, what is the flow over the weir for $\emptyset = 75°$ and crest heights of 2 and 4 in?
solution   The angle α required by Eq. (8-12) equals $180 - 2 \emptyset = 180 - 150 = 30°$, and, from trigometric tables, $\tan \alpha/2 = \tan 15 = 0.2679$. Applying Eq. (8-12) for $h = 2$ in (0.167 ft),

$$Q_v = 0.312 \tan \frac{\alpha}{2} h^2 \sqrt{2gh}$$

$$= 0.312(0.2679)(0.167)^2 \sqrt{64.2 \ (0.167)}$$
$$= 0.0076 \text{ ft}^3/\text{s} = 3.41 \text{ gal/min}$$

For $h = 4$ in (0.333 ft),

$$Q_v = 0.312(0.2679)(0.333)^2 \sqrt{64.2 \ (0.333)}$$
$$= 0.043 \text{ ft}^3/\text{s} = 19.3 \text{ gal/min}$$

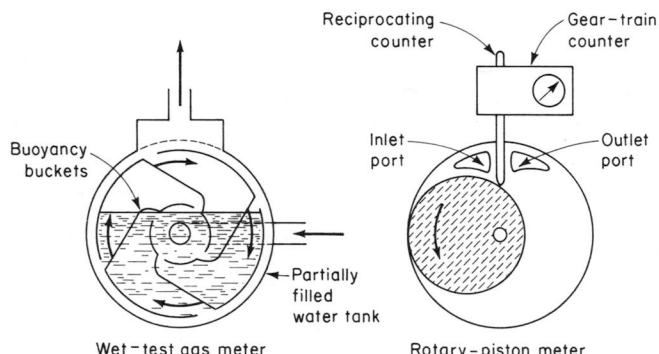

Wet-test gas meter   Rotary-piston meter

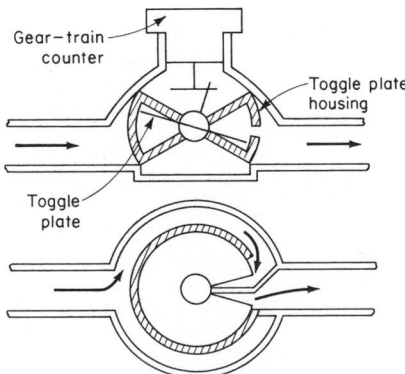

Rotary-disk water meter

Fig. 8-12   Mechanical flowmeters.

8-3 PIPES AND FITTINGS

Piping systems are constructed from straight-run piping, fittings to change direction, and valving to modify or control the flow. A large variety of standard pipes and fittings are available, any specific selection being based on the flow conditions desired (flow, pressure, temperature, etc.) and the chemical requirements of the system (corrosion resistance, chemical stability, etc.). Pipes and fittings are connected by welded, threaded, flanged, flared, compressed, or sweated joints, as illustrated in Fig. 8-15. In general, welded and flanged joints are used for iron pipes of large diameter, i.e., 2 in or larger, while threaded joints are used for small diameters. Flanged, flared, compressed, or sweated joints are used for small-diameter copper and brass tubing.

Piping is specified by a nominal size which, for low-pressure work, is approximately the inside diameter of the pipe. Iron piping comes in a number of *schedules*, or wall thicknesses,

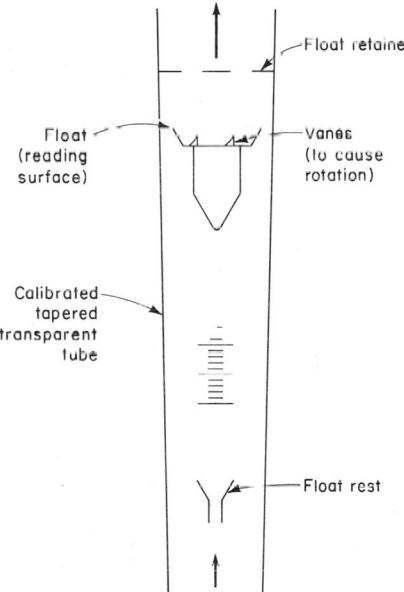

Fig. 8-13 Rotameter. The upper surface of the float is used to read its position on the tapered calibrated tube. The size and weight of the float can be changed to provide versatility.

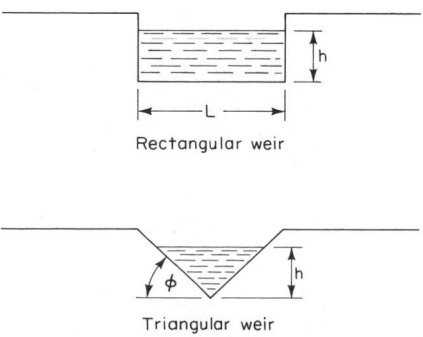

Rectangular weir

Triangular weir

Fig. 8-14 Weirs.

to accommodate various maximum pressures. Table 8-5 gives the schedule designations and the maximum pressure for which each schedule is to be used. Table 8-6 gives the standard physical dimensions of commercial iron pipe. The flow characteristics of iron pipes are also frequently needed, and these are given in Table 8-7. Dimensional data for various types of copper tubing, glass pipe, and glass tubing are given in Tables 8-8 to 8-11.

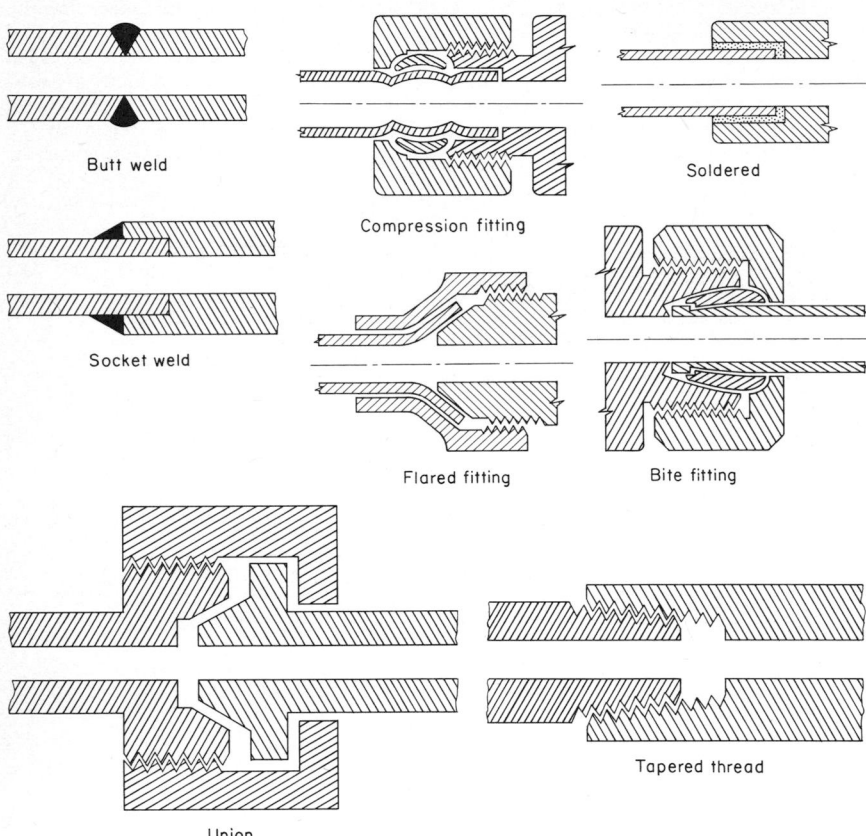

Butt weld

Compression fitting

Soldered

Socket weld

Flared fitting

Bite fitting

Union

Tapered thread

Fig. 8-15 Commonly used pipe joints.

TABLE 8-5   Allowable Pressures for Standard Pipe Schedules

	Wrought iron		Steel	
Schedule	Max allowable pressure (lb/in²)	Normal design max (lb/in²)	Max allowable pressure (lb/in²)	Normal design max (lb/in²)
10	50	25	75	50
20	100	50	150	75
30	150	75	200	100
40	250	125	300	150
60	350	200	450	250
80	450	250	600	300
100	600	300	700	350
120	700	350	900	450
140	800	400	1,000	500
160	900	450	1,200	600

TABLE 8-6 Dimensions of Wrought-Iron or Steel Pipe

Size (in)	Diameters External (in)	Internal (in)	Thickness, in	Circumference External (in)	Internal, in	Transverse areas External, in²	Internal (in²)	Metal (in²)	Length of pipe/ft² External surface (ft)	Internal surface (ft)	Length of pipe containing 1 ft³ (ft)	Weight/ft plain ends (lb)	Wt. of water/ft (lb)
						SCHEDULE 40 (ST—STANDARD WALL)							
1/8	0.405	0.269	0.068	1.272	0.845	0.129	0.057	0.072	9.431	14.199	2533.775	0.244	0.025
1/4	0.540	0.364	0.088	1.696	1.144	0.229	0.104	0.125	7.073	10.493	1383.789	0.424	0.045
3/8	0.675	0.493	0.091	2.121	1.549	0.358	0.191	0.167	5.658	7.747	754.360	0.567	0.083
1/2	0.840	0.622	0.109	2.639	1.954	0.554	0.304	0.250	4.547	6.141	473.906	0.850	0.132
3/4	1.050	0.824	0.113	3.299	2.589	0.866	0.533	0.333	3.637	4.635	270.034	1.130	0.231
1	1.315	1.049	0.133	4.131	3.296	1.358	0.864	0.494	2.904	3.641	166.618	1.678	0.375
1 1/4	1.660	1.380	0.140	5.215	4.335	2.164	1.495	0.669	2.301	2.767	96.275	2.272	0.65
1 1/2	1.900	1.610	0.145	5.969	5.058	2.835	2.036	0.799	2.010	2.372	70.733	2.717	0.88
2	2.375	2.067	0.154	7.461	6.494	4.430	3.355	1.075	1.608	1.847	42.913	3.652	1.45
2 1/2	2.875	2.469	0.203	9.032	7.757	6.492	4.788	1.704	1.328	1.547	30.077	5.793	2.07
3	3.500	3.068	0.216	10.996	9.638	9.621	7.393	2.228	1.091	1.245	19.479	7.575	3.20
3 1/2	4.000	3.548	0.226	12.566	11.146	12.566	9.886	2.680	0.954	1.076	14.565	9.109	4.29
4	4.500	4.026	0.237	14.137	12.648	15.904	12.730	3.174	0.848	0.948	11.312	10.790	5.50
4 1/2	5.000	4.506	0.247	15.708	14.156	19.635	15.947	3.688	0.763	0.847	9.030	12.538	6.91
5	5.563	5.047	0.258	17.477	15.856	24.306	20.006	4.300	0.686	0.756	7.198	14.617	8.67
6	6.625	6.065	0.280	20.813	19.054	34.472	28.891	5.581	0.576	0.629	4.984	18.974	12.51
7	7.625	7.023	0.301	23.955	22.063	45.664	38.738	6.926	0.500	0.543	3.717	23.544	16.80
8	8.625	8.071	0.277	27.096	25.356	58.426	51.161	7.265	0.442	0.473	2.815	24.696	22.18
8	8.625	7.981	0.322	27.096	25.073	58.426	50.027	8.399	0.442	0.478	2.878	28.554	21.70
9	9.625	8.941	0.342	30.238	28.089	72.760	62.786	9.974	0.396	0.427	2.294	33.907	27.20
10	10.750	10.192	0.279	33.772	32.019	90.763	81.585	9.178	0.355	0.374	1.765	31.201	35.37
10	10.750	10.136	0.307	33.772	31.843	90.763	80.691	10.072	0.355	0.376	1.785	34.240	34.95
10	10.750	10.020	0.365	33.772	31.479	90.763	78.855	11.908	0.355	0.381	1.826	40.483	34.20
11	11.750	11.000	0.375	36.914	34.558	108.434	95.033	13.401	0.325	0.347	1.515	45.557	41.20
12	12.750	12.090	0.330	40.055	37.982	127.676	114.800	12.876	0.299	0.315	1.254	43.773	49.70
12	12.750	12.000	0.375	40.055	37.699	127.676	113.097	14.579	0.299	0.318	1.273	49.562	49.00
						SCHEDULE 80 (XS—EXTRA STRONG WALL)							
1/8	0.405	0.215	0.095	1.272	0.675	0.129	0.036	0.093	9.431	17.766	3966.392	0.314	0.016
1/4	0.540	0.302	0.119	1.696	0.949	0.229	0.072	0.157	7.073	12.648	2010.290	0.535	0.031
3/8	0.675	0.423	0.126	2.121	1.329	0.358	0.141	0.217	5.658	9.030	1024.689	0.738	0.061
1/2	0.840	0.546	0.147	2.639	1.715	0.554	0.234	0.320	4.547	6.995	615.017	1.087	0.102
3/4	1.050	0.742	0.154	3.299	2.331	0.866	0.433	0.433	3.637	5.147	333.016	1.473	0.188
1	1.315	0.957	0.179	4.131	3.007	1.358	1.719	0.639	2.904	3.991	200.193	2.171	0.312
1 1/4	1.660	1.278	0.191	5.215	4.015	2.164	1.283	0.881	2.301	2.988	112.256	2.996	0.56
1 1/2	1.900	1.500	0.200	5.969	4.712	2.835	1.767	1.068	2.010	2.546	81487	3.631	0.77
2	2.375	1.939	0.218	7.461	6.092	4.430	2.953	1.477	1.608	1.969	48.766	5.022	1.28
2 1/2	2.875	2.323	0.276	9.032	7.298	6.492	4.238	2.254	1.328	1.644	33.976	7.661	1.87
3	3.500	2.900	0.300	10.996	9.111	9.621	6.605	3.016	1.091	1.317	21.801	10.252	2.86
3 1/2	4.000	3.364	0.318	12.566	10.568	12.566	8.888	3.678	0.954	1.135	16.202	12.505	3.84
4	4.500	3.820	0.337	14.137	12.020	15.904	11.497	4.407	0.848	0.998	12.525	14.983	4.98
4 1/2	5.000	4.290	0.355	15.708	13.477	19.635	14.455	5.180	0.763	0.890	9.962	17.611	6.27
5	5.503	4.813	0.375	17.477	15.120	24.306	18.194	6.112	0.686	0.793	7.915	20.778	7.88
6	6.625	5.761	0.432	20.813	18.099	34.472	26.067	8.405	0.576	0.663	5.524	28.573	11.29
7	7.625	6.625	0.500	23.955	20.813	45.664	34.472	11.192	0.500	0.576	4.177	38.048	14.95
8	8.625	7.625	0.500	27.096	23.955	58.426	45.663	12.763	0.442	0.500	3.154	43.388	19.78
9	9.625	8.625	0.500	30.238	27.096	72.760	58.426	14.334	0.396	0.442	2.464	48.728	25.30
10	10.750	9.750	0.500	33.772	30.631	90.763	74.662	16.101	0.355	0.391	1.929	54.735	32.35
11	11.750	10.750	0.500	36.914	33.772	108.434	90.763	17.671	0.325	0.355	1.587	60.075	39.40
12	12.750	11.750	0.500	40.055	36.914	127.676	108.434	19.242	0.209	0.325	1.328	65.415	46.92
						SCHEDULE 160 (XX—DOUBLE EXTRA STRONG WALL)							
1/2	0.840	0.252	0.294	2.639	0.792	0.554	0.050	0.504	4.547	15.157	2887.164	1.714	0.022
3/4	1.050	0.434	0.308	3.299	1.363	0.866	0.148	0.718	3.637	8.801	973.404	2.440	0.064
1	1.315	0.599	0.358	4.131	1.882	1.358	0.282	1.076	2.904	6.376	510.998	3.659	0.122
1 1/4	1.600	0.890	0.382	5.215	2.815	2.164	0.630	1.534	2.301	4.263	228.379	5.214	0.273

TABLE 8-6   Dimensions of Wrought-Iron or Steel Pipe—(Continued)

Size (in)	Diameters External (in)	Internal (in)	Thickness, in	Circumference External (in)	Internal, in	Transverse areas External, in²	Internal (in²)	Metal (in²)	Length of pipe/ft² External surface (ft)	Internal surface (ft)	Length of pipe containing 1 ft³ (ft)	Weight/ft plain ends (lb)	Wt. of water/ft (lb)
						SCHEDULE 160 (XX—DOUBLE EXTRA STRONG WALL)							
1½	1.900	1.100	0.400	5.969	3.456	2.835	0.950	1.885	2.010	3.472	151.526	6.408	0.42
2	2.375	1.503	0.436	7.461	4.722	4.430	1.774	2.656	1.608	2.541	81.162	9.029	0.77
2½	2.875	1.771	0.552	9.032	5.564	6.492	2.464	4.028	1.328	2.156	58.457	13.695	1.07
3	3.500	2.300	0.600	10.996	7.226	9.621	4.155	5.466	1.091	1.660	34.659	18.583	1.80
3½	4.000	2.728	0.636	12.566	8.570	12.566	5.845	6.721	0.954	1.400	24.637	22.850	2.53
4	4.500	3.152	0.674	14.137	0.902	15.904	7.803	8.101	0.848	1.211	18.454	27.541	3.38
4½	5.000	3.580	0.710	15.708	11.247	19.635	10.066	9.569	0.763	1.066	14.306	32.530	4.36
5	5.563	4.063	0.750	17.477	12.764	24.306	12.966	11.340	0.686	0.940	11.107	38.552	5.61
6	6.625	4.897	0.864	20.813	15.384	34.472	18.835	15.637	0.576	0.780	7.640	53.160	8.16
7	7.625	5.875	0.875	23.955	18.457	45.664	27.109	18.555	0.500	0.650	5.312	63.079	11.75
	8.625	6.875	0.875	27.096	21.598	58.426	37.122	21.304	0.442	0.555	3.879	72.424	16.10

TABLE 8-7   Flow Capacities of Iron Pipe

Nominal size (in)	Capacity at velocity of 1 ft/s ST gal/min	lb H₂O/h	XS gal/min	lb H₂O/h	XX gal/min	lb H₂O/h
⅛	0.179	89.5	0.113	56.6		
¼	0.323	162	0.224	112		
⅜	0.596	298	0.440	220		
½	0.945	472	0.730	365	0.158	79.2
¾	1.665	833	1.345	673	0.460	230
1	2.690	1,345	2.240	1,120	0.878	439
1¼	4.57	2,285	3.99	1,995	1.97	983
1½	6.34	3,170	5.49	2,745	3.02	1,512
2	10.45	5,225	9.20	4,600	5.54	2,772
2½	14.92	7,460	13.20	6,600	7.70	3,852
3	23.00	11,500	20.55	10,275	12.96	6,480
3½	30.80	15,400	27.70	13,850	18.22	9,108
4	39.6	19,800	35.80	17,900	24.34	12,168
4½					31.39	15,696
5	62.3	31,150	57.7	28,850	40.39	20,196
6	90.0	45,000	81.1	40,550	58.75	29,376
7					84.60	42,300
8	155.7	79,800	142.3	71,200	116	57,960
8	152.4	78,100				
9	191.0	97,900	182.1	91,100		
10	248.4	127,300	232.9	116,500		
10	245.5	125,800				
10	240.2	123,100				
11	289.4	148,300	283.5	141,800		
12	290.6	178,900	337.7	168,900		
12	286.5	176,400				

ST = standard wall
XS = extra strong wall
XX = double extra strong wall

TABLE 8-8 Dimensions of Copper Water Tubing

Nominal size	Actual outside diam., in.	Mean outside diam. tolerances, in.		Wall thickness, in.						Theoretical weight, lb./ft.		
		Soft annealed	Hard drawn	Type K		Type L		Type M		Type K	Type L	Type M
				Nominal	Tolerance	Nominal	Tolerance	Nominal	Tolerance			
¼	0.375	0.002	0.001	0.035	0.004	0.030	0.0035			0.145	0.126	0.145
⅜	.500	.0025	.001	.049	.004	.035	.0035	.025	.0025	.269	.198	.204
½	.625	.0025	.001	.049	.004	.040	.0035	.028	.0025	.344	.285	
⅝	.750	.0025	.001	.049	.004	.042	.0035			.418	.362	
¾	.875	.003	.001	.065	.0045	.045	.004	.032	.003	.641	.455	.328
1	1.125	.0035	.0015	.065	.0045	.050	.004	.035	.0035	.839	.655	.465
1¼	1.375	.004	.0015	.065	.0045	.055	.004	.042	.0035	1.04	.884	.682
1½	1.625	.0045	.002	.072	.005	.060	.0045	.049	.004	1.36	1.14	.940
2	2.125	.005	.002	.083	.007	.070	.0045	.058	.006	2.06	1.75	1.46
2½	2.625	.005	.002	.095	.007	.080	.006	.065	.006	2.93	2.48	2.03
3	3.125	.005	.002	.109	.007	.090	.006	.072	.006	4.00	3.33	2.68
3½	3.625	.005	.002	.120	.008	.100	.007	.083	.007	5.12	4.29	3.58
4	4.125	.005	.002	.134	.010	.110	.007	.095	.009	6.51	5.38	4.66
5	5.125	.005	.002	.160	.010	.125	.009	.109	.009	9.67	7.61	6.66
6	6.125	.005	.002	.192	.012	.140	.010	.122	.010	13.9	10.2	8.92
8	8.125	.006	+.002 −.004	.271	.016	.200	.011	.170	.014	25.9	19.3	16.5
10	10.125	.008	+.002 −.006	.338	.018	.250	.014	.212	.015	40.3	30.1	25.6
12	12.125	.008	+.002 −.006	.405	.020	.280	.018	.254	.016	57.8	40.4	36.7

Adapted from Robert H. Perry and Cecil H. Chilton (eds.), "Chemical Engineers' Handbook," 5th ed., McGraw-Hill Book Company, New York, 1973.

TABLE 8-9   Dimensions of General-Service Copper Tubing

·Nominal size	Actual outside diam. (in)	Mean outside diam. tolerances (in)	Wall thickness (in)	Wall-thickness tolerances (in)	Nominal wt. (lb/ft)
$\frac{1}{8}$	0.125	0.002	0.030	0.003	0.0347
$\frac{3}{16}$	0.188	0.002	0.030	0.0025	0.575
$\frac{1}{4}$	0.250	0.002	0.030	0.0025	0.0804
$\frac{5}{16}$	0.312	0.002	0.032	0.0025	0.109
$\frac{3}{8}$	0.375	0.002	0.032	0.0025	0.134
$\frac{1}{2}$	0.500	0.002	0.032	0.0025	0.182
$\frac{5}{8}$	0.625	0.002	0.035	0.003	0.251
$\frac{3}{4}$	0.750	0.0025	0.035	0.003	0.305

Adapted from Robert H. Perry and Cecil H. Chilton (eds.), "Chemical Engineers' Handbook," 5th ed., McGraw-Hill Book Company, New York, 1973.

TABLE 8-10   Dimensions of Glass Pipe

Inside diam. (in)	Outside diam. (in)	Wall thickness (in)	Weight per ft (lb)
1	1.31	0.156	0.55
1.5	1.84	0.171	0.87
2	2.34	0.171	1.13
3	3.41	0.202	1.97
4	4.50	0.264	3.41
6	6.66	0.328	6.30

Adapted from Robert H. Perry and Cecil H. Chilton (eds.), "Chemical Engineers' Handbook," 5th ed., McGraw-Hill Book Company, New York, 1973.

TABLE 8-11   Dimensions of Industrial Glass Tubing

Outside diameter (in)	Medium wall Wall thickness (in)	Weight per ft (lb)	Heavy wall Wall thickness (in)	Weight per ft (lb)
$\frac{1}{4}$	$\frac{3}{64}$	0.029		
$\frac{1}{2}$	$\frac{1}{16}$	0.083	$\frac{3}{32}$	0.11
$\frac{3}{4}$	$\frac{1}{16}$	0.131	$\frac{1}{8}$	0.24
1	$\frac{3}{32}$	0.26	$\frac{5}{32}$	0.40
$1\frac{1}{4}$	$\frac{3}{32}$	0.33	$\frac{5}{32}$	0.53
$1\frac{1}{2}$	$\frac{3}{32}$	0.41	$\frac{5}{32}$	0.64
$1\frac{3}{4}$	$\frac{3}{32}$	0.48	$\frac{5}{32}$	0.76
2	$\frac{1}{8}$	0.72	$\frac{3}{16}$	1.05
$2\frac{1}{4}$	$\frac{1}{8}$	0.82	$\frac{3}{16}$	1.19
$2\frac{1}{2}$	$\frac{1}{8}$	0.91	$\frac{3}{16}$	1.33
$2\frac{3}{4}$	$\frac{1}{8}$	1.01	$\frac{3}{10}$	1.48
3	$\frac{1}{8}$	1.10	$\frac{3}{10}$	1.60
$3\frac{1}{4}$	$\frac{1}{8}$	1.20	$\frac{3}{16}$	1.77
$3\frac{1}{2}$	$\frac{1}{3}$	1.30	$\frac{3}{16}$	1.91
4	$\frac{3}{16}$	2.17	$\frac{1}{4}$	2.85
$4\frac{1}{2}$	$\frac{3}{16}$	2.45	$\frac{1}{4}$	3.20

Adapted from John H. Perry (ed.), "Chemical Engineers' Handbook," 4th ed., McGraw-Hill Book Company, New York, 1963.

TABLE 8-12 Dimensions of Standard (Schedule 40) Pipe Fittings

All dimensions in inches

Nominal pipe size	a	c	e min	h min	j	k	r min	s	p	q	f	Wall thickness min	Wall thickness max
1/8	0.69	0.68	0.200	0.693	1.00	0.84	0.2638		0.97	0.34	0.96	0.405	0.435
1/4	0.81	0.73	0.215	0.844	1.14	0.94	0.4018	1.00	1.19	0.43	1.06	0.540	0.584
3/8	0.95	0.80	0.230	1.015	1.44	1.03	0.4078	1.13	1.43	0.50	1.16	0.675	0.719
1/2	1.12	0.88	0.249	1.197	1.63	1.15	0.5337	1.25	1.71	0.61	1.34	0.840	0.897
3/4	1.31	0.98	0.273	1.458	1.59	1.29	0.5457	1.44	2.05	0.72	1.52	1.050	1.107
1	1.50	1.12	0.302	1.771	2.14	1.47	0.6828	1.69	2.43	0.85	1.67	1.315	1.385
1 1/4	1.75	1.29	0.341	2.153	2.45	1.71	0.7068	2.06	2.92	1.02	1.93	1.660	1.730
1 1/2	1.94	1.43	0.368	2.427	2.69	1.88	0.7235	2.31	3.28	1.10	2.15	1.900	1.970
2	2.25	1.68	0.422	2.963	3.26	2.22	0.7565	2.81	3.93	1.24	2.53	2.375	2.445
2 1/2	2.70	1.95	0.478	3.589	3.86	2.57	1.1375	3.25	4.73	1.52	2.88	2.875	2.975
3	3.08	2.17	0.548	4.285	4.51	3.00	1.2000	3.69	5.55	1.71	3.18	3.500	3.600
3 1/2	3.42	2.39	0.604	4.843	5.05	3.35	1.2500	4.00	6.25	1.85	3.43	4.000	4.100
4	3.79	2.61	0.661	5.401	5.59	3.70	1.3000	4.38	6.97	2.01	3.69	4.500	4.600
5	4.50	3.05	0.780	6.583	6.86	4.44	1.4063	5.12	8.43	2.34	4.22	5.563	5.663
6	5.13	3.46	0.900	7.767	8.03	5.18	1.5125	5.86	9.81	2.66	4.75	6.625	6.725
8			1.125	9.995			1.7125	7.25			5.75	8.625	8.725

90° elbow Tee Cross Coupling 45° ell Y branch 90° street elbow 45° street elbow Street tee Reducing coupling

The types of fittings in general use and their dimensions are given in Table 8-12. The available types of valves are shown in Fig. 8-16, with their corresponding dimensions being the same as for an elbow or tee as given in Table 8-12. It should be noted that fittings and valves come in the same schedules as the pipes with which they are to be used, and that the nominal size of the fitting is the same as the nominal size of the pipe it fits.

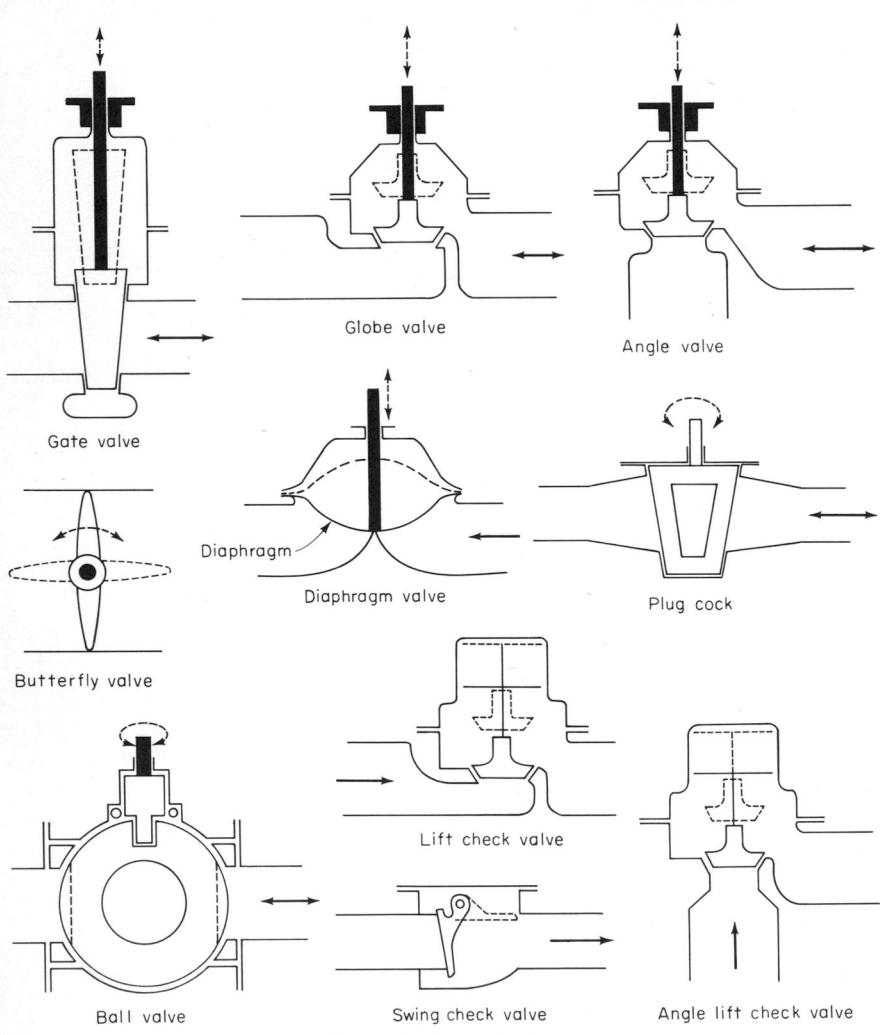

Fig. 8-16 Types of valves.

Chapter 9

Engineering Operations

9-1 CRUSHING AND GRINDING

Crushing and grinding involves the reduction of the particle size of a material by physical means. Usually, at least one of three purposes is served by comminution processes:

1. To increase chemical reactivity
2. To change physical characteristics
3. To improve mixing

Where a reaction is confined to the surface of a solid, the rate of reaction will be a function of the exposed surface. To complete such a reaction in a minimum time, the surface area must be made as large as possible, and this can be accomplished by a grinding process. As an example, powdered coal is used where a high combustion rate is desired. Grinding may also alter the shape of a particle, for example, in the grinding of a pigment, where the grinding may flatten the particle to enable it to cover better or hide the substrate being painted. In the case where the pigment is ground in an oil base, the grinding also serves to provide a better mixture between the oil and the pigment. Although the terms are used interchangeably, crushing is usually used in connection with coarse materials, while grinding is usually associated with the production of fine particles.

The properties of a solid most directly concerned with particle-size reduction are its hardness and grindability. *Hardness* has been described in Sec. 1-5, together with the Mohs system for measuring hardness. *Grindability* is a more general term and refers to the ease with which a particle of a material can be fractured so as to reduce its size. Although there is no well-accepted method for determining grindability, it can be measured by subjecting a sample to a grinding operation in a special grinder in which the energy input for a given size reduction can be measured: the less energy required, the greater the grindability. To describe a particular grinding operation, the reduction ratio is frequently used. The *reduction ratio* is the average particle size of the material before grinding, divided by the average particle size after grinding. Particle-size measurements are difficult to carry out in a reproducible manner because a uniform particle size and shape are rarely achieved and results must therefore be expressed in terms of a *particle-size distribution*. The results are also dependent on the particular particle property actually being measured, and how the results are expressed, since most particles have a highly irregular geometry. The term particle size itself is therefore not precisely descriptive.

Perhaps the most widely used description of particle-size distribution is by *screen analysis*. In a screen analysis, a sample is placed on a series of sieves, called a nest, and the nest is shaken with a circular motion to which is added an occasional bump (Fig. 9-1). The particles trickle down through the coarsest screen on top to the finest on the bottom. Each particle larger than the opening between the threads of the screen is retained by that particular sieve. In this way, a sequence of retentions is obtained and each retention can then be weighed. Thus, the weight percent retained on a particular sieve can be determined.

Standard sieves are made of woven wires that provide essentially square openings. There are three systems for designating sieve sizes:

1. Standard
2. Alternate
3. Tyler

The Tyler designation was promoted by the W. S. Tyler Company, which pioneered in commercial sieve-analysis equipment. The physical sieve dimensions are given for the three systems in Table 9-1. Other types of sieves are available, such as those using silk bolting cloth (usually used for fine sieves, 150- to 400-mesh), electroplated mesh with square openings (usually for fine sieving, 150- down to 400-mesh), and punched sheet-metal sieves (usually used for coarse work, i.e., 0.25-in particles and larger).

During the sieving of fine powders, the materials may have a tendency to ball, a phenomenon in which the particles seem to agglomerate into fairly large spheres which do not pass through the openings in the sieve, even though the particle size of the deagglomerated material is fine enough to pass easily through the sieve. In such a case, the sieving can be accomplished by dispersing the material in a liquid which does not dissolve or swell the particles. Balling frequently is associated with the presence of moisture and can sometimes be avoided by predrying the sample.

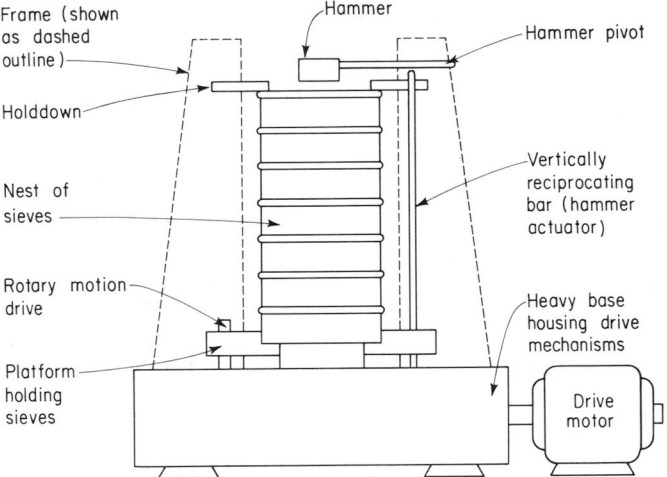

Fig. 9-1 Automatic sieve shaker. The platform holding the sieves has a rotary motion while the hammer on top provides a regular tapping action.

Another important means of determining particle-size distribution is by *elutriation,* a process in which the settling velocities of differently sized particles are used to obtain the particle-size distribution. Where the settling is carried out in a liquid suspending medium, using a horizontal flow in equipment of the type shown in Fig. 9-2, convenient particle-size separations can be made. Where gas is the suspending medium, the elutriation is most often carried out in vertical chambers, with the different particle sizes settling out in sequential chambers, as illustrated by Fig. 9-3. Any number of designs for sedimentation equipment are in use, but these are essentially modifications of the equipment shown in Figs. 9-2 and 9-3.

A *centrifugal classifier* is a special type of sedimentation equipment using centrifugal forces to accelerate the separation. A centrifugal classifier is generally used to make a single particle-size cut and is schematically shown in Fig. 9-4. In a centrifuge, the fine particles leave at the centrally located overflow, while the coarse particles fall to the bottom, where they can be drained. A cyclone type of classifier can be used to remove particles suspended in a liquid or a gas, and the drain can be continuous or intermittent.

Certain specialized procedures can also be used to determine particle-size distribution. In one such procedure, the *air permeability* or resistance to the flow of air can be used. In this type of determination, which is usually used for fine particles, the sample is placed inside a container with fine sieves (or some other filter medium) at each end. The pressure drop required to maintain a particular flow of air through the bed of particles can be related to the particle-size distribution of the sample. Where particle-size measurements on a particular material are made as a routine, the particle-size distribution can be empirically correlated to the pressure drop. An air-permeability particle-size measuring device is schematically shown in Fig. 9-5.

TABLE 9-1   Sieve Dimensions for Standard, Alternate, and Tyler Sieves

Sieve designation		Sieve opening		Nominal wire diam.		Tyler equivalent designation
Standard	Alternate	mm.	in. (approx. equivalents)	mm.	in. (approx. equivalents)	
107.6 mm.	4.24 in.	107.6	4.24	6.40	0.2520	
101.6 mm.	4 in.	101.6	4.00	6.30	.2480	
90.5 mm.	3½ in.	90.5	3.50	6.08	.2394	
76.1 mm.	3 in.	76.1	3.00	5.00	.2200	
64.0 mm.	2½ in.	64.0	2.50	5.50	.2165	
53.8 mm.	2.12 in.	53.8	2.12	5.15	.2028	
50.8 mm.	2 in.	50.8	2.00	5.05	.1988	
45.3 mm.	1¾ in.	45.3	1.75	4.85	.1909	
38.1 mm.	1½ in.	38.1	1.50	4.59	.1807	
32.0 mm.	1¼ in.	32.0	1.25	4.23	.1665	
26.9 mm.	1.06 in.	26.9	1.06	3.90	.1535	1.050 in.
25.4 mm.	1 in.	25.4	1.00	3.80	.1496	
22.6 mm.	⅞ in.	22.6	0.875	3.50	.1378	0.883 in.
19.0 mm.	¾ in.	19.0	.750	3.30	.1299	.742 in.
16.0 mm.	⅝ in.	16.0	.625	3.00	.1181	.624 in.
13.5 mm.	0.530 in.	13.5	.530	2.75	.1083	.525 in.
12.7 mm.	½ in.	12.7	.500	2.67	.1051	
11.2 mm.	⁷⁄₁₆ in.	11.2	.438	2.45	.0965	.441 in.
9.51 mm.	⅜ in.	9.51	.375	2.27	.0894	.371 in.
8.00 mm.	⁵⁄₁₆ in.	8.00	.312	2.07	.0815	2½ mesh
6.73 mm.	0.265 in.	6.73	.265	1.87	.0736	3 mesh
6.35 mm.	¼ in.	6.35	.250	1.82	.0717	
5.66 mm.	No. 3½	5.66	.223	1.68	.0661	3½ mesh
4.76 mm.	No. 4	4.76	.187	1.54	.0606	4 mesh
4.00 mm.	No. 5	4.00	.157	1.37	.0539	5 mesh
3.36 mm.	No. 6	3.36	.132	1.23	.0484	6 mesh
2.83 mm.	No. 7	2.83	.111	1.10	.0430	7 mesh
2.38 mm.	No. 8	2.38	.0937	1.00	.0394	8 mesh
2.00 mm.	No. 10	2.00	.0787	0.900	.0354	9 mesh
1.68 mm.	No. 12	1.68	.0661	.810	.0319	10 mesh
1.41 mm.	No. 14	1.41	.0555	.725	.0285	12 mesh
1.19 mm.	No. 16	1.19	.0469	.650	.0256	14 mesh
1.00 mm.	No. 18	1.00	.0394	.580	.0228	16 mesh
841 micron*	No. 20	0.841	.0331	.510	.0201	20 mesh
707 micron	No. 25	.707	.0278	.450	.0177	24 mesh
595 micron	No. 30	.595	.0234	.390	.0154	28 mesh
500 micron	No. 35	.500	.0197	.340	.0134	32 mesh
420 micron	No. 40	.420	.0165	.290	.0114	35 mesh
354 micron	No. 45	.354	.0139	.247	.0097	42 mesh
297 micron	No. 50	.297	.0117	.215	.0085	48 mesh
250 micron	No. 60	.250	.0098	.180	.0071	60 mesh
210 micron	No. 70	.210	.0083	.152	.0060	65 mesh
177 micron	No. 80	.177	.0070	.131	.0052	80 mesh
149 micron	No. 100	.149	.0059	.110	.0043	100 mesh
125 micron	No. 120	.125	.0049	.091	.0036	115 mesh
105 micron	No. 140	.105	.0041	.076	.0030	150 mesh
88 micron	No. 170	.088	.0035	.064	.0025	170 mesh
74 micron	No. 200	.074	.0029	.053	.0021	200 mesh
63 micron	No. 230	.063	.0025	.044	.0017	250 mesh
53 micron	No. 270	.053	.0021	.037	.0015	270 mesh
44 micron	No. 325	.044	.0017	.030	.0012	325 mesh
37 micron	No. 400	.037	.0015	.025	.0010	400 mesh

Adapted from Robert H. Perry and Cecil H. Chilton (eds.), "Chemical Engineers' Handbook," 5th ed., McGraw-Hill Book Company, New York, 1973.

* The "micron" is also known as the "micrometer," abbreviated µm.

A *cascade impactor* is also a specialized device which is useful where it can be calibrated for a particular type of material. In this equipment, the sample is suspended in a jet of air or liquid which is then made to impinge on a plate. The larger particles are retained by the plate, while the finer ones stay with the airflow. Moistening the impinging plate can assist the particles in sticking to the plate. A cascade impactor and typical plate pattern are illustrated in Fig. 9-6. Cascade impactors can be used at different air velocities to provide a stagewise separation, if desired.

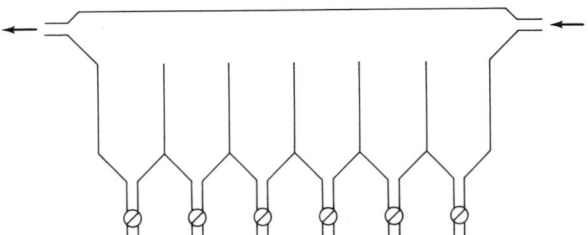

Fig. 9-2 Hydraulic classifier. As the liquid containing the suspended solids moves slowly through the classifier, the larger particles settle out fastest and are collected in the first compartment. The finest particles settle out most slowly and are caught in the last compartment. There can be some overlap between adjacent compartments, and some fines leaving with the effluent, depending on the nature of the solids and the hydraulic-flow conditions.

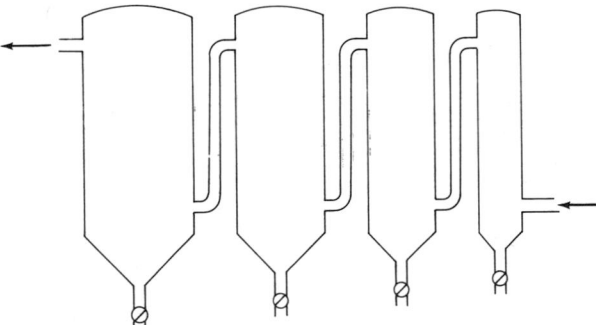

Fig. 9-3 Air elutriation. The air velocity is greatest in the first chamber (having the smallest diameter) and the largest particles settle in this chamber. Sequentially, the air velocity decreases as the chamber diameter increases and the particle size settling in each chamber decreases. The effluent would contain only particles too small to settle in the last chamber.

The results of particle-size-distribution analyses are presented as a fractional distribution or a cumulative distribution. In a *fractional distribution,* the percent of the total sample that is retained on each mesh size is plotted against the mesh size as illustrated in Fig. 9-7. In this figure, powder A has a narrower distribution of particle sizes than does powder B, although both have about the same average particle size. In a *cumulative distribution,* all the particles that are above or below a particular particle size are plotted against the particle size, as shown in Fig. 9-8. In specifying particle sizes, the particle-size distribution can be used, or if such detail is not warranted, the particle sizes can be designated by the screens which pass and retain the particle. For example, a particle size designated as $-40 + 60$ is used to indicate particles which pass through a 40-mesh screen but are retained by a 60-mesh screen.

In some instances, the particle-size distribution may be measured by relating it to the exposed surface area. The total surface area of a particulate material can be determined by gas adsorption in which the amount of nitrogen necessary to cover the surface of the particles with a monomolecular layer of nitrogen is measured, and through the dimensions of the nitrogen molecule related back to the actual surface area. In an analogous procedure, an amount of dye adsorbed by the surface under particular conditions can also be used as a measure of the surface area. Usually, the relationship between surface area and particle-size distribution is empirically determined for any particular material. Table 9-2 gives the cumulative particle-size distribution for a number of particulate materials.

Example 9-1   For a typical filter sand, what percent of the material has a particle size larger than 1 mm? What percent of the material has a particle size smaller than 50 μm?
solution   From Table 9-2 for filter sand, no particles are larger than 1,500 μm, or 1.5 mm, while 10 percent are larger than 770 μm, or 0.770 mm. Interpolating between 0.770 mm and 1.5 mm indicates that about 7 percent of the particles are larger than 1 mm. Similarly, since 90 percent of the particles are larger than 220 μm, that is, 10 percent of the particles are smaller than 220 μm, and 0 percent are smaller than 30 μm, interpolation indicates that about 1 percent of the particles are smaller than 50 μm.

Example 9-2   What would a sieve analysis be for a typical filter sand using 20, 40, 60, 80, and 100 alternate standard screens?
solution   From Table 9-1, the sieve openings (particle sizes) for the screens used are:

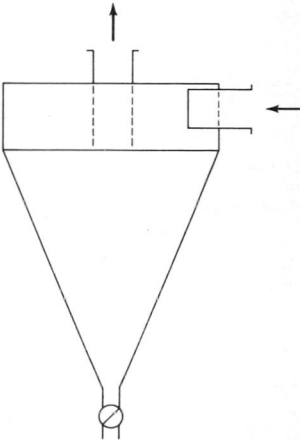

Fig. 9-4 Centrifugal classifier. The fluid (usually air) containing the suspended material (solid or liquid) enters the upper cylindrical section tangentially, imparting a spinning motion to the fluid. The particulate material moves to the periphery by centrifugal action and moves down the conical walls aided by the vortex action of the fluid in the conical section. The particulate material can then be blown out the bottom while the clean fluid passes out through the centrally located exit on top.

Mesh No. (Alternate)	Sieve opening (mm)	Sieve Opening (μm)
20	0.841	841
40	0.420	420
60	0.250	250
80	0.177	177
100	0.149	149

From Table 9-2, the percent particles larger than the above sizes are:

Mesh No.	% Larger particles
20	8
40	56
60	84
80	93
100	96

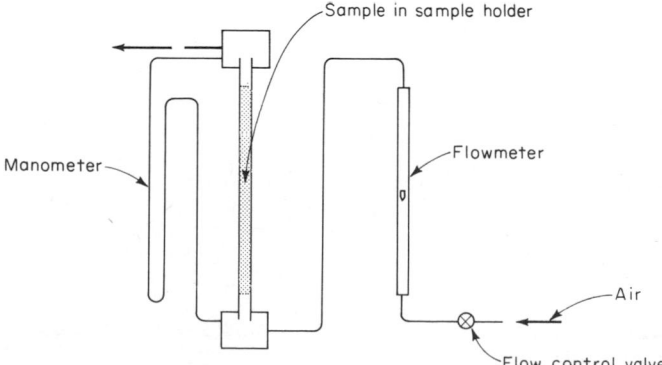

Fig. 9-5 Air-permeability device for measuring particle size. The pressure drop, read on the manometer, is related to the particle size of the sample held between two filters in the sample tube. The measurement is empirical and depends on calibrating the device with samples of known particle sizes and at carefully measured airflows.

While 56 percent of the particles are larger than the openings in a No. 40 screen, 8 percent are also larger than the openings in a No. 20 screen and would have been retained on this screen. Therefore, the percent retained on a No. 40 screen would be 56 − 8 = 48 percent. Similarly, that retained on a No. 60 screen would be the fraction larger than the openings in a No. 60 screen, less that already retained on the Nos. 20 and 40 screens, i.e., 84 − 56 = 28 percent. Thus a sieve analysis for this material would be:

Particle size	Mesh No.	% Retained
+20	20	8
−20 + 40	40	48
−40 + 60	60	28
−60 + 80	80	9
−80 + 100	100	3
−100		4
		100

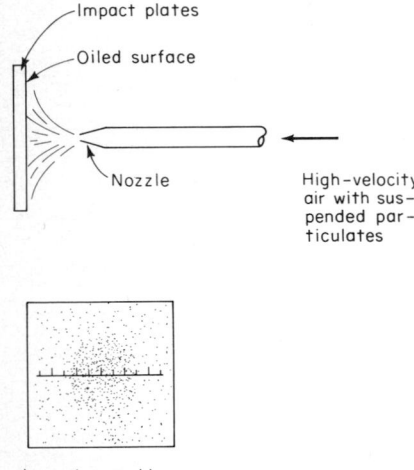

Impactor pattern

Fig. 9-6 Cascade impactor and pattern.

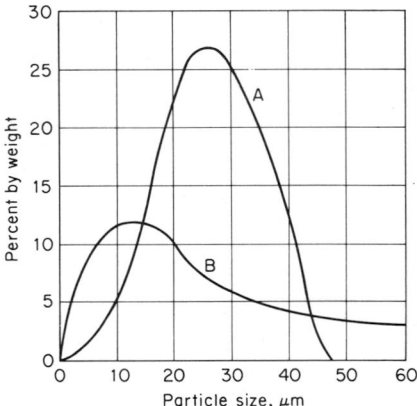

Fig. 9-7 Fractional size distribution. *A* is narrow size distribution; *B* is broad size distribution.

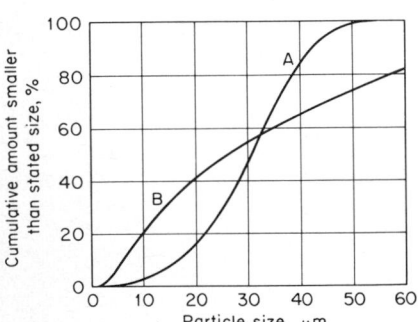

Fig. 9-8 Cumulative particle size distribution. *A* is narrow size distribution; *B* is broad size distribution.

Equipment

The equipment used for crushing and grinding falls into about seven general types:

1. Ball
2. Jaw
3. Cone
4. Pan
5. Roll
6. Hammer
7. Pug

A *ball mill* utilizes a rotating chamber which is loaded with the material to be ground, together with porcelain or steel balls. The rotation of the container causes the balls to roll around and impinge on each other, grinding the material between the balls. A ball mill is commonly used in the laboratory because of its convenience, but it can achieve large dimensions for industrial purposes, as illustrated in Fig. 9-9.

A *jaw crusher,* illustrated in Fig. 9-10, is generally used to break up large particles of hard

TABLE 9-2 Particle-Size Distribution (μm) for Various Common Materials

| | Cumulative oversize (%) | | | | | | | | | | |
Material	0	10	20	30	40	50	60	70	80	90	100
Coarse aggregate	80,000	46,000	32,000	24,000	19,000	16,000	14,000	11,000	9,000	6,000	2,000
Mixed aggregate	80,000	40,000	26,000	18,000	14,000	9,000	5,300	2,800	1,000	280	50
Fine aggregate	6,500	2,800	1,700	1,000	700	480	350	260	190	100	10
Filter sand	1,500	770	580	480	440	400	330	350	290	220	30
Powdered coal	300	180	150	130	100	80	50	45	28	15	3
Portland cement	180	70	48	36	23	20	16	12	7	4	1
Mineral fillers	50	15	9	7	5	5	4	3	2	1.5	0.6
Pigments	2	1.6	1.5	1.7	1.3	1.0	0.9	0.8	0.6	0.4	0.2
Carbon black	0.2	0.09	0.07	0.065	0.05	0.055	0.05	0.044	0.04	0.03	0.01

Illustration: For filter sand, 30% of the particles are larger than 480 μm (and thus 70% of the particles are smaller than 480 μm). All particles are smaller than 1,500 μm, and all particles are larger than 30 μm.

materials such as rocks. In a jaw crusher, a hardened-steel jaw moves back and forth against a stationary surface with the material to be crushed fed in at top where the clearance is greatest, and moving to the bottom, where the clearance is minimum. While jaw crushers in large sizes are widely used in mineral processing, they are also available in small sizes suitable for laboratory operations.

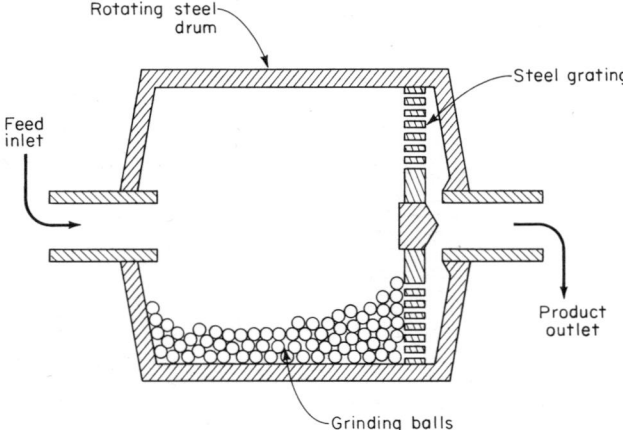

Fig. 9-9 Industrial ball mill.

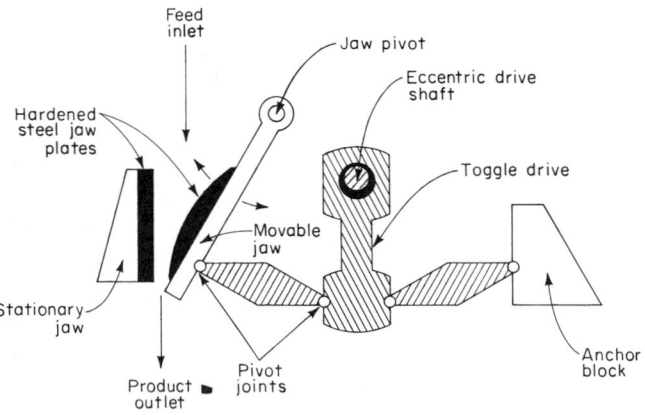

Fig. 9-10 Jaw crusher.

A *cone crusher*, also available in a variety of sizes, involves rotating a hardened conical surface in a stationary conical housing. Since the stationary conical housing has an apex angle different from the rotating conical surface (which may also be grooved), the arrangement is such as to reduce the clearance toward the bottom or discharge end, thereby crushing the particles as they move downward. An industrial-type cone crusher is illustrated in Fig. 9-11.

A *pan crusher* is used not only for crushing but also for mixing liquids and solids. This type of apparatus, illustrated in Fig. 9-12, usually is available only in large sizes. It consists of two heavy wheels, called mullers, which roll around the inside of the pan, crushing the material with which it is filled against the side and bottom of the pan. Doctor blades are suitably positioned to scrape the sides and bottom of the pan as well as the sides of the muller wheels and to distribute the mix in such a way as to present new material continuously to the muller wheels.

A *roll crusher*, illustrated in Fig. 9-13, is also a heavy-duty device. It consists of a roller with heavy bosses mounted on the surface. The roller rotates inside a hardened cylinder with a fixed clearance between the bosses and the cylinder so that any material that is trapped between a boss and the cylinder is crushed. Roll crushers are widely used in the mining industry.

Hammer mills, of the type illustrated in Fig. 9-14, are available in a wide range of sizes. They consist of hammers which are free to swing as the rotor on which they are mounted rotates at relatively high speeds. The hammers swing out toward a grating in large mills, or against a wire-screen mesh in smaller mills. The impact action of the hammers breaks up the particles against the casing, and the openings in the casing automatically size the particles, since large particles stay in the mill until they can pass through the grating or screen. Hammer mills are frequently used in the laboratory to produce very fine powders.

Pug mills are heavy-duty crushers. As illustrated in Fig. 9-15, they consist of two counterrotating cylinders from which heavy-metal bosses protrude. The bosses are located to clear each other and the op-

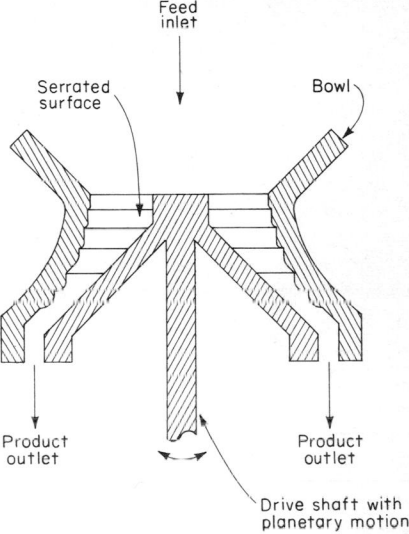

Fig. 9-11 Industrial cone crusher.

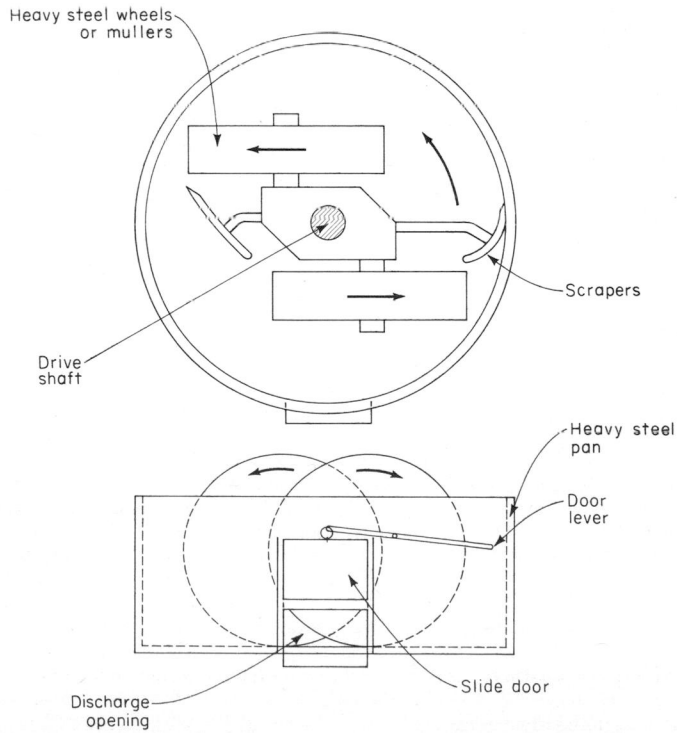

Fig. 9-12 Pan crusher or "muller." Each muller wheel rolls around its own axis as it rotates around the drive shaft.

posing row by the desired clearance, since this is what determines the particle size of the discharge.

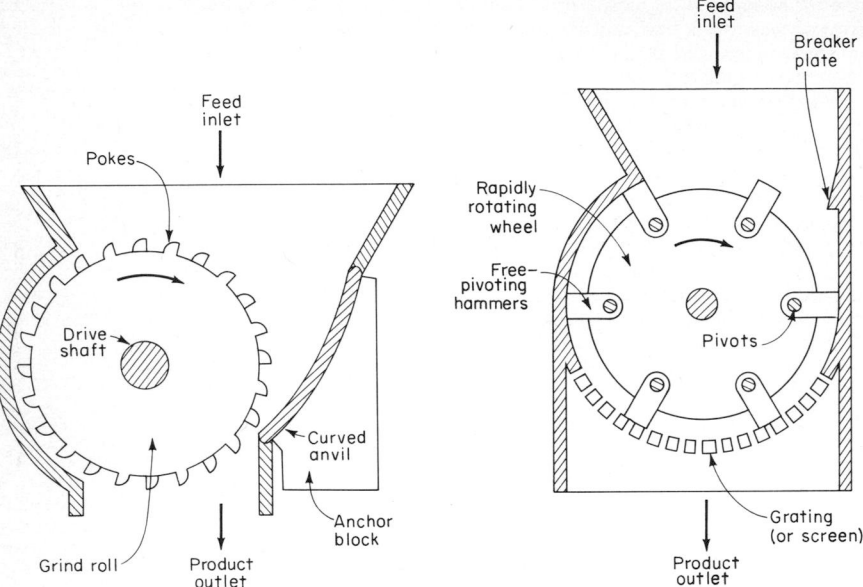

Fig. 9-13 Roll crusher.

Fig. 9-14 Hammer mill.

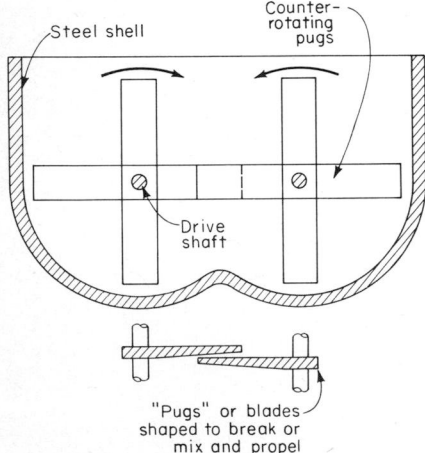

Fig. 9-15 Open-top pug mill. The feed is introduced near the back of the open top. The pugs then propel the mix forward where it is discharged through a door in the front of the shell.

9-2 MIXING AND BLENDING

In mixing and blending operations, the objective is to produce as uniform a material as possible by the physical combination of two or more ingredients. While the two terms are used more or less interchangeably, *mixing* is generally associated with the combination of materials that are substantially different from each other in composition or form, while *blending* is associated with the combination of different varieties of essentially similar materials. Mixing may be used to promote a chemical reaction, such as the polymerization of a plastic, or a physical reaction, such as the dissolution of sugar in water, or to coat one material with another, such as the candy coating of a pharmaceutical. The *homogeneity* of a mixture is the uniformity of composition achieved throughout the batch. Quantitatively, homogeneity can be determined by chemical or physical analyses of several small samples taken from various points in the mixture. In the case of a solids-liquids mixture, the degree of uniformity is referred to as the *dispersion,* a good dispersion being one in which equal volumes of the mixture contain equal numbers of individual solid particles (as contrasted to agglomerated solids).

In the case of solid mixtures, sampling devices called thiefs and rifflers are used. A *thief,* such as that illustrated in Fig. 9-16, generally consists of a tube with openings along its side and a removable rod which fits snugly into the tube. In use, the thief is plunged into the mixture and the rod is carefully removed, allowing the mixture to enter the various segments of the tube. The tube can then be removed and the various segments examined individually. A *riffler* is a device which separates a given batch into several smaller amounts, thus permitting a single analysis to be representative of the entire batch. A riffler is illustrated in Fig. 9-17. Devices such as thiefs and rifflers are useful in obtaining small samples without disturbing the homogeneity.

In mixing solids, the following component properties are of great importance:

1. Particle-size distribution
2. Bulk density
3. True density
4. Particle shape
5. Surface characteristics
6. Flow characteristics
7. Friability
8. State of agglomeration
9. Moisture (or liquid) content

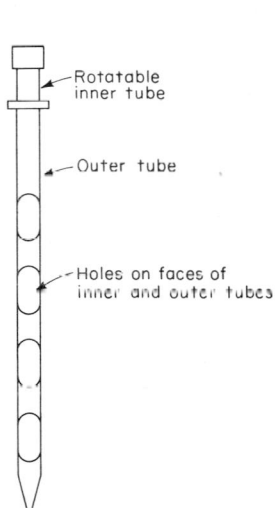

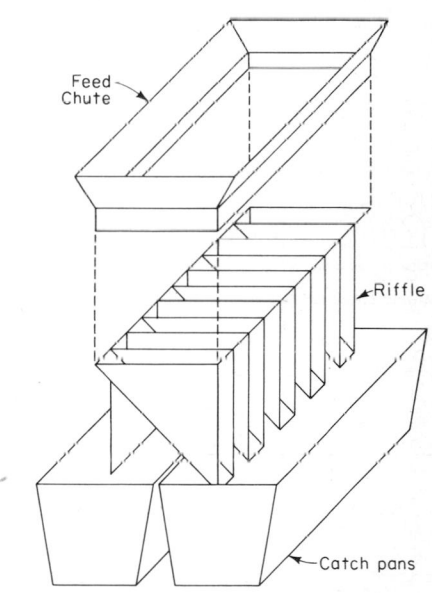

Fig. 9-16 Thief probe sampler. With the holes on opposite faces of the inner and outer tubes, the sampler is plunged into the material being sampled. The inner tube is then rotated so that the holes match up, allowing material to enter the tube at several points. The inner tube is then rotated to shut the holes, and the sampler can be withdrawn.

Fig. 9-17 Riffle for reducing size of particulate samples.

Generally speaking, those materials whose above properties are similar are most easily mixed; conversely, where they differ greatly, a mixture can be made only with great difficulty. In the case of a mixture of two liquids, or a solid and a liquid, two additional properties are important:

10. Viscosity of the liquid(s)
11. Ease of wetting

Equipment

Since mixing and blending are widely used operations, a large variety of mixers and blenders are commercially available. Figure 9-18 illustrates equipment used for mixing solids and liquids, and Fig. 9-19 illustrates equipment used for liquids. Frequently, these basic types are modified to handle special circumstances. For example, double-cone and twin-shell blenders are frequently equipped with means for adding liquids to the solids without interrupting the blending. With suitable modifications, practically all the solids-solids mixing machines can be used to blend a solid with a liquid.

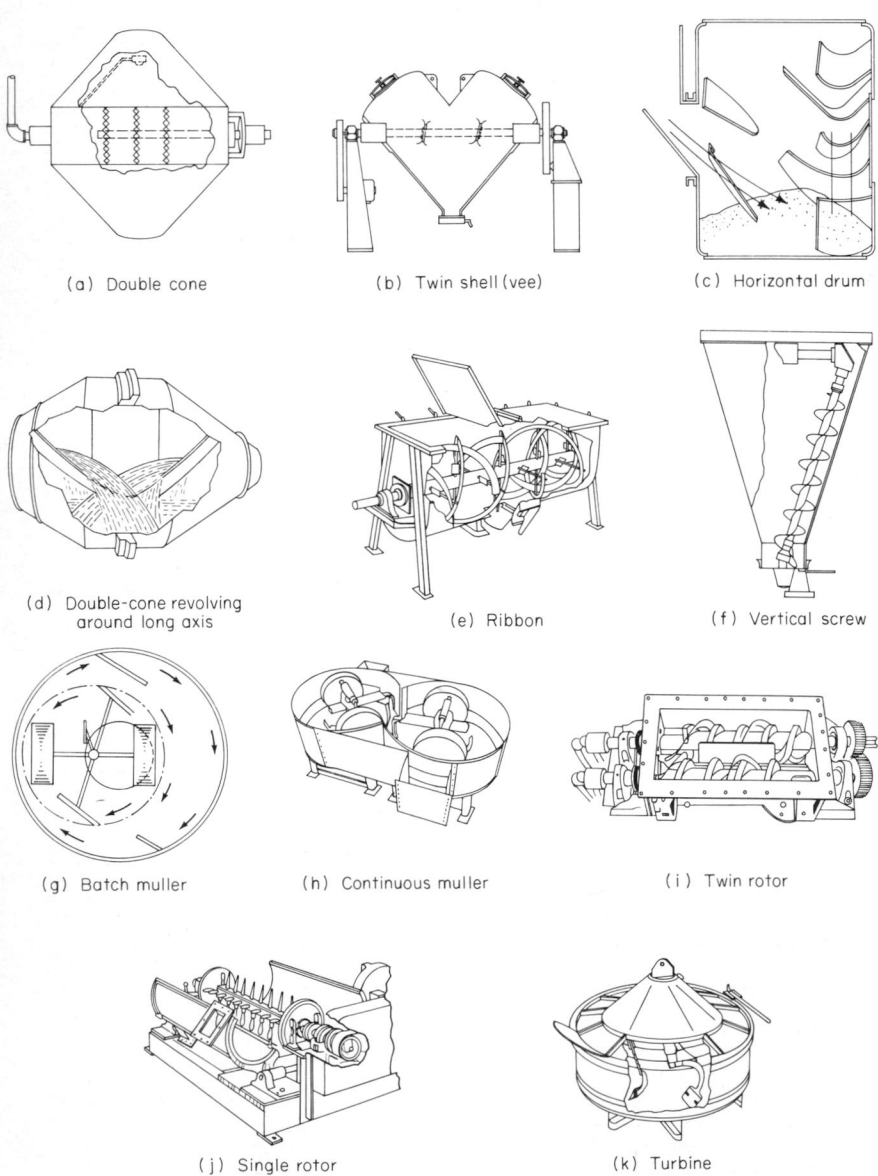

(a) Double cone

(b) Twin shell (vee)

(c) Horizontal drum

(d) Double-cone revolving around long axis

(e) Ribbon

(f) Vertical screw

(g) Batch muller

(h) Continuous muller

(i) Twin rotor

(j) Single rotor

(k) Turbine

Fig. 9-18 Equipment for solids-solids and solids-liquids mixing.

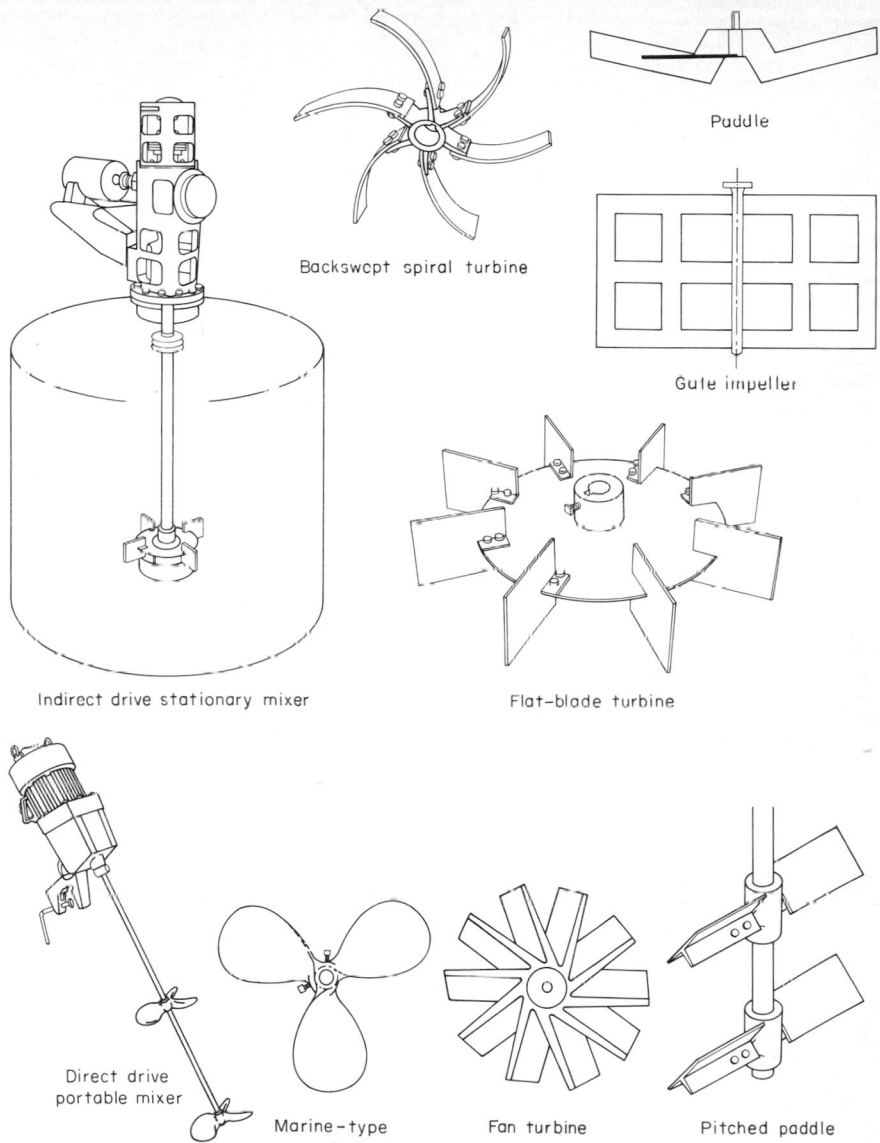

Paddle

Backswept spiral turbine

Gate impeller

Indirect drive stationary mixer

Flat-blade turbine

Direct drive
portable mixer

Marine-type
propeller

Fan turbine

Pitched paddle

Fig. 9-19 Mixing equipment for liquids.

9-3 FILTRATION

The separation of particulate matter from a fluid by passage of the fluid through a membrane that retains the solids is called *filtration*. The original mixture is called a *slurry*, the fluid that passes through the membrane is the *filtrate*, and the retained solid is the *filter cake*. Filters are classified according to the filtering membrane or *filtering medium* used:

1. Loose or granular materials
2. Felted or woven fabrics
3. Rigid porous materials
4. Semipermeable membranes

A filter using a loose or granular filtration medium can be as simple as a box with a perforated bottom covered with successive layers of coarse gravel, fine pebbles, and sand. Such a filter (which is also referred to as a sand filter) is frequently an effective filtration means for use in fairly large installations where a high degree of clarity of the filtrate is not necessary. Water-purification systems, for example, frequently rely on sand filters, although these may be modified by using such materials as charcoal to help absorb some of the organic matter that may be present in the slurry.

The most widely used type of filtration is that using a felted or woven membrane. This type of filtration can range from a very simple laboratory filtration, in which a filter paper is held in a funnel and the slurry is allowed to trickle through, to large specialized equipment, to be described later. The felted or woven membrane is used to retain the solids, the membrane being selected from a wide variety of woven and nonwoven materials on the basis of chemical and physical stability and the size and shape of the spaces between the threads. Woven cotton fabrics are most extensively used. Table 9-3 lists representative cotton filter fabrics in three conventional weaves. In addition to cotton, filter fabrics are made of glass fibers, silk, wool, and many synthetic fibers, including nylon, Dacron, Dynel, Saran, and polypropylene. Usually, the synthetic fibers are used where the slurry would attack the natural fibers. Metal fabrics are also used, particularly where physical strength is required, such as in high-pressure or high-temperature filtrations.

TABLE 9-3   Physical Characteristics of Woven Cotton Filter Cloths

Weave style	Weight (oz/yd²)	Count	Ply	Uses
Duck	11.5	50 × 34	2 × 2	Tight weave, light filtrations
	14.7	50 × 30	3 × 3	Light filtrations
	18.0	45 × 28	3 × 4	Medium weight
	21.3	36 × 26	3 × 3	Heavy weight
	24.5	31 × 24	4 × 4	Fine filtrations
Twill	15.5	38 × 28	4 × 4	Vacuum filters, light service
	15.5	66 × 44	2 × 2	Vacuum filters
	17.5	36 × 25	3 × 3	General use; leaf filters
	18.0	67 × 36	2 × 4	General use; leaf filters
	22.0	34 × 24	4 × 4	High-pressure filtrations
	20.0	58 × 42	3 × 4	High-pressure, fine filtrations
Chain	12.0	56 × 50	2 × 2	Vacuum filters
	15.0	68 × 42	2 × 4	Low-pressure filtrations
	18.0	57 × 37	3 × 3	General use
	22.0	34 × 30	3 × 4	Filter-press use
	20.0	67 × 38	3 × 5	Filter pressure

Count: number of threads per inch in each direction.
Ply: number of small threads twisted together to make final thread.

Porous carbon, graphite, alundum, or silica are also used as filtration media, as well as certain porous plastic materials, including polypropylene and Teflon. Porous media of this type are capable of very fine filtrations (i.e., capable of removing very fine particles) and are usually used as a final filtration to impart a high clarity to the filtrate. Other porous media include porous metals, particularly nickel and stainless steel, which are used in hydraulic systems to maintain the hydraulic fluid free of particulate contaminants. Semipermeable filtration membranes are used where extremely fine particles have to be removed from the slurry, or even those in colloidal solution. Semipermeable filtration membranes are made of collodion, a uniform film of regenerated cellulose. Filtration through semipermeable membranes has found considerable application in the purification of biological materials.

Frequently, it is possible to increase the filtration rate by adding a noncompressible, particulate material to the slurry. Such materials, known as *filter aids*, act by forming a filter bed with relatively large pores but with tortuous paths which entrap the smaller particles. Materials such as diatomaceous earth, in various particle sizes, are frequently used for this purpose, as are sawdust, magnesia, gypsum, carbon, and a variety of other materials. It is also possible, where it is desired to decolorize or deodorize a liquid, to use an activated carbon or activated clay as a filter aid. Such materials are used in sugar decolorization as well as in deodorizing oils, fats, and waxes.

Actually, the filtration membrane usually does not do the filtering at all but rather acts as a support for the buildup of a *filter cake* consisting of particulate matter which has been removed from the slurry. This cake forms a porous bed which serves to trap additional particles as the fluid passes through the cake, as illustrated in Fig. 9-20. Fundamentally, there are two systems for filtration, constant-pressure and constant-volume. In constant-pressure filtration, a given pressure differential is maintained across the filter cake by means of a pressure regulator. The flow rate through the filter will thus decrease as the filter cake is built up and its resistance to hydraulic flow increases. In constant-volume filtration, the slurry is fed by a constant-displacement (piston) pump. The flow will thus be uniform, but the pressure differential across the filter cake will increase as the cake builds up. Equation (9-1) has been developed to describe constant-pressure filtration:

$$\frac{P\theta}{V/A} = m\frac{V}{A} + b \tag{9-1}$$

in which P = filtration pressure (lb/in^2), θ = total time of operation (min), V = total weight of filtrate passing through the filter up to time θ, A = area of filtering surface (in^2), m = a constant depending on the physical conditions of the filtration and the properties of the filter cake, and b = a constant depending largely on the flow conditions through the filter cake.

For constant-rate filtration, Eq. (9-2) is used:

$$\frac{P\theta}{V/A} = 2m\frac{V}{A} \tag{9-2}$$

in which the symbols have the same meaning as in Eq. (9-1). As can be seen, Eq. (9-1) is in the form $y = mx + b$, a plot of which is a straight line of slope m and intercept b. This enables us to correlate and describe filtration operations. If $P\theta/(V/A)$ is plotted against V/A, we should get a straight line whose slope and intercept are easily determined. Thus, by determining the slopes and intercepts, it has been possible to correlate the filtration characteristics of a number of insoluble materials, as shown in Table 9-4.

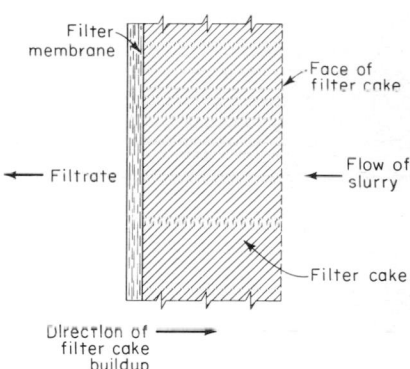

Fig. 9-20 Mechanics of filter-cake formation.

Example 9-3 The following data were recorded for a constant-pressure filtration of Fe(OH)$_3$:
Total filtration area 720 in^2
Filtration pressure 40 lb/in^2

Time θ (min)	Weight of filtrate V (lb)
2.6	14.4
8.8	28.8
18.6	43.2
31.6	57.6

Calculate the filtration constants m and b for Eq. (9-1).
solution V/A is calculated by dividing the above values of V by $A = 720$ in^2. $P\theta/(V/A)$ is calculated by multiplying the above values of θ by $P = 40$ lb/in^2 and then dividing by the calculated value of V/A:

V/A	$P\theta/(V/A)$
0.02	5,200
0.04	8,800
0.06	12,400
0.08	15,800

Plotting $P\theta/(V/A)$ against V/A yields a straight line whose slope, 177,000, is m, and whose intercept, 1,700, is b.

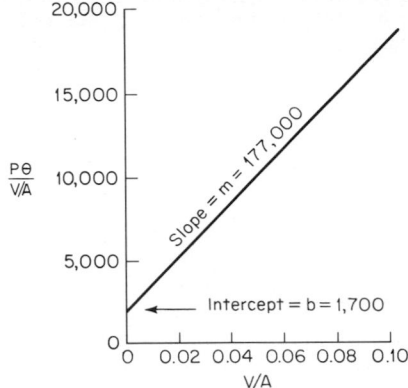

Example 9-4 The filtration described in Example 9-3 is altered so that the filtration rate is kept constant at 3 lb filtrate/min. How long will it take for the inlet pressure on the filter to reach 40 lb/in²?
solution For constant-rate filtration, Eq. (9-2) is used:

$$\frac{P\theta}{V/A} = 2m\frac{V}{A}$$

From Example 9-3, $m = 177,000$.

$$\frac{V}{A} = \frac{3\theta}{720} = 0.0042\theta$$

Note that each quantity must be expressed in the units for which m was developed.

$$\frac{40\theta}{0.0042\theta} = 2(177,000)(0.0042\theta)$$

$$9,524 = 1,487\theta$$
$$\theta = 6.4 \text{ min}$$

Example 9-5 How much filtrate will be obtained after 15 min of filtration of 2 percent slurry of $BaCO_3$ at a constant pressure of 40 lb/in² in equipment having an effective filtration area of 10 ft²?
solution From Table 9-4, the filtration constants for $BaCO_3$ under the stipulated conditions are $m = 6,300$ and $b = 60$. From Eq. (9-1),

$$\frac{40(15)}{V/A} = 6,300\frac{V}{A} + 60$$

$$600 = 6,300\left(\frac{V}{A}\right)^2 + 60\frac{V}{A}$$

This equation is solvable for V/A, but the solution is tedious. A more convenient solution is by trial and retrial: Assume $V/A = 0.30$:

$$6,300(0.30)^2 + 60(0.30) = 567 + 18 = 585$$

The assumed value of V/A is thus too small. Try $V/A = 0.35$:

$$6,300(0.35)^2 + 60(.035) = 771 + 21 = 792$$

Retrial, $V/A = 0.31$:

$$6,300(0.31)^2 + 60(0.31) = 605 + 19 = 624$$

Thus $V/A = 0.30$ is closer to a solution for the equation than is $V/A = 0.31$, and since this is sufficient precision, $V/A = 0.30$ will be used. Since A must be expressed in square inches, $A = 10 \times 144 = 1440$ in². Thus $V = 0.30 \times 1,440 = 432$ lb of filtrate will be obtained.

Note: Because of the diversity of filtration conditions, slurry concentrations, solids compositions, etc., the values cited in Table 9-4 are not intended for design purposes, but rather are used to estimate filtration times and quantities. Where a specific material, slurry

TABLE 9-4   Filtration Characteristics of Insoluble Materials

Material	Slurry concentration (lb solids/lb filtrate)	P (lb/in^2)	m	b
Kieselguhr	0.0123	30	1,000	20
	0.0123	15	1,550	90
	0.0265	15	3,200	25
CaCO$_3$	0.0411	15	2,900	140
	0.0021	10	170	60
	0.0021	20	120	50
	0.0121	50	168	65
BaCO$_3$	0.0219	40	6,300	60
70-mesh marble	0.0048	5	120	70
	0.0048	20	120	70
100-mesh marble	0.0045	25	105	250
MgCO$_3$	0.0022	30	2,500	360
PbCrO$_4$	0.0148	20	390	35
Al(OH)$_3$	0.0046	15	17,300	350
	0.0037	10	21,500	300
	0.0040	20	19,000	200
	0.0040	40	34,000	−300
	0.0009	90	13,600	2,600
	0.0012	60	11,700	3,900
	0.0060	20	33,000	−200
	0.0060	80	102,000	2,600
	0.0005	80	17,000	400
	0.023	80	200,000	500
	0.023	40	128,000	820
Fe(OH)$_3$	0.0028	10	42,000	300
	0.0027	40	110,000	−3,000
	0.0037	10	13,000	−90
	0.0189	40	420,000	200
	0.0173	60	485,000	1,300
	0.0176	20	216,000	500
	0.0028	20	76,000	200
ZnC$_2$O$_4$	0.0400	50	311,000	500
Cr(OH)$_3$	0.0250	5	26,000	400
	0.0250	10	40,000	600
	0.0250	25	96,000	1,000
	0.0250	50	150,000	1,300
	0.0250	75	210,000	1,700
	0.0250	100	240,000	2,000

From William H. Walker, Warren K. Lewis, William H. MacAdams, and Edwin R. Gilliland, "Principles of Chemical Engineering," 3d ed., McGraw-Hill Book Company, New York, 1937.

concentration, or pressure is not cited in the table, a gross estimate can still be made using values for a generally similar material.

Because of the diversity of commercial filtrations, a variety of filtration equipment has been developed. Units generally used for removing solids from a liquid are illustrated in Fig. 9-21, and Fig. 9-22 shows the principal types of equipment used to filter solids from a gas.

It should be noted that in the case of the removal of a solid from a gas, methods other than filtration are available. These include:

1. Gravity settling
2. Impingement
3. Cyclone separation
4. Electrostatic precipitation
5. Water-film washing

9-4   ABSORPTION AND EXTRACTION

Absorption and extraction are processes in which a material is transferred from one phase to another across a phase boundary. Actually, absorption and extraction are really one and the same process, but *absorption* usually refers to the transfer of a gas to a liquid or solid, or the transfer of a liquid to a solid, while *extraction* refers to the transfer of a gas from a liquid or solid to a gas phase, or the removal of one or more components from a liquid or solid. The difference between absorption and extraction is therefore the intent rather than the nature of the process, and consequently they can be treated as one. There are several classifications of absorption and extraction operations:

1. Treatment of gases by liquids
2. Treatment of gases by solids
3. Treatment of liquids by gases
4. Treatment of liquids by liquids
5. Treatment of liquids by solids
6. Treatment of solids by gases
7. Treatment of solids by liquids

By common usage, the treatment of gases by liquids is also referred to as gas *washing* or *scrubbing,* the treatment of liquids by gases is also referred to as *stripping* or *denuding,* and the treatment of solids by liquids is also referred to as *lixivation* or *leaching.*

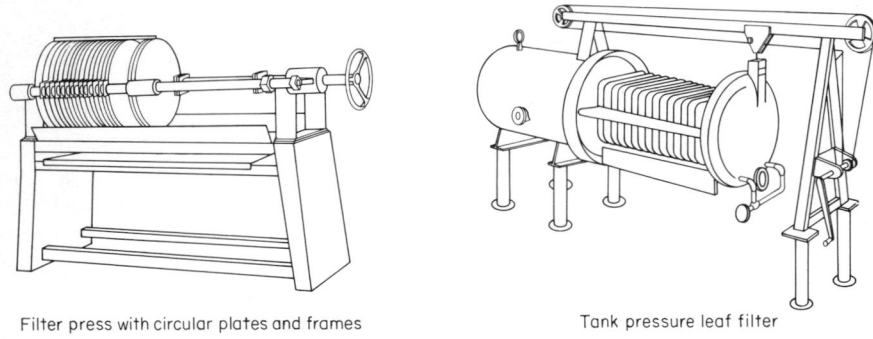

Filter press with circular plates and frames   Tank pressure leaf filter

Vacuum segment for washing

Vacuum segment for drying

Vacuum segment for dewatering

Filter fabric

Scraper blade

Rotating segmented cylinder

Stationary segmented manifold

Filter cake collector

Vacuum segment and filtrate collector

Slurry tank

Continuous drum filter

Fig. 9-21 Equipment for filtering liquids.

The engineering literature on absorption and extraction tends to be very complicated, since it is usually quite difficult to describe the chemical, physical, and flow conditions, all of which have a significant effect on the rate at which the desired material transfer occurs. However, two factors have a preponderant effect, the equilibrium conditions and the rate at which the transferring material diffuses through the system. When two (or more) phases are brought into physical contact, a transfer of materials can occur between the phases until an equilibrium between the phases is established. For example, alcohol exposed to air in a closed vessel will evaporate until the rate of evaporation is equal to the rate at which alcohol vapor from the air reenters the liquid alcohol. At a given temperature this occurs at a particular partial pressure, or *equilibrium pressure,* of alcohol. When the equilibrium pressure is reached, there is no net transfer of alcohol from the liquid to the gas phase or from the gas to the liquid phase.

In the case of a material dissolving from a solid into a liquid, solute continues to dissolve (i.e., material is transferred to the liquid phase), until equilibrium is reached at the limit of solubility of the material. Thus the total amount of material which can be transferred from one phase to another is established by the equilibrium conditions. Table 9-5 gives the solubilities of a large number of common gases in water, the absorption of these gases in water being commercially important operations. Table 9-5 gives the solubilities of a number of gases in terms of the Henry's law constant H defined by the equation

$$H = \frac{p_A}{x_A} \qquad (9\text{-}3)$$

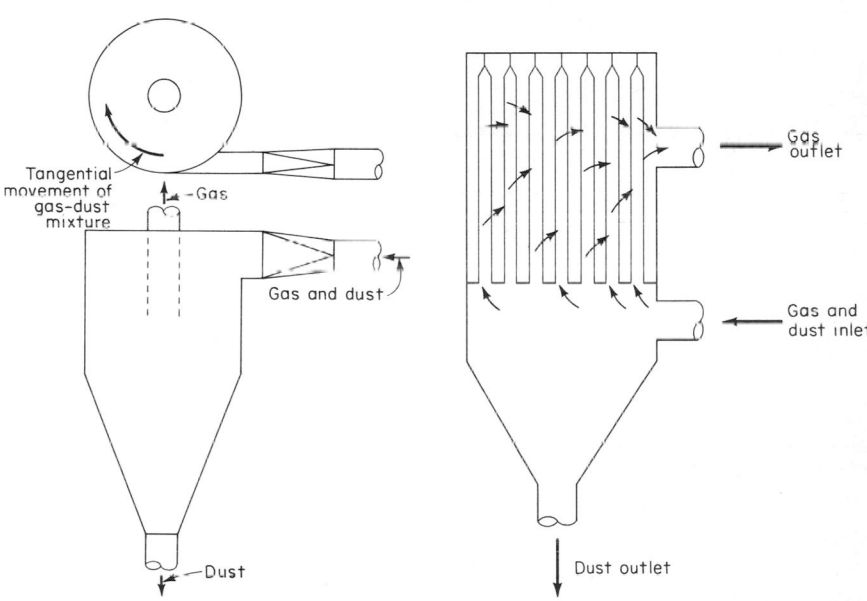

Fig. 9-22 Equipment for filtering gases.

where p_A = partial pressure of A in the gas space above the liquid (atm) and x_A is the mol fraction of A dissolved in the liquid. If a chemical reaction occurs between a gas and its absorbent liquid, such as the absorption of carbon dioxide by sodium hydroxide solution, the equilibrium is greatly displaced in favor of the absorption. Where such a chemical reaction is involved, it is possible to absorb large amounts of gas in relatively small amounts of absorbent, but in this case, it is usually difficult to reverse the process to recover the gas.

Example 9-6   How much oxygen can be dissolved in 100 g of water at 20°C when the total pressure of the gas phase is 760 mmHg and the partial pressure of O_2 is 360 mmHg?
solution   From Table 9-5, $H = 4.01 \times 10^4$ for oxygen at 20°C. (Note that the value given in the table is $H \times 10^{-4}$ in order to have a convenient number of decimal places. Therefore, H = value given in the table $\times 10^4$.)

$$P_{O_2} = \frac{360}{760} = 0.474 \text{ atm}$$

$$x_{O_2} = \frac{0.474}{4.01 \times 10^4} = 0.118 \times 10^{-4} = 11.8 \times 10^{-6}$$

$$100 \text{ g H}_2\text{O} = 5.56 \text{ mols}$$

Let y = number of mols of O_2 dissolved:

$$\frac{y}{5.56 + y} = 11.8 \times 10^{-6}$$

$$y = 65.6 \times 10^{-6} + 11.8 \times 10^{-6}y$$

TABLE 9-5 Solubilities of Gases in Water

ACETYLENE

t, °C	0	5	10	15	20	25	30
$10^{-3} \times H$	0.72	0.84	0.96	1.08	1.21	1.33	1.46

"International Critical Tables," vol. 3, p. 260, McGraw-Hill, 1928.

AIR

t, °C	0	5	10	15	20	25	30	35
$10^{-4} \times H^*$	4.32	4.88	5.49	6.07	6.64	7.20	7.71	8.23

t, °C	40	45	50	60	70	80	90	100
$10^{-4} \times H^*$	8.70	9.11	9.46	10.1	10.5	10.7	10.8	10.7

"International Critical Tables," vol. 3, p. 257.

$^*H$ is calculated from the absorption coefficients of O_2 and N_2, taking into consideration the correction for constant argon content.

BROMINE (Br_2)

t, °C	0	5	10	15	20	25
$10^{-2} \times H$	0.213	0.275	0.366	0.466	0.593	0.737

t, °C	30	40	50	60	70	80
$10^{-2} \times H$	0.905	1.33	1.91	2.51	3.21	4.04

CARBON DIOXIDE (CO_2)

Weight of CO_2 per 100 weights of H_2O

Total pressure, atm.	12°C	18°C	25°C	31.04°C	35°C	40°C	50°C	75°C	100°C
25	7.03	3.86	....	2.80	2.56	2.30	1.92	1.35	1.06
50	7.18	6.33	5.38	4.77	4.39	4.02	3.41	2.49	2.01
75	7.27	6.69	6.17	5.80	5.51	5.10	4.45	3.37	2.82
100	7.59	6.72	6.28	5.97	5.76	5.50	5.07	4.07	3.49
150	....	7.07	....	6.25	6.03	5.81	5.47	4.86	....
200	....	....	....	6.48	6.29	6.28	5.76	5.27	....
300	7.86	7.35	7.54	7.27	7.06	6.89	6.20	5.83	5.08
400	8.12	7.77	....	7.65	7.51	7.26	6.58	6.30	5.84
500	....	....	....	....	....	....	7.58	7.43	6.40
700	....	....	....	....	....	....	....	....	7.61

CARBON MONOXIDE (CO)

Partial pressure of CO, mm. Hg	$10^{-4} \times H$	
	17.7°C	19.0°C
900	4.77	4.88
2000	4.77	4.91
3000	4.77	4.93
4000	4.78	4.95
5000	4.80	4.97
6000	4.82	4.98
7000	4.86	5.02
8000	4.88	5.08

"International Critical Tables," vol. 3, p. 260.

AMMONIA (NH_3)

Partial pressure of NH_3, mm. Hg

Weight NH_3 per 100 weights H_2O	0°C	10°C	20°C	25°C	30°C	40°C	50°C	60°C
100	947							
90	785							
80	636	987	1450					
70	500	780	1170			3300		
60	380	600	945			2760		
50	275	439	686			2130		
40	190	301	470		719	1520		
30	119	190	298		454	1065		
25	89.5	144	227		352	692	825	834
20	64	103.5	166		260	534	596	583
15	42.7	70.1	114		179	395	405	361
10	25.1	41.8	69.6		110	273	247	165
7.5	17.7	29.9	50.0		79.7	167	179	129.2
5	11.2	19.1	31.7		51.0	120	115	94.3
4		16.1	24.9		40.1	76.5	91.1	77.0
3		11.3	18.2	23.5	29.6	60.8	67.1	61.0
2.5			15.0		19.4	45	(55.7)	48.7
2			12.0	15.3	15.3	(37.6)*	(44.5)	36.3
1.6				12.0	11.5	(30.0)	(35.5)	30.2
1.2				9.1		(24.1)	(26.7)	
1.0				7.4		(18.3)	(22.2)	
0.5				3.4		(15.4)		

* Extrapolated values.

"International Critical Tables," vol. 3, p. 260.

CHLORINE (Cl₂)

Partial pressure of Cl₂ mm. Hg	Solubility, g. of Cl_2 per liter					
	0°C.	10°C.	20°C.	30°C.	40°C.	50°C.
5	0.488	0.451	0.438	0.424	0.412	0.398
10	.679	.603	.575	.553	.532	.512
30	1.221	1.024	.937	.873	.821	.781
50	1.717	1.354	1.210	1.106	1.025	.962
100	2.79	2.08	1.773	1.573	1.424	1.313
150	3.81	2.73	2.27	1.966	1.754	1.599
200	4.78	3.35	2.74	2.34	2.05	1.856
250	5.71	3.95	3.19	2.69	2.34	2.09
300	.....	4.54	3.63	3.03	2.61	2.31
350	.....	5.13	4.06	3.35	2.86	2.53
400	.....	5.71	4.48	3.69	3.11	2.74
450	.....	6.26	4.88	3.98	3.36	2.94
500	.....	6.85	5.29	4.30	3.61	3.14
550	.....	7.39	5.71	4.60	3.84	3.33
600	.....	7.97	6.12	4.91	4.08	3.52
650	.....	8.52	6.52	5.21	4.32	3.71
700	.....	9.09	6.90	5.50	4.54	3.89
750	.....	9.65	7.29	5.80	4.77	4.07
800	.....	10.21	7.69	6.08	4.99	4.27
900	.....	.....	8.46	6.68	5.44	4.62
1000	$Cl_2 \cdot 8H_2O$ separates	.....	9.27	7.27	5.89	4.97
1200		.....	10.84	8.42	6.81	5.67
1500		.....	13.23	10.14	8.05	6.70
2000		.....	17.07	13.02	10.22	8.38
2500		.....	21.0	15.84	12.32	10.03
3000	.....	.....	.....	18.73	14.47	11.70
3500	.....	.....	.....	21.7	16.62	13.38
4000	.....	.....	.....	24.7	18.84	15.04
4500	.....	.....	.....	27.7	20.7	16.75
5000	.....	.....	.....	30.8	23.3	18.46

CHLORINE (Cl₂) (continued)

Partial pressure of Cl₂ mm. Hg	Solubility, g. of Cl_2 per liter					
	60°C.	70°C.	80°C.	90°C.	100°C.	110°C.
5	0.383	0.369	0.351	0.339	0.326	0.316
10	.452	.470	.447	.431	.415	.402
30	.743	.704	.671	.642	.627	.598
50	.912	.863	.815	.781	.747	.722
100	1.226	1.149	1.085	1.034	.987	.950
150	1.482	1.382	1.294	1.227	1.174	1.137
200	1.766	1.580	1.479	1.396	1.333	1.276
250	1.914	1.764	1.642	1.553	1.480	1.413
300	2.10	1.932	1.793	1.700	1.610	1.542
350	2.28	2.10	1.940	1.831	1.736	1.661
400	2.47	2.25	2.08	1.965	1.854	1.773
450	2.64	2.41	2.22	2.09	1.972	1.880
500	2.80	2.55	2.35	2.21	2.08	1.986
550	2.97	2.69	2.47	2.32	2.19	2.09
600	3.15	2.83	2.59	2.43	2.29	2.19
650	3.29	2.97	2.72	2.55	2.41	2.28
700	3.44	3.10	2.84	2.66	2.50	2.37
750	3.59	3.23	2.96	2.76	2.60	2.47
800	3.75	3.37	3.08	2.87	2.69	2.56
900	4.04	3.63	3.30	3.08	2.89	2.74
1000	4.36	3.88	3.53	3.28	3.07	2.91
1200	4.92	4.37	3.95	3.67	3.43	3.25
1500	5.76	5.09	4.58	4.23	3.95	3.74
2000	7.14	6.26	5.63	5.17	4.78	4.49
2500	8.48	7.40	6.61	6.05	5.59	5.25
3000	9.83	8.52	7.54	6.92	6.38	5.97
3500	11.22	9.65	8.53	7.79	7.16	6.72
4000	12.54	10.76	9.52	8.65	7.94	7.42
4500	13.83	11.91	10.46	9.49	8.72	8.13
5000	15.25	13.01	11.42	10.35	9.48	8.84

Note: H = Henry's Law Constant

Adapted from Robert H. Perry and Cecil H. Chilton (eds.), "Chemical Engineers' Handbook," 5th ed., McGraw-Hill Book Company, New York, 1973.

TABLE 9-5 Solubilities of Gases in Water—(Continued)

HYDROGEN SULFIDE (H₂S)

t, °C	0	5	10	15	20	25	30	35
$10^{-2} \times H$	2.68	3.15	3.67	4.23	4.83	5.45	6.09	6.76

t, °C	40	45	50	60	70	80	90	100
$10^{-2} \times H$	7.45	8.14	8.84	10.3	11.9	13.5	14.4	14.8

"International Critical Tables," vol. 3, p. 259.

HYDROGEN (H₂)

t, °C	0	5	10	15	20	25	30	35
$10^{-4} \times H$	5.79	6.08	6.36	6.61	6.83	7.07	7.29	7.42

t, °C	40	45	50	60	70	80	90	100
$10^{-4} \times H$	7.51	7.60	7.65	7.65	7.61	7.55	7.51	7.45

Partial pressure H₂, mm. Hg	$10^{-4} \times H$	
	20°C.	23°C.
900	7.42	....
1100	....	7.75
2000	7.42	7.76
3000	7.43	7.77
4000	7.47	7.81
5000	7.56	7.89
6000	7.70	8.00
7000	7.87	8.16
8200	....	8.41
8250	8.17	....

"International Critical Tables," vol. 3, p. 256.

NITROGEN (N₂)

t, °C°	0	5	10	15	20	25	30	35
$10^{-4} \times H$	5.29	5.97	6.68	7.38	8.04	8.65	9.24	9.85

t, °C°	40	45	50	60	70	80	90	100
$10^{-4} \times H$	10.4	10.9	11.3	12.0	12.5	12.6	12.6	12.6

Partial pressure of N₂, mm. Hg	$10^{-4} \times H$	
	19.4°C.	24.9°C.
900	8.24	9.08
2000	8.32	9.15
3000	8.41	9.25
4000	8.49	9.38
5000	8.59	9.49
6000	8.74	9.62
7000	8.86	9.75
8100	9.04	....
8200	....	9.91

° "International Critical Tables," vol. 3, p. 256.

HYDROGEN CHLORIDE (HCl)

Weights of HCl per 100 weights of H₂O	Partial pressure of HCl, mm. Hg						
	0°C.	10°C.	20°C.	30°C.	50°C.	80°C.	110°C.
78.6	510	840					
66.7	130	233	399	627			
56.3	29.0	56.4	105.5	188	535	623	760
47.0	5.7	11.8	23.5	44.5	141	188	253
38.9	1.0	2.27	4.90	9.90	35.7	54.5	83
31.6	0.175	0.43	1.00	2.17	8.9	15.6	28
25.0	0.0316	0.084	0.205	0.48	2.21	4.66	9.3
19.05	0.0056	0.016	0.0428	0.106	0.55	1.34	3.10
13.64	0.00099	0.00305	0.0088	0.0234	0.136	0.39	0.93
8.70	0.000118	0.000563	0.00178	0.00515	0.0344	0.095	0.280
4.17	0.000018	0.000069	0.00024	0.00077	0.0064	0.0245	
2.04	....	0.0000117	0.000044	0.000151	0.00140		

Oxygen (O₂)

t, °C.	0	5	10	15	20	25	30	35
$10^{-4} \times H$	2.55	2.91	3.27	3.64	4.01	4.38	4.75	5.07

t, °C.	40	45	50	60	70	80	90	100
$10^{-4} \times H$	5.35	5.63	5.88	6.29	6.63	6.87	6.99	7.01

Partial pressure of O₂, mm. Hg	$10^{-4} \times H$	
	23°C.	26°C.
800	...	4.79
900	4.58	
2000	4.59	4.80
3000	4.60	4.83
4000	4.68	4.88
5000	4.73	4.92
6000	4.80	4.98
7000	4.88	5.05
8150	4.98	
8200	...	5.16

"International Critical Tables," vol. 3, p. 257.

Sulfur Dioxide (SO₂)

Weight of SO₂ per 100 weights of H₂O	Partial pressure of SO₂, mm. Hg							
	0°C.	7°C.	10°C.	15°C.	20°C.	30°C.	40°C.	50°C.
20	646	857	726					
15	474	637	574					
10	308	417	349	567	698	688		
7.5	198	307	226	419	517	452	665	458
5.0	118	198	165	270	336	216	322	266
2.5	69	92	59	127	161	125	186	172
1.5	38	51	37	71	92	79	121	116
1.0	23.3	31	23.6	44	59	52	87	82.0
0.7	15.2	20.6	15.6	28.0	39.0	36	57	.....
0.5	9.9	13.5	7.9	19.3	26.0	19.7	.....	31.0
0.3	5.1	6.9	4.6	10.0	14.1	11.8	12.9	20.0
0.2	2.8	3.7	3.1	5.7	8.5	8.1	7.5	12.0
0.15	1.9	2.6	1.75	3.8	5.8	4.7		
0.10	1.2	1.5	0.75	2.2	3.2	1.7	2.8	4.7
0.05	0.6	0.7	0.3	0.8	1.2	0.6	0.8	1.3
0.02	0.25	0.3		0.3	0.5			

Here $11.8 \times 10^{-6} y$ is very small compared with y, so that $y - 11.8 \times 10^{-6} y$ is essentially equal to y. Thus

$$y = 65.6 \times 10^{-6}$$

$$= \text{number of mols of oxygen dissolved}$$
$$= 65.6 \times 10^{-6}(32) = 0.0021 \text{ g}$$

Example 9-7   A 250-gal tank contains 125 gal of water containing 4.48 g Cl_2/l at 20°C. What weight of Cl_2 is present in the space above the liquid?

solution   From Table 9-5, the partial pressure of Cl_2 in equilibrium with water containing 4.48 g Cl_2/l at 20°C is 400 mmHg.

$$125 \text{ gal} \times \frac{3.785 \text{ l}}{\text{gal}} = 473.1 \text{ l}$$

$$473.1 \text{ l} \times \frac{400}{760} \times \frac{273}{293} \times \frac{\text{mol}}{22.4 \text{ l}} \times \frac{71 \text{ g}}{\text{mol}} = 735.4 \text{ g} \equiv \frac{735.4}{453} = 1.62 \text{ lb}$$

Example 9-8   A 250-gal tank contains a mixture of air and Cl_2 at atmospheric pressure, the partial pressure of the Cl_2 being 430 mmHg. 125 gal of water at 20°C is then introduced into the tank under suitable pressure. At equilibrium, how much chlorine is dissolved in the water and what is the pressure of the gas phase above the liquid?

solution   The weight of Cl_2 present in the tank is

$$250 \text{ gal} \times \frac{3.785 \text{ l}}{\text{gal}} = 946.3 \text{ l}$$

$$946.3 \text{ l} \times \frac{430}{760} \times \frac{273}{293} \times \frac{\text{mol}}{22.4 \text{ l}} \times \frac{71 \text{ g}}{\text{mol}} = 1,581 \text{ g}$$

The air in the tank will have a partial pressure of $760 - 430 = 330$ mmHg. After the addition of water, the air will have been compressed to half its original volume, and its partial pressure will be $330 \times 2 = 660$ mmHg. The chlorine, however, will have been divided between that remaining in the gas phase and that dissolving in the water. Assume that the partial pressure in the gas phase is 100 mmHg. Then, from Table 9-5, the water in equilibrium with the gas will contain 1.773 g of Cl_2/l. The Cl_2 in the gas phase will be

$$473.1 \text{ l} \times \frac{100}{760} \times \frac{273}{293} \times \frac{71 \text{ g}}{22.4 \text{ l}} = 183.8 \text{ g}$$

The Cl_2 dissolved in the water will be

$$473.1 \text{ l} \times \frac{1.773 \text{ g } Cl_2}{\text{l}} = 838.8 \text{ g}$$

Total Cl_2 present:

$$183.8 + 838.8 = 1,022.6 \text{ g}$$

For the assumed conditions, the total Cl_2 is less than that actually present. Another set of conditions must be assumed, say, a partial pressure of Cl_2 in the gas phase of 150 mmHg. From Table 9-5, at this pressure, the water will contain 2.27 g Cl_2/l. The Cl_2 in the gas phase will be

$$473.1 \times \frac{150}{760} \times \frac{273}{293} \times \frac{71}{22.4} = 275.8 \text{ g}$$

The Cl_2 dissolved in the water will be

$$473.1 \times 2.27 = 1,073.9 \text{ g}$$

Total Cl_2 present:

$$275.8 + 1,073.9 = 1,349.7 \text{ g}$$

The calculated total Cl_2 present is still less than that actually present and a still higher partial pressure of Cl_2 must be tried.

The next trial will be a partial pressure of Cl_2 in the gas of 170 mmHg. Interpolating the values of 150 and 200 mmHg from Table 9-5,

$$2.27 + \frac{170 - 150}{200 - 150} \times (2.74 - 2.27) = 2.27 + 0.19 = 2.46$$

Cl_2 in gas phase: $473.1 \times \dfrac{170}{760} \times \dfrac{273}{293} \times \dfrac{71}{22.4} = 312.5 \text{ g}$

Cl_2 in liquid: $473.1 \times 2.46 = 1,163.8 \text{ g}$

Total Cl_2: $312.5 + 1,163.8 = 1,476 \text{ g}$

This is about equal to the total Cl_2 originally present in the tank (1,581 g). At equilibrium 1,164 g of Cl_2 are dissolved in the water and the pressure of the gas phase will be $660 + 170 = 830$ mmHg. If a more precise answer is desired, another trial at, perhaps, 180 mmHg can be made.

In the case of a solute being extracted from one solvent by another, the *distribution ratio*, or the ratio of the solubilities of the solute in each of the liquids, establishes how much solute can be transferred from one liquid to the other. The transfer of solute between the two liquids will occur until the distribution ratio is satisfied.

Example 9-9   Chloroform ($CHCl_3$), acetic acid (CH_3COOH), and water are mixed in a tank and then allowed to separate, after which the chloroform layer is analyzed and found to contain 20 percent acetic acid. If the distribution ratio of acetic acid for chloroform and water is 0.417 for concentrations expressed as weight percents, what will the concentration of acetic acid be in the water layer?

solution   Distribution ratio $= \dfrac{\text{concentration in } CHCl_3}{\text{concentration in } H_2O}$

$$0.417 = \frac{0.2}{x}$$

$$x = 0.48 \equiv 48 \text{ percent}$$

Example 9-10   How much water will be required to extract half of the acetic acid from 100 lb of a 20 percent solution in chloroform?

solution   Weight of acetic acid in original solution $= 20$ lb; weight of chloroform $= 80$ lb. Weight of acetic acid in final solution $= 10$ lb. Concentration of acetic acid in final solution $= 10/(10 + 80) = 0.111$.

$$\text{Concentration in } H_2O = \frac{\text{concentration in } CHCl_3}{\text{distribution ratio}} = \frac{0.111}{0.417} = 0.266$$

Since the water contains 10 lb of acetic acid,

$$0.266 = \frac{10}{10 + x}$$

$$2.66 + 0.266\,x = 10$$

$$x = 27.59 \text{ lb}$$

$$27.59 \text{ lb} \times \frac{\text{gal}}{8.35 \text{ lb}} = 3.3 \text{ gal}$$

Frequently, nonaqueous liquids can be very effective solvents for the absorption of gases. Table 9-6 lists liquids that are commonly used for gas absorption and the gases with which they are effective.

The rate at which absorption or extraction occurs depends on the rates of diffusion of the material being transferred in each of the two phases. In the case of a gas being absorbed from a gas phase by a liquid, the gas must first diffuse from the bulk of the gas phase through an essentially stagnant gas layer to the interface between the liquid and the gas. At the interface, there is a sudden change in the concentration, and the gas, which has now entered the liquid, must diffuse away from the interface into the bulk of the liquid, again through an essentially stagnant liquid layer, as illustrated in Fig. 9-23. The diffusivities of various gases in air and water are given in Table 9-7. The value of this table lies in providing an indication of the relative rapidity with which an absorption or extraction can occur, since in the absence of direct information, the diffusivities in air and water can serve as a relative guideline for other gases or liquids.

The rate at which an absorption or an extraction occurs will also be influenced by the physical conditions of the operation. For example, it is desired to have as large an interface between the phases as possible. In the case of a gas-liquid interface, this can be achieved by allowing the liquid to flow downward in a vertical column filled with various regular and irregular materials known as *packing*. The liquid wets the packing in a thin film, thereby producing an extended interface with the gas flowing upward through the column. Another factor is the relative velocity between the two phases, since this dictates the thickness of the stagnant films through which the transferring material must diffuse. In the case of a gas-liquid system, it is most convenient to have the gas move at high velocity. Consequently, the packing material must present a large wettable surface without producing a large resistance to the flow of gas. Figure 9-24 illustrates a number of commercial packing materials having this property. The physical characteristics of these packings are given in Table 9-8, and their important operating characteristics in Table 9-9. In the operation of a packed tower, the gas and liquid flow rates are limited by the tendency of the column to flood. *Flooding* occurs when either the liquid or gas velocity is sufficient to have an excessive amount of liquid held up in the packed area, thereby reducing the free volume available for gas flow with a resultant sharp increase in the pressure drop of the gas as it flows through the packed column. Table 9-9 gives the holdup and flooding velocities for commercial packings with air and water as the two phases. The flooding velocity will be different for different liquid and gas combinations, generally in inverse relation to the viscosity of the liquid phase (i.e., a higher gas velocity is required to flood a packed column for a liquid of lower viscosity).

TABLE 9-6 Nonaqueous Solvents Used in Gas Absorption

Solutes	Solvents
Acetylene, C_2H_2	Acetic acid (glacial), $C_2H_4O_2$
Air	Acetic anhydride, $C_4H_6O_3$
Ammonia, NH_3	Acetone, C_3H_6O
Bromine, Br_2	Amyl alcohol, $C_5H_{12}O$
Carbon dioxide, CO_2	Aniline, C_6H_7N
Carbon monoxide, CO	Benzene, C_6H_6
Chlorine, Cl_2	Bromobenzene, C_6H_5Br
Ethane, C_2H_6	Carbon disulfide, CS_2
Ethylene, C_2H_4	Carbon tetrachloride, CCl_4
Hydrogen, H_2	Chlorobenzene, C_6H_5Cl
Hydrogen chloride, HCl	Chloroform, $CHCl_3$
Hydrogen sulfide, H_2S	Ethyl acetate, $C_4H_8O_2$
Methane, CH_4	Ethyl alcohol, C_2H_6O
Methyl chloride, CH_3Cl	Ethylene chloride, C_2H_4Cl
Nitric oxide, NO	Ethyl ether, $C_4H_{10}O$
Nitrogen, N_2	Methyl acetate, $C_3H_6O_2$
Nitrous oxide, N_2O	Methyl alcohol, CH_4O
Oxygen, O_2	Nitrobenzene, $C_6H_5NO_2$
Sulfur dioxide, SO_2	Propyl alcohol, C_3H_8O
Etc.	Propylene, C_3H_6
	Toluene, C_7H_8
	Etc.

From Robert H. Perry and Cecil H. Chilton (eds.), "Chemical Engineers' Handbook," 5th ed., McGraw-Hill Book Company, New York, 1973.

Example 9-11 An absorption column has an internal diameter of 32 in and is filled with $\frac{3}{4}$-in broken-stone packing. What is the maximum allowable flow of air through this column?
solution From Table 9-9, the superficial air velocity at the flooding point for $\frac{3}{4}$-in broken-stone packing is 3.6 ft/s. (The superficial air velocity is the velocity of the air if the column were empty. The actual air velocity will, of course, be much greater as it flows around the packing.)

Diameter of column = 2.67 ft

Superficial area of column = $\dfrac{\pi (2.67)^2}{4}$ = 5.60 ft²

Maximum allowable airflow = 5.60×3.6 = 20.2 ft³/s

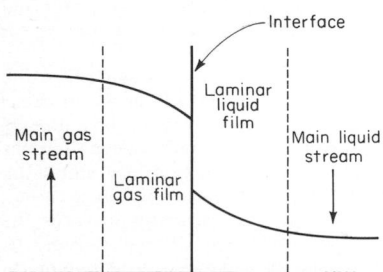

Fig. 9-23 Concentration distribution at gas-liquid interface in absorption.

While packed columns are widely used for absorption and extraction operations, practically any other type of equipment which results in a high surface area between phases can be used. Frequently, where a liquid-liquid or liquid-solid system is involved, simple mixing equipment can be used for carrying out the material transfer between phases.

Because of its commercial importance, ion exchange, a special type of absorption-extraction operation, warrants consideration. In *ion exchange,* particular ions are transferred from an aqueous solution to a particulate solid by having the solution percolate through a bed of the solid. The solid material can be naturally occurring material or, more frequently, a synthetic polymer. In either case, the material has the property of exchanging particular ions with those originally present at certain sites in its molecule.

TABLE 9-7   Diffusivities of Gases in Air and Water

Gas	Diffusivity ($D \times 10^4$)
IN AIR	
Water vapor	0.090
Ammonia	0.0089
Carbon dioxide	0.062
Hydrogen	0.281
Oxygen	0.077
Sulfur dioxide	0.055
Carbon monoxide	0.090
Methanol vapor	0.062
Ethanol vapor	0.048
Benzene vapor	0.035

Illustration: For air—ammonia, $D = 0.0000105$ mol/cm-s.

	IN WATER
Hydrogen	5.94
Oxygen	2.08
Carbon dioxide	1.74
Ammonia	2.04

Illustration: For oxygen—water, $D = 0.0000208$ cm²/s.

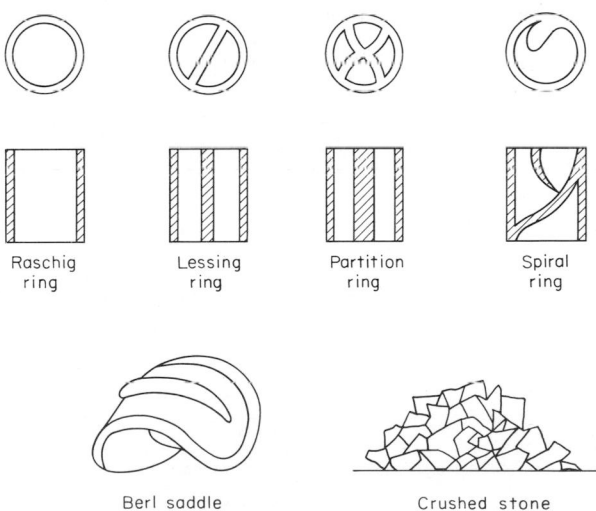

Raschig ring   Lessing ring   Partition ring   Spiral ring

Berl saddle   Crushed stone

Fig. 9-24 Commercial packing materials.

The exchange is reversible, the direction depending on the concentrations of the ions being exchanged, and the system can thus be regenerated as required. Commercial ion-exchange materials are capable of many thousands of cycles.

Those materials which absorb anions in exchange for other anions (usually hydroxyl ions) are said to be *anion-exchange materials*, while those which absorb cations in exchange for other cations (usually hydrogen ions) are said to be *cation exchangers*. Frequently, anion exchangers are mixed with cation exchangers so that a double exchange can occur simultaneously in such a mixed bed. While ion exchange has a variety of applications, its widest use is in water purification, as illustrated by Eq. (9-4):

$$\text{Mg}^{++} + \text{R} - \text{H}_2 \ (s) \rightleftharpoons \text{R} - \text{Mg} \ (s) + 2\text{H}^+$$
$$\text{SO}_4^{--} + \text{R} - \text{OH} \ (s) \rightleftharpoons \text{R} - \text{SO}_4 \ (s) + \text{OH}^- \tag{9-4}$$

in which Mg^{++} and SO_4^{--} are typical hard-water components and R is an insoluble-resin base.

TABLE 9-8   Physical Characteristics of Packings

Material	Bulk density (lb/ft^3)	Surface area (ft^2/ft^3)	Void volume (%)
3-in partition rings	64	18	54
6-in partition rings	80	37	42
Dumped $\frac{1}{2}$-in Raschig rings	49	132	68
Dumped 1-in Raschig rings	58	58	73
Dumped 2-in Raschig rings	35	14	73
Packed 3-in Rasching rings	52	20	64
3-in spiral rings	53	29	64
6-in spiral rings	53	15	58
1-in Berl saddles	50	79	75
2-in hollow spheres	52	19	33
4-in helix blocks	60	20	60

The ion-exchange reactions can be reversed, that is, the bed can be regenerated, by bringing acids (H^+ ions) and alkalies (OH^- ions) in contact with the exchange resins.

Adsorption is a process similar to absorption but is different in that the molecules of the adsorbed material are attached only to the surfaces of the adsorbent. A layer of gas, for example, can be adsorbed on a glass surface, or the coloring matter dissolved in a sugar solution can be adsorbed onto the surface of activated charcoal. The attraction is usually by relatively weak physical forces that are active only at the interface between the two phases, and the capacity of the adsorbent is therefore usually limited. Table 9-10 gives the properties of commonly used adsorbents.

In the case of adsorption, there is a definite equilibrium between the adsorbed material and the surface to which it is attracted. Equation (9-5), known as the Freundlich equation, describes this equilibrium for gas adsorption:

$$q = ap^{1/n} \tag{9-5}$$

where $q =$ quantity of gas adsorbed (mols), $a =$ a constant proportional to the surface area of the adsorbent, $n =$ a constant (greater than 1) that is characteristic of the system, and $p =$ partial pressure of the gas being adsorbed (mmHg).

Example 9-12 What is the value for n in the Freundlich equation for a system in which 38 g of N_2 are adsorbed on mica at $-100°\text{C}$ at a partial pressure of N_2 of 34 mmHg?
solution The constant a in Eq. (9-5) is proportional to the surface area of the adsorbent. We can thus use a value of $a = 1$ for the given system, since a fixed quantity of adsorbent is present and its surface area, whatever it is, will be fixed and can be assigned a relative value of 1.

$$38 \text{ g N}_2 = \frac{38}{28} = 1.36 \text{ mols}$$

From Eq. (9-5),

$$1.36 = (34)^{1/n}$$

This equation is best solved by logarithms, noting that $\log 1.36 = (1/n) \times \log 34$, from which $n = 11.2$.

Example 9-13 How much N_2 would be adsorbed by the system of Example 9-12 if the partial pressure of the N_2 were doubled?
solution

$$q = (68)^{1/11.2}$$

This equation is best solved by logarithms:

$$\log q = \frac{1}{11.2} \times \log 68$$

From which $q = 1.46$ mols $\text{N}_2 = 40.9$ g. (Note that doubling the partial pressure of N_2 increases the amount adsorbed by only 8 percent.)

TABLE 9-9 Operating Characteristics of Packings

Packing	Nominal size (in)	Apparent density lb/ft³	Hold-up with no circulation Water volume as a percentage of tower volume	Pieces per ft³	Mean dimensions (in)	Estimated surface (ft²/ft³)	Percentage voids	Superficial air velocity at flooding point (ft/s)
Broken stone	1/4	95.5	6.1	67,700	0.5 × 0.3 × 0.15	150	41	1.5
Broken stone	1/2	90.5	4.7	14,750	0.87 × 0.43 × 0.25	94	43	2.8
Broken stone	3/4	88.3	3.3	3,560	1.24 × 0.67 × 0.55	52	54	3.6
Clay spheres	1/2	84.9	3.8	27,200	0.457	92	46	2.3
Clay spheres	3/4	81.2	2.6	4,700	0.736	57	51	3.2
Clay spheres	1	77.1	2.0	1,950	0.99	41	50	4.7
Jack chain	3/16	104.4	....	3,900 lin ft	.... · · · · · · · ·	198	79	5.3
Carborundum	1/2	59	12.1	7,360	1 × 0.375	...	60	4.3
Glass rings	1/4	34.5	3.2	77,250	0.27 × 0.27	188	75	4.1
Glass rings	1/2	26.5	1.2	13,000	0.45 × 0.45	133	82	4.2
Bregeat spirals	1/2	57.2	3.3	9,350	0.875 × 0.5	144	83	7
Bregeat spirals	1¼	45.2	1.6	620	1.25 × 1.25	81	91	12
Lessing rings	1	35.0	0.5	1,145	1 × 1	66	92	4
Stoneware rings	1/2	48.8	2.3	11,000	0.5 × 0.5	132	68	4
Twisted fence wire	1/2	.....	....	103	0.5 × 36.0	43	...	25

NOTE: Flooding velocities determined at water flow of 500 lb/(hr)(ft²).
From Thomas K. Sherwood, "Absorption and Extraction," McGraw-Hill Book Company, New York, 1937.

TABLE 9-10 Properties of Commonly Used Adsorbents

Material and typical use	Shape of particles[a]	Size range, U.S. Standard mesh	Internal porosity X, %	Bulk dry density, lb./cu. ft.	Avg. pore diam., angstroms	Surface area, sq. meter/g.	Adsorptive capacity, g./g., dry solid	Trade designation
Aluminas								
Active alumina (transition alumina)	G	Various	25–30	50	35–45	235	0.15[a]	Alcoa F-1; Reynolds R-2101 = RA-1, R2101 = RA-3
Uses: drying gases and liquids; catalyst; catalyst support;	S	3–8, etc.	50–60	47–50	40–50	400	0.22[a]	Alcoa H-151; Kaiser KA-201; Pechiney Alumina A
defluoridation of alkylates; neutralization of lube oils	T	1-1/8 in.[c]	30, 47	~50	136, 99	90, 190		Harshaw Al-0104T, -1404T
CoCl₂-impregnated	G	8–14	30	54	45	200	0.14	Alcoa F-6 (indicator grade)
Desiccant (single-use), CaCl₂-impregnated	G	3–8, etc.	30	57	45	200	0.22[a]	Alcoa F-5
Catalytic alumina, low soda	G	Various	62	47	45	300		Pechiney CR
Dry column grade	S		55	~50				Woelm Dry Column Al₂O₃
Activated bauxite	G	8–20, etc.	35	~53	~50		0.04–0.2[d]	Florite
Chromatographic alumina[f]	G	80–200	30	58	45	225	0.14	Alcoa F-20
	S, P	30–140		~50				Pechiney CBT, CBL
		70–270		54				Woelm Al₂O₃-W200, Al₂O₃-TLC
Siliceous adsorbents								
Aluminosilicates	C, S, P	4–8, 8–12,	40–55	40–55		770		Molecular sieves: Davison 3A, 4A, 5A, 13X, 700 (acid-resist-
Use: selective adsorption based on molecular size and		1/16, 1/8						ant); Linde 3A, 4A, 5A, 10X, 13X, AW-300, AW-500
shape	C, S	1/16 or 1/8 in.	30	44	3, 4, 5	600–700	0.22[f]	Siliporite NK30, NK10, NK20
	P	<400		31	3, 4, 5	700		Siliporite NK10AP, NK20AP
Acid-treated clay	P			30–45		225–300		Clarsil
Uses: refining petroleum fractions; vegetable oils, juices;	S	4–8		53				Filtrol 120
catalyst base								
Magnesia-silica gel	G	Various	33	~30		300		Florisil
Fuller's earth	G	Various	~54	~40		130–250		Cecacite; Clarsil PSC-G, Clarsol ATC
Uses: same as for clay	P	<200		50				Clarsil PCS; Florex
Diatomaceous earth	P			9				Celite; Dicalite (powder and granules); Sorbo-cel (for emulsi-
								fied oil)
	G	10–140[e]		25–27		4		Chromosorb P (straight calcined)
	G	30–140		11–12		1.0		Chromosorb W (flux calcined)
	G	10–140		30–22		0.5		Chromosorb G (flux calcined)
	G	10–80		25–27		2.7		Chromosorb A (flux calcined)
	G	<200		6–25				Clarcel
Silica gel	G	Various	35–50	~27–45	Various	300–800	Various	Davison Silica Gel
Uses: drying of gases, separation of hydrocarbons, catalyst	G	Various		40–48	20–40	650–900	0.4–0.6[f]	Cecagel, Sorbsil
base	S,P	1/16 in., etc.	34–51	41–52	21–28	650–700	0.4–0.6[f]	Cecagel; Mobil Sorbead R, H
	G	70–140		~45				Woelm (adsorption)
	P	<270						Woelm TLC, F-TLC, G-TLC, GF-TLC
	S	4–8	45	46–51	72	250	0.27[f]	Mobil Sorbead W
Other inorganic materials								
Anhydrous CaSO₄	G	Various	38	60		100–110	0.12[d]	Drierite
Calcium silicate (fatty-acid removal)	P			12				Micro-Cel T-49, T-13
Magnesium silicate (decolorizing)	P, G			13		180		Celkate T-21; Woelm TLC
Carbons								
Shell-based	G	Various	50–60	27–34	20	800–1100	0.45[d]	Cochranex FCB; Pittsburgh PCB
Uses (for all carbons): water treatment, gas purification,	G	Various	50–80	25–35	20–30	1000–1600	0.5–0.95[b]	Picactif T.A., T.E.
solvent recovery and purification, decolorizing	G, (P)	Various	60–65	27–36	18–19	1200–1500	~0.4	Acticarbone NC, WNC
	G	Various	~50	27–32	20	800–1100	0.45[d]	Barnebey-Cheney AC, KE, VG, PC, PL
	G	4–6, 8–30	~50	33–35	20	800–900		Girdler 32E (Fe-impregnated), 32W (Cu-Cr-impregnated)
	P		60–80	20–22	~30	1200		Barnebey-Cheney YF, JF, JU (high capacity)

Material	Form[i]	Size, mesh or in.						Typical commercial products
Wood-based	C, P	⅛ ⅓ in.	70–75	12–28	22–24	750–1450	……	Acticarbone AC, etc.; Anticromos
	P	Various	30–50	9–35		600–1200	……	Darco KB, C60, Nuchar Aqua, Nuchar WA, B-100, C-115, CEE, C-190 C, C-1000
	G	5–7	……	24	5–10	1400	……	Supersorbon W
	P	Various	……	21–27	8–30	1000–1500	……	Carboraffin (various grades)
	P	Various	……	27–29	3–10	750–900	……	Brilonit (various grades)
	G	10 × 30, etc.	60	~20	20–40	660–1000	……	Cochranex FCN-1, FCN-2, FCA, FCC
	C, G	Various	40–50	15–20	20–100	800–1200	40–70%	Picatif C. O.
	P	100–30	40–60	27–33		600–1200	80–130%	Picatif CM
Peat-based	C, G, P	Various	~55	15–32	30–60	500–1600	……	Norit (various grades)
	P	5–7	……	20–24	5–20	1300–1400	……	Supersorbon various grades;
	P	<200	……	28–31		700–900	0.	Acticarbone AM, AH
Coal-based	G	Various	65–75	20–30	20–28	500–1200	……	Darco Granular; Permutit Carb-Dur
	G	12 × 40, etc.	60	30	60–65	800–1000	……	Cochranex FCP-1, FCP-2, FCW-V
	G, (P)	Various	55–67	25–30	20	1000–1400	……	Pittsburgh BEL, CAL, SGL, etc. (RB, RC, BL)
	G	……	80	28	22	110	5	Barnebey-Cheney MN
	G	Various	……	27–37		850–350	30–50%	Acticarbone M; Nuchar WV-W, WV-L, WV-G, WV-H
	G	5–7	……	20–24	5–15	1300–1500	……	Contarbon (various grades)
	C, P	……	……	25–30		600–700	……	Darco BG, DC, S-51; Hydrodarco B
Petroleum-based	C, P	Various	65–85	~30	18–22	800–1100	……	Columbia (various grades)
Organic materials								
Porous resin (decolorizing)	S, G	16–50	……	20–45		3	……	Asmit 224, Duolite S-35; Permutit S-360; Wofatit E
	S, (P)	16–50	……	30–40			0.1, (0.2)	Ionex RV (Macro-Ionex RV)
Cross-linked polystyrene	G	50–140	40–45	19–22	500	55–35	……	Chromosorb 01, 103
	S	20–50		39	90	330	……	Amberlite XAD-2
	S	20–50	50–55	18–20	50	750	……	Amberlite XAD-4
Cross-linked polystyrene	G	60–200	……	22–25	35	300–400	……	Chromosorb 102
Phenolic	S	16–50	……	41	80	450	……	Duolite S-30, ES-33, S-37 (general adsorbent)
Acrylic ester	S	20–50	50–55	43	250	450	……	Amberlite XAD-7
	S	20–60	50–54	40–50		140	……	Amberlite XAD-8
Aromatic-amine resin	S	10–50	……	40–50				Asmit 173N
Quaternary amine chloride resin	S	16–50	~65					Asmit 259N, 261; Duolite ES-111, A-140
Copper-amine resin (O₂ removal)	S	10–50	……	30				Duolite S-10
Cellulose	G, (P)	100–200	……			3 × 10⁴	0.1	Whatman CC31 (CF1, 2, 11, 12)

From Robert H. Perry and Cecil H. Chilton (eds.), "Chemical Engineers' Handbook," 5th ed., McGraw-Hill Book Company, New York, 1973.

[a] Water at 60 per cent humidity.
[b] Carbon tetrachloride; test conditions not specified.
[c] Various sizes available within stated range.
[d] Water; test conditions not specified.
[e] Benzene, at 20°C. and 7.5 mm. partial pressure.
[f] Water, at 100 per cent relative humidity.
[g] Accelerated chloropicrin test.
[h] Oxygen.
[i] C, cylindrical pellets; G, granular; P, powder; S, spherical beads; T, tablets.
[j] Separate grades specified for use at pH 4, 7.5 or 10.
[k] Iodine.

Adsorption, as has been indicated, results in an essentially monomolecular layer of the adsorbed material adhering to the surface of the adsorbent. However, when gas molecules impinge on the surface of a liquid, they do not, in general, rebound elastically, but condense on the surface, being held by the attractive forces of the liquid which preceded it. This can occur in an adsorption process, giving rise to higher capacities than would be accounted for simply by the surface area of the adsorbent.

The apparatus in which adsorption is carried out is usually one in which the fluid is passed through a bed of adsorbent. As the adsorbent bed becomes exhausted (i.e., has been loaded with as much adsorbed material as it can hold), the fluid stream can be diverted to a second bed while the original bed is stripped, that is, regenerated by removal of the adsorbed material. A common procedure for removing the adsorbed material is to pass a hot gas or steam through the bed so as to heat the bed and thereby weaken the forces that hold the adsorbed material, as well as to increase its vapor pressure. The heated gas stream, now containing the stripped material, is usually directed to a condenser where the stripped material can be condensed and removed, as illustrated in Fig. 9-25. Adsorption and stripping systems are widely used for purifying a gas or liquid, or for recovering and concentrating a valuable component.

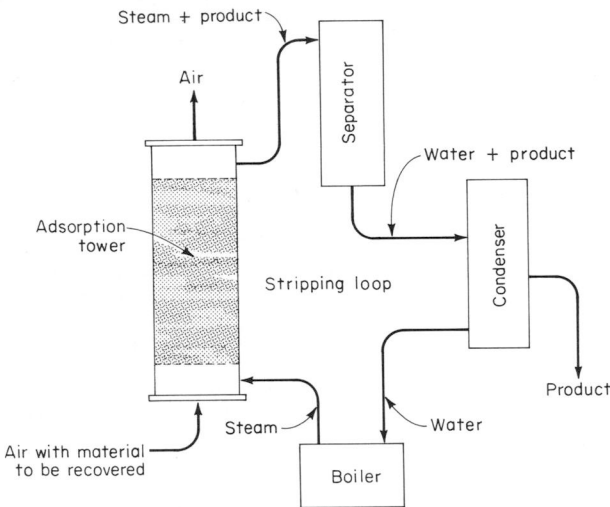

Fig. 9-25 Adsorption-stripping material-recovery system. In operation, the material to be recovered is removed from an airstream by adsorption. The air is then shut off, and the stripping loop is operated.

9-5 DRYING AND HUMIDIFICATION

Drying involves the removal of water from a solid or a liquid, usually by an evaporative process. There are other ways of removing water from a material, such as freezing to crystallize ice, decomposition of water, chemical precipitation, absorption, adsorption, and mechanical separation, but these are used where some special circumstance precludes the use of evaporative drying.

The moisture present in a material can be free or bound. The *free-moisture* content of a material is that water which can be removed without causing a chemical or physical change in the material. *Bound moisture* is that water which is associated with the chemical structure of the material. Free moisture can be removed by simple evaporative heating, while bound moisture may require more complex procedures.

The drying of a solid, illustrated in Fig. 9-26, involves the diffusion of water vapor to the surface of the solid, then through a relatively stagnant air film, and finally into the main air stream. Many materials carry a water content which is sufficient to permit the surface of the solid actually to remain wet for a period of time during the initial phase of the drying operation, the moisture being supplied to the surface by capillary action between the particles of the solid. Under this condition, drying occurs at an essentially constant rate until a moisture content is reached such that the surface is no longer uniformly wet. From this point on, the drying rate decreases, more or less uniformly, as illustrated in Fig. 9-27. The moisture content below which the drying rate will continually decrease is known as the *critical moisture content*. Table 9-11 gives the critical moisture content for a number of common materials.

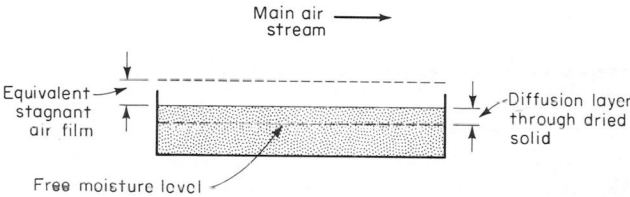

Fig. 9-26 Mechanism of drying of a particulate solid. As drying proceeds, the free-moisture level recedes from the surface of the solid. At the critical moisture, the free-moisture level coincides with the surface of the solid.

The rate of drying is dependent on the manner in which the solid is heated, since the latent heat of evaporation must be supplied before the water can evaporate. Where the material to be dried is sensitive to heat, only low temperatures may be used, and the drying time must therefore be extended, or the drying can be carried out in a vacuum to lower the boiling point.

Several commercial devices are used to improve the uniformity of drying and to decrease the time required. Simple tray driers are used for particulate solids, such as various chemical materials, resins, and natural fibers. Such driers, generally referred to as *cabinet driers,* are essentially intermittent in their operation in that a batch is put into the drier, dried to the desired degree, and then removed. Several other types of driers are used which are intended for continuous operation. These include *tunnel driers* in which individual batches are dried by continuous movement through a tunnel; *drum driers,* in which the material to be dried is spread onto the surface of a heated drum which rotates slowly, permitting the dried material to be scraped from the drum at some later point in the cycle; and *rotary driers,* in which an inclined cylinder, usually of large diameter, rotates while being heated and the material to be dried slowly tumbles and moves down the cylinder. Because drying occurs at the surface of a material, an efficient way to dry a solution is to spray it into a chamber against a countercurrent flow of hot air. The extremely fine particles produced by the spray permit evaporation to take place very rapidly. If the solution is dilute, meaning that most of the particle is water, this method can be used to produce very fine dry powders. Typical drying units are illustrated in Fig. 9-28.

Humidification refers to the addition of moisture to a material, which can be a liquid or a solid but is usually a gas. The humidity, sometimes called the *absolute humidity,* is the weight of water vapor carried per unit weight of gas. The *percent absolute humidity* is the weight of water vapor carried by a unit weight of dry gas at a particular temperature, divided by the weight of water vapor that 1 lb of the dry gas would carry if it were saturated at that temperature. The *percent relative humidity* (RH) is defined by Eq. (9-6):

$$\text{Percent RH} = \frac{p}{P_s} \times 100 \qquad (9\text{-}6)$$

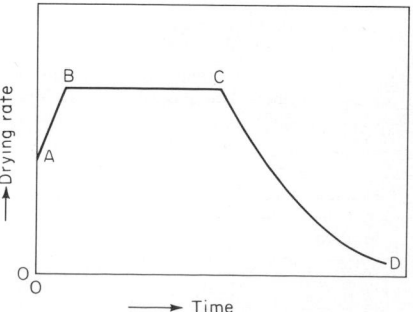

Fig. 9-27 Sequences of drying cycle of a particulate solid. *A* to *B* heating period; *B* to *C* constant rate drying period; *C* to *D* decreasing rate drying period.

Example 9-14 A 1,500-l vessel contains dry N_2 at 20°C and 1 atm. What is the absolute humidity of this gas if 18 ml of water is added and allowed to evaporate?

solution

$$1,500 \ l \times \frac{273}{293} = 1,398 \ l \text{ at STP}$$

$$1,398 \ l \times \frac{mol}{22.4 \ l} \times \frac{28 \ g}{mol} = 1,747 \ g$$

The absolute humidity is therefore $18/1,747 = 0.0103$.

TABLE 9-11   Critical Moisture Content of Various Materials

Material	Thickness, in.	Critical moisture, % water, dry basis
Barium nitrate crystals, on trays...........	1.0	7
Beaverboard..............................	0.17	Above 120
Brick clay................................	.62	14
Carbon pigment...........................	1	40
Celotex...................................	0.44	160
Chrome leather...........................	.04	125
Copper carbonate (on trays)...............	1–1.5	60
English china clay........................	1	16
Flint clay refractory brick mix.............	2.0	13
Gelatin, initially 400% water..............	0.1–0.2 (wet)	300
Iron blue pigment (on trays)..............	0.25–0.75	110
Kaolin....................................	....	14
Lithol red................................	1	50
Lithopone press cake (in trays)............	0.25	6.4
	.50	8.0
	.75	12.0
	1.0	16.0
Niter cake fines, on trays..................	....	Above 16
Paper, white eggshell......................	0.0075	41
Fine book.............................	.005	33
Coated................................	.004	34
Newsprint.............................	....	60–70
Plastic clay brick mix......................	2.0	19
Poplar wood..............................	0.165	120
Prussian blue.............................	....	40
Pulp lead, initially 140% water.............	....	Below 15
Rock salt (in trays)......................	1.0	7
Sand, 50–150 mesh........................	2.0	5
Sand, 200–325 mesh.......................	2.0	10
Sand, through 325 mesh...................	2.0	21
Sea sand (on trays).......................	0.25	3
	.5	4.7
	.75	5.5
	1.0	5.9
	2.0	6.0
Silica brick mix...........................	2.0	8
Sole leather..............................	0.25	Above 90
Stannic tetrachloride sludge...............	1	180
Subsoil, clay fraction 55.4%...............	....	21
Subsoil, much higher clay content..........	....	35
Sulfite pulp..............................	0.25–0.75	60–80
Sulfite pulp (pulp lap)....................	0.039	110
White lead................................	....	11
Whiting...................................	0.25–1.5	6.9
Wool fabric, worsted......................	....	31
Wool, undyed serge........................	....	8

From Robert H. Perry and Cecil H. Chilton (eds.), "Chemical Engineers' Handbook," 5th ed., McGraw-Hill Book Company, New York, 1973.

Example 9-15   What is the percent absolute humidity for the conditions of Example 9-14, given that the vapor pressure of water is 17.54 mmHg?
solution

$$1,500 \ l \times \frac{17.54}{760} \times \frac{273}{293} \times \frac{18}{22.4} = 25.9 \ g$$

This is the amount of water the nitrogen would hold at saturation.

$$\text{Percent absolute humidity} = \frac{18}{25.9} \times 100 = 69.5 \text{ percent}$$

Example 9-16 What is the percent relative humidity for the conditions of Example 9-15?
solution

$$18 \text{ g H}_2\text{O} \times \frac{22.4 \text{ l}}{18 \text{ g H}_2\text{O}} \times \frac{273}{293} = 24 \text{ l}$$

$$p_{\text{H}_2\text{O}} = \frac{24}{1,500} \times 760 = 12.16 \text{ mmHg}$$

$$\text{Percent RH} = \frac{12.16}{17.54} \times 100 = 69.3 \text{ percent}$$

Note that the percent relative humidity is approximately, but not exactly, the same as the percent absolute humidity.

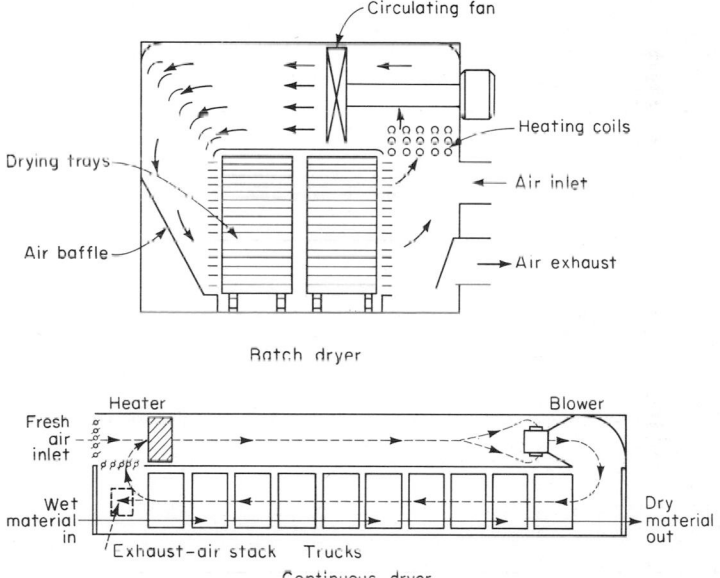

Fig. 9-28 Commercial drying systems.

For a small volume of gas, a convenient way of obtaining a constant and known humidity is to allow the gas to come to equilibrium with certain chemicals that have a definite water-vapor pressure at a particular temperature. A list of such materials is given in Table 9-12.

For larger volumes of gas, moisture may be added by direct injection of steam or atomized water.

The relationship among temperature, humidity, and relative humidity is shown on a *humidity chart* of the type illustrated by Fig. 9-29.

The usual way of measuring the humidity of a gas is by *wet- and dry-bulb thermometry,* using an apparatus called a *psychrometer.* A psychrometer consists of two thermometers, one of which has its temperature-sensitive bulb covered with a wetted wicking material. When air is passed over both thermometers, evaporation of water from the wick covering one of the thermometer bulbs will cause the temperature of that bulb to be reduced in proportion to the rate of evaporation of water. In turn, the rate of evaporation of the water will be a function of the humidity of the air passing over the wick. The difference between the wet- and dry-bulb temperatures can then be used to determine the relative humidity by means of a humidity chart, or from a *psychrometric chart,* that is, a chart relating the humidity to the wet- and dry-bulb temperatures. Such a chart is given in Table 9-13.

TABLE 9-12 Chemicals Used to Maintain Constant Humidities

Solid phase	Use temp., °C.	% humidity
$H_3PO_4.\frac{1}{2}H_2O$	24.5	9
$ZnCl_2.\frac{1}{2}H_2O$	20	10
$KC_2H_3O_2$	168	13
$LiCl.H_2O$	20	15
$KC_2H_3O_2$	20	20
KF	100	22.9
$NaBr$	100	22.9
$CaCl_2.6H_2O$	24.5	31
$CaCl_2.6H_2O$	20	32.3
$CaCl_2.6H_2O$	18.5	35
CrO_3	20	35
$CaCl_2.6H_2O$	10	38
$CaCl_2.6H_2O$	5	39.8
$K_2CO_3.2H_2O$	24.5	43
$K_2CO_3.2H_2O$	18.5	44
$Ca(NO_3)_2.4H_2O$	24.5	51
$NaHSO_4.H_2O$	20	52
$Mg(NO_3)_2.6H_2O$	24.5	52
$NaClO_3$	100	54
$Ca(NO_3)_2.4H_2O$	18.5	56
$Mg(NO_3)_2.6H_2O$	18.5	56
$NaBr.2H_2O$	20	58
$Mg(C_2H_3O_2)_2.4H_2O$	20	65
$NaNO_2$	20	66
$(NH_4)_2SO_4$	108.2	75
$(NH_4)_2SO_4$	20	81
$NaC_2H_3O_2.3H_2O$	20	76
$Na_2S_2O_3.5H_2O$	20	78
NH_4Cl	20	79.2
NH_4Cl	25	79.3
NH_4Cl	30	79.5
KBr	20	84
Tl_2SO_4	104.7	84.8
$KHSO_4$	20	86
$Na_2CO_3.10H_2O$	24.5	87
K_2CrO_4	20	88
$NaBrO_3$	20	92
$Na_2CO_3.10H_2O$	18.5	92
$Na_2SO_4.10H_2O$	20	93
$Na_2HPO_4.12H_2O$	20	95
NaF	100	96.6
$Pb(NO_3)_2$	20	98
$TlNO_3$	100.3	98.7
$TlCl$	100.1	99.7

From Robert H. Perry and Cecil H. Chilton (eds.), "Chemical Engineers' Handbook," 5th ed., McGraw-Hill Book Company, New York, 1973.

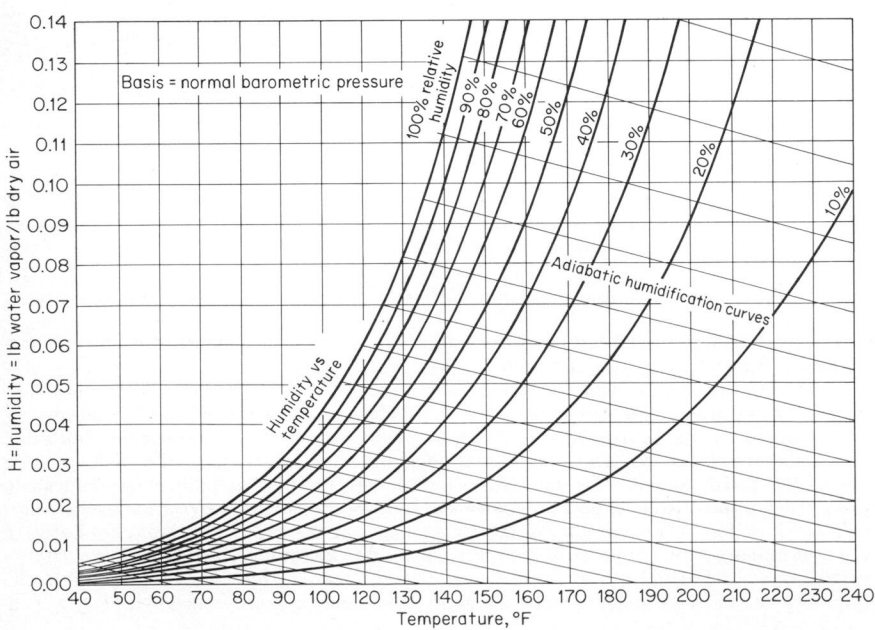

Fig. 9-29 Humidity chart.

TABLE 9-13 Relative Humidity from Wet- and Dry-Bulb Temperatures

Rows = Wet-bulb temp (°F); Columns = Dry-bulb temp (°F)

Wet \ Dry	21	22	23	24	25	26	27	28	29	30	31	32	33	34	35	36	37	38	39	40	41	42	43	44	45	46	47	48	49	50	51	52	53	54	55	56	57	58	59	60
14	1																																							
15	15	4																																						
16	28	17	7																																					
17	42	31	20	10	1																																			
18	56	44	33	22	13	4																																		
19	71	58	46	35	25	16	7																																	
20	85	71	59	47	37	27	18	10	3																															
21	100	86	72	60	49	39	29	21	13	6																														
22		100	86	73	62	51	41	32	23	16	8	2																												
23			100	87	74	63	52	43	34	26	18	11	5																											
24				100	87	75	64	54	44	36	28	20	14	8	2																									
25					100	87	76	65	55	46	37	30	23	16	10	5																								
26						100	88	76	65	56	47	39	32	25	19	13	7	2																						
27							100	88	77	67	58	49	41	34	27	21	15	10	5																					
28								100	88	78	68	59	51	43	36	29	23	17	12	7	3																			
29									100	89	78	69	60	52	45	38	31	25	20	15	10	5																		
30										100	89	79	70	62	54	46	40	33	27	22	17	12	8	4																
31											100	89	80	71	63	55	48	42	35	29	24	19	14	10	6															
32												100	90	81	72	64	57	50	43	37	31	26	21	16	12	8	5	1												
33													100	90	81	73	65	58	51	45	39	33	28	23	18	14	10	7	3											
34														100	91	82	74	66	59	52	46	40	35	30	25	20	16	12	9	5										
35															100	91	83	75	67	60	54	47	42	36	31	26	22	18	14	10	7	4	1							
36																100	91	83	75	68	61	55	48	43	38	32	28	23	19	16	12	9	6	3						
37																	100	91	83	75	69	62	55	49	44	39	34	29	25	21	17	14	10	8	5	2				
38																		100	92	83	76	69	63	56	51	45	40	35	31	27	23	19	16	12	9	7	4	1		
39																			100	92	84	77	70	63	57	52	46	41	36	32	28	24	20	17	14	11	8	6	3	1
40																				100	92	85	77	71	64	58	52	47	42	38	34	29	26	22	19	16	13	10	7	5
41																					100	92	85	78	71	65	59	54	48	43	39	35	31	27	23	20	17	14	11	9
42																						100	92	85	78	72	66	60	54	49	45	40	36	32	28	25	22	18	16	13
43																							100	93	86	79	72	66	61	55	50	46	41	37	33	30	26	23	20	17
44																								100	93	86	79	73	67	61	56	51	47	42	40	34	31	27	24	21
45																									100	93	86	79	73	67	63	57	52	48	45	39	35	32	29	25
46																										100	93	86	80	74	68	63	58	53	50	44	40	36	33	30
47																											100	93	87	80	75	69	63	59	54	50	45	41	38	34
48																												100	93	87	81	75	69	64	59	55	50	45	42	39
49																													100	93	87	81	75	70	65	60	55	51	47	43
50																														100	94	87	81	76	70	65	61	56	52	48
51																															100	94	87	82	76	71	66	62	57	53
52																																100	94	88	82	76	71	67	62	58
53																																	100	94	88	82	77	72	67	63
54																																		100	94	88	82	77	72	68

From Norbert A. Lange (ed.), "Handbook of Chemistry," 10th ed., McGraw-Hill Book Company, New York, 1967.

TABLE 9-13 Relative Humidity from Wet- and Dry-Bulb Temperatures—(Continued)

Wet-bulb temp (°F) — rows; Dry-bulb temp (°F) — columns.

Wet \ Dry	41	42	43	44	45	46	47	48	49	50	51	52	53	54	55	56	57	58	59	60	61	62	63	64	65	66	67	68	69	70	71	72	73	74	75	76	77	78	79	80
41	100																																							
42		100																																						
43		92	100																																					
44		85	92	100																																				
45		78	85	93	100																																			
46		71	78	86	93	100																																		
47		65	72	79	86	93	100																																	
48		59	66	72	79	86	93	100																																
49		54	60	66	72	79	86	93	100																															
50		48	54	61	67	73	79	86	93	100																														
51		43	49	55	61	67	73	80	86	93	100																													
52		39	45	50	56	61	68	74	80	87	93	100																												
53		35	40	46	51	57	63	69	75	81	87	94	100																											
54		31	36	41	47	52	58	63	69	75	81	87	94	100																										
55		27	32	37	42	48	53	59	64	70	75	81	88	94	100																									
56		23	28	33	38	43	49	54	59	65	70	76	82	88	94	100																								
57		20	25	30	34	39	44	50	55	60	65	71	76	82	88	94	100																							
58		17	22	26	31	35	40	45	50	55	60	66	71	77	82	88	94	100																						
59		14	18	23	27	32	37	41	46	51	56	61	66	72	77	82	88	94	100																					
60		11	16	20	24	29	33	38	42	47	52	57	62	67	72	77	83	89	94	100																				
61		9	13	17	21	26	30	34	39	43	48	53	58	63	68	73	78	83	89	94	100																			
62		7	10	14	18	22	27	31	35	40	44	48	53	58	63	68	73	78	83	89	94	100																		
63		4	8	12	16	20	24	28	32	36	41	45	50	54	58	64	69	74	78	84	89	94	100																	
64		2	6	10	13	17	21	25	29	33	37	42	46	50	55	60	65	69	74	79	84	89	95	100																
65			4	7	11	15	18	22	26	30	34	38	43	47	51	55	60	65	69	74	79	84	89	95	100															
66			2	5	9	12	15	18	22	27	31	34	38	44	48	52	56	61	65	70	74	79	84	90	95	100													
67				3	7	9	12	15	19	22	26	31	35	38	42	47	51	55	60	64	69	74	79	84	90	95	100												
68				1	5	7	10	13	16	20	23	27	31	35	39	43	48	52	56	60	65	69	74	79	85	90	95	100											
69					3	5	8	11	14	17	21	24	28	32	36	40	44	48	53	58	62	67	71	76	81	86	90	95	100										
70					1	4	6	9	12	15	19	22	25	29	33	36	41	45	51	55	62	66	72	76	81	85	90	95	100										
71						3	5	8	10	13	16	20	23	26	30	34	38	42	47	51	55	60	65	70	74	80	85	90	95	100									
72						1	3	6	9	12	14	18	21	25	28	32	35	39	43	48	52	56	62	66	71	75	80	85	90	95	100								
73								4	7	10	13	16	19	23	26	29	34	38	42	47	51	56	61	65	70	75	80	86	90	95	100								
74								3	5	8	11	14	16	20	23	27	30	34	38	43	48	52	58	63	68	73	78	82	86	91	95	100							
75								1	4	6	9	12	14	18	21	24	28	32	35	41	45	51	57	61	65	70	74	82	86	91	95	100							
76									3	5	8	10	13	16	18	22	25	29	33	36	40	45	50	55	60	65	71	75	83	87	91	95	100						
77									1	4	6	9	11	14	16	19	22	26	30	34	37	42	48	52	59	63	68	74	78	83	91	96	100						
78										3	5	7	9	12	14	17	20	23	27	31	34	38	43	49	56	60	67	71	79	83	87	96	100						
79											3	5	7	9	12	14	18	20	23	26	32	35	38	41	47	53	60	64	68	75	83	91	96	100					
80															3	5	7	10	12	15	18	20	23	26	29	32	38	44	50	57	61	68	75	79	83	91	96	100		100

Dry-bulb temp (°F)

9-38

Example 9-17   What is the absolute humidity for air at a relative humidity of 30 percent and a temperature of 150°F?
solution   The desired value can be read directly from Table 9-13: 0.052 lb water vapor/lb dry air.

Example 9-18   What is the relative humidity of air at 120°F for an absolute humidity of 0.060?
solution   From Table 9-13, the relative humidity will lie between 70 and 80 percent. By interpolation, the relative humidity for the stipulated conditions would be about 76 percent.

The relative humidity from wet- and dry-bulb readings can also be determined from a humidity chart as shown below. The abscissa corresponding to the wet-bulb temperature t_w is followed upward until it intersects the saturation curve. From that point, an adiabatic-humidification line is followed until it intersects the abscissa corresponding to the dry-bulb temperature t_d. At this point, the relative humidity is read by interpolating from the nearest relative-humidity lines. For the case shown, the relative humidity would be 23 percent.

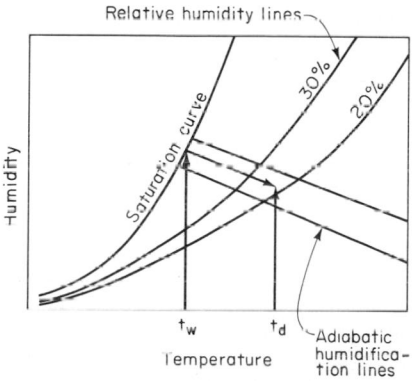

Example 9-19   What are the absolute and percent relative humidities for a dry-bulb reading at 140°F and a wet-bulb reading of 100°F?
solution   At the intersection of the 100°F ordinate and the saturation curve in Fig. 9-29, follow the appropriate adiabatic-cooling line until it intercepts the 140°F ordinate. At this point, read 26 percent RH or an absolute humidity of 0.032 lb H_2O/lb dry air.

Example 9-20   What is the percent RH of air for which the dry-bulb temperature is 80°F and the wet-bulb temperature is 70°F?
solution   From Table 9-13, the percent RH is between 60 and 70 percent. Interpolating between these two values by inspection, the percent RH can be read directly as 62 percent.

Chapter 10

Safety Practices

10-1 FIRE HAZARDS

Because of their continuous exposure to a wide variety of dangerous materials, chemical technicians must have a high degree of awareness of the specific hazards involved. Generally speaking, fire and explosion are the most prevalent hazards, and a good deal of basic information concerning slow and rapid combustion is now available. A combustible material will spontaneously start to burn when its temperature is raised to its *ignition point* in the presence of air. Ignition temperatures for a wide variety of materials are given in Table 10-1.

> **Example 10-1** To what temperature must nitrobenzene be heated in the presence of air to have it ignite? What are the physical changes that occur in the nitrobenzene-air system as it is heated from room temperature to the ignition temperature?
>
> **solution** Table 10-1 indicates that the ignition temperature of nitrobenzene is 924°F. At room temperature nitrobenzene is a liquid. As it is heated, its vapor pressure increases until, at its boiling point (412°F), all the liquid is converted to vapor. Since this occurs in the presence of air, the gaseous mixture can be further heated until a temperature of 924°F is reached, at which point the nitrobenzene-air mixture spontaneously ignites. (For ignition to occur, a proper concentration of nitrobenzene must be maintained, for example, by enclosing the vapor-air mixture to prevent an overdilution of the vapor by air.)

In the case of a liquid, raising its temperature increases its vapor pressure (i.e., increases the concentration of vapor above the liquid) but does not result in combustion until its ignition temperature is reached. With liquids, it is often difficult to reach the ignition-point temperature by simple heating, but the vapors above the liquid can be readily ignited by a spark or flame, after which the heat generated will continue to maintain the liquid at a temperature sufficient to supply the flame with more vapor until the liquid is consumed.

As the temperature of the liquid is raised in the presence of air, a concentration of vapor in air is achieved which will result in ignition, but for which the concentration is too low to result in sufficient heat being generated to sustain the combustion. The vapor simply ignites in a momentary flash. The liquid temperature at which this flash is first observed as the liquid is being slowly heated is known as the *flash point*. As the temperature of the liquid is raised above the flash point, ignition will cause longer and more pronounced flashes until the combustion becomes self-sustaining. Thus flash point is a very important measure of the ease with which a liquid can ignite and explode or burn, the lower the flash point, the more flame-hazardous the material. In determining the flash point, a sample of the liquid contained in a metal cup is slowly heated under carefully stipulated conditions with the temperature of the liquid noted at frequent intervals. Periodically, a small flame is brought in contact with the vapors at the top of the cup, a procedure that is repeated until a flash is observed. Two types of cups are used, open and closed, as illustrated in Figs. 10-1 and

TABLE 10-1 Combustibility Properties of Solids, Liquids, and Gases

Identification of extinguishing agents:
1. Water
2. Foam
3. Carbon dioxide
4. Dry chemical
5. Powdered talc

Name	Flash point (°F)		Explosive limits % by vol. in air		Auto-ignition temp. (°F)	Suscepti-bility to sponta-neous heating	Suitable extin-guishing agents
	Closed Cup	Open Cup	Lower	Upper			
Acetaldehyde	−17		4.0	57.0	365	No	1, 3, 4
Acetanilide	345	345					1, 2, 3, 4
Acetic acid (glacial)	104	110	4.0		1050	No	1, 3, 4
Acetic anhydride	121	130			752	No	1, 3, 4
Acetic ether, see ethyl acetate							
Acetoacetic ester, see ethyl aceto-acetate							
Acetone	0	15	2.15	13.0	1118	No	3, 4
Acetophenone	221					No	1, 2, 3, 4
p-Acetotoluide	335	355					1, 2, 3, 4
Acetyl chloride	40					No	1, 3, 4
Acetylene	gas		2.5	80	635	No	
Acetylene dichloride, see dichloro-ethylene-1, 2							
Acetylene tetrabromide	non-flammable					No	
Acetylene tetrachloride	non-flammable					No	
Adipic acid	385						1, 2, 3, 4
Aldol	181					No	1, 2, 3, 4
Allyl alcohol	70	75	3.0		713	No	3, 4
Allylene	gas		1.74				
Almond oil (artificial), see benzalde-hyde							
Aluminum (powder)						No	5
Amino benzene, see aniline							
Ammonia (anhydrous)	gas		16.0	27.0	1436	No	
Amyl acetate-n	76	80	1.1		714	No	3, 4
Amyl acetate-iso	92	100			715	No	3, 4
Amy alcohol-n	91	120	1.2		621	No	3, 4
Amyl alcohol-rim. iso	109	115			667	No	3, 4
Amyl alcohol-sec. n	94				650–725	No	3, 4
Amyl alcohol-sec. iso	103					No	3, 4
Amyl chloride			1.4			No	2, 3, 4
Amylene-n			1.6			No	2, 3, 4
Amyl salicylate	270						1, 2, 3, 4
Aniline	168				1418	No	2, 3, 4
Aniline hydrochloride		380				No	1, 3, 4
Anthracene	250	385	0.63		881 in O_2	No	1, 2, 3, 4
Anthraquinone	365	485				No	1, 2, 3, 4
Antifebrin, see acetanilide							
Asphalt (typical)	400+	535+			905	No	1, 2, 3, 4
Benzaldehyde	148	165				No	1, 2, 3, 4
Benzene	12		1.4	8	1076	No	2, 3, 4
Benzine, see naphtha V.M. & P.							
Benzoic acid	250					No	1, 3, 4
Benzol, see benzene							
Benzyl acetate	216				862	No	1, 2, 3, 4
Benzyl alcohol	213	220			817	No	1, 2, 3, 4
Benzyl benzoate	298					No	1, 2, 3, 4
Benzyl cellosolve		265				No	1, 3, 4
Benzyl chloride	140		1.1			No	1, 2, 3, 4
Borneol	150					No	1, 2, 3, 4
Bromobenzene	149					No	1, 2, 3, 4
Bromoethane, see ethyl bromide							
Bromomethane, see methyl bromide							
Bronze dust (aluminum free)	No Hazard					No	
Butane-n	−76	gas	1.6	8.5	806	No	
Butyl acetate-n	72	90	1.7	15	790	No	3, 4
Butyl acetate-iso	64					No	3, 4
Butyl alcohol-n	84	110	1.7		693	No	3, 4
Butyl alcohol-iso	82		1.68		825	No	3, 4
Butyl alcohol-ter.	52				901	No	3, 4
Butyl benzene-sec.	126					No	2, 3, 4
Butyl carbinol, see amyl alcohol-prim. iso							

Name	Flash point (°F)		Explosive limits % by vol. in air		Auto-ignition temp. (°F)	Suscepti-bility to sponta-neous heating	Suitable extin-guishing agents
	Closed Cup	Open Cup	Lower	Upper			
Butyl carbitol	172	200			442	No	1, 3, 4
Butyl carbitol acetate		240				No	1, 2, 3, 4
Butyl cellosolve	141	165			472	No	1, 3, 4
Butylene-n	gas		1.7	9.0		No	
Butylene glycol	104					No	3, 4
Butyl formate-n	64					No	2, 3, 4
Butyl lactate	160					No	2, 3, 4
Butyl phthallyl butyl glycollate	385	390				No	1, 2, 3, 4
Butyl propionate-n	90	110			800	No	2, 3, 4
Butyl ricinoleate	230						1, 2, 3, 4
Butyl stearate-n	320	385					1, 2, 3, 4
Butyraldehyde	20					No	3, 4
Butyric acid-n	170					No	1, 3, 4
Butyric anhydride-n	190						1, 3, 4
Butyrone	190					No	3, 4
Camphor	150	200			871	No	1, 2, 3, 4
Camphor oil (light)	117	125				No	2, 3, 4
Carbitol	201	215				No	1, 3, 4
Carbitol acetate	225	230				No	1, 3, 4
Carbolic acid, see phenol							
Carbon, see charcoal							
Carbon disulfide	−22		1.0	50	257	No	1, 3, 4
Carbon monoxide	gas		12.5	74.2	1204	No	
Carbon oxysulfide, see carbonyl sulfide							
Carbonyl sulfide			11.9	28.5			1, 3, 4
Carnauba wax	540	595					1, 3, 4
Castor oil	445	545			840	Yes	1, 2, 3, 4
Cellosolve	104	120	2.6	15.7	460	No	3, 4
Cellosolve acetate	124	135	1.71		715	No	3, 4
Charcoal						Yes	1
China wood oil, see tung oil							
Chlorobenzene	90					No	2, 3, 4
Chloroethane, see ethyl chloride							
Chloroethyl acetate	129						1, 2, 3, 4
Chloroform	non-flammable					No	
Chloromethane, see methyl chloride							
Coal tar pitch	405	490					1, 2, 3, 4
Cocoanut oil	420	510				Yes	1, 2, 3, 4
Colophony, see pine resin							
Corn oil	490					Yes	1, 2, 3, 4
Cottonseed oil	590				650	Yes	1, 2, 3, 4
Creosote oil	165	185			638	No	1, 2, 3, 4
o-Cresol	178					No	1, 2, 3, 4
p-Cresol	187					No	1, 3, 4
Cresylic acid, see p-cresol							
Crotonaldehyde	55		2.95	15.5		No	3, 4
Crotonylene			1.37				2, 3, 4
Cumene, see propylbenzene-iso							
Cyanamide	285	405					1, 3, 4
Cyclohexane	1		1.31	8.35		No	2, 3, 4
Cyclohexanol	154					No	1, 2, 3, 4
Cyclohexanone	147					No	3, 4
Cyclohexanone delta	93					No	3, 4
Cyclohexyl acetate	136					No	2, 3, 4
Cyclopropane	gas		2.41	10.3		No	
p-Cymene	117	145			921	No	2, 3, 4
Decahydronaphthalene	136				504	No	2, 3, 4
Docalin, see decahydronaphthalene							
Decane-n	115		0.67	2.6	>500	No	2, 3, 4
Denatured alcohol—95%	60					No	3, 4
Diacetone alcohol (comm'l)	48	55				No	3, 4
Diamyl phthalate	245	340					1, 2, 3, 4
Dianisidine	403						1, 3, 4
Dibenzyl ether	275					No	1, 2, 3, 4
Dibutyl ether-n	77					No	3, 4
Dibutyl oxalate-n	220					No	1, 2, 3, 4
Dibutyl phthalate-n	315	335				No	1, 2, 3, 4
Dibutyl phthalate-iso	322					No	1, 2, 3, 4
Dibutyl tartrate-n	195	230				No	1, 2, 3, 4
Dicaproate, see triethylene glycol							
o-Dichlorobenzene	151	165				No	1, 2, 3, 4

Name	Flash point (°F)		Explosive limits % by vol. in air		Auto-ignition temp. (°F)	Suscepti-bility to sponta-neous heating	Suitable extin-guishing agents
	Closed Cup	Open Cup	Lower	Upper			
p-Dichlorobenzene	150	165				No	1, 3, 4
Dichlorodifluoromethane	non-flammable					No	
Dichloro ethane-1, 2, see ethylene dichloride							
Dichloroethene-1, 1, see dichloro-ethylene-1, 1							
Dichloro ethylene-1, 1	57		5.6	13	856	No	1, 2, 3, 4
Dichoro ethylene-1, 2	43		9.7	12.8		No	1, 2, 3, 4
Dichloroethyl ether-sym.	131	185			696	No	1, 2, 3, 4
Dichloroisopropyl ether		185				No	1, 2, 3, 4
Dichloromethane, see methylene chloride							
Dichloropropene-1, 2, see propylene dichloride							
Dichlorotetrafluoroethane	non-flammable					No	
Diethanolamine		280			1224	No	1, 3, 4
Diethylaminoethanol		140				No	3, 4
Diethyl carbitol		180				No	1, 3, 4
Diethyl carbonate	77					No	2, 3, 4
Diethyl cellosolve		95				No	3, 4
Diethyl diphenyl urea	302						1, 2, 3, 4
Diethylene glycol	255	290			444	No	1, 3, 4
Diethylene oxide	65		1.97	22.2			3, 4
Diethylene triamine		215					1, 3, 4
Diethyl ether	−20		1.7	48.0	366	No	3, 4
Diethyl glycophthalate	343						1, 3, 4
p-Diethyl phthalate	243	305					1, 2, 3, 4
Diethyl selenide			2.5			No	1, 2, 3, 4
Diethyl sulfate	220					No	1, 2, 3, 4
Diglycol chlorohydrin		225				No	1, 3, 4
Dihexyl, see dodecane							
o-Dihydroxy benzene, see pyro-catechol							
m-Dihydroxy benzene, see resorcinol							
Dimethoxy tetraglycol		285				No	1, 3, 4
Dimethyl aniline	145	170			700	No	1, 2, 3, 4
Dimethyl aniline, see also xylidine							
Dimethyl carbinol, see propyl alcohol,	sec.						
p-Dimethyl cyclohexane	52					No	2, 3, 4
Dimethyl ether	−42					No	3, 4
Dimethyl glycophthalate	369					No	1, 3, 4
Dimethyl ketone, see acetone							
o-Dimethyl phthalate	295	325				No	1, 2, 3, 4
Dimethyl sulfate	182	240				No	1, 2, 3, 4
Dinitro aniline-2, 4	435	510					1, 3, 4
Dinitro benzene	302						1, 3, 4
Dinitro chlorobenzene	382	405					1, 3, 4
Dinitro toluene-2, 4						No	1, 3, 4
Dioxane, see diethylene oxide							
Diphenyl	235	255				No	1, 3, 4
Diphenylamine	307					No	1, 3, 4
Diphenyl ether, see diphenyl oxide							
Diphenyl methane	266					No	1, 2, 3, 4
Diphenyl oxide	239					No	1, 2, 3, 4
Dipropylene glycol		280				No	1, 3, 4
Ditane, see diphenyl methane							
Divinyl ether	<−22		1.7	27.0	680	No	3, 4
Dodecane	165		0.6			No	1, 2, 3, 4
Ester gum	375	555				No	1, 3, 4
Ethane	gas		3.12	15.0	950	No	
Ethanol, see ethyl alcohol							
Ethanolamine		200				No	1, 3, 4
Ether, see diethyl ether							
Ethyl acetanilide	126						2, 3, 4
Ethyl acetate	24	30	2.18	11.5	907	No	3, 4
Ethyl acetoacetate	184					No	1, 3, 4
Ethyl alcohol	55		3.28	19	799	No	3, 4
Ethyl benzene	59	75				No	2, 3, 4
Ethyl bromide			6.75	11.25	952	No	1, 3, 4
Ethyl butyl carbonate	122					No	2, 3, 4
Ethyl butyrate	78	85				No	3, 4

Name	Flash point (°F)		Explosive limits % by vol. in air		Auto-ignition temp. (°F)	Suscepti-bility to sponta-neous heating	Suitable extin-guishing agents
	Closed Cup	Open Cup	Lower	Upper			
Ethyl cellosolve, see cellosolve							
Ethyl chloride	−58	−45	3.6	14.8		No	3, 4
Ethylene	gas		3.02	34	1009	No	
Ethylene chlorohydrin		140				No	1, 3, 4
Ethylene diamine	93	110				No	3, 4
Ethylene dichloride	56	65	6.2	15.9	775	No	1, 2, 3, 4
Ethylene glycol	232	240			775	No	3, 4
Ethylene oxide			3	80	804	No	3, 4
Ethyl formate	−4	10	3.5	16.5		No	2, 3, 4
Ethyl glycol acetate	117					No	3, 4
Ethyl lactate	115					No	3, 4
Ethyl nitrate	50	50	3.8			No	2, 3, 4
Ethyl nitrite	−31	−30	3.01	>50		No	2, 3, 4
Ethyl oxalate						No	1, 3, 4
Ethyl phthalyl ethyl glycollate	365	365					1, 3, 4
Ethyl propionate	54					No	2, 3, 4
Ethyl propyl carbinol acetate	135					No	3, 4
Ethyl *p*-toluene sulfonamide	260	380					1, 3, 4
Ethyl *p*-toluene sulfonate	316						1, 2, 3, 4
F-11, see monofluorotrichloro-methane							
F-12, see dichlorodifluoromethane							
F-114, see dichlorotetrafluoroethane							
Formaldehyde (soln. of gas in water)	130	200			806	No	3, 4
Freon, see dichlorodifluoromethane							
Fuel oil No. 1	100–165				490	No	2, 3, 4
Fuel oil No. 2	110–190				494	No	2, 3, 4
Fuel oil No. 3	125–200				498	No	2, 3, 4
Fuel oil No. 4	150+	250			505	No	1, 2, 3, 4
Fuel oil No. 5	150+					No	1, 2, 3, 4
Fuel oil No. 6	150+	320			755	No	1, 2, 3, 4
Furfural	140	155	2.1		739	No	1, 2, 3, 4
Fusel oil, see *n*-amylalcohol							
Gas, blast furnace			35	74		No	
Gas, coal gas			5.3	31	1200	No	
Gas, illuminating			5.3	31	1094	No	
Gas, natural			4.8	13.5		No	
Gas, oil gas			6.0	13.5	637	No	
Gas, producer			20.7	73.7		No	
Gas, water			9.0	55.0		No	
Gas oil	150+				640	No	1, 2, 3, 4
Gasoline	−50		1.3	6	495	No	2, 3, 4
Glycerine	320	350			739	No	1, 3, 4
Glycerol, see glycerine							
Glyceryl triacetate	280	295				No	1, 2, 3, 4
Glyceryl trinitrate, see nitroglycerin							
Glycol, see ethylene glycol							
Glycol diacetate		220				No	1, 2, 3, 4
Heptane-*n*	25		1	6	452	No	2, 3, 4
Hexahydro-xylol, see *p*-dimethyl cyclohexane							
Hexalin, see cyclohexanol							
Hexalin acetate, see cyclohexyl acetate							
Hexane-*n*	−7		1.25	6.90	477	No	3, 4
Hexanol-1	145					No	3, 4
Hexanol-3	137					No	3, 4
Hexanol-2	114					No	3, 4
Hexanone, see methyl butyl ketone							
Hexyl acetate, see ethyl propyl car-binol acetate and methyl butyl carbinol acetate							
Hydrocyanic acid	0		5.6	40	1000	No	
Hydrogen	gas		4.1	74.2	1076	No	
Hydrogen sulfide	gas		4.3	45.5		No	
Hydroquinone	329						1, 3, 4
Kerosene, see fuel oil No. 1							
Lanolin	460	560			833	Yes	1, 2, 3, 4
Lard oil (commercial)	395				833	Yes	1, 2, 3, 4
Lead tetramethyl			1.8				1, 2, 3, 4
Linseed oil	435	535			820	Yes	1, 2, 3, 4
Lubricating oil, cylinder		535			783	No	1, 2, 3, 4

Name	Flash point (°F)		Explosive limits % by vol. in air		Auto-ignition temp. (°F)	Suscepti-bility to sponta-neous heating	Suitable extin-guishing agents
	Closed Cup	Open Cup	Lower	Upper			
Lubricating oil, light machine	318	370				No	1, 2, 3, 4
Lubricating oil, motor		450				No	1, 2, 3, 4
Lubricating oil, spindle	169	200				No	1, 2, 3, 4
Lubricating oil, turbine		400			700	No	1, 2, 3, 4
Magnesium (powder or chips)						No	5
Manganese (powder)						No	5
Marsh gas, see methane							
Menhaden oil	435				828	Yes	1, 2, 3, 4
Metaldehyde	97	135				No	1, 3, 4
Methanal, see formaldehyde							
Methane	gas		5.3	13.9	999	No	
Methanol, see methyl alcohol							1, 2, 3, 4
Methox, see methoxy ethyl phthalate							
Methoxy ethyl phthalate	275	370				No	1, 2, 3, 4
Methyl acetate	15	20	4.1	13.9	935	No	3, 4
Methyl acetoacetate	180					No	1, 2, 3, 4
Methyl alcohol	54	60	6.0	36.5	878	No	3, 4
Methyl amine	0	10				No	3, 4
Methyl aniline, see toluidine							
Methyl bromide	practically non-flammable		13.5	14.5		No	
Methyl butyl carbinol acetate	113					No	3, 4
Methyl n-butyl ketone		95	1.22	8.0		No	2, 3, 4
Methyl iso-butyl ketone	62	75				No	2, 3, 4
Methyl butyrate	57					No	2, 3, 4
Methyl carbitol		200				No	1, 3, 4
Methyl cellosolve	107	115			551	No	3, 4
Methyl cellosolve acetate	132	140				No	3, 4
Methyl chloride	gas		8.2	19.7		No	
Methyl cyclohexane	25		1.15			No	2, 3, 4
Methyl cyclohexanol	154					No	1, 2, 3, 4
Methyl cyclohexanone	118					No	2, 3, 4
Methyl cyclohexyl acetate	147					No	2, 3, 4
Methylene chloride	practically non-flammable				1224	No	1, 2, 3, 4
Methyl ether, see dimethyl ether							
Methyl ethyl ether	−35		2	10.1	374	No	3, 4
Methyl ethyl ketone	30		1.81	11.5		No	3, 4
Methyl formate	−2		5.0	22.7	840	No	3, 4
Methyl glycol	97					No	3, 4
Methyl glycol acetate	111					No	3, 4
Methyl hexalin, see methyl cyclo-hexanol							
Methyl phenol, see o-cresol							
Methyl phthalyl ethyl glycollate	375	380				No	1, 2, 3, 4
Methyl propionate	28					No	2, 3, 4
Methyl iso-propyl carbinol, see amyl alcohol, sec. iso							
Methyl n-propyl ketone		60	1.55	8.15		No	3, 4
Methyl salicylate	214	225			850	No	1, 2, 3, 4
Methyl sulfate, see dimethyl sulfate							
Mineral seal oil (typical)	170	255				No	1, 2, 3, 4
Mineral spirits, see naphtha, safety solvent							
Monocresyl diphenyl phosphate	450	450				No	1, 2, 3, 4
Monofluorotrichloromethane	non-flammable					No	
Mustard oil	566					Yes	1, 2, 3, 4
Naphtha, coal tar	100–110				900–950	No	2, 3, 4
Naptha, safety solvent	100–110		1.1	6.0	450–500	No	2, 3, 4
Naphtha, V. M. & P.	20–45		1.2	6.0	450–500	No	2, 3, 4
Naphtha (high flash), see naphtha, coal tar							
Naphthalene	174	190	0.9		1053	No	1, 3, 4
Naphthol (β)	307	340				No	1, 3, 4
Naphthylamine (α)	315					No	1, 3, 4
Neatsfoot oil	470				828	Yes	1, 2, 3, 4
Nitric ether, see ethyl nitrate							
p-Nitroaniline	390	390					1, 3, 4
Nitrobenzene	190				924		1, 2, 3, 4
p-Nitrochlorobenzene	261						1, 3, 4
Nitroglycerin	Explodes				518	No	
Nitronaphthalene (α)	327						1, 3, 4
p-Nitrotoluene	223						1, 3, 4

Name	Flash point (°F) Closed Cup	Flash point (°F) Open Cup	Explosive limits % by vol. in air Lower	Explosive limits % by vol. in air Upper	Auto-ignition temp. (°F)	Susceptibility to spontaneous heating	Suitable extinguishing agents
m-Nitro p-toluidine	315	345					1, 3, 4
Nitrous ether, see ethyl nitrite							
Nonane-n	88		0.74	2.9		No	3, 4
Octane-n	56		0.84	3.2	450	No	3, 4
Octyl acetate-n	180					No	1, 2, 3, 4
Octyl alcohol-n	178					No	2, 3, 4
Octyl aldehyde	125					No	2, 3, 4
Oleic acid	372				685	Yes	1, 2, 3, 4
Oleo oil						Yes	1, 2, 3, 4
Olive oil	437				826	Yes	1, 2, 3, 4
Ozokerite	236	315				No	1, 3, 4
Palm oil	421				650	Yes	1, 2, 3, 4
Paraffin wax	390	430			473	No	1, 3, 4
Paraformaldehyde	158	200				No	1, 3, 4
Paranitraniline, see p-nitroaniline							
Peanut oil	540				833	Yes	1, 2, 3, 4
Pentane-n	<−40		1.4	8.0	588	No	2, 3, 4
Pentanone, see methyl propyl ketone							
Perchlorethylene, see tetrachloro-ethylene							
Perilla oil	522					Yes	1, 2, 3, 4
Petroleum, crude	20–90					No	2, 3, 4
Petroleum ether	50		1.4	5.9	475	No	3, 4
Phenanthrene						No	
Phenol	175	185			1319	No	1, 3, 4
Phenyl bromide, see bromobenzene							
Phenyl carbinol, see benzyl alcohol							
Phenyl cellosolve		250				No	3, 4
Phenyl chloride, see chlorobenzene							
p-Phenylene diamine	312	310				No	1, 3, 4
Phenyl ethane, see ethyl benzene							
Phenyl ethyl alcohol	216					No	1, 2, 3, 4
Phenyl methyl ketone, see aceto-phenone							
Phosphorous (red)					500	No	1
Phosphorous (yellow)					86	No	1
Phosphorous sesquisulfide					212	No	1
Phthalic anhydride	305	330				No	1, 3, 4
Picric acid	Explodes				<572	No	1
Pine oil	172	175				Yes	1, 2, 3, 4
Pine pitch	285					Yes	1, 3, 4
Pine resin	370	430				Yes	1, 3, 4
Pine tar	130	255			671	Yes	2, 3, 4
Pine tar oil	144					Yes	3, 4
Potassium						No	5
Potassium xanthate	205			9.5		No	1, 3, 4
Propane	gas		2.37	9.5	871	No	
Propyl acetate-n	58		2.0			No	2, 3, 4
Propyl acetate-iso	43	45	2.0		860	No	2, 3, 4
Propyl alcohol-n	59	85	2.5		812	No	3, 4
Propyl alcohol-iso	53	60	2.5		852	No	3, 4
Propyl alcohol-sec	67					No	3, 4
Propyl benzene-n	86					No	2, 3, 4
Propyl benzene-iso	102					No	2, 3, 4
Propylene	gas		2.0	11.1		No	
Propylene chlorohydrin	125	125				No	3, 4
Propylene dichloride	59	65	3.4	14.5	1035	No	2, 3, 4
Propylene glycol	210	225				No	1, 3, 4
Propylene oxide	<20					No	3, 4
Propyl ether-iso	−18	−15			830	No	2, 3, 4
Propyl formate-n	27					No	2, 3, 4
Propyl formate-iso	22					No	2, 3, 4
Pyridine	68		1.8	12.4	1065	No	3, 4
Pyrocatechol	261					No	1, 3, 4
Quenching oil	365	405				No	1, 2, 3, 4
Racemic acid, see tartaric acid							
Range oil, see fuel oil, No. 1							
Rape seed oil	325	550			836	Yes	1, 2, 3, 4
Red oil, see oleic acid							
Resorcinol	261					No	1, 3, 4
Rosin oil	266				648	Yes	1, 2, 3, 4
Sodium						No	5
Soya bean oil	540				833	Yes	1, 2, 3, 4

TABLE 10-1   Combustibility Properties of Solids, Liquids, and Gases—(Continued)

Name	Flash point (°F) Closed Cup	Open Cup	Explosive limits % by vol. in air Lower	Upper	Auto-ignition temp. (°F)	Suscepti-bility to sponta-neous heating	Suitable extin-guishing agents
Sperm oil, see whale oil							
Stearic acid	385	425			743	Yes	1, 3, 4
Stoddard solvent, see naphtha, safety solvent							
Sulfur	405	440			450	No	1
Sulfur chloride	245	None			453	No	1, 3, 4
Tallow	509					Yes	1, 3, 4
Tallow oil	492					Yes	1, 3, 4
Tannic acid		390				No	1
Tartaric acid		410				No	1
Tetrabromoethane, see acetylene tetrabromide							
Tetrachloroethane, see acetylene tetrachloride							
Tetrachloroethylene	non-flammable					No	
Tetradecane	212		0.5			No	1, 2, 3, 4
Tetraethylene glycol		345				No	1, 3, 4
Tetralin, see tetrahydronaphthalene							
Tetrahydronaphthalene	171	180				No	1, 2, 3, 4
Tin Tetramethyl			1.9			No	2, 3, 4
Toluene	40	45	1.27	7.0	1026	No	2, 3, 4
o-Toluidine	185				900	No	1, 2, 3, 4
p-Toluidine	188	205			900	No	1, 3, 4
Toluol, see toluene							
Tolylacetamide, see p-acetoluide							
o-Tolyl p-toluene sulfonate	363						1, 3, 4
Transformer oil, see transil oil							
Transil oil		295				No	1, 2, 3, 4
Triacetin, see glyceryl triacetate							
Tributyl phosphate		295				No	1, 2, 3, 4
Trichlorethylene	non-flammable					No	
Trichloromethane, see chloroform							
o-Tricresyl phosphate	460	504				No	1, 2, 3, 4
Triethanolamine	355	365				No	1, 3, 4
Triethylene glycol	350	385				No	1, 3, 4
Triethylene tetramine		260				No	1, 3, 4
Triglycol dichloride		250				No	1, 3, 4
Triisopropanolamine		305				No	1, 3, 4
Trimethylene, see cyclopropane							
Trinitrophenol, see picric acid							
Triphenyl phosphate	428					No	1, 3, 4
Tung oil	552				855	Yes	1, 3, 4
Turkey red oil	476				833	No	1, 2, 3, 4
Turpentine	95		0.8		488	Yes	2, 3, 4
Vinyl acetate	18	30			800	No	2, 3, 4
Vinyl chloride			4	22		No	3, 4
Vinyl ether, see divinyl ether							
Whale oil	446					Yes	1, 2, 3, 4
o-Xylene	63	75	1.0		924	No	2, 3, 4
o-Xylidine	206					No	1, 2, 3, 4
Zinc (dust or powder)						No	5
Zirconium (powder)					550	No	5

Adapted from Norbert A. Lange (ed.), "Handbook of Chemistry," 10th ed., McGraw-Hill Book Company, New York, 1967.

10-2, respectively. While both types of cups can be used on almost any combustible liquid, the open cup is usually used for less volatile liquids (i.e., having a high flash point, such as a cylinder lubricating oil), while the closed cup is used for more volatile liquids (i.e., having a low flash point, such as ethyl alcohol). The flash points of combustible liquids are given in Table 10-1.

Example 10-2   What is the flash point of ether?
solution   From Table 10-1, the flash point of ether (diethyl ether) is given as $-20°F$, indicating that this material has the potential for igniting at temperatures well below room temperature. Since it is a volatile liquid, the closed-cup flash point is used.

Under the conditions described, the liquid will flash only momentarily. However, if the concentration of vapor above the liquid is too high before an igniting flame is brought in contact with it, the mixture of vapor and air can burn violently in the form of an explosion. The same would be true of a combustible gas, such as hydrogen or methane, when the concentration of the combustible gas in air is sufficiently high. If the concentration of combustible gas or vapor is too low, either it will not burn or the combustion will take place without violence. At the other end of the scale, if the concentration of vapor or gas is too high, there will be insufficient oxygen in the mixture to sustain an explosive combustion. The explosive concentrations of various vapors and gases when mixed with air are given in Table 10-1. Any concentration of vapor or liquid below the lower explosive limit or above the upper explosive limit is free of explosion hazard. However, the converse is true in that any concentration that lies between these limits represents a mixture which will explode if ignited. These concentrations are therefore referred to as the *explosive limits* for the particular mixture.

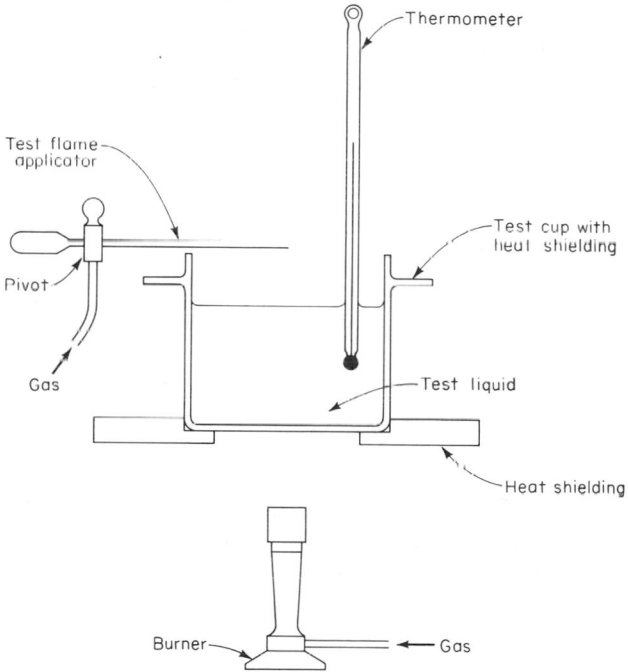

Fig. 10-1   Open-cup flash-point apparatus.

Example 10-3   Are the following gas or vapor concentrations in air explosive?
   10 percent ethyl alcohol
   10 percent *n*-heptane
   3 percent hydrogen
   40 percent methyl alcohol
solution   From Table 10-1, the explosive limits for ethyl alcohol are 3.28 and 19 percent. Since a vapor concentration of 10 percent lies between these limits, such a mixture is explosive. The other three mixtures are not explosive, since they all fall outside their respective explosive limits.

Example 10-4   The partial pressure of ethyl alcohol vapor in a mixture of vapor in air at atmospheric pressure is determined to be 145 mmHg. If the atmospheric pressure is found to be 770 mmHg, is the mixture explosive? If the atmospheric pressure is found to be 750 mmHg, is the mixture explosive?
solution   The concentration of ethyl alcohol vapor at an atmospheric pressure of 770 mmHg is $145/770 \times 100 = 18.8$ percent. Since this is within the explosive limits for ethyl alcohol as given in Table 10-1 (i.e., 3.28 to 19 percent), the mixture is explosive. At a total pressure of 750 mmHg, the concentration is $145/750 \times 100 = 19.3$ percent, which is outside the explosive limits, and the mixture is not explosive.

Example 10-5 Is the gas mixture in a partially filled can of ethyl alcohol explosive at room temperature? (The vapor pressure of ethyl alcohol at room temperature is 59 mmHg.)
solution The vapor concentration is $59/760 \times 100 = 7.8$ percent, which is within the explosive limits for ethyl alcohol, and the mixture is explosive.

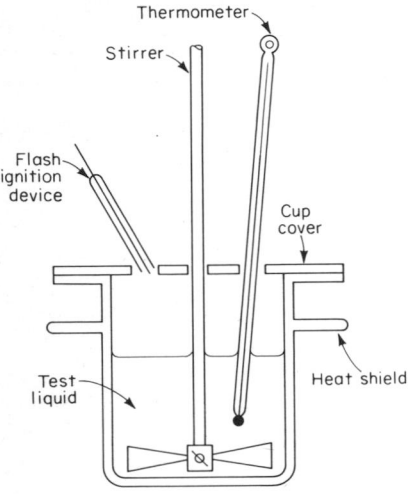

Fig. 10-2 Cup arrangement for closed-cup flash point.

Up to this point, we have considered combustion of a flammable mixture initiated by the application of a flame or spark. However, if such a flammable mixture is simply heated, a temperature is reached at which ignition automatically takes place (i.e., without the application of an ignition agent). This is known as *autoignition,* and the *autoignition temperature* is the temperature at which such a flammable mixture will spontaneously ignite. Autoignition temperatures for various flammable liquids, gases, and solids are also given in Table 10-1.

Under certain conditions, a flammable material can ignite without the apparent application of heat in a process known as *spontaneous combustion.* In spontaneous combustion, the ignition temperature is reached as a result of heat generated by a slow chemical reaction, but because of the physical circumstances, the heat cannot dissipate fast enough and the temperature of the system rises. Corn dust, for example, can undergo spontaneous combustion as a result of the heat generated by the fermentation of various sugars and starches within the corn under conditions such that the heat accumulates until the autoignition temperature is reached. Various substances, as listed in Table 10-1, have a susceptibility to spontaneous heating when exposed to air during storage. These materials are, of course, particularly susceptible to spontaneous combustion.

Table 10-1 also lists suitable extinguishing agents for each material in the event of fire. When using a combustible material on a regular basis, it is wise to post the recommended extinguishing agent prominently in all areas in which the material is used, but particularly in storage areas.

10-2 TOXIC MATERIALS

A *toxic substance* is one that has the potential to produce short- or long-term disease or bodily injury, to affect health adversely, to produce acute discomfort, or to endanger life (human or otherwise) as a result of exposure to the material. Frequently, even small amounts of toxic materials can be dangerous because they have a tendency to concentrate or attack particular organs, as indicated in Table 10-2. Since we are dealing with the effect on individuals who may react to various toxic materials to different degrees, it is difficult to establish an absolute evaluation of toxicity. However, most toxic materials do not have any particularly adverse effect when present in small amounts. Consequently, the concept of the *threshold limit value* (TLV) has been developed and is defined as that concentration of toxic material below which it is believed nearly all individuals may be repeatedly exposed, over a long duration, without adverse effect. Usually, TLV values refer to time-weighted average limits unless preceded by a C to designate a ceiling limit. The *time-weighted average* refers to the average concentration of toxicant over a period of time, while the *ceiling limit* is that concentration above which toxic effects are produced even after only short exposures. The time-weighted average concentration is used for those materials which may be continuously present or are generally slow-acting. For those materials which may be present only intermittently or are fast-acting and potent, the time-weighted average does not reveal the fact that a toxic concentration may actually be present during the exposure. For such substances or conditions, the C limit is more meaningful. Recommended TLVs are published periodically by the American Conference of Governmental Industrial Hygienists. Since TLVs inherently contain an experience factor, they are frequently updated as additional experience becomes available.

Additional definitions used in describing toxic substances are as follows:

Toxic concentration (TXC)—the concentration of a chemical in air (or solution) which has been reported actually to have produced harmful effects in man.

Toxic dose (TXD)—the dose of a chemical which has been reported to have produced toxic but nonfatal effects on man.

Lethal dose 50 (LD50)—a term used to describe the dose of a toxic substance which is expected to kill half (50 percent) of an entire population of experimental animals exposed to the toxic material by other than respiratory means.

TABLE 10-2 Body Organs Affected by Common Chemical Agents

Eyes	Dichloroethyl ether
Cresol	Mica
Quinone	Talc
Hydroquinone	Nitrogen dioxide
Acetic anhydride	
Acrolein	Skin
Benzyl chloride	Butyl alcohol
Butyl alcohol	Nickel
	Phenol
Upper Respiratory	Trichloroethylene
Mucus Membranes	
Ozone	Brain or Central
Dimethylsulfate	Nervous System
Chromium	Benzene
Acetic anhydride	Carbon tetrachloride
Acrolein	Carbon disulfide
Hydrogen sulfide	Butylamine
Butyl alcohol	Hydrogen sulfide
Acetaldehyde	Tetraethyl lead
	Manganese
Liver	Mercury
Cresol	Lead
Dimethylsulfate	Dimethylaniline
Chloroform	Acetaldehyde
Carbon tetrachloride	Nitrobenzene
Trichloroethylene	Thallium
Perchloroethylene	
Toluene	Kidney
	Chloroform
Heart	Mercury
Aniline	Dimethylsulfate
Lungs	Blood
Nickel	Nitrobenzene
Silica	Aniline
Asbestos	Arsenic
Betyllium	Benzene
Chromium	Carbon monoxide
Hydrogen sulfide	Toluene
Allyl chloride	

Note: Where a metal is indicated, it must be in finely divided form, or in the form of an oxide or other compound of metal.

Approximate lethal dose (LDca)—the amount of a toxic material which has been reported to produce a lethal effect or to be the maximum dose which can be tolerated without death. Other similar terms used include *Minimum lethal dose* (MLD), *lethal dose* (LD), and *maximum tolerated dose* (MTD).

Lethal concentration 50 (LC50)—the calculated concentration administered by respiration which is expected to kill 50 percent of a test population of experimental animals as a result of an exposure of 4 h to the toxic fumes.

Approximate lethal concentration (LCca)—that concentration of a toxic substance which has been reported either to produce a lethal effect or to be the maximum concentration that can be tolerated without lethal effect during an exposure of 1 day when administered by respiratory means.

These are the official terms used to describe toxic substances by the U.S. Department of Health, Education, and Welfare. This department, through its National Institute for Occupational Safety and Health, publishes a list of "Toxic Substances." Since it is updated annually, this listing is one of the most authoritative guides to the toxicity of a large number of chemical materials.

Entry of toxic substances is by skin *contact, inhalation,* or *ingestion.* Of these, exposure by skin contact is most frequent, followed by inhalation. Oral ingestion is generally of minor importance, except where exceptionally toxic agents are involved. Generally speaking, the skin is an extremely effective barrier for the protection for the underlying body tissues, and few substances are absorbed through this barrier in toxic quantities. Nevertheless, toxic agents can disturb or injure the skin, in which case penetration to the underlying tissues becomes easier. Alternatively, the agent may react with the skin itself to produce an irritation or sensitization. Of particular importance is contact with sensitive membranes such as those in the eyes or nose. The respiratory tract, however, presents an easy access to body tissues by toxic materials. The lung, which provides a large interfacial area between the air and bloodstreams, also permits foreign material to enter the lung and be readily absorbed into the body system. In addition, inhalation of particulate matter such as dust, mist, and similar substances can in itself result in acute irritations. Frequently, these irritations can lead to still greater toxicological effects. If the dust particle is small enough, it can actually enter the lung proper where, as in the case of a toxic gaseous material, it can be absorbed directly into the tissue of the recipient. Ingestion is usually limited to sporadic incidents, since ingestion is not likely to occur accidentally (except, of course, in the case of children). However, because ingested substances are usually present in high concentrations (even if only in small doses) their effects can be devastating.

Because mercury is widely used, *mercury vapor* warrants special consideration. Mercury vapor hazards arise from breathing mercury vapors, for which the lethal concentrations happen to be very low. Mercury vapors can originate from any source exposing the free metal to the atmosphere. A common source of mercury vapor is from droplets that are spilled (e.g., from a broken thermometer), and which find their way into various cracks and crevices. Fortunately, the vapor pressure of mercury at room temperatures is very low, as illustrated by Table 10-3, but exposure to even small concentrations of mercury vapor can have toxic effects if the exposure is over an extended time. Unfortunately, the vapor pressure of mercury rises very rapidly with increasing temperatures, so that even at modest temperatures, such as 100°C, the vapors emanating from free mercury droplets can reach dangerous levels. The atmospheric concentration TLV for mercury is 0.1 mg/m^3 for a normal work schedule. Since the equilibrium concentration of mercury vapor at 20°C corresponds to a concentration of about 2 mg/m^3, it can be seen that even small amounts of free mercury can be hazardous, particularly in a confined volume or in an area of poor ventilation. The detection and measurement of the concentration of mercury vapor is easily accomplished by ultraviolet absorption, and a number of very sensitive, direct-reading instruments are commercially available. These are portable and (usually) have a self-contained power source for convenient spot checks or continuous monitoring.

Fortunately, the measures necessary to minimize the hazards of mercury vapor can be quite simple. In fact, cleanliness and reasonable ventilation are usually all that is required for safety. Where mercury is used regularly or in quantity, the floor covering should be free of cracks and crevices and easily washed down with large amounts of water. Where mercury droplets appear on the floor, sink, or bench top, they can be picked up by means of a vacuum tube connected to a flask that can act as a trap.

Toxic *radiation* is even more difficult to contend with than mercury vapor, since radioactivity cannot always be swept away by ventilation or contained in a closed vessel. Like mercury vapor, it is completely undetected by the human senses, requiring special detection and precautionary devices to avoid accidental exposure. There are various forms of radiation having various degrees of toxicity, but as far as radioactive materials are concerned, they generally fall into four classes:

    Class 1   Very highly toxic
    Class 2   Highly toxic
    Class 3   Moderately toxic
    Class 4   Slightly toxic

TABLE 10-3 Vapor Pressure of Mercury

Temp (°C)	Vapor pressure (mmHg)	Temp (°C)	Vapor pressure (mmHg)
0	0.000185	120	0.7457
2	0.000228	122	0.8198
4	0.000276	124	0.9004
6	0.000335	126	0.9882
8	0.000406	128	1.084
10	0.000490	130	1.186
12	0.000588	132	1.298
14	0.000706	134	1.419
16	0.000846	136	1.551
18	0.00101	138	1.692
20	0.00120	140	1.845
22	0.00143	142	2.010
24	0.00169	144	2.188
26	0.00200	146	2.379
28	0.00236	148	2.585
30	0.00278	150	2.807
32	0.00326	152	3.046
34	0.00382	154	3.303
36	0.00447	156	3.578
38	0.00522	158	3.873
40	0.00608	160	4.189
42	0.00707	162	4.528
44	0.00820	164	4.890
46	0.00950	166	5.277
48	0.01098	168	5.689
50	0.01267	170	6.128
52	0.01459	172	6.596
54	0.01677	174	7.095
56	0.01925	176	7.626
58	0.02206	178	8.193
60	0.02524	180	8.796
62	0.02883	182	9.436
64	0.03287	184	10.116
66	0.03740	186	10.839
68	0.04251	188	11.607
70	0.04825	190	12.423
72	0.05469	192	13.287
74	0.06189	194	14.203
76	0.06993	196	15.173
78	0.07889	198	16.200
80	0.08880	200	17.287
82	0.1000	202	18.437
84	0.1124	204	19.652
86	0.1261	206	20.936
88	0.1413	208	22.292
90	0.1582	210	23.723
92	0.1769	212	25.233
94	0.1976	214	26.826
96	0.2202	216	28.504
98	0.2453	218	30.271
100	0.2729	220	32.133
102	0.3032	222	34.092
104	0.3366	224	36.153
106	0.3731	226	38.318
108	0.4132	228	40.595
110	0.4572	230	42.989
112	0.5052	232	45.503
114	0.5576	234	48.141
116	0.6150	236	50.909
118	0.6776	238	53.812

TABLE 10-3   Vapor Pressure of Mercury—(Continued)

Temp (°C)	Vapor pressure (mmHg)	Temp (°C)	Vapor pressure (mmHg)
240	56.855	300	246.80
242	60.044	302	257.78
244	63.384	304	269.17
246	66.882	306	280.98
248	70.543	308	293.21
250	74.375	310	305.89
252	78.381	312	319.02
254	82.568	314	332.62
256	86.944	316	346.70
258	91.518	318	361.26
260	96.296	320	376.33
262	101.28	322	391.92
264	106.48	324	408.04
266	111.91	326	424.71
268	117.57	328	441.94
270	123.47	330	459.74
272	129.62	332	478.13
274	136.02	334	497.12
276	142.69	336	516.74
278	149.64	338	537.00
280	156.87	340	557.90
282	164.39	342	579.45
284	172.21	344	601.69
286	180.34	346	624.64
288	188.79	348	648.30
290	197.57	350	672.69
292	206.70	352	697.83
294	216.17	354	723.73
296	226.00	356	750.43
298	236.21	358	760.00

From C. D. Hodgman (ed.), "Handbook of Chemistry and Physics," The Chemical Rubber Co., Cleveland, 1942.

Table 10-4 gives the relative radiotoxicity for a number of radioactive isotopes.

The first line of defense against radiotoxicity is, as with mercury vapor, cleanliness of the working area so as to avoid a buildup of hazardous concentrations. While protective clothing will not stop radiation, such clothing can be worn and destroyed after use to help prevent the dissemination of the toxic materials and to decrease the continuity of exposure.

In addition to these measures, shielding, or confining the radiation by surrounding the source with a radiation-absorbing material, is effective. Shielding, if adequate, can be very effective and relatively inexpensive. The most common shielding materials are lead and concrete blocks. Shielding should be as close to the source as possible to minimize stray radiation.

TABLE 10-4   Toxicity Classification of Radioisotopes

Class 1 (very high toxicity)	$Sr^{90} + Y^{90}$, $Pb^{210} + Bi^{210}$ (Ra D + E), Po^{210}, At^{211}, $Ra^{226} + 55\%$, daughter products, Ac^{227}, U^{233}, Pu^{239}, Am^{241}, Cm^{242}
Class 2 (high toxicity)	Ca^{45}, Fe^{59}, Sr^{89}, Y^{91}, $Ru^{106} + Rh^{106}$, I^{131}, $Ba^{140} + La^{140}$, $Ce^{144} + Pr^{144}$, Sm^{151}, Eu^{154}, Tm^{170}, $Th^{234} + Pa^{234}$, natural uranium
Class 3 (moderate toxicity)	Na^{22}, Na^{24}, P^{32}, S^{35}, Cl^{36}, K^{42}, Sc^{46}, Sc^{47}, Sc^{48}, V^{48}, Mn^{52}, Mn^{54}, Mn^{56}, Fe^{55}, Co^{58}, Co^{60}, Ni^{59}, Cu^{64}, Zn^{65}, Ga^{72}, As^{74}, As^{76}, Br^{82}, Rb^{86}, $Zr^{95} + NB^{95}$, Nb^{95}, Mo^{99}, Tc^{98}, Rh^{105}, $Pd^{103} + Rh^{103}$, Ag^{105}, Ag^{111}, $Cd^{109} + Ag^{109}$, Sn^{113}, Te^{127}, Te^{129}, I^{132}, $Cs^{137} + Ba^{137}$, La^{140}, Pr^{143}, Pm^{147}, Ho^{166}, Lu^{177}, Ta^{182}, W^{181}, Re^{183}, Ir^{190}, Ir^{192}, Pt^{191}, Pt^{193}, Au^{196}, Au^{198}, Au^{199}, Tl^{200}, Tl^{202}, Tl^{204}, Pb^{203}
Class 4 (slight toxicity)	H^3, Be^7, C^{14}, F^{18}, Cr^{51}, Ge^{71}, Tl^{201}

From "Accident Prevention Manual for Industrial Operations," 6th ed., National Safety Council, Chicago, 1969.

Fortunately, monitoring an area for radiation and measuring the degree of exposure of personnel are conveniently accomplished through instruments that indicate the presence of radiation electronically or by photographic film, which in fact, integrates the time and intensity of the radiation it receives. Radiation badges, as shown in Fig. 10-3, are worn by personnel exposed to radiation. They contain a small piece of photographic film the blackening of which (after development) provides a quantitative measure and permanent record of the amount of radiation the wearer has received.

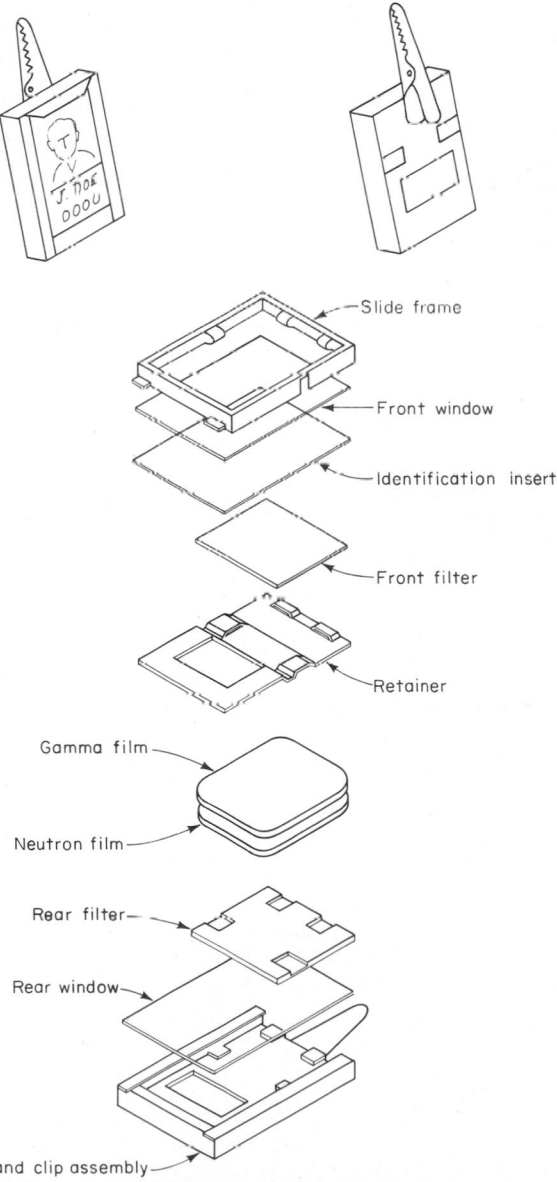

Fig. 10-3 Radiation badge for monitoring exposure of personnel to radiation. (*From "Accident Prevention Manual for Industrial Operations," 6th ed., National Safety Council, Chicago, 1969.*)

As a quick, ready reference to the hazardous nature of a large variety of materials, the National Fire Protection Association (NFPA) issues a guide in which the hazards of the material are rated according to health, flammability, and reactivity. In the NFPA system, a rating of 0 is used to indicate low hazard, ranging to a value of 4 for high hazard. Table 10-5 summarizes the classification system and identification color code used by NFPA, and Table 10-6 gives the NFPA ratings for a large number of materials. Table 10-7 summarizes the toxicity classifications based on standard procedures for determining toxicity relative to eye contact, skin penetration, skin irritation, and ingestion.

10-3   ELECTRICAL HAZARDS

Electrical hazards fall into three categories, shock, burns, and ignition. *Electrical shock* is a sudden disruption of certain body functions as a result of an electric current flowing through the body between two points. The severity of an electrical shock is dependent on the nature of the electrical impulse and the path it takes through the body, and can range from a mild tingling sensation to permanent injury or death. Physiologically, the current can interfere with the contraction and expansion of the chest muscles to an extent that it hinders or even prevents breathing, in which case death can actually occur by asphyxiation if the exposure is of sufficient duration. Alternatively, asphyxiation can occur by a temporary paralysis of nerve centers controlling respiration. This paralysis can continue for some time after removal of the traumatic current. In either case, the application of artificial respiration until the muscular spasm or paralysis passes can be an effective emergency procedure.

The passage of an electric current through the body can also interfere with the electrical signals that set up the normal heart rhythm, thus disorganizing the heart's contractions into an uncoordinated movement that does not accomplish the task of pumping the blood. This condition, known as ventricular fibrillation, can be caused by relatively small currents (i.e., of the order of 100 mA). Emergency treatment is with a defibrillator, a device that applies small, rhythmic shocks to the heart in the hope of getting it to contract in a coordinated rhythmic manner. Early application of a defibrillator under medical supervision provides a good chance for recovery. Complete suspension of heart action by muscular contraction can occur on passage of large currents. In this case, the heart does not go into fibrillation, and the normal rhythm of the heart may resume spontaneously when the victim is freed from the circuit. Finally, hemorrhaging accompanied by tissue, nerve, and muscle destruction can occur from heat effects induced by the passage of large currents.

It is difficult to forecast safe exposures to electric currents, owing to the variability among people. However, Table 10-8 indicates the general current levels needed to induce various physiological conditions, and since it is current that does the damage, it is possible to generalize that a current of 100 mA from a 60-Hz ac (common commercial or household) source will probably be fatal if it passes through a vital organ for only 1 s. However, 60-Hz alternating currents as low as 16 mA can induce muscular contractions from which the victim may not be able to release himself, thus setting up the condition under which the exposure time can be sufficiently large to be fatal. Except for very low-voltage exposures (i.e., below 24 V) and very low body currents (i.e., below 0.2 mA), it is difficult to state safe limits for electrical exposures, and even these low-energy exposures can be traumatic for people with heart diseases. However, several generalizations are useful:

1. Direct current passing through the body is more dangerous than alternating current.

2. Alternating current of low frequency is more dangerous than alternating current of high frequency.

3. Contact with a high-voltage source is more dangerous than contact with a low-voltage source (since the resultant current is proportional to the voltage, and it is the current that causes the primary damage).

4. Contacts that pass current through vital organs are more dangerous than those which bypass such organs, as illustrated in Fig. 10-4.

5. Short current paths through the body that pass through a vital organ are more dangerous than a long path, since the longer path has a higher resistance, thereby tending to reduce the current, as illustrated in Fig. 10-5.

6. Electrical contact with wet skin is more dangerous than with dry skin since the conductivity of a wet skin surface is 600 to 1,000 times greater than that of a dry skin surface, as shown in Table 10-9.

TABLE 10-5 NFPA Hazardous-Materials Classifications and Identification Color Code

Signal	Identification of health hazard, color code blue — Type of possible injury	Signal	Identification of flammability, color code red — Susceptibility of materials to burning	Signal	Identification of reactivity, stability color code yellow — Susceptibility to release of energy
4	Materials which on very short exposure could cause death or major residual injury even if prompt medical treatment is given	4	Materials which will rapidly or completely vaporize at atmospheric pressure and normal ambient temperature, or which are readily dispersed in air and which will burn readily	4	Materials which in themselves are readily capable of detonation or of explosive decomposition or reaction at normal temperatures and pressures
3	Materials which on short exposure could cause serious temporary or residual injury even though prompt medical treatment is given	3	Liquids and solids that can be ignited under almost all ambient-temperature conditions	3	Materials which in themselves are capable of detonation or explosive reaction but require a strong initiating source or which must be heated under confinement before initiation or which react explosively with water
2	Materials which on intense or continued exposure could cause temporary incapacitation or possible residual injury unless prompt medical treatment is given	2	Materials that must be moderately heated or exposed to relatively high ambient temperatures before ignition can occur	2	Materials which in themselves are normally unstable and readily undergo violent chemical change but do not detonate. Also materials which may react violently with water or which may form potentially explosive mixtures with water
1	Materials which on exposure would cause irritation but only minor residual injury even if no treatment is given	1	Materials that must be preheated before ignition can occur	1	Materials which in themselves are normally stable, but which can become unstable at elevated temperatures and pressures or which may react with water with some release of energy but not violently
0	Materials which on exposure under fire conditions would offer no hazard beyond that of ordinary combustible material	0	Materials that will not burn	0	Materials which in themselves are normally stable, even under fire exposure conditions, and which are not reactive with water

From "Accident Prevention Manual for Industrial Operations," 6th ed., National Safety Council, Chicago, 1969.

TABLE 10-6   Hazardous Chemicals Ratings

Display: Based on recommendations of the National Fire Protection Association, hazardous chemicals should bear a diamond-shaped label which is further divided into four smaller diamond, three of which contain colored numerals indicating the degree of hazard for health, flammability, and reactivity, as shown below. The fourth diamond is reserved for special information, such as "radioactive."

Color Coding: Health hazard—Blue
Flammability—Red
Reactivity—Yellow

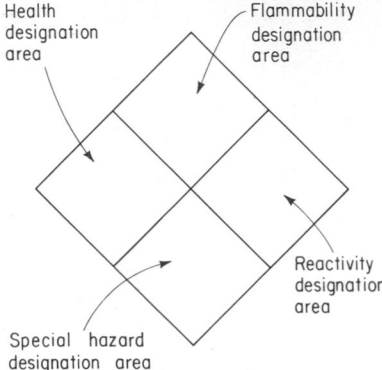

Ratings: Materials are rated from 0 for a completely safe material through 1, 2, 3, and 4, the latter indicating an extremely hazardous material. The relative ratings are based on the following guiding descriptions.

Health hazard rating (Blue):
1. Materials which cause irritation and have only minor residual effects even if no medical treatment is administered.
2. Materials which on intense or prolonged exposure cause temporary incapacitation or possibly residual injury unless prompt medical treatment is given.
3. Materials which on short exposure can cause serious temporary injury or residual injury even though prompt medical treatment is given.
4. Materials which on even short exposure can cause death or serious residual injury even though prompt medical treatment is given.

Flammability hazard rating (Red):
1. Materials which must be heated before ignition can occur.
2. Materials that must be heated or exposed to relatively high ambient temperatures before ignition can occur.
3. Liquids and solids that can ignite under almost all ambient temperature conditions.
4. Materials that rapidly vaporize at atmospheric pressure and normal ambient temperatures, and are readily ignitable.

Reactivity hazard rating (Yellow):
1. Materials that are normally stable but can become unstable at elevated temperatures and pressures; or that may react with water with some, but nonviolent release of energy.
2. Materials that are normally unstable and readily undergo violent chemical changes, but do not detonate; or that may react with water violently, or form a potentially explosive mixture when in contact with water.
3. Materials that are capable of strong or explosive reaction, but require a strong initiating source; or that require heat under confinement; or that react explosively with water.
4. Materials which are readily capable of detonation or of explosive decomposition or reaction at normal temperatures and pressures.

Chemical	Health	Flammability	Reactivity	Chemical	Health	Flammability	Reactivity
Acenaphthene		1		Acetonitrile	3	3	3
Acetal	2	3	0	Acetyl acetone	2	2	0
Acetaldehyde	2	4	2	Acetyl bromide	3	3	2
Acetic acid	2	2	1	Acetyl chloride	3	3	2
Acetic anhydride	2	2	1	Acetylene	1	4	3
Acetone	1	3	0	Acetylene dichloride	2	3	2

TABLE 10-6 Hazardous Chemicals Ratings—(Continued)

Chemical	Health	Flammability	Reactivity	Chemical	Health	Flammability	Reactivity
Acetylene tetrabromide	1	3	0	Auramine			
Acetylene tetrachloride	3			Aziridine	3	3	3
Acetylene trichloride	2	1		Azonaphthalene			
Acridine	2			Azoxybenzene		1	
Acrolein	3	3	3	Barium	2	2	
Acrylic acid	3	2	2	Barium chlorate	1	0	1
Acrylonitrile	4	3	2	Barium chloride	2	0	
Adipic acid	1	1		Barium hydroxide	2		
Adiponitrile	3	2		Barium nitrate	1	0	1
Aldrin	3	1	0	Barium peroxide	1	0	1
Allene		4		Benzal chloride			
Allyl acetate	3	1		Benzaldehyde	2	2	0
Allyl alcohol	3	3	1	Benzene	2	3	0
Allyl amine	3	3	1	Benzene sulfonic acid	3		
Allyl bromide	3	3	1	Benzene sulfonyl chloride			
Allyl chloride	3	3	1	Benzidine	3		
Allyl iodide	3	2		Benzoic acid	1	1	
Alum	1			Benzonitrile	3		
Aluminum	0	1	1	Benzotrichloride	2		
Aluminum borohydride		3		Benzotrifluoride	4	3	0
Aluminum bromide				Benzoyl chloride	3	2	1
Aluminum carbide				Benzoyl peroxide	1	4	4
Aluminum chloride	3	0	2	Benzyl acetate	1	1	0
Aluminum nitrate	2	2		Benzyl alcohol	2	1	0
Amino diphenylene oxide				Benzyl amine	3		
Aminopyridine	3			Benzyl benzoate	1	1	0
Ammonia	3	1	0	Benzyl bromide	3		
Ammonium dichromate	3	2		Benzyl chloride	2	2	1
Ammonium fluoride	3			Benzyl chloroformate	2		
Ammonium hydroxide	2	1		Benzyl cyanide	3		
Ammonium nitrate	2	1	3	Benzyl mercaptan		2	
Ammonium perchlorate	2	1	4	Beryllium	4	1	1
Ammonium persulfate	1	2		Beryllium chloride	4	1	
Ammonium sulfamate	1	1		Biphenyl	2	1	0
Ammonium sulfide		2		Biphenylamine	2	1	0
Ammonium thiocyanate	1			Bismuth chloride	3		
Amyl acetate	1	3	0	Borax	2	3	
Amyl alcohol	1	3	0	Boric acid	2		
Amylamine	3	3	0	Boron	2	2	
Amyl bromide	1	3	0	Boron hydride	3	4	3
Amylene	1	4	0	Boron oxide	2		
Amyl ether	1	2	0	Boron tribromide	2		
Amyl formate	1	3	0	Boron trichloride	2		
Amyl nitrate	1	2	2	Boron trifluoride	3	0	1
Amyl nitrite	1	2	2	Bromic acid	3	2	
Aniline	3	2	0	Bromine	4	0	1
Anisaldehyde	2	1	0	Bromine pentafluoride	3	3	
Anisole	1	2	0	Bromine trifluoride	3		4
Anthracene	1	2		Bromo-acetic acid	2		
Anthramine				Bromobenzene	2	2	0
Anthraquinone	1	1		Bromochloromethane			
Antimony	3	2		Bromoethane	2	3	0
Antimony hydride				Bromoethylene	2	3	
Antimony pentachloride	3			Bromoform	2		
Antimony pentasulfide	3	1	1	Bromotoluene	2	2	0
Antimony trichloride	3			Bromotrifluoromethane			
Antimony trioxide	3			Butadiene	2	4	2
Argon				Butane	1	4	0
Arsenic	3	2		Butene	1	4	0
Arsenic acid	3			Butyl acetate	1	3	0
Arsenic trichloride	3			Butyl alcohol	1	3	0
Arsenic trioxide	3			Butyl amine	3	3	0

TABLE 10-6   Hazardous Chemicals Ratings—(*Continued*)

Chemical	Health	Flammability	Reactivity	Chemical	Health	Flammability	Reactivity
Butyl bromide	2	3	0	Chlorobenzene	2	3	0
Butyl cellosolve	2	2		Chlorobromomethane	2		
Butyl chloride	2	3	0	Chlorocresol	3		
Butyl chromate	3			Chlorodifluoroethane	2	0	
Butyl formate	2	3	0	Chlorodinitrobenzene	3	1	4
Butyl hydroperoxide	1	4	4	Chlorodiphenyl	3	1	
Butyl mercaptan	2	3	0	Chloroethanol	3	2	0
Butyl methacrylate	2	2	0	Chloroform	3	1	0
Butyl methyl ketone	2	3	0	Chloromethyl ether			
Butyl peracetate	2	3	1	Chloronaphthalene	3		
Butyl perbenzoate	1	3	4	Chloronaphthylamine			
Butyl phosphate				Chloronitroaniline	3	1	
Butyl toluene	3	2		Chloronitrobenzene	3	1	1
Butyl vinyl ether	2	3	2	Chloronitropropane	3	2	3
Butyne		2		Chlorophenol	3	2	
Butyraldehyde	2	3	1	Chloropicrin	4	0	1
Butyric acid	2	2	0	Chloroprene	3		
Butyronitrile				Chlorosulfonic acid	3	0	2
Butyryl chloride	3			Chlorotrifluoroethylene	3	4	2
Cadmium	3	2		Chromic acid			
Cadmium chloride	3			Chromium bromide			
Cadmium oxide	3			Chromium chloride			
Calcium	1	1	2	Chromium nitrate			
Calcium arsenate	3			Chromium sulfate			
Calcium carbide	1	4	2	Chromium trioxide	1	0	1
Calcium carbonate	1			Chromyl chloride	3		
Calcium chlorate	2	2		Cinnamaldehyde			
Calcium chloride				Citric acid	1	1	
Calcium cyanide	2	0	0	Cobalt bromide	1		
Calcium hydride	3			Cobalt chloride	1		
Calcium hydroxide	2			Cobalt nitrate	2	0	1
Calcium hypochlorite	2	1	2	Cobalt sulfate	1		
Calcium oxide	1	0	1	Collodion	1	4	0
Caproic acid	2	1	0	Copper bromide	1		
Capryl alcohol	1	2	0	Copper chloride	1		
Caprylic acid	1	2	0	Copper nitrate	1	0	1
Carbazole	1	1		Copper sulfate	1		
Carbitol	1	1	0	Cresol	2	2	0
Carbon	1	1		Crotonaldehyde			
Carbon disulfide	2	3	0	Crotonitrile			
Carbon monoxide	2	4	0	Crotonyl chloride	1		
Carbon tetrabromide	3	0	1	Cumene	0	2	0
Carbon tetrachloride	3	0	0	Cyanamide	2	1	
Carbon tetrafluoride	2			Cyanoacetamide	3	2	
Carbonyl fluoride	2	3		Cyanogen	4	4	2
Cellosolve	2	2		Cyanogen chloride	3		
Cellosolve acetate	2	2		Cycloheptanone	2		
Cellulose nitrate	2	3	3	Cyclohexane	1	3	0
Chloral	3			Cyclohexanol	1	2	0
Chloral hydrate	2	1		Cyclohexanone	1	2	0
Chlordane	3			Cyclohexene	1	3	0
Chloric acid	3	3	3	Cyclohexylamine	2	3	0
Chlorine	3	0	1	Cyclohexylbenzene	2	3	0
Chlorine dioxide	3	3		Cyclopentadiene		2	
Chlorine trifluoride	3	3	3	Cyclopentane	1	3	0
Chloroacetaldehyde	3	2		Cyclopentanone	2	3	0
Chloroacetamide				Cyclopropane	1	4	0
Chloracetic acid	2	2		Cymene	2	2	0
Chloroacetonitrile	3			Decaborane	3	2	1
Chloracetophenone	3			Decahydronaphthanlene	2	0	0
Chloracetyl chloride	3			Decaldehyde			
Chloroaniline	3			Decane	0	2	0

TABLE 10-6 Hazardous Chemicals Ratings—(Continued)

Chemical	Health	Flammability	Reactivity	Chemical	Health	Flammability	Reactivity
Decyl alcohol		2		Dimethylazobenzene			
Diacetone alcohol	1	2	0	Dimethyldiphenylamine			
Diatomaceous earth				Dimethyl butane	1	3	0
Diazomethane	3			Dimethyl carbonate		3	1
Dibenzanthracene				Dimethyl ether	3	4	0
Dibenzcarbazole				Dimethylformamide	1	2	0
Dibenzofluorine				Dimethyl fumarate			
Dibenzylamine	3	1		Dimethylhydrazine	3	3	1
Diborane	3	4	3	Dimethylnaphthalene			
Dibromoethane	3			Dimethyl phthalate	0	1	0
Dibutylamine	3	2	0	Dimethyl propane		4	0
Dibutyl ether	2	3	0	Dimethyl sulfate	4	2	0
Dibutyl oxalate	3	1	0	Dimethyl sulfide	4	4	0
Dibutyl peroxide	2	3	4	Dimethyl sulfoxide	1	1	0
Dibutyl phthalate	0	1	0	Dinitroaniline	3	1	3
Dichloroacetic acid	3			Dinitrobenzene	4		4
Dichloroacetyl chloride	3	2	1	Dinitrocresol	2	2	
Dichloroaniline	3	1	0	Dinitrofluorine			
Dichlorobenzene	2	2	0	Dinitrophenol	3	2	
Dichlorobenzidine				Dinitropiperazine			
Dichlorobutane	2	2	0	Dinitrotoluene	3	1	3
Dichlorobutene	2	3	0	Dioctyl phthalate	0	1	0
Dichlorodifluoromethane	1			Dioxane	2	3	1
Dichloroethane	2	3	0	Diphenylamine	3	1	0
Dichloroethylene	2	3	2	Diphenylmethane	1	1	0
Dichloroethyl ether	2	2	0	Diphenylsulfide			
Dichloromethane	2	0	0	Dodecane	0	2	0
Dichloromonofluoromethane	1			Elon			
Dichloronitroethane	3	2	3	Endrin	3	1	0
Dichlorophenol		1	0	Epichlorohydrin	3	3	2
Dichlorophenylphosphine	3			Epoxy resin			
Dichloropropane	2	3	0	Ethane	1	4	0
Dichloropropene	3	2		Ethanethiol	2	4	0
Dichlorotetrafluoroethane	1	0		Ethanol	0	3	0
Dicyclohexylamine		2		Ethanolamine	2	2	0
Dicyclopentadiene	1	3	1	Ethoxynitroaniline			
Diethanolamine	1	1	0	Ethyl acetanilide	0	2	0
Diethyladipate				Ethyl acetate	1	3	0
Diethylaluminum chloride	3	3	3	Ethyl acetoacetate	2	2	0
Diethylamine	3	3	0	Ethyl acrylate	2	3	2
Diethylaminoethanol	3	2	0	Ethylamine	3	4	0
Diethylaniline	3	2	0	Ethylamyl ketone			
Diethyl carbonate	2	3	1	Ethyl aniline	3	2	0
Diethylene glycol	1	1	0	Ethyl benzene	2	3	0
Diethylenetriamine	3	1	0	Ethyl benzoate	1	2	0
Diethyl ether	2	4	1	Ethylbutyl ketone	1	2	0
Diethyl ketone	1	3	0	Ethyl butyrate	2	3	0
Diethyl malonate	0	1	0	Ethyl chloride	2	4	0
Diethyl phthalate	0	1	0	Ethyl chloroacetate	2	2	0
Diethyl sulfate	3	1	1	Ethyl chloroformate	3	3	1
Diethyl zinc	0	3	3	Ethylene	1	4	2
Dihydroxyanthraquinone		1		Ethylenediamine	3	2	0
Diiosbutyl ketone	1	2	0	Ethylenedichloride	2	3	0
Diisopropyl amine	3	3	0	Ethylene glycol	1	1	0
Diisopropyl ether	2	3	1	Ethylene nitrate	3	2	
Dimethoxybenzidine				Ethylene oxide	2	4	3
Dimethoxyethane		2	0	Ethylene oxide	2	4	3
Dimethoxymethane	2	3	2	Ethyl fluoride	1		
Dimethoxypropane		2		Ethyl fluoroacetate	4		
Dimethyl acetamide	2	1		Ethyl formate	2	3	0
Dimethyl amine	3	4	0	Ethyl hexanol	2	2	0
Dimethylaniline	3	2	0	Ethyl iodide	2	2	

TABLE 10-6 Hazardous Chemicals Ratings—(Continued)

Chemical	Health	Flammability	Reactivity	Chemical	Health	Flammability	Reactivity
Ethyl morpholine	2	3	0	Hydroquinone	2	1	
Ethyl nitrite	2	4	4	Hydroxylamine	1	3	3
Ethyl oxalate	3	2	0	Indanol			
Ethyl phenol		1	0	Iodic acid	3	2	
Ethyl silicate	2	2	0	Iodine	3		
Ethyl vinyl ether	2	4	2	Iodine chloride	3	2	
Fatty acids				Iodine pentafluoride	3		3
Ferric chloride	1			Iodine trichloride	3	1	
Ferrous ammonium sulfate	1			Iodoacetic acid	1		
Ferrous chloride	1			Isoprene	2	4	1
Ferrous sulfate	1			Jet fuel (kerosene)	1	3	0
Fluorine	4	0	3	Lactonitrile	3	2	
Fluoroacetic acid	2			Latex	2		
Fluoroboric acid	3			Lauric acid		1	
Fluoroethane				Lauroyl peroxide	0	2	3
Fluoroethylene	2	4	1	Lead	3	2	
Fluorosilicic acid	3			Lead acetate	3		
Fluorotrichloromethane	1			Lead arsenate	3		
Formaldehyde	2	4	0	Lead carbonate	3		
Formalin	2	2	0	Lead nitrate	1	0	1
Formamide	2	2		Lead oxide	3	1	
Formic acid	3	2	0	Lead thiocyanate	1	1	1
Fumaric acid	1	1		Lindane	3		
Furan	1	4	1	Liquified petroleum gas		3	
Furfural	1	2	1	Lithium	1	1	2
Furfuryl alcohol	1	2	1	Lithium aluminum hydride	3	1	2
Gallic acid				Lithium borohydride1	2		
Gasoline	1	3	0	Lithium carbonate			
Germane	2	3		Lithium hydride	1	4	2
Germanium dioxide	2			Magnesium	0	1	2
Germanium hydride				Magnesium chlorate	2	3	
Germanium tetrachloride	3			Magnesium chloride	2		
Glutaric acid				Magnesium nitrate	1	0	1
Glycerol	1	1	0	Magnesium oxide	2		
Glycolic acid	2			Magnesium perchlorate	1	0	1
Glyoxal	2	1		Malathion	3		
Grease				Maleic acid	2	1	
Heptachlor	3			Manganese	2	2	
Heptane	1	3	0	Manganese sulfate	2		
Heptylamine	2	2	0	Mercaptoethanol	2	2	
Hexachlorobenzene	1	1		Mercuric chloride	3		
Hexachloroethane	2			Mercury	3		
Hexachloronaphthalene	3			Mercury compounds (organic)	3		
Hexafluoroethane		0		Mercury fulminate	3	3	
Hexamethylene tetramine	2	1		Mesityl oxide	3	3	0
Hexane	1	3	0	Methacrylic acid	3	3	2
Hexanediamine	2	1		Methane	1	4	0
Hexanol	1	2	0	Methoxychlor	1		
Hexanone				Methoxyphenol	2	2	
Hexene	1	3	0	Methyl acetate	1	3	0
Hexyl acetate	1	2	0	Methyl acrylate	2	3	2
Hexylamine	2	3	0	Methyl alcohol	1	3	0
Hydrazine	3	3	3	Methyl amyl alcohol	2	2	0
Hydrazoic acid	3			Methyl amyl ketone	1	2	0
Hydriodic acid	3			Methyl benzoate	0	2	0
Hydrobromic acid	3			Methylbenzyl alcohol	1	1	0
Hydrochloric acid	3	0	0	Methyl bromide	3	1	0
Hydrocyanic acid	4	4	2	Methyl butene		4	0
Hydrofluoric acid	4	0	0	Methylbutylamine			
Hydrogen	0	4		Methyl butyl ketone	2	3	0
Hydrogen peroxide	2	0	3	Methyl butyrate	2	3	0
Hydrogen sulfide	3	4	0	Methyl cellosolve	2	2	

TABLE 10-6 Hazardous Chemicals Ratings—(Continued)

Chemical	Health	Flammability	Reactivity
Methyl cellosolve acetate	3	2	
Methyl chloride	2	4	0
Methyl chloroformate	3		
Methyl cyclohexane	2	3	0
Methyl cyclohexanol	3	2	0
Methyl cyclohexanone	3	2	0
Methyl cyclohexene		3	0
Methyl ethyl ether	2	4	1
Methyl ethyl ketone	1	3	0
Methyl formate	2	4	0
Methyl furan	2	3	1
Methyl hydrazine			
Methyl iodide	3		
Methyl isobutyl ketone	2	3	0
Methyl isobutyrate			
Methyl isocyanate			
Methyl isothiocyanate	3		
Methyl mercaptan	2	4	0
Methyl methacrylate	2	3	2
Methylnaphthalene	2	2	0
Methyl nitrosoaniline			
Methyl nitrosourea			
Methyl propyl ketone	2	3	0
Methylpyrrole		3	
Methyl salicylate	1	1	0
Methyl styrene	1	2	1
Methyl tetranitroaniline	2	3	
Methyl toluenesulfonate			
Methyl vinyl ether	2	4	2
Monochloroamine			
Monomethylamine	3	4	0
Morpholine	2	3	0
Naphtha	2	2	0
Naphthalene	2	2	0
Naphthol	2	1	
Naphthylamine	2	1	0
Naphthylisothiocyanate			
Nickel		2	
Nickel carbonyl	4	3	
Nickel nitrate	1	0	1
Nickel sulfate	1	0	
Nicotine	4	1	0
Nitric acid	2	0	1
Nitric oxide	3		3
Nitroacetophenone			
Nitroaniline	3	1	1
Nitrobenzene	3	2	0
Nitrobiphenyl	2	1	0
Nitroethane	1	3	3
Nitrofluorine			
Nitrogen dioxide	3	0	1
Nitrogen trifluoride	3	3	
Nitrogen trioxide	3		
Nitromethane	1	3	4
Nitronaphthalene	1	1	0
Nitrophenol	3		
Nitropropane	1	2	3
Nitrosodiethylamine			
Nitrosomethylaniline			
Nitrosomorpholine			
Nitrosopiperazine			
Nitrosopiperidine			

Chemical	Health	Flammability	Reactivity
Nitrosyl chloride	3		
Nitrotoluene	2	2	
Nonyl phenol	2	1	
Octachloronaphthalene			
Octafluorocyclobutane			
Octafluoropropane			
Octane	0	3	0
Octanoic acid			
Oils (fuel)	0	2	0
Oils (lubricating)	0	1	0
Oils (vegetable)	0	1	0
Oleic acid	0	1	0
Osmium tetroxide	3		
Oxalic acid	3		
Oxalyl chloride	3		
Oxygen	3	0	0
Ozone	3	3	
Paraffin	0	1	
Paraformaldehyde	2	2	2
Paraldehyde	2	3	1
Parathion	4	1	0
Pentaborane	3	4	3
Pentachloroethane	3	2	
Pentachloronaphthalene	3		
Pentachlorophenol	3		
Pentane	1	4	0
Pentanediol	1	1	0
Peracetic acid	3	2	4
Perchloric acid	3	0	3
Perchloromethyl mercaptan	3		
Perchloryl fluoride	2	2	3
Petroleum oil	1	2	0
Phenanthrene		1	
Phenanthreneacetamide			
Phenol	3	2	0
Phenyl acetate	1	2	0
Phenylazonaphthol			
Phenylenediamine	2		
Phenyl ether			
Phenylhydrazine	3	2	0
Phenylisocyanate			
Phenylnaphthylamine		2	
Phenylphenol	2	1	0
Phorone	2	2	0
Phosphine	3	3	
Phosgene	3		
Phosphoric acid	2		
Phosphorous	3	3	1
Phosphorous oxychloride	3		
Phosphorous pentachloride	3	2	
Phosphorous pentasulfide	3	1	2
Phosphorous tribromide	3		
Phosphorous trichloride	3	0	2
Phthalic anhydride	2	1	0
Picoline	2	2	0
Picric acid	2	4	4
Pimelic acid			
Pindone			
Pinene	1	3	0
Piperidine	2	3	3
Polyvinyl acetate			
Potassium	3	1	2

TABLE 10-6 Hazardous Chemicals Ratings—(Continued)

Chemical	Health	Flammability	Reactivity	Chemical	Health	Flammability	Reactivity
Potassium acetate	1			Silica gel (silicic acid)	1		
Potassium borohydride	2			Silicon tetrachloride	3		
Potassium carbonate				Silicon tetrafluoride	3		
Potassium chlorate	1	0	2	Silver nitrate	1	0	1
Potassium cyanide	2	0	0	Sodium	3	1	2
Potassium dichromate	3	2		Sodium acetate	0		
Potassium ferrocyanide	1			Sodium amide	3	2	
Potassium fluoride	3			Sodium azide	3		
Potassium hydroxide	3			Sodium benzoate	0		
Potassium nitrate	1	0	2	Sodium bicarbonate			
Potassium perchlorate	1	0	2	Sodium bisulfate	2		
Potassium permanganate	0	0	1	Sodium borohydride	2	2	
Potassium persulfate	1	0	1	Sodium carbonate	2		
Potassium peroxide	3	0	2	Sodium chlorate	1	0	2
Potassium sulfide	2	1	0	Sodium chloride	0		
Propane	1	4	0	Sodium chlorite	1	1	2
Propanediamine	2	3	0	Sodium chromate	3		
Propanediol	0	1	0	Sodium cyanide	2	0	0
Propenyl acetate	2	3		Sodium dichromate	3		
Propriolactone				Sodium ethoxide	3	3	
Proprionaldehyde	2	3	1	Sodium fluoride	3		
Proprionic acid	2	2	0	Sodium fluoroacetate	3		
Proprionitrile	3	2		Sodium formate	3	1	
Proprionyl chloride	3	3	1	Sodium hydride	3	2	
Propyl acetate	1	3	0	Sodium hydroxide	3	0	1
Propyl alcohol	1	3	0	Sodium hypochlorite	2		
Propylamine	3	3	0	Sodium iodide	2		
Propyl benzene		3	0	Sodium methoxide			
Propyl benzoate	1	1		Sodium nitrate	1	0	2
Propyl bromide	3	3		Sodium nitrite	3	2	
Propyl chloride	2	3	0	Sodium perchlorate	2	0	2
Propylene	1	4	1	Sodium peroxide	3	0	2
Propylene carbonate	1	1	0	Sodium phosphate			
Propylene disulfate				Sodium proprionate			
Propylene glycol				Sodium silicate	1		
Propylene imine	2	2		Sodium sulfide	2	1	0
Propylene oxide				Sodium sulfite	2		
Propyl ether	2	3	1	Sodium tetraborate	3		
Propyl formate	2	3		Sodium thiocyanate	3		
Propyl nitrate	2	3	3	Sodium thiosulfate	1		
Propyne	2	4	2	Stannic chloride	3	1	
Pyrethrum	2			Stearic acid	1	1	0
Pyridine	2	3	0	Stibine	3	2	
Pyrogallic acid	3			Stoddard solvent	1	2	
Pyrrolidine	2	3	1	Strontium	2	2	
Pyruvic acid				Strontium carbonate			
Quinoline	2	1	0	Strontium nitrate	1	0	1
Quinone	3			Strontium peroxide	1	0	1
Resorcinol	2	1		Strychnine	3		
Rotenone	2	1		Styrene	2	3	
Rubber				Styrene monomer	2	3	2
Rubidium	2	3		Succinic acid	2	1	
Ruthenium		2		Succinonitrile	3	1	
Salicylaldehyde		2	0	Sulfamide			
Salicylic acid	1	1		Sulfur	2	1	0
Selenic acid	3	2		Sulfur decafluoride	3		
Selenium	3	2		Sulfur dichloride	3	2	
Selenium dioxide	3			Sulfur dioxide	3	0	0
Selenium hexafluoride	3			Sulfur hexafluoride			
Selenium oxychloride	3			Sulfuric acid	3	0	1
Silane	3	3		Sulfur monochloride	2	1	1
Silica	2			Sulfur trioxide	3		

TABLE 10-6 Hazardous Chemicals Ratings—(*Continued*)

Chemical	Health	Flammability	Reactivity	Chemical	Health	Flammability	Reactivity
Sulfuryl chloride	3			Tricresyl phosphate	2	1	0
Sulfuryl fluoride	3			Tridecanol			
Tall oil		1		Triethyl aluminum	3		
Tallow	0	1	0	Triethylamine	2	3	0
Tannic acid	2	1		Triethanolamine	1	1	1
Tar	3	2		Triethylene glycol	1	1	0
Tellurium hexafluoride	2			Triethylene tetramine	3	1	0
Terphenyl		1	0	Triethyl formate	3	2	
Tetradecane		1	0	Trifluoroacetic acid	3		
Tetraethylenepentamine	2	1	0	Trifluoromethane	2		
Tetraethyl lead	3	2	3	Trimethylamine	3	4	0
Tetrahydrofuran	2	3	1	Trimethyl borate	2	3	1
Tetrahydronaphthalene	1	2	0	Trimethyl cyclohexanone	2	1	0
Tetramethyl lead	3	3	3	Trimethyl pentane	2	3	0
Tetramethyl silane	3			Trimethyl pentene		3	0
Tetranitromethane	3	3		Trinitrobenzene	2	4	4
Thallium	3			Trinitrotoluene	2	4	4
Thallous sulfate	3			Trioxane	2	2	0
Thioacetamide	2			Triphenyl phosphate	2	1	0
Thiodiethanol	1	1	0	Triphenyl phosphine	3	2	
Thioglycollic acid	3			Tripropylamine	2	2	0
Thionyl chloride	3			Trisodium phosphate	2		
Thionyl fluoride	3			Turpentine	1	3	0
Thiophene	2	3		Uranium	3	3	
Thiourea	1			Uranium nitrate	1	0	1
Thorium	1	2		Urea	0		
Thorium nitrate	1	0	1	Valeraldehyde			
Tin	0	1		Valeric acid	1		
Tin compounds (inorganic)				Vanadium dichloride	3		
Tin compounds (organic)	3			Vanadium oxytrichloride	3		
Titanium	0			Vanadium pentoxide		2	
Titanium dioxide	1			Vinyl acetate	2	3	2
Titanium tetrachloride	3	0	1	Vinyl chloride	2	4	2
Toluene	2	3	0	Vinyl ether	2	3	2
Toluene diisocyanate	2	1	2	Vinylidene chloride	2	4	2
Tolyl phosphate				Ucylene	2	3	0
Toluidine	3	2	0	Xylenol	3	1	
Triamyline	2	1	0	Xylidine	3	1	0
Tributylamine	2	2	0	Yttrium		2	
Tributylchlorotin				Zinc	0	1	1
Tributyl phosphate	2	1	0	Zinc acetate	0		
Trichloroacetic acid	3			Zinc chlorate	2	0	2
Trichloroacetonitrile				Zinc chloride	2	0	2
Trichloracetyl chloride	3			Zinc oxide	3		
Trichlorobenzene	2	1	0	Zirconium	1	4	1
Trichloroethylene	2	1		Zirconium acetate	1		
Trichloroanphthalene	2			Zirconium chlorate	1		
Trichloropropane	3	2	0	Zirconium nitrate	1		
Trichlorotrifluoroethane		1		Zirconium oxide	1		

TABLE 10-7   Relative Hazard Classification Based on Standard Toxicity Tests

EYE CONTACT

The most severe injury to the cornea of the eyes of 5 male albino rabbits following instillation of various volumes of undiluted fluid material, or of an excess of solutions of a material in glycol, deodorized kerosene, or water. (Excess, as used below, means the greatest amount which can reach the eye by accident. In the rabbit test an excess is 0.5 ml)

Relative hazard 1.   0.5 ml (an excess) does not cause severe injury
Relative hazard 2.   0.005 ml does not cause severe injury
Relative hazard 3.   0.005 ml or excess 40% solution causes severe injury
Relative hazard 4.   Excess 5% solution causes severe injury
Relative hazard 5.   Excess 1% solution causes severe injury

When solubilities have limited the concentrations applied to the rabbit eye, or when the material was applied only as a solid, the letter y indicates that correction has been applied to make the prediction consistent with the above.

BREATHING VAPORS

Mortality among 6 male albino rats weighing 90 to 120 g, inhaling vapors substantially saturated at room temperature and observed for 14 days thereafter

Relative hazard 1.   8 h inhalation kills 0, 1, 2, or 3 of 6 rats
Relative hazard 2.   2 or 4 h inhalation kills 2, 3, or 4 of 6 rats
Relative hazard 3. $\frac{1}{4}$ to 1 h inhalation kills, 2, 3, or 4 of 6 rats
Relative hazard 4.   2 or 5 min inhalation kills 2, 3, or 4 of 6 rats
Relative hazard 5.   2 min inhalation kills 5 of 6 rats

The test described is strictly a test of hazard: the likelihood that a material free to evaporate at room temperature will produce a vapor concentration which will be injurious to breathe for a short period. The relative hazard of breathing dusts and some volatile materials has been estimated from other tests, believed to be consistent with the above. When the hazard is based on mortality among 6 rats weighing 90 to 120 g inhaling known vapor concentrations for 4 h and observed for 14 days thereafter, the experimental results are:

Relative hazard 1.   4 h inhalation of 128,000 ppm kills 0 to 4 of 6 rats
Relative hazard 2.   4 h inhalation of 16,000 ppm kills 0 to 4 of 6 rats
Relative hazard 3.   4 h inhalation of 2,000 ppm kills 0 to 4 of 6 rats
Relative hazard 4.   4 h inhalation of 250 ppm kills 0 to 4 of 6 rats
Relative hazard 5.   4 h inhalation of 250 ppm kills 5 to 6 of 6 rats

SKIN PENETRATION

LD_{50} is the dosage killing half of a group of male albino rabbits weighing about 3 kg, within 14 days following administration by 24 h contact with about 40% of the body surface

Relative hazard 1. LD_{50}, more than 20 ml/kg body weight
Relative hazard 2. LD_{50}, 2 to 20 ml/kg
Relative hazard 3. LD_{50}, 0.2 to 1.99 ml/kg
Relative hazard 4. LD_{50}, 0.02 to 0.19 ml/kg
Relative hazard 5. LD_{50}, less than 0.02 ml/kg

IRRITATION OF UNCOVERED SKIN

The most severe reaction observed following contact of 0.01 ml of undiluted chemical, or of a 40% solution of a chemical, with the clipped belly of 5 male albino rabbits

Relative hazard 1.   Undiluted causes only capillary injection
Relative hazard 2.   Undiluted causes only slight erythema
Relative hazard 3.   Undiluted causes erythema and slight edema
Relative hazard 4.   Undiluted causes necrosis
Relative hazard 5.   10% solution causes necrosis

When the material has not been applied to the rabbit belly in undiluted form or when a somewhat different test of irritation was relied upon, the letter y is appended to the numerical hazard grade. Correction in such nonstandard tests has been made for the estimated reduction in irritation due to dilution, so that the entries are extrapolated to irritation expected from undiluted material.

It is important to stress the fact that irritation and all other effects on the skin are more severe when a material is covered, as when it contacts the skin on contaminated clothing. This increase is most dramatic when the material is between the skin and impervious gloves, or when it is inside shoes.

TABLE 10-7 Relative Hazard Classification Based on Standard Toxicity Tests—(Continued)

SWALLOWING

LD_{50} is the dosage killing half of a group of male albino rats weighing 90 to 120 g, within 14 days following administration through a stomach tube. In general, a smaller dosage would be required to kill a human.

Relative hazard 1.	LD_{50}, more than 10 g/kg body weight
Relative hazard 2.	LD_{50}, 1 to 10 g/kg
Relative hazard 3.	LD_{50}, 0.1 to 0.99 g/kg
Relative hazard 4.	LD_{50}, 0.01 to 0.099 g/kg
Relative hazard 5.	LD_{50}, less than 0.01 g/kg

From Irving Sunshine (ed.), "Handbook of Analytical Toxicology," The Chemical Rubber Co., Cleveland, 1969.

TABLE 10-8 Effects of Electric Currents on Humans

	Current (mA)					
	Direct		60 Hz		10,000 Hz	
Effect	Men	Women	Men	Women	Men	Women
Slight sensation on hand	1	0.6	0.4	0.3	7	5
Perception threshold	5.2	3.5	1.1	0.7	12	8
Shock—not painful, muscular control not lost	9	6	1.8	1.2	17	11
Shock—painful, muscular control not lost	62	41	9	6	55	37
Shock—painful, let-go threshold	76	51	16	10.5	75	50
Shock—painful and severe, muscular contractions, breathing difficult	90	60	23	15	94	63
Shock—possible ventricular fibrillation effect from 3-s shocks	500	500	100	100		
Short shocks lasting t s			$165/\sqrt{t}$	$165/\sqrt{t}$		
High-voltage surges, W-s or J	50	50	13.6	13.6		

From "Accident Prevention Manual for Industrial Operations," 6th ed., National Safety Council, Chicago, 1969.

Usually, circumstances which are known to be electrically hazardous do not turn out to be responsible for electrical-shock injuries simply because proper precautionary measures are easily established. Frequently the existence of a hazardous condition is not recognized simply because people do not realize that they are practically always connected to one "wire" of an electric circuit, namely, the ground, and that contacting only one other "hot" wire, if it has sufficient electrical potential above ground, can result in an electrical shock. Standing in water simply improves the electrical contact with the ground, while touching a hot wire with wet hands reduces the contact resistance. Either condition alone can increase the intensity of the shock by 100 to 1,000 times. Together, these conditions tremendously increase the chances of a shock being fatal.

Even where the shock condition is minimal, as, for example, for the passage of a shock current through the body without a direct path passing through a vital organ, considerable damage can occur to tissues, nerves, and muscles simply from the heat generated by the passage of the current, the damage being typical of *burns*. Where internal tissues are destroyed, hemorrhaging can occur, still further complicating the traumatic effects. External burns can also occur by contact with the electrical source, which may itself be hot. Usually, electrical shock is complicated by the simultaneous incidence of internal and/or external burns.

The hazard of *electrical ignition* refers to the fact that electrical conditions can cause combustion and explosion. In contact with combustible materials, an overheated circuit can readily cause the combustible to burst into flame. Most electrical-wiring codes are cognizant of this possibility and spell out conditions that effectively control this hazard (such as sheathing the wire in metallic conduit). However, an electrical spark, while it may be small and not carry much energy, nevertheless concentrates this energy in a small space. Thus, even small sparks can ignite an explosive mixture. The combustion so induced, of course, spreads very quickly, resulting in an explosion. This type of ignition can be very serious, since a spark can occur from the simple opening of a switch. (As the contacts of the switch open, current continues to flow between the contacts until they are sufficiently far apart to quench the spark.) Switches on low-voltage circuits, say, up to 3 V, usually are free of this hazards, but those operating at higher voltages are subject to this problem, and the spark can be quite pronounced on 110-V circuits. Also, ac circuits have a greater tendency to spark than dc circuits. Table 10-1 cites combustible (explosive) limits for organic vapors and gases. However, it should also be recognized that fine, combustible dusts, properly mixed with air, can also burn with explosive violence. Table 10-10 lists some of the more common explosive dusts.

Since electrical sparks or other forms of ignition can readily occur, gas and dust explosions are a serious hazard. However, there are convenient and inexpensive methods of preventing an explosion or minimizing its consequences. The best means is to have an

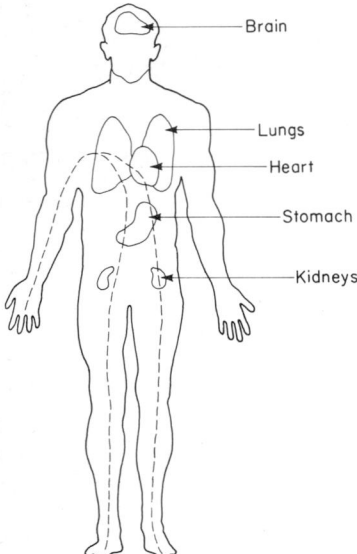

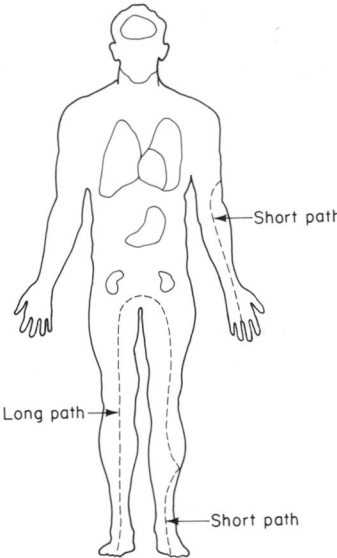

Fig. 10-4 Electric-current paths passing through vital organs. As shown, electric current enters hand and leaves through feet, as would be the case of touching a live wire while the feet are grounded.

Fig. 10-5 Short and long paths for electric currents passing through the body. A short path might occur by brushing a ground connection with the arm while holding a live wire. A long path might occur by brushing one leg against a live wire with the feet on a grounded surface. (Note also that the latter circumstance would also produce a short path.)

TABLE 10-9   Body Resistance to Passage of Electric Currents

Body area	Resistance (Ω)
Dry skin	100,000–600,000
Wet skin	1,000
Internal body—hand to foot	400–600
Ear to ear	About 100

From "Accident Prevention Manual for Industrial Operations," 6th ed., National Safety Council, Chicago, 1969.

TABLE 10-10   Common Combustible Dusts

Type	Example
Carbon	Coal, peat, charcoal, coke, lampblack
Fertilizers	Bone meal, fish meal, blood flour
Food products and by-products	Starches, sugars, flour, cocoa, powdered milk, grain dust
Metal powders	Aluminum, magnesium, zinc, iron
Resins, waxes, and soaps	Shellac, rosin, gum sodium resinate, soap powder, waxes
Spices, drugs, and insecticides	Cinnamon, pepper, gentian, pyrethrum, tea fluff
Wood, paper, tanning materials	Wood flour, wood dust, cellulose, cork, bark dust, wood extract
Miscellaneous	Hard rubber, sulfur, tobacco, many plastics

From "Accident Prevention Manual for Industrial Operations," 6th ed., National Safety Council, Chicago, 1969.

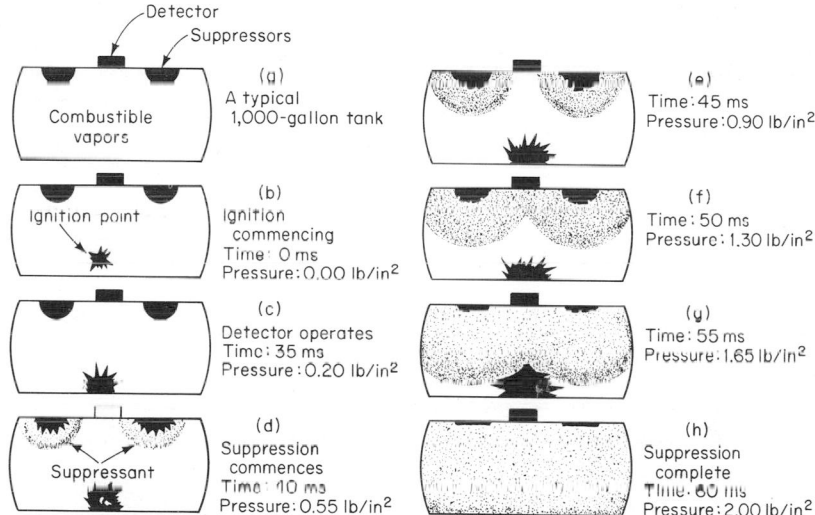

Fig. 10-6 Suppression of an explosion in a large tank. (*From "Accident Prevention Manual for Industrial Operations," 6th ed., National Safety Council, Chicago, 1969.*)

adequate ventilation system so that the concentration of combustible gas or dust never builds up to an ignitable mixture. To protect against sparks from switches or relays, the device can be encased in metal or screening to confine the heat of any combustion to the small space within the device, or a hermetically sealed version can be used. In closely confined spaces, such as inside storage tanks, explosion suppressors have been developed. These devices, activated by pressure waves set up at the instant of detonation, act at split-second speeds to counter the explosion, usually by a rapid venting of the vessel to minimize the destructive force of the explosion or, as illustrated in Fig. 10-6, by releasing a noncombustible gas (usually CO_2) that dilutes the combustible mixture to below combustible levels.

10-4 RADIATION HAZARDS

In recent years, there has been a growing awareness of the hazards inherent in exposure to various types of radiation. There are many types of radiation, and these are generally classified as to whether they are ionizing or nonionizing. *Ionizing radiations* are high-energy forms which can bring about atomic alterations which in turn can result in chemical reactions that do not occur under normal circumstances. X-rays, alpha particles, beta particles, gamma radiation, and neutron particles are typical of ionizing radiation. Because of their high energy content, ionizing radiations are those which present potential hazards. *Nonionizing radiations* are of lower energy content and generally do not bring about chemical changes (unless these involve easily brought about reactions). Typical of nonionizing radiations are light rays, heat waves, and radio waves.

X-rays are produced when high-speed electrons impact a metal target and the energy of the electrons is released as radiation, as illustrated in Fig. 10-7. The nature of the x-rays is determined by the velocity with which the electrons impinge the metal target as well as the nature of the target itself. X-rays can be thought of as being similar to light rays, except that they are of very high energy and short wavelengths. Because of the way in which x-rays are formed, they can be of a variety of wavelengths and intensities. X-rays having high energy contents and short wavelengths are referred to as "hard" x-rays and generally have an energy capable of penetrating at least several inches of steel. X-rays having a lower energy content and longer wavelengths are less penetrating and are referred to as "soft" x-rays. The ability of x-rays to penetrate matter is called their *quality*, while the *intensity* is the energy associated with a particular beam.

A convenient way of describing the combined intensity and quality of x-rays is in terms of the *half-value layer,* or the thickness of material which will reduce the incident radiation intensity by one-half. Concrete is frequently used to shield x-rays, and the half-value layer is generally in the order of several inches of concrete.

It is important to recognize that x-rays can be reflected as well as absorbed by a material on which they are incident. For protection in a room in which x-rays are being generated, it is important to surround an individual as completely as possible with absorbent shielding. As a minimum, such shielding should be disposed between the source of the x-rays and the individual.

Alpha particles are generated by nuclear reactions and consist of protons traveling at very high speeds. Because the proton is a fairly large particle, alpha particles are rather

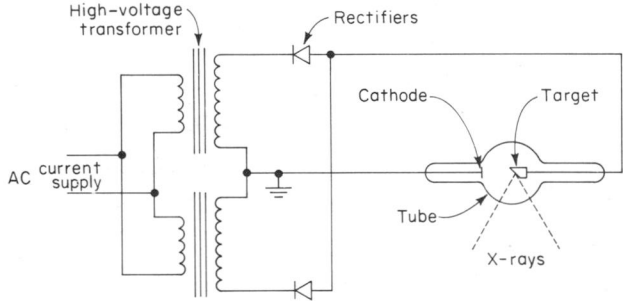

Fig. 10-7 Generation of x-rays.

easily stopped by various materials. For example, the energy of an alpha particle can be fairly completely dissipated by passage through 4 in of air or by thin films of water, paper, and practically any other liquid or solid. Because alpha particles are so easily stopped, exposure of an individual to an external source does not cause very serious physiological damage. If there is a high-intensity beam and parts of the body come close to the source, there can be some surface burning of the skin, but the penetration is not very deep and the damage, unless there is prolonged exposure, is usually not very serious.

The real danger from alpha particles comes from ingesting alpha-particle emitters. Alpha-particle emitters are generally high-atomic-number elements that are undergoing radioactive decay. Ingestion of an alpha-particle emitter, even in small doses, can result in its concentration in certain body components. Some emitters, for example, concentrate in bone, others in the kidneys or lungs; and even though the alpha particle is not very penetrating, it does do serious damage to the organ in which it resides. The main defense, therefore, against alpha particles is to avoid their entry into the body by avoiding ingestion or respiratory entry.

Beta particles also originate from nuclear reactions and differ from the proton in that they have the weight of an electron. A beta particle can be positively charged, in which case it is referred to as a *positron,* or it can be negatively charged, in which case it is called a *negatron.* Since a beta particle is a very small particle, it can have a significant penetration even though it may have a low energy content. For example, beta particles can penetrate wood to an extent of about 1.5 in and can penetrate body tissue from 0.1 to 0.5 in. Light-atomic-weight materials such as aluminum can completely stop beta particles within a distance of about 0.5 in. In air, beta particles are usually stopped in a few inches, but extremely high energy particles can penetrate to as much as 60 ft. Because they are easily stopped, the physiological damage from beta particles is usually restricted to surface burns.

Gamma rays are similar to x-rays but originate from nuclear reactions.

Neutrons are similar to alpha particles but are electrically neutral. Like alpha particles, they are released by nuclear reactions and are easily absorbed by thin layers of materials. Neutrons have an energy content sufficient to produce nuclear reactions in atoms with which they may collide, and these nuclear reactions can cause secondary emissions which can be alpha or beta particles or gamma rays. In addition, neutrons themselves are unstable and may decay to protons with the emission of beta particles. Again, ingestion or respiratory entry can result in major difficulties and should be avoided.

As a source of ionizing radiation, radioactive materials must receive special consideration. *Radioactive decay* involves spontaneous nuclear changes which emit alpha, beta, and gamma radiation. The important thing to remember about radioactive decay of a radioactive material is that it is going on all the time at rates which are inexorably fixed. They can be neither accelerated nor slowed down and continue until there is a complete transformation of the radioactive species to a nonradioactive form. The intensity and quality of the radiation generated by radioactive decay depend upon how much material is present and how much radioactivity has already taken place. It is a law of nature that the amount of radioactive decay decreases by a factor of one-half at equal time intervals. The time interval is different for each radioactive material. The *half-life* is the length of time for the intensity of radiation from a radioactive material to be reduced by one-half. The half-life varies for different radioactive materials from fractions of a second to billions of years. As an example, cobalt-60, a radioactive material used extensively in the treatment of cancer, has a half-life of 5.3 years, meaning that whatever the intensity of the radiation from a cobalt-60 source is at any particular time, it will be half that value in 5.3 years, and half again in another 5.3 years. Decay according to the half-life law is illustrated by Fig. 10-8, in which the quantity of radioactive radiations is expressed in roentgens. A roentgen (R) is a unit of radiation measurement and is equivalent to the amount of radiation emitted

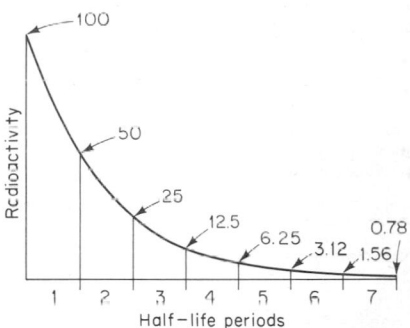

Fig. 10-8 Radioactive decay according to the half-life law. For each radioactive species there is a particular and exact half-life. After seven half-life periods, the radioactivity is less than 1 percent of what it was at the particular time at which the radioactivity of the sample was determined.

by 1 g of pure radium. A roentgen is a very large amount of radiation, and it is more common to express an amount of radiation in milliroentgens (mR), 1/1,000 R.

In recent times, a distinction has been made between the amount of radiation to which a specimen (or individual) has been exposed, and the amount of radiation it has absorbed. The rad, which has the same units as roentgens, is used to measure absorbed radiation. For biological purposes, the rad is more meaningful than the roentgen in measuring radiotoxicity. The physiological effects of radiation are difficult to measure quantitatively, particularly since they do not necessarily show up immediately after an exposure. However, the usual effects are skin redness, dermatitis, loss of hair, and in severe cases, cancer. These effects result from damage to the cell nucleus, which, in turn, results in the cell's being incapable of reproducing itself, or reproduces itself in a faulty manner. The intensity of the radiotoxicity and the permissible doses are extremely difficult to stipulate, since they vary not only with the source and nature of the radiation but with the age and health of the individual and the organs receiving the radiation. The age factor for the permissible accumulated dose is given by the formula

$$F = K(N - 18) \qquad (10\text{-}1)$$

where F = age factor, K = a body-organ factor, and N = age of the individual where $N \geqslant 18$. Body-organ factors are given in Table 10-11. For a person 18 years old or less, there is no permissible exposure to radiation. This is probably due to the decrease in reproductive activity associated with age.

Example 10-6 What are the age factors for the permissible ionizing-radiation dosage for total body exposure for 27- and 54-year-old individuals?
solution From Table 10-11, K for total body exposure = 5. Thus, from Eq. (10-1) for a 27-year-old,

$$F = 5(27 - 18) = 45$$

For a 54-year-old,

$$F = 5(54 - 18) = 180$$

Thus, it is permissible for a 54-year-old individual to have four times the accumulated radiation of a 27-year-old individual.

Example 10-7 For a 54-year-old individual, what are the age factors for permissible ionizing-radiation dosage for exposure of the thyroid and gonads?
solution From Table 10-11, K for thyroid exposure is 30, and for gonad exposure is 5. Thus, the age factors are $F = 30(54 - 18) = 1,080$ for thyroid exposure, and $F = 5(54 - 18) = 180$ for gonad exposure. Thus the permissible exposure for the thyroid is six times that for the gonads.

TABLE 10-11   Body-Organ Factors in Determining the Allowable Accumulated Radiation Dose for Individuals

Body organ	K [for Eq. (10-1)]
Blood-forming organs	5
Total body	5
Head and trunk	5
Gonads	5
Eyes	5
Skin	30
Thyroid	30
Feet	75
Ankles	75
Hands	75
Forearms	75
Bone	30
Other single organs	15

Dose factor-average equivalent dose in one year

From "Accident Prevention Manual for Industrial Operations," 6th ed., National Safety Council, Chicago, 1969.

TABLE 10-12   Radiation Intensity at a Distance of 1 ft from the Source for Common Radioisotopes

Isotope	Radiation intensity at 1 ft (mR/h)
Cobalt 60	14.5
Radium 226	9.0
Cesium 137	4.2
Iridium 192	5.9
Thulium 170	0.027

From "Accident Prevention Manual for Industrial Operations, "6th ed., National Safety Council, Chicago, 1969.

As protection against ionizing radiation, it is necessary not only to avoid contact with the source but even to avoid getting close to it. To facilitate this, indirect handling using mechanical hands, handling materials in contamination hoods, shielding, and distance are the best protective measures. Since radiation, like light, travels in straight lines and in all directions, the inverse-square law applies, meaning that the intensity of the radiation is inversely proportional to the square of the distance from its source. Thus, for cobalt-60, having a radiation intensity of 14.5 mR/h at a distance of 1 ft, the intensity drops to 3.6 mR at 2 ft, 0.9 mR at 4 ft, and 0.145 mR at 10 ft. Table 10-12 lists the radiation intensity at a distance of 1 ft from the source for a number of commonly used radioisotopes.

Fortunately, it is easy to measure radioactivity by the ionizing effects it produces. A wide variety of direct-reading monitoring instruments are available which give not only the total radiation and intensity but, if required, the type of radiation. Thus, it is easy to monitor the degree of radiation hazard that exists in any particular area. However, personnel normally move among areas of varying intensities, and the effects of some ionizing radiation are cumulative. Particularly where a low level of radiation is involved but exposure is likely to be of long duration, it is desirable to measure the total radiation to which an individual has been exposed. This can be conveniently accomplished by means of a radiation-film badge like the one shown in Fig. 10-3.

10-5   PERSONNEL PROTECTION

In addition to special protective devices that are specific for particular operations, chemical operations utilize a variety of devices for *general personnel protection:*

Head protection (safety hats, bump caps, visor caps, hairnets)

Ear protection (insert plugs, muffs)

Eye and face protection (spectacles, goggles, shields, and helmets)

Respiratory-tract protection (air purifiers, dust masks, and hoods)

Miscellaneous devices (safety belts, safety shoes, instep protectors, shoe covers, gloves, and impact shields)

Because of the particular problems associated with toxic materials, flammable materials, radioactive materials, explosion hazards, and electrical hazards, the chemical industries require special attention to personnel safety devices and protective equipment.

Eye protection in chemical operations is generally associated with preventing corrosive liquids, particularly acids, alkalies, and strong oxidizing agents, from entering the eyes. Plastic face shields (Fig. 10-9a) are a good first line of defense against mildly corrosive liquids. Such shields are lightweight and comfortable, but since they are open on the sides and bottom, they may permit some of the offensive liquid to reach the eyes if a large amount is involved. For more positive protection, side-enclosed plastic splash goggles with indirect or shielded vents (Fig. 10-9b) are used. In using splash goggles, care must be taken not to transfer the corrosive liquid from, say, the hands to the frame of the goggles, since such liquids have a tendency to creep to points where they can cause damage.

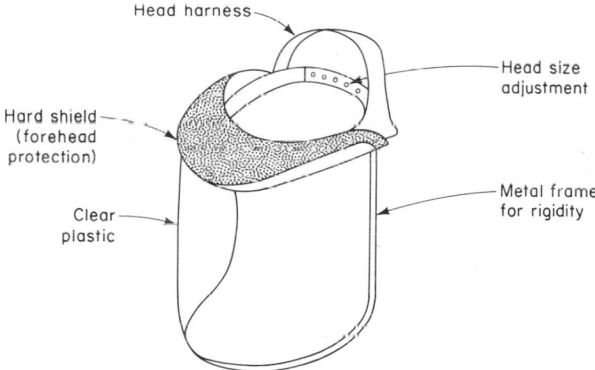

Fig. 10-9a   Plastic face shield.

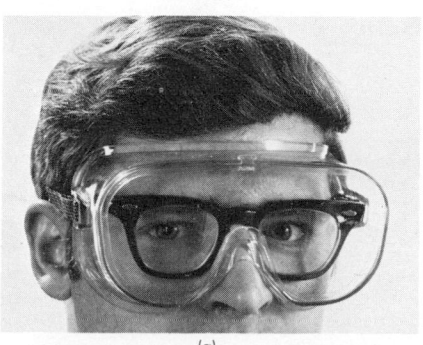

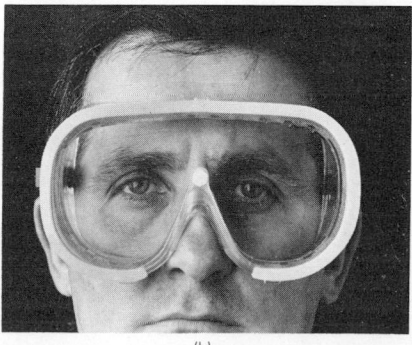

(a)

(b)

Fig. 10-9b   Plastic goggles for splash protection. (*a*) All-plastic, soft-sided cover goggles with shielded vents. (*b*) All-plastic, soft-sided chemical splash goggles with indirect and filtered ventilation for protection against heavy splash and driven mist. (*Mine Safety Appliance Co.*)

A special circumstance relating to infrared radiation should be mentioned. Such radiation, in large concentrations or prolonged exposure, can cause irreversible opacity in the proteinous fluid of the eye. While this is not a common difficulty, those who work around furnaces operating at incandescent temperatures should wear green-colored protective goggles. For low heat intensitites, or where it is also desired to protect facial tissues, a face screen made of metal screen (Fig. 10-10) can be very effective.

In the event of an accident in which a harmful material enters the eye, prompt and extensive flushing with water is perhaps the most effective first action that can be taken. Special eye-flushing faucets (Fig. 10-11) are available, and these should be installed in conveniently located stations in areas where eye-damaging liquids are handled.

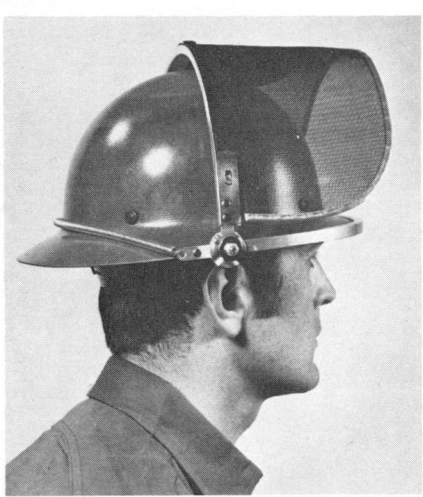

Fig. 10-10 Face screen for radiant-heat protection. (*Mine Safety Appliance Co.*)

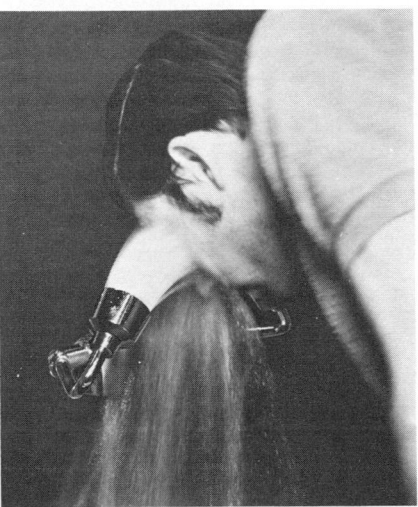

Fig. 10-11 Eye-flushing faucet. Faucet shown in full operation. (*Haws International.*)

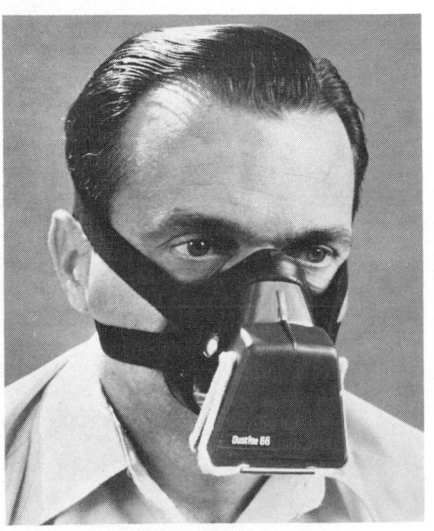

(a)

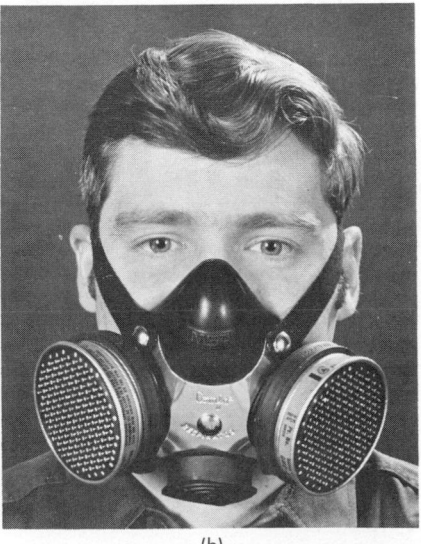

(b)

Fig. 10-12 (*a*) Light-service, single-cartridge respirator. (*b*) Heavy-duty, double-cartridge respirator. (*Mine Safety Appliance Co.*)

Respiratory protection ranges from lightweight filter masks covering the nose and the mouth to gas-absorbent gas masks. Dust respirators may be of the single-cartridge type (Fig. 10-12*a*) for light loads or nontoxic dusts. For finer filtration, a double-cartridge respirator (Fig. 10-12*b*) is generally used. The latter must provide sufficient filtration area to minimize breathing resistance.

For guidance, the U.S. Bureau of Mines defines two schedules. The first of these, designated 21A, is for respirators whose use is limited to pneumococcus-producing dusts, nuisance dusts or mists, acid mists, toxic dusts, and metal fumes. The second schedule, designated 21B, is much more rigid and is related to the threshold limit value (TLV) of the dust or mist. This schedule covers dusts of highly toxic materials such as beryllium, as well as highly radioactive materials.

In particularly difficult areas, respirators can be supplied with an independent source of respirable air. This is usually accomplished by supplying air at a small positive pressure to a full-face respirator mask. The air can be supplied through an air-line hose from a fixed compressed-air source, as in Fig. 10-13*a*, or from a self-contained compressed-air cylinder as in Fig. 10-13*b*.

A gas mask differs from a respirator in that it is equipped for purifying the contaminated air before it reaches the breathing mask. A gas mask is generally equipped with a canister through which the contaminated air passes and which contains chemicals which react with or absorb the contaminant. A diagram of a conventional gas mask is shown in Fig. 10-14. Since there is no universal decontaminant, gas masks are made for specific toxicants and must be carefully selected. The characteristics of canisters and their identifying colors are given in Table 10-13. The capacity of a gas mask, or its period of protection, depends on the type and size of the canister, the concentration of contaminant to be removed, and the air requirements (activity level) of the wearer. It is thus difficult to forecast exactly how long a particular canister can be used. In practice, canisters are often replaced long before they are exhausted, or even after a single use in order to be sure of their effectiveness. Where a

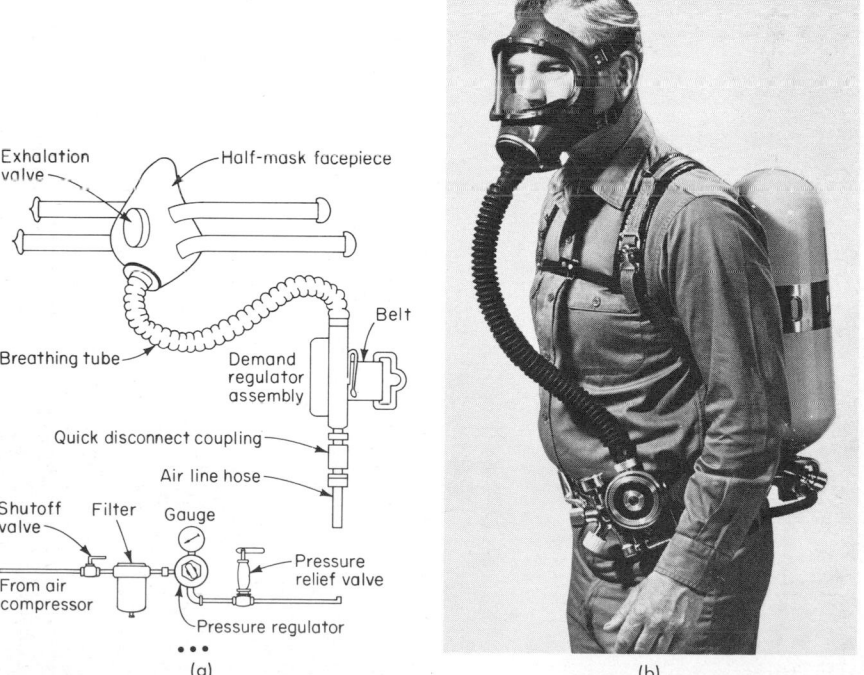

Fig. 10-13 (*a*) Air-pressure mask (air supplied by hose). (*From "Accident Prevention Manual for Industrial Operations," 6th ed., National Safety Council, Chicago, 1969.*) (*b*) Air-pressure mask (air supplied by portable cylinder of compressed air). (*Mine Safety Appliance Co.*)

TABLE 10-13 Gas-Mask Characteristics

Mask	Type designation	Maximum concentration of gas (% by vol.)	Color of canister	Material in canister	Remarks
Acid gas (for protection against gases such as hydrogen sulfide, sulfur dioxide, chlorine, hydrocyanic acid)	A	2	White	Soda lime or soda lime and activated charcoal	The time of protection decreases rapidly as the concentration of gas increases*
Organic vapor (for protection against vapors such as aniline, benzene, ether, gasoline, carbon tetrachloride, chloropicrin)	B	2	Black	Activated charcoal	*
Ammonia gas	C	3	Green	Silica gel or porous granules impregnated with metallic salts such as those of copper or cobalt	†
Carbon monoxide	D	2	Blue	Hopcalite	The air becomes noticeably warmer as the percentage of carbon monoxide increases ‡
Dust fumes, mists, fogs, smokes in combination with any of the above gases or vapors	AE	2	Any of above, plus top gray stripe	Any of above, plus mechanical filter	†
Combination acid gas and organic vapor	AB	2 acid gases, 2 organic vapors	Yellow	Activated charcoal and soda lime	
Combination acid gas, organic vapor, and ammonia gas	ABC	2 acid gas, 2 organic vapor, 2 ammonia	Brown	Soda lime, activated charcoal, and silica gel or impregnated porous granules	As the number of gases increases, the service time of the canister decreases
Combination acid gas, ammonia gas	AC	2 acid gases, 3 ammonia	Green with white stripe at bottom	Soda lime, silica gel, or porous granules impregnated with metal salts such as those of copper or cobalt	
Universal (for protection against combinations of acid gases, organic vapors, ammonia, carbon monoxide, and smokes)	N	Not exceeding 2 gases, 3 ammonia, 2 carbon monoxide, 2 organic vapors; not more than 2 total poisonous gases when more than one gas is present	Red	Soda lime, activated charcoal, calcium chloride, silica gel, or impregnated porous granules, hopcalite, and smoke filter	Stripes indicate that filters are also necessary

From "Accident Prevention Manual for Industrial Operations," 6th ed., National Safety Council, Chicago, 1969.

*Canister or single gas to have ½-in colored stripe (green for HCN, yellow for Cl₂; special markings for NH₃ and CO).

†½-in black or white stripe on canister if dust or mist filter combined with gas filter.

‡Canisters with additional dust or mist filters to have ½-in black or white stripe (near top).

large number of masks are in use, it may be economical to determine canister capacity by direct analysis. With the proper analytical procedure, usually stipulated by the manufacturer, such analyses can be conveniently run and yield reliable results.

In addition to conventional protective devices, chemical plants and laboratories should be capable of being thoroughly and conveniently cleaned; and where an accumulation of toxic materials can build up to a hazardous condition, such as with elemental mercury, cracks and crevices should be avoided or filled. Heavy-dose showers, capable of delivering at least 50 gal of water/min, should also be located at conveniently reached points.

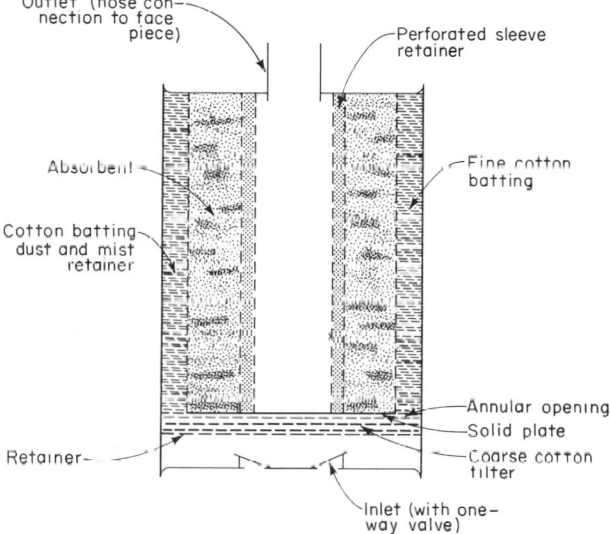

Outlet (hose connection to face piece)

Perforated sleeve retainer

Absorbent

Fine cotton batting

Cotton batting, dust and mist retainer

Annular opening

Solid plate

Retainer

Coarse cotton filter

Inlet (with one-way valve)

Fig. 10-14 Diagram of gas-mask canister.

10-6 WASTE DISPOSAL

Waste disposal is generally concerned with *spills,* or small (usually accidentally generated) quantities, *package lots,* or large bulk quantities, and *process waste,* or continuously generated waste. The disposal of wastes can be accomplished by relatively few methods:

    1. Incineration
    2. Dumping
        a. Into the ground
        b. Into a sewer system
        c. Into waterways
    3. Dispersion into the air

The disposal of wastes by any method has been made more difficult in recent years by expanded and tightened safety and pollution-control measures. This is particularly true where process waste is concerned. Here, the waste-disposal system has to be specific for the particular problem and is usually an integral part of the plant design and operation.

Spills and package lots, however, require special attention, since the selection of disposal equipment and procedures frequently becomes an obligation of the chemical technician. General procedures for the disposal of spills and package-lot wastes are summarized in Table 10-14.

In the case where a combustible, volatile liquid is to be incinerated, sawdust should be used to absorb the liquid before it is introduced into the incinerator, or it can be diluted with a low-volatility, combustible liquid in which it is soluble. Either of these procedures reduces the rate of release of vapor from the volatile liquid.

Special consideration must also be given to radioactive wastes, particularly since the radioactivity cannot be altered or terminated. However, here disposal standards are stipulated by law, based on regulations set down by government regulatory agencies, and must be an integral part of the plant's design and operation.

TABLE 10-14 Protection from and Disposal of Chemicals

Special instructions: Table 10-14 is intended as a guide for handling laboratory quantities of chemicals and provides minimal precautions to be followed for safe handling and disposal. The chemicals are listed alphabetically according to common usage names for ready reference. Occasionally, two names are given where both are in common usage. Since isomers require generally similar handling and disposal procedures, only generic names are given. At the end of the alphabetical listing, classes of compounds are given, to be used in the event a specific chemical is not listed. In any event, it is advisable to become acquainted with the various procedures *before* handling a chemical. In the case of protective devices, manufacturers of the equipment indicated can be consulted for specific models to be used. In the case of particularly hazardous materials, or those having a high recovery value (as is the case for most metals and metallic compounds) reference to the suppliers' recommendations is indicated. In all cases, local disposal ordinances and regulations should be consulted.

Disposal procedures:
1. Add slowly to large excess of dry soda ash or lime.
2. Absorb in vermiculite and collect in combustible boxes.
3. Mix with sand or limestone.
4. Add slowly to large excess of water.
5. Dissolve in acid (or alkali) as required, and dilute with large excess of water.
6. Dissolve in (or mix with) flammable solvent (alcohol, paint thinner, etc.) and spray into incinerator.
7. Mix with reducing agent (or oxidizing agent) as required, and dilute with a large excess of water.
8. Neutralize and flush down drain with large excess of water.
9. Transfer to open pit, add combustibles (paper, wood, etc.), and ignite. (Excelsior moistened with alcohol can be used as a safe "fuse.")
10. Remove to landfill, or special procedure required. Consult with supplier.

Chemical	Rubber gloves	Leather gloves	Laboratory coat	Fireproof clothing	Protective shoes	Safety glasses	Face shield	Body shield	Good ventilation	Fume hood	Respirator	Self-contained breathing apparatus	1	2	3	4	5	6	7	8	9	10
Acenaphthene	√	√				√			√		√			√							√	
Acetal	√	√				√				√								√			√	
Acetaldehyde	√	√								√				√							√	
Acetic acid	√	√				√	√	√	√											√	√	
Acetic anhydride	√	√				√	√	√	√											√	√	
Acetone	√	√				√			√		√							√			√	
Acetonitrile	√	√					√	√		√				√				√				√
Acetyl acetone	√	√				√			√		√			√							√	
Acetyl bromide	√	√								√			√			√				√		
Acetyl chloride	√	√								√			√			√				√		
Acetylene	√	√				√			√		√										√	
Acetylene dichloride	√	√							√	√	√										√	
Acetylene tetrabromide		√			√																	√
Acetylene tetrachloride	√	√							√	√	√										√	
Acetylene trichloride	√	√							√	√	√										√	
Acridine	√	√								√							√				√	
Acrolein	√	√								√								√			√	
Acrylic acid	√	√				√	√	√	√											√	√	
Acrylonitrile	√	√					√	√		√	√			√				√				√

TABLE 10-14 Protection from and Disposal of Chemicals (Continued)

Chemical	Rubber gloves	Leather gloves	Laboratory coat	Fireproof clothing	Protective shoes	Safety glasses	Face shield	Body shield	Good ventilation	Fume hood	Respirator	Self-contained breathing apparatus	1	2	3	4	5	6	7	8	9	10
Adipic acid	√		√			√	√	√	√											√	√	
Adiponitrile	√		√				√	√			√		√			√						√
Aldrin	√		√						√		√		√								√	
Allene	√		√			√		√		√				√							√	
Allyl acetate	√		√			√		√		√				√							√	
Allyl alcohol	√		√			√		√		√				√							√	
Allyl amine	√		√		√			√		√				√							√	
Allyl bromide	√		√					√		√				√							√	
Allyl chloride	√		√					√		√				√							√	
Allyl iodide	√		√					√		√				√							√	
Alum	√		√		√											√				√		
Aluminum		√		√	√												√					√
Aluminum borohydride	√			√		√	√								√							√
Aluminum bromide	√	√					√		√	√	√		√			√				√		
Aluminum carbide	√	√		√												√						√
Aluminum chloride	√	√					√		√	√	√		√			√				√		
Aluminum nitrate	√	√		√												√				√		
Amino-diphenylene oxide	√	√									√				√						√	
Aminopyridine	√	√									√				√						√	
Ammonia	√	√		√			√		√							√				√		
Ammonium dichromate	√	√			√		√	√	√							√						√
Ammonium fluoride	√	√		√												√				√		
Ammonium hydroxide	√	√		√			√		√							√				√		
Ammonium nitrate	√	√		√												√				√		
Ammonium perchlorate	√	√			√		√	√	√							√						√
Ammonium persulfate	√	√			√		√	√	√							√						√
Ammonium sulfamate	√	√			√	√										√				√		
Ammonium sulfide	√	√		√			√		√							√						√
Ammonium thiocyanate	√	√		√												√				√		
Amyl acetate	√	√				√		√		√				√							√	
Amyl alcohol	√	√				√		√		√				√							√	
Amylamine	√	√		√				√		√				√							√	
Amyl bromide	√	√						√		√	√			√							√	
Amylene	√	√				√		√		√				√							√	
Amyl ether	√	√				√					√								√		√	
Amyl formate	√	√				√		√		√				√							√	
Amyl nitrate	√	√						√			√		√			√					√	
Amyl nitrite	√	√						√			√		√			√					√	
Aniline	√	√									√				√						√	

TABLE 10-14   Protection from and Disposal of Chemicals (Continued)

Chemical	Rubber gloves	Leather gloves	Laboratory coat	Fireproof clothing	Protective shoes	Safety glasses	Face shield	Body shield	Good ventilation	Fume hood	Respirator	Self-contained breathing apparatus	1	2	3	4	5	6	7	8	9	10
Anisaldehyde	√	√										√		√							√	
Anisole	√	√				√						√							√		√	
Anthracene	√	√			√		√		√					√							√	
Anthramine	√	√										√			√						√	
Anthraquinone	√	√			√		√		√					√							√	
Antimony		√		√	√													√				√
Antimony hydride	√	√										√										√
Antimony pentachloride	√	√			√				√	√								√				√
Antimony pentasulfide	√	√			√				√	√								√				√
Antimony trichloride	√	√			√				√	√								√				√
Antimony trioxide	√	√			√				√	√								√				√
Argon		√			√																	√
Arsenic	√	√			√				√	√								√				√
Arsenic acid	√	√			√				√	√								√				√
Arsenic trichloride	√	√			√				√	√								√				√
Arsenic trioxide	√	√			√				√	√								√				√
Auramine	√	√										√				√					√	
Aziridine	√	√										√				√					√	
Azonaphthalene	√	√										√				√					√	
Azoxybenzene		√				√	√	√								√				√		
Barium	√	√			√													√				√
Barium chlorate	√	√				√			√	√	√							√				√
Barium chloride	√	√			√													√				√
Barium hydroxide	√	√			√													√				√
Barium nitrate	√	√			√													√				√
Barium peroxide	√	√			√													√				√
Benzal chloride	√	√							√		√	√	√								√	
Benzaldehyde	√	√									√				√						√	
Benzene	√	√				√			√	√											√	
Benzene sulfonic acid	√	√							√	√	√		√				√			√		
Benzene sulfonyl chloride	√	√									√	√	√				√			√		
Benzidine	√	√										√				√					√	
Benzoic acid	√	√				√	√	√	√											√	√	
Benzonitrile	√	√						√	√			√	√				√					√
Benzotrichloride	√	√								√	√	√	√								√	
Benzotrifluoride	√	√								√	√	√	√								√	
Benzoyl chloride	√	√									√	√	√				√			√		
Benzoyl peroxide	√	√				√	√			√				√			√				√	
Benzyl acetate	√	√				√		√			√				√						√	

TABLE 10-14   Protection from and Disposal of Chemicals (Continued)

Chemical	Rubber gloves	Leather gloves	Laboratory coat	Fireproof clothing	Protective shoes	Safety glasses	Face shield	Body shield	Good ventilation	Fume hood	Respirator	Self-contained breathing apparatus	Disposal 1	2	3	4	5	6	7	8	9	10
Benzyl alcohol	✓	✓				✓			✓		✓			✓							✓	
Benzyl amine	✓	✓										✓			✓						✓	
Benzyl benzoate	✓	✓				✓			✓		✓			✓							✓	
Benzyl bromide	✓	✓								✓		✓	✓								✓	
Benzyl chloride	✓	✓								✓		✓	✓								✓	
Benzyl chloroformate	✓	✓								✓		✓	✓								✓	
Benzyl cyanide	✓	✓					✓	✓				✓	✓				✓					✓
Benzyl mercaptan	✓	✓										✓							✓			
Beryllium	✓	✓			✓													✓				✓
Beryllium chloride	✓	✓										✓						✓				✓
Biphenyl	✓	✓				✓			✓	✓				✓							✓	
Biphenylamine	✓	✓										✓			✓						✓	
Bismuth chloride	✓	✓			✓				✓	✓								✓				✓
Borax	✓	✓			✓												✓			✓		
Boric acid	✓	✓			✓												✓			✓		
Boron		✓			✓																✓	
Boron hydride	✓			✓		✓	✓							✓								✓
Boron oxide	✓	✓					✓					✓		✓						✓		
Boron tribromide	✓	✓		✓								✓	✓			✓				✓		
Boron trichloride	✓	✓		✓								✓	✓			✓				✓		
Boron trifluoride	✓	✓		✓								✓	✓			✓				✓		
Bromic acid	✓	✓				✓		✓	✓	✓						✓						✓
Bromine	✓	✓				✓		✓	✓	✓						✓						✓
Bromine pentafluoride	✓	✓		✓								✓	✓			✓				✓		
Bromine trifluoride	✓	✓		✓								✓	✓			✓				✓		
Bromo-acetic acid	✓	✓						✓	✓	✓		✓	✓			✓				✓		
Bromobenzene	✓	✓								✓	✓	✓									✓	
Bromochloromethane		✓			✓																	✓
Bromoethane	✓	✓								✓	✓	✓									✓	
Bromoethylene	✓	✓								✓	✓	✓									✓	
Bromoform		✓			✓																	✓
Bromotoluene	✓	✓								✓	✓	✓									✓	
Bromotrifluoromethane		✓			✓																	✓
Butadiene	✓	✓				✓	✓		✓	✓											✓	
Butane	✓	✓				✓	✓		✓	✓											✓	
Butene	✓	✓				✓	✓		✓	✓											✓	
Butyl acetate	✓	✓				✓	✓		✓	✓											✓	
Butyl alcohol	✓	✓				✓	✓		✓	✓											✓	
Butyl amine	✓	✓			✓				✓		✓			✓							✓	

10-41

TABLE 10-14   Protection from and Disposal of Chemicals (*Continued*)

Chemical	Rubber gloves	Leather gloves	Laboratory coat	Fireproof clothing	Protective shoes	Safety glasses	Face shield	Body shield	Good ventilation	Fume hood	Respirator	Self-contained breathing apparatus	Disposal procedures 1	2	3	4	5	6	7	8	9	10
Butyl bromide	√		√						√			√	√								√	
Butyl cellosolve	√		√			√		√	√				√								√	
Butyl chloride	√		√						√		√		√								√	
Butyl chromate	√		√			√		√	√	√						√						√
Butyl formate	√		√			√		√	√				√								√	
Butyl hydroperoxide	√		√			√	√		√			√				√				√		
Butyl mercaptan	√		√									√						√				
Butyl methacrylate	√		√			√		√	√				√								√	
Butyl methyl ketone	√		√			√		√	√				√								√	
Butyl teracetate	√		√			√	√		√				√			√			√			
Butyl perbenzoate	√		√			√	√		√				√			√			√			
Butyl phosphate	√		√									√			√			√			√	
Butyl toluene	√		√			√		√	√				√								√	
Butyl vinyl ether	√		√			√						√							√		√	
Butyne	√		√			√		√	√				√								√	
Butyraldehyde	√		√									√	√								√	
Butyric acid	√		√			√	√	√	√											√	√	
Butyronitrile	√		√				√	√		√	√					√						√
Butyryl chloride	√		√							√	√					√				√		
Cadmium		√		√		√												√				√
Cadmium chloride	√		√		√				√	√								√				√
Cadmium oxide	√		√		√				√	√								√				√
Calcium		√	√			√		√					√								√	
Calcium arsenate	√		√		√				√	√								√				√
Calcium carbide	√		√		√											√						√
Calcium carbonate		√			√																	√
Calcium chlorate	√		√			√		√	√	√								√				√
Calcium chloride	√		√		√													√			√	
Calcium cyanide	√		√				√	√		√	√					√						√
Calcium hydride	√			√		√	√									√						√
Calcium hydroxide		√			√																	√
Calcium hypochlorite	√		√			√		√	√	√								√				√
Calcium oxide		√			√																	√
Caproic acid	√		√		√													√				√
Capryl alcohol	√		√			√		√	√						√						√	
Caprylic acid	√		√			√	√	√	√										√		√	
Carbazole	√		√	√							√	√									√	
Carbitol	√		√			√						√								√	√	
Carbon		√			√																	√

Chemical	Rubber gloves	Leather gloves	Laboratory coat	Fireproof clothing	Protective shoes	Safety glasses	Face shield	Body shield	Good ventilation	Fume hood	Respirator	Self-contained breathing apparatus		Disposal procedures									
														1	2	3	4	5	6	7	8	9	10
Carbon disulfide	√					√			√	√					√							√	
Carbon monoxide	√		√				√		√		√				√							√	
Carbon tetrabromide	√		√						√			√			√								√
Carbon tetrachloride	√		√						√			√			√								√
Carbon tetrafluoride	√		√						√			√			√								√
Carbonyl fluoride	√		√			√						√		√			√				√		
Cellosolve	√		√				√		√		√				√							√	
Cellosolve acetate	√		√				√		√		√				√							√	
Cellulose nitrate		√	√				√								√								√
Chloral	√		√									√			√							√	
Chloral hydrate	√		√									√			√							√	
Chlordane	√		√							√		√		√								√	
Chloric acid	√		√				√		√	√	√						√						√
Chlorine	√		√				√		√	√	√						√						√
Chlorine dioxide	√		√				√		√	√	√						√						√
Chlorine trifluoride	√		√			√						√		√			√				√		
Chloroacetaldehyde	√		√									√			√							√	
Chloroacetamide	√		√			√			√		√				√							√	
Chloracetic acid	√		√						√	√	√			√			√				√		
Chloroacetonitrile	√		√					√	√			√		√			√						√
Chloracetophenone	√		√							√		√		√								√	
Chloracetyl chloride	√		√									√		√			√				√		
Chloroaniline	√		√									√				√						√	
Chlorobenzene	√		√							√		√		√								√	
Chlorobromomethane		√				√																	√
Chlorocresol	√		√							√		√		√								√	
Chlorodifluoroethane		√				√																	√
Chlorodinitrobenzene	√		√		√							√		√								√	
Chlorodiphenyl	√		√							√		√		√								√	
Chloroethanol	√		√						√			√		√			√				√		
Chloroform		√		√		√									√								
Chloromethyl ether	√		√				√					√							√			√	
Chloronaphthalene	√		√							√		√		√								√	
Chloronaphthylamine	√		√		√							√		√								√	
Chloronitroaniline	√		√		√							√		√								√	
Chloronitrobenzene	√		√		√							√		√								√	
Chloronitropropane	√		√						√			√		√			√				√		
Chlorophenol	√		√		√							√		√								√	
Chloropicrin	√		√						√			√		√			√				√		

TABLE 10-14 Protection from and Disposal of Chemicals (Continued)

Chemical	Rubber gloves	Leather gloves	Laboratory coat	Fireproof clothing	Protective shoes	Safety glasses	Face shield	Body shield	Good ventilation	Fume hood	Respirator	Self-contained breathing apparatus	1	2	3	4	5	6	7	8	9	10
Chloroprene	✓		✓						✓			✓	✓								✓	
Chlorosulfonic acid	✓		✓					✓	✓	✓		✓				✓				✓		
Chlorotrifluoroethylene		✓			✓																	✓
Chromic acid	✓		✓				✓		✓	✓	✓					✓						✓
Chromium bromide	✓		✓		✓											✓				✓		
Chromium chloride	✓		✓				✓		✓	✓	✓					✓						✓
Chromium nitrate	✓		✓		✓											✓					✓	
Chromium sulfate	✓		✓		✓											✓					✓	
Chromium trioxide	✓		✓				✓		✓	✓	✓					✓						✓
Chromyl chloride	✓		✓				✓		✓	✓	✓					✓						✓
Cinnamaldehyde	✓		✓							✓				✓							✓	
Citric acid	✓		✓				✓	✓	✓	✓									✓		✓	
Cobalt bromide		✓		✓	✓												✓					✓
Cobalt chloride		✓		✓	✓												✓					✓
Cobalt nitrate		✓		✓	✓												✓					✓
Cobalt sulfate		✓		✓	✓												✓					✓
Collodion		✓	✓			✓									✓							✓
Copper bromide	✓		✓		✓											✓				✓		
Copper chloride	✓		✓		✓											✓				✓		
Copper nitrate	✓		✓		✓											✓				✓		
Copper sulfate	✓		✓		✓											✓				✓		
Cresol	✓		✓				✓		✓		✓					✓					✓	
Crotonaldehyde	✓		✓									✓				✓					✓	
Crotonitrile	✓		✓					✓	✓			✓	✓			✓						✓
Crotonyl chloride	✓		✓									✓	✓			✓				✓		
Cumene	✓		✓				✓		✓		✓					✓					✓	
Cyanamide	✓		✓					✓	✓			✓	✓			✓						✓
Cyanoacetamide	✓		✓					✓	✓			✓	✓			✓						✓
Cyanogen	✓		✓					✓	✓			✓	✓			✓						✓
Cyanogen chloride	✓		✓					✓	✓			✓	✓			✓						✓
Cycloheptanone	✓		✓				✓					✓							✓		✓	
Cyclohexane	✓		✓				✓		✓	✓						✓					✓	
Cyclohexanol	✓		✓				✓		✓	✓						✓					✓	
Cyclohexanone	✓		✓				✓		✓	✓						✓					✓	
Cyclohexene	✓		✓				✓		✓	✓						✓					✓	
Cyclohexylamine	✓		✓		✓				✓		✓					✓					✓	
Cyclohexylbenzene	✓		✓				✓		✓		✓					✓					✓	
Cyclopentadiene	✓		✓				✓		✓		✓					✓					✓	
Cyclopentane	✓		✓				✓		✓		✓		✓								✓	

TABLE 10-14   Protection from and Disposal of Chemicals (*Continued*)

Chemical	Rubber gloves	Leather gloves	Laboratory coat	Fireproof clothing	Protective shoes	Safety glasses	Face shield	Body shield	Good ventilation	Fume hood	Respirator	Self-contained breathing apparatus	1	2	3	4	5	6	7	8	9	10
Cyclopentanone	✓	✓				✓	✓		✓					✓							✓	
Cyclopropane	✓	✓				✓	✓		✓					✓							✓	
Cymene	✓	✓				✓	✓		✓					✓							✓	
Decaborane	✓		✓			✓	✓								✓							✓
Decahydronaphthalene	✓	✓				✓	✓		✓					✓							✓	
Decaldehyde	✓	✓									✓				✓						✓	
Decane	✓	✓				✓	✓		✓					✓							✓	
Decyl alcohol	✓	✓				✓	✓		✓					✓							✓	
Diacetone alcohol	✓	✓				✓	✓		✓					✓							✓	
Diatomaceous earth		✓			✓																	✓
Diazomethane		✓				✓	✓	✓								✓				✓		
Dibenzanthracene	✓	✓				✓	✓		✓					✓							✓	
Dibenzcarbazole	✓	✓									✓				✓						✓	
Dibenzofluorine	✓	✓				✓	✓		✓					✓							✓	
Dibenzylamine	✓	✓									✓				✓						✓	
Diborane	✓		✓			✓	✓							✓								✓
Dibromoethane	✓	✓							✓	✓			✓								✓	
Dibutylamine	✓	✓			✓		✓		✓					✓							✓	
Dibutyl ether	✓	✓				✓					✓			✓						✓	✓	
Dibutyl oxalate	✓	✓				✓			✓	✓				✓							✓	
Dibutyl peroxide	✓	✓				✓	✓		✓				✓				✓			✓		
Dibutyl phthalate	✓	✓				✓			✓	✓				✓							✓	
Dichloroacetic acid	✓	✓					✓	✓	✓				✓			✓				✓		
Dichloroacetyl chloride	✓	✓									✓		✓			✓				✓		
Dichloroaniline	✓	✓	✓								✓		✓								✓	
Dichlorobenzene	✓	✓	✓								✓		✓								✓	
Dichlorobenzidine	✓	✓	✓								✓		✓								✓	
Dichlorobutane	✓	✓							✓	✓	✓										✓	
Dichlorobutene	✓	✓							✓	✓	✓										✓	
Dichlorodifluoromethane		✓			✓																✓	
Dichloroethane	✓	✓							✓	✓	✓										✓	
Dichloroethylene	✓	✓							✓	✓	✓										✓	
Dichloroethyl ether	✓	✓				✓					✓									✓	✓	
Dichloromethane		✓			✓																✓	
Dichloromonofluoromethane		✓			✓																✓	
Dichloronitroethane	✓	✓							✓		✓	✓	✓			✓				✓		
Dichlorophenol	✓	✓	✓								✓	✓									✓	
Dichlorophenylphosphine	✓	✓	✓								✓	✓									✓	
Dichloropropane	✓	✓							✓	✓	✓										✓	

TABLE 10-14   Protection from and Disposal of Chemicals (Continued)

Chemical	Rubber gloves	Leather gloves	Laboratory coat	Fireproof clothing	Protective shoes	Safety glasses	Face shield	Body shield	Good ventilation	Fume hood	Respirator	Self-contained breathing apparatus	1	2	3	4	5	6	7	8	9	10
Dichloropropene	✓	✓							✓		✓	✓									✓	
Dichlorotetrafluoroethane		✓			✓																✓	
Dicyclohexylamine	✓	✓		✓					✓	✓				✓							✓	
Dicyclopentadiene	✓	✓				✓			✓	✓				✓							✓	
Diethanolamine	✓	✓		✓					✓	✓				✓							✓	
Diethyladipate	✓	✓				✓			✓	✓				✓							✓	
Diethylaluminum chloride	✓	✓							✓		✓	✓									✓	
Diethylamine	✓	✓		✓					✓	✓				✓							✓	
Diethylaminoethanol	✓	✓		✓					✓	✓				✓							✓	
Diethylaniline	✓	✓									✓					✓					✓	
Diethyl carbonate	✓	✓				✓			✓	✓				✓							✓	
Diethylene glycol	✓	✓				✓			✓	✓				✓							✓	
Diethylenetriamine	✓	✓		✓					✓	✓				✓							✓	
Diethyl ether	✓	✓				✓					✓								✓		✓	
Diethyl ketone	✓	✓				✓			✓	✓				✓							✓	
Diethyl malonate	✓	✓				✓			✓	✓				✓							✓	
Diethyl phthalate	✓	✓				✓			✓	✓				✓							✓	
Diethyl sulfate	✓	✓							✓		✓	✓									✓	
Diethyl zinc		✓				✓	✓				✓										✓	
Dihydroxyanthraquinone	✓	✓				✓			✓	✓				✓							✓	
Diiosbutyl ketone	✓	✓				✓			✓	✓				✓							✓	
Diisopropyl amine	✓	✓			✓				✓	✓				✓							✓	
Diisopropyl ether	✓	✓				✓					✓								✓		✓	
Dimethoxybenzidine	✓	✓									✓				✓						✓	
Dimethoxyethane	✓	✓				✓					✓								✓		✓	
Dimethoxymethane	✓	✓				✓					✓								✓		✓	
Dimethoxypropane	✓	✓				✓					✓								✓		✓	
Dimethyl acetamide	✓	✓		✓															✓		✓	
Dimethyl amine	✓	✓		✓					✓	✓				✓							✓	
Dimethylaniline	✓	✓									✓					✓					✓	
Dimethylazobenzene		✓				✓	✓	✓										✓		✓		
Dimethyldiphenylamine	✓	✓	✓								✓	✓									✓	
Dimethyl butane	✓	✓				✓			✓	✓				✓							✓	
Dimethyl carbonate	✓	✓				✓			✓	✓				✓							✓	
Dimethyl ether	✓	✓				✓					✓								✓		✓	
Dimethylformamide	✓	✓		✓															✓		✓	
Dimethyl fumarate	✓	✓				✓		✓	✓					✓							✓	
Dimethylhydrazine	✓		✓								✓								✓		✓	
Dimethylnaphthalene	✓	✓				✓		✓	✓						✓						✓	

TABLE 10-14 Protection from and Disposal of Chemicals (Continued)

Chemical	Rubber gloves	Leather gloves	Laboratory coat	Fireproof clothing	Protective shoes	Safety glasses	Face shield	Body shield	Good ventilation	Fume hood	Respirator	Self-contained breathing apparatus	1	2	3	4	5	6	7	8	9	10
Dimethyl phthalate	✓	✓					✓		✓					✓							✓	
Dimethyl propane	✓	✓					✓		✓					✓							✓	
Dimethyl sulfate	✓	✓							✓		✓	✓									✓	
Dimethyl sulfide	✓	✓									✓								✓			
Dimethyl sulfoxide	✓	✓							✓		✓	✓									✓	
Dinitroaniline	✓	✓	✓								✓	✓									✓	
Dinitrobenzene	✓	✓	✓								✓	✓									✓	
Dinitrocresol	✓	✓	✓								✓	✓									✓	
Dinitrofluorine	✓	✓	✓								✓	✓									✓	
Dinitrophenol	✓	✓	✓								✓	✓									✓	
Dinitropiperazine	✓	✓	✓								✓	✓									✓	
Dinitrotoluene	✓	✓	✓								✓	✓									✓	
Dioctyl phthalate	✓	✓			✓	✓			✓					✓							✓	
Dioxane	✓	✓			✓	✓			✓					✓							✓	
Diphenylamine	✓	✓									✓					✓					✓	
Diphenylmethane	✓	✓			✓	✓			✓					✓							✓	
Diphenylsulfide	✓	✓									✓									✓		
Dodecane	✓	✓			✓	✓			✓					✓							✓	
Elon	✓	✓	✓								✓	✓									✓	
Endrin	✓	✓	✓								✓	✓									✓	
Epichlorohydrin	✓	✓							✓		✓	✓									✓	
Epoxy resin		✓			✓																	✓
Ethane	✓	✓			✓	✓			✓					✓							✓	
Ethanethiol	✓	✓									✓							✓				
Ethanol	✓	✓			✓	✓			✓					✓							✓	
Ethanolamine	✓	✓		✓		✓			✓					✓							✓	
Ethoxynitroaniline	✓	✓	✓								✓	✓									✓	
Ethyl acetanilide	✓	✓	✓																	✓	✓	
Ethyl acetate	✓	✓			✓	✓	✓		✓					✓							✓	
Ethyl acetoacetate	✓	✓			✓	✓	✓		✓					✓							✓	
Ethyl acrylate	✓	✓			✓	✓	✓		✓					✓							✓	
Ethylamine	✓	✓		✓		✓	✓		✓					✓							✓	
Ethylamyl ketone	✓	✓			✓	✓	✓		✓					✓							✓	
Ethyl aniline	✓	✓									✓					✓					✓	
Ethyl benzene	✓	✓			✓	✓	✓		✓					✓							✓	
Ethyl benzoate	✓	✓			✓	✓	✓		✓					✓							✓	
Ethylbutyl ketone	✓	✓			✓	✓	✓		✓					✓							✓	
Ethyl butyrate	✓	✓			✓	✓	✓		✓					✓							✓	
Ethyl chloride	✓	✓							✓		✓	✓									✓	

TABLE 10-14   Protection from and Disposal of Chemicals (Continued)

Chemical	Rubber gloves	Leather gloves	Laboratory coat	Fireproof clothing	Protective shoes	Safety glasses	Face shield	Body shield	Good ventilation	Fume hood	Respirator	Self-contained breathing apparatus	1	2	3	4	5	6	7	8	9	10
													Disposal procedures									
Ethyl chloroacetate	✓	✓								✓	✓		✓								✓	
Ethyl chloroformate	✓	✓								✓	✓		✓								✓	
Ethylene	✓	✓				✓	✓		✓					✓							✓	
Ethylenediamine	✓	✓			✓		✓		✓					✓							✓	
Ethylenedichloride	✓	✓								✓	✓		✓								✓	
Ethylene glycol	✓	✓				✓	✓		✓					✓							✓	
Ethylene nitrate	✓	✓					✓				✓		✓			✓				✓		
Ethylene oxide	✓	✓				✓					✓								✓		✓	
Ethyl fluoride	✓	✓							✓		✓		✓								✓	
Ethyl fluoroacetate	✓	✓							✓		✓		✓								✓	
Ethyl formate	✓	✓					✓	✓	✓					✓							✓	
Ethyl hexanol	✓	✓					✓	✓	✓					✓							✓	
Ethyl iodide	✓	✓							✓		✓		✓								✓	
Ethyl morpholine	✓	✓									✓				✓						✓	
Ethyl nitrite	✓	✓							✓		✓		✓			✓				✓		
Ethyl oxalate	✓	✓					✓	✓	✓					✓							✓	
Ethyl phenol	✓	✓					✓	✓	✓					✓							✓	
Ethyl silicate	✓	✓					✓	✓	✓					✓							✓	
Ethyl vinyl ether	✓	✓					✓					✓							✓		✓	
Fatty acids	✓	✓					✓	✓	✓	✓									✓		✓	
Ferric chloride	✓	✓						✓	✓	✓				✓			✓			✓		
Ferrous ammonium sulfate	✓	✓	✓		✓												✓			✓		
Ferrous chloride	✓	✓	✓		✓												✓			✓		
Ferrous sulfate	✓	✓	✓		✓												✓			✓		
Fluorine	✓	✓					✓	✓	✓	✓							✓					✓
Fluoroacetic acid	✓	✓					✓	✓	✓					✓			✓			✓		
Fluoroboric acid	✓	✓				✓				✓						✓				✓		
Fluoroethane	✓	✓							✓		✓	✓									✓	
Fluoroethylene	✓	✓							✓		✓	✓									✓	
Fluorosilicic acid	✓	✓					✓			✓						✓				✓		
Fluorotrichloromethane		✓			✓																	✓
Formaldehyde	✓	✓									✓				✓						✓	
Formalin	✓	✓									✓				✓						✓	
Formamide	✓	✓	✓																✓		✓	
Formic acid	✓	✓				✓	✓	✓	✓										✓		✓	
Fumaric acid	✓	✓				✓	✓	✓	✓										✓		✓	
Furan	✓	✓				✓			✓	✓					✓						✓	
Furfural	✓	✓								✓					✓						✓	
Furfuryl alcohol	✓	✓				✓		✓		✓					✓						✓	

TABLE 10-14 Protection from and Disposal of Chemicals (Continued)

Chemical	Rubber gloves	Leather gloves	Laboratory coat	Fireproof clothing	Protective shoes	Safety glasses	Face shield	Body shield	Good ventilation	Fume hood	Respirator	Self-contained breathing apparatus	1	2	3	4	5	6	7	8	9	10
Gallic acid	√		√			√	√	√	√									√			√	
Gasoline	√		√			√			√		√			√							√	
Germane		√				√																√
Germanium dioxide	√		√			√			√	√				√								√
Germanium hydride	√			√		√	√								√							√
Germanium tetrachloride	√	√					√		√	√			√			√				√		
Glutaric acid	√	√				√	√	√	√									√			√	
Glycerol	√	√				√			√		√			√							√	
Glycolic acid	√	√				√	√	√	√									√			√	
Glyoxal	√	√									√			√							√	
Grease	√	√				√			√		√			√							√	
Heptachlor	√	√							√		√	√	√								√	
Heptane	√	√				√	√		√		√			√							√	
Heptylamine	√	√			√		√		√		√			√							√	
Hexachlorobenzene	√	√							√		√	√									√	
Hexachloroethane		√				√																√
Hexachloronaphthalene	√	√							√		√	√									√	
Hexafluoroethane		√				√																√
Hexamethylene tetramine	√	√			√				√		√	√									√	
Hexane	√	√				√	√		√		√			√							√	
Hexanediamine	√	√			√		√		√		√			√							√	
Hexanol	√	√				√	√		√		√			√							√	
Hexanone	√	√				√	√		√		√			√							√	
Hexene	√	√				√	√		√		√			√							√	
Hexyl acetate	√	√				√	√		√		√			√							√	
Hexylamine	√	√			√		√		√		√			√							√	
Hydazine	√	√				√	√		√		√			√							√	
Hydrazoic acid		√				√	√	√							√					√		
Hydriodic acid	√	√					√				√			√						√		
Hydrobromic acid	√	√					√				√			√						√		
Hydrochloric acid	√	√					√				√			√						√		
Hydrocyanic acid	√	√					√	√			√		√				√					√
Hydrofluoric acid	√	√					√				√			√						√		
Hydrogen		√				√																√
Hydrogen peroxide	√					√	√						√				√			√		
Hydrogen sulfide	√	√			√				√	√							√					√
Hydroquinone	√	√				√	√		√					√							√	
Hydroxylamine	√	√			√		√		√					√							√	
Indanol	√	√				√	√		√					√							√	

TABLE 10-14 Protection from and Disposal of Chemicals (Continued)

Chemical	Rubber gloves	Leather gloves	Laboratory coat	Fireproof clothing	Protective shoes	Safety glasses	Face shield	Body shield	Good ventilation	Fume hood	Respirator	Self-contained breathing apparatus	1	2	3	4	5	6	7	8	9	10
Iodic acid	√	√						√				√			√							√
Iodine	√	√				√	√	√	√						√							√
Iodine chloride	√	√		√								√		√		√				√		
Iodine pentafluoride	√	√		√								√		√		√				√		
Iodine trichloride	√	√		√								√		√		√				√		
Iodoacetic acid	√	√					√	√	√			√		√		√				√		
Isoprene	√	√				√	√	√		√						√					√	
Jet fuel (kerosene)	√	√				√	√	√		√						√					√	
Lactonitrile	√	√					√	√		√	√				√		√					√
Latex		√			√																	√
Lauric acid	√	√				√	√	√	√											√		√
Lauroyl peroxide	√	√				√	√		√	√						√				√		
Lead		√	√	√														√				√
Lead acetate	√	√		√			√	√										√				√
Lead arsenate	√	√		√			√	√										√				√
Lead carbonate	√	√		√			√	√										√				√
Lead nitrate	√	√		√			√	√										√				√
Lead oxide	√	√		√			√	√										√				√
Lead thiocyanate	√	√		√			√	√										√				√
Lindane	√									√	√	√									√	
Liquefied petroleum gas	√	√				√		√	√					√							√	
Lithium		√	√			√	√			√											√	
Lithium aluminum hydride	√		√			√	√									√						√
Lithium borohydride	√		√			√	√									√						√
Lithium carbonate	√	√		√														√		√		
Lithium hydride	√		√			√	√									√						√
Magnesium		√	√	√															√			√
Magnesium chlorate	√	√				√	√	√	√										√			√
Magnesium chloride	√	√		√															√		√	
Magnesium nitrate	√	√		√															√		√	
Magnesium oxide		√		√																		√
Magnesium perchlorate	√	√				√	√	√	√										√			√
Malathion	√	√									√					√			√		√	
Maleic acid	√	√				√	√	√											√		√	
Manganese		√	√	√															√			√
Manganese sulfate	√	√		√															√		√	
Mercaptoethanol	√	√									√								√			
Mercuric chloride	√	√										√										√
Mercury	√	√										√								√		√

TABLE 10-14 Protection from and Disposal of Chemicals (*Continued*)

Chemical	Rubber gloves	Leather gloves	Laboratory coat	Fireproof clothing	Protective shoes	Safety glasses	Face shield	Body shield	Good ventilation	Fume hood	Respirator	Self-contained breathing apparatus	1	2	3	4	5	6	7	8	9	10
Mercury compounds (organic)	✓		✓									✓						✓				✓
Mercury fulminate	✓		✓									✓						✓				✓
Mesityl oxide	✓	✓				✓		✓	✓					✓							✓	
Methacrylic acid	✓	✓				✓	✓	✓	✓									✓			✓	
Methane	✓	✓				✓		✓		✓				✓							✓	
Methoxychlor	✓	✓							✓		✓		✓								✓	
Methoxyphenol	✓	✓				✓					✓							✓			✓	
Methyl acetate	✓	✓				✓		✓	✓					✓							✓	
Methyl acrylate	✓	✓				✓		✓	✓					✓							✓	
Methyl alcohol	✓	✓				✓		✓	✓					✓							✓	
Methyl amyl alcohol	✓	✓				✓		✓	✓					✓						✓		
Methyl amyl ketone	✓	✓				✓		✓	✓					✓							✓	
Methyl benzoate	✓	✓				✓		✓	✓					✓							✓	
Methylbenzyl alcohol	✓	✓				✓		✓	✓					✓							✓	
Methyl bromide	✓	✓							✓		✓	✓									✓	
Methyl butene	✓	✓				✓		✓	✓					✓							✓	
Methylbutylamine	✓	✓		✓				✓	✓					✓							✓	
Methyl butyl ketone	✓	✓				✓		✓	✓					✓							✓	
Methyl butyrate	✓	✓				✓		✓	✓					✓							✓	
Methyl cellosolve	✓	✓				✓		✓	✓					✓							✓	
Methyl cellosolve acetate	✓	✓				✓		✓	✓					✓							✓	
Methyl chloride	✓	✓							✓		✓	✓									✓	
Methyl chloroformate	✓	✓							✓		✓	✓									✓	
Methyl cyclohexane	✓	✓				✓		✓	✓					✓							✓	
Methyl cyclohexanol	✓	✓				✓		✓	✓					✓							✓	
Methyl cyclohexanone	✓	✓				✓		✓	✓					✓							✓	
Methyl cyclohexene	✓	✓				✓		✓	✓					✓							✓	
Methyl ethyl ether	✓	✓				✓					✓								✓		✓	
Methyl ethyl ketone	✓	✓				✓		✓	✓					✓							✓	
Methyl formate	✓	✓				✓		✓	✓					✓							✓	
Methyl furan	✓	✓				✓		✓	✓					✓							✓	
Methyl hydrazine	✓			✓							✓								✓		✓	
Methyl iodide	✓	✓							✓		✓	✓									✓	
Methyl isobutyl ketone	✓	✓				✓		✓	✓					✓							✓	
Methyl isobutyrate	✓	✓				✓		✓	✓					✓							✓	
Methyl isocyanate	✓	✓				✓		✓	✓					✓							✓	
Methyl isothiocyanate	✓				✓		✓	✓	✓					✓							✓	
Methyl mercaptan	✓	✓									✓							✓			✓	
Methyl methacrylate	✓	✓				✓		✓	✓					✓							✓	

TABLE 10-14   Protection from and Disposal of Chemicals (*Continued*)

Chemical	Rubber gloves	Leather gloves	Laboratory coat	Fireproof clothing	Protective shoes	Safety glasses	Face shield	Body shield	Good ventilation	Fume hood	Respirator	Self-contained breathing apparatus	Disposal procedures 1	2	3	4	5	6	7	8	9	10
Methylnaphthalene	√	√					√		√		√			√							√	
Methyl nitrosoaniline	√	√	√								√		√								√	
Methyl nitrosourea	√	√	√								√		√								√	
Methyl propyl ketone	√	√					√		√		√			√							√	
Methylpyrrole	√	√	√								√		√								√	
Methyl salicylate	√	√					√		√		√			√							√	
Methyl styrene	√	√					√		√		√			√							√	
Methyl tetranitroaniline	√	√		√							√		√								√	
Methyl toluenesulfonate	√	√					√		√		√			√							√	
Methyl vinyl ether	√	√					√				√								√		√	
Monochloroamine	√	√					√	√								√				√		
Monomethylamine	√	√			√				√		√					√					√	
Morpholine	√	√									√					√					√	
Naphtha	√	√					√		√		√			√							√	
Naphthalene	√	√					√		√		√			√							√	
Naphthol	√	√					√		√		√			√							√	
Naphthylamine	√	√									√					√					√	
Naphthylisothiocyanate	√	√									√					√				√	√	
Nickel		√		√	√													√				√
Nickel carbonyl	√	√					√		√		√			√							√	
Nickel nitrate	√	√			√												√			√		
Nickel sulfate	√	√			√												√			√		
Nicotine	√	√									√					√					√	
Nitric acid	√	√						√			√					√				√		
Nitric oxide	√	√						√			√					√				√		
Nitroacetophenone	√	√	√								√		√								√	
Nitroaniline	√	√	√								√		√								√	
Nitrobenzene	√	√	√								√		√								√	
Nitrobiphenyl	√	√	√								√		√								√	
Nitroethane	√	√							√		√		√				√			√		
Nitrofluorine	√	√	√								√		√								√	
Nitrogen dioxide	√	√						√			√					√				√		
Nitrogen trifluoride	√	√				√		√	√	√						√						√
Nitrogen trioxide		√				√																√
Nitromethane	√	√							√		√		√				√			√		
Nitronaphthalene	√	√	√								√		√								√	
Nitrophenol	√	√	√								√		√								√	
Nitropropane	√	√							√		√		√			√				√		
Nitrosodiethylamine	√	√				√			√		√				√						√	

TABLE 10-14   Protection from and Disposal of Chemicals (Continued)

Chemical	Rubber gloves	Leather gloves	Laboratory coat	Fireproof clothing	Protective shoes	Safety glasses	Face shield	Body shield	Good ventilation	Fume hood	Respirator	Self-contained breathing apparatus	Disposal 1	2	3	4	5	6	7	8	9	10
Nitrosomethylaniline	✓		✓	✓								✓	✓								✓	
Nitrosomorpholine	✓		✓	✓								✓	✓								✓	
Nitrosopiperazine	✓		✓	✓								✓	✓								✓	
Nitrosopiperidine	✓		✓	✓								✓	✓								✓	
Nitrosyl chloride	✓		✓				✓		✓	✓	✓							✓				✓
Nitrotoluene	✓		✓		✓						✓	✓	✓								✓	
Nonyl phenol	✓		✓			✓			✓		✓			✓							✓	
Octachloronaphthalene	✓		✓							✓	✓	✓									✓	
Octafluorocyclobutane		✓			✓																	✓
Octafluoropropane	✓		✓							✓	✓	✓									✓	
Octane	✓		✓			✓		✓		✓				✓							✓	
Octanoic acid	✓		✓			✓	✓	✓											✓		✓	
Oils (fuel)	✓		✓			✓		✓	✓					✓							✓	
Oils (lubricating)	✓		✓			✓		✓	✓					✓							✓	
Oils (vegetable)	✓		✓			✓		✓	✓					✓							✓	
Oleic acid	✓		✓			✓	✓	✓	✓										✓		✓	
Osmium tetroxide		✓			✓																	✓
Oxalic acid	✓		✓			✓	✓	✓	✓										✓		✓	
Oxalyl chloride	✓		✓								✓	✓	✓			✓				✓		
Oxygen		✓			✓																	✓
Ozone		✓			✓																	✓
Paraffin	✓		✓			✓		✓	✓					✓							✓	
Paraformaldehyde	✓		✓								✓			✓							✓	
Paraldehyde	✓		✓								✓			✓							✓	
Parathion	✓		✓								✓					✓			✓		✓	
Pentaborane	✓			✓		✓	✓							✓								✓
Pentachloroethane	✓		✓						✓		✓	✓									✓	
Pentachloronaphthalene	✓		✓						✓		✓	✓									✓	
Pentachlorophenol	✓		✓						✓		✓	✓									✓	
Pentane	✓		✓			✓		✓	✓					✓							✓	
Pentanediol	✓		✓			✓		✓	✓					✓							✓	
Peracetic acid	✓		✓			✓		✓	✓	✓						✓						✓
Perchloric acid	✓		✓			✓		✓	✓	✓						✓						✓
Perchloromethyl mercaptan	✓		✓								✓								✓			
Perchloryl fluoride	✓		✓		✓						✓	✓	✓			✓				✓		
Petroleum oil	✓		✓						✓					✓							✓	
Phenanthrene	✓		✓			✓		✓	✓					✓							✓	
Phenanthreneacetamide	✓		✓		✓														✓		✓	
Phenol	✓		✓			✓		✓	✓					✓							✓	

TABLE 10-14   Protection from and Disposal of Chemicals (Continued)

Chemical	Rubber gloves	Leather gloves	Laboratory coat	Fireproof clothing	Protective shoes	Safety glasses	Face shield	Body shield	Good ventilation	Fume hood	Respirator	Self-contained breathing apparatus	Disposal procedures 1	2	3	4	5	6	7	8	9	10
Phenyl acetate	√		√			√		√		√				√							√	
Phenylazonaphthol		√				√	√	√								√					√	
Phenylenediamine	√	√									√				√						√	
Phenyl ether	√	√				√					√								√		√	
Phenylhydrazine	√			√							√								√		√	
Phenylisocyanate	√	√				√		√	√					√							√	
Phenylnaphthylamine	√	√									√					√					√	
Phenylphenol	√	√				√	√	√						√							√	
Phorone	√	√				√	√	√						√							√	
Phosphine	√	√									√					√			√		√	
Phosgene	√	√				√	√	√						√							√	
Phosphoric acid	√	√					√				√				√					√		
Phosphorous	√					√									√							√
Phosphorous oxychloride	√	√		√							√		√		√				√			
Phosphorous pentachloride	√	√		√							√		√		√				√			
Phosphorous pentasulfide	√	√		√							√		√		√				√			
Phosphorous tribromide	√	√		√							√		√		√				√			
Phosphorous trichloride	√	√		√							√		√		√				√			
Phthalic anhydride	√	√				√	√	√	√										√		√	
Picoline	√	√									√					√					√	
Picric acid	√	√		√							√		√								√	
Pimelic acid	√	√				√	√	√	√										√		√	
Pindone	√	√				√	√		√						√						√	
Pinene	√	√				√	√		√						√						√	
Piperidine	√	√		√							√	√			√						√	
Polyvinyl acetate	√	√				√	√		√						√						√	
Potassium		√	√			√	√				√				√						√	
Potassium acetate	√	√		√												√			√			
Potassium borohydride	√		√			√	√									√						√
Potassium carbonate	√	√		√												√		√				
Potassium chlorate	√	√				√	√	√	√							√						√
Potassium cyanide	√	√					√	√			√		√			√						√
Potassium dichromate	√	√				√	√	√	√							√						√
Potassium ferrocyanide	√	√		√												√			√			
Potassium fluoride	√	√		√												√			√			
Potassium hydroxide	√	√		√		√	√		√							√			√			
Potassium nitrate	√	√		√												√			√			
Potassium perchlorate	√	√		√		√	√	√	√							√						√
Potassium permanganate	√	√				√	√	√	√							√						√

TABLE 10-14 Protection from and Disposal of Chemicals (Continued)

Chemical	Rubber gloves	Leather gloves	Laboratory coat	Fireproof clothing	Protective shoes	Safety glasses	Face shield	Body shield	Good ventilation	Fume hood	Respirator	Self-contained breathing apparatus	1	2	3	4	5	6	7	8	9	10
Potassium persulfate	√		√			√			√	√	√					√						√
Potassium peroxide	√					√	√						√			√				√		
Potassium sulfide	√	√			√	√	√									√						√
Propane	√	√				√			√		√			√							√	
Propanediamine	√	√			√				√		√			√							√	
Propanediol	√	√				√			√		√			√							√	
Prophenyl acetate	√	√				√			√		√			√							√	
Propriolactone	√	√				√	√	√											√		√	
Proprionaldehyde	√	√										√		√							√	
Proprionic acid	√	√				√	√	√	√										√		√	
Proprionitrile	√	√					√	√				√	√			√						√
Proprionyl chloride	√	√										√	√			√				√		
Propyl acetate	√	√				√			√		√			√							√	
Propyl alcohol	√	√				√			√		√			√							√	
Propylamine	√	√			√				√		√			√							√	
Propyl benzene	√	√				√			√		√			√							√	
Propyl benzoate	√	√				√			√		√			√							√	
Propyl bromide	√	√							√		√	√									√	
Propyl chloride	√	√							√		√	√									√	
Propylene	√	√				√			√		√			√							√	
Propylene carbonate	√	√				√			√		√			√							√	
Propylene disulfate	√	√							√		√	√									√	
Propylene glycol	√	√				√			√		√			√							√	
Propylene imine	√	√			√				√		√			√							√	
Propylene oxide	√	√				√						√							√		√	
Propyl formate	√	√				√			√		√			√							√	
Propyl nitrate	√	√							√		√	√	√			√				√		
Propyne	√	√				√			√		√			√							√	
Pyrethrum		√			√																√	
Pyridine	√	√										√			√						√	
Pyrogallic acid	√	√				√	√	√	√										√		√	
Pyrrolidine	√	√			√				√		√			√							√	
Pyruvic acid	√	√				√	√	√	√										√		√	
Quinoline	√	√										√			√						√	
Quinone	√	√				√			√		√			√							√	
Resorcinol	√	√				√			√		√			√							√	
Rotenone	√	√				√			√		√			√							√	
Rubber		√				√																√
Rubidium	√		√		√				√	√			√									√

TABLE 10-14 Protection from and Disposal of Chemicals (Continued)

Chemical	Rubber gloves	Leather gloves	Laboratory coat	Fireproof clothing	Protective shoes	Safety glasses	Face shield	Body shield	Good ventilation	Fume hood	Respirator	Self-contained breathing apparatus	1	2	3	4	5	6	7	8	9	10
Ruthenium	√		√			√			√	√		√										√
Salicylaldehyde	√		√								√			√							√	
Salicylic acid	√		√		√	√	√	√												√	√	
Selenic acid	√		√		√				√	√							√					√
Selenium	√		√		√				√	√							√					√
Selenium dioxide	√		√		√				√	√							√					√
Selenium hexafluoride	√		√		√				√	√							√					√
Selenium oxychloride	√		√		√				√	√							√					√
Silane	√			√		√	√							√								√
Silica		√				√															√	
Silica gel (silicic acid)		√				√																√
Silicon tetrachloride	√		√					√	√	√			√			√				√		
Silicon tetrafluoride	√		√		√						√	√	√			√				√		
Silver nitrate	√		√		√				√	√			√									√
Sodium		√	√			√	√					√								√		
Sodium acetate	√		√		√												√			√		
Sodium amide	√		√			√	√										√			√		
Sodium azide		√				√	√	√									√			√		
Sodium benzoate	√		√		√												√			√		
Sodium bicarbonate	√		√		√												√			√		
Sodium bisulfate	√		√			√			√	√	√						√					√
Sodium borohydride	√			√		√	√									√						√
Sodium carbonate	√		√		√												√			√		
Sodium chlorate	√		√			√			√	√	√						√					√
Sodium chloride	√		√		√												√			√		
Sodium chlorite	√		√			√			√	√	√						√					√
Sodium chromate	√		√			√			√	√	√						√					√
Sodium cyanide	√		√				√	√			√		√				√					√
Sodium dichromate	√		√			√			√	√	√						√					√
Sodium ethoxide		√	√			√			√				√								√	
Sodium fluoride	√		√		√												√				√	
Sodium fluoroacetate	√		√							√	√		√								√	
Sodium formate	√		√		√												√				√	
Sodium hydride	√			√		√	√									√						√
Sodium hydroxide	√		√			√			√		√						√			√		
Sodium hypochlorite	√		√				√		√	√	√						√					√
Sodium iodide	√		√			√											√				√	
Sodium methoxide		√	√			√			√				√								√	
Sodium nitrate	√		√		√												√				√	

TABLE 10-14 Protection from and Disposal of Chemicals (Continued)

Chemical	Rubber gloves	Leather gloves	Laboratory coat	Fireproof clothing	Protective shoes	Safety glasses	Face shield	Body shield	Good ventilation	Fume hood	Respirator	Self-contained breathing apparatus	Disposal 1	2	3	4	5	6	7	8	9	10
Sodium nitrite	√		√			√				√												√
Sodium perchlorate	√		√				√		√	√	√					√						√
Sodium peroxide	√						√	√				√		√		√				√		
Sodium phosphate	√		√		√											√				√		
Sodium proprionate	√		√		√											√				√		
Sodium silicate	√		√		√											√				√		
Sodium sulfide	√		√		√			√	√							√						√
Sodium sulfite	√		√		√				√													√
Sodium tetraborate	√		√		√											√				√		
Sodium thiocyanate	√		√		√											√				√		
Sodium thiosulfate	√		√		√					√												√
Stannic chloride	√		√		√											√				√		
Stearic acid	√		√			√	√	√	√										√		√	
Stibine	√		√		√				√	√								√				√
Stoddard solvent	√		√			√		√		√					√						√	
Strontium	√		√		√													√				√
Strontium carbonate	√		√		√													√				√
Strontium nitrate	√		√		√													√				√
Strontium peroxide	√		√		√													√				√
Strychnine	√		√									√			√						√	
Styrene	√		√				√		√		√					√					√	
Styrene monomer	√		√				√		√		√					√					√	
Succinic acid	√		√			√	√	√	√										√		√	
Succinonitrile	√		√				√	√			√		√			√						√
Sulfamide	√		√			√	√									√				√		
Sulfur		√			√																	
Sulfur decafluoride	√	√		√							√	√	√			√				√		
Sulfur dichloride	√	√		√							√	√	√			√				√		
Sulfur dioxide	√		√		√				√													√
Sulfur hexafluoride	√		√		√						√	√	√			√				√		
Sulfuric acid	√		√				√								√					√		
Sulfur monochloride	√		√		√					√	√		√			√				√		
Sulfur trioxide	√		√				√				√				√					√		
Sulfuryl chloride	√		√		√					√	√		√			√				√		
Sulfuryl fluoride	√		√		√					√	√		√			√				√		
Tall oil	√		√			√		√	√					√							√	
Tallow	√		√			√		√	√					√							√	
Tannic acid	√		√			√	√	√	√											√	√	
Tar		√			√																	√

10-57

TABLE 10-14   Protection from and Disposal of Chemicals (Continued)

Chemical	Rubber gloves	Leather gloves	Laboratory coat	Fireproof clothing	Protective shoes	Safety glasses	Face shield	Body shield	Good ventilation	Fume hood	Respirator	Self-contained breathing apparatus	1	2	3	4	5	6	7	8	9	10
Tellurium hexafluoride	√	√				√				√	√						√					√
Terphenyl	√	√					√		√	√				√						√		
Tetradecane	√	√					√		√	√				√						√		
Tetraethylenepentamine	√	√				√			√	√				√						√		
Tetraethyl lead	√	√							√		√		√							√		
Tetrahydrofuran	√	√					√		√	√				√						√		
Tetrahydronaphthalene	√	√					√		√	√				√						√		
Tetramethyl lead	√	√							√		√		√							√		
Tetramethyl silane		√				√																√
Tetranitromethane	√	√							√		√		√			√				√		
Thallium	√	√				√			√	√			√									√
Thallous sulfate	√	√				√			√	√			√									√
Thioacetamide	√	√								√								√				
Thiodiethanol	√	√								√								√				
Thioglycollic acid	√	√								√								√				
Thionyl chloride	√	√			√					√			√			√				√		
Thionyl fluoride	√	√			√					√			√			√				√		
Thiophene	√	√								√								√				
Thiourea	√	√								√								√				
Thorium	√	√				√			√	√			√									√
Thorium nitrate	√	√				√			√	√			√									√
Tin		√	√			√												√				√
Tin compounds (inorganic)	√	√				√												√		√		
Tin compounds (organic)		√				√																√
Titanium		√	√			√												√				√
Titanium dioxide		√				√																√
Titanium tetrachloride	√	√					√		√	√			√			√				√		
Toluene	√	√					√		√	√				√							√	
Toluene diisocyanate	√	√	√				•				√		√								√	
Tolyl phosphate	√	√									√				√			√			√	
Toluidine	√	√									√				√							
Triamyline	√	√				√			√	√				√							√	
Tributylamine	√	√				√			√	√				√							√	
Tributylchlorotin	√	√							√		√		√								√	
Tributyl phosphate	√	√									√				√			√			√	
Trichloroacetic acid	√	√							√	√	√		√			√			√			
Trichloroacetonitrile	√	√					√	√			√		√			√						√
Trichloracetyl chloride																						
Trichlorobenzene	√	√							√		√		√								√	

TABLE 10-14   Protection from and Disposal of Chemicals (Continued)

Chemical	Rubber gloves	Leather gloves	Laboratory coat	Fireproof clothing	Protective shoes	Safety glasses	Face shield	Body shield	Good ventilation	Fume hood	Respirator	Self-contained breathing apparatus	1	2	3	4	5	6	7	8	9	10
Trichloroethylene	✓		✓						✓		✓	✓									✓	
Trichloronaphthalene	✓		✓						✓		✓	✓									✓	
Trichloropropane	✓		✓						✓		✓	✓									✓	
Trichlorotrifluoroethane		✓			✓																	✓
Tricresyl phosphate	✓	✓									✓				✓		✓				✓	
Tridecanol	✓	✓					✓	✓	✓			✓									✓	
Triethyl aluminum		✓	✓				✓	✓				✓									✓	
Triethylamine	✓	✓			✓		✓		✓			✓									✓	
Triethanolamine	✓	✓			✓		✓		✓			✓									✓	
Triethylene glycol	✓	✓				✓	✓		✓			✓									✓	
Triethylene tetramine	✓	✓			✓		✓		✓			✓									✓	
Triethyl formate	✓	✓				✓	✓		✓			✓										✓
Trifluoroacetic acid	✓	✓					✓	✓	✓			✓				✓				✓		
Trifluoromethane		✓				✓															✓	
Trimethylamine	✓	✓			✓		✓		✓			✓									✓	
Trimethyl borate	✓	✓				✓	✓		✓			✓									✓	
Trimethyl cyclohexanone	✓	✓					✓	✓	✓			✓									✓	
Trimethyl pentane	✓	✓					✓	✓	✓			✓									✓	
Trimethyl pentene	✓	✓					✓	✓	✓			✓									✓	
Trinitrobenzene	✓	✓		✓							✓	✓									✓	
Trinitrotoluene	✓	✓		✓							✓	✓									✓	
Trioxane	✓	✓				✓	✓		✓			✓									✓	
Triphenyl phosphate	✓	✓									✓					✓		✓			✓	
Triphenyl phosphine	✓	✓									✓					✓		✓			✓	
Tripropylamine	✓	✓		✓			✓		✓			✓									✓	
Trisodium phosphate	✓	✓		✓													✓		✓			
Turpentine	✓	✓				✓	✓		✓			✓									✓	
Uranium	✓	✓		✓			✓	✓				✓										✓
Uranium nitrate	✓	✓		✓			✓	✓				✓										✓
Urea		✓		✓																		✓
Valeraldehyde	✓	✓									✓			✓							✓	
Valeric acid	✓	✓				✓	✓	✓	✓										✓		✓	
Vanadium dichloride	✓	✓				✓	✓	✓	✓									✓				✓
Vanadium oxytrichloride	✓	✓				✓	✓	✓	✓									✓				✓
Vanadium pentoxide	✓	✓				✓	✓	✓	✓									✓				✓
Vinyl acetate	✓	✓				✓				✓						✓					✓	
Vinyl chloride	✓	✓							✓		✓	✓									✓	
Vinyl ether	✓	✓				✓					✓								✓		✓	
Vinylidene chloride	✓	✓							✓		✓	✓									✓	

TABLE 10-14   Protection from and Disposal of Chemicals (Continued)

Chemical	Rubber gloves	Leather gloves	Laboratory coat	Fireproof clothing	Protective shoes	Safety glasses	Face shield	Body shield	Good ventilation	Fume hood	Respirator	Self-contained breathing apparatus	1	2	3	4	5	6	7	8	9	10
Ucylene	√		√				√		√		√			√							√	
Xylenol	√		√				√		√		√			√							√	
Xylidine	√		√									√			√						√	
Yttrium		√				√																√
Zinc		√		√	√													√				√
Zinc acetate	√		√		√											√				√		
Zinc chlorate	√		√				√		√	√	√					√						√
Zinc chloride	√		√		√											√				√		
Zinc oxide		√			√																	√
Zirconium		√			√																	√
Zirconium acetate	√		√		√											√				√		
Zirconium chlorate	√		√		√											√				√		
Zirconium nitrate	√		√		√											√				√		
Zirconium oxide	√		√		√											√				√		

General classifications

Chemical	Rubber gloves	Leather gloves	Laboratory coat	Fireproof clothing	Protective shoes	Safety glasses	Face shield	Body shield	Good ventilation	Fume hood	Respirator	Self-contained breathing apparatus	1	2	3	4	5	6	7	8	9	10
Acids, inorganic	√	√						√			√					√					√	
Acids, organic	√	√							√	√						√					√	
Alcohols	√	√				√			√	√					√							√
Aldehydes	√	√							√	√												√
Alkalis		√	√			√			√						√							√
Alkali earth metals		√	√			√			√						√							√
Amines, aromatic	√	√							√							√						√
Amides, inorganic	√	√				√	√									√				√		
Amides, organic	√	√			√													√		√		
Azides		√					√									√				√		
Azo compounds		√					√									√				√		
Carbides	√	√			√											√						√
Chlorohydrins	√	√							√		√	√				√				√		
Cyanides	√	√					√	√			√	√				√						√
Esters	√	√				√			√		√				√						√	
Ethers	√	√				√					√							√			√	
Halides, inorganic	√	√					√		√	√				√		√				√		
Halides, organic	√	√							√		√				√							√
Halides, organic acid	√	√							√		√	√		√		√				√		
Halogens, organic	√	√								√		√		√							√	
Hydrazines	√			√							√								√		√	
Hydrides	√			√			√	√								√						√
Hydrocarbons	√		√				√		√		√				√						√	

TABLE 10-14   Protection from and Disposal of Chemicals (Continued)

General classifications	Rubber gloves	Leather gloves	Laboratory coat	Fireproof clothing	Protective shoes	Safety glasses	Face shield	Body shield	Good ventilation	Fume hood	Respirator	Self-contained breathing apparatus	Disposal procedures 1	2	3	4	5	6	7	8	9	10
Inorganic salts	✓		✓		✓											✓				✓		
Ketones	✓		✓			✓		✓		✓				✓							✓	
Metal alkoxides		✓	✓			✓		✓				✓	✓								✓	
Metal alkyls		✓	✓			✓		✓				✓	✓								✓	
Mercaptans	✓	✓										✓						✓				
Niriles	✓	✓					✓	✓				✓	✓			✓						✓
Nitro compounds	✓	✓		✓								✓	✓								✓	
Nitro paraffins	✓	✓							✓			✓	✓			✓				✓		
Organo-metallic compounds	✓	✓			✓							✓	✓			✓				✓		
Oxidizing agents	✓	✓				✓		✓	✓	✓						✓						✓
Peroxides, inorganic	✓					✓	✓						✓			✓				✓		
Peroxides, organic	✓	✓				✓	✓		✓				✓			✓				✓		
Phosphates, inorganic	✓	✓			✓											✓				✓		
Phosphates, organic	✓	✓									✓				✓		✓				✓	
Reducing agents	✓	✓			✓				✓													✓
Sulfides, inorganic	✓	✓			✓		✓	✓								✓						✓
Sulfides, organic	✓	✓									✓							✓				

Index